AF700636

from
THE QUARTERLY REVIEW OF BIOLOGY

42158 LO
4/16/15 # 125.00

VENOMOUS REPTILES AND THEIR TOXINS

VENOMOUS REPTILES AND THEIR TOXINS

Evolution, Pathophysiology and Biodiscovery

Bryan G. Fry

OXFORD
UNIVERSITY PRESS

OXFORD
UNIVERSITY PRESS

Oxford University Press is a department of the University of Oxford. It furthers the University's objective of excellence in research, scholarship, and education by publishing worldwide.

Oxford New York
Auckland Cape Town Dar es Salaam Hong Kong Karachi
Kuala Lumpur Madrid Melbourne Mexico City Nairobi
New Delhi Shanghai Taipei Toronto

With offices in
Argentina Austria Brazil Chile Czech Republic France Greece
Guatemala Hungary Italy Japan Poland Portugal Singapore
South Korea Switzerland Thailand Turkey Ukraine Vietnam

Oxford is a registered trademark of Oxford University Press in the UK and certain other countries.

Published in the United States of America by
Oxford University Press
198 Madison Avenue, New York, NY 10016

© Oxford University Press 2015

All rights reserved. No part of this publication may be reproduced, stored in a retrieval system, or transmitted, in any form or by any means, without the prior permission in writing of Oxford University Press, or as expressly permitted by law, by license, or under terms agreed with the appropriate reproduction rights organization. Inquiries concerning reproduction outside the scope of the above should be sent to the Rights Department, Oxford University Press, at the address above.

You must not circulate this work in any other form and you must impose this same condition on any acquirer.

Library of Congress Cataloging-in-Publication Data
Fry, Bryan (Bryan Grieg)
Venomous reptiles and their toxins : evolution, pathophysiology, and biodiscovery / Bryan Fry.
pages cm
Includes bibliographical references and index.
ISBN 978-0-19-930939-9 (alk. paper)
1. Reptiles—Venom. 2. Reptiles—Chemical defenses. 3. Poisonous snakes—Venom.
4. Poisonous animals—Toxicology. I. Title.
QP632.V46F79 2015
615.9′42—dc23
2014030485

1 3 5 7 9 8 6 4 2
Printed in the United States of America
on acid-free paper

CONTENTS

LIST OF ILLUSTRATIONS

LIST OF TABLES

CONTRIBUTORS

Alagón, Alejandro. Departamento de Medicina Molecular y Bioprocesos, Instituto de Biotecnología, Universidad Nacional Autónoma de México, Av. Universidad 2001, Cuernavaca, Morelos 62210, México; alagon@ibt.unam.mx

Alam, Mehtab. Department of Biochemistry, Dow International Medical College, Dow University of Health Sciences, Ojha Campus, Karachi, Pakistan; promet213@yahoo.com

Ali, Syed. HRJ Research Institute of Chemistry, International Centre for Chemical and Biological Sciences (ICCBS), University of Karachi, Karachi 75270, Pakistan; dr.syedabidali@gmail.com

Arlinghaus, Franziska T. Center for Molecular Medicine, Department of Vascular Matrix Biology, Excellence Cluster Cardio-Pulmonary System, Frankfurt University Hospital, Frankfurt 60590, Germany; f.t@arlinghaus.org

Barten, Stephen. Vernon Hills Animal Hospital, 1260 S. Butterfield Rd., Mundelein, IL 60060, USA; sbartendvm@gmail.com

Barve, Sahas. Ecology and Evolutionary Biology Department, Cornell University, Ithaca, NY 14850, USA; sahasbarve@gmail.com

Bdolah, Avner. Department of Zoology, Tel Aviv University, Tel Aviv 69978, Israel; avnerb@post.tau.ac.il

Bénard-Valle, Melisa. Departamento de Medicina Molecular y Bioprocesos, Instituto de Biotecnología, Universidad Nacional Autónoma de México, Av. Universidad 2001, Cuernavaca, Morelos 62210, México; melitza61@gmail.com

Boyer, Leslie. Venom Immunochemistry, Pharmacology and Emergency Response (VIPER) Institute and Department of Pathology, University of Arizona, Tucson, AZ 85724, USA; boyer@viper.arizona.edu

Brandl, Diane. Melbourne Aquarium, Melbourne, Victoria, Australia; diane.brandl@melbourneaquarium.com.au

Burbrink, Frank. Department of Biology, College of Staten Island, 2800 Victory Blvd., Staten Island, NY 10314, USA; frank.burbrink@csi.cuny.edu

Calvete, Juan J. Laboratorio de Venómica Estructural y Funcional, Jaime Roig 11, Valencia 46010, Spain; jcalvete@ibv.csic.es

Carmichael, Rob. Wildlife Discovery Center, City of Lake Forest Parks, Forestry and Recreation, 1401 Middlefork Dr., Lake Forest, IL 60045, USA; robertcarmichael@comcast.net

Casewell, Nicholas R. Alistair Reid Venom Research Unit, Liverpool School of Tropical Medicine, Pembroke Place, Liverpool L3 5QA, UK; nicholas.casewell@lstmed.ac.uk

Chippaux, J.-P. Mère et Enfant Face aux Infections Tropicales, Paris; Sorbonne Paris Cité, Université Paris Descartes, Faculté de Pharmacie, Paris; 08 BP 841, Cotonou, Bénin; jean-philippe.chippaux@ird.fr

Citron, Diane M. R. M. Alden Research Laboratory, Culver City, CA 90230, USA; d.m.citron@att.net

Clarkson, Myke Herpetological Conservation International, Los Angeles, CA, USA; myke@herpconservation.org

Clemetson, Kenneth J. Haemostasis Research, Department of Haematology, University of Berne, Inselspital, CH-3010, Berne, Switzerland; kenneth.clemetson@tki.unibe.ch

Cochran, Chip. Department of Earth and Biological Sciences, Loma Linda University, Loma Linda, CA 92350, USA; skipc8384@hotmail.com

Cousin, Xavier. Ifremer, Laboratoire d'Ecotoxicologie, Place Gaby Coll, BP 7, L'Houmeau 17137, France; Campus de Beaulieu, Rennes 35042, France; xavier.cousin@ifremer.fr

Cribb, Bronwen W. Centre for Microscopy and Microanalysis and the School of Biological Sciences, The University of Queensland, St Lucia, QLD 4072, Australia; b.cribb@uq.edu.au

Daly, Norelle L. Australian Institute of Tropical Health and Medicine, James Cook University, Smithfield, Cairns, QLD 4878, Australia; norelle.daly@jcu.edu.au

De Sá, Paulo L., Jr. Laboratorio de Genetica, Instituto Butantan, Avenida Vital Brasil, 1500, Butantã, São Paulo SP 05503-900, Brazil; paulsaj2001@yahoo.com.br

Ducancel, Frédéric. CEA, IBiTec-S, Service de Pharmacologie et d'Immunoanalyse, Laboratoire d'Ingénierie des Anticorps pour la Santé, Gif-sur-Yvette F-91191, France; frederic.ducancel@cea.fr

Dunstan, Nathan. Venom Supplies, Tanunda, SA 5352, Australia; nathan@venom-supplies.com

Dziemborowicz, Sławomir. School of Medical & Molecular Biosciences, University of Technology Sydney, PO Box 123, Broadway NSW 2007, Australia; slawomirdziembo-rowicz@gmail.com

Earl, Stephen. UQ Centre for Clinical Research and School of Medicine, The University of Queensland, Brisbane 4029, Australia; s.earl@uq.edu.au

Eble, Johannes A. Institute for Physiological Chemistry and Pathobiochemistry, University of Muenster, Waldeyerstr. 15, 48149 Muenster, Germany - johannes.eble@ uni-muenster.de

Eng, Wai Soon. School of Chemistry and Molecular Biosciences, The University of Queensland, St. Lucia, QLD 4072, Australia; wai.eng@uqconnect.edu.au

Fry, Bryan G. 1. Venom Evolution Lab, School of Biological Sciences, The University of Queensland, St. Lucia, QLD 4072, Australia; 2. Institute for Molecular Bioscience, The University of Queensland, St. Lucia, QLD 4072; bryan@venomdoc.com

Fung, Shin Yee. Department of Molecular Medicine, Faculty of Medicine, University of Malaya, Kuala Lumpur, Malaysia; fungshinyee@gmail.com

Georgieva, Dessislava. Laboratory of Structural Biology of Infection and Inflammation, Institute of Biochemistry and Molecular Biology, University of Hamburg, Hamburg, Germany; octopus_dofleini@yahoo.com

Gibbins, Jonathan M. Institute for Cardiovascular and Metabolic Research, School of Biological Sciences, University of Reading, Reading RG6 6AS, UK; j.m.gibbins@reading.ac.uk

Gillett, Amber. Australia Zoo Wildlife Hospital, Steve Irwin Way, Beerwah, QLD 4519, Australia; drambergillett@hotmail.com

Goldstein, Ellie J. C. R. M. Alden Research Laboratory, Culver City, CA 90230, USA; UCLA School of Medicine, Los Angeles, CA 90095, USA; ejcgmd@aol.com

Guddat, Luke W. School of Chemistry and Molecular Biosciences, University of Queensland, St. Lucia, QLD 4072, Australia; luke.guddat@uq.edu.au

Gutiérrez, José. Instituto Clodomiro Picado, Facultad de Microbiología, Universidad de Costa Rica, San José, Costa Rica; jose.gutierrez@ucr.ac.cr

Haberfield, James. Unusual Pet Vets, Veterinary Services for Reptiles, Birds, and Small Mammals, Murdoch University Veterinary Hospital, 90 South St., Murdoch, 6150, Australia; james@unusualpetvets.com.au

Harrison, Jim. Kentucky Reptile Zoo, 200 L and E Railroad Pl., Slade, KY 40376, USA; kentuckyreptilez@bellsouth.net

Hayashi, Mirian A. F. Departamento de Farmacologia, Universidade Federal de São Paulo, São Paulo, Brazil; mhayashi.unifesp@gmail.com

Hendrikx, Iwan. Venom Evolution Lab, School of Biological Sciences, The University of Queensland, St. Lucia, QLD 4072, Australia; iwanhx@yahoo.com

Herzig, Volker. Institute for Molecular Bioscience, The University of Queensland, St. Lucia, QLD 4072, Australia; v.herzig@imb.uq.edu.au

Hocknull, Scott. Geosciences, Queensland Museum, 122 Gerler Rd. Hendra, QLD 4011, Australia. Email: scott.hocknull@qm.qld.gov.au

Hodgson, Wayne C. Department of Pharmacology, Monash University, Clayton, VIC, 3800, Australia; wayne.hodgson@monash.edu

Jackson, Timothy N. W. 1. Venom Evolution Lab, School of Biological Sciences, The University of Queensland, St. Lucia, QLD 4072, Australia; 2. Institute for Molecular Bioscience, The University of Queensland, St. Lucia, QLD 4072, Australia; tnwjackson@gmail.com

Johnson, Robert. South Penrith Veterinary Clinic, 126 Stafford St., Penrith, NSW 2750, Australia; clinic@reptilevet.com.au

Jones, Rob. Aquarium Vet, P.O. Box 2327, Moorabbin, VIC 3189, Australia; rob@theaquariumvet.com.au

Kerkis, Alexandre. Laboratorio de Genetica, Instituto Butantan, Avenida Vital Brasil, 1500, Butantã, São Paulo SP 05503-900, Brazil; akerkis@usp.br

Kerkis, Irina. Laboratorio de Genética, Instituto Butantan, Avenida Vital Brasil, 1500, Butantã, São Paulo SP 05503-900, Brazil; irina.kerkis@butantan.gov.br

Kochva, Elazar. Department of Zoology, Tel Aviv University, Tel Aviv 69978, Israel; ekochva@post.tau.ac.il

Koludarov, Ivan. Venom Evolution Lab, School of Biological Sciences, The University of Queensland, St. Lucia, QLD 4072, Australia; jcoludar@gmail.com

Kurniawan, Nyoman D. Centre for Advanced Imaging, University of Queensland, St. Lucia, QLD 4072, Australia; nyoman.kurniawan@cai.uq.edu.au

Kwok, Hang Fai. 1. Faculty of Health Sciences, University of Macau, Avenida de Universidade, Macau SAR; 2. CRUK Cambridge Institute, University of Cambridge, CB2 0RE, Cambridge, UK; 3. School of Pharmacy, Queen's University Belfast, BT9 7BL, Belfast, UK; hfkwok@umac.mo

Lavergne, Vincent. Institute for Molecular Bioscience, The University of Queensland, St. Lucia, QLD 4072, Australia; v.lavergne@imb.uq.edu.au

Lock, Brad. Department of Herpetology, Zoo Atlanta, 800 Cherokee Ave. S.E., Atlanta, GA 30315, USA; block@zooatlanta.org

Lomonte, Bruno. Instituto Clodomiro Picado, Facultad de Microbiología, Universidad de Costa Rica, San José, Costa Rica; bruno.lomonte@ucr.ac.cr

Manchadi, Mary-Louise Roy. School of Biomedical Sciences, The University of Queensland, St. Lucia, QLD 4072, Australia; m.roymanchadi@uq.edu.au

Martelli, Paolo. Ocean Park Corporation, Hong Kong; paolo.martelli@oceanpark.com.hk

McCarthy, Sean. Snakehandler Pty. Ltd., P.O Box 262, Hastings, VIC 3915, Australia; sean@snakehandler.com.au

Messenger, Kevin. 1. Alabama A & M University, Department of Biological and Environmental Sciences, 4900 Meridian Street North, Normal, AL 35811, USA; 2. Nanjing Forestry University, Department of Biology and Environmental Science, Nanjing, Jiangsu, China; herpsrule2@aol.com

Mobli, Mehdi. Center for Advanced Imaging, University of Queensland, St. Lucia, QLD 4072, Australia; m.mobli@uq.edu.au

Monagle, Paul. Department of Paediatrics, University of Melbourne, Royal Children's Hospital, Murdoch Children's Research Institute, Melbourne, VIC 3000, Australia; paul.monagle@rch.org.au

Morris, M. Cale. Heritage Academy, 32 South Center St., Mesa, AZ 85205, USA; cale@venomteacher.com

Neri-Castro, Edgar Enrique. Departamento de Medicina Molecular y Bioprocesos, Instituto de Biotecnología, Universidad Nacional Autónoma de México, Av. Universidad 2001, Cuernavaca, Morelos 62210, México; neri@ibt.unam.mx

Nicholson, Graham. School of Medical & Molecular Biosciences, University of Technology Sydney, PO Box 123, Broadway NSW 2007, Australia; graham.nicholson@uts.edu.au

Nouwens, Amanda. School of Chemistry and Molecular Biosciences, The University of Queensland, St. Lucia, QLD 4072, Australia; a.nouwens@uq.edu.au

Oguiura, Nancy. Laboratório Especial de Ecologia e Evolução, Instituto Butantan, Avenida Dr. Vital Brasil 1500, Butantã, São Paulo SP 05503-900, Brazil; nancy.oguiura@butantan.gov.br

Olvera-Rodríguez, Felipe. Departamento de Medicina Molecular y Bioprocesos, Instituto de Biotecnología, Universidad Nacional Autónoma de México, Av. Universidad 2001, Cuernavaca, Morelos 62210, México; folvera@ibt.unam.mx

Paniagua, Dayanira. Departamento de Medicina Molecular y Bioprocesos, Instituto de Biotecnología, Universidad Nacional Autónoma de México, Av. Universidad 2001, Cuernavaca, Morelos 62210, México; dashpame@ibt.unam.mx

Pereira, Alexandre. Laboratorio de Genetica, Instituto Butantan, Avenida Vital Brasil, 1500, Butantã, São Paulo SP 05503-900, Brazil; pereira_alex@butantan.gov.br

Phillip, Terry. Black Hills Reptile Gardens, Rapid City, SD 57709, USA; terry@reptilegardens.com

Pittman, Joe. Florida Snakebite Institute, 18106 Lembrecht Way, Tampa, FL 33647, USA; traumajunkiejoe@juno.com

Prieto da Silva, Álvaro R. B. Laboratorio de Genética, Instituto Butantan, Avenida Vital Brasil, 1500, Butantã, São Paulo SP 05503-900, Brazil; alvaro.prieto@butantan.gov.br

Pyron, R. Alexander. Department of Biological Sciences, George Washington University, 2023 G St. N.W., Lisner Hall 345, Washington, DC 20052, USA; rpyron@colubroid.org

Rádis-Baptista, Gandhi. Laboratory of Biochemistry and Biotechnology, Institute for Marine Sciences, Federal University of Ceara, Av. Abolicao 3207, Fortaleza CE 60165-081, Brazil; gandi.radis@ufc.br

Ragnarsson, Lotten. Institute for Molecular Bioscience, The University of Queensland, St. Lucia, QLD 4072, Australia; l.ragnarsson-mcgrath@imb.uq.edu.au

Rash, Lachlan D. Institute for Molecular Bioscience, The University of Queensland, St. Lucia, QLD 4072, Australia; l.rash@imb.uq.edu.au

Reeks, Timothy. Institute for Molecular Bioscience, The University of Queensland, St. Lucia, QLD 4072, Australia; t.reeks@uq.edu.au

Richards, Renee S. University of Queensland Centre for Clinical Research, Building 71/918, Royal Brisbane and Women's Hospital Campus, Herston, QLD 4029, Australia; r.richards@uq.edu.au

Roelants, Kim. Amphibian Evolution Lab, Biology Department, Vrije Universiteit Brussel, Pleinlaan 2, Brussels B-1050, Belgium; kroelant@vub.ac.be

Rowley, Paul. Alistair Reid Venom Research Unit, Liverpool School of Tropical Medicine, Liverpool L3 5QA, UK; p.rowley@lstmed.ac.uk

Sasa, Mahmood. Instituto Clodomiro Picado, Universidad de Costa Rica, San José, Costa Rica; msasamarin@gmail.com

Scheib, Holger. Venom Evolution Lab, School of Biological Sciences, The University of Queensland, St. Lucia, QLD 4072, Australia; holger@moltalk.org

Shankar, Gowri. 1. Kalinga Foundation, Kalinga Mane, Churrchihakklu, Agumbe Hobli, Hosur Grama, Guddekere, Thirthahalli, Shivamogga -577411, Karnataka, India; 2. Deptartment of Ecology and Genetics, Evolutionary Biology Centre, Uppsala University, Uppsala 752 36, Sweden; gowrishankar.pogiri@gmail.com.

Simpson, Shane. Karingal Veterinary Hospital, 328 Cranbourne Rd., Frankston, VIC 3199, Australia; info@thereptiledoctor.com.au

Sunagar, Kartik. 1. Centro Interdisciplinar de Investigação Marinha e Ambiental, Universidade do Porto, Rua dos Bragas, 177, Porto 4050-123, Portugal; 2. Small RNA and Venom Evolution Lab, Alexander Silberman Institute for Life Sciences, Hebrew University of Jerusalem, Jerusalem 91904, Israel; anaturalist@gmail.com

Takacs, Zoltan. ToxinTech, Inc., P.O. Box 6266, New York, NY 10150, USA; zoltan@zoltantakacs.com

Tan, Nget Hong. Department of Molecular Medicine, Faculty of Medicine, University of Malaya, Kuala Lumpur, Malaysia; tanngethong@yahoo.com.sg

Trabi, Manuela. Centre for Integrative Clinical and Molecular Medicine, School of Medicine, The University of Queensland, St. Lucia, QLD 4072, Australia; manu.trabi@uq.edu.au

Tsai, Inn-Ho. Institute of Biological Chemistry, Academia Sinica; Institute of Biochemical Sciences, National Taiwan University, Taipei, Taiwan; bc201@gate.sinica.edu.tw

Tyrrell, Kerin L. R. M. Alden Research Laboratory, Culver City, CA 90230, USA; k.l.tyrrell@att.net

Undheim, Eivind A. B. 1. Institute for Molecular Bioscience, The University of Queensland, St. Lucia, QLD 4072, Australia; 2. Center for Advanced Imaging, The University of Queensland, St. Lucia, QLD 4072, Australia; e.undheim@imb.uq.edu.au

Utkin, Yuri. Laboratory of Molecular Toxinology, Shemyakin-Ovchinnikov Institute of Bioorganic Chemistry, Russian Academy of Sciences, ul. Miklukho-Maklaya 16/10, Moscow 117997, Russia; utkin@mx.ibch.ru

Vaiyapuri, Sakthivel. Institute for Cardiovascular and Metabolic Research, School of Biological Sciences, University of Reading, Reading RG6 6AS UK; s.vaiyapuri@reading.ac.uk

Van der Ploeg, Harold. 1. Working Group Adder Research Netherlands; 2. Reptile, Amphibian, and Fish Conservation Netherlands; info@eyecreations.nl

Vetter, Irina. 1. Institute for Molecular Bioscience, The University of Queensland, St. Lucia, QLD 4072, Australia; 2. School of Pharmacy, The University of Queensland, Woolloongabba 4102, Australia; i.vetter@imb.uq.edu.au

Vidal, Nicolas. UMR 7205 ISYEB Institut de Systématique, Évolution et Biodiversité, Service de Systématique Moléculaire, Muséum National d'Histoire Naturelle 43, rue Cuvier Paris 75005, France; nvidal@mnhn.fr

Vonk, Freek J. Naturalis Biodiversity Center, 2333 CR, Leiden, Netherlands; freek.vonk@naturalis.nl

Wagstaff, Simon C. Liverpool School of Tropical Medicine, Pembroke Place, Liverpool L3 5QA, UK; simon.wagstaff@lstmed.ac.uk

Weise, Christoph. Institut für Chemie und Biochemie, Freie Universität Berlin, Thielallee 63, D-14 195 Berlin, Germany; chris.weise@biochemie.fu-berlin.de

Whitehead, Darryl L. School of Biomedical Sciences, The University of Queensland, St. Lucia, QLD 4072, Australia; darryl.whitehead@uq.edu.au

Wiley, Kristen. Kentucky Reptile Zoo, 200 L and E Railroad Pl., Slade, KY 40376, USA; kentuckyreptilez@bellsouth.net

Wroe, Stephen. Function, Evolution, and Anatomy Research Lab, Zoology, School of Environmental and Rural Science, University of New England, Armidale, NSW 2351, Australia; swroe@une.edu.au

Wüster, Wolfgang. School of Biological Sciences, Bangor University Environment Centre Wales, Bangor LL57 2UW, Wales, UK; w.wuster@bangor.ac.uk

Yamane, Tetsuo. **1.** National Institute of S&T, Center for Environment and Biodiversity Studies, Universidade do Estado do Amazonas, 69065-001, Manaus, Amazonas, Brazil; 2. Laboratorio de Bioquimica e Biologia Molecular, Centro de Biotecnologia da Amazonia, Ministerio de Desenvolvimento, Industria e Comercio Exterior, Brazil; tetsuo@usp.br

Yang, Daryl C. Department of Pharmacology, Monash University, Clayton, VIC, 3800, Australia; daryl.yang@monash.edu

Young, Bruce. Department of Anatomy, Kirksville College of Osteopathic Medicine, A. T. Still University of Health Sciences, 800 W. Jefferson, Kirksville, MO 63501, USA; byoung@atsu.edu

Zaharenko, Andre J. Laboratório de Genética, Instituto Butantan, Avenida Vital Brasil, 1500, Butantã, São Paulo SP 05503-900, Brazil; a.j.zaharenko@ig.com.br

Zivkovic, Miller. School of Biological Sciences, The University of Queensland, St. Lucia, QLD 4072, Australia; m.zivkovic@uq.edu.au

VENOMOUS REPTILES AND THEIR TOXINS

CHAPTER 1

THE ORIGIN AND EVOLUTION OF THE TOXICOFERA REPTILE VENOM SYSTEM

B. G. FRY, K. SUNAGAR, N. R. CASEWELL, E. KOCHVA, K. ROELANTS, H. SCHEIB, W. WÜSTER, N. VIDAL, B. YOUNG, F. BURBRINK, R. A. PYRON, F. J. VONK, AND T. N. W. JACKSON

1.1 INTRODUCTION

Insight gained into the evolutionary history of the venom system of toxicoferan reptiles (Serpentes [snakes] plus Anguimorpha [anguimorph lizards] and Iguania [iguanian lizards]) in recent years has generated considerable controversy regarding the accepted definition of "venom" and what constitutes a venomous animal. Although this may seem little more than a semantic debate, it is of crucial importance that toxinological researchers apply consistent terminology in order to avoid confusion and to facilitate the efficient formulation of research questions into the evolution, ecology, and medical application of venom and its constituent toxins. The definition of "venom" utilized in the present text is as follows: "A secretion produced in specialized cells in one animal, delivered to a target animal through the infliction of a wound and that disrupts endophysiological or biochemical processes in the receiving animal to facilitate feeding, defense or competition by/of the producing animal" (Fry et al. 2009a, 2012a).

This definition applies to a large number of animal taxa and not solely to toxicoferan reptiles. It incorporates criteria that allow the distinction between "venoms" and "poisons." The definition of "poison" is taxonomically more inclusive (such as "poisonous" plants), but when it is being used to describe an animal, it refers to a defensive secretion passively delivered (e.g., by epithelial absorption) to a potential predator. Conversely, the above-mentioned definition of "venom" rejects the criteria of "rapid prey death" and danger to humans as being necessary for the designation of an animal as "venomous." In many cases (such as parasitoid wasps), venom may be used to subdue, but not kill, intended prey, or it may be used exclusively in defense (such as in bees and most venomous fishes). The criterion of danger to humans has little relevance to the evolution of venom, and it is arguable that the historical anthropocentric bias in toxinological research has hindered our understanding of this phenomenon. It is important to note further that "venom" is a functional definition, and the presence of toxins or toxin genes alone is not sufficient evidence for the classification of an animal as "venomous"; there must also be evidence of a mechanism for creating a wound in order to facilitate the delivery of toxins to either prey (for feeding), apparent predators (for defense), or conspecifics (such as for territorial competition). However, differentiated teeth are not required for delivery, just so long as a wound is created.

Some authors have argued that it must be "proven" through controlled prey-handling experiments that an animal requires venom to subdue prey before a "venomous" classification is granted (Kardong 2012; Weinstein, Keyler, and White 2012). While such experiments are interesting and informative and should continue to be conducted, if they were required for classification, then the vast majority (probably more than 95%) of all species currently classified as venomous (including the families Elapidae [elapid snakes] and Viperidae [viperid snakes]) would no longer qualify as such since the majority of even these front-fanged lineages have not been subjected to such stringent testing. Here we rely on inductive reasoning: if an animal possesses specialized toxin-secreting tissue associated with a specialized venom-delivery apparatus (such as modified dentition or limbs), then this is considered strong evidence that the animal is venomous. If the specialized tissue and delivery apparatus are homologous with those of other, related venomous animals (as in the venom glands and fangs of caenophidian or advanced snakes, then the evidence is further strengthened, and the classification of the animal as venomous is reasonable (Fry et al. 2012a).

Venoms are key evolutionary innovations in the Kingdom Animalia and have evolved independently in multiple taxa (see color plate 1). Deeply grooved and even hollow teeth, strongly suggestive of venom systems, have even been observed in archosauriform dinosaur fossils (Gong et al. 2010; Mitchell, Heckert, and Sues 2010). Venoms allow predators such as snakes, with their lack of limbs, to feed on a far wider range of prey items than may otherwise be possible. Venom can also facilitate effective defense for otherwise vulnerable animals. The level of convergence among venomous lineages in toxin recruitment and functional evolution is striking and is particularly notable in the targeting of the hemostatic and neurological systems (see color plate 2).

1.2 VENOM SYSTEM ORIGIN AND EVOLUTION

1.2.1 PLESIOTYPIC TOXICOFERA REPTILE VENOM SYSTEMS

Reptile venoms perform concerted, multifaceted attacks on a victim's physiology and are capable of acting on any area of the body perfused by the blood. All variables of the venom system have diverged independently, from biochemical variation and specialization of the venoms themselves to variations in the anatomy of the glands and dentition that constitute the delivery system. Because of the paucity of the squamate fossil record, resulting in part from the poor fossilization of soft tissue, the origin and subsequent diversification of the system must largely be inferred from the study of extant species. Early investigation concentrated on morphological characters such as dentition, glands, and musculature associated with the venom-delivery system (Johnson 1956; Kardong 1980; Jackson 2003). In addition, the field of toxinology shows a large anthropocentric bias, as taxa that are broadly recognized as posing a medical hazard, such as snakes, have received most attention. Yet of the approximately 2,850 species of caenophidian snakes, only the approximately 680 front-fanged species are widely recognized as "venomous." It is worth noting that the majority of even these species remained uninvestigated until recently (many continue to be completely neglected) and were considered venomous only because of the similarity of their venom systems to those of species that were known to be dangerous to humans. The toxic oral secretions of the non-front-fanged caenophidian (NFFC) snakes were almost entirely ignored until the past decade or so (Hill and Mackessy 2000; Vidal 2002; Fry et al. 2003a; Fry et al. 2003c; Huang and Mackessy 2004; Fry et al. 2006; Mackessy et al. 2006; Pawlak et al. 2006; Pahari, Mackessy, and Kini 2007; Fry et al. 2008; Pawlak et al. 2009; Weldon and Mackessy 2010; Fry et al. 2012b; Heyborne and Mackessy 2013). Similarly, despite having a state of fame transcending herpetology, the venom system of helodermatid lizards remained poorly investigated, with even the number of venom gland compartments not resolved until recently (Fry et al. 2010a; Fry et al. 2010b). The toxic oral secretions of other lizards and their potential to be functionally venomous had received no research attention until the last few years (Fry et al. 2006; Fry et al. 2009b; Fry et al. 2010b; Koludarov et al. 2012).

The development of molecular systematics provided the phylogenetic framework necessary for a reconstruction of the evolutionary history of the glands and fangs and thus helped to resolve a number of fundamental questions regarding snake and lizard evolution. The first two phylogenetic studies using multiple nuclear genes and broad taxonomic coverage of squamate (amphisbaenians, lizards and snakes) higher-level relationships were in stark contrast to phylogenetic hypotheses based on morphology (Townsend et al. 2004; Vidal and Hedges 2004). Both studies identified a clade composed of snakes, anguimorph lizards, iguanian lizards, Amphisbaenia, Lacertidae, and Teiidae, but the interrelationships among these major lineages remained unresolved. Subsequent analysis of nine nuclear genes provided strong support for a clade composed of: (i) snakes, (ii) iguanian lizards, and (iii) anguimorph lizards (Vidal and Hedges 2005; Fry et al. 2006) This arrangement was given the name Toxicofera due to the shared characteristic of oral glands with distinct protein secreting regions, with the heightened expression of toxin-related genes. Later genetic studies corroborated the clade of toxicoferan reptiles (Vidal and Hedges 2009; Wiens et al. 2012; Pyron, Burbrink, and Wiens 2013). In contrast, phylogenetic studies relying only on morphological data disagreed with molecular data and continued to support the traditional groupings of Scleroglossa and Iguania (Gauthier et al. 2012).

In light of estimated divergence times, the origin of the toxicoferan venom system occurred approximately 170 million years ago (Vidal and Hedges 2005; Fry et al. 2006; Vidal and Hedges 2009). The relative relationships within the toxicoferan reptile clade, however, remain to be elucidated. While the analyses of nuclear gene sequences support a sister-clade relationship between anguimorph and iguanian lizards (to the exclusion of snakes), analyses of small-interspersed nuclear elements (SINEs) instead seem to favor the clustering of snakes and anguimorph lizards (Piskurek, Austin, and Okada 2006). Thus, we consider the arrangement of the three lineages as an unresolved trichotomy (see color plate 3). However, phylogenetic analyses of CRiSP proteins (see chapter 10.3 and figures 10.1 and 10.2) identify a clade composed of all lizard toxins and a separate clade composed of snake toxins. This arrangement lends supports to a closer relationship between anguimorph and iguanian lizards relative to snakes.

Nontoxicoferan reptiles have labial glands that do not have segregated protein and mucus production and histological staining overwhelmingly reflects mucus secretion. In contrast, histological analyses of oral glands in various representatives of snakes, anguimorph lizards and iguanian lizards demonstrated that enlarged glands with newly derived discrete regions for protein and mucus production already existed in the last common ancestor of all toxicoferan reptiles (Fry et al. 2006; Fry et al. 2013). The inferred ancestral toxicoferan venom system consisted of thin, simple seromucous glands in both the mandibular and maxillary regions (Fry et al. 2006; Fry et al. 2013). These glands differed sharply from those of non-toxicoferan lizards in not only being larger and more structured but also staining strongly for protein production in the broader part of the lobule, while the mucus-secreting region near the teeth is thinner and narrower (Fry et al. 2006; Fry et al. 2013). These same histological studies showed that the broadest part of the teardrop-shaped lobules is where protein production occurs, thus amplifying protein production. Thus even if similar proteins are expressed compared to nontoxicoferan reptiles, the relative production would be considerably higher. All toxicoferan lineages were shown to be characterized by having such glands. They were at the thinnest and least developed in the iguanian lizards. The apparent lack of specialization of the venom system among early diverging iguanian lineages is likely because most of their extant descendants are insectivorous or herbivorous. However, toxin genes continue to be expressed by the oral glands of these lizards (Fry et al. 2006; Fry et al. 2013), and some species that feed on vertebrates have larger glands with a higher relative proportion of protein-secreting cells, suggesting a role in predation (Fry et al. 2006; Fry et al. 2013). By contrast, in the snakes and anguimorph lizards, the venom system has been extensively refined (Fry et al. 2003c; Fry and Wüster 2004; Fry et al. 2006; Fry et al. 2009b; Fry et al. 2010a; Fry et al. 2010b; Fry et al. 2013) (see color plates 4, 5, and 6). Since their last common ancestor, the venom systems in these two groups have undergone major structural and anatomical divergence. The maxillary venom glands underwent extensive diversification in snakes, while the mandibular glands were diversified in anguimorph lizards. Differences are notable in the types of secretory epithelia (serous, seromucous, and mucous) and their locations (gland, duct, transition area) and in the number and physical orientation of gland compartments and the relative encapsulation of the glands and the muscles attached to them.

1.2.2 DIVERSIFICATION OF THE ANGUIMORPH LIZARD VENOM SYSTEM

The inferred plesiomorphic anguimorph lizard venom system is represented by the limbless lizard *Pseudopus apodus* (formerly *Ophisaurus apodus*), which has the seromucous mandibular glands present in all anguimorph lizards but also uniquely possesses maxillary seromucous glands (Fry et al. 2010b) (see color plate 3). This plesiomorphic state entails the presence of mixed glands with a ventral serous portion and a dorsal mucous region, which together are encapsulated by a single thin membrane. Even at this stage, they are larger and more complex relative to the variations within iguanian lizards, even the vertebrate-feeding species. Each gland on either side of the jaws is compartmentalized along the anterioposterior axis, with one compartment per tooth and a separate duct for each compartment that leads to the base of the tooth. Within compartments, extensive intralumen drainage channeling is evident, but the major lumen is unstructured. This unstructured, seromucous arrangement is retained in the more robust mandibular venom glands of the Anguidae (anguids), as is the arrangement of one compartment per tooth. However, in all anguimorphs examined to date other than *P. apodus*, the maxillary gland is entirely lost (see color plates 3 and 4). Such an enlargement of the mandibular gland and more refined structure result in a dramatically greater number of protein-producing cells, including the evolution of intercellular lumens, which would retain secretion as a liquid in these new hollow spaces within the gland compartments.

In contrast to the relatively simple and unstructured glands of the other anguimorph lizards, the *Heloderma* and varanoid (*Lanthanotus/Varanus*) mandibular venom glands have independently undergone complete segregation of the protein- and mucus-secreting regions into distinct gland regions, and the entire gland is encapsulated by thick membranes (Fry et al. 2006; Fry et al. 2009b; Fry et al. 2010b; Koludarov et al. 2012; Fry et al. 2013) (see color plates 3 and 4). In both cases, the significantly enlarged protein-secreting serous glands have extensive intercellular lumens, leading to well-structured central lumens. The relative size of the glands varies proportionally with head size. *Heloderma* species have the largest glands proportional to head size, even accounting for the large, broad heads relative to the size of the bodies. In these segregated glands, numerous mucous lobules are located dorsally and are distinct from the protein glands. Convergent fusion of posterior compartments in these two lineages has increased lumen storage space so that in both *Heloderma* and *Varanus* only six compartments remain.

The glands of varanid lizards and Lanthanotidae (lanthanotid lizards) are similar in almost all respects but may differ in the number of compartments. *Lanthanotus* species may have as few as three compartments, but additional investigation is required, as previous studies have been limited to the examination of poorly preserved museum specimens. The physical architecture differs between the helodermatids and the lanthanotids/varanids (see color plates 4 and 5). While the compartments of *Heloderma* glands have a more layered arrangement than those of varanids/lanthanotids, the glands of the latter are thinner and more tubular. In both clades, extensive intralumen drainage channels feed into highly structured lumens. The ducts of helodermatid glands terminate at the base of the thin, deeply grooved teeth, while those of the lanthanotid/varanid glands terminate between the large, bladelike teeth.

Consistent with the anguimorph venom-delivery system being less sophisticated than the high-pressure injection mechanism of the front-fanged snakes, a bite by most anguimorph lizards typically poses only trivial direct medical risks to humans, with the symptoms largely confined to localized pain and swelling, accompanied occasionally by persistent bleeding from the wounds. Recently, however, a death resulting from varanid lizard envenomation has been documented (Vikrant and Verma 2014). In the case of large varanids, the primary danger to humans is posed by the physical damage that may be caused by the large teeth. The potential danger for injury is greatest in the clade of large species with serrated teeth (*Varanus komodoensis, V. salvadorii*, and *V. varius*). Conversely, the effects of envenomation by helodermatid species has been better documented and shown to be clinically complex, with symptoms including extreme pain, acute local swelling, nausea, fever, faintness, myocardial infarction, tachycardia, hypotension, and inhibition of blood coagulation (Bogert and del Campo 1956;

Bouabboud and Kardassakis 1988; Hooker and Caravati 1994; Miller 1995; Strimple et al. 1997; Cantrell 2003).

1.2.3 DIVERSIFICATION OF THE SNAKE VENOM SYSTEM

The mandibular and maxillary glands within the "primitive" Henophidia snakes highlight the dynamic evolution and diversification within toxicoferan reptiles and snakes in particular. For example, the large mandibular and maxillary glands of the early-diverged *Cylindrophis ruffus* (Cylindrophiidae family) are seromucous, with a dominance of protein-secreting cells, significant drainage and evident lumen space (Fry et al. 2013). In contrast, the glands of the more recently diverged lineages characterized by secondarily evolved powerful constriction (the families Boidae [boid snakes] and Pythonidae [pythonid snakes]) possess mandibular and maxillary glands that histologically stain as predominantly (but not exclusively) mucoidal (Fry et al. 2013; see color plate 3 and 6). The differences in gland structures and types between cylindrophiids and boids/pythonids parallel the differences in their predatory ecology: *C. ruffus* is representative of the presumably plesiomorphic snake condition and is a weak "constrictor" that feeds on prey typically not wider than its head (eels, caecilians, other reptiles); while boid and pythonid snakes are powerful constrictors of massive prey items.

Thus, alternative selective pressures operate on the oral glands of these snakes: *C. ruffus* would benefit from chemical assistance in prey capture while the boid and pythonid snakes easily dispatch their prey using powerful constriction but need mucus to lubricate the fur of mammalian or the feathers of avian prey items. It is noteworthy that while *Aspidites melanocephalus* is unusual among pythonid snakes in that it feeds primarily on reptiles (slender prey items not dissimilar to those preyed on by *C. ruffus*), it is probable that its oral secretions are of little to no use in prey capture, as this species relies on constriction to subjugate prey. Despite this, however, it may be problematic for human bite victims as a result of its capacity to produce false positives in the medical diagnosis of snake envenomation.

A test of oral secretions from a wide range of Australian snakes demonstrated a curious cross-reactivity between oral secretions of the pythonid snake *A. melanocephalus* and the venom of the elapid snake *Notechis scutatus* in the Australian snake venom detection kit (sVDK) (Jelinek et al. 2004). Consistent with this false-positive reaction, it has been demonstrated that toxin genes are not only transcribed in the oral glands of *A. melanocephalus* but also continue to be translated and secreted as relics of the venom system that the boids/pythonids have presumably down-regulated (Fry et al. 2013).

In addition to the mandibular and maxillary glands discussed above, snakes are distinguished from the toxicoferan lizards in having additional protein-secreting glands called rictal glands. Until recently (Fry et al. 2013), rictal glands have received little attention except for a few detailed anatomical studies and some general investigations (Gabe and Saint Girons 1969; Kochva 1978; Saint Girons 1988; Saint Girons 1989; Wollberg, Kochva, and Underwood 1998). McDowell carried out a detailed study of the corner of the mouth of a large number of caenophidian snakes and described the rictal glands and muscles of the area together with a reinterpretation of the homology of the involved structures (McDowell 1986). Investigations have also been undertaken on the circumlabial oral glands of a series of Henophidia such as *C. ruffus* and *A. melanocephalus*, including the glands of the rictal region (Underwood 2002; Fry et al. 2013). *C. ruffus* has well-developed maxillary and mandibular glands along with superior and inferior rictal glands, all of which have protein-secreting cells as the dominant type. The lobules of the maxillary and mandibular glands are mixed with serous regions leading to mucous tubules that open into the oral cavity via short ducts along the margin of the lips. The *C. ruffus* superior and inferior rictal glands consist of branched serous tubules opening into mucous tubules that lead into wide mucous ducts opening directly into the corner of the mouth. The well-developed and protein-secreting rictal glands of *C. ruffus* suggest an adaptive role in the life of these serpents (Fry et al. 2013). Of particular note is the expression of the same toxin transcripts in maxillary, mandibular, and rictal glands. These data support not only the hypothesized shared evolutionary history of the mandibular and maxillary glands but also that the rictal glands may be derived from the same system (Fry et al. 2006; Fry et al. 2013). Supporting a potential role in predation, secretions from the rictal glands of *Eryx conicus* and six species of Uropeltidae snakes were shown to be toxic to birds (Phisalix and Caius 1918).

Similar to the mandibular gland segregation observed in *Heloderma* and *Lanthanothus/Varanus*, the maxillary gland in early caenophidian snakes evolved to become segregated into distinct, encapsulated protein-secreting and mucus glands. The significantly enlarged protein-secreting glands (see color plate 1) allowed increased production of venom, and thereby underpinned the extraordinary diversification of the caenophidian snakes in the Cenozoic era (Vidal 2002; Fry et al. 2003a; Fry and Wüster 2004; Fry et al. 2006; Vidal et al. 2007; Fry et al. 2008; Vidal and Hedges 2009; Pyron and Burbrink 2012). All variables have been significantly modified in the evolution of the venom system as a whole, including gland morphology, musculature, skull morphology, dentition, venom composition, and toxin biochemistry and functionality (see color plates 3, 5, 6, and 7). Because of the diversity in dentition and glands among the caenophidian snakes, these morphological characters have been studied extensively and guided much of the early phylogenetic groupings of the front-fanged snakes into elapid and viperid snakes, with the family Colubridae (colubrid snakes) initially being a taxonomical garbage can for the remaining non-front-fanged caenophidian snakes.

Since the end of the nineteenth century, many authors have tried to establish transformation series between the four recognized snake dentition types: aglyphous (nondifferentiated dentition or enlarged without grooving), opisthoglyphous (enlarged, grooved rear fang for venom delivery), proteroglyphous (hollow front fangs on minimally kinetic maxilla), and solenoglyphous (hollow front fangs on kinetic maxilla). The most commonly cited series was a progressive evolution of the dentition from aglyph to opisthoglyph to proteroglyph and then to solenoglyph. Although the solenoglyph type displayed was commonly believed to be the most sophisticated and therefore the most derived system, both a common origin (Cope 1900; Bogert 1943; Marx and Rabb 1972) and an independent evolution of the proteroglyph and solenoglyph systems (Boulenger 1893; Anthony 1955; Kardong 1980; Kardong 1982) had been inferred at various times during the study of these animals. Recent molecular phylogenetic studies (Vidal et al. 2000; Vidal 2002; Vidal and Hedges 2002; Kelly, Barker, and Villet 2003; Vidal and David 2004; Lawson et al. 2005; Vidal and Hedges 2005; Vidal et al. 2007; Vidal et al. 2008; Vidal and Hedges 2009; Vidal et al. 2009; Vidal, Dewynter, and Gower 2010; Burbrink and Crother 2011; Pyron et al. 2011; Pyron and Burbrink 2012; Pyron, Burbrink, and Wiens 2013) revealed that the non-front-fanged caenophidian snakes are not a single, basal monophyletic group as previously supposed but represent numerous independently evolved groups (see color plate 3). It was further revealed that hollow front fangs evolved on at least three separate occasions (see color plate 3). All elapid snakes are characterized by proteroglyphy, while all viperid snakes are characterized by solenoglyphy. In contrast, the *Atractaspis–Homoroselaps* clade within the Atractaspidinae subfamily of the Lamprophiidae family contains both proteroglyphs (*Homoroselaps* species) and solenoglyphs (*Atractaspis* species). Despite the close genetic affinity of *Atractaspis* and *Homoroselaps* their venom glands differ sharply in morphology.

Visualization of the tooth-forming epithelium in snake embryos using the Sonic hedgehog (Shh) gene as developmental marker revealed that all front fangs develop from the posterior end of the upper jaw, with the morphogenesis of rear fangs using the same genes (Vonk et al. 2008) (see color plate 7). In front-fanged snakes, the anterior part of the upper jaw lacks Shh expression, and ontogenetic allometry displaces the fang from its posterior developmental origin to its adult front position, consistent with a plesiomorphic posterior position of the front fang. In non-front-fanged caenophidian snakes, the fangs develop from an independent posterior dental lamina and retain their posterior position. In light of these findings, a major evolutionary transition underlying the massive radiation of caenophidian snakes was identified: the posterior subregion of the tooth-forming epithelium became developmentally uncoupled from the remaining dentition early in snake evolution. This allowed the posterior teeth to evolve independently and in close association with the venom gland and to become highly modified in different lineages.

The evolutionary decoupling of the venom system from remaining tooth development has given rise to an impressive diversity of maxillary dentition (Fry et al. 2008; Vonk et al. 2008; Young et al. 2011) (see color plates 5 and 7). The teeth are extremely variable, showing either: (i), a smooth surface, (ii) a surface with a shallow groove, (iii) a deep groove present on less than half the length of the tooth, (iv) a deep groove running the entire length of the tooth, and (v) a fully enclosed venom canal, as present in proteroglyphous and solenoglyphous caenophidian snakes (Young and Kardong 1991; Young and Kardong 1996; Jackson 2003; Fry et al. 2008; Young et al. 2011). Evolutionary "tinkering" has produced the impressive diversity of dentition among extant species, and the terms "aglyph" and "opisthoglyph" are meaningless in a phylogenetic context, as snakes with these dentition types do not represent monophyletic groups (Vidal 2002; Vidal and Hedges 2002; Fry et al. 2008). The plesiomorphic condition consists of unspecialized and undifferentiated solid teeth lacking grooves (Fry et al. 2008). In modern species in which the posterior teeth are not grooved (aglyphous) or enlarged in relation to other teeth, the presence of ridges on their anterior and posterior surfaces distinguishes them from the anterior teeth (Fry et al. 2008). Multiple lineages have independently evolved enlarged rear teeth (opisthodont) and in these clades the introduction of venom into a bite wound is often facilitated by open channels or grooves along the lateral or anterolateral surfaces of the fangs (opisthoglyphous) (Vidal 2002; Fry et al. 2008; Young et al. 2011). As such enlargement and grooving has occurred convergently on numerous occasions, it must be noted that opisthoglyphous snakes are not a natural group and thus this is not a useful taxonomical character.

Previously, the name "Duvernoy's gland" was given to the venom glands of non-front-fanged caenophidian snakes that lack compressor muscles and associated hollow fangs (Taub 1967). However, the distinction between "venom glands" in front-fanged snakes and "Duvernoy's glands" in non-front-fanged caenophidian snakes is arbitrary and is not supported by evolutionary or developmental research. All reptile venom gland types are derived from the dental glands and develop from a common primordium at the posterior end of the dental lamina (Kochva 1963a; Kochva 1963b; Kochva and Gans 1965; Kochva 1978; Wollberg, Kochva, and Underwood 1998; Vonk et al. 2008; Fry et al. 2012a; Fry et al. 2013). Thus, although the venom glands of caenophidian snakes exhibit variation in all aspects including structure and topography, with countless variants scattered across the phylogenetic tree (Fry et al. 2008; see color plates 3 and 6), they are all homologous in a developmental and anatomical sense. Consequently, in light of the phylogenetic knowledge described above, the term "Duvernoy's gland" has been abandoned, and the term "venom gland" is equally applicable to the toxin-secreting maxillary glands of all caenophidian snakes (Fry et al. 2003a; Fry and Wüster 2004; Fry et al. 2006; Fry et al. 2008; Fry et al. 2012a; Fry et al. 2013).

The secretory epithelium of snakes' venom glands consists of serous, protein-secreting cells with mucous-secreting cells found in some regions, primarily in the ducts. The secretion (venom) is stored in both cells and lumina to varying degrees in different lineages. The venom glands of viperid snakes, for instance, have large centralized tubular lumina in addition to a large central lumen and relatively few secretory granules in the cells. Elapid snakes, on the other hand, have small lumina, and much of the venom is stored intracellularly. The contours range from the plesiomorphic condition of large ovate ducts, to ducts with reduced diameters surrounded by extensive circular connective tissue, through to internal partitions of the venom duct. Vestibules may be absent, present adjacent to the venom gland, present at the fang sheath, or in contact with the oral cavity. Gland ducts may open directly into the oral cavity, near the tooth bases, or into a fang sheath, the last occurring in front-fanged snakes.

Venom gland compressor muscles that increase the efficiency and speed of venom delivery have evolved on at least six independent times in caenophidian snakes (Fry et al. 2008) (see color plates 3 and 7). Three evolutionary origins are represented by the non-front-fanged snake genera *Brachyophis, Dispholidus*, and *Gonionotophis capensis* (formerly *Mehelya capensis*). Species of these genera have superficial muscle fibers connected to the venom gland capsule that appear to constitute rudimentary compressor systems. *Brachyophis* have a purely serous gland (Taub 1967). In contrast, *Dispholidus* have a very intricate venom system: a large venom gland composed of branched tubules covered by a secretory epithelium; tubules opening into a wide duct lined with mucous cells, with the ducts and entire gland surrounded by circular connective tissue; vestibules present adjacent to the venom gland; and enlarged rear fangs with deep grooves running less than half the length of the fangs (Fry et al. 2008). *G. capensis* has ungrooved (aglyphous) teeth and a small venom gland of peculiar structure with a relatively wide lumen (Fry et al. 2008). Three additional evolutionary origins of compressor musculature are represented by the high-pressure front-fang venom-delivery systems (Fry et al. 2008) of the clade combining *Atractaspis* and *Homoroselaps* (Vidal et al. 2008), of elapid snakes and of viperid snakes (Vidal et al. 2008) in the elapid snakes and again in the viperid snakes (Vidal et al. 2007; Fry et al. 2008; Vidal and Hedges 2009). In front-fanged snakes, compression of the glands by the compressor muscle (in concert with other muscles) propels venom along the duct and into the fang sheath and the enclosed channel running through the shaft of the fangs (Young et al. 2001; Young and Zahn 2001; Young et al. 2004; Young and Kardong 2007; Fry et al. 2008).

Within the elapid and viperid snakes the venom system shows little intrafamilial diversity (Fry et al. 2008). The structure of elapid snake venom glands is a synapomorphy of the family. The venom is stored in the cells, with the central lumen being relatively small. This family also has an elongate accessory mucous gland surrounding the venom duct close to the main venom gland, the function of which is unclear (Vonk et al. 2013). There is a vestibule present adjacent to the fang sheath, and the venom duct opens only into the fang sheath. The complex tubular glands of viperid snakes are divided into several lobes that open into a large central lumen that stores considerable amounts of venom. The primary duct of the gland leads to a globular accessory gland, which then connects to the fang via a secondary duct (Kochva and Gans 1965; Kochva 1978; Kochva 1987). There are no mucous regions in either the elapid or viperid snake venom glands except for the accessory glands characteristic of each family.

The venom system of *Atractaspis* is structurally unique among snakes. The venom glands have a wide, elongate lumen surrounded by radially arranged secretory tubules that show some branching at their peripheral ends. Prominent mucoid regions occupy their luminal margins, and there are no separate accessory glands. Vestibules are present adjacent to the fang sheath, and the venom ducts open only into the fang sheath. Finally, it should be emphasized that the venom glands of *Homoroselaps* are of a pattern entirely different from that of both *Atractaspis* and the other species of the Atractaspidinae subfamily within the Lamprophiidae family. Instead, they convergently evolved to resemble those of the elapid snakes (Kochva, Wollberg, and Golani 1985; Underwood and Kochva 1993), which partially explains why this genus was previously assigned to the family Elapidae.

Spectacular elongation of the venom gland, thus increasing the relative venom production, has occurred independently on at least one occasion in each of the front-fanged snake clades: once in *Atractaspis* (the clade made up of *Atractaspis engaddensis, A. microlepidota, A. micropholis,* and *A. scortecci*), twice in the elapid snakes (once in the common ancestor of *Calliophis intestinalis* and *C. bivirgata* and then again in *Toxicocalamus longissimus*), and once in the viperid snakes (the common ancestor of *Causus resimus* and *C. rhombeatus*) (see color plates 3 and 6). In these morphologically intriguing species, the venom glands extend to a quarter of the body length or more. The biological advantage conferred by these elongations remains unknown. In the case of *Causus*, the change in gland length was not accompanied by a change in venom composition or the recruitment of new toxin types (Fry et al. 2008). Rather, these snakes have a venom composition typical of viperids in that the major toxin types are snake venom metalloprotease (SVMP) (see chapter 23), kallikrein (see chapter 14), and group IIA phospholipases A_2 (PLA_2) (see chapter 21). It remains to be investigated whether there are significant differences in venom composition between long- and short-glanded forms within the genera *Atractaspis, Calliophis,* or *Toxicocalamus.* Such investigations may shed light on the evolutionary advantage gained by these species, and it is noteworthy that three of the four known long-glanded genera are fossorial. It may be possible, at least in the cases of *Atractaspis, Calliophis,* and *Toxicocalamus,* that the long-glanded condition is an adaptation to a fossorial lifestyle that allows for a decrease in head size and girth while maintaining a high venom yield.

Evolutionary trends in the venom system also include secondary loss following a shift to a new prey-capture technique (constriction) or prey type ("defenseless" prey, such as eggs, worms, and snails) that does not require venom. Within the basal snakes, the convergent evolution of powerful constriction has resulted in a secondary loss of the venom system. Many caenophidian snakes have also evolved constriction as a new form of prey capture, resulting in loss or degeneration of the venom system: *Acrochordus* (family Acrochordidae) snakes also constrict, wrapping around the gills of fish prey to suffocate them (Lillywhite 1996), and lack venom glands altogether (Fry personal observations); the constricting Lampropeltini ("rat snakes" typified by *Pantherophis guttatus* [corn snake]) have greatly atrophied glands (Fry et al. 2008), as do some African lamprophiids, such as *Pseudaspis cana* (Taub 1967). The African colubrine snake *Dasypeltis scabra*, which feeds exclusively on bird eggs, also has greatly atrophied venom glands (Taub 1967). This "use it or lose it" evolutionary trajectory is paralleled in the egg-eating sea snakes *Aipysurus eydouxii, A. mosaicus* and *Emydocephalus annulatus*: subsequent to switching from feeding on fish to feeding exclusively on fish eggs, the venom glands of these species have atrophied, and significant reduction of the fangs has occurred (Voris and Voris 1983; McCarthy 1987; Gopalakrishnakone and Kochva 1990). Studies of *A. mosaicus* (*A. eydouxii* at the time of study) revealed these changes in the venom delivery system were accompanied by significant accumulation of deleterious mutations in toxins still transcribed, indicating that they are no longer subject to selection (Li, Fry, and Kini 2005a; Li, Fry, and Kini 2005b). Comparable reduction of the venom system has been observed in species of the Australian terrestrial elapid genus *Brachyurophis* that specialize in feeding on lizard eggs (Fry and Jackson, personal observations). The oral glands of the fossorial extremely derived Scolecophidia (Anomalepididae, Gerrhopilidae, Leptotyphlopidae, Typhlopidae, and Xenotyphlopidae) (Vidal et al. 2010) have not yet been specifically examined, but the morphological uniqueness and specialized diet (such as ant and termite eggs and larvae) of these snakes make it seem likely that they may have also undergone a similar reduction of the plesiomorphic venom system. Malacophagous (snail-eating) specialists (Asian members of the Pareatidae and some American Dipsadinae) have undergone a similar reduction in the maxillary venom system. Intriguingly, however, the infralabial/mandibular glands in the "goo-eating" dipsadines *Atractus reticulatus, Dipsas indica*, and *Sibynomorphus mikanii* secrete snail-specific toxins (de Oliveira et al. 2008) that may help in immobilizing their molluscan prey, to facilitate removal of the shells (Salomão and Laporta-Ferreira 1994). The protein families to which these toxins belong to remain to be elucidated.

1.3 VENOM PROTEINS

1.3.1 ORIGIN AND DIVERSIFICATION OF VENOM PROTEINS

Reptile venoms appear to have evolved via a process in which genes typically expressed in normal body tissues were also expressed in oral glands (Fry 2005; Hargreaves et al., 2014a; Reyes-Velasco et al., 2014; Junqueira de Azevado et al., 2014). While some are known salivary proteins, such as CRiSP or kallikrein, oral glands in squamate reptiles appear to be indiscriminate in their secretion of proteins normally restricted to one or only a handful of body tissue types, such as pancreas or liver (Fry 2005; Reyes-Velasco et al., 2014; Junqueira de Azevado et al., 2014). At the base of the toxicoferan reptile clade, the glands were enlarged to have discrete protein- and mucus-secreting regions. Levels of mRNA and protein secretion increased for some but not all protein types previously encoded for in the glands of nontoxicoferan squamates. The existing salivary proteins were the starting substrate for the evolution of reptile venoms, with CRiSP (see chapter 10) and kallikrein (see chapter 14) remaining the most highly expressed in many species (Fry 2005; Fry et al., 2008; Fry et al 2010b).

In some cases, increased salivary protein-encoding gene expression appears to be underlined by the duplication of these pre-existing genes, after which one duplicate evolves to become selectively expressed in the venom gland (restriction) and the other retains a multi-tissue expression profile (Fry 2005; Junqueira de Azevado et al., 2014; Reza et al. 2005; Vonk et al. 2013). In other cases, venom genes appear to have been derived from genes specifically expressed in tissue types (recruitment; Fry 2005). Genes encoding group II PLA_2 toxins for example, are selectively expressed in the venom glands of viperid snakes after having evolved from genes specifically expressed in the liver (Junqueira de Azevado et al., 2014; Reyes-Velasco et al., 2014). Typically, the types of proteins restricted or recruited to the venom gland are those involved in key regulatory processes or bioactivity.

Some authors contest that the multitissue expression profile of some genes related to venom toxins undermine the theory that venom in squamate reptiles had a single early origin at the base of the Toxicofera (Hargreaves et al. 2014a, 2014b). For example, gene expression profiles of salivary glands, scent glands and skin from the geckonid lizard *Eublepharis macularius*, the pythonid snake *Python regius*, the colubrid snake *Pantherophis guttatus*, the colubrid snake *Opheodrys aestivus* and the viperid snake *Echis coloratus* revealed co-expression of many toxin-related genes, and in some cases, at seemingly comparable expression levels to those observed in the venom gland of *Echis coloratus* (Hargreaves et al. 2014b). Unfortunately, interpretations of these results are severely limited by the taxonomic coverage of the species chosen, particularly since only one venom gland was sampled for comparative purposes. Moreover, (i) it is unclear which 'salivary glands' were isolated from the species sampled and, most importantly, (ii) it is apparent that tissue expression data was averaged between multiple samples taken from each tissue type, despite, in some cases, extensive variation in the expression level of genes between those samples. Finally, such studies ignore other robust lines of evidence supporting the toxicoferan venom system, such as the morphological enlargement of oral secreting glands and their switch to primarily protein secretion in many toxicoferans, and the proteomic identification of venom toxins in the oral secretions of previously characterized "nonvenomous" snakes and lizards, including evidence of accelerated duplication and diversification (Fry et al. 2003a, 2006, 2008, 2009b, 2010b, 2013).

Whilst further sampling of non-venom gland tissues will certainly be of critical importance for future analysis of the evolution and functional importance of venom toxins, it is important to note that the co-expression of toxin genes in other tissues does not inherently mean that these genes have no functional role in venom secretions. For example, many pathogenically important snake venom metalloproteinases were found co-expressed in the kidney and pancreas of the viper *Bothrops jararaca* (Junqueira de Azevado et al., 2014)—co-expression in such cases could be explained by either limiting expression to low levels in internal organs to prevent self-toxicity or through post-transcriptional mechanisms that suppress the translation of the co-expressed transcripts into proteins.

The specific regulatory processes governing venom gene restriction/recruitment remain to be elucidated as a result of a fundamental lack of understanding relating to the regulation of transcription and posttranscription of venom toxins (Casewell et al., 2014). However, the sequencing of the *Ophiophagus hannah* genome (Vonk et al. 2013) obtained some evidence for a similarity of genetic regulatory components of the venom system from a pancreatic origin (Vonk et al. 2013), and this may represent a common feature of highly secretory tissue types. Sequencing and analysis of micro RNA (miRNA) libraries made from a range of different tissues showed homologies between the king cobra venom gland and known human and mouse pancreatic miRNAs. The most abundant miRNA in the venom gland library was miR-375, a canonical miRNA in the vertebrate pancreas. Previously, it had been hypothesized that the snake venom gland evolved by evolutionary modification of the pancreatic system, although this hypothesis has since been abandoned because few of the toxins expressed in the venom gland, and none of the core toxicoferan toxins, appear to have evolved from pancreatic highly expressed proteins (Fry 2005). Nonetheless, these data suggest that some of the regulatory components associated with the venom system may be associated with the pancreas, even if the proteins (toxins) are not.

A core set of proteins were highly expressed the common ancestor of all Toxicofera, and following further toxin-amplification events, resulted in the complex venom protein arsenals observed in modern snakes and lizards (Fry et al. 2003a; Fry and Wüster 2004; Fry et al. 2006; Fry et al. 2009b; Fry et al. 2010a; Fry et al. 2010b; Fry et al. 2013) (see color plate 3). Proteins produced by venom glands (see table 1.1) are related to proteins that exhibit diverse normal-body activities (see table 1.2). The plesiotypic toxic activities vary widely (see table 1.3), and some toxin classes have diversified to facilitate numerous novel apotypic toxic activities (see table 1.4). A number of molecular frameworks expressed in the venom glands are known only from mRNA transcripts, and thus corresponding bioactivities remain to be elucidated for these classes (see table 1.5).

Table 1.1: Secretion Locations of Nearest Nontoxin Relations of Reptile Venom Proteins (Fry 2005)

Protein type [toxin class if known by specific name]	Major tissue type for normal secretion (not excluding lower levels of expression/secretion in other tissues)
3FP [3FTx]	Brain
ACN	Muscle
ADAM [SVMP]	Variety of tissues, including epididymis, colon, lung, lymph node, and Thymus
β-defensin [crotamine]	Brain
β-defensin [helofensin]	Highly expressed in skin and tonsils and to a lesser extent in trachea, uterus, kidney, thymus, adenoid, pharynx, and tongue; low expression in salivary gland, bone marrow, colon, stomach, polyp, and larynx; no expression in small intestine
C3 [CVF]	Liver
[Celestoxin]	No near matches to any characterized normal body peptide/protein
Cholecystokinin [cholecystoxin]	Expressed in brain, duodenum, and small intestine
CRiSP	Numerous exocrine tissues including salivary
Cystatin	Restricted to the stratum granulosum of normal skin, the stratum granulosum/spinosum of psoriatic skin, the secretory coils of exocrine sweat glands, with low expression levels also found in the nasal cavity
Endothelin [sarafotoxin]	Endothelium
[Exendin]	Nearest match unresolved
Factor V	Liver
Factor X	Liver
Ficolin [veficolin]	Peripheral blood leukocytes; also detected in spleen, lung, and thymus, possibly because of the presence of tissue macrophages or trapped blood in these tissues; not detected on lymphocytes and granulocytes
Hyaluronidase	Widely expressed
Kallikrein	Variety of exocrine tissues, including pancreas and salivary glands
Kunitz	Wide variety of tissues, including brain, conceptus membrane, lung, ovary, placenta, and uterus

(Continued)

Table 1.1 (*Continued*)

Protein type [toxin class if known by specific name]	Major tissue type for normal secretion (not excluding lower levels of expression/ secretion in other tissues)
LAAO	Variety of exocrine and immune tissues
Lectin	Widespread
Natriuretic peptide B-type	Brain
Natriuretic peptide C-type	Heart
NGF	Wide variety of tissues, including brain, eye, prostate, and salivary glands
Phosphodiesterase	Kidney
PLA_2 type IB	Pancreas
PLA_2 type IIA	Synovial fluid
PLA_2 type III	Expressed in kidney, heart, liver, and skeletal muscle; also present in placenta and peripheral blood leukocytes
Prokineticin [AVIT]	Expressed at high levels in testis and at lower levels in brain, lung, ovary, spleen, thymus, and uterus
Renin-like aspartic protease	Kidney
SPRY [vespryn]	Hemopoietic lineages
Veficolin	Peripheral blood leukocytes; also detected in spleen, lung, and thymus, possibly because of the presence of tissue macrophages or trapped blood in these tissues; not detected on lymphocytes and granulocytes
VEGF	Various tissues, ranging from brain to heart
WAP [waprin]	Variety of tissues, including lactating mammary gland, lung, ovary, and testis
[Waglerin]	Unknown; no homology to any known peptide
YY peptide [goannatyrotoxin]	Colon mucosa

3FP = three-finger peptide; 3FTx = three-finger toxin; ADAM = a disintegrin and metalloprotease; C3 = complement 3; CRiSP = cysteine-rich secretory protein; CVF = cobra venom factor; LAAO = L-amino oxidase; NGF = nerve growth factor; PLA_2 = phospholipase A_2; VEGF = vascular endothelin growth factor; SVMP = snake venom metalloprotease; WAP = whey acidic peptide.

While many of the protein types highly expressed at the base of the tree became major toxin types with extensive neofunctionalization, such as kallikrein (see chapter 14), not all transcripts homologous to known toxin transcripts occurring in the salivary glands of non-toxicoferan reptiles are expressed at the base of the toxicoferan clade. The recruitment of a protein through amplification of a copy of a gene has occurred at many points along the toxicoferan tree (see color plate 1). Sequencing this transcript at low levels in other lineages would provide further evidence of early expression followed by a latent period before amplification in a particular lineage. Consistent with this theory, such transcripts could be at such low levels that they may only be detected through highly sensitive techniques (e.g. qPCR or deep sequencing).

Table 1.2: Bioactivity of Nontoxin Ancestors of Reptile Venom Proteins (Fry 2005)

Protein type [toxin class if known by specific name]	Normal body function
3FP [3FTx]	Binds to the α7 nicotinic acetylcholine receptor
ACN	Rapidly hydrolyses choline released into the synapse, resulting in less neurotransmitter available for neuromuscular control
ADAM [SVMP]	Enzymatic cleavage of the extracellular matrix
β-defensin [crotamine]	Unknown
β-defensin [helofensin]	Antimicrobial but not hemolytic
C3 [CVF]	Central to both classical and alternative complement pathways
[Celestoxin]	No near matches to any characterized normal body peptide/protein
Cholecystokinin [cholecystoxin]	Hypotensive neuropeptide that binds cholecystokinin receptors
CRiSP	Specific actions largely uncharacterized
Cystatin	Inhibits cysteine proteases such as the cathepsins B, L, and S
Endothelin [sarafotoxin]	Potently vasoconstrictive, modulating the contraction of cardiac and smooth muscle
[Exendin]	Nearest match unresolved
Factor V	Blood cofactor that participates with factor Xa to activate prothrombin to thrombin
Factor X	Vitamin K-dependent glycoproteins that convert prothrombin to thrombin in the presence of factor Va, calcium, and phospholipid during blood clotting
Ficolin [veficolin]	Involved in serum-exerting lectin activity; binds GlcNAc
Hyaluronidase	Random hydrolysis of (1- > 4)-linkages between N-acetyl-beta-D-glucosamine and D-glucuronate residues in hyaluronate
Kallikrein	Releases kinins from circulatory kininogen
Kunitz	Inhibits a diverse array of serine proteinases
LAAO	Induces apoptosis in cells by two distinct mechanisms, one rapid and mediated by H_2O_2, the other delayed and mediated by deprivation of L-lysine
Lectin	Hemagglutination activity
Natriuretic peptide B-type	Produces hypotension mediated by the binding to GC-A with subsequent relaxation of vascular smooth muscle
Natriuretic peptide C-type	Produces hypotension mediated by the binding to GC-B with subsequent relaxation of vascular smooth muscle
NGF	Stimulates division and differentiation of sympathetic and embryonic sensory neurons
Phosphodiesterase	Cleaves a variety of phosphodiester and phosphosulfate bonds including deoxynucleotides, nucleotide sugars, and NAD
PLA_2 type IB	Releases arachidonic acid from the sn-2 position of the plasma membrane phospholipids

(*Continued*)

Table 1.2 (*Continued*)

Protein type [toxin class if known by specific name]	Normal body function
PLA_2 type IIA	Releases arachidonic acid from the sn-2 position of the plasma membrane phospholipids, involved in inflammatory processes and diseases, such as rheumatoid arthritis and asthma
PLA_2 type III	Catalyzes the calcium-dependent hydrolysis of the 2-acyl groups in 3-sn-phosphoglycerides; shows an 11-fold preference for phosphatidylglycerol over phosphatidylcholine (PC); preferential cleavage: 1-palmitoyl-2-linoleoyl-phosphatidylethanolamine (PE) > 1-palmitoyl-2-linoleoyl-PC > 1-palmitoyl-2-arachidonoyl-PC > 1-palmitoyl-2-arachidonoyl-PE; plays a role in ciliogenesis
Prokineticin [AVIT]	Constriction of intestinal smooth muscle
Renin-like aspartic protease	Renin is a highly specific endopeptidase, whose only known function is to generate angiotensin I from angiotensinogen in the plasma, initiating a cascade of reactions that produce an elevation of blood pressure and increased sodium retention by the kidney; cleavage of Leu-|-Xaa bond in angiotensinogen to generate angiotensin I
SPRY [vespryn]	Largely uncharacterized
Veficolin	Involved in serum-exerting lectin activity; binds GlcNAc
VEGF	Increases the permeability of the vascular bed
WAP [waprin]	Inhibits leukoproteases
[Waglerin]	Unknown; no homology to any known peptide
YY peptide [goannatyrotoxin]	Gut peptide inhibits exocrine pancreatic secretion, has a vasoconstriction action, and inhibits jejunal and colonic mobility

3FP = three-finger peptide; 3FTx = three-finger toxin; ADAM = a disintegrin and metalloprotease; C3 = complement 3; CRiSP = cysteine-rich secretory protein; CVF = cobra venom factor; LAAO = L-amino oxidase; NGF = nerve growth factor; PLA_2 = phospholipase A_2; VEGF = vascular endothelin growth factor; SVMP = snake venom metalloprotease; WAP = whey acidic peptide.

The toxin diversity inferred for the base of the toxicoferan radiation is particularly intriguing in regard to the enigmatic iguanian lizard glands. In vertebrate-feeding Iguanian species, the glands are enlarged and have obvious protein-secreting regions which are larger and more developed than the mucus secreting regions located proximal to the teeth (Fry et al. 2006; Fry et al. 2013). Protein types expressed include those that have been shown to have neurotoxic activities such as CRiSP (Brown and Carmony 1999; Yamazaki, Brown, and Morita 2002; Yamazaki et al. 2002; Wang et al. 2005a; Wang et al. 2006b) (see chapter 10), vespryn (Pung et al. 2005) (see chapter 24), and coagulopathic activities such as kunitz (see chapter 15), lectin (see chapter 17) (Yamazaki and Morita 2007; Yamazaki and Morita 2009), and likely veficolin (OmPraba et al. 2010). Sequencing of transcripts that are homologous with well-characterized toxin types provides evidence for the continued diversification of the secreted proteins within the vertebrate-feeding Iguanian lizards. There is, however, no known predatory or defensive function for the venom system within herbivorous or insectivorous species (Fry et al. 2013). However, despite being predominantly mucus secreting in the insectivorous and herbivorous specialists, they are still capable of secreting active peptides. In particular, β-defensin

Table 1.3: Basal Toxicity of Reptile Venom Proteins (Fry 2005)

Protein type [toxin class if known by specific name]	Basal toxic activity
3FP [3FTx]	α-neurotoxicity, antagonistically binding to the nicotinic acetylcholine receptor
ACN	Bioactivities uncharacterized
ADAM [SVMP]	Tissue necrosis
β-defensin [crotamine]	Significant neurotoxic activity, modifying voltage-sensitive Na+ channels, resulting in a potent analgesic effect, and myotoxic activities have been reported; which is basal and which is derived remain to be elucidated
β-defensin [helofensin]	Lethal toxin shown to have an inhibitory effect on direct electrical stimulation of the hemidiaphragm
C3 [CVF]	Unregulated activation of the complement cascade, causing rapid and significant problems such as anaphylactic-type problems and/or tissue damage via hemolysis/cytolysis
[Celestoxin]	Hypertension mediated by unknown pathway
Cholecystokinin [cholecystoxin]	Hypotension mediated by binding cholecystokinin receptors
CRiSP	Paralysis of peripheral smooth muscle and induction of hypothermia through blockage of various channels including ryanodine and L-type calcium channels
Cystatin	Inhibition of body defensive enzymes
Endothelin [sarafotoxin]	Potent vasoconstriction resulting in acute hypertension
[Exendin]	Hypotension
Factor V	Combines with toxic form of factor X to potently convert prothrombin to thrombin
Factor X	Potent conversion of prothrombin to thrombin in the presence of factor V (endogenous or venom forms), calcium, and phospholipid
Ficolin [veficolin]	Unknown
Hyaluronidase	Facilitates spread of other venom proteins
Kallikrein	Increase of vascular permeability and production of hypotension in addition to stimulation of inflammation
Kunitz	Inhibition of circulating serine proteinases
LAAO	Apoptosis
Lectin	Platelet aggregation mediated by galactose binding
Natriuretic peptide B-type	Potent induction of hypotension leading to loss of consciousness
Natriuretic peptide C-type	Potent induction of hypotension leading to loss of consciousness
NGF	Bioactivities uncharacterized
Phosphodiesterase	Inhibition of platelet aggregation
PLA_2 type IB	Presynaptic neurotoxicity is either the basal or is basal-derivative
PLA_2 type IIA	Lipase activity resulting in inflammation and tissue destruction
PLA_2 type III	Prevents platelet aggregation mediated through epinephrine pathway
Prokineticin [AVIT]	Potent constriction of intestinal smooth muscle, resulting in painful cramping, and induction of hyperalgesia

(*Continued*)

Table 1.3 (*Continued*)

Protein type [toxin class if known by specific name]	Basal toxic activity
SPRY [vespryn]	Induces hypolocomotion and hyperalgesia; unknown which, if either, is basal activity
Veficolin	Bioactivities uncharacterized
VEGF	Increase of the permeability of the vascular bed and binding of heparin; results in hypotension and shock
WAP [waprin]	Bioactivities uncharacterized other than antimicrobial
[Waglerin]	Elicits tachypnea, ocular proctosis, rapid collapse, and spasms in mice; the primary cause of death is respiratory failure; selectively blocks the epsilon subunit of muscle nicotinic acetylcholine receptor
YY peptide [goannatyrotoxin]	Potently hypotensive through relaxation of vascular smooth muscle

3FP = three-finger peptide; 3FTx = three-finger toxin; ADAM = a disintegrin and metalloprotease; C3 = complement 3; CRiSP = cysteine-rich secretory protein; CVF = cobra venom factor; LAAO = L-amino oxidase; NGF = nerve growth factor; PLA_2 = phospholipase A_2; VEGF = vascular endothelin growth factor; SVMP = snake venom metalloprotease; WAP = whey acidic peptide.

Table 1.4: Derived Toxicity of Reptile Venom Proteins (Fry 2005)

Protein type [toxin class if known by specific name]	Derived toxic activity
3FP [3FTx]	Basal α-neurotoxicity greatly potentiated by the deletion of the C2 and C3 ancestral cysteines; functional derivations include binding to the postsynaptic muscarinic acetylcholine receptors, presynaptic neurotoxic action on the L-type calcium channels, cytotoxic interactions, acetylcholinesterase inhibition, and others
ACN	None currently documented
ADAM [SVMP]	Prothrombin activation a basal derivation; in Viperidae venoms, proteolytic cleavage of C-terminal domains resulted in numerous other activities, including direct-acting fibrinolytic activity; liberated disintegrin domain inhibits platelets via GP IIb/IIIa integrin receptor
β-defensin [crotamine]	Significant neurotoxic activity, modifying voltage-sensitive Na+ channels, resulting in a potent analgesic effect, and myotoxic activities have been reported; which is basal and which is derived remain to be elucidated
β-defensin [helofensin]	None characterized to date
C3 [CVF]	None currently documented
[Celestoxin]	None currently documented
Cholecystokinin [cholecystoxin]	None currently documented
CRiSP	Blockage of cyclic nucleotide gated calcium channels
Cystatin	None currently documented

(*Continued*)

Table 1.4 (*Continued*)

Protein type [toxin class if known by specific name]	Derived toxic activity
Endothelin [sarafotoxin]	None currently documented
[Exendin]	Derived VIP-like form is more potently hypotensive and cardiotoxic than the ancestral glucagon-derived form
Factor V	None currently documented
Factor X	None currently documented
Ficolin [veficolin]	None currently documented
Hyaluronidase	None currently documented
Kallikrein	Derivations affect the blood, particularly targeting fibrinogen
Kunitz	Derivations include inhibition of plasmin and thrombin and the blockage of L-type calcium channels. Structural derivatives form part of neurotoxic complexes with PLA_2 molecules
LAAO	Derivations include hemorrhagic effects, not only by affecting platelet aggregation but also inhibiting blood factor IX
Lectin	Derivations include stimulation of platelet aggregation (binding GPVI, GPIb, GPIa/IIa or VWF), platelet aggregation inhibition (binding GPIb or GPIa/IIa), or anticoagulant actions by binding blood factors IX, X
Natriuretic peptide B-type	Derivatives contain newly evolved helokinestatin domains located upstream that inhibit bradykinin
Natriuretic peptide C-type	Mutants with C-terminal tail have hypotensive activity mediated by GC-A instead of GC-B receptor (elapid venoms and convergently also in *Cerastes* venom); upstream of the natriuretic-peptide-encoding domain, viperid venom forms contain multiple proline-rich bradykinin-potentiating peptides or brakykinin-inhibiting peptides (pit vipers); other forms have newly derived antiplatelet- (*Macrovipera*) or metalloprotease-inhibiting peptides upstream (*Echis*)
NGF	None currently documented
Phosphodiesterase	None currently documented
PLA_2 type IB	Deletion of pancreatic loop facilitates the derivation of a multiplicity of novel, nonenzymatic activities, including antiplatelet and presynaptic neurotoxicity; some derivatives are parts of neurotoxic complexes
PLA_2 type IIA	Derivations include neurotoxic and antiplatelet activity; some derivatives are parts of complexes
PLA_2 type III	None currently documented
Prokineticin [AVIT]	None currently documented
SPRY [vespryn]	None currently documented
Veficolin	None currently documented
VEGF	None currently documented
WAP [waprin]	None currently documented
[Waglerin]	None currently documented
YY peptide [goannatyrotoxin]	None currently documented

3FP = three-finger peptide; 3FTx = three-finger toxin; ADAM = a disintegrin and metalloprotease; C3 = complement 3; CRiSP = cysteine-rich secretory protein; CVF = cobra venom factor; LAAO = L-amino oxidase; NGF = nerve growth factor; PLA_2 = phospholipase A_2; VEGF = vascular endothelin growth factor; SVMP = snake venom metalloprotease; WAP = whey acidic peptide.

Table 1.5: Novel Venom Gland Secreted Proteins with Bioactivities That Remain to Be Characterized or Known Only from Transcripts

Molecular scaffold type	Species recovered from	Tissue type of ancestral protein	Bioactivity of ancestral protein	Representative references
Aminopeptidase	*Bitis gabonica, Gloydius brevicaudus, Echis coloratus, Echis carinatus sochureki*	Expressed in epithelial cells of the kidney, intestine, and respiratory tract; also found in the vascular endothelium, fibroblasts, granulocytes, and monocytes	Broad specificity peptidases, including regulation of blood pressure; release dipeptidases from the N-terminus of peptides such as angiotensin II	Ogawa et al. 2007; Casewell et al. 2009
Dipeptidyl-peptidase IV	*Bothrops jararaca, Gloydius blomhoffi, Lachesis muta, Pseudechis australis, Demansia vestigiata, Tropidechis carinatus*	Widespread expression in most tissues; highest activity identified in the kidney and lung	Regulatory protease; inactivation of bioactive peptides by liberation of dipeptides from the N-terminus	Cidade et al. 2006; Junqueira de Azevedo et al. 2006; Ogawa et al. 2006; St Pierre et al. 2007a
Epididymal secretory protein	*Liophis poecilogyrus, Varanus indicus, Varanus gouldii, Varanus komodoensis*	Epididymis; detected in the epithelial cells of the epididymal duct	Putative role in binding lipids and collagen	Fry et al. 2006; Fry et al. 2012b
Lipocalin	*Azemiops feae, Dispholidus typus, Rhabdophis tigrinus, Trimorphodon biscutatus*	Preferentially synthesized in nonproliferating cells	Preferentially binds long-chain unsaturated fatty; known allergen	Fry et al. 2012b
Lysosomal acid lipase	*Echis coloratus, Micrurus altirostris, Philodryas olfersii*	Widely expressed in the liver and fibroblasts; has since been identified in a variety of tissue and cell types, with the exception of erythrocytes	Degradation of cholesterol esters and triglycerides; modulator of intracellular cholesterol metabolism	Casewell et al. 2009; Correa-Netto et al. 2011
Phospholipase A_2 Type IIE	*Leioheterodon madagascarensis, Dispholidus typus*	Restricted to the brain, heart, lung, and placenta	Progression of inflammatory processes	Fry et al. 2012b

Table 1.5 (*Continued*)

Molecular scaffold type	Species recovered from	Tissue type of ancestral protein	Bioactivity of ancestral protein	Representative references
Phospholipase B	*Drysdalia coronoides, Pseudechis colletti, Crotalus adamanteus*	Detected in white blood cells (granulocytes and neutrophils), intestinal enterocytes, and the epidermis	Removal of fatty acids from both the sn-1 and sn-2 positions of phospholipids	Bernheimer et al. 1987; Chatrath et al. 2011; Rokyta et al. 2011
Renin-like aspartic protease	*Echis ocellatus, Echis jogeri*	Expression restricted to the kidney	Generation of angiotensin I from angiotensinogen; mediator of extracellular volume and vasoconstriction	Wagstaff and Harrison 2006
Ribonuclease	*Celestus warreni, Gerrhonotus infernalis, Liophis poecilogyrus, Psammophis mossambicus*	Expressed predominantly in the pancreas	Pyrimidine-specific C-preferring nuclease	Fry et al. 2010b; Fry et al. 2012b
Snake venom matrix metalloprotease	*Rhabdophis tigrinus, Thamnodynastes strigatus*	Macrophages and granulocytes	Proteolysis of the extracellular matrix	Ching et al., 2012; *Rhabdophis* UniProt B1Q2M9.
Vitelline membrane outer layer protein	*Enhydris polylepis*	Oviduct; component of the outer membrane of the vitelline layer of the egg	Function unknown	Fry et al. 2012b

(see chapter 9) and cystatin peptides (see chapter 24) exhibit the highest expression level and the greatest evidence of duplication and diversification within the insectivorous and herbivorous Iguanian lizards. All other transcript types appear to be expressed in lower levels and typically exist in only one or two isoforms. It is noteworthy that the plesiotypic activity of these two peptide types is antimicrobial (Shah and Bano 2009; Semple and Dorin 2012), an activity retained by at least some isoforms expressed in the venoms of highly toxic front-fanged snakes. It remains to be elucidated whether the β-defensin peptides of Iguanian lizards retain this plesiotypic activity or whether they may have apotypic novel functions (see chapter 9). Another proteinaceous toxin type present in the venom glands of the toxicoferan ancestor, L-amino acid oxidase, also has well-documented antimicrobial activity (Domingues Passero and de Sousa 2010) (see chapter 16). Combined, the same plesiotypic function of these different protein families and the apparent lack of a venom function of the glands in insectivorous/

herbivorous lizards open the possibility that the gland secretion in these lineages underwent an evolutionary reversal to serve an antimicrobial function.

The accumulation of novel secretory frameworks appears to have underpinned the rapid radiation of the other toxicoferan reptile lineages (anguimorph lizards and snakes). In particular, the venom system likely led to the explosive diversification of the caenophidian snakes (Vidal 2002; Fry et al. 2003a; Fry et al. 2003c; Fry and Wüster 2004; Fry et al. 2006; Vidal et al. 2007; Fry et al. 2008; Pyron and Burbrink 2012; Fry et al. 2013). The recovery of novel protein scaffolds from the glands of henophidian snakes, including an entirely new peptide class discovered in pythonid snakes (Fry et al. 2013), reinforces how little is known about the protein composition of reptile venoms or the evolutionary history of the glands. The peptides and proteins recovered so far are unlikely to give a comprehensive picture of the toxin diversity because of the relatively limited sampling conducted so far, which is likely to recover only the dominant forms. More extensive sampling will no doubt recover novel isoforms of types identified to date in addition to entirely new toxin classes. Of interest for future investigations is the secretion of the segregated mucous lobules in the independently apomorphic mandibular systems of lanthanotid/varanid and helodermatid lizards and also those associated with the maxillary system of snakes.

1.3.2 GENERAL CHARACTERISTICS OF VENOM PROTEINS

Despite their diverse origins, certain trends are evident in venom proteins: all are recruited from secretory protein families, many have a preexisting biochemical role in one or more short-term physiological processes, and toxins are recruited more frequently from body proteins with stable tertiary structures maintained by a high degree of cysteine cross-linking.

The precursor peptides of all known reptile venom toxins contain an N-terminal signal peptide required for transport of the precursor peptide to the endoplasmic reticulum and preparation for eventual exocytosis of the mature toxin. This signal peptide is excised during the formation of the functional protein. To date, there are no documented cases of reptile toxin precursors lacking an N-terminal signal peptide. Even if a nonsecretory body protein had a potentially useful toxic side effect, in order to be suitable for recruitment as a toxin, its gene would require the addition of a segment encoding a signal peptide. This addition could occur either through interlocus gene conversion (nonreciprocal recombination) and retrotransposition or through exon shuffling of a signal peptide protomodule (Patthy 2003).

The ancestral body proteins of many toxins have preexisting bioactivities that are highly conserved in a wide range of potential prey items. The long-term evolutionary conservation of certain fundamental biochemical activities and their ubiquitous importance in the Kingdom Animalia provide the necessary functional basis for an endophysiological (physiologically active within the body of the producing organism) protein in one animal to become an exophysiological (physiologically active outside the body of the producing organism) protein when introduced into another. This allows for immediately beneficial (to the producing organism) toxic activities; in a sense, endophysiological proteins are "exapted" for use as toxins.

Consequently, the activity of most toxins is based on one of the following generalized mechanisms: (1) structural damage caused by catalyzing the hydrolysis of a universally present substrate, such as hyaluronidase enzymes hydrolyzing the 1–4 linkages between N-acetyl-β-D-glucosamine and D-glucuronate residues of hyaluronate; (2) physiological imbalance or a short-term response caused by "mimicking" endogenous body proteins as if they were overexpressed, such as the hyperalgesia and intestinal cramps caused by AVIT toxins (see chapter 24) or blood pressure effects of kallikrein toxins destructively cleaving kinins from kininogen (see chapter 14); or (3) "mimicking" endogenous body proteins by acting as a competitive inhibitor to cause a physiological imbalance or disruption of an endophysiological response, such as the α-neurotoxic three-finger toxins (3FTxs) (see chapter 8). The effectiveness of the latter two mechanisms may be amplified if the toxin mimics the action of a pleiotropic protein that governs multiple downstream metabolic pathways or if the mimicked protein plays a central role in a complex protein cascade with feedback systems (such as thrombin-mimicking proteins affecting the blood coagulation process). Mimicking multiple endogenous proteins sharing similar structures and biochemical activities can attain similar multifunctionality. Kallikrein toxins, for example, mimic the different activities of the related kallikrein and thrombin, thereby lysing both kininogen (causing vasodilation) and fibrinogen (anticoagulation) (see chapter 14). As such, the toxicity of many venom proteins that interfere with hemostasis is enhanced by the fact that they are recruited from the same protein (super) families that constitute multiple components of the involved protein cascades.

The effects of each of the three scenarios described above allow the new protein type, now selectively expressed in the venom gland, to provide an immediately useful toxic bioactivity. While some toxins are recruited from endophysiological proteins that play roles in the same pathways with which the toxins themselves interfere, others are not. The snake venom 3FTxs, for example, are derivations of α-neuropeptides similar to Lynx1/SLUR (Ibanez-Tallon et al. 2002; Fry 2005) (see chapter 8). However, other neurotoxins are not necessarily structurally similar or phylogenetically related to the ligands they mimic. Many neurotoxins act according to the second or third mechanism described above: they either activate or deactivate ion channels or other receptors at neuromuscular junctions or in the central or peripheral nervous system (see color plate 2).

Endophysiological proteins involved in one or several short-term physiological processes are the most suitable candidates for recruitment as toxins. Since venoms are used for predation, competitor deterrence, or defense, they are only effective if the effect they produce has a rapid onset. Potentially useful effects of toxins include fast immobilization of prey, instantaneous inhibition of blood coagulation in a host, or quick induction of malaise or pain in a predator or competitor. For example, although the prokineticin ligand-receptor systems govern a large number of long-term physiological processes with potentially negative consequences for the victim (such as predator anorexogenia caused by frog skin defensive AVIT peptides), the toxicity of AVIT peptides (see chapter 24) is largely determined by their short-term (within minutes) physiological effects, which include powerful contraction of smooth muscle and hyperalgesia. Long-term effects on physiological processes, lethal as they might be (such as accumulation of endogenous waste products or secondary systemic processes after abnormal cellular depolarization), are unlikely to contribute much to the adaptive value of toxins since they are irrelevant to the basic ecological purposes of any given venom. This is not to say that some toxins do not have persistent effects, such as the long-term paralysis of spiders induced by the venom of parasitic wasps or, conversely, the long-term paralysis of prey animals induced by spider venoms. In both cases, the victim/prey is kept alive until, respectively, the parasite's eggs hatch and the spider consumes its prey. However, protein families that only fulfill strictly structural functions (such as keratins) or proteins that only govern time-consuming processes (such as cell growth or tissue differentiation factors) are unlikely to be recruited as toxins. For a similar reason, toxin targeting is restricted to body parts that are readily accessible via the bloodstream, which also guarantees an acute effect. Most toxins additionally have a specific extracellular substrate or site of activity, even when they also affect intracellular processes.

The shared cysteine content of toxins and related (homologous) nontoxic body proteins is often remarkably high and often constitutes evolutionarily conserved motifs. Although such motifs lie at the core of the highly conserved tertiary structure that characterizes toxin multigene families, they also underlie extensive neofunctionalization (Froy et al. 1999; Escoubas and Rash 2004; Zhu et al. 2004; Fry 2005). Functionally important toxin types are selected for through adaptive evolution involving explosive duplication and diversification, resulting in the creation of a venom-gland-specific multigene family. The likelihood of neofunctionalization is increased through random mutation, gene conversion, and unequal crossing over (Fry et al. 2003b). Further duplication of the toxin gene is likely to be selectively favored, since additional gene copies may increase the level of toxin expression in the gland, resulting in greater quantities of toxin being available and faster gland replenishment. The tandem duplication of gene copies provides a template, which progressively increases the chances of unequal crossing over and gene conversion. Differential mutations among paralogs within the toxin multigene family allow them to obtain varying levels of potency, new functions, or prey specificity (Kordis and Gubenek 2000; Fry et al. 2003b). The newly created toxin multigene families preserve the molecular scaffold of the plesiotypic protein but modify key surface functional residues to acquire numerous apotypic activities (de la Vega et al. 2003; Fry et al. 2003b; Mouhat et al. 2004; Fry 2005; Casewell et al. 2011a; Casewell et al. 2013; Fry et al. 2013). In addition to changes in functional residues, neofunctionalization may also be facilitated through changes in tertiary structure as a result of variations in the pattern and number of cysteines.

In globular enzymatic proteins, mutations are much more likely to interfere with correct posttranslational folding than they are in the case of extensively cysteine cross-linked proteins. A single mutation, such as the gain or loss of a proline or crucial charge-state residue, may be enough to prevent correct folding of globular proteins. Duplication and diversification of such mutated genes may be negatively selected for, since they could lead to a large proportion of nonfunctional or maladaptive proteins in the venom, increasing the energetic cost of venom production without adding to the fitness of the venom-producing animal. No new activities have been reported thus far for toxins based on globular enzymes such as hyaluronidase (see chapter 24.11), as opposed to disulfide-linked peptides that have been extensively neofunctionalized such as 3FTxs (see chapter 8).

Variation in venom profiles has been shown between species within the same genus (Fry et al. 2002; Fry et al. 2003c; Fry et al. 2005; Sanz et al. 2006; Calvete, Escolano, and Sanz 2007; Angulo et al. 2008; Fry et al. 2008; Gutiérrez et al. 2008; Lomonte et al. 2008; Tashima et al. 2008; Salazar et al. 2009; Wagstaff et al. 2009; Mackessy 2010) and between individuals within the same species, with the intraspecific differences found among different geographic localities (Daltry, Wüster, and Thorpe 1996; Forstner, Hilsenbeck, and Scudday 1997; Fry et al. 2002; Fry et al. 2003c; French et al. 2004; Boldrini-Franca et al. 2010; Castro et al. 2013), and between juveniles and adults (Mackessy 1988; Daltry, Wüster, and Thorpe 1996; Lopez-Lozano et al. 2002; Calvete et al. 2009a). Venom variation has even been reported between separate venom glands of a single snake individual (Johnson, Kardong, and Ownby 1987). Such taxonomic, geographic and ontogenetic variations have been linked to strong natural selection in response to differential prey species (Daltry, Wüster, and Thorpe 1996; Fry et al. 2003a; Fry et al. 2003b;

Fry et al. 2003c; Pawlak et al. 2006; Fry et al. 2008; Gibbs and Mackessy 2009; Fry et al. 2012a; Sunagar et al. 2012; Brust et al. 2013; Casewell et al. 2013; Sunagar et al. 2014).

1.3.3 VENOM PROTEIN DOMAIN UTILIZATION

Toxin diversity is also increased through gene domain mutations including selective domain expression, with deletion of other domains present in the plesiotypic gene; tandem repeats of a plesiotypic domain (encoding a single product); and newly evolved multiple domains (encoding for multiple peptides or proteins that are posttranslationally cleaved from each other).

Selective expression of particular domains from a plesiotypic multidomain precursor is especially well documented for snake venom metalloproteases (SVMP), which have mutated in such a way on numerous occasions (Casewell et al. 2011a; Brust et al. 2013) (see chapter 23). A different form of selective domain expression has occurred in the exendin peptides from helodermatid venoms (see chapter 11). A plesiotypic glucagon tridomain gene was mutated into a new monodomain gene encoding the exendin-3/4 peptide toxin type present in *Heloderma* venoms. This gene was further mutated to acquire a VIP-like sequence (vasoactive intestinal peptide) and targeting, leading to an increase in toxicity (Fry et al. 2010a).

In contrast to domain deletion, tandem repeats of a plesiotypic domain have been utilized in two very different manners. In the helofensin multigene family expressed in helodermatid lizard venom, the gene encodes for a single product constructed from four tandem repeats of a plesiotypic β-defensin domain, with the consequent emergence of a novel protein scaffold (Fry et al. 2010a) (see chapter 24). Another example is the multiple proteolytically liberated sarafotoxin peptides encoded for by single-gene tandem repeats of a plesiotypic endothelin domain (Borgheresi et al. 2001; Hayashi et al. 2004) (see chapter 22). The most extreme case of domain mutation is the conversion of a monofunctional plesiotypic gene encoding a single natriuretic peptide into a multifunctional gene encoding additional posttranslationally liberated peptides that are structurally unrelated to natriuretic peptides. Such secondary peptide-encoding regions have independently evolved and duplicated in the natriuretic peptide genes of multiple reptile lineages (see chapters 18 and 19). For example, they encode the helokinestatin peptides in helodermatid and anguid lizard venoms (Fry et al. 2010a; Fry et al. 2010b; Koludarov et al. 2012), the antiplatelet peptides from *Macrovipera lebetina* venom (Barbouche et al. 1998), the bradykinin-potentiating peptides (BPPs) in viperid snake venoms (Soares et al. 2005), and the metalloprotease-inhibiting tripeptides from *Echis* venoms (Wagstaff et al. 2008). In all of these examples, the peptides are proline-rich and posttranlationally cleaved from the precursor region that is anterior to the mature natriuretic peptide and is often extended by exon duplication. Investigation of gene intron/exon boundaries would be revealing with regard to the molecular diversification histories of variable domain utilizations.

1.3.4 VENOM PROTEIN SUBUNIT UTILIZATION

Another form of neofunctionalization is the utilization of toxins with multiple subunits, and there is tremendous diversity in the composition, number, and interaction of such subunits (see color plate 8 and table 1.6). The multidomain SVMP are unique in having multiple domain-chains encoded by the same gene. Multimeric toxins raise intriguing evolutionary

Table 1.6: Multimeric Snake Toxin Complexes

Functional type	Source	No. of subunits	Type of subunits and uniprot accession	Type of linkage	Specific activity
COLUBRIDAE					
Neurotoxin	*Boiga*	Homologous heterodimer	3FTx (α-ntx) A0S864, 3FTx (α-ntx) A0S865	Disulfide	α-neurotoxic
DIPSADIDAE					
Coagulopathic toxin	*Philodryas*	Homologous heterodimer	Lectin (Q09GK0), lectin (Q09GJ8)	Disulfide	Uncharacterized
ELAPIDAE					
Cardiotoxin	*Ophiophagus*	Homodimer	3FTx (Q69CK0)	Noncovalent	β-blocker
Coagulopathic toxin	Elapidae plesiotypic	Homodimer	Lectin (Q90WI6)	Likely noncovalent	Hemaggluti-nation
Coagulopathic toxin	Australian Elapidae plesiotypic	Monomer cleaved into two chains	fX (Q1L658)	Disulfide	Prothrombin activation
Coagulopathic toxin	*Hemachatus*	Homologous heterodimer	3FTx (hemextin sequences not reported)	Noncovalent	fVIIa inhibition
Coagulopathic toxin	*Naja*	Heterologous heterotrimer	SVMP (P0DJJ4), lectin (sequence not reported), lectin (sequence not reported)	Disulfide	Inhibition of the classical complement pathway
Coagulopathic toxin	*Pseudonaja* & *Oxyuranus*	Heterologous heterodimer	fX (Q58L96), fV (Q58L91)	Noncovalent	Prothrombin activation
Coagulopathic toxin	*Bungarus*	Heterologous heterodimer	3FTx (P00987), PLA_2 (P00617)	Disulfide	Acetylcholine release blocker
Coagulopathic toxin	*Bungarus*	Homodimer or homologous heterodimer	3FTx(α-ntx P15816)	Noncovalent	α3/β2, α4/β2 antagonist
Cytotoxin	*Naja*	Homodimer and oligomers	3FTx (P60301)	Noncovalent	Pore-forming cytotoxin
Neurotoxin	*Dendroaspis*	Homodimer	3FTx (P17696)	Noncovalent	α-ntx enhancer
Neurotoxin	*Micrurus*	Heterologous heterodimer	Kunitz (G9I929), PLA_2 (G9I930)	Noncovalent	ASIC1/ACCN2 agonist
Neurotoxin	*Naja*	Homodimer or homologous heterodimer	3FTx (α-ntx) P01391 or 3FTx (α-ntx P01391), 3FTx (cytotoxin P60305)	Disulfide	α7/β2 antagonist
Neurotoxin	*Ophiophagus*	Homodimer	3FTx (A8N286)	Noncovalent	α7 antagonist
Neurotoxin	*Oxyuranus*	Heterologous heterotrimer	3FTx (P0CJ35), kunitz (B7S4N9) PLA_2 (Q7LZG2)	Noncovalent	Cav1/CACNA1 blocker and KCa2/KCNN blocker

Table 1.6 (*Continued*)

Functional type	Source	No. of subunits	Type of subunits and uniprot accession	Type of linkage	Specific activity
Neurotoxin	*Pseudonaja*	Homologous heterohexamer	PLA_2 in four subunits: A (P23026), B (P23027), C (P30811), D (P23028), in combinations of (2A or 2B) + 2C + 2D or 1A + 1B + 1C + 2D	Noncovalent association except the 2D, which are covalently linked to each other	Presynaptic
Neurotoxin	*Pseudonaja*	Heterologous heterotrimer	PLA_2 α (P00614), β (P00615), γ (P00616)	Noncovalent	Presynaptic
VIPERIDAE					
Apoptosis	*Crotalus*	Homologous heterodimer	SVMP (Q90282), SVMP (A4PBQ9)	Noncovalent	Apoptosis without triggering cell detachment
Coagulopathic toxin	plesiotypic	Homologous heterodimer	Disintegrin (Q6T271, Q6T6T3)	Disulfide	Potent inhibitor of the adhesion of the RGD-dependent integrin α5/β1
Coagulopathic toxin	plesiotypic	Homologous heterodimer	Lectin α (P81508), β (P81509)	Disulfide	Extremely varied
Coagulopathic toxin	plesiotypic)	Homodimer	Plesiotypic lectin (Q9YGP1)	Disulfide	Hemaggluti-nation
Coagulopathic toxin	plesiotypic	Homologous heterodimer	Kallikrein (Q9PRM8)	Unknown	α and β fibrinolysis
Coagulopathic toxin	plesiotypic)	Heterologous heterotrimer	SVMP (P86536),lectin (Q4PRD1), lectin (Q4PRD2)	Disulfide	fX activation
Coagulopathic toxin	*Cerastes*	Homodimer or homologous heterodimer	Kallikrein (Q7LZF4)	Unknown	Platelet aggregation induction
Neurotoxin	plesiotypic	Homologous heterodimer	PLA_2 acidic (P18998) and basic (P62023)	Noncovalent	Blocks acetylcholine release

fX = factor X (see chapter 13); ntx = neurotoxin; Elapidae PLA_2 = Type I PLA_2 (see chapter 20); Viperidae PLA_2 = Type II PLA_2 (see chapter 21); uniprot = www.uniprot.org; for more information: 3FTx, see chapter 8; kallikrein, see chapter 14; kunitz, see chapter 15; lectin, see chapter 17; SVMP/disintegrin, see chapter 23.

questions regarding fundamental geometric combinations and neofunctionalization. For example, multiunit toxins may evolve at slower rates because of structural constraints, as has been noted for the dimeric κ-neurotoxic 3FTxs (see chapter 8) and the dimeric presynaptic neurotoxic PLA_2s that are responsible for the extreme neurotoxicity of certain *Crotalus* species (see chapter 21). Such questions remain largely unexplored and thus represent an exciting, virtually untapped area of research in the molecular evolution of toxins. Multimeric complexes may be selectively favored if they yield more efficient toxins or, alternatively, if multimerization improves their storage lifetime in the venom glands.

Subunits of a multimeric toxin may be homologous in that they are individual representatives of the same protein family (i.e. defined by the same protein domain) and may be either identical or closely related isoforms. Alternatively, they may be heterologous, in that they originate from two or more toxin classes (protein families defined by different domains). In addition, subunits may associate through covalent or noncovalent binding. While such complexes have been isolated from various snake families, elapid snake venoms are particularly diverse in this regard (see table 1.6). As with other areas of venom system evolution, there are areas of convergence. For example, it is intriguing that the majority of heterologous complexes from elapid snake venoms contain a phospholipase A_2 (PLA_2) subunit (see chapter 20) and have a neurotoxic action. In addition, kunitz-PLA_2 neurotoxic complexes have evolved on three separate occasions in elapid snake venom (β-bungarotoxin, MitTx, and taicatoxin) (see chapter 15).

Venom complexes have been particularly impactful in the evolutionary history of snake venoms. For example, a toxin complex unique to *Oxyuranus* and *Pseudonaja* snake venoms resulted from the recruitment of factor Va into the venom, an enzyme involved in blood coagulation (see chapter 12). As another blood coagulation enzyme, factor Xa, was recruited earlier, at the base of the Australian elapid snake radiation (Jackson et al. 2013) (see chapter 13), this allowed for the formation of a factor Xa/factor Va complex, resulting in a complex of devastating toxicity. The recruitment of factor Xa and factor Va would have greatly enhanced the feeding success of these snakes, as in mammals only a small amount of these very active enzymes is needed to cause severe disruption of blood chemistry. However, the activity of factor Xa toxins is rate-limited by the requirement for cofactors, particularly the essential need to form a 1:1 complex with factor Va. Thus, their potency and speed of action are greatly enhanced in venoms that contain both factor Xa and factor Va. The factor Xa/Va toxin complex in the venom of the *Oxyuranus/Pseudonaja* clade is an example of the impact that toxin-recruitment events can have on an animal's feeding ecology.

Subsequent to the recruitment of factor Va, the *Oxyuranus/Pseudonaja* common ancestor split into two forms with highly divergent feeding ecologies. The *Oxyuranus* species represent large snakes with long fangs and high venom yields. Unlike most Australian snakes, which often feed on reptiles (particularly as juveniles in the case of other large species), these snakes feed almost exclusively on mammalian prey throughout their lives, including large rats and bandicoots, dangerous prey items capable of potentially causing serious injury to predatory snakes. The snakes successfully feed on such dangerous prey by minimizing contact time through the use of a snap-release form of striking. A limitation of this prey-capture strategy may be that the snakes need to be very warm in order for this physiologically costly system (including high-speed movement and the rapid replenishment of large quantities of venom) to function optimally. Consequently, their distribution is limited to the tropical north of Australia and New Guinea (*Oxyuranus scutellatus*) and the baking heat of the channel country and central ranges in the Australian outback (*Oxyuranus microlepidotus* and *O. temporalis*, respectively).

In contrast, the *Pseudonaja* species have taken advantage of possessing such toxic venom and have greatly minimized the amount of biological energy expended in venom production. *Pseudonaja* species are typically smaller snakes with very short fangs and much smaller venom yields, and they use body coils to hold prey items in place while envenomating them. *Pseudonaja* species are typically generalists, unlike the mammal-specialist *Oxyuranus*, and may include large numbers of reptiles in their diet, particularly as juveniles. Not only does their prey-handling behavior differ from *Oxyuranus*, but it is likely that the short, stiff fangs of *Pseudonaja* are more suitable for getting between the osteoderms of scincid lizards than the long, flexible fangs of *Oxyuranus*, which are best suited for penetrating mammalian fur. *Pseudonaja* are far more widely distributed than *Oxyuranus*, being found over all of continental Australia (even in the extreme south) and also in New Guinea. These snakes have taken advantage of their extremely toxic venom to become one of the most adaptable and successful genera of snakes in the entire Australo-Papuan region.

Besides providing an adaptive benefit by increasing toxic activity, multimerization of protein domains may also improve the stability of a toxin, thereby facilitating its storage. One example is vipoxin, a neurotoxin found in the venom of the European viper species *Vipera ammodytes*. Vipoxin is a noncovalent heterodimer composed of two type IIA PLA_2 subunits. Subunit B potently catalyzes the cleavage of phospholipids (the standard PLA_2 activity; see chapters 20 and 21) and as a monomer is a strong presynaptic neurotoxin (Mancheva et al. 1987). Instead, subunit A lost phospholipase activity and shows no toxicity. Moreover, its noncovalent binding to subunit B causes the latter to shift from a pre- to a postsynaptic neural target (Tchorbanov et al. 1978) and leads to a five-fold decrease of its toxicity (Aleksiev and Tchorbanov 1976). Although this decrease seems counterintuitive from an adaptive evolutionary perspective, the heterodimer has an improved stability compared with the toxic monomer: while subunit B alone loses its catalytic activity after two weeks at 0° C, the vipoxin can be stored for years (Kang et al. 2011). The two subunits are related proteins and show 62% sequence similarity (Mancheva et al. 1987), indicating that the nonactive PLA_2 subunit diverged from a plesiotypic PLA_2 toxin after gene duplication. Together, these observations led to the hypothesis that the nonactive PLA_2 gene was retained to preserve the activity of the neurotoxin during long periods of storage in the venom gland (Kang et al. 2011).

1.3.5 VENOM PROTEIN EMERGENT PROPERTIES

Molecular evolutionary forces promote synergistic activity between toxins in the absence of complex forms. In many elapid snake venoms, presynaptic neurotoxic PLA_2s block the release of neurotransmitters (see chapter 20), while 3FTx derivatives simultaneously block the postsynaptic receptors (see chapter 8), thus preventing the docking of what little acetylcholine or muscarine is released, leading to a state of flaccid paralysis. Viperid snake venoms acting on the blood coagulation cascade may cause similar collective interactions: fibrinogen is cleaved by kallikrein (see chapter 14); platelets are inhibited by disintegrins (see chapter 23), lectins (see chapter 17), and PLA_2s (see chapter 21); and clotting factors are consumed by prothrombin activators (see chapter 23).

Dendroaspis envenomations produce amongst the most complex neurological effects of any venomous reptile because of the plethora of apotypic toxins in the venoms. Some of the bioactivities may at first consideration appear to be counteractive, such as pre-synaptic toxins that facilitate neurotransmitter release while post-synaptic toxins block the binding of neurotransmitters to their target receptors. However, the relative speed of onset must be considered. Like fish-eating cone snails, mambas prey upon species that may easily escape (e.g. birds and fast-moving arboreal mammals), thus the venoms have been shaped by this extreme evolutionary selection pressure. In both cases the presynaptic toxins rapidly produce excitatory paralysis which results in a rapid incapacitation of prey because of the uncontrollable muscle spasms. Mamba venoms include a second step characterized by flaccid paralysis caused by the postsynaptic-acting toxins that potentiate the effects of neurotransmitter depletion resulting from the presynaptic toxins.

The *Dendroaspis* toxins responsible for the two-step envenomation profile in each case represent a remarkable case of convergence for neofunctionalized targeting of the same receptor using two different peptide scaffolds as the starting substrate: 3FTx (see chapter 8.4.2) and kunitz peptides (see chapter 15.4). Presynaptic-acting 3FTx and kunitz peptides inhibit voltage-dependent potassium channels, resulting in the sustained neurotransmitter release and causing tetanic spasms. Parallel to this, other 3FTxs inhibit the ability of acetylcholinesterase to cleave free acetylcholine, resulting in muscle fasciculations. The slower-onset second step includes blockage of post-synaptic receptors to produce flaccid paralysis. In addition, another set of convergent 3FTx and kunitz peptides block L-type calcium channels to produce depression of cardiac contractility. The action of postsynaptic blockage of nicotinic acetylcholine receptors is uniquely potentiated in *Dendroaspis* venoms by other 3FTxs called synergistic toxins. By themselves, these peptides are virtually nontoxic. However, when bound to α-neurotoxins, they greatly potentiate the α-neurotoxic effect.

1.4 IMPLICATIONS OF THE NEW CONSIDERATIONS OF WHAT CONSTITUTES A VENOMOUS REPTILE

Our growing understanding of the pharmacological diversity of venoms and the morphological diversity of venom-delivery systems (glands, dentition, muscles, and associated structures) present in toxicoferan reptiles constitutes an open invitation to functional morphologists, physiologists, and toxinologists. There is now an unprecedented opportunity to investigate the function of this cornucopia of secretions and delivery systems in the life history of the animals. Furthermore, the realization that venom systems are far more widely distributed among reptiles than previously known greatly increases the research area for bioprospectors in search of novel and useful compounds for drug design and development and medical research. While the majority of these animals are of trivial "medical significance," in that their bites are unlikely to be dangerous to humans, this does not mean that they are "insignificant to medicine." It is precisely the single origin of these glands and their secretions that renders the evolution of the venom apparatus of toxicoferans such a potential treasure trove. The recent discoveries on the origins of reptile venom and associated structures (Vidal 2002; Fry et al. 2003a; Fry and

Wüster 2004; Vidal and Hedges 2004; Vidal and Hedges 2005; Fry et al. 2008; Vonk et al. 2008; Fry et al. 2009b; Vidal and Hedges 2009; Fry et al. 2010a; Fry et al. 2010b; Fry et al. 2013) have led to a paradigm shift in our understanding of the evolution of reptilian venoms.

The fact that we have a great deal left to learn about these animals is further highlighted by the recent discovery of new venomous snake species (see color plates 9 and 10). Part of these new species came forth from the discovery that the most medically important sea snake (*Enhydrina schistosa*) is not a single species but actually represents a remarkable case of convergence between two distantly related sea snakes for the same specialized niche (see color plate 9). Some of the newly discovered species are very large in size (such as the three-meter giant spitting cobra *Naja ashei* or the two and half meter Mangshan pitiviper *Protobothrops mangshanensis*) or likely to possess extremely potent venoms (the western taipan *Oxyuranus temporalis*), and thus their discovery is much more remarkable (see color plate 10). These new species may not only be of medical importance because of the potential that their venoms will be poorly neutralized by currently available antivenoms, but they may also yield novel compounds for use in drug design and development. The long history of reptile venoms in drug design underscores their biodiscovery value. In turn, this potential reinforces the fact that we must conserve all biodiversity as a valuable resource.

CHAPTER 2

SIGNS, SYMPTOMS, AND TREATMENT OF ENVENOMATION

L. BOYER, A. ALAGÓN, B. G. FRY, T. N. W. JACKSON, K. SUNAGAR, AND J.-P. CHIPPAUX

2.1 SUMMARY

A venomous reptile bite is a potentially life-threatening condition that disproportionately affects people who live at a considerable distance from tertiary hospital care. It has been estimated that as many as 1.8 million snakebites occur annually, mainly in tropical and subtropical areas of sub-Saharan Africa, South and Southeast Asia, and Latin America, and that these bites result in approximately 94,000 deaths (Kasturiratne et al. 2008). In Africa, approximately 95% of snakebites and 97% of deaths caused by snakebite occur in rural environments (Chippaux 2008). Permanent sequelae (functional loss and amputation) are more common than mortality following snakebite. The annual incidence of limb amputation in Africa alone is estimated as 5,900 to 14,600, with significant economic consequences for rural agriculture (Warrell et al. 1976; Chippaux 2008; Kasturiratne et al. 2008; Habib 2011).

Treatment of envenomation requires emergency intervention, specialized knowledge, and ready access to antivenoms, which are scarce in many parts of the world and with which most medical personnel are unfamiliar. This chapter is intended as an overview for the biologically knowledgeable reader whose occupational or recreational activities present some level of risk for venomous reptile bites or for hospital staff who seek general information in the context of situation-specific research for management of envenomation. It is not intended to substitute for medical toxinology consultation, which should always be considered when establishing local treatment standards or during emergency response.

2.2 POISON VERSUS VENOM: THE CLINICAL PERSPECTIVE

The basic tenets of clinical toxicology include the concepts of dose and of route of exposure. Acknowledgment of the importance of dose is classically attributed to Philippus Aureolus Theophrastus Bombastus von Hohenheim, more commonly known today as Paracelsus, who said that "all things are poison, and nothing is without poison; only the dose permits something not to be poisonous." In this broad sense, any chemical or mixture can function as a poison, regardless of its origin in biology, geology, or a laboratory. At a very low dilution or dose, a substance may exert no effect at all on an exposed individual, or it may serve a nutritional or perhaps therapeutic purpose; but as dose increases, there eventually comes a level at which harmful effects are observed. At this point, we call it a "poison." "Toxins," including those that make up snake venom, are biological poisons, distinguished from the "nonpoisonous" components of the animal by their capacity to cause harm at relatively low doses. Most venoms are toxic enough at undiluted concentration to cause immediate symptoms at the site of venom injection; many venom injections are of sufficient dose to cause injury to adjacent tissues or to bring about systemic consequences following absorption. Venoms may also contain enzymes that exert cumulative effects over time, in addition to toxins with classic dose-response effects.

Routes of exposure to poisons include oral (such as food poisoning), inhalational (such as carbon monoxide), ocular (such as pepper spray in the eye), parenteral injection (such as a heroin overdose), and cutaneous (such as a splash of acid on the skin). Venom toxins are produced by one creature but have specific targets in another; they are "endophysiological" (physiologically active within the producing animal) molecules that have been adapted for "exophysiological" (physiologically active outside the producing creature, such as within another creature) purposes. They tend to be injected in concentrated form into the dermis, subcutaneous tissues, or other anatomical structures of the recipient. In recognition of this specialized mechanism of toxin delivery, it is common to apply the term "venomous" to animals that possess specialized venom-delivery mechanisms (such as a stinger or modified dentition) to distinguish them from "poisonous" things in general (such as "If you bit it and you got sick, it was poisonous; if it bit you and you got sick, it was venomous"). Rather than refer to "venom poisoning"—although this term is technically correct—many clinical toxinologists use the word "envenomation" (or "envenoming") to describe the pathological processes that follow the injection of venom into a human or veterinary patient.

Knowledge of the potency, specificity, and relative abundance of individual venom components is important to the study of ecology and evolutionary biology and to discovery of novel pharmaceutical products. The venom of one single snake may be separable into more than a dozen component toxin groups, each of which may contain many molecular variants of varying target specificity and potency. These details, however, are never fully understood during management of an individual envenomation. Some toxins exert different effects on reptile, arthropod, or small mammal prey different from those on humans, and the proportion of one component versus another in the combined whole venom varies by orders of magnitude. Thus, the greatest clinical consequences of a bite may not, in many instances, be attributable to the most "dangerous-sounding" toxins present. The total quantity of venom delivered in a bite from a snake may also vary by several orders of magnitude, meaning that a range of clinical consequences from none at all (a "dry bite") to rapid fatality may result. Along this same spectrum, there may be specific clinical effects (such as paralysis, thrombocytopenia, etc.) that manifest when particular toxins are delivered at sufficient dose for effect.

Clinical management of this heterogeneous set of potential outcomes requires that healthcare providers consider each case initially in terms of broad generalities and that ongoing observation of local and systemic indicators serves as a guide to specific therapeutic decisions. A list of the clinical patterns of envenomation that appear to be most important to human health is given in table 2.1. Broad generalities underlie choices of antivenom administration also, including type and dose; these are given in table 2.2. Specific observations and interventions vary with time following the bite and may be divided into the phases of acute envenomation, distribution and clearance of venom, subacute and organ-specific disease development, and chronic recovery (see table 2.3).

Table 2.1: Patterns of Primary Toxicity or Secondary Injury following Reptile Envenomation

Toxicity or injury	Caused by	Diagnosis	Management
Interstitial matrix injury	Hyaluronidase (see chapter 24.11), metalloproteinases (see chapter 23)	Physical examination	Analgesics, position, antivenom
Local cytotoxicity	Cytotoxic 3FTx (see chapter 8), "digestive" phospholipases (see chapters 20 and 21), inappropriate early wound management	Physical examination	Analgesics, position, antivenom, physical or occupational therapy, wound management, later excision or amputation of necrotic parts
Compartment syndrome	Inflammation and hemorrhage within an anatomic compartment	Intracompartmental pressure monitoring (true compartment syndrome is very rare; physical exam may be misleading)	Antivenom, position, expert consultation with medical toxicologist and surgeon, rarely surgical decompression
Acute respiratory distress, wheezing (see also paralysis, below)	Allergy to snake, allergy to antivenom	Physical examination, blood gas measurements	Antihistamines, epinephrine, oxygen, ventilatory support as needed
Pain and early-phase vasodilatation with hypotension	Kallikrein (see chapter 14), bradykinins (see chapter 19)	Physical examination	Intravenous volume support, analgesics
Acute inflammatory response	Nonspecific acute-phase reaction	White blood count, platelet count	General supportive care

Table 2.1 (*Continued*)

Toxicity or injury	Caused by	Diagnosis	Management
Paralysis, with enzymatic nerve damage	Presynaptic (β) neurotoxins (e.g., crotoxin, neurotoxic PLA_2) (see chapters 20 and 21)	Physical examination, spirometry, blood gas measurement	Prevention with early antivenom, ventilatory support as needed
Paralysis, reversible	Postsynaptic (α) neurotoxins (see chapter 8)	Physical examination, spirometry, blood gas measurement	Prevention and treatment with antivenom, temporary early relief with a cholinesterase inhibitor such as physostigmine, ventilatory support as needed
Striated muscle damage, rhabdomyolysis	β-defensin peptides (see chapter 9), myotoxic PLA_2 (see chapters 20 and 21), rarely elevated muscle compartment pressure	Physical examination, CPK, urinalysis	Fluid and electrolyte management to prevent secondary renal injury, antivenom, analgesics
Thrombocytopenia or platelet dysfunction, spontaneous bleeding of mucosae, petechiae	RGD and non-RGD disintegrins, metalloproteinases (see chapter 23)	Physical examination, platelet count, platelet function studies, whole-blood clotting time	Antivenom, minimizing use of platelet inhibitors for analgesia
Hypofibrinogenemia, prolonged bleeding from injury	Kallikrein (see chapter 14)	Fibrinogen level, PT, aPTT, whole-blood clotting time, fibrin-degradation products, d-dimer	Antivenom, avoidance of injury and elective surgery
Complex coagulopathy, hemorrhage either in the bitten part or anatomically distant	Thrombocytopenia plus hypofibrinogenemia plus minor clotting factor inhibitors (see chapters 14 and 23)	Same as for thrombocytopenia and hypofibrinogenemia	Antivenom (blood products should be used only in very severe cases involving inadequate oxygen delivery or hemodynamic compromise; donor blood carries a risk of sensitization and of infectious disease, and it does not correct the underlying pathophysiology)

(*Continued*)

Table 2.1 (*Continued*)

Toxicity or injury	Caused by	Diagnosis	Management
Shock	Secondary effect of vasodilatation and/or volume loss from tissue injury or bleeding	Physical examination	Fluid support, occasionally pressors, antivenom if primary causes are ongoing
Renal, cardiac, and/or central nervous system ischemic injury	Secondary effect of shock, rhabdomyolysis, and/or hypoxemia	Organ-specific markers, imaging, EKG	Organ-specific supportive care, supportive management of proximate cause, antivenom if primary causes are ongoing
Wound infection	Opportunistic flora, poor wound management	Physical examination, culture, and sensitivities	Careful wound hygiene, broad-spectrum antibiotics

2.3 INITIAL ENVENOMATION

Reptile venom toxins suspected to be involved in local injury include hyaluronidase (see chapter 24.11) and metalloproteases (see chapter 23). In addition, many other venom toxins have biochemical activities that have demonstrable interaction with tissue targets. (See table 2.1 in this chapter and, for more information regarding specific toxins, relevant chapters elsewhere in this book.) In clinical practice, the particular mixture of toxins injected is seldom known, and it is likely that multiple mechanisms are involved simultaneously.

The outer layer of human skin, the epidermis, is an effective barrier against most environmental poisons. It is waterproof, flexible, and tough enough to withstand most incidental exposures, including a splash of snake venom, long enough to wash the chemical off before harm is done. But snake fangs are highly specialized devices, and venom delivered by snakebite readily bypasses the epidermis. The separation of tissues, vasculature, and nerves within dermis and subcutaneous structures and within muscle results in faster spread of toxins laterally than across tissue boundaries. Depending on the angle and depth of fang penetration, signs of envenomation and speed of systemic uptake may therefore vary. A shallow puncture, such as that delivered by a Gila monster or a snake with short fangs (or a shallow bite generally), may result in venom delivery directly into the dermis, which is a living layer of skin richly supplied by small blood and lymphatic vessels and containing sweat glands, hair follicles, and nerve endings. Most people are familiar with intradermal injections, from experience with "skin tests" (such as for tuberculosis). Initially, fang marks may be the only visible evidence of injury (see color plates 11A–C).

Minutes to hours after the event, as excess fluid is trapped in the dermis, the skin may stretch tight, or a blister may form (see color plate 11D). When blood leaks out of dermal vessels, the red color is visible because of its proximity to the surface. When dermal tissue dies, it

Table 2.2: Antivenom Characteristics of Importance in Clinical Use

Characteristic	Evaluation concept	Interpretation and limitations
Specificity and potency	Immunizing snake species	Qualitative predictor that the product will probably neutralize venom of the same and closely related species. In theory, a single-species (monospecific) product will have greater potency against that target, but this is not always the case (Chippaux, Williams, and White 1991; de Roodt et al. 2011). True single-target antivenom (monovalent, monoclonal) is not currently in commercial use.
	Potency bioassay with target snake venom (such as ED_{50} in mice)	Quantitative confirmation that the product, as manufactured, neutralizes a particular amount of venom of selected snake species (may be same as or different from immunizing species). Minimum potency standards are defined differently around the world and must be interpreted in the context of the amount of venom expected in a bite by the target species. The animal model used may be important for venoms that have disproportionate effects relative to those in humans, such as α-neurotoxins that bind cholinergic receptors in animal models but not in humans (Hart et al. 2013) (see chapter 8).
	Antivenomic cross-neutralization studies	Particularly useful for identifying gaps in neutralizing coverage of selected toxins; neutralization in this *in vitro* assay does not necessarily correlate with functional neutralization *in vivo*.
	Specific functional assays (such as hemorrhagic assay)	May correlate with neutralization of particular toxic effects of venom; interpretation depends on method and validation of the particular model.
	Affinity purification	May enrich the potency of a serum product against target venom; may also cause loss of antibodies with highest affinities. By itself not an indicator of final product potency or specificity.
	Clinical trials, efficacy	If adequately powered and well controlled, the "gold standard." Historically, most antivenom clinical trials have not met this standard (Chippaux, Stock, and Massougbodji 2010).
Purity and antigenicity	Plasma source (such as horse, goat, sheep)	For patients who are allergic to one animal, an alternative (if available) may be preferable.
	Immunoglobulin derivative: Fab, $F(ab')_2$, whole IgG	Whole IgG includes the Fc portion of the molecule, which is involved in pathophysiology of serum sickness. If well purified, Fab and $F(ab')_2$ antivenoms minimize this risk.

(*Continued*)

Table 2.2 (*Continued*)

Characteristic	Evaluation concept	Interpretation and limitations
	Total protein and albumin content	For a given neutralizing potency, the lower the total protein and albumin, the better. These are lower in products with (a) higher titers against venom in the original hyperimmune plasma and (b) cleaner separation of immunoglobulin from the rest of the plasma during manufacturing.
	Pyrogen and microbial assays	These are not always stated on package information, but regulatory authorities commonly require them as part of Good Manufacturing Practices (GMPs). Presence of pyrogens, an indication of microbial contaminants at some stage of manufacturing, may be the cause for rejection of a product batch.
	Clinical trials, safety	Most clinical trials report the rate of acute adverse reactions occurring at the time of antivenom infusion. Very few report on serum reactions after hospital discharge. Pharmacovigilance data (such as MedWatch in the USA) may add information after a product is approved for marketing.
Kinetic profile	Immunoglobulin derivative: Fab, $F(ab')_2$, whole IgG	The small Fab molecule is cleared faster than $F(ab')_2$, which is cleared faster than whole IgG. If antivenom clearance is faster than venom clearance, repeated dosing may be required in order to prevent recurrent venom toxicity.
	Clinical trials, pharmacokinetics	Very few antivenom products have been subjected to kinetic studies. In the absence of product-specific data, it is reasonable to consider comparators of the same molecular type, such as Fab, $F(ab')_2$, whole IgG).
Stability	Liquid or lyophilized product	Typically, liquid products require storage at low temperature, and "cold chain" problems are an issue particularly in tropical climates. Shelf life is longer for lyophilized products. Also, liquid products require greater amounts of preservatives (such as thiomersal, phenol derivatives).
	Preservatives	Preservative chemicals are used to prevent microbial and chemical degradation of the product, but when large doses, particularly of liquid products, are administered, the toxicity of the preservative itself may be of concern.
	Storage conditions	Prolonged exposure to heat and humidity or intermittent freeze/thaw cycling shortens the effective life of the product.
	Quality of seal and cake, residual humidity, excipients	These are signs of quality of freeze-drying that are not always described in product information. A well-lyophilized product in a well-sealed container lasts longer and reconstitutes faster.

Table 2.2 (*Continued*)

Characteristic	Evaluation concept	Interpretation and limitations
	Manufacturer's stated expiration date	Potency of antivenom decreases over time. For those manufacturers adhering to GMPs and subject to regulation by strict government standards, the expiration date reflects the period of time during which data exist showing that the minimum potency is retained. Expiration dates are sometimes extended if later testing shows that the product continues to retain its potency.
Reliability	Good Manufacturing Practices (GMPs)	GMP standards are variably enforced, depending on the regulatory body overseeing the affected factory. Under strictest standards, these involve audit and inspection of manufacturing methods and quality-control measures that ensure the batch-to-batch quality and consistency of antivenom.
	Regulatory approval status	The meaning of "approval" varies depending on the country issuing it. Most countries lack the infrastructure to assess and enforce GMP standards and clinical trial quality. The United States has a two-tiered process consisting of a strictly regulated investigational phase that must be completed before licensure for marketing is granted. Many developing countries allow antivenom sales with little or no proof of safety and efficacy.
	Expert endorsement	Particularly in settings where regulatory infrastructure is lacking, scientific and medical consultative expertise is crucial for selection of appropriate products, and supplemental local research may be useful. Recognized experts may be identified through professional toxinology organizations or via academic centers with established research in envenomation management. Resolution of conflicts of interest, review by properly constituted local ethics committees, and documentation of GMP compliance are essential considerations during this process.
	Source of purchased product	Depending on whether a product is purchased directly from the manufacturer, through a licensed wholesaler, via the Internet, or on the black market, what is actually in the vial does not always reflect what the label says. The high price of many antivenoms attracts many frauds. Caveat emptor.

Table 2.3: Observations and Interventions in Snakebite Management

Phase	Process	Intervention
Prevention and preparation	Local and global epidemiology	The distribution and accessibility of emergency medical facilities and of region-specific antivenoms should be anticipated by appropriate epidemiological studies (Chippaux 2008). Logistical and economical optimization of antivenom distribution may require collaboration among government, manufacturing, and distribution networks.
	Antivenom preparation	Where economically feasible, antivenoms specifically targeted to local needs should be locally tested for safety and efficacy. Where this is not practical, local health authorities should seek expert consultation before purchase or collaborative development of appropriate antivenom products.
	Minimizing risk of snakebite	Expertise in public health and herpetology should be applied to public education and to training of captive collections managers, with a view toward minimization of risk for humans in potential contact with snakes. Region-specific approaches must take into consideration the variety of endemic snake species along with local occupational risks and cultural practices.
	Emergency preparedness	Hospitals and emergency personnel should be aware of local herpetofauna and should ensure that pharmacies are stocked with appropriate antivenoms. Keepers of exotic reptiles should establish affiliations with local medical experts and together should ensure local stocking of exotic antivenoms as needed. First aid, communication, and transport guidelines should be established and understood. Ethics committee approval and regulatory documentation should be in place in advance for use of experimental therapy.
Acute (first 24 hours after a bite)	First aid	In regions (or captive collections) where pressure immobilization is indicated to slow uptake of neurotoxins/hemotoxins, appropriate measures should be employed. In other cases, most first aid interventions should be avoided, in favor of speedy transport to the site of definitive care. Cryotherapy, electric shock, incision, suction, and tourniquet use are contraindicated. *Above all, do no harm.*
	Communication	Telephone notification in advance of the patient's arrival at the hospital may enable faster administration of antivenom. Contact with poison control centers facilitates the mobilization of medical toxicology expertise and/or the search for antivenom if local supplies are not available. The physician responsible for experimental antivenom use should be located as quickly as possible. Photos of the snake may be sent to poison centers or to herpetologists for identification; the animal itself should be either left alone or removed by experts, depending on circumstances.

Table 2.3 (*Continued*)

Phase	Process	Intervention
	Transport	Emergency medical transport via ambulance or air should be considered, whenever it is available and expedient, to the nearest facility able to administer antivenom. Intravenous access should be initiated before or during transport, using either normal saline or lactated Ringer's solution. In regions where α-neurotoxins are endemic, use of cholinesterase inhibitors such as physostigmine may postpone paralysis. If emergency transport is not available, private transport should commence without delay, with a driver other than the bitten individual. *Time is tissue.*
	Medical assessment	The major purposes of initial medical assessment are to stabilize vital functions and to determine need for antivenom. Initial management of any shock and ventilatory impairment should be immediately followed by careful assessment of the local injury with attention to the margins and any lymphangitic streaking or tenderness. In regions where rapid antigen testing is important for antivenom selection, this should be done immediately. In unknown cases or those where coagulopathy is a concern, blood should be tested immediately for indicators of platelet and fibrinogen adequacy. Indicators of early neurotoxicity (such as ptosis, fasciculations) should be noted. Any history of atopy or allergy to antivenom components should be noted.
	General treatment	General supportive care in the acute phase of envenomation includes tetanus prophylaxis, fluid support, analgesia as needed (such as with opiates), and positioning of the limb to minimize restriction of fluid movement. Preventive measures against the possibility of early adverse events to antivenom may be taken in cases of atopy, allergy, or high-risk product formulation.
	Antivenom	Antivenom should be administered as quickly as possible, repeatedly if necessary, in cases where it is desirable (1) to halt progressive local spread of injury during the first 24 hours after a bite, (2) to reverse medically significant coagulopathy or other systemic effects, or (3) to prevent prolonged paralysis after a bite by a snake with venom expected to contain neurotoxins. Physicians who are unfamiliar with or inexperienced in the treatment of snakebite should consult medical toxicology specialists, by phone if necessary, before using antivenom, to ensure selection of the proper product and for assistance with prevention or management of acute reactions. Whenever feasible, antivenom should be administered intravenously, diluted and at a rate commensurate with product-specific guidelines determined from clinical trials.

(*Continued*)

Table 2.3 (*Continued*)

Phase	Process	Intervention
Subacute (two days to three weeks after a bite)	Assessment	Assessment 24 hours after snakebite is primarily for minimization of secondary injury, although ongoing direct venom effects sometimes require additional attention, particularly in severe cases with coagulopathy. Repeated assessments of coagulation and hemoglobin/hematocrit are useful during this phase. Observation of renal, cardiac, neurological, and gastrointestinal indicators may reveal sequelae of shock or of hemorrhage. Paralysis from either α- or β-neurotoxins may take days to weeks to resolve. Ongoing evolution of the local tissue injury may take the form of redistribution of edema and subcutaneous hemorrhage, often in a lymphatic pattern; this does not imply ongoing venom effect so much as the secondary effects of already-injured matrix and vasculature.
	General treatment	Supportive care during the subacute phase may involve ongoing management of fluids and electrolytes along with maintenance of ventilatory effort. In cases where local injury has involved cytotoxic effects in the region of a joint, physical and occupational therapy should be initiated for edema reduction and active and passive range-of-motion exercises. Openly necrotizing lesions should be observed closely and gently managed with the minimum necessary debridement, to allow for local healing so as to minimize any eventual surgical loss.
	Antivenom	Short-half-life products (Fab in particular) require repeated dosing in order to maintain neutralization of toxins during the absorption and distribution phases of envenomation. Supplemental dosing of $F(ab')_2$ and whole-IgG antivenoms is sometimes indicated, where systemic effects such as coagulopathy are of ongoing medical concern. In cases where patients present more than 24 hours after a bite for treatment and have previously received no antivenom, there is little evidence to support the use of antivenom for local injury, but medically significant coagulopathy or neurological manifestations may warrant a trial of antivenom regardless.
Chronic (one to 12 months after a bite)	Assessment for disability	It is common that envenomed patients do not seek medical follow-up after the subacute phase has resolved. Ongoing medical problems may be present, however, including late-phase resolution of local tissue injury requiring surgical intervention, loss of function from contractures affecting movement of fingers or limbs, or chronic intermittent edema, particularly in lower-extremity bites.
	Physical and occupational therapy	Physical therapy may be necessary for restoration of power and range of motion at the ankle or knee. Occupational therapy may help with recovery of hand and digit function. A compression stocking may help with dependent edema during daytime activities, when elevation is impractical.

Table 2.3 (*Continued*)

Phase	Process	Intervention
	Wound management	Chronic open wounds near the bite site are a significant concern, particularly where snakebite folk remedies may encourage infection or otherwise complicate primary wound healing. In such cases, careful patient education and follow-up with a skilled wound team familiar with tropical parasitology or with diabetes care may reduce the severity of late tissue and limb loss.

can slough off as visible necrosis (see color plates 11F–H). Blisters form when the integrity of the dermis is damaged by venom just below the skin's surface (see color plates 11D and 12C). Clear or serous blisters contain lymph; darker, serosanguinous ones are made up of lymph into which blood has been mixed following damage to adjacent vessels. Hemorrhagic bullae do not necessarily mean that underlying tissue is necrotic, as hemorrhage is commonly present in coagulopathic snakebite regardless of tissue damage. Occasionally, bullae on fingers may spread circumferentially, and a tourniquet-like effect may be observed on distal structures. Ischemia of the fingertip is an indication for decompression of the responsible bulla. SVMPs are responsible for many of these effects, and the concentration of such toxins is high in the venoms of most vipers, particularly pit vipers, and of *Atractaspis* species (see chapter 23).

Toxins deposited into the dermis may bind to tissue and remain near the site of injection, may spread among the cells locally, or may be taken into circulation via blood or lymphatic vessels, depending on the quantity injected, chemical characteristics of the particular venom, and how much damage has been rendered to the local vasculature. These phenomena spread laterally within the dermis, and related processes may extend into the deeper, subcutaneous tissues, resulting in a combination of erythema and edema following the combined distal and proximal spread (see color plate 11D 11I, and 11J). Deeper muscle tissues are commonly spared from the effects of intradermal or subcutaneous venom injection. An occasional result is that venoms containing myotoxins of significant *in vitro* or small animal *in vivo* potency may fail to provoke muscle injury in large human victims.

The layer beneath the epidermis and the dermis is subcutaneous (or subdermal) tissue. Depending on anatomic placement, this layer may be thin or thick and may contain abundant fat in addition to intracellular matrix, blood and lymphatic vessels, and nerves (see figure 2.1). The margin between dermis and subcutaneous tissue may be indistinct, with various structures such as hair follicles, vessels, and nerves spanning the boundary. In most snakebites, fangs penetrate into the subcutaneous tissues, releasing venom predominantly into fat and the local tissue matrix. Occasionally, a vessel or nerve may be mechanically damaged by the action of a fang. Similarly to intradermal injection, subcutaneous injection may result in local injury or uptake into blood and lymph. Hemorrhage into subcutaneous tissue is perceived by the eye as a "bruise," often darker in color than dermal bleeding, and edema in this layer may cause massive swelling but at relatively low pressure, because there are few boundaries within this layer to stop the accumulation of fluid (see color plates 11E, 11I, and 11J). Color changes over time reflect the breakdown and clearance of hemoglobin in a process similar to that of a simple

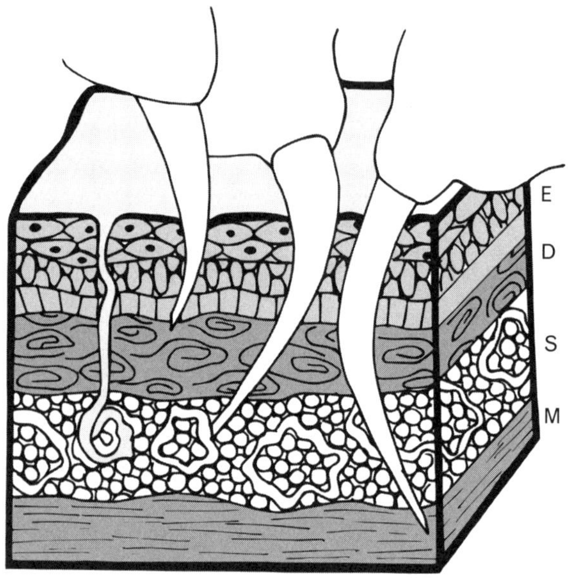

FIGURE 2.1: Fang penetration into different skin and soft tissues. E = epidermis, D = dermis, S = subcutaneous tissue, M = muscle. (Illustration: Genevie Jackson.)

bruise (see color plates 11E, 11I, and 11J); this appearance is commonly mistaken for necrosis, because the dark color appears to increase as overlying edema improves in the days following envenomation (see color plates 11E, 11I, and 11J). SVMPs (see chapter 23) are responsible for many of these effects, and the concentration of such toxins is high in the venoms of most vipers, particularly pit vipers, and of *Atractaspis* species. Local effects in subcutaneous tissue tend to spread laterally rather than to underlying muscle and bone, because tissue planes and vascular anatomy favor movement within the tissue matrix rather than across tissue boundaries. The spread of edema and hemorrhage within the subcutaneous tissues is subject to external physical forces, including gravity, constrictive clothing or jewelry, and compression bandages. True necrosis may occur following a pattern of prominent limb edema with a discolored central zone that continues to necrose, as evidenced by surface blistering and spontaneous sloughing of tissue (see color plates 11F–H). Cytotoxic 3FTxs are responsible for these effects in bites from certain African cobras, such as *Naja nigricollis* (see chapter 8.4.3).

Rarely, when circumstances of fang and human anatomy coincide to allow it, a fang may penetrate past the subcutaneous tissues and into a muscle, joint, or other specialized structure (see figure 2.1). In such cases, the effects of venom are specific to the particular surroundings, and edema and hemorrhage have site-specific consequences. A tough fascial capsule surrounds many muscles (the most familiar example of such fascia may be found in the kitchen, as the surface layer on a chicken breast). This fascial capsule tends to contain the movement of toxins and of fluid, such that edema within a muscle compartment does not produce high-volume swelling the way subcutaneous edema does. Because expansion is limited, pressure in an injured muscle compartment may, rarely, rise to dangerous levels, compressing associated blood vessels and nerves, with consequent injury from loss of blood and restriction of oxygen flow. Hemorrhage and edema within a deep muscle compartment are generally not visible on superficial examination, but signs of injury may be recognized as anatomically specific tenderness and loss of motor nerve function. Compartmental pressure measurement is essential if compartment syndrome is suspected following snakebite, because tenderness and decreased motor activity are more commonly caused by subcutaneous envenomation, and these may closely mimic compartment syndrome on simple physical examination. Fasciotomies should only be undertaken after diagnosis through the measurement of intracompartmental pressure. Massive subdermal external swelling is unrelated to compartment pressure and thus is a poor indicator for surgery.

Venom in the eye is a special case that has been described from surface exposure to aerosolized venom. In this situation, the portion of the eye's anatomy that is most immediately and profoundly affected is the conjunctival surface, both the blepharal (lining of the lid) and orbital (coating of the globe itself) regions. Conjunctival edema may be accompanied by intense vasodilatation and is apparent from an anatomic pattern that spares the central zone of the cornea itself. Corneal injury is less likely but of potentially major concern, because proteolytic toxins have the potential to cause both toxic and inflammatory injury to the cornea, with subsequent (potentially permanent) impairment of vision. Immediate treatment of ocular exposure to venom consists of flooding the eye with running water for 15 minutes, followed by ophthalmologic examination for corneal ulceration or inflammation. While some specialized snakes are notorious for their "spitting" behavior (although "squirting" is a better description), the venom of any snake may affect the tissues of the eye if it somehow comes in contact with them (see color plates 12A, 12B, 13). The venoms of the African spitting cobras are particularly rich in cytotoxic 3FTxs (see chapter 8.4.3), which are responsible for the clinical effects resulting from this defensive use of venom.

2.4 UPTAKE AND DISTRIBUTION OF VENOM FOLLOWING ENVENOMATION

Injected venom disperses away from the injection site by a variety of mechanisms that are a function of venom biochemistry and site anatomy, and also of deliberate or inadvertent secondary processes such as limb position, muscle activity, and first aid activities. Snake venom becomes detectable in blood and in lymph within minutes of experimental injections, and the spread of obvious injury (edema, erythema, hemorrhage, tenderness) typically occurs in bitten patients at a pace much more rapid than is seen with inflammation from infection or simple injection trauma.

In dermal and subcutaneous tissue, the interstitial space through which venom spreads consists basically of two phases: the extracellular matrix, which is a macromolecular scaffold made up of collagen, elastin, proteoglycan, and glycoproteins; and lymph, which is the fluid that circulates among the tissue structures and exits via the lymphatic vasculature. The initial lymphatic vasculature is essentially open to the interstitial space and is made up of endothelial cells with central anchoring filaments attached to the surrounding matrix. The unattached edges of these cells separate when the matrix moves or expands, and they seal together when tissue is compressed, effectively closing off a lumen and pumping its contents forward into larger and more proximal lymphatic vessels. Initial lymphatic vessels merge into collecting ducts, which combine into larger ducts responsible for drainage of the entire anatomic region. Ultimately, the merged lymphatic flow empties into the central blood circulation, mostly at the thoracic duct. Now part of the blood plasma, substances that formerly constituted lymph are carried via the blood vascular system for metabolism or clearance in kidneys, liver, and the reticuloendothelial system, in addition to which, they circulate to organs such as the heart, brain, and lungs. Lymphatic vasculature is responsible for the clearance of waste and foreign material of any molecular size, in addition to particulates, bacteria, and immune cells. This is because the open initial lymphatic vessels, particularly under circumstances of edema, allow entry of anything suspended or dissolved in lymph (Skobe and Detmar 2000; Paniagua et al. 2012).

Hyaluronidase, present in the secretory arsenal of the ancestral toxicoferan reptile (see chapter 24.11 and color plate 3), has direct effects on the interstitial scaffold, and the loss of matrix integrity it causes enables influx of fluid leaked from local cells or vessels. Serous or hemorrhagic fluid may accumulate rapidly as toxin-laden lymph diffuses into adjacent tissue, and in the early phases of envenomation, a palpable leading edge of local edema is sometimes recognized. A clinically useful marker of local injury is measurement, over time, of the proximal movement of this leading edge, which presumably represents the simultaneous migration of multiple toxins and may in theory be interrupted by timely administration of antivenom. Estimation of the volume of edema by serial measurement of limb circumferences is another common approach to lesion assessment. However, this is of less value in guiding therapy, because ongoing hemorrhage and fluid loss into tissues, once vascular injury has taken place, are not directly reversible by neutralization of toxins.

Microcirculatory uptake-control measures may be either useful or harmful in the first hours of snakebite management. In Australia, guidelines for first aid following elapid snake bites that are neurotoxic as a result of venoms rich in 3FTxs (see chapter 8) and presynaptically active group I phospholipase A_2 enzymes (SV-G^I-PLA_2s) (see chapter 20) are based on the work of Struan Sutherland, who demonstrated that systemic neurotoxicity could be delayed in primates by measures to minimize lymphatic uptake of toxins. Delayed neurotoxicity, in turn, should minimize the risk of ventilatory failure while en route to medical care. With this in mind, early compression bandaging is commonly advocated, in order to retard lymphatic uptake of elapid snake neurotoxins that might lead to ventilatory arrest prior to arrival at a hospital (see figure 2.2), although the effectiveness of these guidelines under real-life circumstances is not clear (Norris et al. 2005; Currie, Canale, and Isbister 2008; Canale, Isbister, and Currie 2009). On the other hand, management of pit viper bite in North America normally does not involve early compression, because clearance of tissue-injurious toxins, particularly SVMPs (see chapter 23), is considered desirable, while the risk of acute ventilatory paralysis is negligible in most cases. A joint statement issued by the American College of Medical Toxicology, the American Academy of Clinical Toxicology, the American Association of Poison Control Centers, the European Association of Poison Control Centres, the International Society on Toxinology, and the Asia Pacific Association of Medical Toxicology in 2011 (ACMT-AACT-AAPCC-EAPCC-IST-APAMT 2011) read: "Given that the primary toxic effect of envenomation is local tissue injury, mortality is not an ideal outcome measure to extrapolate to human crotaline envenomation. Available evidence fails to establish the efficacy of pressure immobilization in humans, but does indicate the possibility of serious adverse events arising from its use. The use of pressure immobilization for the prehospital treatment of North American Crotalinae envenomation is not recommended." The position statement leaves out North American elapid snakes (coral snakes, *Micrurus/Micruroides*), because, similar to that of most elapid snakes, coral snake venom is predominantly neurotoxic (German et al. 2005).

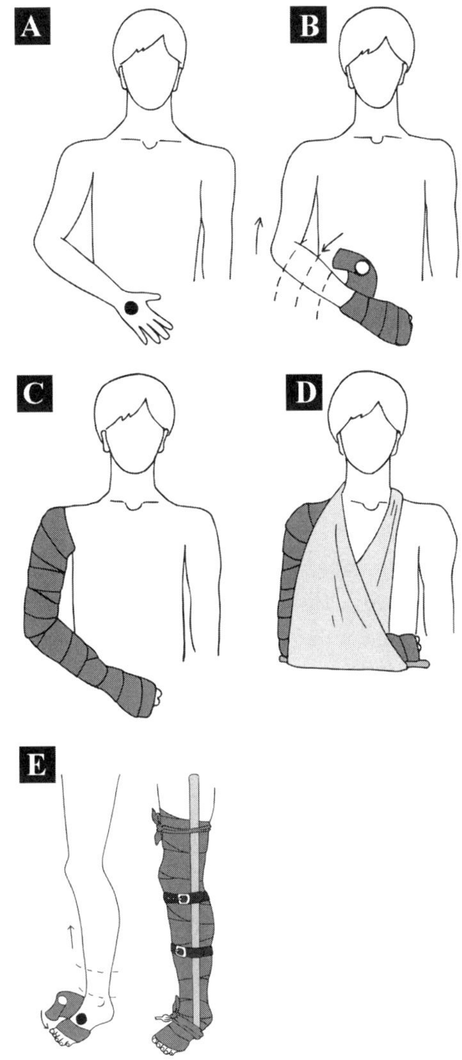

FIGURE 2.2: Pressure-immobilization bandaging of (A–D) an arm or (E) a leg. (Illustration: Genevie Jackson.)

For elapid snakes from regions other than Australasia and North America and all viper species, the use of measures to reduce local venom uptake is not uniformly agreed upon, and first aid measures actually taken vary widely. In Africa, where bites by most vipers (particularly SVMP-rich [see chapter 23] large species of *Bitis* such as *B. arietans*) and certain elapid snakes (especially African spitting cobras such as *Naja nigricollis* with venoms rich in cytotoxic 3FTxs [see chapter 8.4.3]) are associated with tissue injury (Warrell et al. 1976), tourniquet use frequently complicates wound progression, probably by causing tissue hypoxemia while delaying clearance of venom from the affected area. Many snakebite images that show disfiguring necrosis represent the results of a combination of envenomation and tourniquet use, and it is not possible to distinguish the contribution to injury from venom from that caused by the treatment (see color plate 14). In general, compression should probably be avoided with such envenomations, using the same reasoning as applied in the North American consensus statement. Tourniquets are not to be used for any envenomation, including those of neurotoxic elapid snakes, for which the pressure-immobilization method is indicated.

On the other hand, lab and field studies in Burma have shown that potently coagulopathic venom effects resulting from bites by *Daboia siamensis* may be retarded through the use of pressure pads and that localized necrosis was not significantly different from that in bites when such first aid measures were not employed (Tun-Pe et al. 1987; Tun-Pe et al. 1995). Thus, pressure-immobilization first aid may help delay severe systemic disruption of the blood coagulation cascade caused by venoms rich in coagulopathic toxins (see chapter 23), as is the case for the coagulopathy caused by the venoms of some Australian elapid snakes because of the presence of high concentrations of accelerated mutant forms of factor Va and factor Xa (see chapters 12 and 13).

Pressure-immobilization should be considered for envenomation by any lethally potent coagulopathic or neurotoxic species. However, pending further research, it should not be utilized for species known to cause serious local tissue injury. Practitioners should develop and teach methods consistent with the local herpetofauna of any given region, taking into consideration local cultural practices and likely transport times to medical care. Those occupationally at risk should consider the individual reptile types in their collections using expert consultation if necessary, and first aid protocols and specific notes on cage cards should reflect the expected characteristics of the envenomation by the animals contained within (see chapter 5).

Following initial management, it is common to take advantage of gravity (limb elevation) and motor activity (active and passive range of motion) to enhance edema clearance. Neither cold temperature nor antihistamine administration, both of which are advocated for other causes of limb edema, is appropriate in snake venom injury; cryotherapy (packing the limb in ice) may contribute to tissue damage by reducing blood supply to envenomated tissues, and it increases the risk of necrosis and eventual amputation (see chapter 4). Antihistamine administration does not address the underlying toxic etiology and only has benefit in cases of venom allergy.

During the first hours of envenomation, it is common to observe lymphangitic streaking in the superficial tissues of the arm or leg, proximal to the bite site. These red streaks represent injured lymphatic ducts, which become inflamed in the process of carrying venom and products of tissue injury and inflammation. Such materials may leak secondarily out of the injured lymphatic vessel or node, causing an accumulation of fluid and hemorrhagic material in the adjacent tissues. Lymphatic drainage removes venom and products of tissue degradation via ducts that pass through the medial side of the upper arm or thigh, on their way to the axillary or inguinal lymph nodes. Injury to these ducts and nodes may cause tenderness as early as 30 minutes after the bite. Secondary accumulation of hemorrhagic or inflammatory material becomes apparent hours later and is probably related to injury and leakage from the lymphatic structures themselves. It is not clear in such cases how much this secondary lesion represents the effect of active venom and how much it is simply a consequence of dumping previously damaged material into a new region of subcutaneous tissue. Lymphatic damage from active venom effect may explain "skip lesions" sometimes described in necrotizing envenomations, where tissue loss occurs in a patch proximal to the original point of venom entry (see color plate 14D).

Simultaneous with the uptake of injected venom and products of tissue injury into lymphatic vessels, some venom toxins may be taken up into regional blood capillaries. Blood circulation, unlike lymph, is a closed system, with high-pressure arterial flow distributed into smaller arterioles, thence via capillaries into venules and veins. The blood circulation contains formed elements (red blood cells that carry oxygen, white blood cells that participate in the immune response, and platelets that participate in hemostasis) along with high-molecular-weight substances (proteins, mostly albumin), all of which tend to stay within the vascular lumen. Smaller molecules, including oxygen, peptides, glucose, and salts, cross the blood vascular endothelium readily, however, and smaller venom toxins similarly may enter the blood circulation. For small molecules, uptake via the blood vascular system takes place at a relatively high rate (20–40 liters/day) relative to that by the lymphatics (2–4 L/day). It is reasonable, therefore, to expect that low-molecular-weight toxins may exert their effect (and be cleared) more rapidly than larger proteins which must follow the slower lymphatic route; however, comparative toxicokinetics of mixed venom toxins have not been well studied in this sense.

Interacting with the dynamics and kinetics of toxin uptake are the facts that structures of the interstitium and of the vessels themselves may be damaged by venom and that the rate of lymphatic clearance of large molecules increases when edema forces the separation of the distal lymphatic duct cells. Delivery of oxygen, inflammatory cells, and antivenom to the injured limb may be affected acutely by damage to capillary endothelium. In addition, clearance via the blood and the lymphatic vasculature may be impeded by loss of endothelial integrity. In some cases, particularly among older patients bitten on the lower extremities, injury to local vasculature develops a chronic component, with long-term edema in the bitten limb that worsens with inactivity, temperature change, and dependent position.

2.5 INTERACTION OF ABSORBED VENOM TOXINS WITH DISTANT TARGETS

2.5.1 ABSORPTION AND DISTRIBUTION OF TOXINS

In the minutes, hours, and days following snakebite, venom toxins continue to be absorbed from local tissue. Some small venom molecules, after reaching the bloodstream, may be cleared rapidly by the kidneys; others, particularly if they are larger and/or bound by antivenom molecules, may be broken down by the reticuloendothelial system. Depending on the type and quantity of venom injected (and the dose and timing of binding by antivenom), however, toxins with specific molecular targets gradually exert effects at a distance from the original bite site. Broadly speaking, these specific venom effects are sometimes categorized as neurotoxic, myotoxic, and coagulopathic.

2.5.2 NEUROTOXINS

Snake venom neurotoxins generally affect the cholinergic nerve junction, where, depending on the toxin, there may be presynaptic (β-neurotoxic) or postsynaptic (α-neurotoxic) effects, or both, with the net impact of diminishing conduction of nerve signals via the neurotransmitter acetylcholine (see chapters 8, 10, 15, 20, 21, 23.4.4, and 24.25 and color plate 2B). In many ways, pre- and postsynaptic neurotoxicity are clinically indistinguishable; fewer nerve impulses transmitted across to receptors on striated muscle results in less muscle contraction, which translates to weakness, the most severe form of which is flaccid paralysis of affected muscles. Clinical manifestations become apparent over minutes to hours and may include ptosis, external ophthalmoplegias, limb weakness, and, ultimately, ventilatory failure (Ranawaka, Lalloo, and de Silva 2013). In severe cases, mechanical ventilation and intensive supportive care are necessary for maintenance of life (see color plates 11M, 11N).

Many presynaptic neurotoxins (see chapters 8, 10, 15, 20, and 21) damage the presynaptic membrane and subcellular organelles, effectively blocking the release of acetylcholine and therefore preventing signal transmission from nerve to muscle (Sampaio et al. 2010) (see color plate 2B). Damage incurred during this process is not reversible by antivenom, implying that to be maximally effective, antivenom must be administered before neurotoxicity occurs. Intensive supportive care, including mechanical ventilation, may be required for days to weeks in patients who suffer flaccid paralysis prior to effective antivenom neutralization.

Many postsynaptic neurotoxins (see chapters 8, 23.4.4, and 24.25) function as competitive inhibitors of acetylcholine at the receptors on the muscle side of the neuromuscular junction (Arias 2000) (see color plate 2B). They block recognition of the neurotransmitter, preventing signal completion. The neurological effects of this process may be indistinguishable from those of presynaptic toxins, but because the cell itself is not damaged, the toxicity can sometimes be reversed. One means of overcoming the postsynaptic receptor blockade is by the use of anticholinesterase agents (such as neostigmine or pyridostigmine), which interfere with the natural breakdown of the neurotransmitter in the gap between nerve and muscle, thereby overwhelming the postsynaptic receptors during signaling (Brazil, Prado-Franceschi, and Laure 1979; Watt et al. 1986; Watt et al. 1989). Although treatment with anticholinesterases will not overcome presynaptic neurotoxicity, it may successfully be used as a temporizing measure in management of predominantly postsynaptic neurotoxicity, effectively buying time until toxins can be eliminated or neutralized by antivenom. Signs of neurotoxicity may include ptosis (eyelid droop caused by cranial nerve paralysis) (see color plate 11L), diminished expiratory flow rate (weak breathing caused by paralysis of the diaphragm and chest muscles), or complete ventilatory failure (see color plates 11M, 11N).

2.5.3 MYOTOXINS

Snake venom myotoxins that cause injury through direct action on muscle targets are present in many viperid and elapid snake venoms (see chapters 8, 9, 20, and 21). Several types of myotoxin have been described. The "small" myotoxins, nonenzymatic toxins found in rattlesnake venom, cause depolarization of skeletal muscle cells, perhaps via a sodium channel interaction (see chapter 9). Cytotoxic 3FTxs increase membrane permeability through a nonenzymatic interaction with surface proteins (see chapter 8). PLA_2s also contribute to myotoxicity (see chapters 20 and 21). From a clinical standpoint, the syndrome provoked by these toxins may be mild, with muscle aches and slight elevation of creatine phosphokinase (CPK); or it may be dramatic, with fasciculations or gross myokymia of skeletal muscle apparent within hours after envenomation. In some cases, muscle weakness may be profound, with the syndrome complicated by the concurrent presence of paralytic neurotoxins. Muscle breakdown may be accompanied by rising blood urea nitrogen, creatinine, and CPK, plus myoglobinuria, with attendant risk of secondary renal injury. Monitoring of serial CPK values in such cases characteristically shows a slow decline following treatment and may be useful for decision-making in fluid and electrolyte management. It should be noted, however, that antivenom affects CPK only indirectly, through reduction in the ongoing process of muscle injury; changes in CPK have not been proven to have predictive value relative to the use of secondary doses of antivenom. As in the case of neurotoxins, the effects of myotoxins may or may not be mitigated by antivenom use, depending on the extent to which tissue damage has taken place before initiation of treatment. Physiological markers of muscle injury, such as CPK or myoglobin, may therefore continue to be abnormal well after antivenom has been administered.

2.5.4 COAGULOPATHIC TOXINS

Snake venom coagulopathic toxins exert their effects via a variety of mechanisms (see chapters 8, 12, 13, 14, 15, 17, 20, 21, and 23 and color plate 2A), more than one of which may occur in any one envenomation. These mechanisms include altered platelet function and number, fibrinogen consumption or degradation, vascular damage, and interference with clotting factors (Hutton and Warrell 1993; Markland 1998). Some toxins bind with the surfaces of blood platelets, altering membrane function and causing dysfunctional aggregation. Clumped platelets have been described in the peripheral blood smears of envenomed patients, and formal platelet counts (such as those measured by automated Coulter counters) may indicate dramatic drops (by as much as three orders of magnitude) in total numbers of platelets in the first hours after a snakebite. However, it is not clear how much of this apparent thrombocytopenia is present *in vivo* versus how much occurs *in vitro* (such as in the citrated purple-top blood-collecting tube). Clinical correlation is evident, however, in hemorrhagic consequences similar to those observed in severe thrombocytopenia from other causes: spontaneous mucosal hemorrhages,

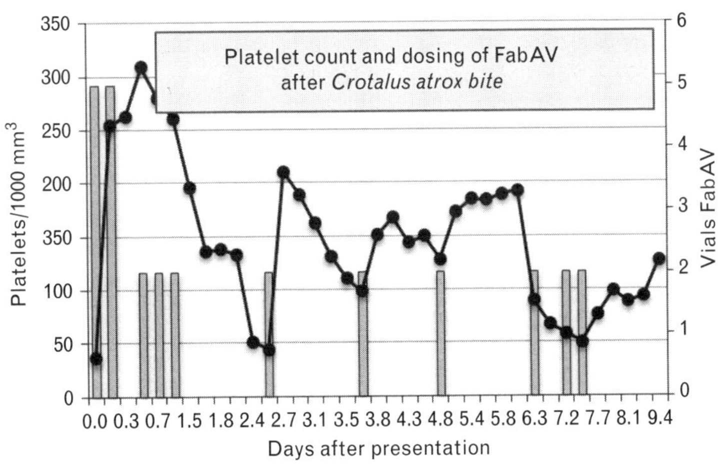

FIGURE 2.3: Response of thrombocytopenia to treatment with Fab ovine antivenom, following envenomation by *Crotalus atrox*. The use of short-acting Fab antivenom in this case made it possible to observe the change in platelet responsiveness over 10 days following envenomation. Note the change in responsiveness with successive antivenom doses over one week's time. Recurrent thrombocytopenia of this type has been described in the United States, where Fab antivenom is in common use for hemotoxic envenomation (Seifert et al. 1997; Boyer et al. 1999; Boyer, Seifert, and Cain 2001; Boyer et al. 2013b).

for example, and petechiae that appear at sites anatomically distant from the point of entry of venom. Remarkably, the apparent thrombocytopenia that is measured early in the course of viperid envenomation appears to be rapidly reversed (often within an hour) by specific antivenom. Late or recurrent thrombocytopenia, on the other hand, may be recalcitrant to treatment, suggesting that different mechanisms may be in effect over the course of time. These effects are most vividly apparent in the treatment of pit viper envenomation in the United States, where the use of short-acting Fab antivenom makes it possible to observe the change in platelet responsiveness with successive doses administered during the one to two weeks following envenomation (see figure 2.3) (Seifert et al. 1997; Boyer et al. 2013b). It is possible that damaged platelets are rapidly replaced during the acute-phase response, in cases where antivenom is applied quickly, but that exhaustion of reserves or damage to megakaryocytes complicates longer-term exposure to these toxins, resulting in a several-day recovery period, when antivenom is provided later, during the subacute phase of injury. Based on what is known from other medical conditions involving severe thrombocytopenia (such as leukemia), it is likely that thrombocytopenia is responsible for the spontaneous hemorrhages that occasionally cause stroke or other serious consequences following envenomation.

A second set of coagulopathic effects that may result from snake envenomation is related to the quantity and function of fibrinogen. Hypofibrinogenemia has been observed during the acute and subacute phases of a variety of viper envenomations (see color plate 2A), and it appears to result from several mechanisms including the action of kallikreins (see chapter 14) and SVMPs (see chapter 23). Depletion of fibrinogen, once it has occurred, is not directly reversible by antivenom. A relatively rapid return toward normal levels is commonly observed, however, in the first hours following antivenom administration, indicating that the ongoing destruction of fibrinogen is halted once adequate neutralization is accomplished and that replenishment of circulating fibrinogen from hepatic sources can then take place. As in the case for platelet counts, the quantitative fibrinogen level is observed to rise relatively rapidly when treatment takes place during the first hours after envenomation; however, consistent with overall substrate depletion, late treatment may have very little immediate impact. Based on what is known from other disease states (such as congenital dysfibrinogenemia or afibrinogenemia), hypofibrinogenemia from snakebite is very likely responsible for the tendency toward unrelenting bleeding from sites of injury that is observed in patients who undergo surgical procedures during the subacute phase of incompletely treated envenomation. Dental extractions, surgical procedures (Boyer et al. 1999; Boyer et al. 2013b), and secondary trauma may be accompanied by greatly increased bleeding under such circumstances. Death has been reported in patients with residual hypofibrinogenemia following recovery from acute pit viper envenomation (Kitchens and Eskin 2008).

In addition to thrombocytopenia and hypofibrinogenemia, snake venom toxins provoke a broad variety of other effects on the blood clotting cascade, both procoagulant and anticoagulant (see chapters 8, 12, 13, 14, 15, 17, 20, 21, and 23 and color plate 2A). Most of these are minor in impact, relative to the apparent effects on platelets and on fibrinogen, however. Overall, although it is possible to demonstrate quantifiable effects on prothrombin time, partial thromboplastin time, thrombin time, bleeding time, and specific clotting-factor levels, these details are seldom clinically useful. Even in the case of Russell's viper envenomation, the net effect of absorbed toxins is generally anticoagulant rather than procoagulant, perhaps because of fibrinogen consumption during the formation of microthrombi as toxins are diluted during absorption. In parts of the world where specific blood assays are readily available, platelet count and fibrinogen (or prothrombin time [PT]) determinations may be sufficient for patient-management purposes; or where these are not available, the whole-blood clotting time (WBCT) alone is commonly a sensitive enough measurement with which to guide care (see color plate 11K).

2.6 SECONDARY CONSEQUENCES OF VENOM INJURY

2.6.1 GENERALITIES OF SECONDARY CONSEQUENCES

In addition to the effects of direct interaction of toxins with tissues, envenomation carries a risk of secondary consequences. These effects vary depending on a variety of factors, including the amount of time between the bite and medical intervention, vulnerability because of preexisting medical conditions, adequacy of fluid resuscitation, duration and severity of impaired oxygenation, severity and anatomic site of hemorrhage, safety and adequacy of antivenom used, and injury caused by first aid or healthcare providers. Although, to some extent, the duration and severity of secondary effects may be minimized by use of antivenom, in general, these phenomena require symptomatic and supportive management, sometimes including prolonged intensive hospital care. Untreated, the secondary effects of envenomation may lead to shock, respiratory failure, organ failure, death, or long-term disability.

2.6.2 SHOCK

Shock results from a combination of factors that varies depending on circumstances. Significant edema fluid loss ("third spacing") may be associated with the local injury itself. Hemorrhage (see chapter 23) may become severe enough to cause hemodynamic compromise within hours of a severe viper bite, both from the volume of blood loss and from acute anemia. Vasodilation (see chapters 11, 14, 18, 19, and 22) may accompany inflammatory processes (see chapters 14 and 18) or anaphylactic phenomena related to direct venom effects (see chapter 24.6), venom allergy, or antivenom administration. The same processes may compromise ventricular function, or vulnerable coronary arteries may be affected by increased myocardial demands, with resultant ischemia and injury. Management of shock includes volume support and pressors, with antivenom for correction of any concurrent coagulopathy.

2.6.3 RESPIRATORY FAILURE

Respiratory failure may be the result of either pulmonary (such as from hemorrhage, edema, or anaphylaxis-associated airway edema and spasm) or mechanical (ventilatory failure in the context of paralysis from neurotoxins or myotoxins) actions. In either case, hypoxemia and altered intrathoracic fluid dynamics complicate the management of concurrent shock. Management of respiratory failure includes oxygen and mechanical ventilation, with antivenom or perhaps anticholinesterase drugs for correction of paralysis (although the efficacy of these may be limited). Elapid snake venoms are rich in the apotypic neurotoxic form of 3FTx which is mutated to be much more potent on mammals (see chapter 8.4.2). Thus, human envenomations from elapid snakes are much more likely to result in such effects than envenomations from the non-front-fanged caenophidian species that feed on lizards or birds. *Boiga* species, for example, are extremely potent to birds and lizards but are orders of magnitude less toxic to mammals (Pawlak et al. 2006; Heyborne and Mackessy 2013).

2.6.4 ORGAN INJURY

Shock and hypoxemia from envenomation come with the added risk of complications from hemorrhage and tissue breakdown. Renal function is especially vulnerable to injury in cases with rhabdomyolysis, but myocardial infarction, stroke, and gastrointestinal hemorrhage may also occur. Management of organ injury is context-specific and mainly supportive. SVMP-rich venoms, such as those of most vipers (true and pit), in particular cause such effects (see chapter 23).

2.6.5 DEATH

Although direct venom effects may lead to fatality from intracranial hemorrhage, as, for example, in the case of envenomations by those Australian elapid snakes that have venoms rich in TV-fVa (see chapter 12) or in TV-fXa (see chapter 13), shock and respiratory failure are the leading causes of death from other reptile envenomations. Timely supportive care is therefore of great importance, regardless of the specific etiology and sequence of events for any particular patient and snake. Whenever possible, patients with severe snakebite should be managed in hospital units equipped with mechanical ventilators, pressor drugs, advanced monitoring capabilities, and personnel skilled in critical care.

2.6.6 LONG-TERM DISABILITY

Permanent injury following reptile envenomation is most often related to damage to tissues in the bitten limb. Necrotizing effects of venom toxins, chronic lymphedema from local vascular injury, contractures related to connective-tissue damage, ischemic damage from tourniquets or cryotherapy, infection, and delayed wound healing from scarification or surgery all affect long-term functional outcome. Management of subacute and chronic injury involves a combination of skilled wound care, physical or occupational therapy, and occasionally surgical repair (such as late debridement, tendon repair, skin grafting).

2.6.7 ALLERGY

A special concern for occupationally or avocationally exposed reptile handlers is venom allergy (see chapter 5). Hypersensitivity to venom may occur following repeated exposure to snake proteins, particularly (but not exclusively) those present in venom itself. There appears to be elevated risk following exposure to spitting cobras, presumably because of respiratory-tract contact with aerosolized proteins. Immediate hypersensitivity reactions (so-called type 1 or anaphylactic reactions) are more likely to develop among people who have a medical history of atopy (asthma, hay fever, or eczema). The development of urticaria (hives) or wheezing while handling animals or venom should be taken seriously by venom professionals, because subsequent exposures have the potential to cause very severe reactions. Cardiovascular collapse and respiratory failure from venom-induced anaphylaxis may occur from a very low-dose exposure (such as with cage cleaning) or at the time of an envenomation. Management of simultaneous anaphylaxis and envenomation is a daunting diagnostic and therapeutic challenge, balancing the use of epinephrine and antihistamines with the often-simultaneous use of serum products (antivenom) that increase the risk of type 1 reactions in atopic individuals. Reptile handlers who develop signs of possible allergy should minimize their exposure to animals and consult with an experienced allergist regarding risk minimization, use of antihistamines and epinephrine, and notification of potential rescuers via medical documentation or allergy-warning bracelets or tags. Venom-desensitization therapy, although established for the use of beekeepers, has not been well established for reptile handlers but is worthy of further investigation.

2.7 MEDICAL MANAGEMENT OF REPTILE VENOM INJURY

Ongoing envenomation concerns fall broadly into two categories of physiology and treatment: (1) tissue injury for which supportive care must be provided, to sustain life while healing can occur, and (2) processes amenable to neutralization by antivenom. Processes in the second category include venom distribution and absorption, and for this reason, timely use of antivenom may alleviate the ongoing injury for which supportive care is required. In theory and under ideal circumstances, where the perfect dose of perfectly safe, perfectly targeted antivenom was administered moments after envenomation, most injury from envenomation could be prevented entirely. Efforts toward improvement of snakebite treatment have therefore emphasized the development of higher-quality antivenoms and minimizing the "fang-to-needle time" between bite and antivenom administration. In practice, however, the "perfect antivenom" does not exist, and some delay is inevitable before treatment of any bite, with a consequent suboptimal course of injury. Supportive care is therefore essential to envenomation management in nearly all cases. In regions and at times of severe antivenom shortage, supportive care may be the only care available for snakebite. Table 2.3 divides snakebite management into four phases, and it suggests specific interventions at each stage.

2.7.1 PREVENTION AND PREPARATION

In many parts of the world, snakebite is an endemic public health issue. Improved housing, occupational safety training, public education, availability of trained "snake-removal services," and culturally specific interventions teaching respect for the environment and avoidance of harmful first aid measures all have a role in minimization of morbidity and mortality (see chapter 5).

Hospitals, captive collections such as zoos, and research facilities should be prepared in advance with appropriate antivenoms and treatment protocols for emergency use. "Antivenom" (also "antivenin") is a general term for any of hundreds of biological agents manufactured from the immune serum of vertebrate animals and used to interrupt the process of envenomation by venomous creatures. Source animals, potency, specificity, purity, safety, and molecular composition vary widely across different antivenom products. Table 2.2 details the antivenom characteristics of importance in product selection. The World Health Organization maintains a list of antivenoms available on the world market (www.who.int/bloodproducts/snake_antivenoms/en/). Commercial antivenom products are not available for all species, although in some cases (Gila monsters, for instance), there is precedence for the preparation and use of experimental antivenom in occupational-exposure settings.

2.7.2. ACUTE ENVENOMATION MANAGEMENT: THE FIRST 24 HOURS

Basic measures for early care are given in table 2.3. Medical tests of potential use are in table 2.4. In essence, the goals of the first day of therapy are to prevent death, to minimize injury from overzealous first aid or surgery, and to administer appropriate antivenom as early as possible. General guidelines for antivenom use are given in table 2.5. Under emergency circumstances, physicians and patients rarely have much choice of which antivenom product to use, because suppliers, distributors, costs, and government regulations tend to limit the local availability of all but a handful of products. In some locations, however, such as the parts of the Americas where coral snakes are endemic, an informed choice must be made between two very different products (in this case, polyvalent pit viper antivenom versus a coral snake specific [*Micrurus*] antivenom) (de Roodt et al. 2013). The risks of antivenom use must be carefully considered (Boyer et al. 2013a; Leon et al. 2013) and appropriate precautions taken to minimize the risk of injury from early adverse reactions. Worldwide, a particularly dangerous emergency may occur if the owner of an exotic (nonnative) venomous snake has failed to acquire appropriate antivenom in anticipation of a potential bite (see chapter 5). It is common in such cases that a delay of more than 24 hours may take place before antivenom can be acquired, if at all.

2.7.3. SUBACUTE INJURY MANAGEMENT: THE FIRST THREE WEEKS

In a severe snake envenomation, venom may remain in the body for many days. If venom remains sufficiently neutralized by antivenom during this time, then priorities in medical care are wound therapy, pain control, and (if required) mechanical ventilation and intensive fluid, electrolyte, and nutritional care. Guidelines for subacute care are summarized in table 2.3. Sometimes venom is not sufficiently neutralized during the subacute phase of envenomation.

Table 2.4: Medical Tests in Management of Acute and Subacute Snakebite

Test	Utility
Clinical grading scale (such as minimal, moderate, severe)	Communication among clinicians; valuable only if reassessed periodically during acute and subacute phases of care; not a validated research tool.
Snakebite severity scale (SSS)	Research tool for ensurance of comparability between groups in *Crotalus* envenomation; not a clinical management scale (Dart et al. 1996).
Serial skin marking to show contiguous spread at different times; may be combined with serial photography	Assessment of rapidity of development of local injury; especially useful after antivenom administration and when case is transferred between care providers. It is essential to mark the edge of any hint of venom effect (tenderness, edema) and not just of discoloration.
Limb circumference measurement	Assessment of volume of fluid lost with edema or hemorrhage; qualitative comparison between limbs if envenomation is uncertain; not generally useful for antivenom dosing.
Venom detection, qualitative	Used to guide choice of antivenom, in some geographic settings (Isbister et al. 2013). Has documented issues for cross-reactivity (Jelinek et al. 2004), false negatives, and false positives caused by the hook effect (Steuten et al. 2007), in addition to false positives from relictual 3FTx still expressed in forensic levels in the oral secretions of pythons (Fry et al. 2013).
Venom detection, quantitative	Research, potentially useful for decisions about repeated antivenom dosing but not available for routine clinical application in most settings.
Antivenom quantitation	Research, potentially for decisions about dosing but not available for routine clinical application in most settings.
Serial hemoglobin and hematocrit	Management of severely hemorrhagic cases.
Platelet count and function	Risk of bleeding, decisions on antivenom dosing in regions where thrombocytopenia is a risk.
PT, aPTT, TT, FSPs, fibrinogen	Risk of bleeding, decisions on antivenom dosing in regions where hypofibrinogenemia is a risk and test is readily available.
Whole-blood clotting time (see color plate 11K)	Risk of bleeding, decisions on antivenom dosing in settings where other coagulation tests are not readily available.
Urinalysis, myoglobin, CPK, renal panel	Secondary-effects monitoring in supportive care of shock, rhabdomyolysis, or secondary renal injury, especially where myotoxic venom is a concern.
ABGs, spirometry, EKG, serum chemistry panels, X-rays	Secondary-effects monitoring in supportive care in cases of ventilatory failure and severe volume loss.
Compartment pressure measurement	Distinguishing rare compartment syndrome from more common severe subcutaneous edema; prevention of unnecessary surgery.

Table 2.5: Use of Antivenom

Step	Process
Assessment for need of antivenom treatment	Antivenom administration carries with it a risk of early acute reactions and delayed serum sickness. It should be given promptly but only when there is a clear medical indication. Signs and symptoms of systemic venom effect (such as abnormal blood coagulation, indications of neurotoxicity) or of progressive local injury are indications for antivenom.
Emergency selection of product (see also table 2.2, for considerations by hospital pharmacies in advance of an event)	In many cases, different local snake venoms require different antivenom products, and exotic snakebites (from captive animals held at a distance from their native range) normally require specialty products not stocked at local hospitals. Physicians in the Americas, Africa, and Asia should be alert to the differences between elapid and viper bites in their areas, because these commonly require treatment with entirely different products. Regional differences in venom within genera, between species, may similarly lead to serious errors in antivenom selection (Visser et al. 2008; Warrell 2008; Boyer et al. 2013a; Leon et al. 2013). Physicians in Australia may choose to use qualitative antigen-detection kits when identity of the snake is in doubt. When the identity of the animal is not known or when a specific monovalent antivenom is not available, polyvalent antivenom is commonly the best choice, but potency against the particular envenomation (if known) must then be considered carefully in dosing. "In date" products are always preferable, but in their absence, well-preserved antivenom is sometimes used years after its original expiration date. Whenever possible, regional poison control centers or medical toxicology consultants should be contacted for advice on specific product selection and use (Warrell 2009).
Dose	Antivenom is provided in single-use vials or flasks. Typical starting doses are in the range of one to 10 vials, depending on the product and the circumstances. Medical toxicology consultation is recommended for determination of starting dose, or guidelines on the manufacturer's package insert may be used.
Reconstitution	Liquid antivenom products do not require reconstitution. Lyophilized (freeze-dried) products may come with a diluent in a separate vial, or they can be reconstituted by swirling with 5–10 ml of normal saline. Well-lyophilized products will go into solution within a few minutes, producing a clear, watery solution that may be slightly yellow in color. Some products take much longer to reconstitute; filling the vial to remove air space and use of a mechanical agitator may speed the process (Hill, Bogdan, and Dart 2001; Quan, Quan, and Curry 2010).
Dilution	In some regions, intravenous infusion is limited by availability of materials for sterile dilution and administration, or high-potency products may minimize the need for dilution. In general, however, to minimize the risk of early adverse reactions, antivenom may be diluted such that the total dose (one to 10 vials) will be given in 250–1000 ml of normal saline.

Table 2.5 (*Continued*)

Step	Process
"Test dose"	In the past, intradermal or conjunctival "test doses" of antivenom or of serum have been used, to determine the risk of allergy before administering antivenom products, but the negative predictive value of these was too low for practical value. Test doses are no longer recommended in favor of the adoption of universal serum-reaction precautions.
Premedication	Premedication, with the aim to reduce the severity and frequency of early adverse reactions, carries its own risks of drug side effects. Epinephrine, H1 and H2 blockers, and corticosteroids have all been used for this purpose. Premedication is based partly on the particular antivenom's adverse event rate (highest for unrefined whole-IgG products, lower for highly purified $F(ab')_2$ antivenoms) (Hill, Bogdan, and Dart 2001; Quan, Quan, and Curry 2010; Boyer et al. 2013a; Leon et al. 2013). Also important is the treating facility's familiarity with antivenom treatment and readiness to manage allergic reaction (Habib 2011).
Administration	Antivenom has greatest bioavailability when administered intravenously. The risk of early adverse reactions is reduced by slow infusion of diluted material, over 30 to 60 minutes. In settings where this is not feasible, liquid or reconstituted antivenom is sometimes administered by slow intravenous push (over 5 minutes), but this technique carries a higher risk and should be avoided when possible.
Observation and treatment for early adverse reaction	Ongoing monitoring during and immediately after antivenom infusion should include frequent observation of the skin (for itch, redness, or urticaria) and cardiorespiratory status (for cough, throat clearing, difficulty breathing, or hypotension). At any sign of reaction, the infusion should be stopped. Treatment with epinephrine and H1 and H2 blockers may be followed by cautious reinitiation of infusion at a lower rate. Expert consultation is very helpful in this setting.
Reassessment	Over the hours following initial treatment, local injury should be reassessed frequently. Blood coagulation should be reassessed one to four hours after antivenom. An expanding area of local injury or ongoing coagulopathy is an indication for additional antivenom.
Secondary dosing	Secondary antivenom dosing is appropriate if reassessment suggests the ongoing effects of unneutralized venom. If a short-half-life antivenom is used (such as Fab), then secondary "maintenance" dosing is sometimes administered every four to six hours initially and every one to two days thereafter, during the subacute phase of envenomation.
Observation and treatment for delayed serum sickness	Delayed type 3 immune reactions, or serum sickness, classically begin seven to 10 days after treatment with antivenom. Serum sickness is very common (50%–90%) with whole-immunoglobulin serotherapy, less common with refined products, and uncommon (< 5%) with highly refined Fab and $F(ab')_2$ antivenoms. The presence of urticaria, vasculitis, arthralgias, myalgias, or fever one to two weeks after antivenom treatment should be evaluated for possible treatment with antihistamines or with a short course of oral corticosteroids.

This may be a result of antivenom supply shortages, underdosing, low-quality antivenom, clearance of short-half-life antivenom, lack of financial resources, venom variation between the species responsible for the envenomation and the species (singular or multiple) utilized in the immunizing mixture of the antivenom (see chapter 3), or other delays such as considerations regarding serious antivenom allergy (it must be noted that true antivenom allergy is far rarer than is conventionally perceived). Regardless of the cause, in such situations, ongoing or recurrent venom effects may be present, particularly those involving coagulopathy. Administration of late or additional antivenom may result in improved coagulation during this time. Other venom effects, such as neurotoxicity, may not reverse quickly when there is a delay in the administering of antivenom, for reasons discussed previously in this chapter. This is particularly the case for presynaptic neurotoxicity such as that caused by bites from *Oxyuranus* species, as presynaptically active PLA_2s present in the venoms of these species physically destroy nerve endings (see chapter 20). This is in contrast to the postsynaptically active acetylcholine receptor-blocking peptide neurotoxins such as the 3FTxs often present in high expression levels in elapid snake venoms (see chapter 8); neurotoxicity resulting from the activity of these latter toxins is more rapidly reversible with appropriate antivenom. Seven to 10 days after initial antivenom treatment, signs of a type 3 immune reaction to antivenom therapy may complicate recovery. In classical serum sickness, which is common after serotherapy with relatively unrefined whole-IgG antivenoms, patients may develop an urticarial or vasculitic rash, followed by arthralgias and myalgias severe enough to prevent normal activity. Treatment with antihistamines and a short burst of oral corticosteroids appears to shorten the syndrome.

2.7.4. CHRONIC DISABILITY MANAGEMENT: MONTHS FOLLOWING THE BITE

Disability from snakebite is common, particularly in the bitten extremity. Amputation of a finger or foot, deep ulceration at the site of a necrotic lesion, chronic lymphedema, muscle atrophy, or contractures at joints near the bite site may require long-term follow-up by physical and occupational therapists, plastic surgeons, or wound clinics (see color plates 11–14). Depending on the region, wound management may be facilitated by diabetes clinics, burn centers, or tropical parasitology programs. There is no role for additional antivenom during the chronic stage of injury from snakebite.

CHAPTER 3

ANTIVENOM RESEARCH AND DEVELOPMENT

M. BÉNARD-VALLE, E. E. NERI-CASTRO, B. G. FRY, L. BOYER, C. COCHRAN, M. ALAM, T. N. W. JACKSON, D. PANIAGUA, F. OLVERA-RODRÍGUEZ, I. KOLUDAROV, K. SUNAGAR, AND A. ALAGÓN

3.1 SUMMARY

To date, antivenom is the only specific therapy that has been proven effective for the treatment of envenomations. Antivenom production and optimization require a multidisciplinary approach, including knowledge of venom composition and pharmacokinetics along with various insights on immunology and purification technologies. An ideal antivenom should have the following characteristics: high neutralization potency against toxicologically relevant venom components, a volume of distribution that allows it to reach toxins that have already been absorbed, and high half-life in systemic circulation to ensure neutralization of toxins that reach the bloodstream after prolonged absorption periods and to promote redistribution of toxins from extravascular space to the bloodstream (Gutiérrez, Leon, and Lomonte 2003). Much research, however, remains to be done in order to improve antivenom production and affordability in addition to distribution and training for its efficient use.

Antivenom production utilizes the basic principles elucidated in the late 1800s, when Emil von Behring and Shibasaburo Kitasato developed the first hyperimmune serum. Their great breakthrough consisted in the observation that the serum of an animal immunized with diphtheria or tetanus toxins could be used in the treatment of these diseases (WHO 2010). This was the beginning of serotherapy. Subsequently, Cesaire August Phisalix at the Natural History Museum of Paris and Albert Calmette at the Pasteur Institute developed the first hyperimmune sera, targeting cobra and viper venoms (Calmette 1894a; Phisalix and Bertrand 1894b; Phisalix and Bertrand 1894a). By 1896, the serum developed by Calmette was already being used in human patients and was involved in the first successfully reported use of antivenom serum therapy (Calmette 1897). Since then, the process called hyperimmunization, defined as the generation of a larger than normal number of antibodies to a specific antigen (Abbas, Lichtman, and Pillai 2011), has provided the sole successful form of envenomation treatment. When injected into an envenomed patient, antibodies present in the hyperimmune serum bind the venom molecules and can prevent their toxicity (see chapter 2).

During hyperimmunization, an animal is exposed for a prolonged period to a foreign substance, in this context known as an antigen, and its adaptive immune system responds by producing antibodies that bind specifically to the substance. When venom is used as the antigen and the resulting hyperimmune serum is later injected into an envenomed patient, antibodies in the serum bind the venom molecules and can prevent their toxicity. The antibodies involved in this neutralization process are called immunoglobulins (Ig), particularly immunoglobulins G (IgGs) produced in mammals. IgGs are globular proteins of about 150 kDa, composed of two heavy chains and two light chains bound by at least 15 disulfide bridges. Functionally, the IgG molecule is divided in two main regions: the crystallizable fragment, or Fc, mainly involved in cellular response to antigens and in complement fixation; and two antigen-binding fragments, or Fab, which recognize foreign substances and bind to them. At the tip of the Fab are the hypervariable regions that directly interact with the antigen (see figure 3.1A) (Abbas and Lichtman 2006). An antigen that has the ability to initiate an immune response is designated an immunogen.

In addition to mammalian IgG, avian IgY from chicken eggs has been explored as an alternative source of antivenom because of the simplicity of its production and management (Costa de Almeida et al. 1998; Devi et al. 2002; Paul et al. 2007; Criste et al. 2008; Meenatchisundaram et al. 2008). However, research has indicated that IgY has an unfavorable pharmacokinetic profile and is a stronger mediator of adverse reactions such as anaphylaxis (Sevcik, Diaz, and D'Suze 2008; Sevcik et al. 2012).

3.2 ANTIVENOM PRODUCTION

3.2.1 HYPERIMMUNIZATION

All antivenoms in current clinical use are developed through hyperimmunization, typically of large mammals, including horses (Sjostrom et al. 1994; Herrera et al. 2005; Guidlolin et al. 2010) and sheep (Smith et al. 1992a; Dart and McNally 2001). Camels (Cook et al. 2010a; Cook et al. 2010b) and llamas (Harrison et al. 2006) have been used to produce experimental antivenoms. The specific hyperimmunization process varies largely depending on the selected animal model and on the characteristics of the immunogen (such as crude venom or pure toxin) (Chippaux and Goyffon 1998). Hyperimmunization may be conducted anywhere from 3 to 15 months and totaling anywhere from 10 to 50 injections. A protocol used to produce equine hyperimmune plasma in Mexico is depicted in color plate 15. In this protocol, the immunogen is injected subcutaneously at various points along the neck of the animal, increasing the area of exposure to the antigen and favoring its absorption (Sriprapat et al. 2003).

In order to monitor the hyperimmunization process, periodic small blood samples are collected, and the specific antibody titer, defined as the dilution at which half of the colorimetric response is obtained, is measured using an enzyme-linked immunosorbent assay (ELISA)

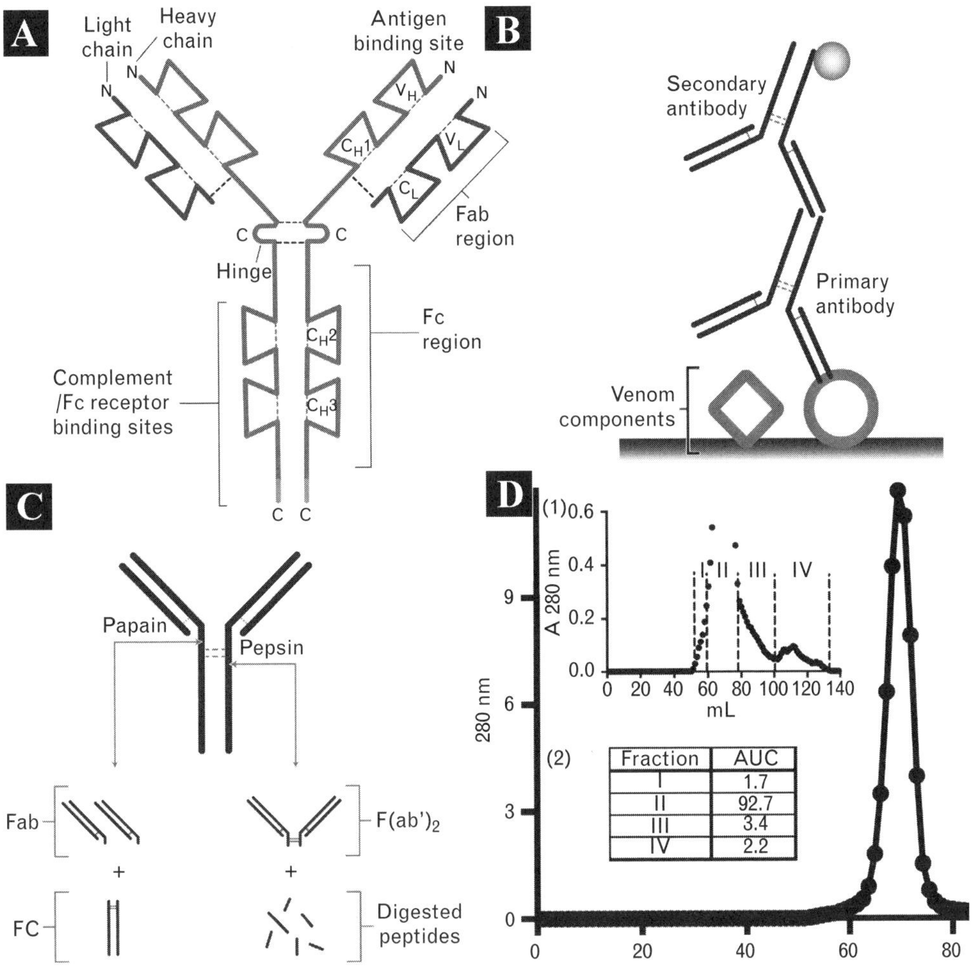

Fraction	AUC
I	1.7
II	92.7
III	3.4
IV	2.2

FIGURE 3.1: (A) Structure of an IgG molecule (Abbas and Lichtman 2006). (B) Schematic of an enzyme-linked immunosorbent assay (ELISA). Primary antibody: sample of hyperimmune plasma. Secondary antibody: IgG against primary antibody, covalently linked to an enzyme such as horseradish peroxidase, which reacts to a substrate to give a colorimetric response. (C) IgG enzymatic digestion to produce Fabotherapeutics (Abbas and Lichtman 2006). (D) Composition of Inoserp MENA as determined by gel filtration. Elution profile of 55 mg of antivenom separated in a column (0.9 x 195 cm) filled with Sephacryl S-200. Column developed with PBS at 12 mL/hr. Insert (1) shows the elution profile in an expanded mode and displays the resolved fractions: I, II, III, and IV. Insert (2) displays the area under the curve (AUC) of the resolved fractions. I: F(ab')$_2$ dimers; II: F(ab')$_2$ monomers; III: medium-mass-weight components; IV: low-molecular-mass components. (Illustration: Paulina Bénard.)

(see figure 3.1B). The neutralizing potency of selected samples is also measured. Once the desired titer and neutralizing potency are achieved, a greater amount of blood, depending on the size of the animal, is collected. The duration and structure of immunization protocols used to raise neutralizing antibodies to the level required for antivenom production are often empirical and depend on the experience of the manufacturers. An example of an immunization protocol is given in table 3.1.

Table 3.1: Example Protocol for Equine Hyperimmunization

Day	Injected venom (μg)	(μl)	Injectable saline (ml)	Adjuvant Type	Vol. (ml)	Total volume (ml)	Inoculation route	Blood sample (ml)
1	100	5	0.9	CFA	1.3	2.2	ID	5
7	100	5	0.5	IFA	0.7	1.2	SC	5
14	200	10	0.6	ALUM	0.4	1	SC	5
21	250	13	0.5	IFA	0.7	1.2	SC	5
28	500	25	0.6	ALUM	0.4	1	SC	5
35	1000	50	1	—	0	1.1	SC	5
42	2000	100	0.4	IFA	0.7	1.2	SC	5
49	4000	200	0.4	ALUM	0.4	1	SC	250
56	5000	250	0.8	—	0	1	SC	250
63	6000	300	0.2	IFA	0.7	1.2	SC	250
70	6000	300	0.3	ALUM	0.4	1	SC	250
77	6000	300	0.7	—	0	1	SC	250

CFA = complete Freund's adjuvant; IFA = incomplete Freund's adjuvant; ALUM = alumina; ID = intradermal; SC = subcutaneous.

Historically, blood serum of hyperimmunized animals was used in the production of antivenoms, and treatment was accordingly termed serotherapy. Today, however, the process of plasmapheresis—where the cellular fraction of blood is reinjected into the donor animal within 24 hours following blood collection—is generally preferred to the use of crude blood serum (see color plate 15). This process reduces the probability of anemia, increasing the animal's productive time and improving its quality of life. Therefore, it is now blood plasma, instead of serum, that is normally used in antivenom production (Chippaux and Goyffon 1998; WHO 2010). During plasmapheresis, blood is collected in bags with an anticoagulant such as sodium citrate. Afterward, the bags are suspended in order to allow the cellular fraction of blood, mainly erythrocytes, to sediment under the effect of gravity. When using horses as production animals, this process is especially efficient, owing to the morphology of equine erythrocytes (see color plate 15B) and their tendency to align like stacked coins (Rouleaux formation) (Shoemaker 2009). The high sedimentation rate of stacked red blood cells permits the collection of the cellular fraction of the blood in a new bag containing isotonic-glucose injectable solution (see color plate 15D) and its reinjection into the horse within two hours following bleeding (see color plate 15D).

3.2.2 ADJUVANTS

During hyperimmunization, substances known as adjuvants can be used to promote or potentiate the animal's immune response to the antigen. Although their mechanisms of action are poorly understood and their use by vaccine and antivenom manufacturers has been largely empirical, some literature has shed some light on this subject (Cox and Coulter 1997). This review described five action mechanisms: immunomodulation, antigen presentation, cytotoxic T-lymphocyte (CTL) induction, antigen targeting, and depot generation. Antivenom production favors adjuvants that act through antigen targeting/presentation and those that generate antigen depots. Substances that are most commonly used for this purpose are Freund's adjuvant (complete and incomplete), bentonite, aluminum hydroxide, and sodium alginate (Chippaux and Goyffon 1998). Complete Freund's adjuvant (CFA) is a water-in-oil emulsion composed of heat killed *Myobacterium tuberculosis* bacteria, paraffin oil, and mannide monooleate. In some hyperimmunization protocols, CFA is combined with the antigen in the first inoculation. Because of CFA's aggressive nature, subsequent inoculations are combined instead with incomplete Freund's adjuvant (IFA), the same substance but lacking the bacteria, or with aluminum salts (see table 3.1).

3.2.3 VENOM DETOXIFICATION

Because whole venoms may be poorly tolerated, the selected animal is sometimes inoculated with a biologically detoxified venom, known as a toxoid, that maintains its ability to raise an immune response but reduces the stress caused to the animals (Chippaux and Goyffon 1998). A variety of detoxification protocols have been used on snake venoms, with varying outcomes. The most common of these are chemical methods using substances such as formalin, but production of toxoids using gamma radiation has also been suggested (Hati, Mandal, and Hati 1990; de la Rosa 2011). Unfortunately, these techniques can significantly increase production costs and complicate manufacturing processes. Generation of toxoids requires some degree of understanding of the immunological properties of venoms. For a toxoid to be useful in antivenom production, epitopes that are relevant for neutralization must be maintained (de la Rosa 2011). Therefore, detoxification methods that significantly alter protein structure are often useless.

3.2.4 ANTIVENOM PROCESSING AND ADVERSE REACTIONS

Adverse effects related to the use of antivenom can be divided into two main categories: early and late reactions. The first include true IgE-mediated anaphylactic reactions, which generally occur within the first minutes of antivenom administration and are characterized by a rapid loss of blood pressure (anaphylactic shock), airway constriction, and swelling of the epiglottis, which can cause suffocation. Anaphylactic reactions can be lethal and can be controlled by halting antivenom administration and by the use of parenteral epinephrine (Janeway et al. 2001). Early adverse events also include non-IgE-mediated anaphylactoid reactions that occur in the presence of pyrogens, complement activation by IgG aggregates in antivenoms, or direct toxic action of preservatives (phenol and thimerosal, especially in liquid antivenoms) (Garcia et al. 2002; Lalloo and Theakston 2003). Late adverse reactions include mainly serum sickness, mediated by IgM or IgG antibodies against antivenom proteins, and the consequent formation of immune complexes in the patient's blood (Leon et al. 2013). Symptoms usually start between one and two weeks after antivenom application and can last for several days. They involve diffuse rash, fever, hematuria, and intense urticaria, among others (Dart and McNally 2001) and usually resolve favorably. (See also chapter 2.)

3.2.5 FIRST-GENERATION ANTIVENOMS

The first therapeutic antivenoms, now referred to as first-generation antivenoms, were crude serum from hyperimmunized animals, injected essentially unmodified into the bitten patient. In addition to the specific and nonspecific IgGs, these products contained high concentrations of irrelevant proteins such as albumin, complement, and close to 200 other serum proteins (see figure 3.2). As a consequence, adverse reactions were very common (Currie et al. 1907).

3.2.6 SECOND-GENERATION ANTIVENOMS

Second-generation antivenoms consist of whole IgG molecules purified from hyperimmune serum or plasma (see figure 3.2). There are two main methods in current use for separation of immune proteins from other components of animal plasma, involving precipitation with ammonium sulfate or caprylic acid (Rojas, Jimenez, and Gutiérrez 1994). The former is a "salting out" protocol involving the precipitation of IgGs using ammonium sulfate. After separation from the supernatant, precipitated IgG is resuspended and submitted to other purification methods such as ultrafiltration (WHO 2010). In contrast with the ammonium sulfate method, caprylic acid causes the precipitation of nonimmune proteins, leaving IgGs in soluble form and therefore eliminating the need to redissolve it from the precipitated fraction. Because of this difference, protein yields and recovery of neutralizing activities are usually better when using the caprylic acid technique. The use of this method also decreases the content of protein aggregates (Otero et al. 1999; WHO 2010) and has been shown to have some antiviral activity (Burnouf et al. 2004).

Safety of second-generation antivenoms depends greatly on the effectiveness of purification, which determines the presence of protein aggregates, albumin, and other contaminants in the final product. A randomized blinded clinical trial comparing the two precipitation protocols

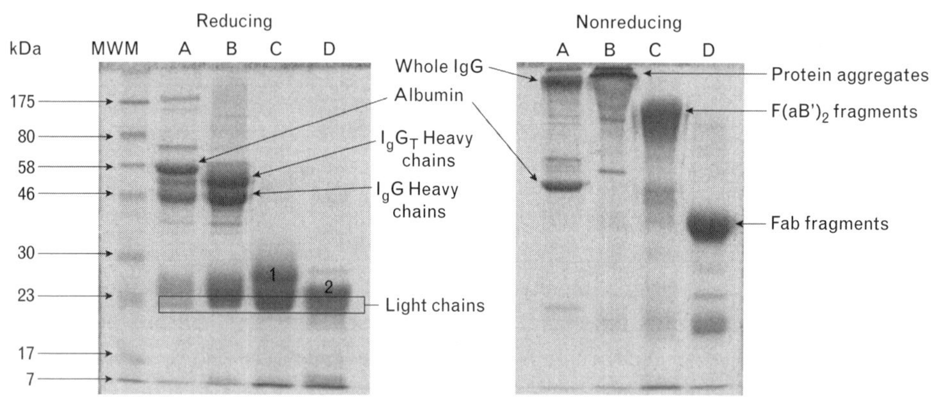

FIGURE 3.2: Comparison of antivenom composition in a 12.5% 1D SDS-PAGE gel, 15 μg of protein loaded per lane. MWM = molecular weight markers; A = Merck anti-*Latrodectus* antivenin (an example of an equine whole-serum antivenom [first-generation antivenom]); B = Suero Antiofídico Polivalente from the Clodomiro Picado Institute (an example of an equine whole-IgG antivenom [second-generation antivenom]); C = ViperStat North America from Veteria (an example of an $F(ab')_2$ antivenom [third-generation antivenom]); D = Crofab (an example of an ovine Fab antivenom [third-generation antivenom]); 1 = pepsin-digested equine IgG/IgGT heavy chains; 2 = papain-digested ovine IgG heavy chains. (Photo: Felipe Olvera.)

demonstrated that the material purified by caprylic acid had a lower content of protein aggregates and albumin and produced significantly fewer early adverse reactions (incidence of 25%) than the antivenom purified by ammonium sulfate (incidence 52%) (Otero et al. 1999). Regardless of the purification method used, adverse reactions caused by second-generation antivenoms are fewer overall than those caused by therapy with first-generation antivenom (crude animal serums).

3.2.7 THIRD-GENERATION ANTIVENOMS

The Fc region is involved in cellular response and what are generally called effector functions of antibodies, and thus it is responsible for some of the more commonly encountered adverse reactions (Alagón 2002). In third-generation antivenoms, the Fc region is removed through enzymatic digestion. Pepsin digestion produces F(ab')$_2$ fragments, whereas papain digestion produces Fab fragments (Chippaux and Goyffon 1991; Guidlolin et al. 2010) (see figure 3.1C). Thus, the term "fabotherapeutic" (or "fabotherapic" in Latin languages), proposed by Alejandro Alagón in the late 1990s, is now generally accepted to refer to antivenoms produced with these procedures. In clinical trials, fabotherapeutics have been proven to be even safer than second-generation antibodies, with rates of immune reactions to an F(ab')$_2$ antivenom sometimes two orders of magnitude lower than to whole-IgG products (Boyer et al. 2013a). Mexican epidemiology data show rare, and usually mild, early adverse reactions or serum sickness among more than 250,000 patients treated with scorpion and viper fabotherapeutics (Alagón 2002). Although lack of the Fc is in part responsible for this increase in safety, the greater purity of the active component and lower overall protein content are also of great importance (Boyer et al. 2013a). Given that the Fc is not involved in the interaction of Fab with venom components, its removal has little impact on the neutralizing capabilities of the antivenom. However, it is important to note that pharmacokinetic parameters, including absorption, residence time, and elimination, are very different among complete IgGs, Fab, and F(ab')$_2$ fragments (see section 3.5).

3.2.8 PURIFICATION

The improvement of purification methodologies has steadily increased the safety of antivenoms over time. Regardless of their active compound (IgG, Fab, or F(ab')$_2$), antivenoms must be submitted to thorough purification, generally involving at least two manufacturing processes before use in human patients. Refinement usually includes an ultrafiltration step to remove low-molecular-weight components and agents used in plasma fractionation such as ammonium sulfate. In some cases, a nanofiltration step is performed to remove potentially present equine viruses, as required by WHO 2010, 5386. Sterilization by filtration through a 0.2 µm membrane is typically the last step before filling and closure of liquid antivenoms or filling for lyophilization of dried antivenoms.

3.2.9 ANTIVENOM STABILITY

In this context, stability refers to maintenance of antivenom safety and neutralizing properties for long periods of time. Many types of antivenom are currently commercialized in liquid form, which requires the use of preservatives such as phenol or thimerosal. For liquid antivenoms, temperature is a particularly crucial variable in stability, with aggregation precipitation occurring above refrigeration temperatures (Rojas et al. 1990). Given that protein aggregates are likely to cause adverse reactions, liquid antivenoms are recommended to be stored between 4° C and 6° C. In these conditions, they can maintain their beneficial properties for up to five years. Thus, cold-chain documentation is an essential aspect (Gutiérrez et al. 2009) and is necessary for the distribution of liquid antivenoms, especially in tropical and warm climates. Freeze-drying or lyophilization is a separation process used to extract the water molecules from a sample, based on sublimation (Boss, Maciel, and Vasco de Toledo 2004). It consists of applying a strong negative pressure to a sample that has been rapidly frozen at very low temperatures, and it is used to conserve its bioactive properties. When used in antivenom processing, lyophilization increases shelf life and reduces or eliminates the need for potentially harmful preservatives and storage at low temperatures. Disadvantages of this technique include an increase in production costs and the fact that solubilization of the antivenom can take more than an hour if lyophilization is not performed correctly. Properly lyophilized antivenom takes less than five minutes to solubilize.

3.3 ANTIVENOM EVALUATION

Evaluation of antivenom has two main objectives. The first is to determine its neutralizing properties against the homologous venom and other relevant venoms. The second is to judge the safety of the antivenom. (See also chapter 2, table 2.2.)

3.3.1 NEUTRALIZATION

The first step in the evaluation of the neutralization properties of antivenom is to use ELISA (enzyme-linked immunosorbent assay) to measure the presence of antibodies that specifically recognize the venom. These assays not only monitor the development of the immune response in hyperimmunized animals, but they may also be used to determine the cross-reactivity of the antibodies to venoms that were not used in the hyperimmunization process (see section 3.4). Western blotting allows the qualitative determination of antibody recognition of particular venom components and is particularly useful for the identification of poorly immunogenic toxins. Lack of recognition of clinically relevant toxins logically suggests lack of useful clinical effect. Recognition of venoms by serum antibodies does not necessarily mean that the antivenom will confer effective neutralization of the desired venom, however. Evaluation of the neutralization potency of antivenoms is therefore necessary. Available tests used for this purpose are numerous, but all consist of the measurement of the decrease of venom effects in relation to increasing amounts of antivenom (Chippaux and Goyffon 1998). The most common *in vivo* test is the neutralization of the venom's lethal activity. This is usually performed in mice because of their size and availability. It is performed by inoculating groups of mice with a determined amount of venom (usually between three and six LD_{50}) preincubated for 15 to 30 minutes with increasing volumes of antivenom. Survival rate is registered 24 to 48 hours after inoculation. The median effective dose (ED_{50}) is defined as the volume of antivenom that prevents the death of half of the injected population. Protocols that involve preincubation of venom with antivenom can be used to analyze the ability of antivenom to neutralize particular venom activities, including hemorrhage, edema, and necrosis, which may be challenging to ascertain otherwise (de Roodt 2002). Even though preincubation protocols, such as previously described, are currently the main tool used to evaluate the neutralization properties of an antivenom, they don't reflect what happens in a real envenomation, where venom and antivenom are often administered through different routes and with varying periods of time between them. Several methodologies with independent inoculation of venom and antivenom have been described for this reason.

3.3.2 SAFETY

Evaluation of the safety of antivenoms involves the verification of the absence of severe adverse effects or ideally any secondary reaction whatsoever. To date, chemical analysis of antivenoms can contribute greatly to the prediction of antivenom safety. An antivenom with the following characteristics can be predicted to cause only mild, if any, adverse reactions: undetectable albumin contamination, low pyrogenicity, lowest possible protein content, high specific neutralization potency, low protein aggregates, and minimized or absent toxic preservatives (Ramos-Cerrillo et al. 2008; Lanari et al. 2014). The WHO recommends the following tests as the minimum necessary for determination of the safety of an antivenom (WHO 2010):

- Neutralization potency against most relevant venoms.
- Protein concentration.
- Purity of active substance.
- Presence of protein aggregates.
- Presence of non-IgG contaminants.
- Pyrogen test.
- Sterility test.
- Concentration of excipients and osmolality.
- pH.
- Concentration of preservatives.
- Determination of traces of agents used in plasma fractionation.
- Visual inspection.
- Stability tests.

In terms of purity, the active substance should constitute more than 90% of the final antivenom product, and albumin content should be less than 1%. Methods such as reducing and non-reducing SDS-PAGE are useful to determine antivenom protein composition (see figure 3.2). Also, methodologies such as size-exclusion HPLC or regular gel filtration (see chapter 7.2.) can be used to analyze the sample for protein composition and for protein aggregates (see figure 3.1D).

3.4 CROSS-REACTIVITY AND VENOM VARIATION

Generally speaking, snake venoms within a particular genus or from closely related genera have similar antigenic and toxicological properties. This allows the production of antivenoms that are able to neutralize most species from a defined group despite using only select venoms in the hyperimmunization process. However, while the gross-level protein profile tends to be maintained between closely related species, the relative proportions and particular isoforms present may vary widely. Given that these variations are reflected in the clinical syndromes developed in patients, it is necessary to characterize venoms in order to predict the course of envenomations and also the relative antivenom cross-reactivity. Significant variation in venom profiles has been shown between species within the same genus (Fry et al. 2002; Fry et al. 2003c; Sanz et al. 2006; Calvete, Escolano, and Sanz 2007; Angulo et al. 2008; Fry et al. 2008; Gutiérrez et al. 2008; Lomonte et al. 2008; Tashima et al. 2008; Wagstaff et al. 2009) and between individuals within the same species, with the intraspecific differences found among geographic locales (Daltry, Wüster, and Thorpe 1996; Forstner, Hilsenbeck, and Scudday 1997; Fry et al. 2002; Fry et al. 2003c; French et al. 2004; Salazar et al. 2009; Boldrini-Franca et al. 2010; Mackessy 2010; Castro et al. 2013; Sunagar et al. 2014), between sexes (Daltry, Wüster, and Thorpe 1996; Menezes et al. 2006), and between juveniles and adults (Mackessy 1988; Daltry, Wüster, and Thorpe 1996; Lopez-Lozano et al. 2002; Calvete et al. 2009a). Venom variation has also been reported between venom glands of a single individual (Johnson, Kardong, and Ownby 1987). In extreme contrast is the extraordinary level of venom conservation within sea snakes as a consequence of their fish-specific diet (Fry et al. 2003c). Thus, the sole sea snake antivenom has an extraordinary level of cross-reactivity (Chetty et al. 2004). Further, it is able to neutralize the unrelated sea kraits (Chetty et al. 2004), which are an independent marine lineage but also specialize in fish prey.

The ability of an antivenom to neutralize a heterologous venom, that is, a venom from a species that is not used during hyperimmunization, depends on the presence of neutralizing epitopes that are shared with the homologous venom. Therefore, similarities and differences in venom composition must be considered during antivenom design. Usually, more than one venom is used as an immunogen during antivenom production, aiming to obtain antibodies that can recognize and neutralize components in homologous and heterologous venoms. The phenomenon of recognition and neutralization of a heterologous venom is referred to as cross-reactivity.

An important consideration is that the abundance of particular toxins within a venom used for hyperimmunization can influence an antivenom's neutralizing properties. Extremely lethal toxins, such as α-neurotoxins, can be clinically relevant even when they are present in very small proportion within a venom. When venoms of this type are used as immunogens, the immune response can be "distracted" by other, more abundant venom components. Such an antivenom would therefore perform poorly against α-neurotoxins (see Bénard-Valle 2009; Carbajal-Saucedo et al. 2013). Also, the smaller the toxin type, the less immunogenic it is. Thus, immunization with venoms composed of different toxin types may result in an antivenom with a stoichiometry skewed toward the larger or more abundant toxins.

In addition, variations in the composition of the venoms used as immunogens can result in heterogeneous neutralizing properties of different batches of antivenom (see table 3.2). *Crotalus simus*, for example, is one of the species utilized as an immunogen to produce Antivipmyn, one

Table 3.2: Comparison of Neutralization Potency between Batches of Antivipmyn

		Batch of Antivipmyn			
		B-9G-21	B-IH-15	B-1C-08	B-1E-09
	$3LD_{50}$ (μg)	EC_{50} (μlAV/$3LD_{50}$)			
C. s. culminatus[1]	487.2	> 400*	> 300*	>460*	>460*
C. s. culminatus[2]	199.6	> 450*	> 400*	115.2*	120*
C. s. simus	11.4	17.5	35.9	11	15

LD_{50} = medium lethal dose.

EC_{50} (μlAV/$3LD_{50}$) = neutralizing potency expressed as microliters of antivenom that neutralize 3 LD_{50}.

* Antivenom was resuspended in half the usual volume of saline (5 mL); still, 100% survival was not obtained with volumes equal to or lower than these values.

[1] Individual from Morelos, Mexico.

[2] Individual from Guerrero, Mexico.

of the commercially available antivenoms in Mexico. Because of this, and given the variation previously observed in the venom of this species (Saravia et al. 2002), further characterization of the venom from adult individuals of this species along its distribution revealed that individuals from the subspecies *C. s. simus* have very similar venoms, while the venoms of the other two subspecies, *C. s. culminatus* and *C. s. tzabcan*, had important individual variations, even when they were from animals collected in the same locality (Neri-Castro 2013). These differences were primarily in components between 4 and 16 kDa and thus are most likely in β-defensins (see chapter 9) and PLA_2 (see chapter 21).

The production of efficient antivenoms requires a great deal of knowledge about the biochemical, toxinological, and immunochemical characteristics of venoms. Much work in this area remains to be done.

3.5 VENOM AND ANTIVENOM PHARMACOKINETICS

Pharmacokinetics (PK) is the science that studies the process of absorption, distribution, and elimination (excretion and/or metabolism) of a substance in the body (Shargel et al. 2005). The previously described heterogeneity of venoms and the mechanism of action of their components give them a range of PK properties with implications in their pharmacodynamics. This last term refers to a venom's interaction with target molecules as well as the consequences of these interactions (Shargel, Wu-Pong, and Yu 2005). The analysis of PK properties of antivenoms is also relevant, since a mismatch between venom and antivenom PK has been shown to be responsible for recurrence of envenomation symptoms after treatment with antivenom (Boyer, Seifert, and Cain 2001; Shargel, Wu-Pong, and Yu 2005). When performing antivenom

PK studies, the nature of the antivenom and the animal model chosen must be considered carefully to avoid discordance in PK profiles. Research has shown that antibody clearance and half-life are influenced by the Fc interaction with specific receptors, which exhibit some variation in specificity between species. For example, murine antibodies have shorter half-life in humans than humanized or chimeric antibodies. Conversely, human IgGs have long half-lives in mice, because murine FcRn (neonatal Fc receptor) binds human IgGs (Tabrizi, Tseng, and Roskos 2006).

Given that antivenom infused intravenously usually reaches its target venom proteins in plasma, interaction with endogenous proteins that could modify an antivenom's PK profile is generally avoided. For this reason, molecular mass becomes the dominant factor influencing antivenom distribution and elimination. Differences in antivenom PK profiles have been studied because of their implications in treatment of envenomations, showing that their distribution volumes and half-life times are crucial. In their review, Gutiérrez and collaborators (Gutiérrez, Leon, and Lomonte 2003) showed that in clinical studies, Fab and $F(ab')_2$ fragments exhibit larger distribution volumes than IgG antivenoms. They also observed that Fab fragments have shorter half-lives than whole IgG and $F(ab')_2$. The large distribution volumes exhibited by Fab fragments suggest distribution in extravascular space, which gives the advantage of reaching and neutralizing toxins in deeper tissues, allowing fast neutralization of toxins. On the other hand, the short half-life exhibited by these fragments suggests that small size leads to rapid clearance, too. This implies that Fab antivenom leaves the body faster than venom, giving place to the recurrence phenomena well described for this type of antivenoms (Warrell et al. 1986; Boyer et al. 1999; Bogdan et al. 2000; Bush et al. 2013). Whole IgG molecules have small volumes of distribution, suggesting that the antivenom only reaches plasma and interstitial fluids of highly perfused organs and limiting the neutralization of venom that has already reached its target tissue. On the other hand, IgG clearance seems to be slow enough to have a long half-life that keeps the neutralization properties of antivenom in the organism for many hours. $F(ab')_2$ seems to be intermediate between IgG and Fab, with a volume of distribution closer to that exhibited by Fab and a half-life long enough to avoid most recurrence phenomena (Boyer et al. 2013b). Selection of the best antivenom for treatment of envenomation should take antivenom and venom PK and pharmacodynamic profiles into account.

CHAPTER 4

INEFFECTIVE TRADITIONAL AND MODERN TECHNIQUES FOR THE TREATMENT OF SNAKEBITE

M. BÉNARD-VALLE, E. E. NERI-CASTRO, L. BOYER, T. N. W. JACKSON, K. SUNAGAR, M. CLARKSON, AND B. G. FRY

4.1 SUMMARY

Snakes and their venoms have fascinated people for thousands of years. Many intriguing and unusual features of snakes have contributed to the mystical aura that surrounds these animals, transforming them into symbols of great importance in almost every society. Because of the shedding of their skins, they have been used to represent rebirth, fertility, and immortality in ancient cultures such as that of the Aztecs, who honored the feathered snake Quetzalcoatl as the creator of the present era and a powerful symbol of life and renovation. Defensive displays and the venoms that back them up in many species have also resulted in fear and aversion, turning snakes into incarnations of evil in various cultures. Thus, they have often been characters in stories representing the battle between good and evil, such as the Nordic myth of Jörmungandr, the world serpent and archenemy of the god Thor. Various shamanic societies, such as the Lacandonians from Chiapas, Mexico, regard snakebites as a punishment for inadequate behavior or sins committed (Leyton-Ovando 2001).

Snakebites involve complex pathophysiology characterized by a heterogeneous set of signs and symptoms. In addition, they are often associated with deeply rooted social practices and religious beliefs. In order to deal with them, traditional medical practitioners from around the world have developed an incredibly diverse array of remedies and emergency protocols, many of which are embedded in strict and complicated rituals to reinforce the central importance of the traditional healer in village life. The Culebreros of Mexico, the Sangomas of South Africa, and the Ozhas of West Bengal are just a few among the great number of traditional healers who claim to have the ability to cure snakebite. This alleged power to heal grants them great status in their communities, and its acquisition requires a great deal of study and numerous initiation rituals.

To date, the only specific therapy that has scientifically validated efficacy for treating snakebites is the application of specific antivenom. However, certain social, religious, and economic issues have made this therapy unavailable or have discouraged its use in many places. A high percentage of people involved in snakebite accidents around the world address a local medical practitioner or traditional healer before reporting to an established medical facility. For example, between 1980 and 1989, in the district of Burdwan, West Bengal, 65.5% of snakebite patients sought traditional healers, and only 22.1% received hospital treatment (Hati et al. 1992). In Africa, traditional practitioners were consulted far more frequently than modern health centers, with 50% to 90% of people seeking traditional treatment prior to attending hospital (Chippaux 2009). This is a serious situation, particularly in developing countries where medical facilities are less accessible and reliable information is not always available. Regrettably, hospitals with well-established treatment protocols stocking specific antivenom are scarce at best. Arrival at these facilities, especially when bites occur in remote areas, can take several hours or even days, and once there, treatment can be extremely expensive (Chippaux 2009). Ironically, when snakebite treatment has suboptimal outcomes resulting from the delay in obtaining modern medical treatment caused by initially seeking traditional remedies instead, the traditional healer is not typically held accountable. More often, modern medicine is held up as the failure.

The development of modern medicine is not causing traditional practices and local recommendations to disappear, even though the optimal course of action in terms of prehospital and hospital care surrounding snakebite accidents has received considerable attention in the medical community. Even industrialized countries, where people have better access to information and medical facilities, are no strangers to this problem. The popular practice of maintaining venomous snakes as pets may result in accidents of varying severity (Schaper et al. 2009). Most of these reptile enthusiasts are not professionals and may base their emergency protocols on unreliable or incomplete information from various sources. A variety of recommendations based mainly on anecdotes from both medical and nonmedical sources are now available through the Internet and specialized literature, adding to an already rich collection of unproved therapies. A recent review of 48 websites found using two major search engines showed that 54.1% contained inappropriate treatment recommendations (Barker et al. 2010). This highlights the fundamental principle that people who keep such animals as pets, which can be done quite safely (see chapter 5), should only do so under a rational permitting system that includes significant training requirements in order to obtain the license.

Characteristics inherent in snakebite accidents may promote the use of unproven remedies, whether traditional or more modern. For example, a high percentage of bites by venomous snakes result in either weak envenomation or no envenomation at all (dry bites). Together with the misidentification of nonvenomous snakes as dangerous species, these cases can result in the incorrect conclusion that the remedy applied was effective, thus anecdotally increasing its credibility. These nonvenomous bites may represent some 50% of all snakebites around the world (Chippaux 1998; Kasturiratne et al. 2008). The employment of unproven remedies has many disadvantages, some of which are not readily apparent. One of them is the generation of a false sense of security, postponing, or, in the worst cases, replacing, the search for adequate treatment. Use of multiple-step pathways to treatment, such as prior consultation of traditional health practitioners, significantly delays presentation at a hospital (Sloan, Dedicoat, and Lalloo 2007). Delays in treatment have been related to increases in the severity of injuries and increased mortality rates in snakebite accidents (Chippaux, Ramos-Cerrillo, and Stock 2007). Another more obvious risk is the utilization of remedies that cause harm, such as scarification or exposure to potentially toxic substances. Multiple cases of damage resulting from inadequate handling of wounds and illnesses by traditional practitioners are reported in medical literature (Dada et al. 2009; Unuigbe et al. 2009). In addition, some alternative measures, such as tourniquets or cryotherapy, can increase the damage caused by snakebite.

The number of people who apply some type of traditional remedy or unproven therapy before going to a hospital is surprisingly high. A study that included 147 snakebite victims treated in the Hlabisa hospital in KwaZulu-Natal, South Africa, between November 1995 and October 1996 showed that 90% had used some form of traditional remedy prior to hospital admission, 89% had used two or more each, and 49% had used at least three each. One patient had used six different remedies (Newman et al. 1997). In some cases, traditional practitioners specifically advise against attendance at a modern medical facility. An example that illustrates this situation can be found in an article from the *Wall Street Journal* in which a local priest from the province of Mushari, India, discourages attendance at hospitals, claiming that "if you don't listen to God and instead go to a hospital, it's at your own risk" (Trofimov 2007). The healing practice employed by this priest consists of bathing in a sacred pool and the application of a special type of mud to the snakebite wound while chanting prayers.

There are very few systematic studies that address the efficacy of alternative methods. In this chapter, we present a critical analysis of the available scientific evidence, evaluating recommended treatments and traditional remedies in order to assess their viability as first aid measures or additional therapies during a snakebite accident. In addition, we evaluate the possible dangers of the application of each unproven therapy.

4.2 MECHANISMS FOR VENOM REMOVAL

4.2.1 SCARIFICATION, INCISION, AND AMPUTATION

The practice of enlarging the wound is one of the oldest first aid treatments for snakebite. It is an intuitive measure and has been favored in many different places; scarification is a common treatment in indigenous cultures, and healing by bleeding was a feature of Western medicine until relatively recently. Inappropriate cutting may result in severing of muscles, tendons, and blood vessels. Such mechanical damage may result in severe bleeding, which may be greatly potentiated in cases involving venoms causing disruption of blood clotting. Cutting may also allow venom remaining on the surface of the skin direct access to the tissues beneath. Furthermore, the likelihood of secondary infection is high. This practice is not beneficial, and the risk of causing damage and increasing tissue loss is very high; no type of incision should be done acutely to any snakebite wound.

Scarification has long been used in ceremonies and religious rituals in places as diverse as Africa, Papua New Guinea, and Australia. Numerous cuts, often in intricate patterns, are used as part of cleansing and healing rituals (Klauber 1956). In addition to promoting bleeding, this is done as a way of administering substances with apparent healing effects into the wound. In addition to the bleeding risk, there is also the possibility of introducing additional toxic components to an already complex pathology or of seeding infections.

In early Western medicine, healing through bleeding was a common medical practice used to treat a variety of diseases including snakebite. Nicander of Colophon first described the use of cutting wounds in Western medicine in medical literature in the second century B.C. (Gow and Scholfield 1953; Klauber 1956). The number and shape of the incisions and the tools used for cutting vary, with X-shaped cuts over the fang marks and suction being the typically recommended snakebite treatment. Other, less common practices include performing multiple cuts of varying depth around the fang marks. Some people have been reported to bite the wound in the absence of tools to make incisions (Wingert and Chan 1988). Suction and a tourniquet are often recommended as accompaniments to incision. Commercial first aid kits have been designed to facilitate incision and suction in the field, and some are still available. They typically include a suction device, a scalpel or sharp blade, sterile swabs, and some kind of tourniquet or ligature. Although it is still performed in many parts of the world, the medical consensus is against this measure (Hultgren 1994). This practice contradicts the first principle of first aid treatment, "first, do no harm," as a lot of additional damage may result (Hall 2001). Sterile surgical blades are rarely available in the field, where most snakebite occurs. Rather, incisions are performed with a variety of nonsterile tools, such as sewing needles, kitchen or field knives, or broken glass. In traditional healing practices, a diverse collection of tools is used to incise snakebites. Culebreros from Veracruz, Mexico use the fangs of nonvenomous snakes, the thorns of orange trees, and sharpened pieces of deer antler (Leyton-Ovando 2001). The likelihood of secondary infection is greatly increased in these cases (Jorge et al. 1990; López, Lopera, and Ramírez 2008). When mouth suction is applied or when the wounds are bitten, risk of infection is even higher because of bacteria present in the human mouth (Goldstein et al. 1978). Even though some experiments performed in the 1920s appeared to show an improvement in survival rates of envenomed dogs as a result of this practice, more recent studies failed to observe any improvement of outcome in envenomed patients treated this way (Wingert and Chan 1988; Hall 2001). In a study of surgical excision of subcutaneous tissue around the bite area in 54 patients, 16% developed complications (Huang et al. 1974). Percentages of complications are much lower in patients treated with antivenom (Hall 2001). A study on human volunteers demonstrated that blood spontaneously oozing from a mimicked snake-fang wound contained a very small amount of the injected "mock venom" (Alberts, Shalit, and LoGalbo 2004). Even if the wound is enlarged and allowed to bleed profusely, the amount of recovered venom is unlikely to be significant.

The most extreme form of incision is amputation of a finger or toe in order to avoid what is believed to be an otherwise certain death. This practice is extremely painful, incapacitating, unnecessary in the case of dry bites, and potentially ineffective because of rapid absorption and transmission of the venom away from the affected digit.

The use of cutting as a snakebite treatment has not been limited to the confines of traditional remedies or misguided amateur attempts at medicine. Fasciotomies have long been a standard practice in the treatment of *Crotalus* envenomations in parts of the United States, and this archaic practice persists today. Purported justification for this radical surgery has been to alleviate intramuscular compartmental pressure (Cumpston 2011). This is despite a conspicuous absence of evidence of venom actually penetrating the fascia and thus causing pressure sufficient to restrict blood flow within the muscle (Holstege et al. 1997). Compartment syndrome as a result of raised intracompartmental pressure is mistakenly diagnosed when, in fact, the true pathology is superficial prefascial envenomation (Cumpston 2011). There is a growing body of evidence that not only are fasciotomies generally unnecessary, but they also cause great harm (Cumpston 2011). This is not to say that legitimate cases of intracompartmental pressure have not been reported (see Roberts, Csencsitz, and Heard 1985), but in these cases, the appropriate treatment is antivenom, and fasciotomy (see color plate 12E) is of unproven value (Tunget-Johnson et al. 1998; Rosen, Leiva, and Ross 2000; Gold et al. 2003).

4.2.2 MECHANICAL VENOM EXTRACTORS

The use of suction is, without a doubt, the most popular of nonmedical attempts at first aid and the most familiar to anyone who has seen snakebite in movies. After the dangers of incision and scarification were made evident in most parts of the world, mechanical suction without incision became widely advocated. Commercial venom-extraction kits usually consist of a negative pressure pump with cups of different sizes or other types of suction devices such as rubber bulbs. Most also include razors to remove hair. These kits are promoted as effective first aid treatment for snake and insect bites or stings, with statements that they can "safely and quickly remove significant amounts of venom" (Sawyer Products 2014). The purpose of this method is to extract venom from the bite area in order to reduce the damage and symptoms caused by envenomation. The use of suction is based on two primary assumptions: that injected venom remains in the bite site long enough to be removed and that it is possible to remove the venom from the bite site through the use of suction.

In snakebite, venom may be delivered to a wide range of tissue depths, from subcutaneously through to intramuscularly, with numerous variables affecting the outcome (see chapter 2). The subsequent speed of dispersion of venom is also highly variable, being affected not only by bite site and depth but also by the diverse natures of the injected toxins. When venom reaches the interstitial space under the skin or in the muscle, it interacts with the molecules of the extracellular matrix (ECM). These interactions favor the spreading of venom components through surrounding tissues, hindering their physical removal. Furthermore, enzymes present in many types of venom, such as hyaluronidase (see chapter 24.11), and can act like spreading factors, degrading the ECM to accelerate venom diffusion and absorption (Kemparaju, Girish, and Nagaraju 2010). The lower the molecular weight of the toxins, the greater the likelihood of absorption through blood capillaries and consequent rapid systemic dissemination. Conversely, the higher the molecular weight of the toxins, the greater the likelihood of slow transport in the lymphatic fluid along the affected limb until a lymph node is reached, after which the toxins are absorbed into the blood and quickly transported to the rest of the body. Given that snake venoms are complex mixtures containing proteins of varied molecular weight, some toxins will reach the bloodstream earlier than others. Neurotoxins, typically of low molecular weight, will reach systemic circulation faster than enzymatic, high-molecular-weight toxins such as metalloproteases. Studies have shown that measurable quantities of venom appear in the bloodstream appears a few minutes after the bite (Audebert et al. 1994; Paniagua et al. 2012). However, significant venom levels can be measured at the bite site several hours after the accident (Paniagua et al. 2012). Hence, there will be at least some venom at the bite site in the minutes after the accident, but it will decrease over time as the absorption process takes place. Given that absorption starts immediately, it is very unlikely that 100% of any venom is available for extraction even in the first minutes after inoculation.

The first controlled experiments for analyzing the amount of venom that can be extracted from tissues were made in the 1960s and reported positive results. In one of these, rabbits injected with the venom of *Daboia palestinae* were used as a study model (Shulov et al. 1969). The authors reported that when the venom was injected subcutaneously, as much as 46% of it could be extracted if the suction device was applied immediately after injection. When the venom was injected intramuscularly, the amount of extractable venom was reduced to 11%. Almost two decades later, Bronstein and collaborators reported the extraction of as much as 30% of *Crotalus* venom injected subcutaneously in rabbits (Bronstein, Russell, and Sullivan 1986). However, this work was published only as a non-peer-reviewed abstract and doesn't detail its precise methodology. Despite these significant shortcomings, it is still cited as evidence of the efficacy of venom extractors by product manufacturers.

The initial success in the extraction of venom was probably a result of the selected animal model. Rabbits, unlike humans, have a very small subcutaneous layer that may make venom extraction easier (Bush 2004). Research using a porcine model (in which skin anatomy is similar to human) analyzed the effect of a commercial extractor device on local tissue injury and swelling. It was concluded that the extractor did not reduce swelling or tissue injury and even resulted in further injury in some cases because suction-induced tissue ischemia increased tissue loss during envenomation and resulted in lesions (Bush et al. 2000; Bush 2004). Lesions associated with suction devices have also been reported in human patients (Bush and Hardy 2001; Holstege and Singletary 2006). A study on human volunteers administered subcutaneously with radioactively labeled "mock venom" reported that the extractor removed only between 0.04% and 2% of the injected venom (Alberts, Shalit, and LoGalbo 2004).

Thus, despite the considerable debate in the medical community concerning the benefits and safety of these first aid devices (Gellert 1992; Forgey 1993), no proper scientific research or systematic clinical trial has demonstrated that using a venom-extractor device will improve the outcome of envenomation. Indeed, not only are such devices now revealed to be ineffective, but they have also been shown to potentiate local damage. Because of this recent evidence, the latest first aid recommendations no longer advocate mechanical suction as a useful measure in snakebite cases (Boyd et al. 2007; Gil-Alarcón, Sánchez-Villegas, and Reynoso 2011).

4.2.3 BLACK STONES OR SNAKE STONES

"Black stones" are pieces of charred bone, usually from a domestic animal, that are applied to the bite site. It is claimed that they stick to the wound until they have absorbed all the venom, after which they fall off. It is also claimed that subsequently boiling and then soaking the stone in milk will recover its healing properties (REAP 2013). Black stones came to the attention of European medicine in the 17th century, when they were carried back by merchants and missionaries from India, China, and Southeast Asia. It was said that expert snake charmers obtained the stones from the skulls of venomous snakes. Proposed as a cure for snakebites and many other poisonings, they generated strong debate concerning their real therapeutic value. Francesco Redi and Athanasius Kircher made the first tests of black stones on animal models while participating in this debate (Redi 1671). Even though these experiments yielded negative results, anecdotes of their success prevented black stones from being discarded as perceived medical tools (Baldwin 1995). Today black stones are still used as a traditional treatment in many parts of the world, particularly in Africa and India. As with many other traditional practices, evidence of their efficacy is based solely on oral tradition (Rasquinha 1996).

There has been only one systematic study to date concerning the therapeutic potential of black stones as treatment for snakebite (Chippaux, Ramos-Cerrillo, and Stock 2007). In this work, the authors used mice as a study model to perform *in vitro* and *in vivo* experiments with the venom of three of the most medically important African snakes, *Bitis arietans, Echis ocellatus*, and *Naja nigricollis*. In four strictly controlled experiments, they made the following observations: (1) *In vivo* application of a black stone at the injection site did not increase the survival rate of mice inoculated with three medium lethal doses (LD_{50}) of venom, even when applied immediately after injection. (2) *In vitro* incubation of venom proteins with black stone powder is not a clinically relevant test. Thus, based on the weight of the evidence, they concluded that black stones were an ineffective form of treatment for snakebite.

4.3 PHYSICAL TREATMENT OF THE BITE AREA

4.3.1 ELECTRIC SHOCK

In 1986, a letter was published describing the use of high-voltage, low-current electric shocks as treatment for snakebites in a population of Waorani Indians from Ecuador (Guderian, Mackenzie, and Williams 1986). The method consisted of the application of four to five high-voltage direct-current (HVDC) shocks (20–25,000 volts; < 1 mA) directly to the bite site. The current was originally delivered using an insulated probe connected to the spark plug of an outboard or car motor. In a further refinement, the motor was replaced by a hand-held electrical shock device, or "stun gun," that delivered a pulsating direct current of similar characteristics. The letter reported 34 snakebite patients on whom the described method was successfully used within the first 30 minutes after the accident. The use of this novel treatment spread rapidly, particularly in the United States, where close to 50 local newspapers and four popular magazines published articles promoting its use as a potentially lifesaving first aid measure (Altman 1986; Johnson 1988; Bucknall 1991). Despite a lack of supporting evidence, it was advertised as useful for virtually every envenomation caused by an animal, including bee stings and bites from spiders or snakes. Most of these articles cited individual case reports and provided only circumstantial evidence of the treatment's value, as commented on by David L. Hardy in the *Tucson Herpetological Society Newsletter* (Hardy 1988). One of the most widely cited articles was an editorial in the magazine *Outdoor Life*, which enthusiastically recapitulated the cases reported by Guderian and cited other cases of allegedly successful treatments utilizing variations of the original HVDC protocol (Mueller 1988b; Mueller 1988a). Unfortunately, this article did not provide evidence of a causal connection between the application of HVDC and positive outcomes of cited envenomations. Following the media excitement concerning the use of electric shock in snakebite treatment, a company called J & K Industries developed a special model of stun gun that delivered close to 25,000 volts of direct current (Bucknall 1991). It promoted this device as a "field tested and proved" cure for snakebite, carelessly implying that seeking professional medical attention was of secondary importance (Mackey 1988). In 1990, the US Food and Drug Administration banned the promotion in that country of electrical devices as having any therapeutic value (Bucknall 1991).

The biological basis of the therapeutic value of HVDC was never clearly established, although various mechanisms were proposed. One of these suggested that the current could directly inactivate the venom proteins (Guderian, Mackenzie, and Williams 1986). Independent of its veracity, this theory did not provide a logical argument that could explain why bodily proteins (belonging to the same families as venom proteins) were not affected by the electric current. Another hypothesis proposed that the electric shock produced an intense vascular constriction that arrested the venom in the bite zone until it broke down and ceased to be active (Guderian, Mackenzie, and Williams 1986). This theory does not account for venom proteins that act on local tissue such as myotoxins and metalloproteases, which are abundant in venoms of snakes such as *Bothrops* and *Crotalus*. More recent research has shown that venom proteins remain active in bodily tissues and the bloodstream for several days after the initial envenomation (Audebert et al. 1994; Riviere et al. 1998).

One study tested the hypothesis that venom proteins were selectively affected by evaluating the effect of an electrical current from a commercial stun gun directly applied to a *Crotalus adamateus* venom solution in an electrolysis cell (Davis et al. 1992). Even though the venom was shocked for a significantly longer time than recommended by stun gun manufacturers, no effect was observed on the potency of the venom as determined by murine LD_{50} testing. In order to test the therapeutic value, various controlled studies were conducted using different animal models. Injecting mice with the venom of *Crotalus oreganus* (Johnson, Kardong, and Mackessy 1987) assessed its lethal effect. The effect on local tissue damage was analyzed in rats using *Bothrops atrox* venom (Howe and Meisenheimer 1988) and in rabbits using *C. adamanteus* venom (Stoud et al. 1989). None of these studies recorded an improvement in the outcome of experimental envenomation related to the application of HVDC. Two literature reviews on the matter also concluded that there was no scientific evidence to support the use of HVDC in the treatment of envenomation by snakes or other animals (Bucknall 1991; Welch and Gales 2001). Other case reports in the literature suggest that HDVC is of little, if any, therapeutic value and that it is actually potentially dangerous (Dart and Gustafson 1991). To date, there is no scientific evidence that supports the use of this method, and, like any unproven therapy, it may be more harmful than beneficial. Cases of mild to severe burns caused by stun guns have been reported (Burdett-Smith 1997), and these could be worsened by the use of an uncontrolled power source such as an outboard motor or a car's spark plug. There is also the possibility of stimulation of cardiac muscle by stun guns and similar devices (Nanthakumar et al. 2008). Finally, lack of standardization of nonmedical protocols such as this one may increase the risks of an already hazardous measure; for example, bite victims may mistakenly utilize high-current power sources that could cause fatal electrocution.

4.3.2 CRYOTHERAPY

Cold therapy was initially projected for treating snakebites in the early 1900s. It involves freezing with ice packs or ethyl chloride sprays and prolonged submersion in ice water (Klauber 1956). All of these methods are based on the hypothesis that reducing temperature reduces venom activity. However, even if this actually worked on enzymatic toxins, it would not affect other toxin types such as peptide neurotoxins. A critical review of the available evidence in favor of and against cryotherapy was conducted and concluded that the method was ineffective and should not be used (Frank 1971). A more recent study that experimentally analyzed this practice failed to demonstrate any benefit of the application of cold (or heat) on local tissue damage induced in rats by *Agkistrodon piscivorous* venom (Cohen, Wetzel, and Kadish 1992).

Treating snakebite envenomation with cold may worsen the state of the wound in several different ways. Initial exposure to cold temperatures on the skin activates thermoregulatory mechanisms and causes vasodilation, which may contribute to the spreading of venom. Additionally, the severe edema caused by many snake venoms can compromise circulation in distal parts of an affected extremity, leading to ischemia and possible tissue loss. Cold-induced vasoconstriction, which takes place when tissues are cooled between 10° C and –2° C, would definitely worsen this effect (Mohr, Jenabzadeh, and Ahrenholz 2009). The same vasoconstriction could also increase the risk of local tissue damage by arresting venom proteins in a single zone. Even without envenomation, prolonged exposure of the whole body or body parts to low temperatures causes severe damage (Kiss 2012). Cold exposure can cause injuries of varying severity, including frosting, chilblain, and frostbite. The last is caused by the slow freezing of tissues and in severe cases leads to tissue loss and amputation (Mohr, Jenabzadeh, and Ahrenholz 2009).

Current snakebite treatment protocols strongly advise against the use of cryotherapy (Boyd et al. 2007; Gil-Alarcón, Sánchez-Villegas, and Reynoso 2011).

4.3.3 CAUTERIZATION

Cauterization is an ancient remedy utilized to stop hemorrhage and sterilize wounds. Different methods, such as the application of a hot iron or burning gunpowder, have been used to treat snakebite according to the theory that intense heat can destroy venom components. Various chemical caustics, such as silver nitrate, nitric acid, and potassium hydrate, have also been used (Klauber 1956). This practice was very common as a complement to other therapies in the 1800s, before antivenoms therapy was an option and when very little information was available concerning the nature of envenomation. In two fatal cases of snakebite, lunar caustic (silver nitrate) was applied to the bite site in order to stop the bleeding caused by previous treatment attempts, but no beneficial effects were recorded (Powell 1853). In 1871, a note was published in *Nature* reporting the successful treatment of 70 snakebite patients using melted wax over the fang marks (*Nature* 1871). In this note, the treating physician stated that the therapeutic effect came from the prevention of air entering the wound.

This is an extremely painful and damaging practice and should not be attempted for any reason. No benefit has ever been shown from cauterization of snakebites in clinical or experimental studies.

4.3.4 TOURNIQUETS

As mentioned earlier in this chapter, numerous first aid practices are aimed at eliminating or delaying the effects of venom. Application of tourniquets or ligatures is one of the most long-standing and common practices around the world (Tun-Pe et al. 1987). The theory is that by tying a tourniquet to the affected extremity, the injected venom will be retained in a small area around the bite site, affording the bite victim a longer period of time to get to a hospital. Unfortunately, it is not that simple, as a lot of aspects have an influence on the possible benefits or damages resulting from this practice (see color plate 14). These aspects include (but are not limited to) the material used to make the tourniquet, the pressure it applies, and its location. Another important feature is which species of snake is responsible for the bite, given that absorption and distribution rates vary greatly between different venom components. Depending on the experience of the person tying the tourniquet and the available materials, many variants of this method have been proposed. All of these have generated contrasting experiences and experimental results, which have led to great controversy regarding this method (see Currie 2006). The pressure applied and the amount of time it is maintained are some of the most relevant variables influencing the outcome of the treatment and the possibility of damage caused by the tourniquet itself. Unfortunately, most tourniquets are applied incorrectly, utilizing too much pressure and maintaining it for too long, which can lead to tissue damage through ischemia. Damage caused by tourniquets can hide the symptoms caused by snakebite and complicate hospital treatments (Tun-Pe et al. 1987; Currie 2006). Epidemiological studies reveal this practice to be an extremely common pre-hospital measure around the world. In two cities in Myanmar, 152 snakebite accidents were reported, of which 139 (91%) victims had had a tourniquet applied (Tun-Pe et al. 1987). In Belo Horizonte, Brazil, 97 of 114 patients bitten by *Crotalus durissus* applied tourniquets (Amaral et al. 1998). This study reported that the application of tourniquets resulted in no therapeutic advantage. Paresthesia, palpebral ptosis, respiratory distress, and high concentrations of creatine kinase enzyme were observed in patients with and without tourniquets. No local tissue damage was observed in these cases, probably because the main component of this venom is crotoxin, a potent presynaptic neurotoxin. Another study showed that 78% of the patients bitten by *Daboia russelii* who applied tourniquets did not achieve any significant delay in venom distribution to the rest of the body (Tun-Pe et al. 1987). Similarly, no efficacy of tourniquets to delay systemic envenomation was observed in patients bitten by *Echis carinatus* (Bhat 1974). In Mexico, there are no statistics available, but the authors' experience has been that this first aid measure is extremely common in the whole country and that no therapeutic benefits are gained from its application.

Far from being beneficial, tourniquets have been shown to cause severe harm to the ligated extremity. In cases of bites by snakes whose venoms cause local damage, such as hemorrhage, inflammation, and necrosis, any type of tight ligature may arrest venom proteins in the vicinity of the bite, accelerating and increasing damage to local tissues (Amaral et al. 1998). In Brazil, ligatures are strongly contraindicated, given that 90% of snakebites in that country are caused by species of the genus *Bothrops*, whose venoms have very potent, locally acting components.

It is clear that in most cases, tourniquets offer virtually no benefits to snakebite patients. Their incorrect application can lead to tissue damage by itself, and even when they are tied correctly restriction of venom proteins to a single area can increase tissue damage.

4.4 BOTANICAL AND CHEMICAL REMEDIES

4.4.1 BOTANICAL REMEDIES

For centuries, medicinal plants have played a fundamental role in traditional treatments, being used for thousands of pathologies around the world. Herbal medicine is one of the oldest sciences, and it doubtless predates the first recognized civilizations. Throughout the centuries, humans have used plants to cure illnesses. Through empirical testing, societies around the world have accumulated great knowledge of their surroundings, including the identification of plants that can be eaten, are dangerous, or have medicinal properties (Otero et al. 2000a).

The use of plants is the most widely distributed remedy for snakebite in traditional medicine, and a great number of plant species are used for this purpose (Morton 1981; Duke 1985). While 578 vascular plants have been reported in the literature to have some type of antiophidian activity, the actual number of plant species used is probably much higher (Martz 1992). Treatments usually consist of a single plant or a mixture of plants that can be used to prepare decoctions or alcoholic extracts, administered orally or used to bathe the patient or the bitten body part. In some cases, the affected zone is exposed to vapors for a determined period of time, and the process is repeated several times. In cases of inflammation or necrosis, traditional practitioners tend to use poultices on the affected zone (Otero et al. 2000a). Plants are selected for the treatment of snakebite accidents for very different reasons in different parts of the world. Most of the species used were originally tested empirically, and they change depending on their local availability and abundance. Interesting cultural aspects such as the "similarity law" operate in plant selection, with many plants used for snakebite having some physical similarity to snakes present in the region (Piojan 2008). An example is the genus *Dracontium*, members of which have trunks that resemble snakes of the genus *Bothrops* in color and texture and are used as snakebite antidotes in all of tropical America.

The species of medicinal plant and the method of administration are determined by a local medical practitioner, whose knowledge is normally transmitted from generation to generation and modified depending on personal experience (Otero et al. 2000a; Leyton-Ovando 2001). Information is often jealously protected and only shared with the initiated, making it very hard to find literature concerning the methods employed. Like many traditional practices, the use of plant remedies is imbued with ritual and ceremony, usually involving the combination of various remedies. In the state of Oaxaca, Mexico, members of the Chinanteca tribe use cutting and mouth suction after the snakebite. Next, they apply a poultice composed of the leaves of seven species of plants and finally a fresh leaf of *Dorstenia contrajerva*. The rest of the treatment depends on symptoms and ends with a ceremony called Limpia (literally, "cleaning") accompanied by several prayers (Piojan 2008). Across the country, along the Gulf of Mexico, the Culebreros have been observed to use close to 120 species of plants for this purpose, 10 of which are very common. Plant harvesting involves a variety of rituals and can only be done by certain people on specified days in March (Leyton-Ovando 2001). Even though this practice is widely spread in Mexico, there are no scientific studies to refute or prove the presence of any type of antiophidian activity. In Swaziland, in southern Africa, the Sangomas treat snakebite with a paste called Sibiba made from various herbs, bark, parts of snakes, and venom. Hundreds of small cuts are made around the wound, and *sibiba* is rubbed in. The paste is also eaten to prevent snakebite. Other Swazi cures include rubbing aloe vera leaves or drinking their juices. Sangomas from Zululand, also in southern Africa, use thorn bushes and monkey fruit (Myke Clarkson, personal observation).

Colombia is one of the few countries where trustworthy data on traditional remedies are available (Otero et al. 2000a). In northwestern Colombia, 60% of patients suffering from snakebite are initially treated with some type of medicinal plant. In this region, 20 medical practitioners, called Chamane, treated 454 patients, 20 of them with lethal outcomes. The administration route of the plants depends on the time and severity of the envenoming determined by the local Chamán. Of the 105 plants used by the Chamanes in the region, 85 have been collected and identified (Otero et al. 2000a).

The experiments that have been used to assess the possible beneficial effects of plants in snakebite accidents can be grouped as follows: (1) preincubation, by which venom or venom fractions are incubated with varying doses of plant extracts and subsequently injected into an animal model or tested for activity; or (2) *in vivo*, which includes protection or rescue experiments where plants are administered through various routes before or after experimental envenomation. Careful interpretation of the results of these experiments is very important in order to understand their real predictive potential. In most cases, experiments are far from reflecting the application of botanical remedies in a real case of envenomation (Houghton et al. 2007). Preincubation experiments are the most common in the scientific literature. In one of these, groups of mice were injected with 1.5 times the predetermined LD_{50} of *Bothrops asper* venom preincubated with extracts of *Dracontium croatti*. The results showed that this group presented a 45% survival rate, while a control group injected with the same amount of non-preincubated venom had a survival rate of 10% (Otero et al. 2000a). This type of experiment in no way reflects what happens in a clinical envenomation, because such an exclusive and direct interaction between venom and plant components is not possible in normal circumstances. Factors such as absorption, dose, and metabolism are not considered in this *in vitro* testing. However, it can be used to select plant species with therapeutic potential for further analysis.

Other studies focus on the neutralization of particular venom activities using very similar approaches. One study in Costa Rica also analyzed the venom of *B. asper*, one of the most medically significant snakes in the region, the bites of which cause severe local and systemic hemorrhage (Castro et al. 1999). Hours can go by before a patient bitten by *B. asper* in this region is treated with antivenom; therefore, delaying or stopping the hemorrhagic activity of the venom is of crucial importance. The ethanolic extracts of 48 species of plants were preincubated with 10 minimum hemorrhagic doses (MHD), and results showed that 10 of the 48 completely neutralized this activity. Neutralizing extracts were subsequently characterized, and the main components identified were flavonoids. The authors propose that these components may be acting as zinc-chelating agents, inactivating hemorrhagic metalloproteases. This is an example of a mechanism that affects neutralization when tested *in vitro*, but given the lack of specificity for venom components, this neutralization would most likely not be maintained on *in vivo* experiments.

Another study evaluated the capacity of some of the previously described plants to neutralize the lethal and enzymatic activity of *Bothrops atrox* venom, this time using *in vivo* rescue and protection experiments (Otero et al. 2000b). After preincubation experiments with the alcoholic extracts of 82 plant species, 10 venom samples were selected for analysis and tested for lethality. In the next set of experiments, extracts of selected plants were administered by different routes (oral, intravenous, intramuscular, and intraperitoneal) before or after inoculation of 1.5 LD_{50} doses of venom, also inoculated through different routes (intravenous, intramuscular, and intraperitoneal). Extracts administered orally and intraperitoneally 60 minutes before intramuscular inoculation of venom provided a low protective effect. Also, two of the extracts were partially effective when administered intravenously 15 minutes after intramuscular inoculation. One of them resulted in a 50% survival of mice when inoculated 5 minutes after intramuscular administration of venom. The authors also reported that the extracts delayed time of death in most cases.

In China, various products based on plant extracts are used to treat snakebite. An example of this is the "Yuang snakebite drug," created using the soluble extracts of 11 plant species. It was reported that the oral administration of the drug 40 minutes before the injection of *Naja atra* and *Gloydius halys* venom into mice resulted in 66.1% and 52.4% survival rates, respectively (Yuliang et al. 1979). A number of similar works have been done claiming the efficacy of various other plants with similar methodological and practical limitations. Table 4.1 shows a few of the plant species that have been studied for their effect on snakebites. Unfortunately, the observation of a positive effect in rescue or preventive experiments does not prove that the tested plant is able to neutralize snake venom, and most performed experiments have obtained partial results like those of the previously discussed examples. However, these results do suggest that some plants may have a positive effect when treating certain envenomations. Isolation of the components responsible for the apparent therapeutic effects and the characterization of their mechanisms of action could further understanding of the treatment of bites from specific snake species. Whereas none of these should be expected to replace the use of specific antivenom, some compounds could be found that help to reduce the damage caused by snakebite in cases where arrival at a clinic with available antivenom may not be possible for several hours following a bite.

To conclude, no plant remedy has been systematically proven to have a beneficial effect on snakebite envenomation in humans, and no clinical study has shown an improvement of outcome in patients treated with a specific plant. Still, much research remains to be done in this field, and some of the discussed studies have hinted at the presence of components that may have therapeutic benefits for the treatment of snakebite. On the other hand, the dangers of application or ingestion of unknown substances are considerable. Many plant-derived substances are extremely toxic. The monkey fruit used by Zulu Sangomas for snakebites contains strychnine, an extremely toxic alkaloid (Philippe et al. 2004). Application of poorly washed leaves and parts of plants over damaged tissue could also cause infection. No remedy should be applied without full understanding of its risks and the extent of its therapeutic potential.

4.4.2 VITAMIN C

Ascorbic acid, or vitamin C, has been recommended for the treatment of many diseases, ranging from burns and the common cold to diabetes (Klenner 1971). High doses of intravenous vitamin C (> 1 g/kg) have been claimed to be effective treatment of envenomations in both humans and animals (Stone 1972). Some physicians and veterinarians have even suggested its use as an alternative to specific antivenom. Much of this information is published on Internet blogs, and evidence in favor of it typically consists of case reports and references to its use in treatment of diseases completely unrelated to envenomation (Kumar et al. 2004). No viable mechanism of neutralization of any venom component by vitamin C has ever been proposed. It has been suggested that the ingestion of antioxidants, such as vitamin C, can have a beneficial effect in the treatment of snake envenomations based on the observation of a decrease in antioxidant levels in snakebite cases (Suat et al. 2013). Even if this is the case, the beneficial effect would probably be minimal and far from a direct neutralization of venom components. No adverse effects of administration or ingestion of high doses of vitamin C have been observed, even when taken periodically for long periods of time, because the excess of this compound is

Table 4.1: Plant Species Analyzed for Antivenom Effect

Plant species	Part used[a]	Preparation	Administration route[b]	Showed beneficial effect experimentally	Experimental method[c]	Reference
Allamanda cathartica	LE	Unknown	Unknown	No	P	Castro et al. 1999
Aristolochia ovalifolia	RO	Pulverization	Oral	Not tested	—	Hernández, Bello, and Mávil 2007
Citrus limon	RF	Juice	Oral, external bath, poultice	Yes	P, Prot, R	Otero et al. 2000a; Otero et al. 2000b
Clinacanthus mutans	LE	Decoction	Oral	No	Prot	Cherdchu et al. 1977
Dorstenia contrajerva	LE	Poultice	Poultice	Not tested	—	Piojan 2008
Dracontium croatti	RH	Decoction	Oral, external bath	Yes	P	Otero et al. 2000a; Otero et al. 2000b
Hemidesmus indicus	RO	Alcoholic extract	Oral	Yes	P	Alam, Auddy, and Gomes 1994
Hibiscus tiliaceus	SE	Alcoholic extract	Oral	Not tested	—	Hernández, Bello, and Mávil 2007
Trichomanes elegans	WP	Decoction	External bath	Yes	P, Prot	Otero et al. 2000a; Otero et al. 2000b

[a] Plant part used as remedy: RH = rizomes; LE = leaves; RO = roots; RF = ripe fruits; SE = seed; WP = whole plant.

[b] Administration of plant species by traditional medical practitioners.

[c] Method used to test effect of plant: P = preincubation; Prot = protection; R = rescue.

excreted in urine (Levine et al. 1996). However, unfounded belief in this practice could delay the use of an effective therapy.

4.4.3 ALCOHOL

The use of various drugs, particularly alcohol, is an old folk cure for snakebites. In the United States, alcoholic beverages, particularly whiskey, were given to snakebite patients in the belief that venom and alcohol were antagonists and would neutralize each other. Laypersons, along with physicians, thought that snakebite would have no effect on a patient intoxicated with alcohol. Many cases of such treatments are available in the literature. For example, two envenomed horses were "successfully treated" in 1867 by oral administration of half a pint of whiskey and half an ounce of ammonia together with the use of tobacco leaf poultices (*Nature* 1870). In 1855, a physician reported the "successful treatment" of an *Agkistrodon contortrix* bite with alcohol. The latter clearly illustrates the use of several remedies at the same time. Saltwater washing, excision of the area, scarification and suction, tourniquet, cauterization with silver nitrate, and oral administration of strong brandy were all used on the same patient during the course of three days (Kunkler 1855). The fact that he fully recovered is probably more related to the characteristics of the venom than to any of the applied remedies. The venom of this species usually causes severe edema and pain, but severe or fatal envenomations are uncommon (Bhakta, Morgan, and Borys 2009). The case of alcohol resembles that of many other folk remedies in that belief in its efficacy is mainly based on such misinterpretations as those previously described. Alcoholic beverages have also been used in snakebites to take the patient's mind off the situation. Ingestion of a depressant chemical while the patient is under a multidimensional chemical attack will at the least mask symptoms and at the worst potentiate effects. Alcohol is also a vasodilator and thus may potentiate the spread of venom. Therefore, the ingestion of alcohol is not only ineffective but may make matters worse.

4.5 SNAKE REPELLENTS

Motivated by fear, people have hunted and killed snakes in order to keep them away from their homes and families. Snake repellents are, of course, a less drastic alternative, and thus hundreds of measures have been proposed for this purpose. These include an incredibly diverse array of practices that range from charms made of snake parts to patented chemical mixtures and electronic pulse emitters. The identification of a substance that could keep snakes away from human dwellings, minimizing undesired encounters, would probably prove beneficial for both parties.

Various patented mixes with claimed snake-repellent properties are presently available on the market; none has been proven effective. One, called Dr. T's Snake-A-Way, was tested for its effects on the behavior of a nonvenomous species (*Pituophis melanoleucus*) and venomous species (*Crotalus viridis*). The experimental design consisted of dividing a room into four quadrants with two 12-inch-wide lines of the granular repellent and measuring the number of times the snake crossed over the lines. It was observed that gopher snakes were "completely undaunted by the repellent," given that all of the tested individuals crossed the repellent line at least once. Rattlesnakes appeared to dislike the smell, but one of the two tested individuals was reported to have crossed the lines twice (Marsh 1993). Several toxic chemicals, such as DDT, dieldrin, and toxaphene, were considered snake repellents in the 1950s and 1960s. Subsequently, it was recognized that their apparent repellent nature was just a consequence of their high lethality (Savarie and Bruggers 1999).

An experiment was performed to test numerous substances that were claimed to repel snakes (San Julian 1985). Among them were mothballs, sulfur, gourd vines, cedar oil, and king snake musk. None of them was observed to cause a reaction in the tested snakes, and therefore no evidence was found that any of them acted as a repellent. Most of the recent scientific research on snake attractants and repellents has been done in an attempt to control populations of *Boiga irregularis* on Guam, where it has been introduced and has caused numerous ecological and medical difficulties (Chiszar, Rodda, and Smith 1997; Savarie and Bruggers 1999). Unfortunately, a suitable repellent has not been identified thus far.

After these and other experiments made the effectiveness of chemical snake repellents very questionable, some manufacturers claimed that their products were only useful when snakes were in their natural environments. Manufacturers did not discuss the reasons for these discrepancies, and supposed evidence of its effectiveness is given through testimonials (Chemical Solutions 2014). Another article addressing the matter of discrepancies between laboratory and field experiments concluded quite the opposite. It noted that repellents that appear effective in laboratory tests are not necessarily useful for snakes in the wild and drew attention to the methodologies used to test attractant and repellent substances that are to be used in the field (Chiszar, Rodda, and Smith 1997).

In Australia, the use of pulse-emitting electronic snake repellents has become popular. Unfortunately, no evidence exists that these devices repel snakes; on the contrary, several professional snake catchers claim that they remove more snakes from gardens that have pulse emitters than from those that do not.

No substance or electronic device has been scientifically validated as capable of preventing snakes from approaching a certain area in field tests. However, research on potentially repellent substances continues.

4.6 CONCLUSION

After an extensive analysis of the available literature concerning traditional remedies for snakebites, it appears evident that very few such measures provide any therapeutic advantage. Many traditional remedies and alternative treatments for snakebite, in fact, pose serious risk of worsening envenomation or causing damage by themselves. Furthermore, these techniques can significantly delay the seeking of effective treatment.

CHAPTER 5

MAINTAINING VENOMOUS REPTILE COLLECTIONS

Protocols and Occupational Safety

B. G. FRY, I. HENDRIKX, P. ROWLEY, T. N. W. JACKSON, H. VAN DER PLOEG, R. JOHNSON, M. SASA, N. DUNSTAN, S. BARVE, B. LOCK, T. PHILLIP, M. ZIVKOVIC, L. BOYER, K. WILEY, J. HARRISON, R. CARMICHAEL, M. C. MORRIS, P. MARTELLI, D. BRANDL, G. SHANKAR, S. MCCARTHY, K. SUNAGAR, J. PITTMAN, AND C. COCHRAN

5.1 SUMMARY

This chapter is written with the intent that it may be used as a citable reference for applications to institutional animal ethics or biosafety committees and that the protocols contained within represent standards of "best practice." However, we recognize that the guidelines may not be practical or economically feasible for all facilities. Thus, allowances must be given to smaller endeavors in this regard, and suitable alternatives may be instituted. For herpetologists, toxinologists, venom producers, and zookeepers, maintenance of a healthy collection of animals for research, venom-extraction purposes, and educational outreach is crucial. Proper husbandry practices are a must for ensuring the health of any institution's collection. In addition to concerns associated with animal health, numerous daily activities involving the routine care and maintenance of a venomous collection can pose significant risks to employee safety. Not only must proper safety precautions be taken to minimize the risks associated with collection maintenance, but steps must also be taken to minimize stress placed on the specimens themselves, thus promoting a healthy collection that will sustainably yield the venom required for research. Although such situations are avoided in "best practice," in cases where staff must work alone with venomous reptiles, it is absolutely essential that there is an on-person panic-button alarm system in place, so that communication in the event of a bite does not rely on mobile-phone contact.

5.2 INSURANCE

Full workers' compensation and public liability insurance must be maintained at all times. All participants or workers need to make sure that they qualify for full workers' compensation and public liability insurance for the activities they will be performing. The level of insurance for these activities may vary between workers and volunteers. Workers should be covered under their employers, but volunteers and external collectors may only have limited accident insurance coverage. Volunteers may need to take out additional personal insurance to ensure that they are covered against the possible consequences of accidents, particularly loss of wages because of time off after an injury. Volunteers also need to be registered with the establishment in control of the fieldwork or activity in which they are taking part, so that there is a record of their qualifications, medical considerations, next of kin, and emergency contact details.

5.3 WORKPLACE SAFETY

5.3.1 SAFETY METHODOLOGY AND RISK ASSESSMENTS

Risk assessments must be in place for tasks. A risk assessment is a risk-management approach that should be adopted for all work activities and must be documented. There are three basic safety categories to be considered:

1. *Hazard:* A source of physical harm or a source of potentially damaging energy.
2. *Risk:* The likelihood of the hazard resulting in damage to people or property.
3. *Control:* The action required to eliminate a hazard or reduce the risk to an acceptable level.

Risk management is hazard identification and assessment and the controlling of risks resulting from a particular hazard. The risks to be controlled are those that a person, an organization, or the environment may be exposed to. All participants must have a good understanding of all tasks that need to be performed. Knowledge- and competency-based training tasks to be accomplished must be signed off on by both the trainer and the trainee. Areas of emphasis for risk-management training include the following:

- Approach handling and restraint each time as a novel encounter. Complacency or a casual approach, no matter how experienced the handler, can result in serious accidents.
- Only interact with venomous animals when in prime mental and physical condition, never when sick, fatigued, mentally stressed, distracted, drowsy, or under the influence of drugs, including alcohol and medications, that may affect alertness or reaction time.
- Always work at least in pairs, never alone.

- Be prepared for a bite, maintain current first aid training, have a stocked first aid kit, have a bite protocol in place and easily accessible for a hospital trip, and arrange a bite-management plan with the local hospital.
- Possess (or have access to) appropriate antivenom, with sufficient stock levels for two severe bites (see section 5.3.7).
- Never manipulate venomous reptiles unless absolutely necessary.
- Do not work with venomous species during periods of severe weather when the possibility of power loss is elevated or rapid transport to the hospital is restricted by poor road conditions.
- Avoid, if possible, manipulating venomous species when professional and reliable emergency aid may not be readily available such as after normal business hours or during holidays.
- Avoid answering phone calls, emails, SMS messages, using radios/TV, or other devices that might distract the handler.

5.3.2 STAFF TRAINING

Staff training must be standardized, frequently updated, and flexible so that the training process is tailored to the individual. For example, consideration must be given to whether the person is right- or left-handed. All staff must be fully versed in, and follow, all aspects of the safety protocol. Yearly refresher courses are required for all employees on topics relating to animal handling, management, and health and safety issues. It is also imperative that all employees be trained to respond properly in the event of an envenomation. Training must include the following:

- Every staff member must hold a current occupational first aid certificate.
- PPE (personal protective equipment) must be available to facilitate the safety of the staff member and the welfare of the animal (see chapter 5.3.3).
- Staffers must have knowledge of basic husbandry needs in addition to species-specific and individual specimen requirements (see chapter 5.4).
- Venom-extraction SOP (standard operating procedures) for delivering a consistent product must be followed (see chapter 5.10).
- Training must cover emergency measures, what to do in the event of an escape or an envenomation. (For more details regarding envenomation treatment, including first aid and anaphylaxis treatment, see chapter 2.)

Additional aspects may warrant coverage, depending on the institution, and these may be delegated to a small subset of employees:

- Knowledge of local or international laws regarding field collection of specimens and/or venom or DNA samples; knowledge of importation/exportation laws if additional animals are to be brought into the collection from captive or wild-caught sources or if there are arrangements to send animals to another institution (see chapter 5.10).
- Quarantine protocols and preventive care (see chapter 6).
- Proper registration of new specimens (see chapter 5.10).

During annual training refreshers for staff, the modules themselves must be reviewed and updated to include potential new tools that have become available or to implement new ideas or methods.

5.3.3 PERSONAL PROTECTIVE EQUIPMENT (PPE)

As Benjamin Franklin said, "An ounce of prevention is worth a pound of cure." With this in mind, species-appropriate PPE is mandatory at all times while working with venomous reptiles. It is as vital as proper handling tools in maintaining handler safety (also see appendix 5.1). The following equipment is a must for any institution maintaining venomous reptiles:

- Sturdy, bite-resistant boots with slip-resistant soles and long pants must be worn at all times while working with venomous reptiles. It must be stressed that at no time should boots, gloves, or other PPE be considered "bite-proof" and thus providing total safety.
- Safety glasses must be worn at all times while working in any room containing venomous reptiles. Full-face shields must be worn at all times while working with *Hemachatus* and spitting species of *Naja*. A lab coat or gown must also be worn while working with any spitting species in order to avoid having clothing covered in venom. Once any protective device becomes soiled with venom, special attention must be given to cleaning it, as dried venom easily becomes airborne dust and is a significant health and safety risk involving allergy induction. Washing on a hot cycle will remove the venom. For cleaning of face shields and goggles, only nonabrasive soft cloths or sponges must be used, as brushes will damage the plastic or Perspex. Certain chemicals may also damage plastic or Perspex, so only appropriate cleaners must be used.
- Disposable gloves are to be worn while cleaning venomous reptile enclosures. Double-gloving is useful for replacing gloves, which can become difficult with inevitably damp hands. The outer glove must be replaced once it is damaged. Remove it by grasping the top of the glove and pulling along the fingers; then remove the second glove by grasping the top of it with the removed glove and folding the gloves into each other so as not to touch the exposed surfaces.
- After the specimen is removed, when cleaning enclosures of spitting species or large species such as *Pseudechis australis* that leave significant amounts of dry venom in the cage while feeding, a double-cartridge particulate-filter mask, sealed eye protection (such as clear swim goggles), and disposable latex gloves must be worn. In addition, the enclosures of these species must be immediately sprayed down thoroughly, in order to lightly saturate all substrate, walls, and enclosure décor before cleaning. This reduces the risk of venom particles becoming airborne.
- Species-appropriate handling tools, gloves, hooks, tongs, tubes, pin boxes, trap boxes, or tweezers are to be used at all times while manipulating venomous reptiles.

(For risk assessment and safe operating procedures, see appendices 5.1–5.4.)

5.3.4 STAYING ALERT AND FOCUSED

Alertness and focus are two vitally important factors of individual safety that may be more difficult to maintain in an institutional environment than in a private setting. Staff scheduling and cross-training should be at such a level that if one employee is sleep-deprived (such as from having a newborn child at home) or recovering from a virus or other illness, that this person can be assigned to a task not requiring direct interaction with venomous reptiles. Venom extraction in particular is to be restricted until the staff member is back to full capability. A staff member should take a five-minute break at least once an hour in order to maintain focus. Under no circumstances should any work with venomous reptiles occur after alcoholic intake during the same day, including a single beer with lunch, or excessive consumption the night before. The same goes for medications that can produce side effects that affect alertness (particularly those that cause drowsiness or nausea), such as strong antihistamines. It is also essential to budget the necessary time to perform each task, as one should never be rushed while working with venomous reptiles. As in any other scientific workplace, personal food and liquids must not be consumed in the facility. Similarly, smoking around the animals will negatively affect their health and should thus be prohibited within 15 meters of any building containing live animals.

5.3.5 WORKING IN PAIRS

Working in pairs is mandatory at all times. In addition, other staffers must be immediately contactable by means such as alarm buttons, radio communication, intercom systems, or mobile phones. Both persons must be well trained in handling and emergency procedures involving venomous reptiles.

5.3.6 BITE PREPAREDNESS AND PROTOCOLS

All staff members must annually retake their certification in an occupational first aid course. If regionally appropriate, or if the facility possesses applicable species, this should include a subsection on the use of pressure immobilization bandages (PIB) for species that are potently neurotoxic or coagulopathic but without severe local tissue effects (see chapter 2.4). Administration of adrenaline injections to treat anaphylactic shock must be included as part of the course. There must be at least one clearly marked first aid station in every room that contains a venomous reptile. These kits must include elastic crepe bandages (if applicable) and injectable adrenaline, preferably in ready-to-use design such as an EpiPen or Ana-Kit, along with injectable antihistamine. Inhalers such as salbutamol/Ventolin may be useful in helping to reduce breathing restrictions. There must also be an eyewash station in each room housing a venomous animal, with the first aid kit located near this. EpiPens must be stored at the temperatures stated in the package insert.

Where an institution has venomous reptiles housed in more than one room or where employees are required to work with venomous reptiles in an isolated area, a suitable communication system must be in place for emergency notification. This may include installed emergency switches (connected to a siren or monitoring system) in each room or intercoms, radios, or mobile phones. Each species must have a laminated card that includes the scientific name, signs and symptoms of envenomation, and antivenom requirements (including the storage location of the antivenom if it is not stored on site or at the local hospital). In the event of an envenomation, this card must be placed around the bitten person's neck on a loose cord.

If antivenom is stored on site, a printout of the treatment protocol is to be stored within the antivenom's ready-to-use travel container (with each antivenom type in a different, clearly labeled box). This protocol can then accompany the staff member to the hospital. Members of the local ambulance service and designated emergency room staff should annually be given a tour through the facility so that they are familiar with the location and with the differing first aid treatments for bites from various species, particularly whether or not to use PIB. An emergency-coordination meeting, with personnel from local medical facilities in attendance, must be held yearly and include training in how to properly treat a bite. The importance of building a positive working relationship with all medical personnel who may be involved in the event of an envenomation cannot be overstated. In addition, a formal medical management plan should be in place with the designated receiving hospital, and a copy of this plan should accompany the bitten patient to the hospital. Depending on local regulations, a management plan may also need to include investigational review board (or local equivalent) approval for emergency use of foreign antivenoms for the treatment of bites from exotic species.

(For more details regarding envenomation treatment, including first aid and anaphylaxis treatment, see chapter 2.)

5.3.7 ANTIVENOM CONSIDERATIONS

For nonnative snakes where local hospitals do not stock appropriate antivenom, enough antivenom to manage at least two severe envenomations must be obtained and stored in an easily accessible location. Holding antivenoms involves significant cost and burden because of import permits, expiration dates, and storage provisions. Thus, it is acceptable to collaborate with other nearby institutions in a network to distribute the cost and storage of antivenom holdings. Cooperating facilities should be no farther than four hours' travel time apart or from the common hospital to which a bite victim would be transported.

Antivenom stored on site must be kept in a clearly marked, dedicated refrigerator (containing no food or drink or laboratory supplies). The inventory (including expiration dates) must be updated biannually. Each type of antivenom is to be stored in a separate plastic container. Each bin should be a different color, matching the color code on the emergency card on each enclosure. In an emergency, this reduces the chance of selecting the wrong antivenom. In order to accommodate color blind staff, bold text should also be written on each bin label detailing the contents.

Medical practitioners should be notified of governmental policy for administration of foreign antivenoms. For instance, in the United States, some antivenoms have received marketing approval from the FDA, meaning that hospitals may stock them for routine use, like any other drug, when treating envenomation by native snakes such as *Crotalus atrox*. In the United States, only CroFab currently has marketing approval for treatment of envenomations by pit vipers. For a physician to use an alternative antivenom for pit viper bite, or a foreign product for an exotic envenomation, compliance with federal law requires that they proceed under the terms of an Investigational New Drug Application (INDA). Unless an INDA is in place with advance approval by the hospital's IRB, doctors may be reluctant to administer foreign antivenom, even if research data have been published suggesting that other products, such as Instituto Bioclon's Antivipmyn used for American *Crotalus* species, may actually be more effective. If an IRB-approved INDA product is not already in place, then the doctor may administer an appropriate antivenom (such as SAIMR polyvalent for bites from *Bitis gabonica*), using the insert that comes with the antivenom serum. Under these circumstances, however, it is necessary for the physician to obtain emergency IND approval from the FDA (301-827-2000 or 301-827-1800 days, 301-796-8240 or 866-300-4374 nights and weekends), with after-the-fact reporting to the hospital IRB and to the FDA. It must not be assumed that an attending emergency physician will know this information or will accept the burden of legal and reporting requirements on short notice. Thus, formal meetings must be held between the institution maintaining the venomous animals and the ER director or other designated INDA Principal Investigator (PI) of the local hospital in order to discuss these details ahead of time. A record of the contact information for the hospital PI should be included in the emergency antivenom kit; and a memo can be posted on the bulletin board where all physicians will see it.

5.3.8 VENOM BIOSAFETY CONSIDERATIONS

Standard chemical PPE must be worn while working with any quantity of venom; this includes but is not limited to closed-toed shoes, eye protection, laboratory coat, and gloves. Extra care must be taken when using a syringe for the injection of venom into a laboratory animal. However, the entire quantity in the syringe used in such experiments is typically sublethal for humans, and the amount coating the needle is typically capable of causing only trivial symptoms. For this reason, antivenom is not required to be stored by institutions working only with venom (rather than with venomous animals), as the likelihood of a severe envenomation is almost zero.

From an institutional biosafety committee perspective, the primary risks to consider in the occupational handling of venom are allergy related issues. Thus, large quantities of dried venom (50 mg or more) must be manipulated only in a fume hood to avoid contaminating the rest of the laboratory with powdered venom. Smaller quantities may be worked with outside the fume hood, but when quantities greater than 5 mg are being manipulated, it is mandatory that the researcher or technician wear a particulate filter in addition to sealed laboratory goggles. Allergies are readily developed through nasal or ocular exposure to venoms, and this may lead to anaphylactic shock following envenomation (see chapter 2). Other than this allergenic risk, venoms are inert by inhalation, ingestion, or touch and thus do not count as "dangerous goods" from the perspective of the International Air Transport Association with regard to transport or storage. However, as they are lethal substances upon injection, special precautions must be taken when significant quantities of venom are in a syringe, such as during antivenom production. In addition, considering their potential for misuse, all venom stocks must be stored in locked freezers if they are kept in rooms that are not otherwise locked at all times. (See also appendix 5.2.)

5.3.9 ESCAPES

Upon the initial entry (each morning) into a room in which venomous snakes are being housed should be conducted with a view to the possibility that an escape has occurred and a snake is loose in the room. Walk in cautiously and slowly, being observant of surroundings. Make an initial scan of the room to look for signs of an escape such as an askew lid, a broom tipped over, or a bowl on the floor. If an escape is discovered, leave the room, lock the door, and call all staff on hand to assist in finding the escaped individual. The department or facility head must be notified immediately. All nonessential personnel are to be escorted from any room containing specimens, and all doors are to be locked until the animal is found and secured.

5.4 HANDLING TECHNIQUES

Streamlined protocols and procedures facilitate safety (see color plate 16). Essential handling equipment includes gloves, hooks, tongs, and clear tubes. Specialized equipment such as full-face shields may be employed while working with spitting species, while heavy polyvinyl aprons may be employed for trunk protection from large, agile species such as *Oxyuranus* species. During enclosure cleaning and maintenance, animals need to be properly secured in holding containers. Appropriately sized and secure (lockable) containers or bins should be obtained for this purpose. These holding containers should also be appropriately labeled: "Venomous Animal."

Practice makes perfect, and it is highly recommended that persons with limited experience in handling venomous animals become comfortable with the techniques by first handling aggressive, nonvenomous reptiles that behave in a similar manner to their venomous counterparts. *Pseustes, Ptyas, Spalerosophis*, or *Stegonotus* are examples of suitable trainer snakes for developing the skills required in handling typical elapid snakes. Handling *Acanthophis*, a nontypical elapid snake, requires experience with trainer snakes similar to those appropriate for developing skills for handling viperid snakes; in this case, *Candoia* or the various species of blood pythons are examples of suitable trainer snakes. In all cases, the chosen species are snakes that are not averse to biting, thus discouraging complacency while handling.

5.4.1 HOOKS

Hooks may be custom-made or purchased commercially from numerous vendors. The hook-and-tail method for handling many venomous snakes is probably the most commonly used technique worldwide because of its effectiveness, versatility, and relative safety (for expert handlers). It involves using a reptile hook that is longer than the snake's strike range to support the cranial third of the snake while the handler controls the caudal aspect of the animal by grasping the snake's tail base in the pericloacal area. It should be noted that holding the tail below the cloaca provides less control for the handler and can cause injury to the snake. Most larger venomous snakes can be handled using this method. However, extreme caution must be exercised with elapids that have prehensile-tail capabilities such as *Dendroaspis viridis* and other similar species. Do not let the area posterior to the cloaca make a full loop around the free hand with this method. In addition, this technique should never be attempted with small, short-bodied venomous snakes (such as *Acanthophis*, dwarf *Bitis*, or *Atractaspis*).

Large, heavy-bodied vipers (such as *Bitis* species) should be handled with two hooks (with wide, flat bases) at once, rather than hooked and tailed, so that the animal's weight can be evenly distributed. Utilizing this method ensures that the animal will feel more comfortable and therefore will more likely want to remain on the hooks during transport to the holding bin or permanent enclosure. Additionally, this method of handling large-bodied vipers minimizes the risk of rib injuries caused by too much pressure being placed on a small area.

5.4.2 SNAKE TONGS

A large range of commercially available snake tongs provide a means of grasping a snake securely at a distance. Midwest Tongs are the highest-quality equipment currently on the market. Better-quality snake tongs allow the snake to be grasped securely while minimizing the chance of damage to the snake's spine by limiting the amount of pressure that is applied. Appropriately sized tongs must be used; otherwise, more slender specimens may be able to twist out of the tongs' grasp. Many snakes do not tolerate the use of tongs very well; the restrictive grasp causes some snakes to respond violently. For this reason, tongs are not generally recommended as the main form of equipment for everyday handling and should be utilized only in certain difficult situations (such as extracting an animal from a tree) or when restraining a particularly agile or aggressive specimen. Modification through the addition of high-friction surfaces and padding may improve performance.

5.4.3 PINNING STICKS

Pinning sticks, or "jiggers," can be very effective when used appropriately. They are often crucial pieces of equipment for venom extraction, where they are employed to facilitate the securing of the head of the specimen to be venom extracted. In addition, they may be necessary in general husbandry, when the head of the specimen must be secured in order to allow examination of the cranial end of the reptile or so that veterinary procedures such as stomach tubing, removal of retained spectacles, and tracheal intubation can be performed. Double-pinning so that the second pinner acts as a shock absorber for body movement reduces the chance that the snake will be able to jerk its head free.

Using tubes and pinning tools together in some procedures (i.e. eye cap removal, cleaning caught debris in snake's mouth, etc.) can prove very effective. This requires two people; one for holding the snake/tube and the other person using the pinning device and securing the head using the proper grip technique (see Section 5.4.6).

5.4.4 NECK BRACES

The techniques mentioned above have some major disadvantages. Particularly for species with long, delicate necks, there is the potential for the snake to damage itself by struggling while being pinned. As the main aim is to support the snake's neck prior to actually grasping the animal behind the head, a neck brace facilitates both increased safety of handling and decreased risk of damaging the snake. The size used will vary by species; for example, for *Crotalus durrissus*, the brace could be a 24-cm length of 55-mm-diameter plastic drainage pipe cut in half lengthwise, attached to an 85-cm handle by a pair of angle brackets with flat-headed coach bolts. All edges have been smoothed off and the inside surface padded with high-density foam. With the handler holding the hook in one hand, the snake is encouraged to straighten out sufficiently for the tool to be placed over the back of the neck, approximately 3 cm behind the head. Once positioned correctly, downward pressure is applied, and the head is then pinned with a jigger.

5.4.5 HOOP BAGS WITH SEWN CORNERS

Hoop bags are essential equipment for anyone working with or intending to treat venomous snakes. The sewn corners provide a "safer" area to hold or secure a bag containing an animal. The corners can be grasped with forceps or tongs to further minimize the chance of envenomation through the bag. There is a range of commercially available bags, but they can also be custom-made. It is important that the seams of the bag are tightly woven and double-stitched. The material chosen should allow both good ventilation and ease of knot tying. Bags must be made of dark material, as snakes will enter them more readily than those made from material of a lighter shade. Bags must be washed after each use and then closely inspected, as wear and tear and some disinfectants can degrade the material. Bags should be inspected before each use for holes or loose threads along the seams. Loose threads may entangle a snake in the bag and create an unnecessarily risky scenario for both animal and handler. If bags attached to rigid frames are not used, there is an increased envenomation potential; bags must be held open with a pair of tongs by a second person to reduce this risk.

It is recommended that bags used for venomous snakes be retired every 2–3 years. Retired bags can be used for non-venomous species as long as no holes or ripped seams are present.

5.4.6 CLEAR PLASTIC RESTRAINT TUBES

Clear plastic circular tubes are very useful for restraining venomous snakes for examination, administration of medication, and induction of anesthesia. It is vital that a correctly sized tube is selected. The general aim is to have a snug-fitting tube that is wide enough to allow free forward movement but slim enough that it is impossible for the snake to double back and reverse out of the tube. As some snakes are hesitant to enter a tube initially, it can be helpful to present a larger-diameter tube for the snake to crawl into and then feed tubes of narrower diameter down until the desired snugness is reached. Tongs may be used to gently bring the snake to the edge of the tube. The tongs also prevent the snake from reversing once inside, allowing the snake's body to be safely grasped.

Clear tubes are best, as they allow for easy visual inspection and observation; however, some snakes may become wary of entering the tube as time progresses. It is advantageous to paint the last 15 to 20 cm of the tube end black; this encourages the snake to enter the tube and also provides a visual indicator of when it is safe to pick up the tube (once the snake's head is past the painted portion).

An effective and safe way to use restraining tubes is to take the tube in one hand and a pair of snake tongs in the other. Use the tongs to gently hold the snake while the tube is presented to the snake. Usually, the snake will crawl quickly into the tube, but it may take a few tries. Once the snake is in the tube, hold the snake with the mouth of the tongs butted up against the end of the tube so the snake cannot back out. Then slide the hand that is holding the tube down the full length of the tube toward the snake. The last step is to carefully transfer the tube and the snake by grabbing the end of the tube and the snake while releasing pressure on the snake tongs. Pressure should be on the snake through the entire transfer process so the snake does not have an opportunity to pull free or back out.

Another technique is to place the snake in a bin and then slide the tube over the snake's head as it climbs the bin's vertical wall. Alternatively, vipers can be placed along a wall or other barrier. As the snake moves along the surface of the wall, place the tube in its path. It is critical that the keeper moves slowly and exercises extreme patience. Oftentimes, it takes a few tries to get the snake safely into the tube. If you try to rush the snake, you will most likely not get the desired results. Work slowly but deliberately.

Regardless of the method used, start with a larger tube at first, and feed narrower tubes down through the larger tube, removing the larger tube each time a narrower one is properly in place. Restraint tubes can be modified to the caretaker's preference. Many herpetologists fit a lockable entrance hole to the venomous snake's enclosure or shift box to which a tube can be connected. The snake can then be gently encouraged into the tube without the need for removal from its enclosure, further minimizing the risk of envenomation.

It is mandatory to clean and disinfect tubes after each use. Not only is this important for disease control, but snakes will smell other snakes that have been in the tube previously, and this may make them reluctant to enter.

5.4.7 BITE-RESISTANT GLOVES

Commercially available thick gloves are very useful when attempting to restrain venomous lizards. The use of these gloves for handling venomous snakes is more controversial. The choice often depends on personal preference, and many clinicians and herpetologists prefer thinner gloves (or none at all) to allow for more dexterity. When used with venomous snakes, gloves must not be relied on as a sole means of defense, but when used properly, they may be considered a good secondary line of defense. They are particularly useful for elapids with fang lengths less than the thickness of the gloves. They are also useful for pit vipers in masking the heat signature given off by hands.

The material chosen will vary by intended use. Puncture-resistant material (such as that of the Hexarmor gloves) is useful for terrestrial venomous snakes, although prolonged chewing by elapids may puncture the gloves and result in an envenomation. While neoprene material of at least 5 mm is appropriate for most sea snakes, very large species of *Hydrophis* (such as *H. stokesii* [formerly *Astrotia stokesii*]) may have fangs approaching 1 cm in length, and this must be taken into account.

The bladelike teeth of varanid lizards are best defended against with heavy leather gloves, while helodermatid lizards may be handled with similar gloves or with gloves made of needle-stick-resistant material.

5.4.8 ENCLOSURE ACCESS FOR HANDLERS

Institutions employing a large number of handlers who have varying levels of handling experience should enact a tiered system utilizing shape- and color-coded stickers on each enclosure to identify what level a person must be at to work with the animal within a given enclosure. It is important to use shapes, as some staffers may be color-blind. In addition, as clone locks are typically used to reduce the number of keys needed, all enclosures of a particular tier should have locks requiring the same key but with different lock-and-key combinations for each tier. Only staffers trained for a particular tier are issued the relevant key. This prevents accidental or intentional access to animals that staffers are unqualified to interact with.

The lowest tier will contain all nonvenomous reptiles. Requirements to work with nonvenomous reptiles will vary by institution, but universities require completion of Institutional Animal Care and Use Committee (IACUC) training before work with vertebrates is allowed. Tier 1 animals, with the exception of constrictors more than 2 meters in length, can be handled alone by anyone with training.

Venomous reptiles can be categorized into three levels, with individuals being allowed to work with any level that is at or below the highest level for which they are cleared. The classification of level 1 venomous reptiles will vary by institution and country, but as a general guideline, they should include the easier-to-handle genera (such as smaller *Crotalus, Vipera*) that are native to the region. Once again, all handling should be done with two qualified handlers present in the room. Level 2 genera require that two people be present, at least one of whom has level 3 clearance. Examples of level 2 genera are the larger *Crotalus*, larger *Bitis*, most elapids, and non-front-fanged genera such as *Dispholidus, Rhabdophis*, or *Thelatornis.* Level 3 animals require two level 3 handlers to be present. Examples of level 3 genera include fast-moving, agile elapids (such as *Dendroaspis, Oxyuranus*, and *Pseudonaja*) and the larger *Bothrops.*

5.4.9 PROCEDURES FOR HANDLING DEAD SPECIMENS

While odor can indicate the presence of a dead snake, it may be impossible to differentiate between a dead snake and a regurgitated prey item. Under no circumstances must a staff member assume that a venomous reptile is dead without first ensuring that the smell is not from a dead prey item. In addition, a venomous snake that is suspected of being dead must still be handled with great caution, as some snake species are known to "sham" death. It is critical to note that dead snakes still have the capacity to envenomate for an hour or two after death because of nerve reflexes. To confirm that the animal is indeed dead, carefully turn the snake over or pinch the tail with a pair of tongs. Care must be still taken, as some species (such as *Hemachatus*) have been known to continue to sham death even after these manipulations.

Once the reptile is determined to be dead, it should be removed from the enclosure with a hook or tongs and placed in a bag that is labeled with the animal's scientific name, institutional identification number, sex, and date of death. Clearly mark all packaging holding the deceased animal as "venomous."

As freezing causes cellular damage from crystallization and thus may negatively affect pathology studies, the specimen should be stored at between 0° C and 4° C and a necropsy performed within 24 hours. If a necropsy cannot be performed within this period, then the specimen should be frozen for a more limited necropsy at a future date. The individual's record sheet and the institution's digital database should be updated.

The risk of being pricked by a fang is present even when the animal has been confirmed to be dead. Therefore it is imperative that the head is immobilized inside a plastic specimen jar; either while still attached to the snake or post-decapitation (see chapter 6.2.11 and color plates 24G/H).

If the specimen was captured in the wild, it should be deposited as a museum specimen after the necropsy. In preparation of this, the specimen is fixed in 10% neutral-buffered formalin and then stored in a jar containing 70% ethanol. A label indicating the specimen or museum number should be sewn into the dorsal region of the specimen.

5.5 HUSBANDRY

5.5.1 THE BUILDING

The building should be of sturdy construction, built under the construction norms of the country. If the venomous reptile collection is housed in a region known to have earthquakes or climatic events such as cyclones, this must be taken into consideration in the building and enclosure designs. Entrance to the building should be limited to facility employees. The primary access to the building must be alarmed and locked at all times. In addition, the entrance to any room housing venomous reptiles must have escape-proof doors that are locked at all times, and only employees cleared to handle venomous species should have access to these rooms. Access gates and emergency exits must be in place and fully operational. Ceilings and finished surfaces in rooms must be smooth to prevent escapes. If possible, rooms should have their own sinks and water faucets to facilitate enclosure cleaning and provide a readily accessible source of water for the animals. Each sink must have a permanent mesh filter over the drain to prevent escapes via the plumbing. Similarly any air vents on the floor or near the floor should be given extra mesh covering so that an escaped snake can't get access into the ductwork of the room.

A complete set of handling instruments (including hooks, tongs, and forceps) must be kept in each husbandry room, to avoid contamination and the transfer of potential pathogens throughout the facility. Each room containing venomous reptiles must have a printed record and a digital record, kept separately from each other, detailing the species and number of specimens within for use during catastrophic events such as earthquakes, cyclones, and tornadoes and for use by first responders. Knowing the contents of each room is vital in the event of an escape or any other emergency. This record must be stored as a hard copy with the institutional or departmental head and must also be digitally accessible off site.

Emergency lighting that remains operational in the event of a power outage is an absolute must, multiple high-quality rechargeable flashlights (with indicator lights for finding them in the dark) should be available just inside all rooms, and emergency snake hooks and/or tongs should be stored by each door. In collections maintained in cold climates, backup power sources must be available for use in the event of electrical-system failure.

Temperature alarms must be installed in each room containing live specimens so that an alarm is activated and notifications (such as text messages to multiple mobile phones) are automatically sent if temperatures go either above or a below a predetermined acceptable level. In order to reduce heat from solar radiation, windows should have solar-reflective coating, and in order to avoid heat loss in cold climates (or decrease in air-conditioning efficiency in hot climates), windows should be double-glazed. Any rooms containing specimens must have daylight-spectrum fluorescent tubes connected to a time-control mechanism. Doors leading to rooms housing animals should, if possible, have a glazed panel that allows for the floor area to be viewed prior to entering. Double entry systems with sliding glass doors can also be employed to reduce the risk of escape.

When keeping snakes or other venomous animals, the rule of three barriers to prevent the animal's escape from the building must be followed. In the majority of facilities, those three barriers are the enclosure, the room, and the building. Thus, keepers should observe that each barrier is properly designed to fulfill its role. For example, to prevent any possible escape, all doors in the building must have no more than a 5-mm gap from door bottom to floor, ideally no gap at all, bearing in mind that small snakes may be able to squeeze through a gap of only a few millimeters. All floor drains must also have screen-mesh coverings with holes no larger than 5 mm. All rooms and areas must be kept free of clutter. The bottoms of enclosures and racks should be at least 250 mm above the floor so that a quick scan of the entire room can be made upon entering. Areas under sinks must be examined for holes or other escape routes.

It is absolutely imperative for all ovoviviparous species that all cage ventilation openings be additionally covered with fly screen in order to prevent escapes by the very small live young. Such species are notorious for sometimes not displaying any sign of pregnancy until babies are noticed. Cages containing ovoviviparous species should always be approached as if there are freshly born neonates within. While this is particularly important for cages containing mixed-sex groups or containing arrivals within the last six months, this must be standard practice even for cages containing only females, as parthenogenesis has been documented for such snakes (Booth et al. 2012).

5.5.2 DISSEMINATION OF CODES OF PRACTICE

Copies of the emergency procedure to be followed in the event of a bite should be posted in an obvious location in all rooms that contain venomous reptiles and in any room where a venomous animal will be handled. The code of practice (COP) for the handling of venom must be posted on the laboratory fume hood and also in any room where venom is to be handled in any form. Copies of all COPs also must be stored in the office of the institutional or departmental head.

5.5.3 ENCLOSURES

5.5.3.1 Enclosure Basics

In addition to a removable laminated enclosure card, all enclosures should be separately labeled with the specimen's scientific name (genus, species, and subspecies), common name, locality (if known), arrival date, and specimen identification number. Such cards must also address whether pressure immobilization has been contraindicated as treatment for a bite because the venom is rich in tissue-destroying SVMP (see chapters 2.4 and 23 and color plates 12C, 12D, 14G, and 14H) or cytotoxic 3FTx (see chapter 8.4.3 and color plates 13 and 14A–F). Genera and species with venoms rich in SVMP include most viperids, *Atractaspis* species, and some elapid snakes such as *Demansia* species. Species with venom rich in cytotoxic 3FTx include African spitting cobras, with some populations of *Naja nigricollis* being particularly rich.

Provided that their "minimum requirements" are addressed, most species will thrive in a basic enclosure setup (see color plate 17). Particularly for large collections, the more basic and generic an enclosure setup is, the more efficiently husbandry and maintenance procedures can be carried. Some species, such as sea snakes, require extremely specialized enclosures, while others (such as large elapid snakes or large varanid lizards) require very large enclosures. Determining which aspects of a species' habitat and behavior need to be addressed in a captive situation requires sound knowledge of the species' ecology.

Generic setups must be modified appropriately to suit the particular requirements of the species kept. Variables such as temperature, humidity, air flow, provision of hides, and climbing substrate are some of the easiest things to identify and modify. For example, species that utilize an ambush-hunting strategy (most viperid snakes) can generally be kept in smaller enclosures than those suitable for similarly sized active-foraging species (most elapid snakes). The minimum sized to facilitate a proper thermal gradient is an enclosure that is at least 1.5 times body-length long and 0.5 times body-length deep. We do recognized that this is "best practice" ideal and that such spacious cages may not be practical or feasible for all institutions.

Some species require more particular microhabitats. For example, arboreal species are accustomed to drinking during rain and therefore must be given water via spray bottles or misting machines. Others require extremely specialized habitats. *Azemiopis feae* is an extreme example demonstrating the need of some montane species for cool, moist microhabitats with ambient temperatures of 22° C to 24° C and brief warm periods of 26° C in the hot zone. These snakes are from high-elevation cloud forests and are prone to dehydration via cutaneous water loss; they require regular cool misting in order to avoid critical fluid loss. Screen lids should not be used, as they not only provide poor insulation but are also a safety hazard. For species requiring well-ventilated enclosures, such as *Atheris* species, these needs can be met with the use of small fans (such as computer fans) mounted on the outside of the cage over vents.

The venomous reptiles that require the most specialized care are sea snakes and sea kraits. Sea snakes in particular require very large enclosures containing thousands of liters of seawater. Only generalist species such as *Hydrophis curtus* (formerly *Lapemis curtus*), *Hydrophis hardwickii* (formerly *Lapemis hardwickii*), or *Aipysurus laevis* adapt well to captivity and then only if collected as juveniles. Specialists such as *Hydrophis elegans* fare extremely poorly in captivity. Sea snakes have a natural lifespan of only four to five years because of the high metabolic rates that result from the energy expenditure of swimming. Consequently, they need to feed every one to two days. The stress of capture and the failure to provide unique microhabitat requirements mean that sea snakes will not readily feed in captivity and often starve to death in as little as two weeks. Sea snakes should be placed into fresh water containers overnight once a month to drink as they rely on fresh water in the wild for hydration rather than entirely hydrating through the filtering salt out of saline water (Lilywhite et al. 2008).

Sea kraits also require specialized environments but are much more robust than sea snakes. Unlike sea snakes, sea kraits lay eggs on land and also come onto land to bask. As a result, they are harder-muscled and adapted to physical disruption, as they are tossed and turned by waves as they come and go from the land. Sea kraits do not stress out as much from handling, and wounds can be treated out of the water (see chapter 6). It is crucial, however, that humidity be maintained at high levels to avoid cutaneous water loss. As with sea snakes, sea kraits should be monthly placed in fresh water overnight in order to properly hydrate.

5.5.3.2 Plastic Tubs

These are a cost- and space-efficient option for housing large venomous reptile collections and have the additional attractiveness of being time-efficient for hygienic maintenance and readily sterilizable. Another benefit of tubs is that they can be sourced in a wide range of sizes and varieties, making them suitable for the maintenance of a wide variety of species. In addition, unlike with fixed enclosures, staffers have the ability to move a specimen around the facility without having to come into physical contact with the animal. Such systems are commonly arranged in tight-fitting rack systems, which secure the lid in place, significantly reducing the likelihood of escapes. The rack system should be built so that each column or row in the system can be locked. A simple method is to use pipe fitting brackets and PVC pipe to make a "catch bar"; holes can be drilled to allow a padlock to lock the entire column). In areas prone to earthquakes, the racks must be secured to the wall to prevent them from toppling over during such an event.

5.5.3.3 Fixed Enclosures

These can be made of a wide variety of materials, generally wood, but plastic is becoming more popular (such as PVC foam boards). These types of enclosures allow for more specialization in design in order to meet the particular needs of certain species and can facilitate safer interactions for the handler. Fixed enclosures come in a variety of designs (such as front-opening or top-opening) that have different functional advantages, disadvantages, and applications. Such enclosures are generally harder to keep clean (more time-consuming, with sharper corners or joints) and generally take up more space in comparison with tubs. Another drawback is that some models also come prefabricated with an upper and lower lip (above and below the entrance to the cage) that the animal can wedge itself into, posing a significant hazard to an unsuspecting caretaker. In addition, extraction of an animal that has decided to wedge itself in the lip can be a long and dangerous process. To solve this problem, expandable foam can be used to seal off the lip. For very large or active species, fixed enclosures are the preferred option in order to provide the minimum enclosure space needed to satisfy husbandry requirements. Exposed wood must be waterproofed with marine polyurethane. Postconstruction, such enclosures must be heated for a minimum of two weeks in a well-ventilated area to allow for all solvents to evaporate.

5.5.3.4 Outdoor Enclosures (Pits)

Outdoor enclosures give reptiles better exposure to seasonal environmental fluctuations (essential for reproduction in some species) and extremes in husbandry parameters (mainly temperature) that cannot easily be replicated inside. However, such enclosures are generally harder to maintain hygienically and may interfere with husbandry observations and feeding regimes. This option is also not practical for most large-scale facilities, as space is a constraint.

5.5.3.5 Trap Boxes and Hide Boxes

Trap boxes must be able to be closed from outside the enclosure (or easily from within the enclosure through the use of a long hook or tong), isolating the reptile inside in order to facilitate safe enclosure cleaning and transport of the individual. Trap boxes should not be permanently fixed to the enclosure. This allows the box to be removed for maintenance and also utilized for other enclosures (sterilized between uses). Additionally, many snakes are generally ambush predators, with thick, slow-moving bodies, and subsequently appear to relax when they are able to get into a tight hiding place (thigmotaxic behavior). In nature, they will back themselves into "corners" or up against rocks, since these areas provide some security against an attack. Regardless of species, hide boxes give any reptile a feeling of security in its environment. Reptiles are more likely to feed naturally in captivity if this method is employed, as stressed reptiles often refuse food. No physiological measurements or venom composition can be expected to be "normal" when the animal providing the data is in a constant state of stress. At the same time, when the lights are out, the reptile may want to get out of its hide and move around the rest of the enclosure a bit, and it should be able to do this if it demonstrates the desire to do so. The physical and "psychological" well-being of reptiles in captivity will lead to longer life spans and increase the likelihood of captive propagation. Additionally, for large enclosures, sliding dividers are essential, allowing for an individual to be isolated on the other side while husbandry procedures, cage cleaning, and maintenance are carried out. Isolation of the reptile in a confined area of the enclosure will also allow close examination without using physical restraint. Large, agile snakes (such as adult *Dendroaspis polylepis*) can be easily and safely bagged using a modified trap box. In such an apparatus, a cover with a clear tube attached to a black bag can be affixed while the snake is secured inside. When the tube is in place, the door to the box can be removed, allowing the snake to leave the box and enter the tube and thus the bag.

5.5.3.6 Substrates

Substrates should not be abrasive and should be easily cleaned. Newspaper is used in many facilities and is a cheap alternative. However, once wet it becomes quite messy. In addition, some newspaper inks contain toxic chemicals. A preferable alternative is butcher's paper. Such substrates are ideal for tubs heated from below but should be used only for species that occupy dry microhabitats. They are unsuitable for fossorial or leaf-litter-dwelling genera such as *Calliophis, Denisonia, Micrurus*, or *Vermicella*. For such types, the setup should be moderately damp peat moss with a layer of less damp sphagnum moss on top. These substrates are rich in tannins and other antifungal or antimicrobial chemicals, which help to reduce the cleaning burden. For such setups, heating is more complicated and is best accomplished by keeping the enclosure in a room set at a constant 26° C to 28° C. These species are typically active above ground only at night and thus are not exposed to the same temperature extremes as diurnal species. For sea snakes and sea kraits, with their high metabolic rates and high frequency of defecation, very large filters should be used, with a turnover rate of one hour.

5.5.3.7 Heating

The first consideration for heating is the room itself, in which the ambient temperature should not drop below 18° C and should preferably not exceed 28° C in order to allow the animals to thermoregulate. Thus, thermostat-guided heating and cooling must be utilized, particularly during annual extremes. As thermostat failure is a common cause of mass death in such systems, the heating systems must be regulated by at least two thermostats in case one fails. Further, backup systems must be in place to cover for the event of electricity loss.

Specific temperature requirements will vary greatly between species, and each cage must be measured using a temperature gun on a regular basis before an animal is placed inside it to ensure that proper temperature conditions are being maintained. If prolonged periods are spent directly under the heat lamp (more than 30 minutes), then modifications must be made so that this hot spot is able to reach a higher temperature, as this signifies that the heating needs of the animal are not being met. With this in mind, varanid lizards must be provided with a basking spot of 50°C to 60°C, which is best accomplished by having a large raised area, allowing for a proper temperature gradient. Sea snakes must be maintained in water 28°C to 30°C. Species very sensitive to warm temperatures (such as *Azemiops feae*) must not be kept in rooms with ambient temperatures exceeding 22°C. Similarly, heat-sensitive fossorial species such as *Calliophis, Micrurus, Sinomicrurus*, and *Vermicella* are best kept at relatively cool ambient temperatures no higher than 22°C, with a gradient to a maximum of 25°C to 26°C.

As reptiles benefit from exposure to UV, which boosts the immune system (Gillespie et al. 2000), UV-emitting bulbs, such as mercury vapor, should be used where possible. Small fans (such as computer-cooling fans) mounted on the outside of the cage above the heat bulbs will drive the warm air down through the vent. This increase in heating efficiency will dramatically reduce electricity use. Having small vents at one-third of the level from the bottom will allow warm air to escape, creating a significant cool gradient below this, facilitating the ability of the reptile to thermoregulate. The heating of very large enclosures may require several heat sources such as heat panels in combination with heating bulbs that also emit ultraviolet light (such as mercury vapor bulbs). This will allow for heating but without the overprovision of UV light. Outdoor enclosures may require supplemental heating or an air-conditioned refuge. Even burrowing species (such as *Heloderma suspectum*) may spend an appreciable amount of time basking, either daily or during seasonal fluctuations, and must also be provided with basking areas. For rack systems, heating is accomplished through the use of heat tape, which is placed on only one end of the enclosure. This creates a thermal gradient, allowing the reptile to thermoregulate.

Care must be taken to avoid burns; protective cages over the bulbs are recommended for snakes. For varanid lizards, such barriers are absolutely required, as there have been many cases of lizards starting fires by pulling out lighting wiring.

5.6 CLEANING

5.6.1 CLEANING BASICS

It is important to understand the difference between cleaning and disinfection. Cleaning an enclosure is the process of removing unwanted matter and does not result in a sterilized enclosure. For example, sifting out a sand substrate may remove some unwanted matter but will not kill bacteria or fungi present in the sand. Disinfecting is the process of killing or removing microorganisms (germs). The best practice is to clean the enclosure frequently and also to disinfect it regularly using a safe and appropriate disinfectant. Proper maintenance will not only keep the animals healthy but will also cut down on potentially unpleasant odors, an important concern if the animals are on public display. Proper maintenance includes cleaning all surfaces of each enclosure and all pieces of equipment, using a suitable disinfectant that will not harm the animals but will still kill microorganisms. F10 veterinary-grade super-concentrate disinfectant diluted to a ratio of 1:125 is an effective option, as are Quatricide PV by Pharmacal and Nolvasan Solution. To minimize the likelihood of transferring diseases, handling and feeding tools and temporary animal holding bins should be room-specific and never transferred between rooms.

5.6.2 ENCLOSURE CLEANING

Remove waste material from the enclosure, and place it in the appropriate receptacle. Spray all surfaces with disinfectant, and allow 15 minutes for the product to take full effect, then wipe down all surfaces using warm water. Clean all glass surfaces in the same manner. Clean water bowls with disinfectant, rinse thoroughly, and then refill (or replace if using disposable bowls made out of recyclable plastic). Place clean substrate, hide areas, and water bowls back in the enclosure. It is important to note that many species will immediately defecate upon being put back into a clean cage. However, if a small piece of shed skin or urate is left in the cage, the snakes are less likely to exhibit this territory-marking behavior.

5.6.3 EQUIPMENT CLEANING

Between uses in different enclosures, spray all equipment surfaces with disinfectant, or wipe them down with a washcloth that has been dipped in disinfectant and wrung out. This is crucial for disease management.

5.7 FEEDING AND WATERING

Hungry reptiles actively anticipate food and often become more alert and active at feeding time; therefore, it is vitally important that handlers exercise extreme caution during feeding. Food items should be introduced quickly with the aid of long forceps or tongs (as appropriate), and the enclosure should then be immediately secured. Venomous reptiles housed in large vivaria may first be isolated in a trap box that can be secured by pulling a wire from outside the enclosure. The enclosure can then be safely entered, the food delivered, and the enclosure once again exited and secured before the trap box is reopened to release the reptile.

Two of the most important requirements for feeding are proper temperature and hydration. Most reptiles will not eat outside their optimal temperature range. A temperature gradient must be present in the enclosure so that the specimen may thermoregulate. As a general principle, the gradient includes a "hot zone" of no more than 40° C maintained by heat tape or heat bulbs and a "cool zone" with an ambient temperature that must not drop below 20° C at night or exceed 30° C during the day.

Water requirements for all species vary considerably. Some desert species require only light misting once or twice a year, as they obtain most of their water metabolically through consumption of prey items, while species from tropical and subtropical environments often require a constant supply of clean water to help maintain adequate levels of humidity, particularly in buildings with refrigerated air-conditioning, as this removes water from the air. Keeping the water source clean at all times will help to stave off opportunistic infections and keep reptiles in good health. In addition, if the reptile is not adequately hydrated, it will not eat. Dry air is very detrimental to some species. Drying of the mucosal membranes lining the airways may lead to inflammation, and subsequently to serious infection (see chapter 6). Adequate humidity levels must be maintained as necessary with the use of misters, damp substrate such as sphagnum moss, or placing water near a heat source. However, great care must be taken not to overwater, as this may promote the growth of fungus. Even some desert dwelling species need a microclimate of high ambient humidity within their cage. This can be easily accomplished by making a small humidity box that has slightly dampened spaghnum moss and an entrance hole. In enclosures with high humidity, it is essential to have moss (peat or sphagnum) as substrate, as it is naturally rich in tannins and other chemicals that are potently antimicrobial.

5.7.1 FOOD SUPPLY

It is important to be creative and flexible when working with an individual that will not feed. Snakes will refuse to feed if their health is compromised or husbandry requirements have not been appropriately met. Sometimes using a different food item, such as a juvenile rabbit instead of a rat, or even a different size class (fuzzy versus hopper mouse) can induce a reluctant snake to feed. Taking a previously thawed rodent and then warming it up under a warm heat lamp for 10–15 minutes will elicit a strong feeding response with pit vipers. An effective method, primarily for sit and wait predators, is to slowly show the dead prey via long hemostats of tongs. Once close to the snake, pull the prey away and then repeat by slowly bringing the prey back to the snake and pulling away again. Typically, after the third or fourth time the snake will not want to take a chance on that prey leaving for good and a strong feeding strike typically ensues.

The natural dietary preference of each species must be taken into consideration (see color plate 18). Venomous lizards are generalists and will readily take a variety of prey items, including insects, week-old chickens, and rodents. Similarly, most species of vipers and elapids accept rodents and other small mammals as prey, and even those snakes that do not feed frequently on mammals can learn to do so in captivity. The newborn of many venomous species can be fed easily with pinky mice (less than one week old). However, for smaller snake species

(*Bothriechis, Porthidium*), geckos such as *Hemidactylus* (or their tails dusted with vitamin powder) can be offered to the newborns. Research facilities that maintain venomous snakes often produce their own lab rats and/or mice, to be used as laboratory animals or for feeding purposes. It is important to keep track of the quality of the rodents produced as food supply and never to feed snakes with mice that are sick, have parasites, or have been used previously in toxinological assays.

Some snake species (such as in the genera *Atractus, Geophis*, and *Ninia*) can be bred as food supply for strictly ophiophagus (snake-eating) snakes; likewise, fish (such as *Poecila* and *Gambusia*) can be used to feed piscivorous (fish-eating) snakes. Many coral snakes, genus *Micrurus*, include fish eels, genus *Synbranchus*, as minor prey items, and in captivity, they learn to feed on other species of fish (Chacon et al. 2012). The health of species that are specialist feeders in the wild may be compromised by a rodent diet, as the fat levels and fiber content of rodents differ considerably from those of their natural prey. Thus, under the "best practice" ideal, species such as *Ophiophagus hannah* should ideally be fed snakes, not rodents. We do recognize that this may not always be practical or feasible. Forcing rodents down into the stomach of a dead snake and sewing the mouth shut with dissolving stitches is one method for increasing the amount of food taken by a snake that will only eat other snakes. Stitching together several dead, shaved rodents using absorbable sutures is an efficient manner of feeding *Ophiophagus hannah* in the event that snakes are not available as a prey item (see color plates 18F/G). Other specialist feeders such as young *Echis* will also feed on locusts and other invertebrate prey such as scorpions. Sea kraits will readily feed on eels presented on forceps or tongs. It is important to note that feeding live, wild-caught prey comes with significant risk of introducing parasites or other diseases. Thus, any such prey must be deep frozen (–80°C) to kill such pathogens.

5.7.2 FEEDING REGIMES

Feeding regimes will vary greatly according to species, age, reproductive cycle, time of year, venom-extraction demands, type of prey (nutritional value), and prey size. Activity levels are a major variable, with active sight feeders such as elapid snakes requiring one appropriate-sized food item every 7 days (adults) or one feed every 4 days (juveniles), while ambush feeders such as viperid snakes (or elapid snakes that have convergently evolved the same hunting strategy, such as *Acanthophis*) require one feed every 14 to 21 days (adults) or one feed every 10 days (juveniles). Monitor lizards should be offered small items daily or every other day and large items on a weekly basis. Helodermatid lizards are less active than monitor lizards, so they require less food volume and on a less frequent basis. Care must be taken not to overfeed individuals, as this is also detrimental to their health.

An important consideration for animals having venom extracted on a regular basis is that they will use more energy to continually replenish their venom stores and therefore need more food and more frequent feedings than a similar-sized counterpart that is not being extracted from regularly.

Sea snakes in particular have special food needs. They will take dead fish such as whiting offered with long poles or tongs. It must be stressed that oily fish are never to be fed, as, like crocodilians and turtles, sea snakes have difficulty processing the oil, resulting in lethal metabolic toxicity (Fry, personal observations). It is recommended to supplement the diets of captive sea snakes with vitamins once a week. Offering an appropriate-sized fish every second day is an appropriate feeding regime for sea snakes.

5.7.3 ASSIST-FEEDING

There are several gradations of assist- or tease-feeding, ranging from "normal feeding" on one side and "liquid-diet feeding" on the other. Assist- or tease-feeding can be used at the curator's discretion. A specimen may be alert and healthy but may not be able to recognize the offered prey type. This is particularly the case for ophiophagus species that are offered rodent prey items.

Three basic methods can be used for assist-feeding

- The animal remains in the enclosure (tree dwellers, ambush hunters) while forceps or a thin, straight wire are used to present a small prey item, with the length chosen to maximize both safety for the keeper and calmness for the specimen. The specimen is now "teased" with the prey item. Watch closely for a response. Is there a tongue flick? Does it assume a strike position? If so, tease it some more. Maybe gently tap it on the head, body, or tail (the last will often work well with species that display caudal luring). Be careful, though; if this method is used too aggressively, the snake will retreat. Observe what happens when the snake does strike. If it hangs on to the prey item, do not move; just sit and observe. The snake may "switch on" and suddenly recognize that the piece of meat it is holding is actually food. Wait until the snake has swallowed the prey completely before you move or close the lid or the sliding door.
- In the case of fossorial snakes (such as *Atractaspis, Calliophis*), the animal is removed from the enclosure and tubed. A prey item is placed on a clean table (or paper or plastic sheet). The head of the snake is allowed to exit the tube on one side and is carefully manipulated toward the prey. Hovering the snake over the prey item or slightly touching its head to the prey item may trigger a feeding response. If the snake holds on to the prey item and starts swallowing, you can gently let go of the snake and the tube. Supervise the snake while it eats, and transfer it carefully into a bin or enclosure once it's done.
- Some snakes are reluctant to strike, but once there is something to chew on, they will start swallowing. In such cases, the snake can be tubed and the prey item then placed in the snake's mouth with forceps.

5.7.4 LIQUID-DIET FEEDING

Force-feeding should be done with liquid diets via a catheter tube and not with solid objects. Liquid force-feeding is faster, resulting in less stress to the specimen and also less contact time, thus reducing risk while also allowing the effective deliverance of medication and rehydration (see chapter 6). In contrast, force-feeding involving the manual pushing down of a prey item (such as a mouse) subjects the specimen to great stress and can result in bruising or tissue damage. The liquid-diet steps are as follows:

- Commercially produced liquid diets, such as Carnivore Care, are well balanced and easy to use. Alternatively, prepare the liquid diet by blending at high speed a 1:1 mixture of veterinary-quality puppy food with water. Cat food is not to be used, as it often contains large amounts of fish oil, which reptiles have difficulty processing, leading to death from metabolic toxicity. Strain the liquid through a fine mesh to remove large particulates that may clog the catheter.
- Place a 1-cc syringe with the plunger removed and the end where the needle is attached cut off (resulting in a smooth barrel) in the snake's mouth. This ensures not only that the feeding catheter does not enter the trachea but also that when the animal bites down, it is not biting on the catheter, which would interfere with the threading into the stomach and may also result in punctures to the catheter, allowing liquid to enter the trachea (potentially resulting in the animal drowning because of fluid in the lungs).
- Use the thinnest round-ended urinary catheter practical for the volume of food being delivered. Premeasure the catheter, and place a guide mark at the spot that would correspond to at least one-third but no more than one-half of the length of the snake (or about one-third of the way into the trunk of a venomous lizard). Attach a liquid-diet-filled syringe to a lubricated catheter, and prefill the catheter with liquid to remove air contained inside the catheter. Alternatively, a tube designed for the artificial insemination (AI) of dogs can be used. This type of tube is rigid but has a rounded end. It can be attached directly to the syringe filled with food and pushed down the throat of the snake. The advantage is that a tube can be used and the snake can be fed without having to hold it by hand, as the AI tube is rigid enough to be pushed down the esophagus even if the specimen is struggling. Extreme care must be used not to puncture the esophagus. Lubrication, along with a gentle touch, helps prevent injury to the snake.
- Gently thread the catheter or AI tube through the syringe barrel, and continue until the guide mark is reached. Slowly depress the plunger to deliver the liquid diet. As a guide, if feeding liquid diets such as Hill's a/d, use 1% of body weight (10 mL/kg) and then dilute 1:1 with water.
- Feeding frequency will depend on the metabolism and general health of the specimen, on average, elapids every 14 days and viperids every 21 days. For very debilitated and cachectic specimens, feed smaller amounts, less concentrated and more often.

5.8 BRUMATION

For captive reproduction of some species of venomous reptiles (generally those from temperate or desert localities), a winter brumation period is required to facilitate gametogenesis (ovulation and sperm production). The duration and magnitude of the temperature reduction required for successful brumation is variable; however, a six-to-eight-week gradual reduction (and subsequent increase) of temperature to 18° C is typically sufficient. Prior to beginning the brumation cycle, animals should be fed more regularly than normal to mimic the late-summer "feast" season and then not fed so that they will defecate their last meal at a normal temperature. Otherwise, undigested meals will rot in the stomach, resulting in the death of the animal. The period of fasting will depend on the relative metabolism of the species. For species with fast metabolisms, such as elapid snakes and varanid lizards, three weeks is sufficient, while for species with slow metabolisms, such as viperid snakes and helodermatid lizards, a minimum of six weeks of fasting is required. At the end of the brumation cycle, warm in waves over two weeks, with each day a bit warmer and longer than the previous. Only once normal temperature parameters are reached should the animals be fed.

Commercial grade wine chillers come in a variety of sizes that make great brumation centers. Animals can be placed in ventilated and secured plastic boxes with several inches of substrate (making sure it is never damp) and a small water dish. These wine chillers allow the keeper to gradually cool the animal to the desired temperatures over several weeks. Refrigerated rooms or rooms with air conditioning can also be utilized depending on how many animals need to be cooled.

5.9 VENOM EXTRACTION

During the extraction process (see color plate 19 and appendix 5.1), the handler is exposed to the highest risk of envenomation because of the close proximity of the person's hands to the fangs of the snake. Because of this, extreme caution must be taken, and extraction should never be undertaken by anyone not in optimal mental and physical condition (such as when sick or sleep-deprived). The handling techniques required also vary from those used in standard husbandry. Regardless of whether venom is extracted for antivenom production or for research purposes, it must be obtained under hygienic conditions, including the use of sterile containers and the immediate deep-freezing of venom samples with liquid nitrogen or dry ice or in a -80° C freezer.

5.9.1 TRAINING

Employees should undergo intensive training before taking part in the extraction of venom. This training is beyond the minimum training standard detailed above, as there are special risks involved. Training records and a competency assessment must be kept for each employee. Safety assessments should occur for each employee on a yearly basis.

5.9.2 EXTRACTION TECHNIQUES

A venom-extraction safety checklist must be completed before starting each extraction. Extraction from sick animals or those undergoing treatment is to be avoided. In order to minimize stress, venom extraction should not occur at intervals of less than six weeks for any animal. Water should be removed from the enclosure 3 to 24 hours before extraction to ensure that the snake does not drink just prior to being handled, thus avoiding any water entering the trachea during the extraction process. If not already housed in the extraction room, the venomous reptiles should be transferred in a secure container to the collection area. After venom extraction, the specimen is returned to a secure transport container and transported back to its enclosure.

To avoid any cross-contamination between samples and to decrease the risk of exposure to venom particles, any pipette, cup, or piece of parafilm should be immediately placed in a sealable plastic bag and discarded after a single use. Reusable glassware or plasticware needs to be properly cleaned and treated between uses. An example of a suitable process is as follows:

1. Soak with bleach (to denature any remaining venom).
2. Wash with soapy water (to remove any particulate matter).
3. Rinse with alcohol (to remove residue of bleach and water).
4. Heat sterilization at 230° C for 60 minutes.
5. Keep glassware in a sealed sterile environment until use during extraction.

5.9.2.1 Sedation

For front-fanged species, venom extraction may be facilitated by light sedation. Nonhelodermatid lizards and non-front-fanged caenophidian snakes should always be sedated, followed by the subcutaneous injection of pilocarpine, adjacent to the venom gland, to stimulate venom production (see chapter 6.2.4.4). For a discussion of cooling to 16° C to 18° C as a form of behavior management, see chapter 6.2.4.5. The use of carbon dioxide gas (CO_2) to render the animal unconscious comes with significant animal-welfare issues (see chapter 6.2.4.5).

5.9.2.2 Venom Extraction Using Head Pinning

The "classic" approach for head pinning (see color plates 16C and 16D) is to remove the animal from its container with an appropriate-sized hook and place it on a hard surface covered by a material that provides a cushioning effect (such as a piece of carpet). The reptile's head is pinned with the flat part of the U-shaped part of a hook or by using a specially constructed pinner. The animal is carefully grasped behind the head, then presented with and induced to bite a plastic container covered by an artificial membrane (such as parafilm) (see color plates 19A–C).

Some front-fanged caenophidian snake species are difficult to extract venom from, such as those that produce smaller volumes of venom (such as small elapids), larger elapids with short fangs (such as *Pseudonaja*), or snakes that like to chew and are reluctant to release a beaker (such as *Pseudechis, Ophiophagus*). An alternative form of venom extraction is to use a pipette tip slipped directly over the fang (see color plates 19D and 19E).

Helodermatid lizards are venom extracted by allowing them to chew on soft rubber tubing over a sterile container and then collecting the venom that drips into the container below (see color plate 19F). Varanid lizards however are most efficiently venom extracted using the same protocols developed for venom extraction of non-front-fanged snakes: anesthesia followed by pilocarpine stimulation, with the secretion pipetted off (see color plate 19G).

5.9.2.3 Venom Extraction Using Tubes

Snakes may be tubed and their heads allowed to exit the opposite end and then induced to bite a parafilm-covered cup that is presented by the use of tongs (see color plate 19C). This works exceptionally well for species that are easily stressed (e.g. dwarf *Bitis* such as *B. atropos*), for species with very short necks and very mobile heads (such as *Aspidelaps*), or for *Atractapis* species (which cannot be safely pinned because of their highly mobile fangs [see color plate 5F]). Since pinning is avoided, not only is stress to the animal reduced, but so is the risk of envenomation to the handler.

5.9.2.4 Venom Handling and Storage

As venom is highly allergenic, with induced hypersensitivity resulting in the risk of anaphylactic shock in the case of envenomation, working with dry venom must be done either in a fume hood or with the user wearing a double-cartridge particulate-filter mask, eye protection, and disposable latex gloves, with paper placed on the working surface (and disposed of immediately afterward). Surfaces of nondisposable implements are then to be sprayed with 70% ethanol to denature venom proteins. Venom extraction waste should then be treated with strong bleach solution. (For handling of venom post-extraction, see chapter 7.2; see also appendix 5.2.)

5.10 FIELD COLLECTION

If possible, all attempts should be made to acquire captive-bred individuals before resorting to field collection. This is not only an issue of environmental ethics but also because captive-born animals adapt better to maintenance in a research or production facility. It is also beneficial from the perspective of disease management, as captive-born animals will be relatively disease- and parasite-free. However, as venom profiles are well documented to vary within geographical regions of a single species, field collection may be necessary to obtain species or geographic variants not already present in captivity or to obtain additional individuals to increase the genetic diversity of captive breeding colonies. Relevant data to be collected at the time of capture include but are not limited to GPS coordinates, vegetation type and canopy cover, microhabitat selection, substrate, 1.5 m temperature, 1 cm and 1.5 m relative humidity, time of day, and moon cycle. If field collection is deemed necessary, steps must be taken to observe appropriate rules and regulations as well as the safety of field collectors. (See also appendix 5.3.)

5.10.1 PERMITS

A complex web of regulations surrounds the establishing of a professional collection of venomous reptiles, including institutional, city, state/provincial, federal, and international regulations regarding the keeping and selling of native and exotic animals, collecting native animals from the wild, and approvals for movement of animals across state lines. Approval from an institutional animal ethics committee and a biosafety committee is also required. In addition to permits from wildlife-control agencies, obtaining land-use permits and permission from government and private individuals may also be necessary. Approvals may also need to be obtained for the importation of antivenom or venom from other countries. Any research involving live animals requires approval from the relevant local animal ethics committee.

5.10.2 FIELD SAFETY

Additional safety concerns arise during fieldwork beyond those discussed in sections 5.3 and 5.4 above and also in appendices 5.3 and 5.4. Every possible safeguard must be employed when working with venomous snakes in the field. This is an uncontrolled situation. Transparency between employees, supervisors, managers, and external experts ensures that there is meaningful and effective communication about matters that may affect health, safety, and welfare. This is critical for the creation of a culture in which safety is the first consideration and zero harm to both animal and person is acceptable. What is considered professional in this area of specialty might not translate to others with less knowledge or confidence or those facing a language barrier.

Poor communication may result in the following:

- A lack of proper preparation.
- Losing time as tasks are repeated when not done correctly the first time.
- An unclear message that may lead people to the wrong conclusion about institutional policy.
- Misguided assumptions in a number of project-related areas.
- A lack of evaluation with regard to the communication needs of differing social cultures (such as language barriers).
- An organizational culture of "shooting the messenger."

A detailed fieldwork plan needs to be developed and acknowledged by all fieldworkers before participation. This plan needs to include the following:

- Destination GPS coordinates (if possible).
- Main activity.
- Weather conditions.
- Trip participants, including their contact details and contact details of next of kin in case of an emergency.
- Travel details, dates, times, location of accommodation, if applicable.
- Details and registration of vehicle used and whether it has a built-in GPS locator or panic button.
- Site destinations; if national parks or state forests, name of local ranger and contact details; if in the wilderness, name of closest emergency services or police station and contact details.
- Risk assessments relating to the activity, signed acknowledgments that all persons in the field party have read and understand the risks involved with the activity.
- Communication devices, satellite phone including number, distress beacon (EPIRB, PLB) ID number and confirmation that it has been registered and is ready for use and charged and that its usage date has not expired.
- Whether personal GPS trackers for remote locations are available.
- Communication plan: how to keep in contact while out in the field; if at night, whether light beacons are displaying the locations of the vehicle and individuals on foot.
- Checklist of all PPE required on field trip.
- Other safety considerations, such as water, first aid kits, torches, insect repellent, vaccinations, certificate of medical fitness if required on long adventures, evacuation plan in case of bush fire or other catastrophic emergency.
- Reviewing the level of understanding of any communication to check that the intended meaning has been understood; this may involve a signature to validate receipt of information and understanding of content.

- Emergency contacts back at base and security phone numbers in protocols provided to designated safety officer.

Each team member must be familiar with the shortest route to a local hospital (if available). Never manipulate a venomous snake in the field unless it is absolutely imperative. Envenomations in remote locations are truly a worst-case scenario. Field collection of reptiles often occurs in remote locations and during the wet season, so it is essential that vehicles used are equipped for difficult driving conditions, with a winch, a snatch kit, and tire-repair kits. It is mandatory that all drivers have completed an occupational four-wheel-driving course, including sand driving. In addition, a satellite telephone or EPIRB should be taken to facilitate emergency extraction. As cellular telephone service is often not available in remote locations, long-range walkie-talkies should be used to facilitate communication between field team members. Field collecting has an added difficulty: working in the dark. Headlamps should always be used when collecting venomous reptiles in the field at night. Each person must carry spare batteries and a secondary light source. A binder containing the bite protocol should be kept in the vehicle, with bite procedures, hospital locations, phone numbers, and medical sheets. Making a predeparture checklist of all necessary materials is essential so that important tools and supplies are not left behind.

Other than special difficulties posed by low-light, uncontrollable situations, safety considerations for the handling of venomous reptiles during field collection largely match those detailed above for husbandry purposes. In short, proper handling tools (such as hooks and tongs) must be utilized, alcohol and medications that may cause drowsiness must be avoided at all costs, and field collecting should be conducted in pairs or groups. In addition, the location the collectors are visiting must be given to colleagues, along with the expected date of return, and daily telephone check-ins are essential. If cellular telephone service is not available, then the team must carry at least one satellite telephone. These safeguards will help protect field collectors in the event that an issue arises that renders them unable to return under their own power.

5.10.3 FIELD COLLECTION TECHNIQUES

For many snake species, driving slowly along quiet roads at night is an effective technique. The sealed roads remain warmer longer than the soil. A snake encountering the road may view it as a large flat rock on which to warm itself. Unfortunately, this results in high snake mortality rates on roads with appreciable amounts of traffic. Such specimens, however, are not only useful indicators of target species being in the area, but if they are freshly hit and still alive (even if mortally wounded), the specimens can be venom extracted and euthanized, with the venom glands removed for transcriptome studies (although this may provide suboptimal mRNA levels). In addition, even long-dead specimens may be useful for genetic studies.

For off-road collecting, in addition to visual spotting by walking, turning over suitable rocks and logs and utilizing traps may be efficient collection methods for some species. Be mindful to return all overturned items back to their original resting positions so that microhabitat conditions may remain as close to preturned levels as possible. For snakes and small varanid lizards, drift-line and funnel-trap combinations are extremely efficient (see color plate 20D). Heavy fly screen supported by stakes is used for the drift-line material, and the bottom is partially buried to prevent animals from easily passing underneath. Pairs of funnel traps are clamped onto the drift line using bulldog clips. It is important to have soil leading up the funnel edge to the opening. This not only masks human scent but also discourages the animal from going under the trap. It is essential to cover the trap with heavy shade cloth and weigh this down with sticks to keep it from blowing off. This will prevent stress or death from heat, which is particularly important for nocturnal species present in the trap before the morning trap checks are completed. Laying out cover boards for snakes to conceal themselves underneath is also useful for sites to which repeated visits are expected.

5.10.4 COLLECTION AND TRANSPORT

During transport to the facility, animals must be placed inside porous black bags that allow sufficient air flow. The animals must be double-bagged, with one bag tied or taped securely and then placed inside a second bag, which is also tied or taped securely. These bags must then be placed inside a rigid lockable container, which is clearly labeled as containing a venomous reptile. Temperatures within the transporting vehicle must be monitored constantly so that they remain within acceptable limits for the animal being transported. Containers must be kept out of direct sunlight and the air-conditioning run to keep the entire vehicle at no warmer than 24° C. In the event that the species in question requires elevated humidity levels, this may be managed by placing a damp cloth inside a plastic container that has holes drilled in it and placing this container inside the bag containing the specimen. The bag itself must not be wetted, as this may block air flow and result in suffocation of the specimen.

Transport of sea snakes requires special modifications. They must be in an enclosed container, bucket, or polystyrene box with a wet towel at the bottom. The snake must be in a mesh bag; as a cotton bag (for example) would saturate with water, drowning the animal. This arrangement will allow the snakes to stay hydrated. If the animal is being transported during the day time in temperatures greater than 24°C then a frozen iceblock should be placed underneath the towel.

It is of paramount importance to keep good records of each specimen, not only for management purposes but also to enhance knowledge of the species. Arrival data linked to the institutional registration number must include (but are not limited to) locality of origin, date of arrival, collector or contact person, size and weight, and health status. In addition, quarantine protocols must be followed (see chapter 6.4.1).

5.11 VENOMOUS REPTILE CONTROL (CALLOUTS) AND TRANSLOCATION

Often, institutions become the first phone call for locals dealing with an unwanted reptilian visitor (see color plates 21 and 22). Similarly, industrial sites such as mines may approach institutions seeking risk-management protocols and training for on-site personnel who may encounter venomous reptiles. In any interaction between a venomous reptile and a human, there are several options:

- Leave the animal alone.
- Translocate the animal within its estimated home range.
- Translocate the animal outside its estimated home range.
- Euthanize the animal, and submit it as a voucher specimen.
- Maintain the live animal for educational purposes at an approved institution.
- Kill the animal (putting the person at great risk of envenomation and likely in violation of wildlife laws).

For those of us who love venomous creatures (such as the authors of this book), leaving the animal where it was found seems to be the obvious choice. However, mitigating factors such as an ingrained fear of snakes, the proximity of children and pets, or the need to conduct work in the area (such as clearing of vegetation) often necessitate the animal's removal or relocation. Which of the above options (excluding the last one) is most appropriate therefore needs to be determined on a case-by-case basis. Organizing an educational outreach program in order to build strong ties with the local community is the best way to obtain the most positive outcomes for people, venomous reptiles, and the environment.

Callouts often include the resident wanting the scaly "intruder" removed regardless of whether it is venomous or not. Numerous factors for both the reptile and the human residents must be taken into account in deciding whether a reptile can be or should be safely and ethically relocated or if it must be subsequently maintained in captivity (rendering it ecologically dead). Factors to be considered during any reptile relocation are training, safety, community perceptions, effects on the relocated animal, and whether the reptile belongs to a protected species.

5.11.1 TRAINING

Reputable courses that provide appropriate education and training in handling ensure optimal occupational wildlife control and rescue operation safety (see color plates 16–22). In many countries, wildlife authorities require the completion of an accredited course for relocators before issuing the appropriate permits. In some cases, the institution or facility is requested to train emergency and first-response personnel (fire fighters, ambulance, police) in the removal of venomous animals. The learning environment must be large enough to handle venomous reptiles safely and also include graded instruction in reptile handling, starting with nonvenomous reptiles and gradually progressing to venomous reptiles. Such courses must include the following as key competency modules:

- Identification of local venomous reptile species and their mimics.
- Identification of a suitable reptile habitat, whether natural (such as nearby woodlands) or artificial (such as flat pieces of plywood), and factors that may attract venomous reptiles (such as climatic conditions, junk piles, spilled feed from chicken farming facilitating an increase in the local rodent population, backyard aviaries).

- Conformation to general occupational safety and health requirements in venomous reptile handling and transport (see section 5.3). In addition, there are unique hazards in removing venomous reptiles from industrial areas because of concurrent hazards such as poor lighting, heavy machinery, toxic chemicals, glass, and restricted access. Such hazards may involve additional PPE requirements, specific to each site, beyond the minimum for working with venomous reptiles (see section 5.3), for example, steel-cap boots and hard hats.
- Conformation to specific local legislation regarding removal and translocation.
- The necessity of working in teams of at least two, so that there are spotters not only for the venomous reptile being removed and any possible additional snakes but also to keep an eye out for industrial hazards while the venomous reptile is being secured.
- Training in first aid and the effects of snakebite (with a particular focus on local species).
- Handling equipment and techniques (see section 5.4).

5.11.2 SAFETY

Safety for both human and reptile must be the first aspect taken into account during any call-out. As previously mentioned, additional hazards may be present not only in industrial settings but also in residential areas (such as cars). In the event that the species of reptile captured is harmless, there is no safety issue for residents that would preclude release of the animal on site. There are, however, potential safety issues for the reptile. If the property owners suffer from ophidiophobia, no amount of reasoning is likely to get them to agree to have a snake (or often even a lizard) released back on their property. Refusal to remove a reptile may lead homeowners to take matters into their own hands, likely resulting in the death of the reptile and potentially (if the reptile in question is venomous) the envenomation of the person.

5.11.3 COMMUNITY PERCEPTION

How well are venomous reptiles perceived in the community? Are they seen as valuable members of the local ecosystem, or are they considered vermin? This is the key in determining whether residents will accept short-distance relocations being performed in their area for venomous species. With this in mind, it would be a good idea to establish a working relationship with the local community. Annual or biannual talks about how to identify local species of venomous and nonvenomous reptiles and the importance of venomous reptiles to the local ecosystem are a great start for gaining acceptance of venomous reptiles as cohabiters and reducing the number of calls (and therefore travel expense) for nonvenomous species. For long-term success, it is beneficial to give talks at local schools, as children are much more receptive to new information than adults who have had their prejudices ingrained for years. If compassion for all of the animals in the ecosystem can be kindled in children, they may successfully pressure their parents to be more accepting of reptiles. Reach out to organizations such as Boy Scouts and Girl Scouts, which do many camping trips, and teach them how to avoid negative interactions with venomous snakes, how to properly identify native species, and, most important, how to enjoy observing them from a safe distance. Local emergency-service departments also benefit from additional training; the public often contacts them to request that they remove "nuisance" venomous reptiles.

Only through education will barbaric practices such as "rattlesnake roundups" be made illegal on animal-cruelty and also conservation grounds (see color plate 23). The stated goal of venom collection for research purposes is fallacious, as venom collected at such roundups falls well short of accepted best practice as detailed in chapter 7.2. Since the venom is pooled from many specimens, it becomes at the very least a mixture of regional variations within that species. Such variation has a direct and immediate impact on venom action and antigenicity, and therefore the source of a sample must be identified in venom research. In some cases, the venom collected is a mixture of entirely different species. Furthermore, the venom is often not snap-frozen immediately after collection, rendering it useless for research purposes or antivenom manufacture. Roundups are also unethical for a number of reasons. First, the snakes are kept alive for months without food, water, or shelter. This is in violation of all best-practice animal-welfare legislation. The animals are decapitated while alive, which is not a humane form of reptile euthanasia (see chapter 6.2.10). This is because the reptile brain remains conscious for 15 minutes or longer after the head is severed. Not only is this horrendous animal cruelty, but it also creates a dangerous situation with a high potential for accidental envenomation. University researchers must receive approval from animal-welfare and biosafety committees, both of which, in addition to considering the statutory legislation, take into consideration the ethics at all levels within the project. Thus, work on samples originating from rattlesnake roundups would not be approved on scientific grounds (poor-quality material) and also on animal-ethics grounds.

5.11.4 EFFECTS ON RELOCATED ANIMALS

A detailed understanding of the ecology of a given species is necessary in order to be able to make an informed decision about whether to relocate an individual. Unfortunately, this information is not known for many species. Translocation of animals may be carried out on two different distance scales. Short-distance translocation (SDT) is defined as transport of an animal to a location near or within its normal or estimated home range (less than 1.6 km for most species of snake). Long-distance translocation (LDT) is defined as transport of an animal to a location outside its normal or estimated home range (Hardy et al. 2001). While large, active species such as *Ophiophagus hannah* may have a home range of up to 6 sq. km, some ambush vipers have been shown to have home ranges of roughly a hectare or less (Glaudas and Rodriguez-Robles 2011; Maritz and Alexander 2012).

LDT does not appear to be an appropriate solution for resolving reptile-human conflict, as it has been found to negatively affect reptiles in terms of altered movement patterns (Reinert and Rupert 1999; Plummer and Mills 2000; Nowak, Hare, and McNally 2002; Sullivan, Kwiatkowski, and Schuett 2004; Butler, Malone, and Clemann 2005a; Butler, Malone, and Clemann 2005b) and increased mortality rates (Reinert and Rupert 1999; Plummer and Mills 2000; Nowak, Hare, and McNally 2002). Translocated *Crotalus horridus* that managed to survive their first overwintering period after LDT exhibited more extensive movement patterns than resident snakes during the following active season. However, they appeared to establish a more resident-like pattern of movement during this second year (Reinert and Rupert 1999). Additional factors that must be taken into account for LDT are the potential for disease transfer (Cunningham 1996; Hardy et al. 2001; Shine and Koenig 2001; Seigel and Dodd 2002; Sullivan, Kwiatkowski, and Schuett 2004) and the impact on the receiving population's genetic structure (Dodd and Seigel 1991; Stockwell, Mulvey, and Vinyard 1996; Whiting 1997; Hardy et al. 2001; Shine and Koenig 2001). However, LDT may prove viable for repatriation attempts of sensitive species (King, Berg, and Hay 2004).

SDT, on the other hand, avoids the risk of disease transfer and potentially detrimental genetic exchange for the receiving population; however, SDT is not without its own potential drawbacks. Although body condition, behavior, mortality rate, and home-range size were shown not to vary after SDT in two rattlesnake species (Sealy 1997; Brown, Bishop, and Brooks 2009), recapture rates of translocated animals remained high (Sealy 1997; Hardy et al. 2001; Sullivan, Kwiatkowski, and Schuett 2004; Brown, Bishop, and Brooks 2009). A high recapture rate has the potential to upset residents and could eventually lead to their deciding to kill the snakes in hopes of preventing future encounters. Not all studies had high recapture rates for SDT individuals (Hare and McNally 1997; Sealy 1997), but this may have been a result of methodological differences.

5.11.5 PROTECTED SPECIES

Occasionally, the animal to be removed may belong to a species with protected conservation status. This adds another dimension to the callout, as there may be federal, state, county, or local laws that stipulate how the animal must be dealt with regardless of the landowner's personal feelings. Additionally, unless the respondent has the proper permits, they may not be legally allowed to touch the animal in question. It is advised, therefore, that an institution acquire such permits for employees charged with going on callouts in order to remain compliant with all regulations. An unfortunate consequence is that if the homeowners realize that their desired result does not align with legislation, they are likely to stop calling for help and instead dispatch the animal in a way they see fit, which in most cases means the death of the animal.

Each institution must ensure that employees who go on callouts are educated about local protected species and any special rules pertaining to such species. For example, *Heloderma suspectum* often wander into urban areas, and because of public fear and misconceptions surrounding this species, translocation is often the only option. The Arizona Game and Fish Department requires that *Heloderma suspectum* be translocated no more than 400 meters from where they were found. However, research has shown that SDT of *H. suspectum* like this is ineffective in keeping them out of populated areas (Sullivan, Kwiatkowski, and Schuett 2004). Thus, education is the best way to help homeowners coexist with the species. If a *H. suspectum* is found in an urban environment in Arizona and there is no suitable desert within 400 meters, then it cannot be released and must spend the remainder of its life in captivity.

APPENDIX 5.1

RISK ASSESSMENT FOR VENOM COLLECTION, CAPTIVE MAINTENANCE, AND COLLECTION OF VENOMOUS REPTILES

The following is applicable to all interactions with live venomous reptiles. The institutional safety officer must be informed of the location and duration of field trips before the trips begin. Venomous reptile species will be collected from various habitats, both terrestrial and aquatic. Collection areas may be subject to extreme weather conditions such as flash floods and cyclones. Thus, the team must be aware of and keep updated about extreme weather warnings posted by local weather services. In the event of an emergency, the relevant local emergency services will be contacted by phone. A satellite phone must be available in case of an emergency in areas without mobile-phone coverage.

HAZARDS

Hazard identification:

Envenomation of personnel.
Allergy induction.

Other hazards:

Injury from tripping over random objects (such as logs and rocks).
Animal bites other than envenomations.
Vehicle accident traveling to and from collection site.
Injury while scuba diving.

Biological hazards:

Substance	MSDS	Conc.	Physical form	Hazardous
Venom	No	Dilute	Liquid	Yes (only by injection)

Conc., concentration.

Health effects								
Acute toxicity	Chronic toxicity	Corrosive	Irritant	Sensitizer	Carcinogen	Mutagen	Teratogen	Asphyxiant
Yes (injection only)	No	No	No	No	No	No	No	No

Hazardous reactions				
Explosive	Flammable	Spontaneous reactivity	Water reactive	Oxidizer
No	No	No	No	No

Routes of exposure				
Inhalation	Ingestion	Skin absorption	Eye contact	Injection/needle stick
No	No	No	Yes	Yes
Evidence of Exposure				
Presence of dust/odors	Leaks/spills/residues	Worker symptoms	Direct contact	Duration of exposure (per procedure)
Yes	No	Yes	No	< 4 hrs

CONTROLS

Hazards are reduced by using proper handling equipment (see section 5.4). In the particular case of sea snakes, hazards are reduced by wearing 7 mm neoprene gloves and arm cuffs during collection activities, as sea snake fangs are typically less than 3 mm (except for large species such as *Aipysurus laevis, Hydrophis elegans, Hydrophis (Astrotia) stokesii*, etc.)

Biological hazards:

Hazard	Associated risk
Envenomation	If victim is untreated, death is a possibility. Any bite requires observation in a hospital. Signs that snake antivenom is required include severe myotoxicity (evidenced by presence of myoglobin in urine and also indicated by severe muscle pain), neurotoxicity (evidenced by loss of control of skeletal or eye muscle control), or coagulopathy (evidenced by changes in clotting times). Satellite phone must be available in case of emergency in areas without mobile-phone coverage. All team members must possess a current occupational first aid certificate. In the event of an envenomation: • Pressure immobilization will be applied for appropriate species such as any elapid, neurotoxic viperids or coagulopathic non-front-fanged genera such as *Dispholidus*, *Rhabdophis* species, and *Thelatornis* species. • Transport will be brought to the victim. • The nearest hospital (which will have already been contacted with notification of the field research being undertaken and to confirm antivenom stocks) will be contacted by phone or radio. • The victim will be transported to the hospital promptly but without undue haste, as a car wreck from rushing would only further complicate the situation.
Envenomation by other organisms while in the field	The risk of fish envenomation will be minimized by wearing appropriate clothing such as dive boots or reef-walking shoes and displaying caution and awareness of the surroundings. The only octopus species of medical importance to humans is the blue-ring octopus. As these are small and with very small beaks, they are not able to penetrate dive gloves and may thus be safely handled with gloves. Deaths have been only very rarely recorded.

Allergy induction	While envenomation is often the primary concern when working with venomous species, development of an allergy is a much higher risk. This may only cause sneezing and other "hay fever"-like symptoms in lab workers and thus be nothing more than an inconvenience; however, in staff interacting with snakes in the field or in captivity, an allergy could result in catastrophic anaphylactic shock in the event of an envenomation. Thus, chronic exposure to dried venom must be avoided, either through the use of a P1 half-face particulate filter or opening containers in the fume hood when working with quantities > 50 mg of powdered venom.
Bites by various organisms while in the field	Bites other than envenomations, such as bites by mosquitoes and sand flies, may occur during collection, with the accompanying possibility of disease transmission such as dengue and malaria. The risk of disease is minimized by wearing appropriate clothing (such as enclosed shoes, long pants, and long-sleeved shirts), vaccination (where possible), and use of appropriate bug spray.

Other hazards:

Hazard	**Associated risk**
Injury caused by tripping over ground terrain while in the field	Injury may be sustained by accidental tripping over objects on the ground such as logs and rocks. This risk may be minimized by being cautious and aware of surroundings.
Injury caused by boat/car/plane accident while in the field	Injury may occur during a boat, car, or plane trip to the collection location if there is an accident. Risk can be minimized by driver caution, defensive driving, and not driving while tired.
Extreme weather	Collection areas may be subject to extreme weather conditions such as flash floods, cyclones, and heat. Be aware of and keep updated about extreme weather warnings posted by the local Bureau of Meteorology and sign up for an automated warning system (if available). Always hydrate properly (avoid coffee in hot conditions).

Control measures:

Elimination or substitution controls	
Potential elimination or substitution of hazards	Risk of envenomation will be minimized by using appropriate live-animal containers (suitable in rigidity and size), proper collecting technique, and appropriate equipment. Allergy induction is eliminated through proper handling of powdered venoms.
Administrative controls	
Administrative measures (such as health monitoring)	Collecting must never be carried out alone. All team members must have current occupational first aid training, including (if relevant) the application of pressure-immobilization bandaging. The collector should carry a first aid kit at all times while collecting, which includes (if relevant) a pressure bandage and sling. Any worker who displays symptoms of allergy induction must be restricted from working with live snakes in the field or in captivity without having injectable adrenaline and injectable antihistamine present at all times.
Training	
Required training	Staff should be experienced in the collection of venomous animals or, alternatively, be trained by staff experienced in the collection of venomous animals. Furthermore, all research staff involved in fieldwork must possess a current occupational first aid certificate. Documentation must be kept of trained people and their first aid certificate details. Training SOPs will cover: • Protective clothing. • Methods for capture. • Methods for transport.

Risk analysis after controls are implemented:

Possibility	Frequency	Impact	Score
Remotely possible	Very rare	Very serious	Low

APPENDIX 5.2

RISK ASSESSMENT FOR WORKING WITH VENOMS IN THE LABORATORY

The following is applicable to all processes (such as proteomics or bioactivity testing) in which crude venom, purified toxins, or synthetic variants are handled. The animal venom/toxin may be in liquid or solid (lyophilized) form.

HAZARDS

Hazard identification:

Envenomation of personnel.
Allergy induction from chronic exposure to powdered venom.

Chemical hazards:

Substance	**MSDS**	**Conc.**	**Physical form**	**Hazardous**
Venom	No	Dilute	Liquid	Yes but only by injection

MSDS, Material Safety Data Sheet; Conc., concentration.

Health effects

Acute toxicity	Chronic toxicity	Corrosive	Irritant	Sensitizer	Carci-nogen	Mutagen	Teratogen	Asphy-xiant
Yes but injection only	No	No	No	No	No	No	No	No

Hazardous reactions

Explosive	Flammable	Spontaneous reactivity	Water reactive	Oxidizer
No	No	No	No	No

Routes of exposure

Inhalation	Ingestion	Skin absorption	Eye contact	Injection/needle stick
Yes	No	No	Yes	Yes

Evidence of exposure

Presence of dust/ odors	Leaks/spills/ residues	Worker symptoms	Direct contact	Duration of exposure (per procedure)
Yes	No	Yes	No	< 4 hrs

CONTROLS

- Avoid inhaling aerosolized venom; use only in well-ventilated areas. Quantities > 50 mg must be used only in a fume hood or while wearing a P1 half-face particulate filter.
- All personnel handling animal venoms/toxins should attend a biosafety seminar and receive appropriate training in the handling of animal venoms/toxins from experienced supervisors.
- Caution must be used at all times when handling animal venoms/toxins.
- Personal protective equipment, including nitrile gloves, lab coat, and appropriate enclosed footwear, must be worn when handling animal venoms/toxins.
- Creating aerosols and fine dust with animal venoms/toxins must be avoided.
- Animal venoms/toxins must be stored in clearly labeled and sealed plastic containers (never glass).

Biological hazards:

Hazard	**Associated risk**
Envenomation	Envenomation will only occur through a needle stick and then depression of the syringe to deliver dilute venom. Thus, this is an extremely unlikely event. Therefore, the storage of antivenom is not necessary, as these are extremely unlikely events. Venom in syringes should not be at concentrations such that an estimated human lethal dose is in a volume < 100 μl (such as 100 mg/mL for highly toxic venoms such as *Naja naja*, which has an estimated human lethal dose of 10 mg). As needles are only used in situations such as *in vivo* testing, the amounts being worked with at any given time are less than the estimated human lethal dose (e.g., a typical rat injection would be only 10–50 μg). Venom, however, should always only be stored in plastic containers, never in glass. Thus, if the container is dropped onto the floor, there is no combined hazard of glass fragments and venom.
Allergy induction	While risk of envenomation when not working with live animals is a minuscule possibility, development of an allergy is a much higher risk. This may only cause sneezing and other "hay fever"-like symptoms in lab workers and thus be nothing more than an inconvenience; however, in staff interacting with snakes in the field or in captivity, such an allergy could result in catastrophic anaphylactic shock in the event of an envenomation. Thus, chronic exposure to dried venom must be avoided either through the use of a P1 half-face particulate filter or opening containers in the fume hood when working with quantities > 50 mg of powdered venom.

Control measures:

Elimination or substitution controls	
Potential elimination or substitution of hazards	Risk of envenomation will be minimized by using personal protective equipment including safety glasses, nitrile gloves, lab coat, and appropriate enclosed footwear and by following protocols. Allergy induction is eliminated through proper handling of powdered venoms/toxins in quantities > 50 mg.
Administrative controls	
Administrative measures (such as health monitoring)	Any worker who displays symptoms of allergy induction must be restricted from working with live snakes in the field or in captivity without having injectable adrenaline present.
Training	
Required training	All handlers of animal venoms/toxins are required to undertake training sessions led by experienced personnel.

Risk analysis after controls implemented:

Possibility	**Frequency**	**Impact**	**Score**
Remotely possible	Very rare	Very serious	Low

APPENDIX 5.3

TEMPLATE SAFE OPERATING PROCEDURES FOR BOATING AND DIVING FIELD TRIPS

STAFF, VOLUNTEERS, AND STUDENTS

- Conduct mandatory discussion session with person in charge regarding field trip procedures, activities, risks, and risk assessments.
- Complete and acknowledge risk assessment.
- Complete and submit authorized volunteer forms if required.
- Fill out appropriate paperwork for working away from the facility.

NOTIFICATION OF LOCAL SCIENTIFIC PERMITTING AGENCIES

- E-mail notification prior to each field trip, with information regarding dates, times, sampling type, locations, activities, vessel(s), and all personnel on board.
- Upon return, e-mail report for each field trip to relevant authorities within mandated time frame.

WEATHER CHECK AND FINAL CONFIRMATION OF TRIP

- Two days before intended trip, check local weather and boating forecasts via Bureau of Meteorology or local weather information supplier, and either confirm or postpone trip.
- Make final weather check on morning of field trip. If weather has changed and looks poor, cancel trip by phoning all participants as early as possible.

RESEARCH STATIONS AND VESSELS

- Complete station form on arrival; information recorded includes all staff, students, and volunteers; room number for each person at station; signature of each person; daily contact detail of trip leader and signature.
- If relevant, complete boating and safety record form online each day before departure from research station; information recorded includes dates, times, destinations, vessel names, all personnel on board each vessel, boat skipper, and activities including whether diving, swimming, or snorkeling is involved. Record confirmation of checks of weather and boating conditions for the day, and supply contact details (phone and radio channel).

- Complete research station whiteboard, if applicable, before departing the station each day; this includes date, departure time, ETA back at research station vessel, destinations, and number of people on board.
- Complete boating safety gear form; this checklist list must be completed and signed off by master of the vessel each day. This is a detailed form on which the presence of each safety item is confirmed, the amount of fuel carried is recorded, and confirmation that vessel and all gear are in excellent working order is established, including expiration dates of flares and EPIRBs.
- Complete boat log at end of each day.

RADIO CHECKS

- Conduct radio check back to base each day.
- Make radio call-in within 30 minutes of vessel launch, to check that radio is working and confirm that all is well and boating conditions are satisfactory.
- Make radio call toward end of day to notify base of ETA.
- Have extra radio call during the day to inform base or research station of progress, conditions, and intentions.
- Keep radio on all day in order to receive calls from base, research station, Coast Guard, and weather alert.

APPENDIX 5.4

TEMPLATE CHECKLIST OF HAZARDS/RISKS AND OTHER ITEMS WHEN PLANNING DIVING WORK

Risk factor	Lower risk	Moderate risk	Higher risk
Boat traffic	Infrequent traffic Depth > 5 m	Some traffic depth > or < 5 m SSBA dive	High traffic/shipping lane depth < 5 m SSBA dive
Dive (duration)	At least 2 repetitive groups less than no-decompression limit	1 repetitive group less than no-decompression limit	Dive to no-decompression limit
Dive (multiple ascents)	None	1 or 2	3 or more
Dive (profile)	"Ideal" profile	"Square" profile	"Reverse" or "sawtooth" profile
Dive (repetitive)	No more than 2 dives/day	3–5 dives/day	> 5 dives/day
Dive (time since last)	< 3 months	3–6 months	> 6 months
Hazards (entrapment)	Unobstructed ascent		Obstructed ascent
Hazards (marine life)	General observation	Observation of dangerous animals	Spearfishing or manipulation of dangerous or venomous animals
Hazards (task-related)	Use of equipment such as slates or cameras	Use of handheld pneumatic tools and/or small lift bags	Use of heavy tools/frames and/or large lift bags
Local knowledge	Know site well	Some experience with or knowledge of site	Little knowledge of site
Medical assistance	< 30 mins away	0.5–2 hrs away	> 2 hrs away
Personnel	2–3 divers (multiples) 1 boat handler per team	2–3 divers (multiples) No boat handler	Single diver
Personnel (experience)	> 50 hrs	20–50 hrs	< 20 hrs
Site (current)	Nil to weak	Moderate	Strong
Site (depth)	< 18 m	18–30 m	> 30 m
Site (exposure)	Sheltered		Exposed
Site (location)	Location not remote Sheltered embayment Uniform bottom profile	Exposed site Close location	Exposed site Remote location
Time of day	Start and finish in full daylight	Start at/before dawn and finish near/after dusk	Night diving
Weather and sea	Calm, settled weather pattern	Calm, unsettled weather pattern	Rough

SSBA dive = surface-supplied breathing apparatus.

CHAPTER 6

VETERINARY CARE OF VENOMOUS REPTILES

J. HABERFIELD, P. MARTELLI, R. JOHNSON, S. BARTEN, A. GILLETT, B. LOCK, R. JONES, S. SIMPSON, T. N. W. JACKSON, C. COCHRAN, N. DUNSTAN, K. SUNAGAR, AND B. G. FRY

6.1 SUMMARY

This chapter has been written with the intent that it may be used as a citable reference for applications to institutional animal ethics or biosafety committees. It contains a description of "best practice" procedures under optimal conditions, and thus, the guidelines may not be practical or economically feasible for all facilities. This is particularly likely to be the case for smaller endeavors, and it is recognized that there may be alternatives to the practices described herein that are more suitable for certain circumstances. The challenges presented in the examination of venomous reptiles are not in the general principles but in the specifics. In recent years, the number of species documented to produce venom has been vastly expanded to include the majority of snakes and numerous lizards. However, many of these species are unlikely to pose a threat to humans in the event of an accidental bite. For the veterinary practitioner, to consider all reptiles that possess venoms as potentially dangerous is not realistic or practical. On the other hand, certain species possess exceptionally potent venoms and are more than capable of delivering a potentially life-threatening bite. Veterinary care of these species presents specific challenges requiring special considerations, techniques, and skills. Therefore, in addition to the usual veterinary concerns of any zoological collection, work with venomous animals presents special considerations and difficulties to the veterinarian. However, while it is imperative that veterinarians be aware of the potential risks of envenomation when conducting an examination on a venomous reptile, it need not compromise the examination. A set of protocols and principles guide "best practice" in the management of exotic and potentially dangerous wildlife for which veterinary care is being given. These general protocols are presented here, and in keeping with the theme of this book, this chapter will also address specific aspects of veterinary medicine that are uniquely applicable to the care of venomous reptilian patients, including safety considerations for the veterinarian.

6.2 BASIC VETERINARY TASKS AND PROCEDURES

6.2.1 GENERAL PRINCIPLES

The same principles apply to the treatment of both venomous reptiles and nonvenomous reptiles. Repetitive handling should be avoided where possible, both to reduce stress for the animal and to minimize risk to staff. After making a diagnosis, a well-thought-out treatment plan should be developed with the following considerations (Johnson 2011):

- Where possible, utilize medications with longer duration of action to reduce the frequency of administration.
- Use suture materials and wound-closure techniques that reduce the need for handling postsurgery, such as self-dissolving or subdermal sutures that do not require removal.
- Whenever possible, treat the animal at its place of residence rather than at the veterinary practice. In the veterinary setting, this is often preferable for minimizing staff risk and animal stress. In order to achieve this, the owner of the animal needs to be able to administer the prescribed medications safely.
- Choose medications based on ease of administration for the caretaker or snake owner. For example, if an oral medication is needed, is it possible to provide the medication in a food item?

6.2.2 GENERAL EXAMINATION

A general examination of any ill or injured patient often includes a physical examination in conjunction with diagnostic procedures such as blood collection, radiography, and possibly ultrasound. Beginning with a thorough physical examination as part of any diagnostic investigation is essential. Observation of the patient's movement and behavior prior to handling or anesthesia can greatly assist with identifying abnormalities such as paralysis or neurological deficits that may be overlooked if the practitioner proceeds straight to restraint or anesthesia. For example, when examining sea snakes, it is imperative that these animals are placed in a large tub of water or a pool to observe swimming and diving behavior before any restraint. As they are not adapted to movement on land, they will often appear uncoordinated or even lethargic when out of the water.

A standard stepwise method of examination ensures that all areas of the animal's body are examined, and the order of examination often comes down to personal preference. Universally, the most adopted method for examination is to start at one end of the animal (either cranial or caudal) and work toward the other end. The eyes, nostrils, oral cavity, vent, and entire body (including joint manipulations) should be thoroughly inspected for abnormalities before moving on to diagnostic sample collection. It is important to have at least two people conduct the examination; at least one person is required to adequately restrain the specimen, while another performs the examination. In case of an accidental envenomation or escape, it is also imperative to have assistance close at hand.

6.2.3 RESTRAINT

When examining a venomous reptile, accidental envenomation can occur if adequate safety measures are not in place (see chapter 5.4 and color plates 16, 24A and 24B). It is paramount that all appropriate equipment necessary for restraint, diagnostic sampling, and induction of anesthetic and all medications that may be required are on hand and in place *before* an examination begins in order to minimize the risk of envenomation. Clear plastic restraint tubes are an efficient, safe, and humane means of restraint for venomous snakes. Most examinations include diagnostics such as blood collection, radiography, and ultrasound, in addition to sex identification, all of which can be performed with the snake safely restrained in the tube. Further procedures, such as anesthesia, radiography, microchipping, tagging, and humane euthanasia, can also be performed at this time. Large venomous lizards may require multiple persons to restrain them adequately and safely. It is important to have control of the tail, hind legs, and head at all times. If the animal must be restrained for longer than 10 minutes, sedation is recommended.

6.2.4 ANESTHESIA

Veterinarians treating venomous reptiles must be comfortable with all manner of restraint, including chemical restraint (Read 2004; Mosley 2005; Schumacher and Yelen 2006), before embarking on an examination. Minor procedures may be performed under manual restraint alone but may be less stressful for the patient if sedation or light anesthesia is used. Delivery methods for sedatives or anesthetics vary greatly and may include gaseous inhalation and intramuscular or intravenous injections. Because of the tendency many reptiles have of holding their breath when gaseous anesthetics are used for induction, administration is more common via intravenous or intramuscular routes.

Some procedures, such as the removal of a retained spectacle or venom extraction, may be facilitated by light sedation (Read 2004; Mosley 2005; Schumacher and Yelen 2006). However, handling techniques that utilize methods such as cooling animals in ice water or in a refrigerator should never be used, as they are painful and can be extremely detrimental to the animal's health (Schumacher and Yelen 2006). Close attention must be paid to temperature exposure in any examination; overheating or thermal burns from tabletop heat mats can occur inadvertently when animals are sedated.

6.2.4.1 Snakes

If the examining veterinarian is familiar with the delivery of intravenous injections to snakes, this method provides rapid immobilization and is relatively well tolerated by most snakes. It does, however, require manual restraint of the animal (often in a Perspex tube), and the duration of anesthesia can be relatively short, depending on the drug selected. Despite this, the duration is usually more than sufficient to perform short procedures such as blood collection, radiography, sexing, cloacal swabs and washes, and ocular and oral examinations. Intravenous drugs commonly used include alfaxalone (10 mg/kg), propofol (1–5 mg.kg), or tiletamine/zolazapram (3 mg/kg) administered via the ventral coccygeal vein. Alfaxalone IV has a short duration of action and is thus the preferred anesthetic for use in sea snakes, as these patients are aquatic and drugs with longer durations of action may lead to death by drowning. It is always recommended that sea snakes having undergone any sort of anesthesia or sedation be given time to recover in a confined enclosure lined with moist towels and not placed in any water. These animals should be left in this state for a minimum of 12 to 24 hours and be assessed for full mobility before being placed back in water.

Intramuscular administration may be an easier option for those not familiar with intravenous access, but this route does take significantly longer to take effect, particularly if the snake is not at optimal body temperature at the time of injection. Depending on the species, intramuscular injection of tiletamine/zolazapram/Telazol at 10 mg/kg will provide a long-lasting state of indifference or torpor adequate for safe handling for procedures such as blood drawing, radiographs, oral and eye examination, manipulation of joints, sexing, cloacal swabs or washes, tracheal washes, or suture removal. The advantage of injectables such as tiletamine/zolazapram is that they can be intramuscularly injected while the animal is still secured in a bag, eliminating unnecessary stress or risk to animals and handlers. When a large number of snakes must be examined (such as following confiscations of shipments or in research settings where a large number of animals is amassed for morphometrics and sampling), sedating all the animals concurrently with injectable drugs allows safe and efficient processing. Aside from tiletamine/zolazapram, other drugs that can be administered intramuscularly include alfaxalone (10 mg/kg) or combinations including alpha-2-agonists and ketamine, which produce similar results (xylazine, detomidine, medetomidine, or dexmedetomidine combined with ketamine). Benzodiazepines (diazepam and midazolam) used alone have proven unreliable in the anesthesia of venomous reptiles. However, they can be used to potentiate dissociative agents such as ketamine or opioids such as butorphanol.

When procedures may be painful, appropriate injectable analgesic agents should be administered. Administration immediately before or shortly after induction is considered acceptable and ensures a plane of analgesia during and after the procedure. Alternatively, injectable local anesthetics such as lignocaine or bupivacaine can be utilized at the site of preserved pain or surgery. When using local anesthetics, the total dose should be calculated based on body weight to prevent overdosage, which is a particular problem in small patients.

Gaseous anesthesia can be used as an induction agent but is more commonly used for prolonged anesthesia of already anesthetized or sedated patients by delivering measurable amounts of vaporized anesthetic and oxygen via a non-rebreathing circuit. Gas anesthesia also is indicated for painful procedures, as injectable sedatives often provide inadequate analgesia or anesthetic depth. Gaseous anesthesia for induction can be delivered via the restraint tube, in an induction box, by mask, or by direct intubation. If the snake is already restrained in a tube, the anesthetic tubing can be taped to the end of the tube to facilitate induction. This method has been used with reasonable success for many venomous species. Induction in a chamber using precision vaporizers may be accomplished with isoflurane (up to 5%) or sevoflurane (up to 8%). Cotton wool soaked in isoflurane and dropped into the induction box will greatly speed up the process, and many field biologists use this method. Clearly, these high concentrations are detrimental to the health of the patient if exposure is prolonged, and once the animal is anesthetized, it must be placed on an anesthetic circuit for continued anesthesia and ventilation. Direct intubation with assisted ventilation provides the fastest induction but requires direct handling of the patient. In order to avoid discomfort to the animal and risk to the practitioner, it should only be done if the animal is already sedated and lignocaine is applied to the epiglottis.

If the patient has had an endotracheal tube inserted, it can be ventilated manually or with an electronic ventilator for the duration of the procedure if the animal is not breathing on its own. In mammals, spontaneous respiration is stimulated by high CO_2 levels. Conversely, in reptiles spontaneous respiration is stimulated by low O_2 levels. Thus reptiles placed on 100% oxygen commonly cease to breathe and manual or mechanical ventilation often is required.

Ceasing anesthetic delivery shortly before finishing the procedure, while continuing ventilation with oxygen or air, will speed up recovery from anesthesia considerably.

6.2.4.2 Lizards

The principles and precautions for anesthetizing venomous lizards are similar to those for snakes, as are anesthetic dosages. Because of the likelihood of breath holding, gaseous induction is not a suitable anesthetic technique for this group. However it must be noted that inter-species differences exist even within a genus. For example *Varanus komodoensis* have proven to be readily amenable to gaseous anesthesia but other varanids such as *V. bengalensis* and *V. salvator* breathhold (Martelli personal observations). Lizards possess a caudal coccygeal vein similar to that in snakes, but intravenous access can also be gained through the jugular and ventral abdominal veins in some species. For prolonged anesthesia, lizards should be intubated and maintained on gaseous anesthesia. Some individuals may spontaneously ventilate; however, ventilation support often is required.

6.2.4.3 Monitoring

Reptiles are challenging anesthetic patients to monitor. Most techniques used to monitor anesthesia in mammals or birds are not applicable to reptiles. A reptile's heart may continue to beat for hours after the animal has died, even in specimens with crushed skulls or pithed brains. Thus, a Doppler flow device, ECG, or esophageal stethoscope may produce a feedback of dubious meaning to the anesthetist as these devices may indicate continuing functional activity when in fact the animal may have expired.

Regardless, it is important to audibly monitor heart rate—a Doppler flow device placed over the heart of any reptile species provides audible monitoring of heart rate and is recommended for any anesthetic procedure. In addition to this, there should be regular checking of cloacal tone, jaw tone, and pinch reflex of the toes in lizards, and the animal must be constantly observed for any movement that may indicate lightening of anesthesia. Generally, in reptiles, a lighter plane of anesthesia is most desirable (particularly where adequate analgesia has been provided) and will ensure good recovery of the patient at the end of the procedure.

6.2.4.4 Use of Sedatives and Pilocarpine to Extract Venom from Nonhelodermatid Anguimorph Lizards and Non-Front-Fanged Caenophidian Snakes

The use of anesthesia such as tiletamine/zolazapram followed by periglandular injection of pilocarpine (10 mg/kg) is a safe and effective manner of facilitating venom extraction from anguimorph lizards and non-front-fanged caenophidian snakes (Hill and Mackessy 2000; Fry et al. 2003c; Fry et al. 2009b).

6.2.4.5 Use of Low Temperatures and Carbon Dioxide (CO_2) Gas prior to Venom Extraction

"Behavioral management" of snakes used in venom production must strike a balance between operator safety and animal welfare. The role of cooling and the use of CO_2 in routine venom extraction are contentious issues. Both have been used by large venom-harvesting laboratories such as Venom Supplies in Australia and Instituto Butantan in Brazil. Anecdotally, the snakes have not suffered distress or health effects. But it must be stressed that neither technique has been subjected to controlled trials, and thus, the mid- to long-term health effects are unknown.

At the time of the writing of this book, Venom Supplies has used this cooling method (with animal-ethics approval) for 8,000 venom extractions over a ten-year period. Records kept by the consulting veterinarian do not suggest any detrimental health effects. Thus, there appears to be no evidence-based objection to overnight cooling to 16° C to 18° C. The caveats are that the snake has not been fed within the previous six days, the cooling period does not exceed 12 hours, and the frequency of cooling events is not greater than monthly. It is reasonable to assume that species with wide temperature variation in their natural environments may be more tolerant to this practice.

The use of CO_2 gas to induce unconsciousness for the purposes of venom extraction, however, carries with it a much greater risk of adverse health effects and/or avoidable distress or suffering.

6.2.5 DRUG ADMINISTRATION ROUTES AND PRINCIPLES

It is important to remember that for optimal drug uptake and utilization, reptiles must be at their preferred body temperature. Administration of any medication to a cold reptile is unlikely to have the desired effect. Drug administration routes are very similar to those of nonvenomous species, with the intramuscular, subcutaneous, intravenous, and topical routes commonly used. Occasionally, intraosseous (lizards only), intracoelomic, or oral routes are used.

The epaxial musculature in snakes and lizards provides an easily accessible area for injection. For many years, it was thought that injection into the caudal third of a reptile's body should be avoided because of the effect of the renal portal system. The two potential effects include a "first pass" effect, in which some of the drug could be carried to the kidney and excreted before entering systemic circulation, and the potential for increased toxicity when nephrotoxic drugs such as amikacin are carried at high concentrations directly to the kidney by the renal portal system. Some studies (Holz et al. 1997; Holz et al. 2002) have suggested that this is not of clinical concern and is more of an academic phenomenon. Further research into this area is needed to truly evaluate its clinical relevance. While Holz's work suggests little clinical relevance, until more is known, it is best to err on the side of caution and give injections in the cranial two-thirds of the body.

The subcutaneous route of injection is not ideal in most species of snakes, because the subcutis adheres tightly to the skin and muscle fascia and does not distend easily. If this route is chosen in snakes, multiple injection sites may be needed. The ventral lateral body wall at the costo-abdominal junction offers a better site for injecting most species of snakes. The dorsolateral body wall of other venomous reptiles often provides a more easily obtainable site for subcutaneous injection. In addition, many lizards have a lateral skin fold between the front and rear legs where the skin is loose, and this site is also suitable for subcutaneous fluid injection. In all venomous reptiles, the intravenous administration of medications can be difficult to perform on a repeated basis unless a venous catheter is placed. Even with the use of a catheter, restraint of the animal is typically required for administration. This is not practical to perform on a regular basis and may have a detrimental effect on the patient; therefore, this route of drug and fluid administration is not commonly used in clinical practice except in severely debilitated patients where other methods of administration are unlikely to be effective. Oral administration is an alternative route for the delivery of fluids and medication, and these may also be inserted in food.

Topical medications (creams, pastes, gels, and other solutions) work well in some situations; however, they often require regular application to be effective, and this may result in stress to the animal or increase in risk to staff. It is important to consider each case on an individual basis, comparing the pros and cons of repetitive handling to those of administration of topical medications. Some topical medications (such as betadine and acaricides) may be administered via baths or paint brushes with minimal disturbance to the animal or risk.

6.2.6 FLUID THERAPY

It can be assumed that most sick reptiles are suffering from some level of dehydration. In general, reptiles have a high tolerance for dehydration, so its presence often indicates chronic disease or inappropriate husbandry. Reptiles that are severely dehydrated often present with sunken eyes, dry and loose skin folds, depression, and anorexia (Donoghue 2006). In these cases, diagnosing dehydration is easy, although the degree of dehydration may be challenging to quantify. The parameters used to quantify dehydration in reptiles include mucous membrane moisture, skin elasticity, and position of the ocular globe (Martinez-Jimenez and Hernandez-Divers 2007). Mucous membrane moisture level can be a helpful indicator of dehydration. Skin elasticity in most reptile species is a poor indicator because of scale cover and the thickened and highly keratinized integument. In principle, the use of sequential assessment of packed cell volume (PCV) and total protein (TP) can aid in quantification of dehydration; however, the interpretation of the results can be quite subjective. It is critical to keep in mind that lymph dilution is common and unpredictable. A 5% to 10% dilution by lymph will not be noted when drawing blood, but in the laboratory, it may lead to the diagnosis of hemodilution resulting in the clinician concluding (wrongly) that rehydration has taken place. Changes in PVC and TP over time will reflect changes in hydration status, and trends may be more indicative than specific numbers, but excessively high PVC and TP may indicate dehydration.

If available, data from a clinically normal conspecific kept under similar conditions should be used for comparison.

As quantification is difficult, many veterinarians will apply the following general rules:

- Patients with suspected or mild dehydration: Shallow warm-water baths daily. This provides rehydration by offering a chance for the animal to drink. Water temperatures should be measured and adjusted to mimic cage temperature and prevent hyperthermia. From a practical standpoint, use water between 25° C and 32° C.
- Moderately dehydrated patients: Subcutaneous or oral fluid therapy using either hypo or isotonic replacement fluid at 30–50 mL per kg of body weight per day. The same fluid rate can also be given intracoelomically. It is often advisable to spread the total daily dose over 2–4 smaller doses. For example rather than giving 30 mL once daily, instead give 10 mL three times daily. This is important to minimize patient pain from the skin stretching from large volumes of subcutaneous fluids and also to avoid over-distending the gastrointestinal system if oral fluids are given.
- Severely dehydrated patients: Rehydration by intravenous or intraosseous fluid therapy where possible. This is often beneficial in unresponsive moribund patients; note that reptiles can remain moribund for weeks. The rate of the fluid therapy varies on the degree of dehydration, however, the general principles are similar to other species: replace the maintenance requirements (1–3% of body weight) plus 25–33% of the dehydration deficit per day (Mader 2006). This allows the animal to be rehydrated slowly over 3–4 days and minimize the risk of over-hydration.

A multimodal approach to fluid therapy often yields the best results.

6.2.7 VENOUS ACCESS

Reptilian hematology and biochemistry is complex, and reported results are rarely reliable as a standalone indicator of disease. For this reason, they should always be interpreted with care and used in conjunction with other diagnostic findings and a thorough physical examination when formulating a treatment plan or to confirm a diagnosis.

The right palantine-pterygoidean vein provides reasonable venous access and rapid induction in large snakes but requires a cooperative patient and excellent restraint or general anesthesia. Therefore, this vein is not recommended for the administration of intravenous injections to venomous species. Complications can include postvenipuncture bleeding and hematoma formation. Pressure should be applied with a cotton-tipped applicator after the blood draw, and some practitioners recommend soaking the applicator in dilute epinephrine to assist with vasoconstriction. In most venomous species, blood can be collected easily and safely from the ventral coccygeal vein, which runs horizontally along the midline of the tail, just ventral to the coccygeal vertebrae. The ventral abdominal vein can also be used in larger venomous lizards, and both vessels are suitable for the intravenous delivery of drugs or fluids. Cardiocentesis can be performed in larger snakes (> 300 g) by immobilizing the heart between two fingers. Sedation is required for cardiocentesis, both to prevent the patient struggling and to provide analgesia. Because of the potential of iatrogenic damage such as heart muscle laceration and hemorrhage, a risk/benefit analysis should always be performed prior to using this technique, and it is not recommended for animals that are to be immediately released into the wild.

Long-term IV access may be needed for some treatments; however, catheterization can be challenging in reptiles and may require ultrasound guidance and/or a surgical approach in many species. The jugular vein is intracoelomic in snakes and generally requires a surgical "cut-down" procedure to access it; in lizards, it is generally lateral and can be located on a line drawn between the tympanum and the shoulder. In many lizards, the lateral saphenous vein is readily accessible via an incision parallel to the tibia. In lizards, the tissue around the vessel can be dissected to provide access to approximately 2 cm of vein. In lizards, catheterizing with a normal IV catheter is optimal. It is necessary to apply a bandage to protect the catheter, as most lizards will try to scratch and pull it out. For long-term venous access in snakes, the jugular vein is accessed cranial to the heart. The cut-down technique and catheterization procedure is similar to that described above for lizards; however, in snakes, a long segment of the polyethylene catheter is directed cranially when inserted and advanced toward the head. The point of entry of the catheter is very small and allows the vein to seal by itself. The free end of the catheter is then fed a short distance subcutaneously before being pulled through a separate opening in the dorsal costal wall a short distance away from the incision point. The catheter is managed as above. While snakes do not try to pull the catheter out and generally a bandage is not needed, stay sutures may be utilized to hold a portion of the catheter tube to the snake's body to prevent its becoming caught on something and being pulled out. In addition, furniture in the enclosure should be evaluated with regard to its potential to snag the catheter. If a sufficient length of polyethylene tubing is left protruding, the IV line can be managed without disturbing the patient.

6.2.8 ULTRASOUND AND RADIOGRAPHY

Radiographs can often be performed without the need for restraint by leaving the animal in a bag or transparent container. Radiographs also can be taken through clear plastic restraint tubes. Lateral views are best obtained by using horizontal beams or with animals under anesthesia, as they may require unnatural positioning of the animal, which is potentially stressful. It is worth noting that some species can be sexed by radiograph (through the visualization of the hemibacula or the presence of eggs), thus avoiding unnecessary handling. For ultrasounds, gel should be liberally applied. In some cases, it may be necessary to warm the gel up slightly to allow it to penetrate between the scales and minimize the degree of artifacts created by air. Alternatively animals can be placed in water baths during the ultrasound procedure to eliminate air between the transducer head and the patient. Snakes are best scanned along the ventral surface of the body to avoid interference from the ribs. Lizards are also examined through their ventral surfaces. While holding the animal on its back, the ultrasonographer can get very good images. In some animals with heavy scalation, the image quality achieved may be poor because of attenuation of the ultrasound beam.

For all radiographic and ultrasound assessment, a minimum of two orthogonal views should be taken.

6.2.9 MICROCHIPPING

The microchipping of venomous reptiles has been used as a means of identification in many wild populations and in some large collections. The techniques are the same for equivalent-sized nonvenomous reptiles. There are many differing opinions on where to insert microchips, with the subcutaneous route (often dorsolaterally along the body wall) being the most commonly used. Try to be as caudal as possible, and insert the microchip from the cranial direction; this will help prevent migration of the microchip out of the skin. It can often be helpful to place a small drop of surgical-grade tissue glue over the microchipping site to prevent migration of the microchip outward before the skin has had adequate time to heal. There is a range of different-sized microchips commercially available, allowing reptiles of almost any size to be microchipped.

6.2.10 EUTHANASIA

Euthanasia is an important part of veterinary medicine and, if done well, is quick, easy, and peaceful for all involved. As for all animals, a "two-step" euthanasia technique is recommended, in which the animal is first sedated or anesthetized. Once adequately sedated or anesthetized, the animal is then given an intravenous or intracardiac injection of a barbiturate. One of the most commonly used barbiturates is pentobarbitone with a dose of 150 mg/kg. It is important to remember that the reptilian heart may continue to contract for a considerable period after death. For this reason, it is often advisable to destroy the brain by pithing or blunt trauma. An extremely effective method to both pith and chemically inactivate the brain in sedated snakes has been described (Martelli, Luz, and Meyer 2001). Care must be taken to prevent accidental envenomation when handling the carcasses of venomous animals (see color plates 24G and 24H). The fangs or entire heads should be removed and placed in sharps containers. Decapitation without prior deep sedation is not considered humane in reptiles. If decapitation is done, the brain should be immediately pithed with a probe or by blunt trauma to rapidly end brain function.

It must be emphasized that freezing reptiles is not recognized as a humane method of euthanasia and is not a method that would be approved by any animal-welfare committee or authority. Also, decapitation alone of a conscious reptile is not considered humane, as the head will remain fully conscious for 15 minutes or longer, during which the animal is assumed to be feeling great pain and distress. Again, this is not a method that would be approved by any competent animal-welfare committee or authority.

Two useful resources are AVMA Euthanasia Guidelines (www.avma.org/kb/policies/documents/euthanasia.pdf) and the Central CITES guideline for the humane euthanasia of reptiles (www.blv.admin.ch/themen/tierschutz/04013/index.html?lang=en).

6.2.11 NECROPSY

Necropsy is an essential health-management tool for all collections. A thorough necropsy is generally needed to reveal or confirm cause of death and to help identify disease processes that are difficult to diagnose premortem and may threaten the rest of the collection. It is important that if tissue samples are to be analyzed, the animal should be refrigerated after death and not frozen. Freezing causes ice crystals to form in the tissue, destroying their microscopic structure and rendering them useless for histopathological evaluation.

Care must be taken to prevent accidental envenomation when handling the carcasses of venomous animals. It is always advisable to use appropriate safety equipment (such as a snake hook) to handle the carcass until death can be confirmed, as, particularly in the case of snakes, very sick animals or even healthy specimens of certain species can lie dead still or mimic death. Specific necropsy occupational health and safety recommendations vary in each state or country, but general principles are often the same. Gloves, protective eyewear, and an appropriately fitting face mask are recommended to prevent potential spread of zoonotic disease. If the oral cavity needs to be examined, great care must be taken to minimize the risk of accidental envenomation. If the brain is of interest for histopathology, it can be removed from the skull of large specimens by careful dissection; alternatively, the whole head of a small specimen can be removed and placed in 10% buffered formalin. A pathology laboratory can then decalcify the specimen in order to examine the fixed brain. It is advisable to contact the preferred laboratory first and confirm that this method is an option before sending; some labs will refuse to process heads from venomous species. The specimen jar should be clearly labeled with a warning of the venomous nature of its contents. If the oral cavity, eyes, brain, or other head structures are not of concern, it is often advisable to decapitate the animal and place the head in a crush-proof container for disposal (Johnson 2011). Alternatively, an empty syringe case or urine-specimen cup can be securely taped over the head or upper jaw to shield the fangs during postmortem examination and/or disposal of the body. (For additional discussion of the handling of deceased specimens, see chapter 5.4.9.)

6.3 DISEASE CONTROL AND PREVENTION

Most reptile pathogens are spread by direct contact, fomites, or aerosolization (Chitty 2011). There is a suspected association between the presence of mites and respiratory disease in snake collections, although this remains to be conclusively proven in controlled studies. All newly acquired reptiles should be housed in a quarantine facility completely separate from the main collection. In larger collections where turnover of reptiles is high, it is advisable to have multiple rooms in the quarantine facility to allow further segregation of ill and apparently healthy animals (Stahl 2001). Any newly acquired reptiles should be seen to after all routine tasks with resident reptiles have been completed. The best practice is to have a keeper who deals solely with the reptiles in quarantine (Miller 1996). All newly acquired reptiles should be kept separate in individual caging to allow individual identification of disease and to minimize any intraspecies aggression or dominance (Pasmans et al. 2008). Newly acquired animals are often more susceptible to infection because of stress and subsequent immunosuppression associated with handling, shipping, and new environments (Schumacher 2006). Simple, easily cleaned enclosures and equipment are important. Plastic containers or tubs that can be easily cleaned and disinfected are preferred for quarantine purposes. The stress of being in a new environment can be minimized by ensuring that the enclosure is appropriately set up for the individual species being kept and that the reptile is not just kept in a plain plastic container or tub without any hides or cage "furniture."

In addition, there should be cage- or section-specific sets of supplies and equipment that come into direct contact with the animals or cage furnishings (such as handling tools, surfaces, and holding tubes). Such dedicated equipment should be disinfected before and after each use. Routine disinfection may be accomplished with a 1:125 dilution of F10 SC or by using potassium monopersulfate spray (such as Virkon), while 5% sodium hypochlorite must be used on enclosures between changing cage inhabitants or before storage. It is important that bleach is thoroughly rinsed away before new inhabitants are introduced to an enclosure.

6.3.1 QUARANTINE

6.3.1.1 Quarantine Principles

The importance of good quarantine practices cannot be overstated. Many people have "learned the hard way," losing a valuable animal or even their entire collection by overlooking this critical component of husbandry. Despite this being a regular occurrence, many herpetologists and institutions still fail to implement effective quarantine strategies. One of the reasons for this is the conservative nature of protocols found in the literature, coupled with the difficulty of testing for diseases of primary concern. Keeping detailed quarantine records is essential, regardless of the size of the collection. Basic data such as feeding records, shedding dates, medical concerns and treatment, weights, and waste production (feces or urates) should all be documented (Stahl 2001).

Any animal that dies during the quarantine period requires a full necropsy. This is extremely important and may provide diagnoses that were elusive in the premortem animal. Histopathological analysis of all organs should be performed and appropriate bacterial cultures undertaken. Frozen samples should be saved for viral testing. Wild-caught or delicate animals will also require more attention during the quarantine period in order to monitor their stress levels during adaptation to life in captivity. Therefore, staff time devoted to quarantine animals will be higher than the relative staff time necessary for an equivalent number of animals postquarantine. Husbandry duties should proceed through the collection in chronological order, with the most recently added animals serviced last and with changes of gloves (latex or nitrile) between each cage. This can more easily be accomplished by wearing double gloves and changing the outer glove, as changing a single pair of gloves repeatedly can be difficult with sweaty hands.

The general principles of quarantine are as follows:

- Sound acquisition practices.
- Simple and clean enclosures.
- Isolated quarantine area.
- Thorough physical examinations, fecal analysis, and general screening tests at multiple points in time.
- Detailed and accurate record keeping.
- Servicing the quarantine area last.
- Necropsying animals that die during quarantine.

6.3.1.2 Quarantine Time Frame

Recent advancements in our understanding of reptile virology indicate the necessity of quarantining newly acquired specimens for a period of at least six months (Pasmans et al. 2008). The six-month period is calculated from the date of the most recent arrival to the quarantine room. All animals are considered to be of the same quarantine age, with the time remaining in quarantine based on the most recent arrival. For example, if a venomous reptile had already been in quarantine for five months but a new animal is added to the same quarantine room, then that specimen has its quarantine age reset to zero. This demonstrates the value of having multiple rooms within the quarantine facility. We recognize that this may be impractical in some situations; the intent of this chapter is to provide "best practice" standards.

For some countries, international importation (such as into Australia) requires six-months prequarantine in full isolation (full gown in/gown out barrier style), followed by a 24-month surveillance period where separation is maintained and any health issues trigger full isolation protocols again. Such protocols should also be implemented for animals acquired from high-quarantine-concern sources, for instance, facilities that have had a history of ferlavirus (paramyxovirus) or that maintain high concentrations of animals prone to such diseases.

6.3.1.3 Quarantine Veterinary Care

A good admission protocol is useful to minimize problems during quarantine. It is essential to weigh, hydrate, and deworm animals upon arrival and to provide a safe hiding box for snakes or suitable perching facilities for lizards to minimize stress and promote immunocompetence. A standardized screening by an experienced reptile veterinarian of all newly acquired reptiles is required, with subsequent treatment protocols implemented as necessary. The specific veterinary testing recommended depends greatly on the species of reptile, the local area, and where the reptile was sourced. General testing often includes repeated fecal examinations, complete blood counts, serum biochemistry panels, infectious disease screening (serologic and/or PCR), and thorough physical examination.

6.4 COMMONLY ENCOUNTERED DISEASES

6.4.1 PARASITIC INFESTATIONS

The stress of being collected from the wild may result in the suppression of an animal's immune system, which can make the animal vulnerable to infection by opportunistic parasites. It is important to identify any parasites that are present and treat accordingly. Using a blanket treatment of all animals coming into captivity can lead to problems with parasite resistance in addition to a higher risk of life-threatening drug reactions. However, acquisition of a large number of specimens in a short period of time may make testing of individuals impractical, and in such situations, blanket treatment may be appropriate.

The two broad groups of parasites are ectoparasites (outside the body) and endoparasites (within the body). Ectoparasites are generally discovered during thorough clinical examination, whereas many endoparasites require fecal analysis to identify.

6.4.1.1 Endoparasites

Fecal examination provides a means to identify endoparasites and is an essential part of routine checkups. Wild-caught animals typically have greater parasite burdens than captive-bred reptiles. Exotic species of parasite may present unique treatment difficulties.

Many different endoparasites cause disease, and there is no single treatment that will be successful for all endoparasites. For this reason, thorough and repeated fecal examinations to identify the type and level of endoparasitic infestation present are vital, as they allow an appropriate treatment plan to be implemented. At least four samples, two to four weeks apart, must be taken. As the parasites normally stay in the gastrointestinal tract of the reptile, looking at a gross fecal sample and not seeing visible worms does not demonstrate their absence. A fecal exam involves the concentration of a fecal sample in a hypertonic salt or sugar solution followed by microscopic examination for the presence of the ova of parasites. Multiple samples are recommended in case the parasites fail to pass ova in every sample. It is important that a sample of fresh feces, ideally less than three hours old (Pasmans et al. 2008), is analyzed to improve accuracy and to aid in successful pathogen identification, especially entamoeba. For laboratory submissions, it is advisable to place samples in a fecal container or zip-lock bag on ice within an insulated container and send them as soon as possible.

It is important to repeat fecal examinations after appropriate treatment has been completed to ensure that pathogens have been eliminated. The general rule for the confirmation of successful elimination is two negative fecal samples taken at two to four weeks apart (Pasmans et al. 2008). Most treatments kill adult nematodes but not always larvae. Thus, additional time must be taken to allow any remaining larvae time to mature into adults, producing ova that are detected in the fecal test. A second dose is recommended in two to four weeks. Ivermectin and fenbendazole are commonly used anthelminthics for the control of nematodes. If intestinal cestodes are identified, praziquantel is typically effective. Three once-daily doses of toltrazuril oral formulation are appropriate. Flagellate infestations can be treated with metronidazole (Flagyl), administered either by stomach tube or in a food item.

The intermediate or immature stage of the *Spirometra* tapeworm occurs in snakes that are frog- or reptile-eaters in the wild. Larval tapeworms (spargana) may appear as small distinct swellings under the skin but rarely cause disease. This condition does not occur in captive-bred animals that eat frozen/thawed mammalian prey items. Fecal testing is of no use in diagnosing this condition as the worms are in their larval stages—immature and not egg-layers. Regular worm treatments are ineffective (Johnson 2012).

6.4.1.2 Ectoparasites

Ectoparasite burdens are common in many reptiles and should also be considered a potential vector of infection and disease and in some cases a zoonotic risk (see color plate 24F.) A complete and thorough examination of the animal's external surfaces is necessary to identify ectoparasites. The most common implantation sites for ectoparasites are external ear canals, around the eye, in heat-sensing pits, between scales, in the crevices or folds of the inguinal and axilla regions of lizards, and in the mental groove and between the scales of the ventral chin of snakes. Magnification and good lighting aid detection. The most common reptile ectoparasite seen in captivity is the blood-sucking mite *Ophionyssus natricis*; however, additional species of mites, ticks, and leeches may also be found. Ticks are generally larger than mites and are usually firmly attached to the skin. Mites are very mobile and will move along the reptile and also spread between cages and even rooms.

Adult mites are visible to the naked eye, often seen on paper substrate, reptile bags, or other equipment. Inspection of the head and eyes of the animal may reveal their presence. Snakes infested with mites may exhibit dysecdysis (difficulty sloughing skin) and spend long periods soaking in their water bowls. Mite infestation is directly related to poor hygiene and quarantine practices (Johnson 2012). Mites are very mobile, feeding on the reptile and then returning to a crevice to hide and breed. Mites can easily travel between rooms. Placement of white double-sided tape outside enclosures and examination of the water surface and feces for dead mites will assist the keeper in diagnosing an infestation. Mite treatment involves treating the enclosure with an appropriate antiparasite spray such as D-phenothrin (Reptile Enclosure Spray, Prevent-A-Mite). The animals themselves may be sprayed with fipronil (Frontline); excess spray should be wiped away after a few minutes. Desiccating agents and other products with "natural ingredients" are widely commercially available for snake mites but can dehydrate small reptiles. Ticks may be removed manually by gentle traction with forceps; care should be taken that the mouth parts are removed and that the tick is not crushed. An older treatment that has toxicity considerations is the use of triclorfon (Neguvon® or Masoten®) at a dose of 2 g/L (0.2%) sprayed on the snake, in the cage, outside the cage and along the plinths and false ceilings is very effective and totally safe for reptiles of all ages and size. However, being a pesticide it is very toxic to invertebrates including spiders and also to fish. Thus it may be considered as a treatment of last resort.

6.4.2 CRYPTOSPORIDIOSIS

All *Cryptosporidium* species have direct life cycles. The two most significant in snakes are *C. serpentis*, which has gastric tropism, and *C. varanii* (formerly *C. saurophilum*), which has small intestinal tropism (Jacobson 2007a; Wellehan and Stahl 2013).

Cryptosporidiosis results in gastrointestinal disease and wasting. Reptiles can be infected with two species, one detrimental to the reptile and the other to humans. *C. parvum*, usually obtained from infected prey items, can pass through the digestive tract of reptiles without causing illness but poses a zoonotic risk to humans. *C. serpentis* infection results in debilitating illness in reptiles without any zoonotic potential. Clinical signs include regurgitation one to three days after feeding, a firm mid-body swelling caused by inflammation and swelling of the lining of the stomach, chronic weight loss, and poor growth.

Diagnosis is based on clinical signs and direct fecal examination. Antemortem diagnosis is made by detection of the oocysts in mucus-coated, regurgitated food, in gastric lavage samples, in impression smears of biopsy samples, or in feces by using acid-fast stains. Nested PCR with sequence analysis is (at the time of this book writing) available at the University of Florida or University of Tennessee for all *Cryptosporidum* species and should be used for identification to the level of species, as morphology alone cannot distinguish pathogenic from nonpathogenic (mouse) species. During quarantine, acid-fast screening of fecal samples with PCR sequencing of positive results should be considered. Shedding of the organism is transient, so multiple fecal samples may be required for diagnosis.

The course of the disease is often long and almost always fatal. There is no effective treatment available commercially. Treatment is little studied and inconsistently effective but in important cases can be considered. In experimentally infected *Pogona vitticeps*, treatment with paromomycin in saline (500 mg/5 mL) at 100 mg/kg every 24 hours for 7 days, then every 84 hours for 72 days did not control shedding of oocytes. Subsequently, the lizards were treated with 360 mg/kg every 48 hours for 10 days, and histological examination confirmed the elimination of the infection from all specimens (Grosset et al. 2011). For animal-welfare reasons and because of the costs associated with control measures, supportive care, and quarantine facilities, serious consideration must be given to euthanasia of infected snakes and cage mates in the event of an outbreak of Cryptosporidiosis. The extensive reorganization of husbandry procedures that may be required in order to prevent the disease from being transmitted to all specimens in the facility is also an impediment to effective treatment (Johnson 2012).

6.4.3 FERLAVIRUS (OPHIDIAN PARAMYXOVIRUS) AND SUNSHINE VIRUS

Ferlavirus is an enveloped RNA virus 146 to 321 nm in diameter. It grows optimally at 30° C and has restricted growth at 37° C (Lunger and Clark 1979). Ferlavirus infections are associated with neurologic, respiratory, and acute immunosuppressive disease and have caused significant mortality in snake collections. The disease has been identified in snakes in Europe, the Americas, and the Canary Islands but is likely distributed worldwide. Anecdotally, the disease is likely present in Australia, but this has not been confirmed. The shedding patterns of ferlavirus are unknown (Jacobson 2007b). There are no controlled studies to support any claims regarding the existence of persistently infected shedding state or of snakes that mount an appropriate immune response and have cleared the infection (Hyndman, Shilton, and Marschang 2013). It is reasonable to assume that ferlavirus can be transmitted between snakes by both oral secretions and cloacal excretions. Ferlavirus has not yet been isolated or detected by PCR from fomites or ectoparasites (Hyndman, Shilton, and Marschang 2013). There are currently no reports concerning vertical transmission of ferlavirus (Pasmans et al. 2008). The incubation period of ferlavirus in naturally acquired infections is unknown. There are claims that the incubation period for ferlavirus may be as short as 21 days (Hernandez-Divers 2006) but will generally exceed 90 days (Hernandez-Divers 2006; Ritchie 2006). However, these claims are not supported by controlled studies (Hyndman, Shilton, and Marschang 2013). Diagnostic tests may be unclear. The presence of circulating antibodies indicates exposure but not necessarily infection. Reverse transcription polymerase chain reaction (RT-PCR) has the potential to produce false positives, and any resultant products must be sequenced to confirm the presence of the virus.

While belonging to the paramyxoviral group of viruses, “sunshine virus” is only distantly related to ferlavirus. First identified in a collection of pythons on the Sunshine Coast of Queensland, Australia, in 2008, the virus causes neurorespiratory disease in affected snakes. Clinical signs in infected animals may be neurological, neuro-respiratory, or non-specific in nature. Some infected snakes may show no overt signs of disease at all. Neurological signs include head tremors, opisthotonus, incoordination, diminished righting reflexes, uncoordinated movement of the cranial and caudal body, and erratic mouth gaping. Respiratory signs include a mild discharge of clear viscous fluid from the mouth and dyspnea. Non-specific signs include anorexia, stomatitis, weakness, lethargy, regurgitation, and inappetence (Hyndman et al. 2012). More recently dermatitis has been reported in pythons infected with Sunshine virus (Johnson and Hyndman 2014).

Little to nothing is currently known about its incubation period, carrier status, transmission, or survival in the environment. The virus is detected using PCR, and at the time of this publication, no venomous species of snake has been found to be infected (Hyndman et al. 2012).

As with Cryptosporidiosis, outbreaks of paramyxoviruses such as the two described above pose a clear and present danger to the entire collection, and infected animals and cage mates should be immediately euthanized.

6.4.4 FUNGAL DISEASE

All reptiles have fungi present on their skin (Paré et al. 2003). It is thus imperative to correlate histopathological and morphological characteristics of fungal elements in tissue sections in order to assign pathogenicity to these isolates or to dismiss them as contaminants. A critical review of the literature pertaining to fungal disease in reptiles indicates that only a very few species of fungi have been implicated in multiple cases of disease in squamates (Pare and Sigler 2002; Pare and Jacobson 2007). Some of the fungi capable of causing disease in squamates under the right conditions include common environmental fungi such as *Purpurocillium lilacinum, Fusarium solani, Metarhizium anisopliae, Beauveria bassiana*, various aspergilla, and yeast such as *Candida* or *Cryptococcus*. Infection with these opportunistic fungi is usually associated with overwhelming exposure or, more commonly, with stress and immune compromise from substandard husbandry conditions.

The Ascomycetous fungi in the order Onygenales are the most important fungal pathogens of squamates worldwide. Interestingly the most significant fungal pathogens of mammals, including *Blastomyces, Histoplasma, Coccidiodes*, and *Microsporum* are contained within this order. While recent reptile isolates have been called the *Chrysosporium* anamorph of *Nannizziopis vriesii* (CANV), molecular genetic evidence shows that very diverse organisms are represented in reptile disease (more than 30 isolates of CANV have been cataloged and are now known to represent an assemblage of closely related fungi) and that CANV does not seem to be a common reptile pathogen (Pare et al. 2006; Lock and Wellehan 2014; Pare 2014).

The species of this group of related fungi most often identified as pathogenic in snakes is *Ophidiomyces ophiodicola* (formally *Chrysosporium ophiodicola*) (Sleeman 2013). Infection with *O. ophiodicola* begins with cutaneous lesions, most commonly on the face, and often progresses to severe necrogranumatous dermatitis. The infection eventually disseminates and is often fatal. *O. ophiodicola* has likely been underdiagnosed, especially in earlier reports, due to unfamiliarity and changing taxonomy. *O. ophiodicola* has caused dermatomycosis in *Pantherophis obsoleta* (Rajeev et al. 2009) and *Hoplocephalus bungaroides* (McLelland, Johnson, and Reuter 2010). *O. ophiodicola* has been detected in wild snakes, with the majority of cases described to date in species of *Crotalus* (Allender et al. 2011; Allender et al. 2012), although it has also been detected in a wide variety of other caenophidian snakes. Facial disfiguration syndrome occurs in snakes with distorted heads caused by the facial lesions characteristic of *O. ophiodicola* dermatitis (Allender et al. 2012). The infection progressively hinders feeding, and snakes become emaciated before death. This disease has been suggested as a possible reason for the substantial decline in timber rattlesnake populations in the northeastern United States. Intraconazole, voriconazole, and terbinafine are all valid drug options for treatment.

6.5 VETERINARY ISSUES OF PARTICULAR NOTE IN VENOMOUS REPTILES

6.5.1 RETAINED FANGS AND FANG SHEATH INFECTION

Captive specimens of long-fanged species will occasionally accumulate fangs and develop inflammation or sepsis of the sheath of gum covering the fangs (see color plates 24C, 24D, and 24E). Some species, such as *Bitis arietans*, seem particularly prone to inflammation of the fang sheath. Trauma to the fangs of wild sea snakes, with associated mucosal abscessation and necrosis, has also been noted. Infections and trauma like this will eventually cause the snake to stop feeding, and infected material in the oral cavity can enter the blood and the airways. Surgical removal of retained fangs and debridement of the surrounding tissue are indicated. Clearly, this is best achieved under anesthesia for safety reasons and for optimal care of delicate oral structures. Dental radiography is useful to image the extent of the damage to the short and mobile maxillary bone. Use of appropriate antimicrobials, based on culture and sensitivity testing, is indicated.

6.5.2 VENOM GLAND ADENITIS

Venom gland adenitis is not uncommon in snakes used for venom extraction when operators are too forceful in massaging the glands. It can also occur spontaneously. Marsupialization of the gland for drainage is tempting but ineffective because of the fibrinous caseous nature of reptilian pus. Surgical removal is preferable. The surgery consists of removing the gland as close as possible to the point of fang insertion. To perform this procedure, the skin over the gland is incised between scales. This exposes the thick, pearl-white capsule of the principal gland. Ideally, the ejector muscles are lifted at the point of their insertion over the gland capsule in order to minimize bleeding and postsurgical deformity of the face. Care must be taken to minimize damage to the large vascular bundle that lies medial to the mid-body of the gland, which will bleed profusely if ruptured.

6.5.3 FREE-LIVING (WILD) ANIMALS

Free-living venomous reptiles are occasionally presented to private veterinary practices and wildlife veterinary facilities for treatment. These animals have often suffered trauma (such as injuries sustained during interaction with household pets and netting) and present with fairly obvious injuries (sometimes requiring immediate euthanasia); but others may appear to be suffering from more chronic conditions that require thorough investigation before a diagnosis can be reached. The same diagnostic techniques, handling methods, anesthetic regimes, and precautionary measures as those for captive specimens can be applied to wild patients.

Patients suitable for treatment will require some form of housing before returning to the wild, and enclosures and methods of feeding should be appropriate for the species in question. Additionally, if captive reptile patients are kept in the same facility, quarantine measures, as detailed above, should be employed. Use of simple, easily cleaned enclosures and equipment is important. Plastic containers or tubs that can be easily cleaned and disinfected are preferred. The stress of being in a new environment can be minimized by ensuring that the enclosure is appropriately set up for the individual species being kept and that the reptile is not just kept in a plain plastic container or tub without any hides or cage "furniture."

Injuries sustained by free-living venomous reptiles differ considerably from those in captive collections. Animals are often injured on roads, by domestic dogs or cats, or by being entangled in fruit netting or fishing nets, or they may present with injures where the cause is unknown. In sea snakes, the etiology is often undeterminable, but injuries such as a fractured spine or jaw are likely attributable to blunt-force trauma, perhaps a result of impact with watercraft or from trawling activity (Gillett et al., unpublished data). Trauma to the fangs of wild sea snakes has also been noted, with associated mucosal abscessation and necrosis, and could be associated with trauma during feeding.

When wild sea snakes are temporarily maintained in captivity, rostral abrasions may occur from rubbing against rough surfaces. This condition is difficult to treat in the aquatic environment. When rostral damage occurs, the protocol is initial treatment with betadine and then ceftazidime (Fortum, Fortaz) 20 mg/kg intramuscular injection once every three days. Restricting handling will reduce stress and increase the chances of the snake feeding. However, the combination of the stress of handling for treatment and the discomfort from infected oral wounds typically results in the sea snake refusing to feed, often leading to anorexia and death.

Disease in free-living reptiles is not a common reason for admission, but it is an important factor to consider when investigating the health of a population or investigating mortality of unknown etiology. Many of the diseases and parasites mentioned in the section above on captive collection can occur in wild reptiles, but there the impact to the individual is often minimal. For example, the presence of internal parasites is considered normal in wild specimens and generally does not appear to affect the animal's health. Small numbers of external parasites are also often encountered on wild reptiles but are kept in check by the natural process of ecdysis. The presence of heavy epibiotic infestations in sea snakes has been noted (Saravanakumar et al. 2012). In stranded animals, it may be an indication of debilitation or interference with the normal shedding process and has been noted in animals with fractured spines where natural ecdysis may be interrupted (Gillett et al., unpublished data). Neoplasia has also been identified in a number of stranded sea snakes, and in such cases, is considered a primary reason for stranding (Gillett et al., unpublished data) (see color plate 24F).

6.5.4 VENOMOID SNAKES

Venomoid snakes are snakes that have had their venom glands surgically removed or the venom duct sectioned (see color plate 24E). Unfortunately, the popularity of this practice is increasing among reptile keepers, and unqualified practitioners, in direct violation of animal-welfare legislation, often perform such procedures. Venom-gland surgery by amateurs is frequently performed without using analgesia or anesthesia, and such incidents constitute clear-cut cases of animal cruelty. These operations typically have poor outcomes resulting from the nonsterile manner in which the surgeries are performed. Venomoid surgery in snakes is recognized as a cruel, disfiguring, and unethical procedure and should not be condoned (www.ava.com.au/policy/31-surgical-alteration-natural-state-animals). Removal of venom glands should only be performed for valid medical reasons to improve the welfare outcome of an animal, as in the case of infection, trauma, or neoplasia. Not only does the practice of venomoid surgery create serious concerns regarding animal welfare, as such surgery is often carried out inexpertly, but some venomoid snakes may still produce venom from residual secretory tissue. The false sense of security provided by the venomoid surgery therefore constitutes a considerable risk to human welfare in these cases. Incomplete gland removal may present a threat to the life of anyone handling the snake. In addition, use of venomoid snakes by exhibitors and educators only creates confusion in the identification of potentially dangerous snakes and hinders the promotion of safe handling procedures. Further, any progeny will be fully venomous.

CHAPTER 7

RESEARCH METHODS

B. G. FRY, E. A. B. UNDHEIM, T. N. W. JACKSON, D. GEORGIEVA, I. VETTER, J. J. CALVETE, H. SCHEIB, B. W. CRIBB, D. C. YANG, N. L. DALY, M. L. ROY MANCHADI, J. M. GUTIÉRREZ, K. ROELANTS, B. LOMONTE, G. M. NICHOLSON, S. DZIEMBOROWICZ, V. LAVERGNE, L. RAGNARSSON, L. D. RASH, M. MOBLI, W. C. HODGSON, N. R. CASEWELL, A. NOUWENS, S. C. WAGSTAFF, S. A. ALI, D. L. WHITEHEAD, V. HERZIG, P. MONAGLE, N. D. KURNIAWAN, T. REEKS, AND K. SUNAGAR

7.1 INTRODUCTION

Venoms are complex concoctions consisting of peptides, proteins, salts, and other compounds. As such, they present a unique set of challenges for the venom researcher. This chapter provides a broad overview of the techniques useful in venom research, ranging from gross characterization of the venom's protein content to purification and characterization of individual toxins, bioactivity testing, and evolutionary assessment of toxin-encoding genes. Special attention is paid to unique challenges posed by reptile venom or techniques that have proved to be particularly useful.

7.2 VENOM STORAGE AND HANDLING

As discussed in chapter 5.9, appropriate venom extraction techniques vary for different types of venomous reptile. However, mucus and precipitates should be removed from all venoms by centrifugation followed by filtration through 0.45 µm filters with a low-protein binding polyvinylidene fluoride (PVDF) membrane (such as Durapore) prior to any use in purification or assay techniques. Centrifugation and filtration are particularly important to prevent contamination with blood, as blood cells rupture during freezing and thus contaminate the venom with intracellular proteins. In anguimorph lizards, a variable amount of bleeding occurs naturally during envenomation. For this reason, it is particularly important that lizard venoms be centrifuged at 12,000 RCF (relative centrifugal force) immediately following venom extraction. The resultant supernatant should then be filtered and frozen. Centrifugation and filtering are also critical for freshly collected venoms of non-front-fanged snakes because of the ubiquity of blood and mucus contamination in these secretions. For front-fanged caenophidian snakes, centrifugation and filtering may be done postfreezing, or even postlyophilization, as there is negligible contamination by mucus or blood if venom extraction techniques are humane and efficient. However, as repeated freeze-thaw cycles should be avoided because of the possibility of denaturing proteins, filtering prior to freezing in liquid nitrogen in anticipation of lyophilization removes one freeze-thaw cycle from the equation. This may prove to be crucially important, particularly for globular toxin types, as such non-covalently folded proteins are notably unstable outside of natural conditions.

Ideally, venoms should always be stored frozen or in powder form and in the dark to prevent degradation by moisture and radiation. Stored correctly, venoms are stable for at least eight decades (Jesupret et al. 2014). Of particular concern, especially for snake venoms, is the presence of a number of highly active proteolytic enzymes, which may quickly have a profound impact on the apparent venom composition by digesting other venom components (Tashima et al. 2012). Venom stored as liquid should be aliquoted to avoid multiple freeze-thaw cycles. Adding protease inhibitors to venom prior to analysis may help prevent enzymatic degradation but should be avoided if the venom is to be used in bioactivity testing where such inhibitors may interfere with functional characterization.

Sample treatment and preparation for further steps of analysis are also important considerations. Buffer and solvent selections should be made in consideration of downstream applications. Ammonium, for example, may remain with purified toxins in the form of counterions and cause issues with protein modification experiments such as N-hydroxysuccinimide esters that react with ammonium ions (such as isobaric tag for relative and absolute quantitation [iTRAQ] or difference gel electrophoresis [DIGE]). Even smaller, disulfide-rich peptides, generally considered to be highly stable, can be denatured by factors such as temperature (too high a column temperature during reversed-phase high-performance liquid chromatography [RP-HPLC]) and pH (high-pH RP-HPLC). High pH (> 7.5) may also induce disulfide shuffling. This is a very important consideration, as maintaining correct disulfide pairing is critical for accurate determination of both activity and structure. In addition, lyophilization creates elevated local salt concentrations at the outer part of the solid sample, which in some cases may result in denaturing conditions, such as in the case of buffers containing ammonium acetate in concentrations greater than 100 mM. Steps using such buffers should be followed by a technique that also desalts, such as RP-HPLC, before lyophilization.

The type of downstream analysis the venom will be subjected to will be broadly guided by the intended outcomes. However, regardless of whether the questions to be answered are evolutionary or applied (such as relative neutralization by antivenom or more esoteric biodiscovery studies), there is a core set of basic techniques for analysis, purification, and functional characterization. These techniques are the fundamental starting point, downstream of which more specialized protocols may be employed.

7.3 PROTEOMICS

In reptile venoms, proteins and peptides account for most of the molecular and functional toxin diversity. Hence proteomics has become a central approach in both initial and later analytical stages of research (Calvete 2011b). However, with molecular masses spanning two orders of magnitude and isoelectric points ranging widely across the pH scale, no single proteomic approach is suitable for characterization of the full spectrum of toxins present in a given reptile venom. Multidimensional analytical approaches are therefore essential for meaningful proteomic analysis of reptile venoms and vital for successful downstream characterizations of toxins (such as bioactivity) (see figure 7.1). Three proteomic techniques are generally used in venomics: liquid chromatography (LC), polyacrylamide gel electrophoresis (PAGE), and mass spectrometry (MS). All three methods are used to separate and detect analytes, but each method achieves this based on differing physiochemical properties, and each has a number of advantages and disadvantages, as discussed in detail below.

7.3.1 SIZE SEPARATION BY CENTRIFUGAL FILTRATION

This technique utilizes a selectively permeable membrane with pores of a particular diameter as a physical barrier for any components with a maximum diameter greater than that of the pore. It can be used to concentrate or desalt samples or to separate venom components according to molecular mass. This is particularly useful when low-molecular-weight components are not present in high concentrations in a venom (Fry et al. 2005). However, attention must be paid to the solvent in which the sample is diluted, as some filter types are not compatible with certain solvents.

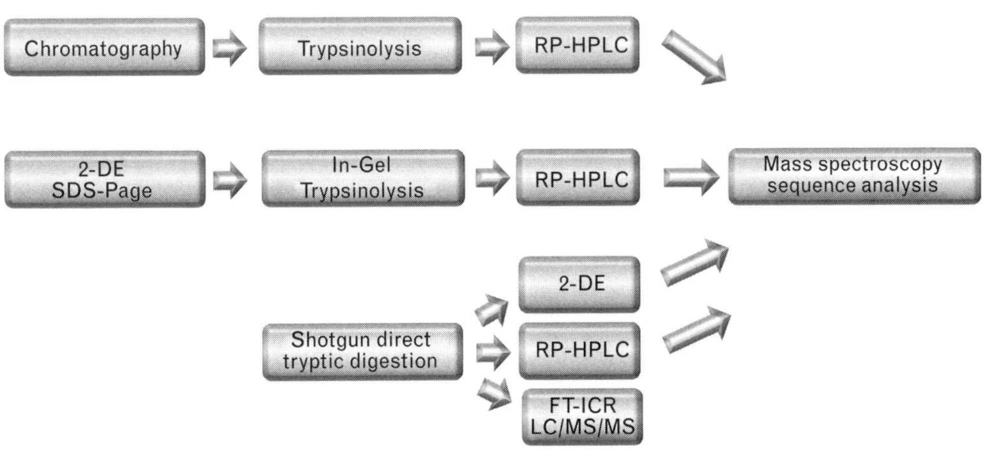

FIGURE 7.1: Proteomics pathways leading to mass spectrometry analysis of digested peptide fragments.

7.3.2 LIQUID CHROMATOGRAPHY (LC)

A wide variety of LC techniques are available for the separation of dissolved components based on several physicochemical properties. A guard column should be employed for any method, especially for crude venom samples. However, it must again be stressed that all venoms should be precleaned by centrifugation and filtration to remove precipitates (see section 7.2). Otherwise, insoluble material may block the guard column, resulting in pressure elevations. Each purification method has inherent advantages and disadvantages. Typically, any single method, even one with high resolving power, will not yield pure proteins. Thus, a combination of orthogonal separation methods based on different principles is often necessary. An efficient combination of techniques is concentration of a particular molecular weight range by centrifugal filtration or size-exclusion chromatography (SEC), followed by ion-exchange chromatography (IEC), then RP-HPLC. The choice of buffers to be used at each step must take the subsequent chromatography step into consideration. Gel filtration typically involves a more concentrated buffer (0.05–0.1 M) than IEC (10–20 mM), and thus the buffer must be appropriately diluted prior to IEC. In contrast, samples in high-salt buffers are of no consequence for RP-HPLC, which is why this technique is the last in the series and provides a desalting step, which is important for subsequent pharmacological testing. While such a multistep approach leads to appreciable loss of material, the end result is extremely pure individual venom components.

Basic chromatography considerations are:

- Desalting. Is this necessary before or after a particular technique?
- Elution techniques. Isocratic versus stepwise versus gradient.
- Packing of columns.
- Porosity and capacity of matrix.
- Column dimensions, sample size, and flow rate.
- Choice of mobile or stationary phase.
- Choice of wavelengths used for monitoring and what each one detects.
- Potential problems. These include adsorption, dissociation of non-covalently linked protein complexes, multiple peaks containing a homogeneous protein, denaturing (may be a particular problem for RP-HPLC because of the use of harsh mobile phases and often very low pH).
- Cleaning and storage of chromatography media.

Liquid chromatography separates analytes dissolved in a liquid mobile phase based on their differential interactions with a stationary phase. Column-based LC is the most popular form of LC. In this method, the stationary phase is packed in a column, typically in the form of small porous particles, through which the solvent (mobile phase) carrying the analytes flows. The retention of analytes by the stationary phase is essentially determined by two factors: (1) the effective volume of the column available to the analyte and (2) the adsorption of the analyte to the stationary phase. This is the most versatile approach for the separation of proteins and peptides because of the vast number of combinations of stationary and mobile phases available. Separation can therefore be based on a wide range of physicochemical properties.

The number of LC techniques can appear rather daunting, but most can be classified into just a few broad categories based on the mechanism of separation. Different techniques can be used at different stages of venom fractionation and toxin purification. Furthermore, with some experimental planning, these can be combined without additional sample treatment steps, enabling fast and convenient multidimensional protein separation.

In most laboratories involved in venom research, LC is carried out on an analytical or semipreparative high-performance liquid chromatography machine (HPLC). These typically operate at pressures less than 60 MPa (600 bar, 8700 psi) and flow rates of 0.4–5.0 mL/min and use columns with an inner diameter (i.d.) of 2.0–10.0 mm and a length of 20–300 mm. Although this is a somewhat arbitrary classification, columns with an i.d. > 4.6 mm are usually termed semipreparative, whereas those with an i.d. < 4.6 mm are referred to as analytical. For detection of analytes, HPLCs are generally equipped with detectors that measure absorbance in the far-UV range. Alternatively or additionally, they may be fitted with other devices such as multiangle laser light scattering (MALLS) detectors, which determine absolute molar masses of proteins. There are other customizable parts of HPLCs, including various software capabilities. Fortunately, the majority of chromatographic techniques can be performed on almost all basic HPLC machines, and the approaches detailed below are therefore applicable to almost any standard HPLC setup.

7.3.2.1 Size-Exclusion Chromatography (SEC)

SEC separates analytes based on their molecular size (specifically, Stokes radius) (Hong, Koza, and Bouvier 2012). SEC is unique among LC techniques in that there is ideally no influence of adsorption of analytes to the stationary phase (Hong, Koza, and Bouvier 2012). Instead, the size range of the pores present on the stationary-phase particles and the sizes of the analytes, such as how much of the interparticle volume is available to an analyte, are the primary determinants of elution time. This means that smaller molecules that are able to enter the pores of the stationary-phase particles travel through a greater volume and elute later than larger molecules that only have access to a smaller portion of the total column volume.

The resolution at which these analytes are separated largely depends on three factors: column length, pore volume, and linear velocity. Longer columns, larger pore volume, and lower velocity yield better resolution (Hong, Koza, and Bouvier 2012). In addition, because the separation of analytes occurs within a single column volume, minimizing dispersion volume of the system is essential for maximizing resolution by ensuring that all analytes start as equal as possible. Keeping the sample concentration high and the loading volume low will aid in achieving this objective. Finally, it is also important to minimize the volume available for mechanical dispersion, such as solvent-mixing effects that may cause significant peak broadening (Kostanski, Keller, and Hamielec 2004). This can be accomplished by minimizing the length and inner diameter of tubing or by reducing the particle size of the stationary phase to allow for more efficient packing of the column. Because of the length requirements for optimal resolution, SEC is generally performed with specialized columns packed with cross-linked dextran gel (such as Sephadex). However, it can also be performed using hydrophilic silica-based stationary phases, such as poly(2-hydroxyethyl aspartamide). These are intended for hydrophilic interaction chromatography (HILIC) with mobile phases lacking organic solvents but yield somewhat lower resolution and have a smaller loading capacity (Alpert 1990; Alpert 1999).

While using long columns and dedicated machines can increase the resolution of SEC, by itself this is generally not sufficient for the purification of protein and peptide toxins from complex venoms. However, when fractionating venoms, even lower-resolution SEC represents an excellent first experimental step. The size of the proteins being investigated is among the most important determinants of optimal proteomic tools, and SEC therefore provides a very useful sample-preparation step by separating venom components into groups with different experimental requirements. SEC also has the advantage that nondenaturing mobile phases can be used, allowing the separation of large enzymes or protein complexes prone to denaturation (such as by hydrophobic solvents).

7.3.2.2 Ion-Exchange Chromatography (IEC)

IEC separates analytes based on ionic interactions of the stationary phase with opposite charges on the analyte. In the case of peptides and proteins, samples are typically solubilized in a low-ionic-strength buffer with a pH that induces the desired charge (net positive charge < pI < net negative charge) and adsorbed onto the stationary phase by replacement of phase counterions. These can be either positive or negative, and the name of each technique refers to the type of stationary-phase counterion that is replaced by the analyte upon adsorption: cation exchange for replacement of cationic counterions with positively charged analytes or anion exchange for replacement of anionic counterions. These two types of IEC can further be grouped into strong or weak IEC, based on the pH range in which the stationary phase retains its charge. Thus, strong cation exchange (SCX) and strong anion exchange (SAX) phases remain fully negatively and positively charged, respectively, over a wide pH range, while weak cation exchange (WCX) and weak anion exchange (WAX) phases are only partially charged over a narrow pH range (Cummins, Dowling, and O'Connor 2011).

The strength of the IEC employed is an important consideration when contemplating the appropriate IEC stationary phase as it affects the ionic strength required for the elution of analytes. Peptides and proteins can be eluted from IEC columns by a solvent gradient of increasing ionic strength (more salt), a pH gradient resulting in loss of net analyte charge, or a combination of the two (Winnik 2005; Cummins, Dowling, and O'Connor 2011). However, because of the high proportion of charged functional groups in strong IEC phases, proteins may adsorb to several exchanger sites, with the result that the ionic strengths required for elution often result in denaturing conditions. For the fractionation and isolation of larger proteins, weak IEC is therefore recommended, as these can be carried out under close to physiological conditions (Cummins, Dowling, and O'Connor 2011). However, SCX is recommended for structurally stable peptides, such as typical medium-to-low-MW disulfide-bonded toxins. Compared with SAX, SCX provides better overall retention of peptides at lower pH, which is desirable for most venom peptides, as higher pH may induce alterations in disulfide connectivity (Kopaciewicz and Regnier 1983). For example, in 5 mM potassium phosphate at low pH (pH 3), even peptides lacking basic residues but containing a free N-terminus can be retained (Alpert and Andrews 1988). Despite the excellent adsorption rates of most peptides, SCX also generally provides very good peptide recovery when elution is conducted with a gradient of intermediate to strong displacing salt such as potassium (such as 300–500 mM potassium chloride [KCl]). High resolution is also achievable with SCX, and this may be further improved by the inclusion of some organic solvent in the mobile phase (such as 25% acetonitrile [ACN]). Although sample-dependent, this may also induce mixed-mode effects, allowing for the separation of peptides with identical charges (Alpert and Andrews 1988). Furthermore, SCX columns can also be run in HILIC mode by including > 60% organic solvent in the mobile phase (Alpert 1990), illustrating the versatility of this technique.

7.3.2.3 Reversed-Phase High-Performance Chromatography (RP-HPLC)

RP-HPLC is probably the most utilized form of LC in proteomics (Jandera 2011). It separates analytes in order of increasing hydrophobicity. These are adsorbed onto polyhydrocarbon chains (usually 4, 8, 10, or 18 carbons in length) on the stationary phase and eluted with a gradient mobile phase containing an increasing concentration of organic solvent. For the analysis of peptides and proteins, the gradient usually consists of water and ACN acidified with trifluoroacetic acid (TFA) or formic acid (FA) (Howard et al. 2012). TFA and FA act as counterions to positively charged residues at low pH, which increases the hydrophobicity and solubility of peptides and reduces electrostatic interactions with the residual silanols of the stationary phase (Štulík et al. 1997). In addition, this acid-base property gives RP-HPLC versatility as a chromatographic technique, enabling experimentation with different ion-pairing reagents along with reversed ion-pairing conditions (Young and Wheat 1990). The latter occur at high pH, where the carboxyl groups rather than amides become charged, and cations such as ammonium can be used as counterions in bicarbonate-buffered mobile phases. This process should only be carried out using columns with either stable-bond or polymer-based stationary phases, as higher pH tends to be detrimental to standard silica-based particles.

Although the effective volume of the column is a less critical factor than adsorption in influencing the retention of analytes, pore size of the particle stationary phase is nevertheless an important consideration in RP-HPLC and also other adsorption-based stationary phases. This is because pores greatly increase the surface area of the stationary phase, and thus pores large enough to allow the movement of analytes will increase the loading capacity of the column. Conversely, pores that are too large will restrict the loading capacity of the column because of reduced surface area as compared with a greater number of smaller pores.

Unlike that of SEC, the effect of adsorption means that the separation volume of RP-HPLC and other adsorptive stationary phases is much greater than the column volume. Higher flow rates result in a lower effective gradient (volume * time / percent) and therefore result in less resolution but sharper peaks. Although RP-HPLC is generally carried out on an HPLC, flow rates restricted by system pressure mean that long gradient times may be required for sufficient resolution, which can be problematic in high-throughput settings. Higher flow rates on normal HPLC systems can be achieved using monolithic columns, where the stationary phase consists of a single porous rod. These are commercially available and allow greater flow rates by largely eliminating the effects of particle compression (Novakova and Vlckova 2009). Another solution has been increasing the temperature of the column, thereby reducing fluid viscosity and allowing for greater flow rates. This, however, poses problems with thermolabile venom components, which may unfold and become denatured at elevated temperatures. Alternatively, ultra-high-pressure liquid chromatography (uHPLC) can be employed, which typically operates at higher pressures (up to 100 MPa) and uses smaller particle stationary phases with only peripheral pores (Howard et al. 2012).

Regardless of the instrumentation used, RP-HPLC offers excellent, reproducible separation of peptides and medium-MW proteins (< 20 kDa). The use of volatile mobile phases means eluted compounds (eluates) can be lyophilized for downstream applications without the need for desalting. It is also well suited for desalting samples fractionated either by other LC techniques or from other experiments. However, because of the denaturing effects of higher organic solvent concentrations, separation of larger proteins and enzymes without loss of activity may be problematic.

7.3.2.4 Hydrophobic Interaction Chromatography (HIC)

HIC is a chromatographic method appropriate for the fractionation of high-MW venom proteins. In contrast to RP-HPLC, HIC stationary phases are characterized by low hydrophobicity (such as C2–C4, phenyl, CH_4) and ligand density of the stationary phase, coupled with a mobile phase of high ionic strength (Szepesy and Rippel 1994). Proteins are adsorbed onto the stationary phase in high-salt-content buffers and then eluted off by a decreasing salt concentration, in order of increasing hydrophobicity at the surface of the tertiary structure. This tends to preserve protein tertiary structure and activity and is also suitable for extremely hydrophilic peptides that do not bind to RP-HPLC phases and extremely hydrophobic peptides that are difficult to elute from RP-HPLC phases (Alpert 1988). In the case of most peptides, however, RP-HPLC is a better option because of better column efficiency and fewer baseline artifacts (Alpert 1988).

7.3.2.5 Hydrophilic Interaction Chromatography (HILIC)

Although the mechanisms of HILIC remain poorly understood, the principle is similar to that of normal-phase chromatography (Alpert 1990). Here analytes are adsorbed onto a hydrophilic stationary phase from a more hydrophobic mobile phase and eluted in order of increased hydrophilicity. Although most frequently used for the separation of small molecules (Jandera 2011), HILIC is also well suited to the separation of peptides, particularly involving differences in polar groups. For this, volatile mobile phases can be used, such as 15 mM ammonium formate, pH 3, with a gradient of 85% to 5% ACN (Novakova and Vlckova 2009). HILIC suffers from the same general limitations as RP-HPLC in terms of the risk of denaturation of large proteins because of the high initial concentrations of organic solvent. Therefore, the technique is recommended mainly for the fractionation of venom peptides.

7.3.2.6 Mixed-Bed Columns

Mixed-bed columns provide an additional set of possibilities for the separation of peptides and proteins. These contain more than one chromatographic functional group, usually HILIC or IEC paired with a hydrophobic group, and are designed to provide specialized loading conditions and unique selectivity (Zhao, Dong, and Sun 2009). One example of a mixed-bed column that provides an excellent additional means of separation of venom proteins is the use of mixed-bed WAX-WCX. This approach enables the retention of a greater number of proteins than either WAX or WCX in isolation at near-neutral pH because of the retention of both acidic and basic proteins (Havugimana, Wong, and Emili 2007). It can also be used with non-denaturing mobile phases, such as a gradient from 0 to 0.5 M NaCl in 10 mM phosphate buffer, pH 6.5, with 5% ACN.

7.3.2.7 Multidimensional LC

While it is possible to achieve high-resolution fractionation of venom components by several of the aforementioned chromatographic methods, each method in isolation is rarely sufficient for the isolation of individual toxins. This may be of particular importance during fractionation of samples prior to analysis by MS, particularly when detection of low-abundance components is desired. Instead, extensive fractionation can be achieved by the combination of multiple methods that separate peptides and proteins based on nonoverlapping, or orthogonal, properties. Such an approach also enables the separation of venom components into fractions that have different requirements for optimal further fractionation by LC.

An example of this is the initial fractionation of venom by SEC, which enables the separation of large venom components prone to denaturation from peptides that are more robust structurally. In addition to minimizing loss during further separation steps, the partitioning of high-, medium-, and low-MW venom components may be of benefit to downstream analytical and pharmacological applications. The limited loading capacity of SEC compared with the remaining chromatographic methods, however, means that this step may have to be repeated multiple times to obtain enough material.

Subsequent to or even without SEC, isolation of proteins can be carried out by weak IEC followed by HIC. The relatively flexible mobile phase composition and small elution volume of SEC mean that it can be transferred either directly or after dilution to the subsequent IEC step. Similarly, the eluted protein fraction from IEC requires only salt addition to promote binding in HIC. IEC-HIC thus represents a convenient means of separating proteins by both polarity and charge without much sample treatment.

For the purification of medium-to-low-MW peptides, several combinations provide highly orthogonal LC techniques (Gilar et al. 2005). Among these, the use of SCX followed by RP-HPLC combines the convenience of the high loading capacity of SCX with the volatile mobile phases and desalting capability of RP-HPLC. Although the use of excessive concentrations of organic solvent necessitates a drying step prior to RP-HPLC fractionation, this can be kept to a minimum for convenience. The combination of SCX-RP-HPLC is also used as a two-dimensional liquid chromatography (2D-LC) fractionation step prior to analysis by MS, a technique referred to as multidimensional protein identification technology (MudPIT) (Zhang et al. 2013). This may be done online or offline, but better resolution, peak capacity, and as a result more protein identifications will result from offline fractionation (Nagele et al. 2004). Another highly orthologous combination that can be used for MudPIT-like separation of peptides is RP-HPLC-HILIC, and this can also be used for the offline isolation of peptides (Gilar et al. 2005; Boersema, Mohammed, and Heck 2008; Xu et al. 2012). HILIC is also suitable for a convenient final third-dimensional isolation step after SCX-RP-HPLC fractionation which yields peptides of high purity solubilized in volatile solvent (Faull et al. 1998).

7.3.3 POLYACRYLAMIDE GEL ELECTROPHORESIS (PAGE)

PAGE is a commonly used analytical technique in proteomic investigations of venoms. Here separation is facilitated by the movement of charged particles in a uniform electric field through polyacrylamide gel. Two orthogonal separation methods of PAGE are typically used to separate proteins based on different properties, namely, sodium-dodecyl sulfate (SDS) PAGE by MW and isoelectric focusing (IEF) by isoelectric point (pI).

7.3.3.1 Sodium Dodecyl Sulfate Polyacrylamide Gel Electrophoresis (SDS-PAGE)

SDS-PAGE is the most commonly used PAGE technique. It separates proteins by MW based on their rate of movement through a cross-linked acrylamide gel, with larger proteins moving more slowly than smaller proteins. To eliminate the effects of variable sizes (Stokes radii) and surface charges of proteins in order to better correlate rate of movement with MW, proteins are linearized and bound with SDS. SDS is an anionic detergent that binds to the polypeptide backbone of proteins, resulting in proteins of similar shape and negative charge and hence similar electrophoretic properties. The most commonly utilized SDS-PAGE method is that of Laemmli (Laemmli 1970), which uses a Tris-glycine-based buffer system to separate proteins that are first loaded and concentrated in a high-porosity (stacking) gel that lies directly above a higher-concentration low-porosity (separating) gel where separation occurs. The rate at which proteins migrate through the gel is not linearly correlated with MW. Therefore, polyacrylamide concentrations must be optimized for the MW range of interest, typically ranging from around 5% (100–400 kDa) to 18% (< 50 kDa) for nongradient gels. Gels containing an acrylamide concentration gradient can also be used to attain a wider MW window of resolution, and these range from 4%–12% (30–300 kDa) to 10%–20% (5–100 kDa). For peptides < 5 kDa, a Tris-tricine buffer system is recommended, as this facilitates the separation of small SDS peptides from the SDS micelles that move at the front of the separation during SDS-PAGE and normally prevent the separation of these low-MW peptides (Schagger and von Jagow 1987). The major limitation is that isoform diversity cannot be visualized. Thus, while, for example, a venom may be seen to be rich in PLA_2, it cannot be determined if these are acidic, neutral, or basic (Ali et al. 2013a; Ali et al. 2013b) (see figure 7.2). These important variations may have profound impact on bioactivity, and thus, isoelectric focusing (see section 7.3.3.2) coupled with gel electrophoresis (see section 7.3.3.3) is a powerful combination to resolve by size and by charge diversity through this two-dimensional process.

7.3.3.2 Isoelectric Focusing (IEF)

IEF separates proteins and peptides based on their pI by making use of the effect of pH on their net charge states. Samples are typically suspended in acrylamide gel strips with immobilized pH gradients and subjected to an electric field. This causes negatively charged molecules to migrate toward the anode at the low-pH end and positively charged molecules to migrate toward the cathode at the high-pH end of the strip. However, because proteins attain a negative charge at pH values above their pI and vice versa, their movement through the strip on either side of the point corresponding to their pI becomes opposite. Given sufficient time, this results in "focusing" at the pI of each protein. Since the pI of proteins varies extensively, separation of proteins in a mixture can be achieved via this method.

Because of the importance of the electric field (volts) in generating movement of charged molecules in IEF, it is sensitive to the presence of conductive molecules that increase the current (amps) between the two electrodes. This reduces the voltage of the electric field and hence

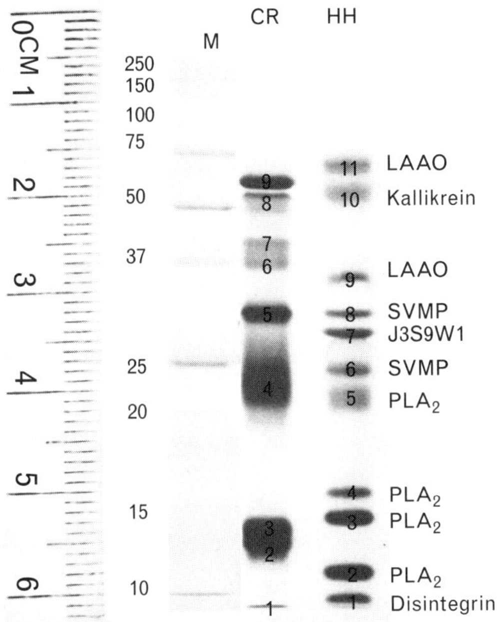

FIGURE 7.2: One-dimensional 12% acrylamide SDS-PAGE analysis of *Calloselasma rhodostoma* (CR) and *Hypnale hypnale* (HH) venoms carried out under dissociating and denaturing conditions. Gel was stained in 0.2% colloidal Coomassie brilliant blue G250. M = molecular mass standard markers. (Ali et al. 2013a).

reduces resolving power in the IEF strip, leading to a phenomenon known as streaking, the result of poor focusing. IEF is therefore usually carried out in a stepwise voltage protocol; low voltage is first applied to allow the migration of conductive molecules out of the strip, and voltage is then gradually increased to effective focusing levels (such as 8000 V) and left for an adequate number of voltage hours (such as 98,400 Vh). The total number of Vh should typically not exceed 100,000 in order to avoid overfocusing effects that may also result in streaking. In venoms, troublesome conductive molecules are usually inorganic salts and glycosidic chains, which may be present in high concentration. Removal of these prior to IEF may therefore be necessary for successful focusing of venom proteins and can be achieved using commercially available IEF cleanup kits, protein precipitation protocols, or LC-based desalting steps.

7.3.3.3 Two-Dimensional Polyacrylamide Gel Electrophoresis (2D-PAGE)

Although it is a powerful separation technique, IEF tends not to be used in isolation. One application of IEF is the fractionation of samples before separation by RP-HPLC. Known as Off-Gel (Ros et al. 2002), it provides a powerful alternative to 2D-LC combinations such as the most commonly used SCX-RP-HPLC. In the vast majority of cases, however, it is coupled to SDS-PAGE, which allows orthogonal separation of proteins based on both pI and MW (Ali et al. 2013a; Ali et al. 2013b; Low et al. 2013) (see figure 7.3). This extremely powerful technique, known as 2D-PAGE, is able to resolve thousands of unique proteins in a single run and thus provides an important tool for fingerprinting venoms based on their higher-MW components. In addition to visual comparison, spots can be excised, digested, and identified by MS sequencing, as outlined below.

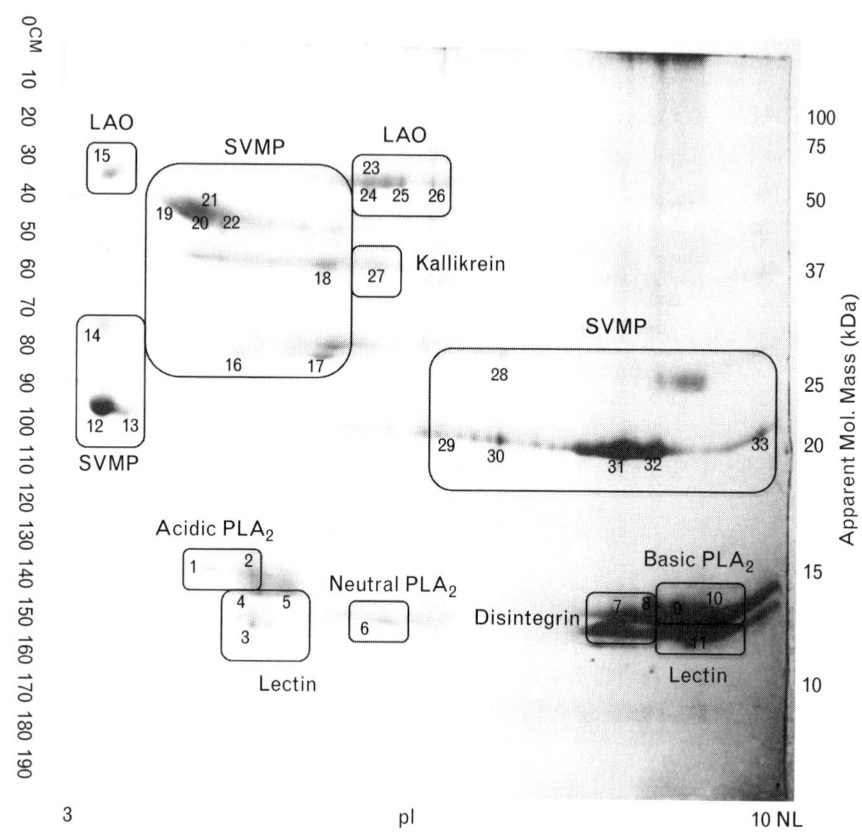

FIGURE 7.3: Two-dimensional gel of the *Calloselasma rhodostoma* crude venom stained with colloidal Coomassie brilliant blue G250. First dimension: isoelectric focusing (pH 3–10 nonlinear gradient); second dimension: 12% SDS-PAGE. The pH gradient and the MW marker positions are shown. (Ali et al. 2013a).

7.3.3.4 Detection and Quantification

Proteins separated by PAGE can be detected through either staining or fluorescent tagging. There exists a wide range of stains and staining techniques, of which Coomassie brilliant blue (R-250) dissolved in water/methanol/acetic acid (Meyer and Lamberts 1965) is probably the most commonly used. Colloidal Coomassie brilliant blue (G-250) in phosphoric acid/ammonium sulfate can also be employed, with the advantage of less background staining and greater sensitivity under visible light (Neuhoff, Stamm, and Eibl 1985; Neuhoff et al. 1988). For the highest sensitivity among nonfluorescent stains, silver staining can be employed. There are several drawbacks, however, including complex and time-consuming preparation, poor dynamic range, and issues with reproducibility (Rabilloud et al. 1994). In comparison, fluorescent dyes, such as SYPRO ruby-red fluorescent dyes for total protein detection, offer greater sensitivity, reproducibility, and dynamic range, although UV or blue-light transilluminator gel imagers are required. If such imagers are available, colloidal G-250 can also be used for detection in the near-infrared range, with equal or better sensitivity, detection, and dynamic range compared to ruby-red stain (Butt and Coorssen 2013). Hence, the most convenient staining technique for total protein detection is probably colloidal G-250. Staining by alcian blue (Wardi and Michos 1972) can not only detect and quantify proteins but also detect posttranslational modifications

(PTMs) of glycoproteins. Fluorescent dyes are also available for the specific staining of PTMs, and these allow subsequent staining with compatible total protein staining dyes for simultaneous detection.

While quantitative comparisons of proteins can be done with noncomplex samples in a one-dimensional PAGE (1D-PAGE) using multiple lanes, this quickly becomes problematic when comparing complex samples such as crude venoms, where bands may contain multiple proteins. A solution to this is the use of 2D differential in-gel electrophoresis (2D-DIGE), where samples are labeled with dyes fluorescing at different wavelengths, mixed, and then separated on the same 2D-PAGE (Birrell et al. 2006; Ali et al. 2013a) (see color plate 25). Fluorescence of each spot can then be compared at each wavelength to quantify the contribution from each sample in each spot, reducing the influence of between-gel experimental variability in 2D-PAGE.

7.3.4 MASS SPECTROMETRY (MS)

Unlike LC and PAGE, MS is a destructive analytical tool; analytes are not available for additional analyses after separation and detection. MS is also not particularly well suited for the separation of complex mixtures and is therefore usually preceded by fractionation through at least one step of LC and/or PAGE (see figure 7.1). The resolution of MS, however, means it is ideally suited for the precise identification of toxins. Hence, MS has become an indispensable tool in proteomic characterization of venoms, as a means of both fingerprinting venoms and identifying toxins and their sequences (Escoubas 2006). While MS can be used for the analysis of large proteins and protein complexes (van den Heuvel et al. 2006; Kaltashov, Bobst, and Abzalimov 2013), it has gained popularity in proteomics because it is ideally suited to the analysis of peptides. These are generally separated according to their mass-to-charge ratio (m/z) and the molecular mass consequently calculated as a function of their charge. Although a simplification, this is achieved by the three main parts of the mass spectrometer: the ionization source produces multiple different gas-phase ions, the mass filter/analyzer separates these ions, and then the ion detector amplifies and transfers the signals.

Although a large number of ionization modes exist, only two are commonly used in proteomics: electrospray ionization (ESI) and matrix-assisted laser desorption/ionization (MALDI). In ESI, ions are formed directly from solution, which means it is amenable to online high-throughput systems. In addition, multiply charged ions are dominant, and this allows for both a wide range of molecular weights to be accurately measured within a narrow m/z and the optimal utilization of certain fragmentation techniques (see section 7.3.3.2). Although the details of the ion-formation mechanisms in MALDI are uncertain, the principle is quite simple: a sample is co-crystallized with a matrix and irradiated with an ultraviolet or infrared light laser. This causes the charged analytes to desorb into the gas phase, and these are then accelerated into the mass analyzer. While the co-crystallization step renders MALDI unsuitable for high-throughput sample analysis, the importance of the matrix in the ionization of analytes and the ability to investigate mixtures of analytes mean that MALDI can greatly increase the versatility of an MS platform.

Along with the type of ionization source, the mass analyzer and detector dictate the utility of a mass spectrometer. To achieve the best performance in speed, sensitivity, resolution, and accuracy, most modern MS instruments have hybrid mass analyzers that combine the different performance characteristics offered by various types of analyzers (such as Time of Flight [TOF], Quadrupole [Q], Trap, and Fourier Transform [FT] types) (Glish and Burinsky 2008). These combinations greatly affect the type of fragmentation, and hence sequence information, that can be obtained from a sample (see section 7.3.3.2). Ultimately, the range of combinations of sources and mass analyzers available means that MS is an approach ideally suited to the wide spectrum of analytical challenges encountered in venom proteomics.

7.3.4.1 Liquid Chromatography Mass Spectrometry (LC-MS)

As mentioned previously, samples analyzed by MS are normally fractionated beforehand by at least one dimension of LC or PAGE. The most commonly used combination is a fractionation by RP-HPLC followed by MS. LC-MS can be either online (ESI) or offline (MALDI), but in either case, it provides a convenient way to reduce the number of analytes to be analyzed simultaneously by the mass spectrometer. It also provides a rapid high-resolution descriptive tool that can be used to examine and compare both general venom characteristics and changes in individual medium-to-low-MW venom components (Fry et al. 2003a; Fry et al. 2003c; Fry et al. 2005; Ali et al. 2013b; Jesupret et al. 2014). Because of this, analysis of venoms by LC-MS provides a natural first step in the proteomic investigation of venoms.

7.3.4.2 Mass Spectrometry Sequencing

Although a highly accurate measurement of the molecular mass of a toxin could in theory enable its identification from a list of sequences (such as a transcriptome assembly), this is rarely sufficient because of the large number of potential PTMs that may occur. Venom glands also tend to express a large number of isoforms of certain toxin types, further complicating the matter. Thus, for confident identification of toxins, sequencing is usually necessary. The traditional proteomic approach to sequencing toxins largely relied on the use of automated Edman degradation and amino acid composition analysis followed by confirmation of MW by MS (such as the method used in the discovery and characterization of exendin-4 (Eng et al. 1992). This approach, however, suffers from being low-throughput and incurring prohibitive costs when characterizing a large number of toxins, and it also requires the complete isolation of toxins prior to sequencing (Fry et al. 2003a; Fry et al. 2005). Fortunately, the advent of highly sensitive, high-resolution MS instruments allowing for novel fragmentation techniques has provided a method to circumvent these issues (Fry et al. 2003a; Ali et al. 2013a; Ali et al. 2013b; Low et al. 2013; Jesupret et al. 2014).

In MS, sequencing is based on fragmenting the polypeptide backbones of peptides and proteins into sufficiently small fragments so that the MW of individual amino acids can be confirmed in order to provide the amino acid sequence. There are two main approaches that can be taken to achieve this: "bottom-up" (fragmenting cleaved proteins) and "top-down" (fragmenting intact proteins) proteomics. For bottom-up proteomics, also often referred to as shotgun sequencing, whole-protein mixtures or isolated proteins (such as by 2D-PAGE) are fragmented typically by digestion with endoproteases. Trypsin is most commonly used, as it is a robust enzyme, it has good specificity, and cleavage at its preferred sites (C-terminal side of the basic amino acids Arg and Lys) results in basic residues at C-termini, which makes MS-fragmentation spectra easy to interpret (Olsen, Ong, and Mann 2004).

The resultant peptides are then separated on an LC column before being passed through multiple MS steps in a process referred to as tandem MS (MS/MS, or MS^n for higher-order MS experiments). During an MS/MS experiment, the mass spectrometer first conducts a survey scan (MS) to measure all ions present at that time, then selects, sequentially isolates, fragments, and measures the fragment spectrum (MS^2) of peptide ions (precursor ions) that fit within a range of specified properties, such as m/z and number of charge. This cycle is repeated throughout the LC gradient so that MS/MS spectra are obtained from as many different peptide ions as possible.

There are two main groups of peptide-fragmentation methods, separated based on the underlying mechanism of breakage of the polypeptide backbone: thermal and radical-driven fragmentation. Because of the popularity of Q/TOF, triple-Q, and TOF/TOF mass analyzers, thermally mediated fragmentation techniques that are frequently employed include post-source decay (PSD), collision-induced dissociation (CID), and higher-collision-energy dissociation (HCD) (Olsen et al. 2007). Cleavage occurs primarily at the amide bond of the peptide backbone and is caused either by dissociation during flight in a unimolecular process (PSD) or by collision into molecules of an inert gas (such as N_2 or Ar) at high energies (CID and HCD). Although both fast and sensitive, these techniques tend also to "knock off" labile PTMs, leaving them undetected. They are also largely restricted to use with relatively small peptides (< 4 kDa) for the generation of informative spectra, and a bottom-up approach is therefore usually necessary for larger proteins.

In contrast to thermally mediated fragmentation, radical-driven fragmentation is not necessarily restricted in terms of precursor size and is therefore used in both bottom-up and top-down proteomics (Ueberheide et al. 2009; Fornelli et al. 2013; Guthals et al. 2013). Here an electron capture (electron-capture dissociation [ECD]), loss (electron-detachment dissociation [EDD]), or transfer (electron-transfer dissociation [ETD] and negative-electron-transfer dissociation [NETD]) results in the formation of a radical, which then induces cleavage of the N-Cα (ECD, ETD) or Cα-CO (EDD, NETD) bond (Anusiewicz et al. 2005; Kim and Pandey 2012). This mode of fragmentation also differs in that even labile PTMs are usually retained, allowing for precise identification of these modifications. It is also worth noting that although the aforementioned fragmentation events are induced in the mass analyzers of specialized mass spectrometers, analogous techniques can be performed on commonly accessible MALDI-TOF machines. These can be done by inducing a fragmentation event known as in-source decay (ISD) through the use of matrices such as 1,5-diaminonapthalene (ETD-like) or 5-nitro-salicylic acid (NETD-like but in positive-ion mode) (Brown and Lennon 1995; Takayama 2001; Asakawa and Takayama 2011; Asakawa et al. 2013). However, because fragmentation occurs before any separation of precursor ions, ISD normally requires pure sample for interpretation of spectra and therefore does not represent a high-throughput strategy. Nevertheless, the ability to induce multiple, complementary types of fragmentation using the same source and mass analyzer illustrates the versatility of MALDI-MS.

For interpretation of fragmentation data, MS-based sequencing can be roughly divided into two general types: true *de novo* sequencing and search-based sequencing. Traditional *de novo* sequencing efforts focus on sequencing either individual or a few functionally interesting toxins. These were often combined with Edman degradation to account for ambiguities in fragmentation spectra and incomplete C-terminal sequences from Edman degradation. However, technological and computational advancements mean that *de novo* shotgun sequencing of whole venoms is now possible (Bandeira, Clauser, and Pevzner 2007). Similarly, full-length peptide toxins can now routinely be sequenced *de novo* on certain mass spectrometers (Ueberheide et al. 2009).

While technological development has enabled rapid true *de novo* sequencing of whole venoms and peptides, this is a daunting task on most mass spectrometers. The alternative, search-based MS sequencing, is a much more commonly adopted approach. In this approach, experimentally derived fragment spectra are searched against theoretically derived fragment spectra calculated from a sequence template using computational tools such as Mascot (Matrix Science), Spectrum Mill (Agilent), or Protein Pilot (ABSciex). Although it is not possible to obtain full sequences of entirely novel toxins using this method, since the successful identification of proteins depends on the availability of known sequences, in combination with a venom-gland transcriptome, it enables both accurate and high-throughput characterization of novel venoms (Wong et al. 2012; Dutertre et al. 2013b; Fry et al. 2013; Low et al. 2013).

7.3.4.3 Quantitation by Mass Spectrometry

Two types of MS-based quantitative comparisons are commonly used: stable isotope labeled and unlabeled. Although there are several approaches to isotopic labeling of peptides for quantitative MS (Bantscheff et al. 2007), only strategies making use of chemical labeling are practical for quantitative studies of venom composition. Among these, use of isobaric tags for relative and absolute quantitation (iTRAQ) is probably the most common method, as it allows for simultaneous monitoring of as many as eight samples (Ross et al. 2004; Choe et al. 2007). An equal amount of protein from each sample is digested, and the N-termini and primary amines of the proteolytic fragments are each labeled with a unique label. The samples are then combined, analyzed by LC-MS/MS, and searched for protein identifications. During MS/MS, the iTRAQ labels dissociate at an equal rate, and the abundance of each identified protein in each sample can then be compared on the basis of the relative intensity of each label in the MS/MS spectra. While labeled quantitative approaches such as iTRAQ have been found to be more reproducible between analyses (Bantscheff et al. 2007), this is a rather costly approach compared with unlabeled quantitative MS experiments.

Label-free quantitative MS is usually done in one of two ways: extraction of ion chromatograms or counting of MS/MS spectra for each protein (Wong and Cagney 2010). In the first, the integrated area under the peak of each peptide is used to estimate its ion intensity from the MS spectrum of each sample, and this is used to estimate the protein ratio in the sample. The second approach estimates abundance based on the number of MS/MS spectra obtained from each peptide, with the assumption that the more abundant a peptide is, the more it will be sent to analysis by MS/MS. While the reproducibility of labeled quantitative MS has been found to be greater (Bantscheff et al. 2007), there are also noneconomic benefits of label-free quantitative MS, such as higher accuracy and fewer sample-preparation steps (Trinh et al. 2013).

7.3.5 STRUCTURE DETERMINATION

The two major methods for elucidating the structure of proteins at atomic resolution are X-ray crystallography and nuclear magnetic resonance (NMR) spectroscopy (Garbuzynskiy et al. 2005; Sikic, Tomic, and Carugo 2010; Feng, Pan, and Zhang 2011). The two techniques

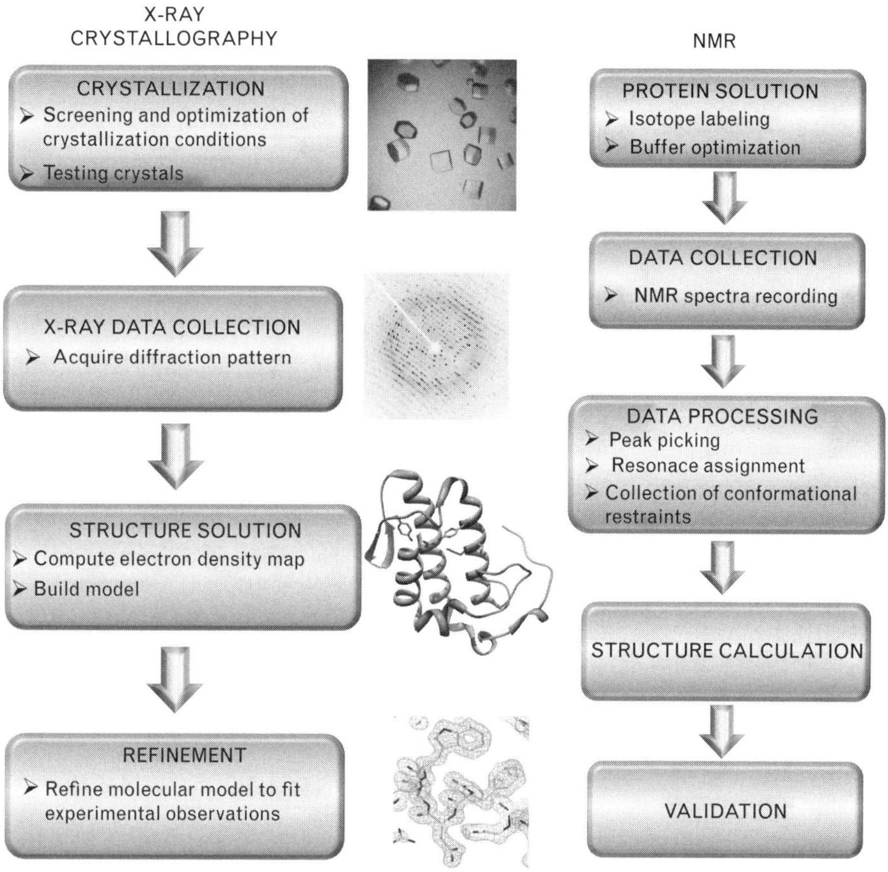

FIGURE 7.4: Comparison of methodological stems of X-ray crystallography versus nuclear magnetic resonance.

are highly complementary, each with distinct advantages and limitations (see figure 7.4 and table 7.1). While X-ray crystallography provides more accurate atomic resolution and has been used to determine the majority of structures in the protein data bank (www.rcsb.org), NMR provides information on the dynamics of the protein in solution. Another advantage of NMR is the possibility of elucidating the binding surfaces between a protein and a ligand without the need to determine the entire structure of the complex. This is particularly pertinent in cases where the binding is too weak to form a stable complex.

7.3.5.1 X-Ray Crystallography

A major hurdle in solving protein structures by X-ray crystallography is obtaining single well-ordered crystals large enough to diffract X-ray beams. Protein crystallization is a trial-and-error procedure, in which a wide variety of conditions are screened with the help of commercially available kits. The availability of high-throughput robotic systems makes it possible to save time, effort, and protein during this step.

Table 7.1: Comparison between NMR and X-ray Crystallography.

	NMR	XRC
State	In solution	Crystals needed
Protein concentration	Variable	Up to 100 mg/mL
Purity	> 95%	High
Homogeneity	Monodispersity	Low polydispersity
Recovery of the sample	Yes	No; radiation damage
Resolution	Lower	Highest reported; 0.44Å
Sensitivity	Low inherent sensitivity	Maximum model fidelity and accuracy
Molecular mass	Applicable for proteins < 50 kDa	No size limitations
Information obtained	Dynamic properties of proteins, molecular interaction of the protein with a ligand in addition to stability and folding-unfolding kinetics	Static structure
Sample preparation time	Short	Long

Garbuzynskiy et al. 2005; Sikic, Tomic, and Carugo 2010; Feng, Pan, and Zhang 2011.

Crystallization is a process in which crystals are grown by slow, controlled precipitation, usually from an aqueous solution of the protein. Factors that affect the crystal growth are protein concentration and purity, pH, temperature, concentration of the precipitant, ionic strength, organic compounds (polyethylene glycol), detergents, volume of the crystallization drop, gravity, vibration and handling by the investigator, magnetic fields, additives, ligands, substrates, inhibitors, and protein structure and dynamics.

Sample requirements for protein crystallization are as follows:

- High purity of samples—this can be checked using SDS-PAGE or MS.
- Low polydispersity—this can be confirmed with dynamic light scattering (DLS) or small-angle X-ray scattering (SAXS).
- Stable in the buffers/additives used.
- High solubility.
- Sufficiently concentrated sample.
- Clear solution—use a Vortex-System or filter the sample.
- Correctly folded—check with CD spectroscopy.
- In case of enzymes—check activity.

In X-ray crystallography, the three-dimensional structures of macromolecules are determined by X-ray diffraction. The protein crystal is targeted by an X-ray beam which is diffracted by the electrons in the crystal; the diffraction pattern is based on the crystal lattice structure.

The crystal is gradually rotated in the beam, and a number of images are recorded. The next step is the calculation of the electron density map, which can be done in several ways:

- Molecular replacement—in cases where the structure of a highly homologous protein (> 30% sequence homology) is available, this can be used as a model to calculate the phases and build the model of the unknown structure. This is the quickest and easiest method for solving crystal structures.
- Multiple isomorphous replacement (MIR)—a heavy atom is introduced into the structure without perturbing the crystalline lattice. At least one native crystal and two crystals soaked in two heavy atom solutions must be available. This is the most common method for ab initio phase determination.
- Multiwavelength anomalous dispersion (MAD)—multiple data sets are recorded at different wavelengths from a single crystal after incorporation of anomalous scattering.

The calculation of an electron density map allows the generation of a model, which is further refined until it fits the experimental observations. The quality of the model (in terms of steric clashes) can be assessed using programs such as MolProbity (Davis et al. 2004).

7.3.5.2 Nuclear Magnetic Resonance (NMR)

NMR is a physical phenomenon exhibited when atomic nuclei that possess magnetic moment are placed in a static magnetic field and simultaneously exposed to an additional, oscillating magnetic field. NMR spectroscopy exploits the NMR phenomenon to study physical, chemical, and biological properties of macromolecules. It gives detailed information about molecules and their environment in solution, without the need to obtain crystals and in nearly physiological conditions. In contrast to X-ray crystallography, NMR can be used for folding or denaturing studies of proteins or for studying the dynamics of protein-protein or protein-ligand interactions in real time.

There are two general approaches through which NMR structures can be determined, and the choice between them is determined essentially by the molecular mass of the protein being investigated. The first approach, which was pioneered in the laboratory of Kurt Wüthrich (Wüthrich 1986) and which led to his being awarded the 2002 Nobel Prize in Chemistry, involves the use of 2D ^{1}H-^{1}H (homonuclear) NMR experiments (such as correlation spectroscopy [COSY], nuclear Overhauser effect spectroscopy [NOESY], and total correlation spectroscopy [TOCSY]). This approach is still widely used today but only for proteins smaller than 5 kDa that cannot be isotopically labeled. Nevertheless, the homonuclear NMR strategy is suitable for studying peptides purified from natural sources or where unnatural amino acids have been introduced synthetically.

The homonuclear strategy relies primarily on 2D experiments, and as the protein size increases, the number of resonances on the associated spectral plots also increases. This leads to extensive overlap between different signals, which in turn results in lower confidence in the assignment procedure and ultimately in poor structural definition. The natural isotopes of nitrogen and carbon are not suitable for NMR studies; however, the nonradioactive ^{13}C and ^{15}N isotopes of these atoms have very favorable NMR properties. These isotopes are most commonly introduced by recombinant expression of the peptide in *Escherichia coli* grown in media enriched with ^{13}C-labeled glucose and ^{15}N-labeled ammonium chloride (Klint et al. 2013). Proteins that have been ^{13}C- and ^{15}N-enriched can be used to acquire additional heteronuclear NMR experiments. These are often acquired as 3D experiments, although 4D experiments are sometimes used (Mobli et al. 2010). The added dimension provides a means by which the ambiguity found in 2D spectra can be resolved, and once again, the NMR approach can be used to provide high-resolution structures (Kwan et al. 2011). The 3D experiments typically take much longer to acquire, and they contain much more specific information. Many more 3D experiments are required to solve a structure using the heteronuclear approach than 2D experiments are required via the homonuclear approach (King and Mobli 2010). The heteronuclear approach therefore naturally requires more instrument time and ultimately longer analysis time, although significant progress has been made toward addressing these constraints (Hoch et al. 2014). The experiments typically acquired include 3D ^{15}N- and ^{13}C-edited NOESY experiments (HSQC NOESYs) along with a suite of experiments for backbone (such as HNCO, HNCACB, and HN(CO)CACB) and side-chain assignments (such as HCCH and HCC[CO]NH TOCSYs) (King and Mobli 2010).

Sample requirements for NMR are as follows:

- MW—currently limited to MW < 30 kDa.
- Purity of samples > 95%—check with SDS-PAGE or MS.
- Monodisperse—check with DLS or SAXS.
- Stable at room temperature for days.
- High solubility.
- Concentrated—ideally 0.5–1.5 mM.
- Clear solution—use a Vortex-System or filter the sample.
- Correctly folded—check with CD spectroscopy.
- In case of enzymes—check activity.
- Buffer—0–50 mM, usually 20 mM.
- pH—usually 5–7.
- Ionic strength such as total salt concentration < 150 mM when using a cryoprobe system.
- Paramagnetics—should be avoided, as they broaden NMR signals.
- Isotope labeling—not required for small peptides (< 5 kDa), ^{15}N required for small (< 10 kDa), ^{15}N and ^{13}C required for medium (< 20 kDa), and ^{15}N, ^{13}C, and ^{2}H required for large (> 25 kDa) proteins.

When the protein is available in sufficient amounts (traditionally 0.5 mL of a 1-mM solution, although with cryoprobes and higher-field-strength spectrometers, lower concentrations are routinely used), the first step is acquisition of the NMR spectra. A set of multidimensional NMR spectra are recorded for the assignment of all chemical shifts (^{1}H,^{15}N,^{13}C). The next step—data analysis—is the most critical and time-consuming. This step includes peak picking (the NMR spectra are analyzed and the important signals identified), resonance assignment (the picked peaks from the NMR spectra are assigned to their corresponding nuclei, allowing determination of the secondary structure of a protein), and NOE (nuclear Overhauser effect) peak assignment (generation of distance restraints). A family of structures that satisfy as many conformational restraints as possible is iteratively generated. The high prevalence of disulfide bonds in toxin structures can introduce additional complexity to the structure-calculation process, as it can be difficult to determine connectivity using only NMR data (Rosengren et al.

2003). In this case, determination of the disulfide connectivity through chemical methods is required. Similarly to X-ray crystallography, programs such as MolProbity can be used to assess the quality of NMR-derived structures.

7.4 MOLECULAR BIOLOGICAL TECHNIQUES

7.4.1 TRANSCRIPTOMICS

As mentioned previously, search-based MS sequencing represents a commonly available, rapid, and economic means of identifying large numbers of venom components. However, reliance on a sequence template for the identification of peptides and proteins hinders effective sequencing of novel toxins. Depending on the sequencing depth and coverage available in public databases, it may also be poorly suited for the detection of isoform variants, low-expression-level toxins, or full sequence determination. A more accurate and faster way of obtaining toxin sequences is therefore to directly sequence the toxin-encoding messenger ribonucleic acid (mRNA) transcripts and subsequently match these to the proteomic data (Fry et al. 2006; Fry et al. 2008; Fry et al. 2010b; Fry et al. 2013). Several approaches exist for this purpose, and they can be broadly categorized as nontargeted and targeted (see below). While nontargeting approaches are not focused on specific toxin types, targeted transcriptome methods do, and typically require prior information regarding the target toxin's sequence.

7.4.1.1 Sample Preparation

Unlike DNA, RNA molecules are mostly single-stranded and thereby far more sensitive to degradation by nucleases. Hence, preventing RNA degradation is of critical importance during extraction. Venom glands should therefore be dissected immediately after euthanasia and, if possible, flash-frozen in liquid nitrogen promptly thereafter. Alternatively, the dissected glands may be stored in a preservative such as RNAlater. Preventing RNase contamination during extraction is paramount, and working quickly and keeping samples cold during the process will help reduce sample degradation. While commercial kits exist for the direct extraction of poly-A RNA (mature mRNA molecules are typically characterized by a 3′ poly-A tail), it may be convenient to first extract total RNA (such as using a TRIzol protocol) followed by a poly-A RNA enrichment step. Alternatively, non-poly-A RNA (e.g. ribosomal RNA, which typically makes up a large proportion of total RNA) may be removed from the sample using a depletion step. Many kit-based protocols use spin columns, and these may become clogged by samples containing large amounts of mucus if utilized at the first step of extraction.

7.4.1.2 Nontargeted Sequencing

Nontargeted transcriptomics represents more accurately what the term "transcriptome" encompasses, in that the outcome strives toward producing as complete a representation as possible of all transcripts present at the time of sample extraction. In the context of venoms, a nontargeted trasncriptome study is meant to provide a comprehensive overview of all genes that participate in the venom gland's functioning. This obviously includes all toxin genes and their splicing variants, but also metabolic genes involved in the synthesis, posttranslational processing, and secretion of toxins. The inherent advantage to this approach is the generation of complementary DNA (cDNA) libraries that may form the foundation for numerous studies and experiments beyond simply attaining the sequences of a few toxins of immediate interest (Fry et al. 2006; Fry et al. 2008; Fry et al. 2010b). This requires high-throughput sequencing of extracted mRNA, and the field has consequently experienced a surge of activity with the continuous development of faster and cheaper DNA-sequencing platforms. Such next-generation sequencing (NGS) platforms have made the acquisition of venom-gland transcriptomes efficient either through the use of benchtop sequencers, collaborations, or outsourcing to one of the numerous sequencing facilities available. This was previously an incredibly expensive task, and the development and adoption of NGS technology has revolutionized many fields in biology in a similar fashion to the revolution previously instituted by the "traditional" Sanger sequencing method (Rogers and Venter 2005; Schuster 2008; Wilhelm 2009).

A number of NGS platforms are suitable for transcriptome characterization, and these differ primarily in the number and length of reads (stretches of sequenced DNA) obtained and also in price (Glenn 2011). Transcriptomes sequenced by NGS are typically constructed from shotgun assemblies, such as mRNA fragmented to a suitable length prior to sequencing. The length of each read is therefore an important factor in determining the ability to subsequently reassemble sequences into the contiguous sequences (contigs) that make up each original transcript. The presence of highly homologous multigene families typical of venoms quickly complicates the assembly of contigs, particularly because of the frequent lack of a reference genome sequence to use as an assembly template. This necessity for *de novo* assembly of data sets therefore means that the short read lengths associated with some sequencing platforms (first developed by Illumina) may give rise to significant issues with data assembly and analysis. Platforms enabling longer reads have therefore been preferred, and 454 pyrosequencing in particular has been the platform of choice for the investigation of venom-gland preparations from various taxa, including centipedes (Undheim et al. 2014); cephalopods (Ruder et al. 2013b), cone snails (Hu et al. 2011; Hu et al. 2012; Terrat et al. 2012; Violette et al. 2012; Dutertre et al. 2013a), remipedes (von Reumont et al. 2014), reptiles (Durban et al. 2011; Rokyta et al. 2011; Durban et al. 2013; Fry et al. 2013; Jackson et al. 2013; Terrat et al. 2013; Sunagar et al. 2014), scorpions (Rendón-Anaya et al. 2012; Sunagar et al. 2013c), spiders (Undheim et al. 2013; Wong et al. 2013), and vampire bats (Low et al. 2013). In contrast, only a few venom-gland preparations have been sequenced using the Illumina HiSeq or MiSeq platforms (Wong et al. 2012; Francischetti et al. 2013; Rokyta, Wray, and Margres 2013). However, with Illumina platform read lengths now exceeding 250 base pairs in length, thereby facilitating sequence assembly, the greater number of reads (up to billions) generated by this technology provides an advantage over the 454 platform.

7.4.1.3 Targeted Sequencing

In targeted sequencing of toxin-encoding genes, partial protein or nucleotide sequences are used to construct primers that amplify the DNA of interest, typically the complementary cDNA sequence to reverse-transcribed mRNA (see Li, Fry, and Kimi 2005a; Li, Fry, and Kimi 2005b). This approach, termed 'rapid amplification of cDNA ends' (RACE), is useful for the acquisition of complete sequences from partial sequence data or for detection of particular protein or peptide families. The former has been the primary motivation for RACE experiments because of the use of traditional Edman-degradation-based protein-sequencing efforts, which often yield incomplete C-terminal sequences. RACE also has the advantage that it requires very little starting material. However, while RACE is a powerful and inexpensive means of using a transcriptome to acquire full-length sequences of toxins of interest and their homologs, there are obvious drawbacks. The process can be quite labor-intensive, the number of obtained sequences is typically relatively small compared to nontargeted sequencing, and their diversity may be biased by primer design. In addition, since prior knowledge on at least part of the sequence (either RNA- or protein-based) is required for primer design, RACE is not the most suitable technique for providing an overview of the full toxin diversity in venoms. Finally, unlike nontargeted methods based on NGS platforms, RACE cannot provide any information on relative levels of expression of different toxin genes.

RACE may be conducted in either direction of the transcript depending on which part (5′ or 3′) is unknown. In either way, the reaction involves a PCR using a universal RACE primer (often supplied by commercial RACE kits) and a user-designed primer that specifically targets the transcript of interest. Criteria for the design of this specific primer may vary according to the used RACE protocol or commercial kit, but typically provide restrictions on primer length (approx. 20–28 bp), GC content (50–70%), GC positioning secondary structure formation or dimerization, and annealing temperature.

Although the user-designed primer as a "gene-specific primer" or "GSP," a single RACE reaction can be used either to complete the sequence of a single gene transcript or of multiple related transcripts (e.g. transcripts of paralogous toxin genes and/or alternative splicing variants of a single gene). In case of the latter, the primer is preferably designed in a conserved region of the transcripts, to optimize the range of potential target transcripts. Either a single primer (e.g. representing the consensus sequence across all available transcripts) or a degenerate primer can be designed (a mixture of primers showing nucleotide variation at specific nucleotide sites). The number of ambiguous nucleotide sites (often called "wobbles") in a degenerate primer should be limited, since it comes at the cost of primer specificity and may yield the amplification of unrelated (e.g. nontoxin) transcripts.

To design a gene-specific primer based on a protein sequence, the latter first needs to be reverse-translated into a nucleotide sequence, which is done using a codon usage table. As most amino-acids are encoded by multiple codons, it is often impossible to obtain an unambiguous primer sequence. If so, the primer can be designed as to represent the most likely codon sequence (if information on relative codon usage is available) or again, a degenerate primer can be designed (see above).

Depending on which end of the transcript is unknown, one of two types of RACE reactions are conducted. A 3′ RACE reaction is applied to obtain the 3′ region of a transcript, e.g. to derive the C-terminal part of a toxin protein, or the 3′ UTR. First-strand (antisense) cDNA is synthesized using an adapter oligonucleotide that targets the 3′ poly-A tail of mRNA molecules and contains the sequence of an (antisense) universal RACE primer. The antisense cDNA is then used as starting material in a normal PCR using the (antisense) universal RACE primer and the user-designed (sense) primer.

A 5′ RACE reaction is applied to complete the 5′ sequence of transcript, e.g. to characterize the N-terminal part of a protein, the signal peptide, or the 5′ UTR. Unlike 3′ RACE, it requires ligation of a (sense) RACE adapter oligonucleotide to the 5′ (cap) side of mRNA molecules prior to first-strand (antisense) cDNA synthesis. As a result, the first-strand cDNA will be terminated by the complement of this RACE adapter. This allows a subsequent PCR with a universal (sense) RACE primer targeting the adapter's complement and a user-designed (antisense) primer complementing the original transcript's sequence.

7.5 BIOINFORMATICS

7.5.1 NEXT-GENERATION SEQUENCING (NGS) DATA ASSEMBLY AND CURATION

The large volumes of data generated by NGS pose unique challenges in assembly and curation that must be addressed prior to the data being used in downstream applications such as search-based MS sequencing, phylogenetic analyses, or molecular modeling. Bioinformatics has therefore become a crucial part of not just analysis of NGS data but also the early steps of data generation and processing. Typically, this happens by passing the millions of reads through a "pipeline," a series of curation and filtering steps to eventually retain only transcripts of interest. In the context of venom toxin identification, such pipeline may include: (1) cleaning of sequence reads; (2) contig assembly; (3) prediction of open-reading frames; (4) identification of (candidate) toxin precursor transcripts; and (5) prediction of the mature (cleaved) toxins and posttranslational modifications.

The first step of a pipeline, cleaning of the reads, involves removing any barcodes used during multiplexed sequence runs and trimming adapter sequences. It is also desirable to remove low-quality reads. Functions for these tasks are often included in bioinformatics software suites such as CLC Genomic Workbench (CLC Bio) or NGS platform software packages, or they may be conducted by running open-source stand-alone programs such as Cutadapt (Martin 2011) or Trimmomatic (Lohse et al. 2012), depending on the NGS platform used.

Second, because read lengths achieved by NGS are generally relatively short, these need to be stitched together into contigs that ideally represent full-length transcripts. Since the vast majority of venomous organisms lack a reference genome sequence, this assembly process is almost invariably *de novo*. However, *de novo* assembly of reads from toxin-encoding transcripts is complicated by the presence of large numbers of isoforms (closely related paralogues or splice variants), which may differ by only a few nucleotides. Due to this high degree of sequence similarity, reads that originated from different transcripts may be erroneously assembled together, resulting in chimeric contigs. As a result, contig variation, apart from reflecting true molecular diversity, may to some extent be biased by errors in sequencing or assembly. In addition, the extremely variable size and expression level of reptile toxins means that there is great heterogeneity in precursor size and the number of reads that will be mapped to each precursor.

Assembly algorithms operate via two general approaches: overlap-based and de-Bruijn-graph-based. Modern short-read assembly programs, such as Trinity (Grabherr et al. 2011), SOAPdenovo-Trans (http://soap.genomics.org.cn/SOAPdenovo-Trans.html), or Oases (Schulz et al. 2012), are based on de Bruijn graphs, where cDNA reads are split into shorter sequences of fixed length k (k-mers), which are used as overlapping sequence tags in order to assemble contiguous reads (Miller, Koren, and Sutton 2010; Compeau, Pevzner, and Tesler 2011). Although this method considerably reduces the redundancy present in short-read data sets and speeds up the reconstruction of the original transcriptome sequences, it also has limitations when working with highly repetitive sequences and can lead to the generation of chimera sequences. The latter is also an issue for overlap-based assemblers, such as Mira (Chevreux et al. 2004) or Newbler (Margulies et al. 2005), which are generally used for platforms such as Roche 454 that yield longer but more error-prone reads (Nagarajan and Pop 2013) and where there is a tradeoff between high and low stringency, the former yielding shorter contigs and the latter more chimera sequences.

Third, after obtaining cDNA sequences of suitable length and quality, *in silico* translation of coding DNA sequences (CDS) into amino acid sequences may yield full toxin precursor sequences, containing an N-terminal signal peptide, propiece (spacer) regions, and mature toxins. Rather than translating each contig along its full length to all six reading frames, sets of open reading frames (ORF) are predicted. This extracts and translates interstop regions (coding sequences in between inferred stop codons) in all defined reading frames that are longer than a specified threshold and (usually) contain a start codon (ATG; encoding methionine).

The fourth step involves screening of the translated ORF for candidate toxin precursors. Depending on the taxon under study, and the available side-information at hand, there are several possible analyses, either in series or in parallel, to filter out candidate toxin proteins. These can be generally divided in analyses that identify ORF with typical toxin-like features (see further), and analyses that involve comparison of the set of ORF to genome, transcriptome, or proteome data. Such data may be obtained in parallel to the venom gland transcriptome study (e.g. a proteome study of the same gland), or may be available through online sequence databases (e.g. NCBI or Uniprot).

Conducting a BLAST search with the set of ORF as query sequences against the protein databases of Uniprot or NCBI is one of the foremost most straightforward and most frequently applied first steps to identify venom gland transcripts that encode proteins homologous to previously described toxins or toxin precursors. Obviously, however, the efficiency of this approach depends on the availability of structurally closely related toxins or toxin precursor sequences in public databases.

Alternatively (preferably additionally), proteomic data obtained via MS-based sequencing (see section 7.3.3.2) can be used to identify toxin-encoding sequences in a transcriptome library. There are several pieces of search software available for this purpose, some of the more popular being Mascot (Perkins et al. 1999); X! Tandem (Craig and Beavis 2004); Spectrum Mill (Agilent Technologies, Santa Clara, CA); and Protein Pilot (AB Sciex, Framingham, MA). While employing slightly different algorithms, the underlying principle of these programs is generally the same (Eng et al. 2011). The search-based MS sequencing strategy relies on an experimentally (transcriptome) or informatically (downloaded from sites such as UniProt or NCBI) acquired sequence database to produce theoretical peptide sequences, based on the experimental conditions (such as enzyme or fragmentation method). These are then searched against the MS-derived peptide-fragment spectra. Such a database is also commonly supplied with nontarget proteins (such as reversed amino acid sequences) that are used as a measure of false positives. Given the enormous number of theoretical peptides that can be generated from just a single transcriptome and the computational expense of searching all of these, candidate peptides are generally selected by being within a set range (peptide mass tolerance) of the experimental precursor ion mass (Shilov et al. 2007). The sequences of these are then matched against the fragment spectra and ranked according to a score calculated by the search software (such as least probable random match), and the peptide sequence with the best score is deemed the most probable candidate sequence. A second algorithm of the search program then groups the identified peptides into proteins based on the parent protein in the sequence base and again ranks these based on scores calculated from variables including number of unique peptides and sequence coverage.

While search-based MS sequencing utilizing high-quality data is a very powerful tool, there are some inherent limitations. First, only proteins present in the supplied sequence database will be identified. This means that the approach is sensitive to sources of biological variation such as sequence variation between transcriptome source and venom source and differences between protein content in venom and expression profile of the venom gland at the time of RNA extraction. Second, the sensitivity of NGS technologies is currently much greater than that of MS sequencing technologies. While it is easy, or even desirable, to enrich nucleic acid molecules by PCR during the preparation of cDNA libraries, enrichment of proteins is dependent on the amount of starting material available and on the organism studied (such as the venom yield). As a consequence, low-abundance proteins are not always detectable by MS compared to cDNA reads, which will exist even for components present at monomolecular level.

7.5.2 PHYLOGENETICS

Like any gene-encoded inheritable trait, toxin proteins are continuously subjected to various evolutionary forces (like DNA mutations in their genes), and will change over the course of evolutionary time. These changes are caused my mutation of their underlying gene sequences. Hence, genes that originate from a common ancestral gene (through speciation or duplication) diverge over the course of time. The resulting sequence variation is inevitably translated into the proteins they encode, and reptile toxins, being gene-encoded peptides and proteins, are no exception. Apart from explaining difference in bioactivity (allowing comparative structure-function analyses), sequence variation between multiple homologous toxins can be used to reconstruct their evolutionary relationships. Such analyses require that their amino-acid sequences (or the DNA sequences of their underlying genes) are compared in an evolutionarily meaningful way, exposing all shared (pleisotypic) and newly derived (apotypic) characters.

Therefore, sequences need to be arranged alongside each other in order to optimize homology, with individual characters (amino-acids or nucleotides) aligned as having a common evolutionary origin. Construction of a multiple sequence alignment (MSA) of nucleotide/amino acid sequences is almost invariously the first step in molecular evolutionary studies (see figures 8.2, 9.2, 11.1, 14.1, 15.2, 18.1, 19.1, 20.2, 24.1-24.7). For smaller data sets (around 100 sequences) of short-to-medium-length sequences (around 500 nucleotides) that share at least 70% sequence identity, the state-of-the-art algorithms and applications for constructing MSA can create accurate alignments in a matter of minutes. However, aligning larger data sets of longer and more diverse sequences can be computationally expensive.

Depending on the research question, molecular evolutionary studies of venom are typically conducted to infer relationships among homologous toxins of one or several venomous taxa, or between toxin proteins and non-toxic members of the same protein (super)family. In either case, it is likely that homologous sequences will have to be retrieved from public databases. An efficient way to do so is via BLAST using the set of toxins of interest (e.g. those newly recovered from a transcriptome study) as query.

The compiled sequences are then aligned using applications such as Clustal (Thompson et al. 1997), MUSCLE (Edgar 2004), or MAFFT (Katoh and Standley 2013). Although the resulting MSA often seem fairly accurate, manual editing may be required to rectify obvious alignment errors. This is particularly true if the aligned sequences show major length differences, or contain multiple tandem-duplicated domains. Aligning protein-encoding DNA sequences, which evolve in triplets or codons, may introduce non-threefold insertions and deletions that would imply highly unlikely frame-shift errors upon translation. Moreover, amino acid sequences typically are more evolutionarily conserved than their underlying nucleotide sequences, and may thereby produce a more accurate alignment. If phylogenetic analyses are to be conducted on nucleotide sequences, it may be sensible to use an amino-acid alignment as a reference for guiding a subsequent nucleotide alignment.

Phylogenetic analyses have elucidated the molecular evolutionary histories of various toxin protein-encoding genes (Fry et al. 2003b; Fry and Wüster 2004; Fry 2005; Fry et al. 2006; Fry et al. 2008; Fry et al. 2009a; Fry, Roelants, and Norman 2009; Fry et al. 2009b; Fry et al. 2010a; Fry et al. 2010b; Fry et al. 2012b; Fry et al. 2013) (see figures 8.1, 11.2, 15.1, 17.1, 17.2, 17.3, 19.2, 20.1, and 23.4). Several methods of tree reconstruction are available. Distance-based methods, like Unweighted Pair Group Method with Arithmetic Mean (UPGMA) and neighbor-joining, involve the computation of pairwise genetic distances between all sequences of the MSA, and using those to construct a tree via a hierarchical clustering algorithm. Distance-based methods represent a phonetic approach; they only take into account overall sequence (dis)similarity, and do not consider the evolutionary trajectories of individual characters. As a typical result, a paraphyletic grade of plesiotypic sequences, by sharing more similarity to each other than to more apotypic sequences, may be erroneously recovered as a monophyletic group. In organismal phylogenetics, distance-based methods have therefore been largely been superseded by cladistic approaches that do take individual character evolution into account. However, neighbor joining has remained a frequently used method in the construction of gene or protein phylogenies, because of its computational ease and its application in a wide range of software packages.

Cladistic approaches include both nonparametric (maximum parsimony tree searching) and parametric methods (maximum likelihood tree searching and Bayesian inference). The maximum-parsimony (MP) method searches the tree (or set of trees) that implies the least number of evolutionary steps (DNA or amino-acid substitutions) to explain the data. Since this method aims to minimize the number of inferred mutations, MP is prone to overlooking recurrent substitutions at a single site, and is considered more sensitive to errors caused by convergent evolution or reversals.

Instead, parametric methods implement substitution models to reconstruct the evolution of DNA or protein sequences and may consider character state frequencies, relative substitution rates, and heterogeneity of substitution rates across the sequences. While nucleotide-based substitution models (considering four character states) allow estimation of optimal substitution parameters during the phylogenetic analysis, this is computationally unfeasible for amino-acid models (considering 20 character states). As a result, parametric analyses of protein sequences require empirical substitution models, incorporating generalized amino acid substitution rate matrices inferred from large protein data sets (e.g., Dayhoff, JTT, WAG or LG). Selection of an appropriate model for the data set at hand, whether composed of nucleotides or amino-acids is essential and can typically be done via a series of likelihood-ratio tests, or information criteria as implemented in applications like jModeltest (Darriba et al. 2012) and ModelGenerator (Keane et al. 2006).

Two commonly used model-based methods in molecular phylogenetics are maximum likelihood tree searching and Bayesian phylogeny inference. Maximum-likelihood (ML) methods implement a heuristic search for the tree that maximizes the probability of observing the data (the sequences) given an evolutionary model (the substitution model and rate parameters). For years, the application of ML in phylogenetics was largely limited by the computational expensiveness of model-based tree searching, especially for larger data sets. The advent of highly efficient tree searching algorithms, as implemented in the applications PhyML (Guindon et al. 2010) and RAxML (Stamatakis 2014) and the availability of online portals like CIPRES Science Gateway (http://www.phylo.org/sub_sections/portal/) now allows the analysis of extremely large (up to several thousands of sequences) data sets. In addition, the original objective of ML analyses has shifted from reconstructing a single most optimal tree (most likely to contain errors) to estimating the credibility of individual clades of sequences, often assessed by inferring clade support values via nonparametric bootstrapping. A typical bootstrap analysis involves the generation of 500–1000 data sets by resampling the characters of the original data set and analyzing those data sets using ML. Clades recovered by over 75% of those replicate analyses are generally considered to be well supported.

Bayesian methods of phylogeny inference estimate the posterior probability of trees given the data (the sequences), an evolutionary model (the substitution model and rate parameters) and some user-defined prior probabilities (priors). Although this objective differs in principle of that of ML, both methods often converge on similar phylogenetic trees. Bayesian methods make use of Markov chain Monte Carlo (MCMC) algorithms to approximate the posterior probability of trees and model parameters. In practice, this process represents one, or multiple parallel chains of millions of calculations (generations), during which trees and parameter values are continuously changed and sampled (saved to an output file) in function of their prior probability and their likelihood given the data. The posterior probability of a tree is determined by the frequency by which it is sampled through the MCMC process and can be inferred by analyzing the output file. In addition, calculation of a majority-rule consensus tree of the sampled trees gives the posterior probabilities of individual sequence clades. Typically, clades represented by over 95% of the sampled trees are considered to be well supported.

MrBayes (Ronquist et al. 2012) is the most commonly used application for Bayesian inference of phylogenetic trees. Rather than specifying a single model of amino acid substitution, MrBayes allows the use of a mixed prior for the amino-acid substitution model (set by the command "prset aamodelpr = mixed"), which enables the MCMC chain to switch among nine different amino acid substitution matrices and utilize them in function of their posterior probability. A single analysis by MrBayes typically implements four parallel MCMC chains. These chains should be run long enough to allow adequate sampling of the posterior tree set after passing through a "burn-in" phase (an initial period during which suboptimal parameters and trees are sampled). The required duration is largely dependant on the size and nature of the data set and is difficult to predict beforehand, but in many cases, a chain length of 1×10^7 generations (with a burnin of 2×10^6 generations and a sampling frequency of 100–1000 generations) will yield a good approximation of clade-specific posterior probabilities. In addition, it is advised to run a Bayesian analysis at least twice to verify if both runs converge on similar posterior sets of parameters and trees. Determination of an appropriate burnin phase, and assessment of adequate sampling and chain convergence is best assessed using the program Tracer (http://tree.bio.ed.ac.uk/software/tracer/). Subsequently, posterior probabilities of clades can be obtained by compiling the sampled trees of both runs and calculating their majority-rule consensus tree.

7.5.3 RATE OF EVOLUTION

Like any molecular adaptive trait, toxins and their genes are subjected to both random (mutation) and deterministic (selective) evolutionary forces that may dramatically change over time and across lineages. A major objective of molecular evolutionary research is to understand how various regimes of evolution affect the fitness of an organism. Signatures of natural selection in genes and proteins in particular reflects an organism's adaptation to constantly changing ecological variables, such as diet, temperature, habitat, predatory pressures, and prey abundance. The simplest way of identifying the mode of selection using genetic sequence data is by calculating the "summary statistics" from the MSA of the gene in question. Ideally, the sequences should represent different individuals from a single population or multiple individuals from different populations or species. However, the sequences should belong to the same gene or homologous genes. By utilizing the given alignment, summary statistics summarize information regarding the relative frequency of polymorphic sites. These observed data are now compared with the results that would be expected under neutral evolutionary processes (null model). Any significant deviation of the observed data from the null model results in rejection of the null hypothesis, indicating the role of natural selection in shaping the evolution of the gene being evaluated. The standard null hypothesis makes several important assumptions: a constant effective population size, a constant mutation rate, absence of recombination, absence of migration, random mating, infinite sites (such as each mutation occurring at a different site or absence of reverse mutations), and that all mutations are either lethal or neutral in nature. These assumptions are often violated, and the rejection of the null model may also result from this.

Mutations that accumulate in nucleotide sequences that encode for proteins can be categorized as either synonymous (silent or dS) or nonsynonymous (replacement or dN) substitutions. The relative rate of accumulation of these mutations (denoted dN/dS or ω) reflects the nature of selection governing the evolution of the gene in question. The hallmark of neutral evolution is that the rate of replacement (dN) is exactly the same as the rate of silent mutations (dS), hence dN/dS (ω) being exactly 1 (Yang et al. 2000). Any significant statistical deviation from $\omega = 1$ denotes the influence of natural selection. If the accumulation of replacement changes is slower than the accumulation of silent substitutions, then dN/dS < 1, which indicates the influence of negative selection (Yang et al. 2000). In contrast, a dN/dS ratio of more than 1 indicates a greater accumulation of nonsynonymous than synonymous mutations, the phenomenon of positive Darwinian selection (Yang et al. 2000). It is important to remember that while the ω value is extremely informative in terms of indicating the nature of the selection regime in operation, it does not, however, equate to the absolute "strength" of selection.

In molecular evolutionary studies, the influence of natural selection on various toxin-encoding genes is often evaluated by employing maximum-likelihood models (Goldman and Yang 1994; Yang 1998) implemented in Codeml of the PAML package of programs (Yang 2007). Several site, branch, and branch-site models are available to assess the relative accumulation of dN and dS across sites, branches, and sites and branches (Yang and Nielsen 2002;

Zhang, Nielsen, and Yang 2005). Amino acid sites under positive selection are identified using the Bayes empirical Bayes (BEB) approach implemented in site model 8 (Yang, Wong, and Nielsen 2005), which calculates the posterior probability that a particular amino acid belongs to a given selection class (neutral, conserved, or highly variable). Sites with greater posterior probability (PP ≥ 95%) of belonging to the "ω > 1 class" are inferred to be positively selected. Additionally, there are several useful models in the HyPhy package (Pond, Frost, and Muse 2005), and the choice of model often depends on the question being addressed and the size of the data set (Kosakovsky Pond and Frost 2005): (i) single likelihood ancestor counting (SLAC) can be used for large data sets (40 or more sequences) (Kosakovsky Pond and Frost 2005); (ii) random effects likelihood (REL) can be used for data sets of intermediate size (20–40 sequences) (Kosakovsky Pond and Frost 2005); (iii) fixed effects likelihood (FEL) can be used for intermediate to large datasets (more than 50 sequences) (Kosakovsky Pond and Frost 2005); and (iv) FUBAR can be used for analyzing very large data sets (the authors tested this model on a data set of 3142 sequences, which is not feasible when using other methods; Murrell et al. 2013), where sequences evolve under the influence of pervasive diversifying and purifying selection pressures. When the forces of natural selection act in an episodic manner, identification of the precise regions that have undergone adaptive evolution is often difficult (Murrell et al. 2012). Moreover, when there is a strong influence of purifying selection on a majority of lineages in the phylogenetic tree, the signal of positive selection is often masked (Murrell et al. 2012). Hence, in order to address these shortcomings and to reliably identify sites that are influenced by both episodic and pervasive influence of positive selection at the level of an individual site, the mixed-effects model evolution (MEME) method was proposed (Murrell et al. 2012). This state-of-the-art method for identifying episodic influence of selection has become popular in recent times and has been utilized in the study of a number of toxin types (Sunagar et al. 2012; Brust et al. 2013; Fry et al. 2013; Jackson et al. 2013; Low et al. 2013; Sunagar et al. 2013a; Sunagar et al. 2013b; Sunagar et al. 2013c; Sunagar et al. 2014) (see color plate 26).

7.5.4 HOMOLOGY/COMPARATIVE MODELING OF VENOM PROTEINS

The sequences of venom proteins can vary in length from a few dozen to hundreds of amino acid residues. A protein's sequence may provide clues to its activity. However, the three-dimensional structure (fold) of a protein is what ultimately determines its function. It is possible for a non-conserved sequence to have a conserved fold. Proteins with a conserved fold are termed structural homologs and are commonly observed in the toxin components of venoms. Structural homologs often share key residues, usually cysteine residues that form disulfide bridges with one another, which are responsible for stabilizing the protein's architecture. The overall protein architecture is usually critical for activity, along with particular residues that are directly responsible for bioactivity. These residues are therefore often conserved, while less critical residues are subject to variation.

Determining the sequence of a protein is much faster than determining its three-dimensional structure, and this is why there are substantially more sequence data available than there are three-dimensional structures of proteins. However, with the use of homology modeling, which predicts the structure of a protein based on previously determined structures of proteins with similar sequences, the number of predicted three-dimensional structures is growing rapidly. This has proven to be particularly useful in the study of protein types that are extensively disulfide-bonded. Ideally, homology modeling would perfectly replicate an experimentally determined structure. Unfortunately, this is not the case, and the quality of the model of the unknown structure (target) corresponds to the degree of sequence similarity with the known structure (template) (Chothia and Lesk 1986). It is thus imperative that there be absolute conservation in cysteines involved in disulfide bonding and that the spacing between cysteines be highly conserved. In such cases, high-quality models may be generated (Fry et al. 2006; Fry et al. 2008; Fry et al. 2009a; Fry et al. 2009b; Fry et al. 2010b) (see color plates 8 and 27).

7.5.4.1 Template Selection

The first step in protein-structure modeling is to identify a suitable template structure for a target sequence with an unknown structure. The best results are obtained when the sequences of target and template are highly homologous. If homology is high, a simple BLAST (Altschul et al. 1990) or FASTA (Lipman and Pearson 1985) search usually yields at least one suitable candidate. If target-to-template homology is low, PSI-BLAST with its position-specific scoring matrix is better suited to finding templates than BLAST or FASTA (Altschul et al. 1997). Another option for identifying distantly related template structures is protein threading (Peng and Xu 2011). For the modeling of venom proteins, BLAST or PSI-BLAST queries against the Protein Data Bank (PDB) (Berman et al. 2000) usually provide a sufficient number of possible templates. Sometimes, however, the highest-scoring structure does not possess the required cysteine pattern and has to be discarded. If several similar template structures exist, which is often the case for smaller venom proteins, more than one template may be selected for the model-building process. The resulting model (see section 7.5.4.3) is then based on several templates. This can be advantageous in the case that loops have to be built and one of the templates may already contain a suitable scaffold. On the contrary, it may also be disadvantageous, since the resultant model is a hybrid of existing structures. Therefore, templates for venom protein modeling are selected for the following:

- Sequence similarity of target to template.
- Number of residues in aligned regions.
- Cysteine pattern.
- Functional residues (if function is preserved).
- Quality of template structure (such as RMSD, missing residues).
- Additional information from template file (such as mutations, co-crystallized ligands, or hybrid proteins).

7.5.4.2 Target-to-Template Alignment

Homology modeling can be used for targets that share at least 20% homology with the template. Central to this approach is the alignment of the target and template sequences. It is likely that structurally conserved regions (SCRs) occur where amino acids in the target sequence are aligned to residues in the template sequence, while structurally variable regions (SVRs) occur where target residues do not find a match in the template (insertions) or vice versa (deletions). The greater the percentage of SCRs shared between the target and the template, the higher the quality of the resultant model. Good-quality homology models possess greater than 70% sequence identity.

The core of the protein structure is more highly conserved between homologs than the exterior loops (Chothia and Lesk 1986). This is taken into consideration when creating a homology model, as mistakes in alignment at this stage are irreversible at later stages of the modeling process.

The target-to-template alignment of venom proteins is guided initially by the number and position of conserved cysteine residues. Subsequently, insertions and deletions are moved into surface loops, if applicable. Swiss PDB Viewer (SPDBV) is useful for aligning a target sequence to one or more templates (Guex and Peitsch 1997). The alignment window in this software is interactively linked to the graphical user interface (GUI), which facilitates the placement of insertions and deletions.

7.5.4.3 Model Generation

In a third step, the model is built. This can be done either locally or via a Web service. Two main approaches exist:

1. Modeling by satisfaction of spatial restraints.
2. Fragment assembly.

Modeling by satisfaction of spatial restraints applies an approach derived from NMR spectroscopy. Modeling by satisfaction of spatial restraints is a global approach in which the entire protein is modeled. MODELER server or software applies this method (Sali and Blundell 1993). The target-to-template alignment is used to define geometrical restraints that are subsequently converted into electron density functions. Electron density functions describe the space where protein heavy atoms (carbon, oxygen, nitrogen, sulfur, not hydrogen) are likely to be found. Into these electron density functions, the protein structure is modeled. The degrees of freedom for the model are significantly limited by the number of geometrical (spatial) restraints that are either derived from the structure of the template or provided by the researcher. Model building following this approach is an automated iterative process of building and refinement until the global (or best possible) energy minimum is reached.

An alternative to this global modeling approach is the fragment-assembly method (Greer 1981). Here a distinction is made between SCRs and SVRs in the first step of modeling. SCRs are first aligned, followed by modeling of SVRs. Several programs exist that implement this approach, such as Swiss-Model (Guex and Peitsch 1997; Arnold et al. 2006), 3D-JIGSAW (Bates et al. 2001), nest (Petrey et al. 2003), and Builder (Koehl and Delarue 1994; Koehl and Delarue 1995). Swiss-Model server together with the locally installed SPDBV software (Guex and Peitsch 1997) provides a powerful modeling suite for target to template alignment. The venom proteins generated are modeled using SPDBV and Swiss-Model or, alternatively, MODELER.

7.5.4.4 Loop Modeling

Modeling loop regions generally decreases the accuracy of a model. This is the main reason for the lower quality of models resulting from low homology between target and template. If loops consist of only a few residues, such as turns (one to five amino acids), the resulting loops of the model may still be accurate enough. However, accuracy decreases for longer unmatched regions. Loops of venom proteins are then modeled by software such as Swiss-Model (Guex and Peitsch 1997; Arnold et al. 2006) or MODELER (Sali and Blundell 1993). In rare cases, these loops have to be replaced manually by choosing a different loop template from a fragment library.

7.5.4.5 Side-Chain Refinement

Placing side chains correctly is of less consequence than template selection and target-to-template alignment, in which errors lead to poor or incorrect models. However, correctly placed side chains not only play a role in connecting sulfhydryl groups to disulfide bonds, but they also are essential for the arrangement or presentation of functional residues. The side chains of a template structure may arrange in suboptimal states during experimental structure determination, such as from crystal packing in the case of X-ray crystallography. These errors may be exacerbated in a model structure if the modeling software transfers side chains without optimizing placement. To prevent such suboptimal placement of side chains, libraries of common side-chain rotamers such as SCWRL may be used (Canutescu, Shelenkov, and Dunbrack 2003).

7.5.4.6 Model Assessment

The quality of every 3D model must be assessed. Both Swiss-Model and MODELER, but also other modeling tools, provide energetic evidence for the quality of the model. Swiss-Model implements atomic nonlocal environment assessment (ANOLEA), a method that calculates the mean force potentials of the nonlocal environment of all atoms in a protein structure or model (Melo and Feytmans 1997). The nonlocal environment contains all heavy atoms within a sphere of 7 Å. Heavy atoms of the same chain that are more than 11 amino acids away from the atom under consideration, along with heavy atoms from different chains, are used for calculating mean force potentials. Atoms that belong to closer sequence neighbors are considered local and thus are not used in calculations. Mean force potentials are pseudoenergy values that are calculated for all atom pairs in a nonlocal environment. They take into account the number of observations of each atom pair in a nonlocal environment and also the relative frequency of each atom pair with respect to all atom pairs. ANOLEA results can, among other things, be used to identify errors in the target-to-template alignment.

What-Check was originally programmed to assess experimentally determined protein structures, but it can also be used for models (Hooft et al. 1996). What-Check analyzes and calculates numerous parameters of a 3D structure, such as amino acid nomenclature, bond lengths and angles, bumps, amino acid packing, and many others. Although What-Check is overcritical when applied to models instead of experimental protein structures, the provided information is still valuable for the modeling scientist.

Verify3D (Bowie, Luthy, and Eisenberg 1991; Luthy, Bowie, and Eisenberg 1992) is a Web server that was originally designed to refine X-ray structures before submission to the PDB (Berman et al. 2000). Like What-Check, it can also be applied to model structures. Verify3D analyzes how well a 3D model represents its protein sequence (1D structure). Depending on its location and environment (such as secondary structure element, amino acid type), each residue is assigned a structural class. A score is calculated for every amino acid in this structural class, with a database of good-quality structures used as a reference. PROCHECK analyzes the stereochemistry and local geometry of amino acid residues in a protein structure (Laskowski et al. 1996). ProSA-web (Wiederstein and Sippl 2007) is an interactive Web server that identifies errors in 3D structures and models using the ProSA (Protein Structural Analysis) algorithm (Sippl 1993). ProSA also calculates a mean force potential using Cα-atoms as a base. This allows ProSA to be applied to low-resolution models where at least a Cα-trace exists.

7.6 BIOACTIVITY TESTING

7.6.1 NEUROMUSCULAR TARGETS

Neuromuscular transmission is a complex process that underlies locomotion and the mechanics of physiological respiration in vertebrates. Thus, it is not surprising that many venom components target a wide range of the receptors, channels, and enzymes critically involved in this process. In order to cause paralysis in the prey, toxins can, for example, act presynaptically or postsynaptically by interfering with voltage- or ligand-gated ion channels and enzymes such as acetylcholinesterase (see color plate 2). Because of the complexity of the process, different and complementary techniques must be used to fully explore the mechanisms of neurotoxic action. Isolated nerve-muscle preparations, in particular the chick biventer nerve-muscle preparation, are useful as initial screening assays for the examination of neurotoxic effects because of the simplicity of their preparation and the large amounts of information they can provide. More precise information regarding molecular targets can be gleaned from electrophysiological or high-throughput assays on cells expressing specific ion channels or receptors.

7.6.1.1 Organ Bath Assays

Organ baths have been employed by physiologists and pharmacologists for more than 100 years and are considered to be the classical preparation for the tissue-level assessment of agonist or antagonist activity. This technique, applied to nerve-muscle preparations, has played a pivotal role in studying the neuromuscular activity of animal venoms, particularly those from snakes. Tissues from rodents, chickens, and toads have been used for this purpose, with the chick biventer cervicis nerve-muscle preparation being formalized as the standard method of assessing neurotoxicity of snake venoms in 1994, following a WHO-commissioned study (Harvey et al. 1994). Such preparations are extremely useful for the initial characterization of crude venoms or isolated toxins and determining the pre- or postsynaptic nature of their paralytic activity (Hodgson and Wickramaratna 2002; Fry et al. 2003a; Ali et al. 2013b). The use of other, nonneuromuscular tissue preparations (such as ileum, vas deferens, and uterine horn) to study venom bioactivity is discussed in section 7.6.2).

7.6.1.1.1 Setting Up the Tissue in the Organ Bath

When using isolated tissues, it is important that the integrity of the preparation is maintained and that optimal physiological conditions are provided in order to maximize the longevity of the tissue and the repeatability of the experiment. An organ bath consists of a chamber of defined volume encased in a temperature-controlled water jacket. It contains a separate coil system for preheating the buffer in order to keep all solutions in contact with the tissue at appropriate physiological temperatures. The organ bath and its reservoir are filled with a physiological saline solution (such as Krebs', Tyrode's, Ringer's solutions) appropriate to the tissue being used. The organ bath is continuously bubbled with carbogen (95% O_2, 5% CO_2), which oxygenates the tissue, maintains buffering, and assists in mixing of test compounds.

Following dissection of the desired tissue for study, the appropriate section is suspended within the organ bath chamber. In the case of the chick biventer cervicis muscle, the entire muscle, with the intact nerve running along the tendon, is mounted in the bath. Tissues such as guinea pig ileum can be cut to a length suitable to the size of the organ bath being used. One end of the tissue is anchored by a thread to a tissue holder near the bottom of the chamber, and the other end is anchored by a thread to a calibrated isometric force transducer. The analog signal is digitized using an A/D converter (such as PowerLab), and the data are visualized, recorded, and analyzed using appropriate software (such as LabChart). The tissue is placed under a small amount of tension and equilibrated until the desired resting tension is stable. Subsequent contractions of the tissue can then be measured. In order to study neuromuscular activity, the nerve/tissue needs to be electrically stimulated. Thus, during the setup of nerve-muscle preparations, the nerve is placed in contact with an electrode that is connected to a stimulator capable of delivering a defined intensity and frequency of electrical pulses (such as a Grass S44 or S88 unit).

7.6.1.1.2 Testing of Venom or Other Compounds

Prior to the addition of a venom sample, the viability of the tissue should be established using an appropriate agonist and, if required, electrical stimulation. For example, in the chick biventer nerve-muscle preparation, a supramaximal stimulation voltage is determined that results in stable twitch responses. Subsequently, in the absence of electrical stimulation, acetylcholine is added to elicit a contraction. It is very important that if contracture is observed, sufficient time is allowed to determine if the effect has reached a plateau (usually within one minute). The agonist is removed from the tissue by draining and refilling the chamber with buffer a minimum of three times, thus "washing" the tissue. If done with care, this will preserve tissue integrity, which will be reflected in reproducible responses.

Venom or pure toxin is added to the organ bath chamber via a pipette, by positioning the pipette tip just below the fluid surface and slowly administering to the bath. Care must be taken to avoid disturbing the string connecting the tissue to the transducer. The bubbling of carbogen is sufficient to quickly and uniformly mix the venom and the physiological solution to assist in ensuring rapid equilibrium between the venom and receptor population being studied. The volume of venom or toxin added to the bath should be as small as possible, as changes in osmolality can affect tissue behavior. Thus, controls using the same total volume of vehicle in which the venom sample is diluted are used to identify artifacts, especially when performing cumulative concentration-response curves.

7.6.1.1.3 Chick Biventer Cervicis Organ Bath Assay

Since its introduction in 1960 (Ginsborg and Warriner 1960), the isolated chick biventer cervicis nerve-muscle preparation has become a valuable tool to study neurotransmission at the skeletal neuromuscular junction. It has been frequently used in toxinological research to test for neurotoxic activity of whole venoms or isolated venom components in vertebrates (see figure 7.5). The chick biventer cervicis muscle contains the fast-twitch, focally innervated alpha fast-twitch fibers and the two types of slow-tonic, multi-innervated beta fibers, with the fast-twitch alpha fibers making up the vast majority (70%–80%) (Toutant, Rouaud, and Douarin 1981).

Studies using snake venoms or toxins account for more than half of all publications using the chick biventer cervicis nerve-muscle assay. This can be partially explained by the large venom yields of snakes, which are therefore usually not a limiting factor for the assay (see list of limitations below). Another reason for snake venom researchers to choose this assay is the fact that snakes mainly target vertebrate prey, and the chick biventer nerve-muscle assay has advantages over other similar vertebrate assays when it comes to seeking ethical approval.

This assay has several advantages that account for its increasing popularity, as follows:

- Quick and easy dissection of the superficially located muscle and associated tendon which includes somatic nerve.
- Two muscles present per chick, enabling reduction in the number of animals used and allowing for "within animal" match control tissues.
- Pre- and postsynaptic effects can be discriminated.

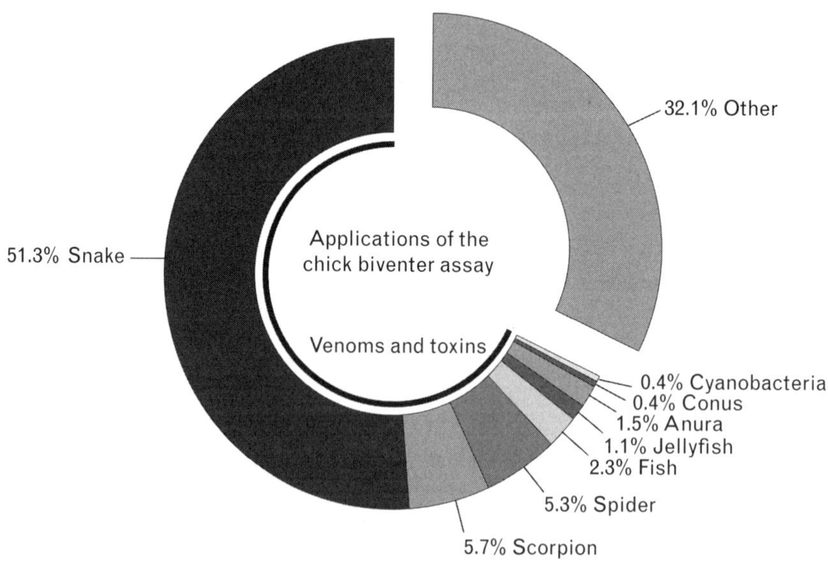

FIGURE 7.5: Approximately two-thirds of all chick biventer assays were used to study the effects of venoms or toxins from a range of sources, whereas most of the remaining studies aimed more generally at understanding muscle function or to evaluate the effect of small-molecule drugs on muscle function. Data are based on 265 publications resulting from entering the search term "chick biventer" in PubMed (www.ncbi.nlm.nih.gov/pubmed, data extraction on January 1, 2014.

- Both direct (direct depolarization of the muscle fibers, bypassing the neuromuscular junction) and indirect (via stimulation of the motor nerve, thereby incorporating neuromuscular transmission) stimulation of the muscle are possible.
- Provides information on changes in both baseline muscle tension and twitch height.
- Increase in baseline tension and inhibition is indicative of myotoxic activity (Kuruppu et al. 2007).
- Ethical benefit of using surplus male chickens from commercial chicken breeders.

The chick biventer cervicis nerve-muscle assay has been previously used for the characterization of a number of snake toxins (see Fry et al. 2003a; Wickramaratna et al. 2003; Lumsden et al. 2004a; Lumsden et al. 2004b; Wickramaratna et al. 2004; Lumsden et al. 2005; Blacklow, Escoubas, and Nicholson 2010; Blacklow et al. 2011; Marcon et al. 2012; Brust et al. 2013; Herzig et al. 2013; Marcon et al. 2013). The first step, after the muscle has been set up in the organ bath, is to confirm whether any observed muscle contraction is caused by electrical stimulation of the nerve or the muscle. To achieve this, a few minutes of electrically induced muscle twitches are recorded before adding d-tubocurarine (dTC), a competitive antagonist at postsynaptic nicotinic acetylcholine receptors (nAChR), to the bath solution. If dTC is able to completely abolish electrically induced twitches, it can be concluded that the twitches were mediated by stimulation of the nerve (i.e. indirect stimulation) and subsequent acetylcholine (ACh) release at the neuromuscular junction. In the majority of chick biventer assays, indirect stimulation is used.

After washing off the dTC and waiting until the twitches have returned to their initial levels, the next step is to measure the contraction response of the muscle to exogenous ACh, carbachol, and KCl in the absence of electrical stimulation. The muscle response to the same agents is again tested at the conclusion of the experiment (after testing the venoms or toxins), and based on the comparison of the responses before and after the experiment, a conclusion on the mode of action of the active toxin can be drawn. A toxin that acts postsynaptically on nAChR should result in a decreased muscle response to the nAChR agonists ACh and carbachol, whereas presynaptically acting toxins leave these responses unaltered. KCl induces membrane depolarization of the muscle that is independent of both pre- and postsynaptically acting toxins, and its response should be same before and after the experiment. Response to KCl is therefore an indicator of how "healthy" the muscle is at the conclusion of the experiment as myotoxins (in contast to toxins interfering with excitation-coupling) will cause a decrease in response to the application of KCl.

The myotoxic effects of venom or toxin can be examined using direct stimulation of the biventer cervicis muscle (see figure 7.6). Muscles are directly stimulated using electrodes and contractions elicited using higher voltages and a longer duration of stimulation (as compared to indirect stimulation assays). Muscle contractions caused by any nerve-evoked release of ACh are blocked by the addition of 10 μM d-tubocurarine. Venoms or toxins are considered to be myotoxic if they inhibit twitches elicited by direct stimulation and/or cause an increase in baseline muscle tension (Harvey et al. 1994). Furthermore, an absence of response to exogenous KCl following addition of the toxin will be observed. Therefore, a decreased response to all three exogenous agonists following addition of the toxin should be seen. An example of this is a study in which the venom of a *Naja naja* from the Sindh desert in Pakistan notably exhibited the classical signs of *in vitro* myotoxicity: a significant increase in baseline muscle tension in addition to a decreased response to exogenous KCl (Ali et al. 2013b) (see figure 7.6). It must

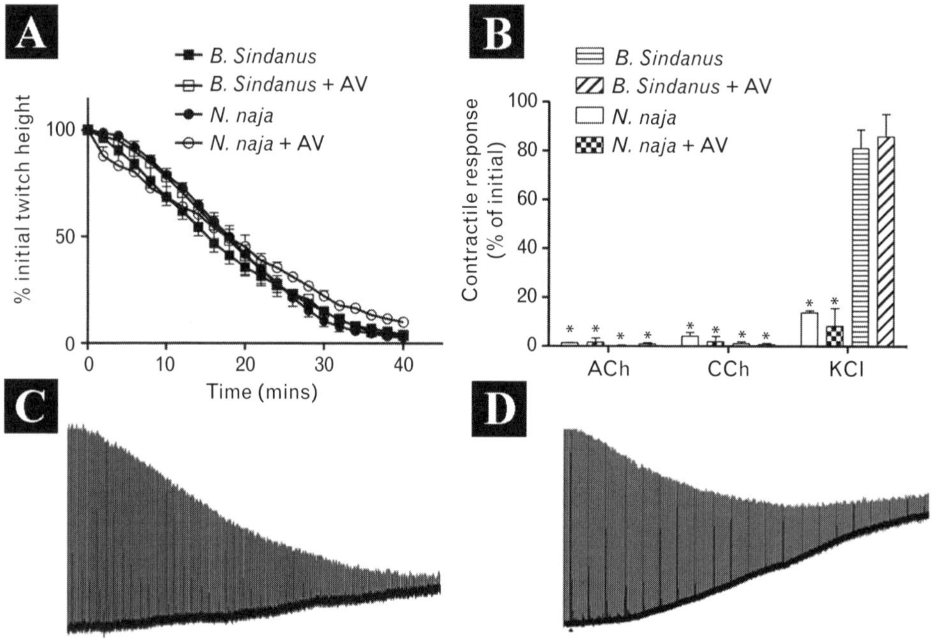

FIGURE 7.6: Effects of *Bungarus sindanus* and *Naja naja* crude venom (10 µg/mL) in the absence and presence of 10x dose of Bharat Polyvalent Antivenom (Ali et al. 2013b) on (A) nerve-mediated twitches of the chick biventer nerve-muscle preparation and (B) responses to exogenous ACh (1 mM), CCh (20 µM), and KCl (40 mM). * $P < 0.0001$, two-way ANOVA, compared to initial response, $n = 3$. Relative induced rise in baseline tension indicative of myotoxicity for (C) *Bungarus sindanus* and (D) *Naja naja* crude venom.

be noted that confirmation of such results requires histological examination of the tested tissue (see section 7.8.1).

Limitations of this assay are as follows:

- Taxon specificity—the relative potency of venoms/toxins may be different between avian and mammalian tissues (Pawlak et al. 2006), and thus the more difficult mammalian preparation must also be undertaken in order to investigate the functional evolution of neurotoxins in relation to the diet of venomous organisms.
- Determination of some molecular targets is not efficiently accomplished with the chick biventer cervicis nerve-muscle preparation; other techniques such as electrophysiological or neurochemical assays may be required.
- Not high-throughput, although multiple organ baths can be run in parallel if available.
- Organ bath volume could be a limiting factor for venoms or toxins that are only available in small amounts.

7.6.1.2 High-Throughput Methods

In recent years, high-throughput methods traditionally reserved for the screening of extensive small-molecule-compound libraries have been adapted for pharmacological profiling and activity-guided isolation of toxins and venom peptides across a range of ion channels and receptors. The advantage of such approaches over more traditional *in vivo* or *ex vivo* approaches lies in the ability to rapidly screen the specific action of crude venoms, partially purified venoms or isolated venom components, on pharmacological targets of interest.

High-throughput fluorescence assays are particularly well suited for the measurement of ion-channel activity based on the availability of a range of fluorescent dyes able to detect intracellular ions such as calcium (Ca^{2+}) (such as Fluo-4, Fura-2, Calcium Green), sodium (Na^+) (such as SBFI, CoroNa Green, Sodium Green), potassium (K^+) (such as PBFI), and chloride (Cl^-) (such as SPQ) (Johnson and Spence 2010). In addition, fluorescence detection of membrane potential or voltage-sensitive dyes can be used to monitor changes in membrane potential. These latter assays are particularly useful for ion channels that are impermeable to Ca^{2+}, as the relatively small-concentration gradient of ions such as Na^+ or K^+ across the cell membrane often leads to a small signal when using fluorescent Na^+ or K^+ dyes, but changes in membrane potential are still detectable.

Because of the large Ca^{2+} gradient across cell membranes and the superior spectrophotometric properties of Ca^{2+} dyes (Vetter 2012), high-throughput Ca^{2+} assays are particularly useful for isolating, identifying, and characterizing venom components with activity at a range of voltage- and ligand-gated ion channels (such as TRP channels, nAChRs, or $Ca_v2.2$) and G protein-coupled receptors (Vetter and Lewis 2010; Vetter 2012; Sousa et al. 2013). Indeed, the utility of Ca^{2+} assays can even extend to ion channels not generally permeable to Ca^{2+}. Functional coupling of Na_v to Ca_v endogenously expressed in the human neuroblastoma cell line SH-SY5Y has been exploited to generate a $Na_v1.7$ assay in which membrane depolarization by the $Na_v1.7$-selective scorpion toxin OD1 elicits a downstream increase in intracellular Ca^{2+}. This assay, which has yet to be used in the investigation of reptile toxins, thus benefits from the superior signal-to-noise inherent to Ca^{2+} assays (Vetter et al. 2012).

Although a range of high-throughput platforms, including Cell Lux (Perkin Elmer) and FDSS 6000 (Hamamatsu), are available, the FLIPR^{TETRA} (Molecular Devices) platform is generally considered the industry standard high-throughput fluorescent plate reader. This platform is based on sophisticated integrated liquid handling and use of the EMCCD or ICCD cooled charge-coupled device (CCD) camera for fluorescence and/or luminescence detection. The camera can acquire kinetic data from 96, 384 or 1534 wells simultaneously (Vetter 2012).

Now in its fourth generation, the FLIPR^{TETRA} incorporates high-precision user-exchangeable pipettor heads that simultaneously dispense test compounds, agonists, or antagonists to all wells of the cell plate. This allows fluorescent readings to be taken both before and after compound addition and simultaneous detection of agonists or antagonists in a single experiment. Such functional information can provide important insights into the molecular identity of the primary pharmacological target of the venom component. In addition, high-throughput FLIPR assays often negate problems associated with conventional imaging that arise from uneven dye loading, extrusion, intracellular compartmentalization, or photobleaching, as loading and imaging conditions are constant across all wells. Thus, FLIPR assays benefit from relatively high assay sensitivity, accuracy, robustness, and reproducibility, defined by a high Z' score (Zhang, Chung, and Oldenburg 1999; Vetter 2012; Sousa et al. 2013). In addition, high-throughput fluorescence assays are inherently amenable using small amounts of material, which is of particular benefit for assay-guided isolation of components from venom samples that are only available in limited quantities.

Activity of crude venoms at specific pharmacological targets can be assessed with as little as 1–10 μg of venom per well in 384-well format. Typically, a two-addition protocol consisting of addition of crude venom, venom fractions, or toxins, followed by stimulation of the target of interest with a known pharmacological activator (see figure 7.7), is most informative and can aid in verifying or delineating toxin pharmacology. Specifically, as functional kinetic responses are measured, it is possible to characterize the effect of full or partial agonists and of competitive and noncompetitive antagonists on a range of targets. In addition, for venom components with agonistic activity, a second stimulation with a known pharmacological activator can identify heterologous desensitization, which can provide information on activation mechanisms. Unfortunately, interpretation of these responses can be difficult, because plate-based fluorescence assays operate via successive additions of compounds, without the opportunity to remove or wash off previous stimulants.

Many venoms contain salts, small molecules, or neurotransmitters which can elicit nonspecific effects or false positive responses or otherwise obscure activity at the target of interest. Thus, for crude venoms that elicit addition responses, separation of venom components using RP-HPLC is generally needed before detailed assessment of pharmacological activity is possible. Similarly, separation of venom components is also necessary for successive rounds of bioactivity screening and activity-guided purification. This generally requires removal of organic solvents used in purification steps such as methanol and acetonitrile, as most fluorescence

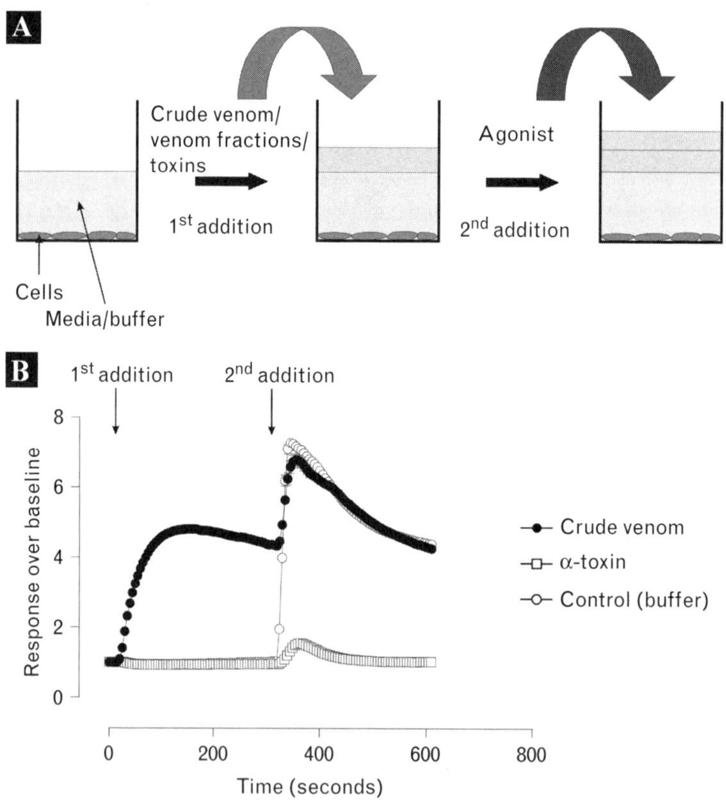

FIGURE 7.7: Basic FLIPR protocol. (A) Cells expressing the target of interest are plated on 96-, 384-, or 1536-well plates and loaded with fluorescent dye. Crude venoms, venom fractions, or toxins are added first, followed by stimulation with a suitable agonist. (B) Typical fluorescence responses of SH-SY5Y human neuroblastoma cells loaded with Calcium 4 No Wash dye containing PNU120596 (10 μM). Shown are responses to the α7 nAChR agonist choline (25 μM, 2nd addition) after addition of (1st addition) crude venom (10 μg/well), an α-toxin, and control (physiological salt buffer).

assays can only tolerate small amounts of mobile phase. The additional handling steps arising from lyophilization and subsequent reconstitution of venom fractions in suitable assay buffers eliminate some of the inherent advantages of plate-based assays, in particular the ability to screen very large numbers of venom components. An additional drawback is that assays which are designed to assess pharmacological activity at a specific molecular target, are inherently ill suited for the delineation of unknown activity or the identification of novel molecular targets.

To assess such activity, more traditional assays, including organ bath, nerve conduction, or *in vivo* studies, can provide immediate insight into the physiological consequences, and thus the likely mechanisms of action, caused by venoms or venom components. Nonetheless, FLIPR assays have the potential to dramatically increase the rate of discovery of novel bioactive venom components and can significantly accelerate the detailed pharmacological characterization of known toxins.

7.6.1.3 Electrophysiological Methods

The molecular target of a venom component may be analyzed using a variety of electrophysiological techniques. A classic one uses intracellular microelectrode recording of nerve-evoked end-plate potentials and spontaneous synaptic potentials. This technique has long been useful in identifying the postsynaptic actions of snake α-neurotoxins on muscle and ganglionic nAChRs and the effects of presynaptic PLA_2 neurotoxins (see chapters 20 and 21) on neurotransmitter release (Datyner and Gage 1973; Wilson et al. 1995a; Hodgson, Dal Belo, and Rowan 2007). However, more recent electrophysiological studies have used novel approaches to determine the target and mode of action of snake toxins on ionic currents mediated via a variety of voltage-gated (such as cyclic-nucleotide gated, K_V, K_{Ca}, and Ca_V channels) and ligand-gated ion channels (acid-sensing ion channel and various nAChR subtypes) (Wu, Wang, and Shi 2002; Takacs, Wilhelmsen, and Sorota 2004; Wang et al. 2005b; Zhang et al. 2009; Bohlen et al. 2011; Diochot et al. 2012; Peigneur et al. 2012).

7.6.1.3.1 Voltage-Clamp Method

The voltage-clamp method can measure the ionic current flowing through the ion channels spanning the cell membrane. Voltage clamping can be achieved with either a single electrode (patch clamp) or two microelectrodes (two-electrode voltage clamp, TEVC). In these experiments, the membrane potential is fixed ("clamped") at a predetermined value by injecting current into the cell through the electrodes. The amount of current required to maintain the membrane potential is equivalent to the current passing through the ion channels and is recorded with high temporal resolution, which allows for the separation of these currents from capacitive currents that activate when the membrane potential changes. The voltage-clamp technique is the leading method for the study of the voltage dependence of channel opening and closing ("gating") in response to changes in membrane potential, such as activation of Ca_V channels by depolarizing test potentials (Zhang et al. 2009), or the transient application of neurotransmitters, such as ACh (Takacs, Wilhelmsen, and Sorota 2004), and neuromodulators, such as H^+ for acid-sensing ion channels (Diochot et al. 2012). The voltage-clamp technique may also determine state-dependent or voltage-dependent binding of toxins that may lead to use-dependent effects as seen with other animal toxins (Yamaji et al. 2009).

7.6.1.3.2 Patch-Clamp Technique

The patch-clamp method for voltage clamping of excitable cells was developed and refined in the 1970s and '80s by Neher and Sakmann, earning them the 1991 Nobel Prize in Physiology or Medicine (Hamill et al. 1981). Its success lies in the use of a single blunt electrode (patch pipette) to first establish a high-resistance seal (> 10 GΩ) with the cell membrane and subsequently to measure membrane potential and inject current into the cell. This technique is now widely used to examine ionic currents flowing through single ion channels or the ensemble "whole-cell" current from all ion channels in small cells. The latter configuration is the most popular and the only one enabling measurement of action potentials using current-clamp protocols (Yamaji et al. 2009). However, various other "inside-out" or "outside-out" configurations allow the measurement of single-ion-channel gating and kinetics (see figure 7.8).

Patch pipettes are usually pulled from borosilicate glass capillary tubes (Rae and Levis 1992). This type of glass provides a good compromise between electrical noise level (related to its dielectric constant) and ease of pipette fabrication in relation to its low softening temperature. Thick-walled tubing is preferable for single-channel recordings, because it has a lower noise level, while thin-walled tubing is preferred for whole-cell recordings because of its lower access resistance. To minimize access resistance, the pipette is made as short as possible, with a tip of 1–2 μm after fire-polishing (see below). Sylgard silicone elastomer can also be used to coat the electrode in order to reduce the pipette capacitance, a key source of noise in single-channel recordings (Labarca, Lindstrom, and Montal 1984). The hydrophobic properties and high resistivity of Sylgard prevent the spread of solution up the outer surface of the electrode and create the same benefits as those of thicker-walled glass. Fire-polishing, using commercially available microforges, can also be used to smooth the surface of the tip and help burn off excess Sylgard or dust particles on the tip. For successful formation of a GΩ seal, filtering both internal and external solutions through a 0.22–0.45 μm filter can minimize the presence of particulates and other contaminants that may impede seal formation.

The most popular approach to isolating a specific voltage-gated ionic current is to employ one or more of the following procedures: (i) block other ion-channel currents using specific inorganic or organic channel blockers (such as tetrodotoxin to block Na_V channels), (ii) eliminate ions that carry unwanted channel currents by replacement with nonpermeant ions (such as replacement of Na^+ with choline to eliminate Na_V channel currents), (iii) use of ions to buffer intracellular calcium (such as F^-), and/or (iv) use of voltage protocols that inactivate unwanted currents or only activate the current to be investigated (such as depolarizing prepulses to inactivate "A-type" K_V channel currents). Occasionally, the channel type under investigation cannot be recorded in isolation without contamination by the current of another channel type because of the absence of selective blockers. In these situations, an alternative "current-subtraction" approach must be employed, in which macroscopic currents are first recorded followed by the selective blocking of the target channel. The residual current recordings are then digitally removed offline from the macroscopic current recordings to isolate the target channel currents (Gunning et al. 2008).

The composition of the internal and external solutions must be carefully considered, especially in whole-cell and outside-out patch configurations. This is because the internal pipette solution dialyzes with the intracellular compartment of the cell, and intracellular components can diffuse from the cell into the patch pipette. This can quickly lead to the phenomenon of current "rundown" caused by dephosphorylation of the channel as the result of dilution of a

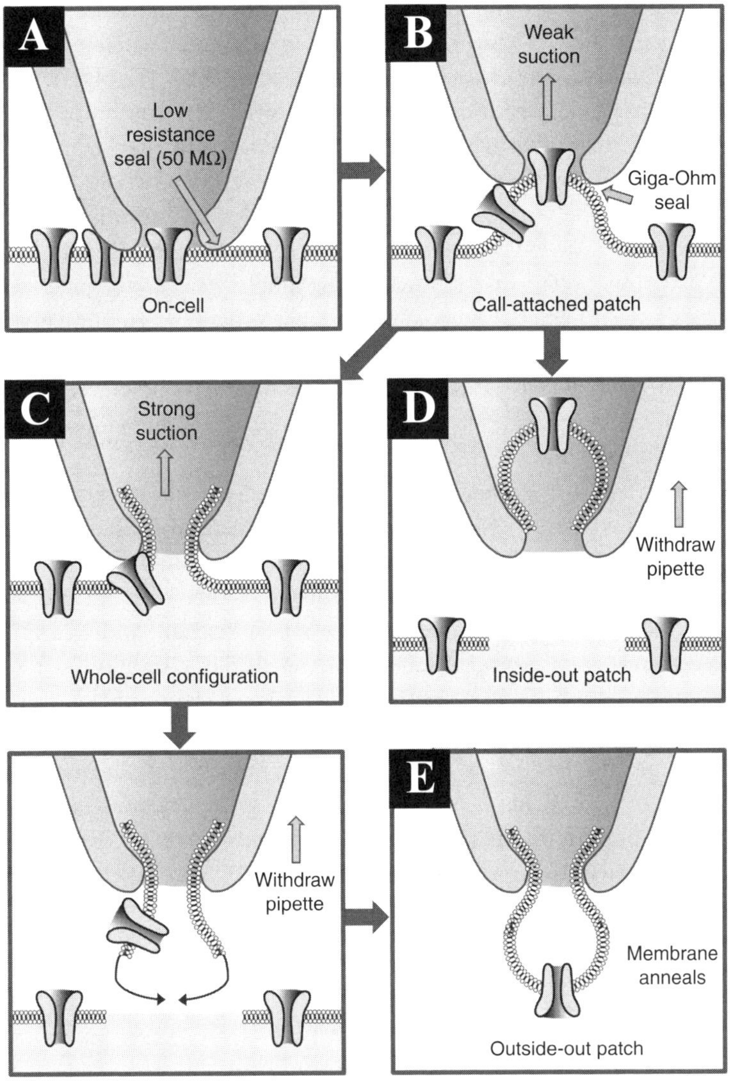

FIGURE 7.8: Schematic representation of the procedures that lead to the four patch-clamp recording configurations on isolated single cells. (A) "On-cell" configuration with patch pipette in simple mechanical contact with the cell. (B) Application of slight negative pressure to the inside of the recording pipette leads to an increase in seal resistance to form the giga-Ohm seal, creating the "cell-attached" patch configuration. The cell-attached configuration allows measurement from single ion channels located within the area of the tip of the recording pipette. (C) Application of stronger negative pressure or a short-duration, high-voltage pulse to the pipette ruptures the patch of membrane inside the tip of the electrode and creates the "whole-cell" configuration. In this configuration, the inside of the patch pipette is in direct contact with the intracellular compartment of the cell. (D) The "inside-out" patch is formed by quickly withdrawing the patch pipette from the cell after creating the cell-attached patch. The patch of membrane can be torn from the cell while maintaining a giga-Ohm seal. (E) The "outside-out" patch configuration is formed by very slowly withdrawing the pipette from the cell in the whole-cell configuration. A small patch of membrane separates from the cell after the tip has been pulled away from the cell by 50–200 μm, which then seals, as depicted, and forms a patch on the tip of the pipette. In all four cases, a giga-Ohm seal is formed between the patch pipette and the cell membrane.

secondary messenger. Current rundown can be slowed by use of higher-resistance pipettes (the narrower-diameter tip slows the rate of diffusion) or even reversed by adding ATP-Mg^{2+} or cAMP to the internal pipette solution (McDonald et al. 1994). The pipette solution should also be isosmotic with the external bath solution to prevent the membrane from bursting or deforming because of differences in osmotic pressure. Some researchers make the pipette solution hyposmotic (~10%), claiming that it improves the stability of whole-cell recordings. The liquid junction potential between the internal pipette and external bath solutions also needs to be taken into consideration. This arises because of the varying mobility of ions in solution. This potential is typically offset on the patch-clamp amplifier when the pipette is in the bath solution but will disappear when the whole-cell or outside-out configuration is achieved as the pipette solution equilibrates with the intracellular environment. This must be compensated for if an accurate voltage clamp is to be achieved. Fortunately, the liquid junction potential is easily determined for internal pipette and external bath solution combinations using the JPCalc software package (Barry 1994).

Many types of cells are amenable to patch clamping. These include primary cells obtained by tissue dissociation following enzyme treatment and used either immediately or maintained in primary cell culture (such as DRG, trigeminal, spinal or hippocampal neurons, chromaffin cells, and skeletal muscle fibers). Other cells include immortal cell lines expressing the desired ion channel (such as SH-SY5Y, N18, and A7R5). The primary issue associated with these nontransfected cells is the need to eliminate currents carried by voltage-gated ion channels not under investigation. This may be a difficult proposition given the numerous ion-channel subtypes, overlapping selectivity of ion-channel blockers, and similar activation potentials. A more appealing option is to express the ion-channel protein in cells deficient in other ion channels by transiently or stably transfecting the DNA encoding the ion channel protein into mammalian somatic cells (such as HEK293, CHO-K1, CV 1, COS-1, or COS-7 cells). In transient transfection, the DNA is taken up by the cell using calcium-phosphate-mediated, electroporation, or liposome-mediated transfection. The DNA will enter the nucleus and persist episomally for a few days. Stably transfected cells, in which the DNA encoding the ion channel is covalently integrated into the DNA of the cell, causing the channel to be permanently expressed, are ideal for longer-term, more in-depth studies of a single ion-channel target.

Caveats for patch-clamp experiments are as follows:

- The whole-cell technique is limited, in most cases, to cells that are small, round, clean of debris, and lacking branched processes. This is because of the necessity of control voltage over the entire cell surface using a single electrode (space-clamp problems lead to inadequate voltage clamp).
- Large cells, or cells with high channel densities, can generate large currents in the whole-cell configuration that can cause series resistance errors and saturate patch-clamp amplifiers (> 10–20 nA), necessitating a reduction in the driving force of the charge carrier by adjusting the concentration in either the external or internal solution.
- Institutional Animal Ethics Committee approval is required for primary-cell tissue harvesting.
- Acutely dissociated cells are likely to have similar levels of expression of ion channels and receptors to those present *in vivo*. However, enzyme treatments involved in the

dissociation can affect ion-channel function (for example, pronase, papain, and trypsin all remove Na_V channel inactivation if they access the intracellular compartment).

- It can be difficult to identify neuronal cell types once tissues have been dissociated.
- Transient transfection often leads to low levels of expression in only a small percentage of the cells (< 1%–50%) and for only a short time. This necessitates co-transfection of the channel with a reporter gene (such as green fluorescent protein, GFP) to assist in identifying cells that have undergone successful transfection.
- The creation of a stably transfected cell line is time-consuming and labor-intensive, as cells have to be grown for several passages in selective medium, subcloned, screened for expression levels, and then maintained using standard tissue-culture techniques. However, some stably transfected cell lines are commercially available, albeit at considerable cost and often with commercial IP limitations.
- Immortal cell lines or stably transfected cells maintained in culture for long periods of time are likely to have altered levels of expression or express nonnative or mutant channels and receptors, particularly after many passages.

7.6.1.3.3 Two-Electrode Voltage-Clamp (TEVC) Technique with *Xenopus* Oocytes

TEVC is an alternative to the patch-clamp technique for studying effects of venoms toxins on voltage- and ligand-gated ion channels. In the TEVC variant of the voltage-clamp technique, two electrodes impale a cell, with one intracellular microelectrode measuring the cell membrane potential (voltage electrode) and the second injecting current (current electrode). The separate current electrode allows for greater control of the membrane potential than patch clamping, especially if low-resistance blunt electrodes are used. Given the need to impale the cell with two electrodes, this method is mainly reserved for investigations using very large cells such as *Xenopus* oocytes (1–1.3 mm in diameter).

Xenopus oocytes, stage V or VI egg progenitor cells of the South African clawed frog *Xenopus laevis*, are commonly employed for heterologous expression and investigation of an extensive variety of receptors and ion channels. The first TEVC experiments to use *Xenopus* oocytes to investigate receptors were carried out in 1982 using mRNA encoding for nACh receptors (Miledi, Parker, and Sumikawa 1982). Prior to injection, oocytes are removed from adult female frogs and defolliculated by incubation with collagenase. Oocytes are then injected with either cDNA into the nucleus or complementary RNA (cRNA) into the cytoplasm, usually with a nano-injector (Goldin 1992; Bossi, Fabbrini, and Ceriotti 2007). Subsequently, they are incubated to allow transient widespread expression of the exogenous ion channel under relatively simple conditions. This typically occurs within one or two days and can last up to several weeks. Levels of expression can easily be adjusted by varying the amount of RNA or DNA injected.

One advantage of using *Xenopus* oocytes is that many ion channels and receptors are difficult to express in mammalian cells resulting in low (or no) expression. This occurs partly because the channels can desensitize, endocytose, or simply cause cell death from overactivation by components within the culture media or sera (Boeckman and Aizenman 1996). The use of *Xenopus* oocytes as a transient expression system can minimize this problem, as the oocytes achieve high transfection levels following expression in a simple inorganic buffer (ND96 solution).

Additionally, oocytes are particularly useful when co-expression of several ion-channel subunits is required. This is significantly more difficult to achieve in mammalian cells. Therefore, oocytes are particularly well suited for TEVC because of their large diameter, relatively facile expression of ion channels and receptors, and relatively few endogenously expressed channels that might interfere with TEVC recordings (apart from calcium-activated chloride channels; Bowie and Smart 1993). *Xenopus* oocytes are also particularly hardy cells and can withstand impalement by comparatively blunt, low-resistance electrodes for injecting current. This enables large currents (> 1 μA) to be faithfully recorded without loss of voltage-clamp integrity.

In the context of studying the neuromuscular activity of reptile venoms, TEVC of *Xenopus* oocytes has been particularly useful in studying the potency, subtype selectivity, and structure-activity relationships of snake α-neurotoxins at nACh receptors. The system allows the isolated expression of a series of ion-channel or receptor subtypes or site-directed mutants on the proteins and a simple method of measuring the activity of toxins at these receptors. This can be applied to the in-depth characterization of known toxins as illustrated by early work in this field determining the subtype-dependent activity of α-bungarotoxin (Couturier et al. 1990; Luetje et al. 1990).

More recently, the utility of the TEVC oocyte assay as a discovery platform for novel toxins was highlighted by the identification of the analgesic ASIC1 antagonists, Mambalgin 1 and 2, from the venom of *Dendroaspis polylepis polylepis*. The activity and responsible molecule were identified, isolated, and characterized using TEVC of *Xenopus* oocytes expressing acid-sensing ion channel 1a and other subtypes (Diochot et al. 2012). A subsequent study that reported the synthesis and 3D structure of Mambalgin 2 also employed TEVC of *Xenopus* oocytes expressing a site-directed mutant of ASIC1a (F350A) to identify the binding site of this novel family of three-finger toxins (Schroeder et al. 2014).

Caveats for *Xenopus* oocytes experiments are as follows:

- Institutional Animal Ethics Committee approval is required for long-term maintenance of and surgery on *Xenopus* frogs.
- The large size of the *Xenopus* oocyte requires constant perfusion, with requiring larger volumes of toxins during application. This necessitates larger supplies of toxins than patch-clamp experiments.
- Unlike mammalian cells, practically every oocyte that is injected will express the exogenous channel (Grinevich et al. 2005). However, *Xenopus* oocytes are prone to exhibiting batch-to-batch variations in viability and expression levels often related to seasonal variations (Stühmer and Parekh 1995). As with all transient transfection techniques, it is a relatively robust approach but is highly dependent on the condition of the cells and quality of the RNA/DNA.
- There are altered biophysical characteristics and ligand-binding affinities for some channels in *Xenopus* oocytes compared with mammalian cells because of differences in the cell membrane, different posttranslational modifications (Stühmer and Parekh 1995), or ion-channel subunit stoichiometry, such as differing α4 and β2 nAChR subunit ratios between *Xenopus* and mammalian cells (Nelson et al. 2003).
- Lipophilic compounds can bind nonspecifically to the yolk and vitelline membrane of the oocyte. This can lead to a rightward shift in the concentration-response curve and widely disparate IC_{50} values between oocytes expressing a mammalian ion channel and patched mammalian cells (Kiehn et al. 1996).

- Endogenous calcium-dependent chloride channels in *Xenopus* oocytes can cause problems in the study of current-voltage relationships or ionic selectivity of other ion channels (Dascal 1987).

In summary, the above electrophysiological methods, particularly patch clamping, can provide extremely detailed information concerning the effects of toxins on ion-channel gating and kinetics, pharmacology, and desensitization, all with ultrahigh (μs, pA) resolution. Accordingly, manual patch clamping of single cells represents the gold standard in electrophysiology. However, these methods are very time-consuming and require a high level of user skill and knowledge to yield good-quality data. Typically, only a handful of ligands can be studied on one cell and only one ion channel subtype at a time, with substantial user input. Automated TEVC and patch-clamp systems have gone some way to alleviate some of these concerns; however, both platforms are viewed as less desirable than high-throughput indirect methodologies such as fluorescence-based cellular assays (such as FLIPR) and cellular efflux assays (such as Rb^+) when a large number of compounds, such as large chemical libraries, need to be screened (for a review, see Gonzalez et al. 1999). They are thus considered better suited as methods for providing detailed functional information at the "hit" validation stage of drug screening, for screening of more focused compound libraries, for drug safety screening against voltage-gated ion channels (such as hERG channels), or when targets are not compatible with fluorescence-based or efflux assays.

The development of the planar-array patch-clamp electrophysiology technique overcame many of the drawbacks inherent in manual or automated patch clamping (Fertig, Blick, and Behrends 2002). Since then, a range of dedicated medium- to high-throughput commercial platforms have been developed, including CytoPatch (Cytocentrics), IonFlux HT (Fluxion), IonWorks HT/Quattro and Barracuda (Molecular Devices), PatchXpress (Molecular Devices), PatchLiner and SynchroPatch 96/384PE (Nanion Technologies), and Qpatch HT/HTX (Sophion Bioscience) (for a review, see Farre and Fertig 2012). The details vary in each system, but essentially planar-array patch clamping employs a single-use 16- to 384-well disposable plate system. A vacuum is used to place cells on top of micron-sized holes at the base of each well of a multiwell planar patch plate made of silicon rubber, glass, plastic, or silicon oxide/nitride substrate. In some cases, the population patch-clamp method is also used (Finkel et al. 2006). In this approach, recording wells are perforated with multiple holes such that recordings are made from 64 cells at a time within each well. This reduces the intercell current variability observed with single-hole-per-well formats. Some platforms then apply further suction to rupture the membrane patch covering the aperture and create a high-resistance seal (up to 1 GΩ) between the cell and the edges of the holes. With other platforms, a solution containing a pore-forming agent (such as amphotericin-B) is perfused under the hole, allowing access to the intracellular compartment. This "perforated patch" voltage-clamp configuration is similar to the standard whole-cell configuration using glass pipettes but prevents many of the intracellular components controlling channel modulation from dialyzing out of the cell.

Rather than manipulating a glass pipette onto a single cell, cells are pulled onto the holes in each plate to form seals. As a result, the whole method of patch clamping is far simpler and massively more parallel, decreasing the amount of time and the skills required to perform the assay. At this stage, however, automated planar patch-clamp platforms are typically expensive devices mainly available only to pharmaceutical companies or high-throughput screening laboratories.

7.6.2 SMOOTH-MUSCLE ORGAN BATH PREPARATIONS

In contrast to the chick biventer assay, which is used to assess the neurotoxic effects of venoms, the smooth-muscle configuration of the organ bath allows for the study of direct action of venom components on smooth muscle itself (Ruder et al. 2013a). Most of the targets of interest are G-protein coupled receptors (GPCR), and their activation by venom may lead to tissue contraction or relaxation within the organ bath, quantitated as a change in smooth-muscle tension. The concentration-dependent relationship between venom and tissue behavior provides evidence of some specificity of action and a potential means to begin the process of understanding receptor affinity. As in the *in vivo* condition, excessive concentrations of any compound may lead to collateral nonspecific effects mediated by other receptors and even possible perturbation of membrane phospholipids, the effects of which may be harder to reverse once the test compound is removed.

The advantages of using this technique on smooth-muscle tissue with an accessible lumen, such as ileum, uterus, and vas deferens, are as follows:

- Depending on the preparation, several tissue pieces (4–10) can be obtained from a single animal.
- Potencies (such as EC_{50} values) and effects of inhibitors (IC_{50} values) can be determined on the whole-tissue level by measuring the contraction or relaxation of the tissue.
- Functional GPCRs such as muscarinic and adrenergic receptor subtypes are present in ileal tissue, as are histaminergic receptors within guinea pig ileum and oxytocin receptors within rat uterus.
- Tissues are generally very hardy in the organ baths and can last more than six hours, allowing for the effective generation of behaviors across a range of concentrations if desired.
- If compounds to be tested are available in short supply, the volume of the organ bath can be adjusted, or smaller organ baths can be designed or obtained.

While there are numerous advantages to this technique, there are disadvantages, some of which have already been mentioned. These include the following:

- If response is a result of effects on GPCRs, they obviously must be present in the chosen tissue for the assay to work. For many tissues, especially nonmammalian ones, specific receptor populations present may not be known.
- Receptor populations may not be expressed uniformly in a smooth-muscle preparation.
- There may be damage to tissue during the dissection and hanging process, which may affect results. Generally, normalization of data removes this factor.
- Receptor populations may, in some cases, vary in size as a function of developmental age.
- The size of tissue used may require a significant volume of buffer, which may also require large amounts of venom to reach a desired concentration. If the venom is hard to obtain or in scarce supply, careful thought should be given to minimizing venom use.

Regardless, measurement of whole-venom effects on smooth-muscle contractile behavior in tissue organ baths can serve as a powerful screening tool to identify potential mechanisms of action by which the venom exerts its effects on prey species. An example of this is the characterization of natriuretic peptides via their action on aortic smooth muscle, followed by screening on cells overexpressing the receptor of interest (Schweitz, Bidard, and Lazdunski 1990; Li et al. 2001; Joseph et al. 2004; Lumsden et al. 2004b; Fry et al. 2005; Lumsden et al. 2005; Chaisakul et al. 2012; Chaisakul et al. 2013).

7.6.3 TESTING FOR OTHER PATHOLOGICAL EFFECTS

The majority of viperid snake venoms, and some elapid and colubrid snake venoms, inflict pathological alterations at the locale of venom injection—so-called local effects. The most significant of these are hemorrhage, myonecrosis, dermonecrosis, and edema. Hence their study in experimental models is necessary to understand the mechanism of action of locally acting toxins and to assess the efficacy of neutralizing agents such as antivenoms and other natural and synthetic inhibitors. Experimental models have been developed, primarily in rodents. A concern in this field is the pain and discomfort that these animals suffer in these assays. Following the general ethical framework of "3 Rs" (replacement, reduction, and refinement of laboratory methods performed in or on animals), attempts have been made to find surrogate *in vitro* assays for these tests. In addition, the use of precautionary (prophylactic) analgesia is recommended when studying these effects.

7.6.3.1 Hemorrhage

Disruption of the integrity of microvessels, mainly capillaries, is one of the main actions of snake venom metalloproteases (SVMP) (see chapter 23). Some SVMP degrade components of the basement membrane that surrounds endothelial cells and provides mechanical support to microvessels (such as type IV collagen, nidogen, laminin, and the proteoglycan perlecan). Hydrolysis of these components therefore results in the weakening of the mechanical stability of these vessels. As a consequence, the hemodynamic biophysical forces that normally operate in circulation, such as hydrostatic pressure and shear stress, distend and eventually disrupt the integrity of endothelial cells, with consequent extravasation (Gutiérrez et al. 2005; Escalante et al. 2011). Venom-induced hemorrhage can also occur systemically, contributing to hypovolemia and cardiovascular shock. Enzymes that affect hemostasis contribute to bleeding by provoking incoagulability, thrombocytopenia, and platelet hypoaggregation (Gutiérrez et al. 2005).

Hemorrhagic activity of venoms and toxins can be readily studied by a simple method originally described in rabbits (Kondo et al. 1960) and later on adapted for use in rats (Theakston and Reid 1983) and mice (Gutiérrez et al. 1985). It is based on the intradermal injection of solutions of venoms/toxins (diluted in physiological saline) into experimental animals, most often rats or mice. After a time interval, usually two to three hours in mice and 24 hours in rats or rabbits, animals are sacrificed by an overdose of anesthetic or by CO_2 inhalation, the skin is removed, and the diameter of the hemorrhagic halo in the inner side of the skin is measured as a quantitative index of hemorrhagic activity. The minimum hemorrhagic dose (MHD) corresponds to the venom or toxin dose that induces a 10-mm-diameter hemorrhagic spot (Theakston and Reid 1983).

The advantage of this procedure lies in its simplicity and ease of performance. It has the limitation that the intensity of the hemorrhagic lesion is not evaluated, only its extension. To circumvent this problem, the hemorrhagic skin can be dissected out, placed in a tube containing 1 mL of distilled water, and left at 2° C to 8° C for 24 hours. Subsequently, the tubes are centrifuged and the supernatant collected. The absorbance of the supernatant at 540 nm is a quantitative indication of the amount of hemoglobin present in the tissue and, hence, of the intensity of the hemorrhagic lesion.

7.6.3.2 Effects on Hemostasis

Most reptile venoms affect hemostasis and do so in a variety of ways. Some of them contain coagulant and procoagulant toxins which induce blood clotting *in vitro* and provoke defibrinogenation and incoagulability *in vivo*. In addition, many venoms affect circulating platelet numbers and/or platelet function. Components responsible for this effect include disintegrins (see chapter 23), kallikrein (see chapter 14), lectin (see chapter 17) and SVMP (see chapter 23).

The coagulation system is a complicated interaction between three separate major components. The first component is the plasma proteins that constitute what is often referred to as the clotting system. The second is the platelets, small anuclear cells that when activated release a multitude of proteins into the immediate environment and through a complex system of surface receptors and ligands can facilitate many cellular and acellular interactions. The third component of the coagulation system is the blood-vessel walls, highly dynamic and interactive surfaces playing a critical role in the balance between clotting and bleeding. There are many other cell surfaces within the blood, including red-cell membranes and a variety of white cells that almost certainly play an additional role in this complex system.

Thus, when considering the interaction of complex biological toxins such as snake venoms with the coagulation system, one needs to consider the activity of the venom across all relevant components. Testing this activity is not easy. There is no validated mechanism to test the impact of any therapy or toxin on the blood-vessel wall, other than animal studies to document *in vivo* clinical effect. The impact of toxins on platelets can be measured by assessing change in platelet function, in terms of aggregation responses, although such tests are poorly reproducible and require dedicated skill and equipment. Flow cytometry is increasingly being used to document specific aspects of platelet activity (up-regulation or down-regulation of surface proteins) in response to variable agonists or antagonists. This again requires specific expertise and equipment, and one must ensure that adequate controls are in place and a relevant spread of surface proteins is examined.

In general, a range of *in vitro* assays are utilized to test the plasma clotting system, including the following:

- Clot-based: functional assays observing the time taken to clot formation after a specific stimulus. These assays may be general in that they assess the overall activity of the system, or they may be specific in that they are manipulated to identify the function of a single component protein within the system.
- Chromogenic: functional assays based on the ability of activated enzymes to cleave specific chromogenic substrates.
- Immunological: quantitative assays based on amount of a specific protein.

General assays of coagulation (such as activated clotting time [ACT], activated partial thromboplastin time [APTT], prothrombin time [PT], thromboelastography [TEG], and endogenous thrombin potential [ETP]) respond differently depending on the initiator used, conditions of the test, and detection methods. Specific functional assays of individual proteins are usually based on similar methodology to general assays but often involve manipulation of the test conditions. Thus, specific assays suffer from all the same technical flaws as the general assays. In summary, coagulation testing to determine the impact of venoms is fraught with difficulties, and careful planning and consultation with coagulation experts are required to understand the limitations of assays and ensure that interpretation of results is reasonable.

7.6.3.2.1 Coagulant Activity *in vitro*

Blood is collected from healthy human donors, with sodium citrate used as an anticoagulant. The plasma is separated and collected by centrifugation at 1000–1300 RCF in a swing-bucket centrifuge. The clotting time of plasma by varying concentrations of venom is quantified by adding 0.1 mL of varying venom concentration, diluted with physiological saline, to aliquots of 0.2 mL of plasma, previously incubated at 37° C. The minimum coagulant dose (MCD) corresponds to the dose of venom or toxin that induces clotting of plasma in 60 seconds (Theakston and Reid 1983).

7.6.3.2.2 Defibrinogenating Activity *in vivo*

Groups of mice or rats are intravenously injected with various doses of venoms or toxins, dissolved in physiological saline. Controls are injected with saline solution alone. One hour after injection, a blood sample is collected by cardiac puncture, and 0.2 mL of blood is placed in a dry glass tube and left undisturbed for 20 minutes. Then the tubes are gently tilted, and the formation of a clot is observed. In the case of blood samples from controls injected with saline solution alone, a consistent clot should be observed, whereas no clotting occurs in the blood of defibrinogenated animals. Defibrinogenating activity of venoms or toxins is expressed as the minimum defibrinogenating dose (MDD), defined as the minimum dose of venom or toxin in which all injected animals have blood incapable of clotting (Theakston and Reid 1983).

7.6.3.2.3 Thrombocytopenic Activity

Groups of animals (usually mice or rats) are injected intravenously with various doses of venoms or toxins that have been dissolved in physiological saline. Controls are injected with saline solution alone. One hour after injection, a blood sample is collected by cardiac puncture and immediately mixed with sodium citrate solution as anticoagulant. A platelet count is performed using an automated hematological analyzer. Thrombocytopenic activity is expressed as the median thrombocytopenic dose (TD_{50}), which is the dose of venom or toxin that reduces the number of circulating platelets by 50% as compared with control animals (Rucavado et al. 2005).

7.6.3.3 Myotoxicity

Local and systemic acute muscle damage, myonecrosis, is a common consequence of snakebite envenomings (Lomonte and Rangel 2012). The vast majority of myotoxins in snake venoms belong to the PLA_2 families (see chapters 20 and 21), although the β-defensin peptides are another major class (see chapter 9).

A widely used method to assess myotoxicity is based on quantification of the activity of enzymes, such as creatine kinase (CK), released from muscle fibers when cell damage occurs. There are diagnostic kits commercially available for the quantification of CK activity in plasma or serum (Gutiérrez, Arroyo, and Bolaños 1980). The method is based on the injection of solutions of venoms or toxins (diluted in physiological saline) intramuscularly in mice or rats. A typical muscle for injection is the gastrocnemius, although other muscles are also used, such as the soleus or the muscles of the thigh. After a time interval (usually several hours), a sample of blood is collected, either with anticoagulant or not. Then plasma or serum is separated by centrifugation, and the CK activity is quantified. Depending on the kind of myotoxic agent, a preliminary time-course survey for CK activity increase is useful to determine the optimal sampling time, generally between two and six hours. Control animals are injected with physiological saline solution alone. The minimum myotoxic dose (MMD) is the dose of venom or toxin that increases the plasma, or serum, CK activity four times as compared with CK activity of animals injected with saline solution alone.

Myotoxicity can also be studied by histological assessment of muscle damage. For this, animals are injected as described above, and 24 hours after injection, they are sacrificed by an overdose of anesthetic or by CO_2 inhalation. The injected muscle is dissected out and immediately placed in a fixative solution (for example, 10% formalin). It is recommended that the orientation of the muscle is maintained in such a way that transverse sections are obtained. After fixation, tissue is processed routinely for embedding in paraffin, and sections of 5–8 μm are cut, placed on a slide, and stained with hematoxylin and eosin for light microscopic observation. Necrotic cells are identified on the basis of hypercontraction of myofibrils. The total numbers of muscle cells and necrotic cells are counted, and myotoxic activity can be expressed as the Myonecrotic Index, the ratio of necrotic to total muscle cells (Teixeira et al. 2003).

7.6.3.4 Dermonecrosis

When injected intradermally in the skin of animals, snake venoms may induce necrosis of the skin (dermonecrosis). This is caused by the action of typical tissue-damaging toxins, such as SVMP (see chapter 23) and PLA_2 (see chapters 20 and 21). In the case of cytotoxic *Naja* species venoms, this effect is caused by "cardiotoxins," which are cytotoxins belonging to the three-finger family of toxins (3FTx) (see chapter 8). The method used to quantify dermonecrosis is basically the same used for the quantification of hemorrhage. Mice or rats are injected intradermally with venom or toxin solutions, and after a more prolonged time interval, usually 72 hours, animals are sacrificed, their skin removed, and the diameter of the necrotic lesion in the inner side of the skin is measured as a quantitative index of dermonecrosis (Theakston and Reid 1983). The minimum dermonecrotic dose corresponds to the dose of venom or toxin that induces a necrotic lesion of 5 mm diameter (Theakston and Reid 1983).

7.6.3.5 Edema

One of the most characteristic effects of snake venoms, particularly of viperid snake venoms, is an increase in vascular permeability leading to accumulation of fluid in the interstitial space, resulting in edema. This effect is induced by diverse venom components, such as SVMP (see chapter 23), PLA_2 (see chapters 20 and 21), kallikrein (see chapter 14), VEGF (see chapter 24.21), and others. The action of these venom components in the tissues results in the synthesis and/or release of a large variety of inflammatory mediators, such as histamine, eicosanoids, bradykinin, complement anaphylatoxins, nitric oxide, and cytokines.

Edema is often studied using the rodent footpad model, mainly in rats or mice. Solutions of venoms or toxins (dissolved in physiological saline solution) are injected subcutaneously into the posterior footpad of animals. Controls are injected with saline solution alone. At various time intervals after injection, the edema can be evaluated by measuring the thickness of the footpad using a low-pressure spring caliper (Lomonte, Tarkowski, and Hanson 1993) or, alternatively, by measuring the volume of the footpad by plethysmography (Chacur et al. 2001). An alternative procedure to assess the increase in vascular permeability is based on use of the dye Evans blue, which binds to plasma albumin. In this protocol, animals are injected with venoms or toxins either in the footpad or intradermally in the abdominal region. After a time interval, a solution of Evans blue is administered intravenously and, after a period of time, animals are sacrificed and the injected tissue (footpad or abdominal skin) dissected out and placed in a solution of formamide at 4° C, protected from light. Afterward, the supernatant of this preparation is collected, and the absorbance is recorded at 600 nm as a quantitative index of extravasation of albumin, such as of plasma.

7.6.3.6 Cytotoxicity

Several kinds of animal venom components have the ability to induce a rapid cytotoxic effect on a variety of cell types in culture, such as cytotoxic 3FTx (see chapter 8) and myotoxic PLA_2 (see chapters 20 and 21). Their rapid cytotoxic effect is related to their ability to disrupt the integrity of the plasma membrane, therefore causing cytolysis, which can be measured by a variety of assays. This effect can be more pronounced on certain types of cells, and therefore, selection of an optimal cell line to be used as a target is relevant. A particularly useful cell line for the study of venom cytolysins is the C2C12 myogenic line (ATCC CRL-1772; Yaffe and Saxel 1977). These murine myoblasts are readily maintained in culture, grow rapidly, and have the ability to fuse into multinucleated, elongated myotubes. Assays for the cytolytic activity of toxins can be used in many cases as a replacement for *in vivo* experiments (for example, myotoxicity) and are therefore attractive from the perspective of ethics. Experimental conditions can also be varied and controlled in more detail than in an animal. Nevertheless, caution should be taken when extrapolating the results to an *in vivo* setting.

Cytotoxicity assays tend to be more useful when studying purified toxins than for crude venoms, since the complex composition of the latter may complicate interpretation of the results. For example, venoms that contain a mixture of a cytolytic agent and a proteolytic enzyme will simultaneously induce cell death and cell detachment, the latter not necessarily being an indicator of cell damage. Depending on the principle on which the final measurement of cell death is based, detachment can introduce errors.

Among the alternatives for measuring rapid cytolysis, the release of lactate dehydrogenase (LDH; EC.1.1.1.27) from the cytosol represents a simple, sensitive, and reproducible marker.

Cells, at near confluence in 96-well plates, are exposed to the toxic agent for a given time period, and then an aliquot of the medium (40–60 mL) is assayed for LDH activity, using any commercially available reagents, preferably based on a UV kinetic principle. Essential reference points for interpreting the results include cells exposed only to the culture medium (0% cytolysis) and cells exposed to a detergent such as Triton X-100 (0.1% final concentration) in order to solubilize all intracellular contents (100% cytolysis). Test samples can be easily interpolated between these two controls to obtain a % cytolysis (or, if preferred, % survival) value. Depending on the type of toxic agent, preliminary time-course experiments can be used to select an optimal sampling time, for example, three hours. An important consideration concerns the presence of low amounts of LDH in the serum supplements routinely used in cell-culture work; in order to minimize this factor, improve sensitivity, and avoid its "masking" effect, the assays can be performed in the presence of a minimal amount of serum proteins, for example, 1% fetal calf serum (Lomonte et al. 1999).

7.6.3.7 Lethality

Venom lethality is typically assessed using an *in vivo* model. These studies involve calculating the lethal dose 50 (LD_{50}), the amount of venom required to kill 50% of the animals used for testing. This assay therefore produces data that can be used for comparison among species to assess the relative potencies (within a given model organism and via a specific route of administration) of the venoms of interest. The majority of snake venom LD_{50} studies are undertaken using a mouse model (Theakston and Reid 1983; Casewell et al. 2010), although nonmodel organisms have also been used to assess the relative potency of venom to different prey types (Richards, Barlow, and Wüster 2012). The murine assay involves injecting varying doses of venom (in 100 μl of phosphate-buffered saline) intravenously into the tail vein of laboratory mice. At least five mice should be used in each experimental group, and a minimum of four groups (such as different venom doses) should be used to ensure robust data analysis. As an alternative to the intravenous route, venom can also be delivered intraperitoneally, although different injection routes result in incomparable LD_{50} data because of differences in venom pharmacokinetics. Following injection, mice are observed for a period of 24 hours, the number of deaths observed in each group are recorded, and the LD_{50} and 95% confidence intervals are calculated using probit analysis (Finney 1971). The advantage of such lethality studies is that they provide an overview of the action of all venom components in an *in vivo* system. The murine assay is validated by the European Pharmacopoeia and, in combination with effective dose 50 (ED_{50}) assays that determine the venom-neutralizing potential of an antivenom, is a prerequisite for the preclinical assessment of antivenom efficacy. However, since LD_{50} studies simply measure the lethality of venom, external pathology is often absent, and therefore, postmortem studies are typically required to provide an indication of the cause of death.

7.7 ANTIVENOMICS

More than a century after the development of the first *serum antivenimeux* by Calmette (Calmette 1894b; Calmette 1894a) and Phisalix and Bertrand (Phisalix and Bertrand 1894b; Phisalix and Bertrand 1894a) (see chapter 3), the timely parenteral administration of appropriate antivenom remains the only effective treatment for snakebite envenomings. Antivenoms are generated by the immunization of large animals, usually horses, with venoms of one or several snake species, thus generating monospecific or polyspecific antivenoms, respectively (Gutiérrez, Leon, and Burnouf 2011). Although there are many important considerations in producing a polyvalent antivenom, a key technical issue in the generation of novel antivenoms is the design of optimized immunization venom mixtures that ensure that the resulting antidote will be effective against the highest number of venoms from snakes of medical concern across the geographical range in which it will be used. The design of venom mixtures should be based on a rigorous analysis of epidemiological, clinical, proteomic, immunological, and toxicological information. Such methodologies include biochemical and proteomic analysis of venoms, pharmacological profiling of venom activities, and analysis of immunological cross-reactivity of antivenoms toward homologous and heterologous venoms.

Antivenomics is a proteomics-based protocol (Lomonte et al. 2008) developed to complement *in vitro* and *in vivo* preclinical tests (Williams et al. 2011; Gutiérrez et al. 2013) in the qualitative and quantitative characterization of the immunological profile and extent of cross-reactivity of antivenoms against homologous and heterologous venoms. The combination of antivenomics and neutralization assays provides a powerful toolbox for analyzing the preclinical efficacy of antivenoms at the molecular level. First-generation antivenomics (Lomonte et al. 2008) consisted of the physical depletion of antivenom-binding toxins by incubating the whole venom with antivenom. This was followed by (i) precipitation of antigen-antibody complexes (Ag-Ab) out of solution via the addition of a secondary antibody (such as rabbit antihorse IgG) or (ii) depletion of Ag-Ab from the reaction mixture using IgG-binding protein A coupled to Sepharose beads. Antigen-antibody (Ag-Ab) complexes contain the toxins against which antibodies present in the antivenom are directed. The fraction of nonimmunodepleted molecules (% NR) can be estimated as the relative ratio of the chromatographic areas of the toxin in the supernatant and in the whole venom (WV): % NR = (NR/WV) x 100. Based on the extent of immunoprecipitation, toxins were classified as (i) nonimmunoprecipitated (proteins present in the supernatant that failed to raise antibodies in the antivenom or that triggered the production of very low-affinity antibodies), (ii) toxins exhibiting variable degree of immunoprecipitation (proteins that generated low- to high-affinity antibodies), and (iii) toxins quantitatively immunodepleted from the venom (highly antigenic proteins bearing antigenic determinants for very high-affinity antivenom antibodies). These components can be easily identified by comparison of an RP-HPLC separation of the nonprecipitated fraction (supernatant of immunoprecipitation) with that of the whole venom (the toxin composition of which had been previously characterized by a venomics approach) (Calvete, Juarez, and Sanz et al. 2007; Calvete 2011b; Calvete et al. 2011).

The combination of immunoaffinity chromatography and proteomic analysis of non-immunocaptured (flow-through fraction) and immunocaptured fractions is the basis of the "second-generation antivenomics" protocol (Williams et al. 2011; Pla, Gutiérrez, and Calvete 2012). Second-generation antivenomics was designed to overcome some shortcomings of the in-solution immunodepletion protocol. In particular, because of the absence in the market of molecules able to specifically and efficiently immunoprecipitate $F(ab')_2$ and Fab antivenoms produced by many manufacturers, first-generation antivenomics was not useful to investigate the immunological profile of these antibody fragments. In addition, the immunoprecipitated fraction contains predominantly IgG molecules from primary and secondary antibodies and is thus not suitable for RP-HPLC. Consequently, first-generation antivenomics provided only indirect evidence of the immunoreactivity of IgG antivenom; antivenom immunoreactivity was inferred indirectly through proteomic characterization of the toxin fraction remaining in solution after immunoprecipitation. In the immunoaffinity-based second-generation method, the fraction of nonimmunodepleted molecules (% NR) can be determined as 100 – ([R/(R + NR)] x 100), where R corresponds to the area of the same protein in the chromatogram of the fraction eluted from the affinity column. The inclusion of R in the calculation compensates for possible losses during sample handling and chromatographic analysis. In addition, experiments in which venom is incubated with mock matrix and matrix-coupled preimmune antibodies are run in parallel to the immunoaffinity antivenomics analysis as matrix and immunospecificity controls, respectively.

Antivenomics is translational venomics. Immunoaffinity-based antivenomics provides qualitative and quantitative information on both the set of toxins bearing antivenom-recognized epitopes and those exhibiting poor or no immunoreactivity. The superior chromatographic performance results, particularly evident in the case of low-abundance peaks, become masked in the more noisy chromatograms of supernatants of immunoprecipitation assays. Similarly, in the high-percentage acetonitrile region of the elution gradient of the RP-HPLC run some venom proteins (mainly SVMP) and IgG molecules (such as primary and secondary antibodies used for immunoprecipitation) coelute, causing additional interference. A further advantage of second-generation antivenomics is the reusability of affinity columns, contributing to the economy and the reproducibility of the method.

Its ease of use, reproducibility, sensitivity, low cost, quantitative character, and molecular resolution make second-generation antivenomics a viable option to supersede the commonly employed immunoassays and Western blots for assessing the immunoreactivity of antivenom antibodies. The immunochemical detection of electroblotted proteins provides a yes/no response: a given protein is either recognized or not by the antivenom, and it is essentially a nonquantitative technique. On the other hand, quantification of antibody immunoreactivity by ELISA-type methods is compromised if the immobilized fractions do not contain pure proteins.

7.8 VENOM SYSTEM MORPHOLOGICAL CHARACTERIZATION

Research on the evolution of toxicoferan reptiles, or any other venomous group, would not be complete without an examination of the venom production and delivery system (Fry et al. 2006; Fry et al. 2008; Fry et al. 2009b; Fry et al. 2010b). Such studies are undertaken not only on the venom-producing glands and associated compressor muscles (if present) but also on the wound-creating dentition.

7.8.1 HISTOLOGY

The principles of histology are (i) the chemical treatment of tissues with fixatives, (ii) a method of tissue processing and sectioning, and (iii) the usage of histochemical dyes in solution that react specifically with targeted cell structures. Histological stainings are complex, in that each method and chemical stain targets the morphological and chemical characterization of specific cell structures. In each situation, both natural and synthetic dye stains may be results of (a) chemical reactions, (b) physical adsorptions or absorption, and (c) other physicochemical processes.

The initial stage of the histological process is fixation, the critical period in which autolysis (self-digestion) is arrested. Autolysis causes cell proteins to divide into amino acids, which can then diffuse out of the cells and thus break down the tissues. For the purposes of histology, fixation will modify a tissue in a way that is insoluble during further treatment and shields the tissue from shrinkage or distortion as it is subjected to processing chemicals and conditions, such as ethanol, xylene, heat, and paraffin wax. An additional aim for the process of fixation is to improve the conditions for histochemical staining of specified features of the tissue. The introduction of an incorrect chemical at this point may cause breakdown of the target tissue.

Upon the completion of the fixation process, excess fixative solution is removed by running the sample under water for short periods of time. Some investigators maintain that tissues processed in Bouin's fixative will lose soluble picrates if washed in water; therefore, they suggest placing the tissues in 50% ethanol for a short period of time before transferring to 70% ethanol. Samples may be stored in 70% ethanol for extended periods, but in general, the preference is always to dehydrate and embed tissues in wax as rapidly as possible.

At a primary stage in the processing, tissues must be dehydrated for histological preparation. During sectioning, if moist cells are penetrated with a blade, they can collapse or easily deform. In order to avoid this, intracellular fluids are replaced by a medium that hardens to form an easily sectioned material. Paraffin waxes are a generally suitable medium that acts to structurally support tissues for sampling. This medium may be easily cut to thicknesses of 1–15 μm with either glass or metal blades.

Before tissue samples are mounted in paraffin wax, the tissue must first be dehydrated via a series of stepwise increases in alcohol concentration. Second, in a clearing action, traditionally xylene or a safer modern alternative, histolene, is used to remove the alcohol from the tissue, as it does not mix with paraffin wax. If at any time the clearer agent becomes clouded or turbid, this is evidence of water being present within the fluid. All samples should be returned to a new sample of ethanol in order to completely dehydrate the tissues before clearing using fresh clearing agent.

The final chemical stage of tissue processing involves immersing it in heated paraffin wax. Commonly, a second change of wax is required, and this is performed in a vacuum oven; the use of a vacuum during infiltration will remove chemical residues and air adjacent to or within tissues. The pressure will eliminate holes within the paraffin blocks adjacent to the tissues.

The procedure of fixation and chemical processing specifically aims to arrest postmortem changes in the tissues and thus protect cells and tissues from shrinkage and distortion before sectioning and mounting of the samples onto slides. Once tissue samples are mounted on slides, the use of specific dyes and other chemicals will improve the refractive indices of intracellular components of these tissues. This targeting allows for the establishment of evidence of the presence of chemicals or structures within the tissues. This can be used in order to provide evidence for the presence, density, and position of subcellular elements within the cells and/or tissues.

Typically, tissues and organs are too large for surveying with optical microscopy. This requires general techniques to be followed that yield thin, translucent histological sections which can be stained for examination and compared with control sections. A common procedure for viewing biological tissues is via transillumination with the use of optical or modern computer-assisted microscopes. The most useful stain for the visualization of protein-secreting cells is Masson's trichrome, a three-color staining protocol, with protein-secreting cells staining strongly (see color plates 4D2, 4E, 4F, 6I, 6J). A complementary stain is Periodic Acid Schiff's (PAS), which detects cells in mucus.

7.8.2 MAGNETIC RESONANCE IMAGING (MRI)

Microimaging using MRI is perfect for providing high-resolution three-dimensional images (> 6 micron) with high-quality soft-tissue contrast for visualization of venom-gland structures (see color plates 4A1, 4A2, 4A3, 4B3, 4C1, 4D1, 6A–6H). MRI has an advantage compared with other radiology modalities such as computer tomography (CT), which can only produce very high-resolution images (> 2–3 micron) of hard tissues such as bone, cartilage, and ligaments. The advantage of MRI compared with the histological imaging method is that the sample can be kept intact, while the images can be resliced for viewing in oblique planes, and 3D rendering of the venom-gland structures can be reconstructed (Fry et al. 2008; Fry et al. 2009b).

In general, high-field microimaging MRI scanners (> 9.4 T) with strong gradients (> 2.5 G/cm/A) and optimized radiofrequency (RF) coils are required for small venom-gland samples (1–2 mm in size). Higher-field magnets are preferable because they provide higher sensitivity for signal detection, which is critical for high-resolution imaging up to 10-micron isotropic 3D resolution. Larger samples (up to 34 mm in diameter) can be accommodated using larger coils; however, the achievable resolution will generally decrease to 100–200 micron because of the lower sensitivity of the RF coils and gradient performance.

At 16.4 Tesla, a 30–50 micron 3D-resolution image will typically be acquired for one to two hours. The length of MRI acquisition will increase to 10 hours or more for a 10-micron-resolution image. The image signal-to-noise ratio (SNR) and contrast-to-noise ratio (CNR) can be improved by a small addition of MRI contrast agent and increasing the number of scan averages. It is therefore often useful to acquire high-resolution MR images overnight (16 hours). To maintain tissue stability for long acquisition periods, samples are normally prefixed in a solution of 4% paraformaldehyde (PFA) in phosphate-buffered saline (PBS). PFA is favored over other fixatives (alcohol or oxidizing agents), as the tissue contrast is easily recovered to native conditions upon washing with PBS. Prior to MRI, PFA will need to be removed by incubation with PBS, normally over four days of incubation. Incomplete removal of PFA will result in poor tissue contrast because of short T2 (spin-spin or transverse) relaxation. A small concentration of MRI contrast agent, typically 0.1% to 0.5% gadolinium diethylene triamine penta-acetic acid (GD-DTPA), can be included in the PBS wash (Ullmann et al. 2010). The MRI contrast agent is used to reduce T1 (spin-lattice or longitudinal) relaxation, in order to enable faster recycle time delay of MRI sequence, allowing more signal averaging and improving SNR. The contrast agent and incubation length should be optimized for various types of samples in order to obtain optimal T1 and T2 (spin-spin or transverse) relaxation times. High concentrations of contrast agent and prolonged incubation may be detrimental, as they can dramatically shorten T2 relaxation resulting in loss of signals.

For MRI, venom-gland samples can be submersed in perfluoro-ether solution (Fomblin, Solvay Solexis, Italy). This medium is used to provide a black background and uniform local-field homogeneity. Size-specific containers (typically 5–20 mm NMR tubes or 50 mL plastic tubes) are used to hold the sample snugly, avoiding vibrations during MRI scanning. Subsequently, the whole sample container is placed under vacuum to remove air pockets from the tissues. This step is critical, as micro air bubbles can produce severe image artifacts. MRI radiofrequency coils must be chosen with a maximum filling factor (such as fitting as close as possible to the sample) in order to obtain maximum sensitivity.

For high-resolution 3D imaging (10–50 micron), a gradient echo sequence, such as Bruker's 3D FLASH (fast low angle shot), is preferable as it produces better image quality in comparison with spin-echo sequences. For imaging at 16.4 T, the following parameters can be used: TR (recycle delay)/TE (echo time) = 40/8 ms, flip angle 20°, 4–8 excitations. The field of view and matrix will need to be varied to fit the individual samples, with the resulting voxels having a range of between 20 and 40 μm isotropic resolution. Total scan time is 8 to 15 hours per sample, depending on size and resolution (Fry et al. 2008).

The MRI data can be exported in DICOM format. The 3D structures of venom glands can be analyzed using various visualization software such as OSIRIX (www.osirix-viewer.com) or Medical Imaging Processing, Analysis, and Visualization v6.0.0 (MIPAV, mipav.cit.nih.gov). For more complicated 3D image segmentation, surface rendering, and volumetric measurements of the glands will need to be performed manually using programs such as ITK-SNAP (www.itksnap.org) or AMIRA (www.fei.com).

7.8.3 ELECTRON MICROSCOPY

Electron microscopy may be used to image the morphology of venom systems at higher resolution and at a finer scale than available with the use of light microscopy (Fry et al. 2009b). An electron microscope produces an image by focusing a beam of electrons at a sample in a vacuum and then detecting the electrons that return or pass through. There are two main types of electron microscopes: scanning (SEM) and transmission (TEM). SEM provides information about the surface of a sample and renders it as an image in real time in 3D. Such an image, while monochromatic, is easy to interpret. TEM involves sending the electron beam through the sample. Since electrons are stopped by all but the thinnest of sample slices, it is only possible to view sections of around 50 to 90 nm in thickness with TEM. It is also necessary to introduce heavy metals into the sections to provide image contrast. These metals (usually osmium, lead, and uranium) interact with known cellular components. Because the metals scatter electrons, they produce dark regions of contrast, showing up organelles and their boundaries in the resultant images.

Beyond deciding whether images of surface morphology or internal detail are required, the major challenge is to determine the best protocol for a particular tissue in order to avoid autolysis, extraction, shrinkage, expansion, and possible destruction of cell membranes. The choices are complicated by the presence of soft-tissue components adjacent to or surrounding hard-tissue components in many venom systems. Hard tissues are more tolerant of preparation protocols than soft tissues but can cause challenges during sectioning if thin sections are required. They can easily break or tear away from softer structures. Since sectioning is not required for SEM, it may be advisable to view whole systems with SEM first and subsequently undertake TEM on separated structures excluding the hard components. There is an alternative in some cases: calcified tissues can be decalcified (such as by using ethylene diamine tetra-acetic acid, EDTA) in order to section through both hard and soft tissues.

The conventional approach to tissue preparation for morphological assessment by both SEM and TEM involves chemical fixation. A number of sample and fixative factors affect fixation. These are the osmolarity of the sample tissue (freshwater versus marine); need for additional ions (such as Ca^{2+} ions to bind to negatively charged membrane proteins and stabilize them); pH of the tissue (usually 7.2–7.4); temperature of applied chemicals (4° C versus room temperature); size of the tissue (excision < 1 mm^2 blocks versus in situ treatment perfusion); and duration of fixation (30 minutes to 24 hours). Choices depend on where the animal is when the sample is taken. If the sample can be obtained from a live organism in a laboratory, then ideal conditions can be met, but sampling out at sea or on a terrestrial field trip introduces constraints. Initial fixation is undertaken using an aldehyde, such as glutaraldehyde, or a mixture of aldehydes: PFA (formalin) and glutaraldehyde, which stabilizes the cellular structures, in particular the proteins. The sample is then washed in buffer before a secondary fixative is applied, to prevent interactions between the two fixatives. The secondary fixative (such as osmium tetroxide) interacts with lipids and stabilizes cell membranes.

It is important for the tissue to be fixed as soon as possible, as samples rapidly decay after dissection from the animal. This constraint extends to tissue size, as fixatives penetrate slowly through tissues, traveling at rates as slow as 0.2 mm per hour (osmium tetroxide at 4° C), unless injected or perfused. Glutaraldehyde penetrates more quickly, but it is advisable to cut tissue into blocks of 1 to 0.5 mm^2 for full penetration to occur in a few hours. Protocols differ, and while guidelines can be provided for general processing, it is best to consult published papers that deal with the animal of interest and display good-quality images. Because so many factors are involved, few researchers have the time or resources to undertake methodological experimentation and choose instead to follow published protocols in their area.

Concentrations of primary fixatives, buffers, and additives can differ substantially. Assessing a few protocols used for the same organism will quickly establish the likely role osmolarity will play. For example, when comparing two protocols for the study of snake venom-gland morphology (Mackessy 1991; Sakai, Carneiro, and Yamanouye 2012), it is apparent that two aldehydes are mixed in one protocol, whereas a single aldehyde (glutaraldehyde) is used in the other; however, the overall percentage used (4.5%–5% aldehyde) is similar, and the addition of sucrose indicates the need for a high osmolarity. Cultured cells require different treatment even if they are from the same animal, such as snake venom-gland cells from *Bothrops jararaca* (Carneiro et al. 2006).

Fixation is followed by removal of the water component through dehydration with either ethanol or acetone. Ethanol is preferred for SEM preparations, because it is compatible with equipment and chemicals used in the later stages of drying the sample. Acetone is preferred for TEM processing, since it removes water more completely and is highly miscible with the resin used to infiltrate the tissue. The resin is polymerized to provide a hard block for sectioning; any water left in the sample will inhibit the process for the commonly used resins such as Epon and Spurr's resin.

Fixation chemicals are dangerous and toxic. They are nondiscriminatory and will affect both the handler and the sample, so care must be taken with them. Their main function is to stop autolytic processes and preserve cellular structure. It is also important to consider the toxicity of the venom from the organism being sampled. Anesthetizing the animal and wearing suitable protective equipment are required.

For conventional TEM, after samples have been fixed, dehydrated, infiltrated with a chosen resin, and polymerized (usually at 60° C), the resulting block is sectioned. Initially, thick sections are taken from the trimmed block face, mounted on slides, and contrasted using toluidine blue. When an area of interest is found, then ultrathin sections are taken using a glass or diamond knife. The sections are mounted on a mesh copper grid. Additional contrasting metals are added, usually by application of uranyl acetate and lead citrate solutions, and the grids are viewed at 60–100 kV (usually 80 kV).

Alternatives exist to conventional fixation and processing approaches. Using a microwave designed for laboratory fixation (such as Pelco BioWave) can reduce fixation, dehydration, and resin-embedding times from a few days down to half a day in total. Application of microwaves is thought to increase membrane permeability to fixatives and solvents. It is, however, important to avoid heating the tissue.

If the equipment is available, for small enough samples, rapid freezing is possible followed by cryosubstitution. High-pressure freezing is the method of choice if available. This yields superb results, although it can take some trial and error to exclude ice-crystal formation and subsequent distortion of tissue. Cryosubstitution that follows freezing involves introduction of the fixative and contrasting metals into the tissue block at temperatures below that at which water freezes (Kheyri, Cribb, and Merritt 2013). A benchtop protocol is now available that provides a simple, cost-effective, and rapid cryosubstitution process (McDonald and Webb 2011).

For SEM imaging, the sample simply needs fixation, dehydration, and drying. There are a number of possible approaches. The first is to follow the conventional protocol used for TEM, stopping before resin infiltration and drying the sample at that stage. It is not recommended to simply air-dry from ethanol or acetone, because tissue collapse often occurs if mineralization is not present. The conventional approach is to replace an introduced solvent (ethanol or acetone) with liquid CO_2 under pressure. The CO_2 is then heated while under pressure to change phase from liquid to gas, instantly drying the sample without damaging it but causing around 10% shrinkage. This process is referred to as critical-point drying (CPD). Hexamethyldisilazane (HMDS) can be used after ethanol dehydration as an alternative method of drying. The HMDS is left to evaporate overnight, rendering the sample completely dry and usually undamaged, although a trial is necessary to ensure that the process will work with the tissue being studied.

Freeze-drying is another form of sample drying. It is undertaken after plunge-freezing a sample into liquid nitrogen. The result is a dry sample that can be mounted for standard SEM imaging. Samples can be fractured cleanly while frozen (using a cooled razor blade and protective gloves), providing an added advantage, since this allows a view into the interior of the venom-gland system rather than just an image of its surface.

A less commonly used technique is freezing the sample and imaging it directly while still frozen. This approach is referred to as cryo-SEM. It is also possible to store the sample in buffer (after fixation) and freeze it immediately before viewing. The advantage of using cryo-SEM is that the water component in a sample does not need to be removed, and so the dehydration and drying step can be eliminated, avoiding the potential of shrinkage or collapse of delicate tissues. Freezing samples by plunging into liquid nitrogen slush (–210° C) results in a more rapid freezing rate than that which results from using the liquid form alone. The sample is then transferred to the SEM and held at a low temperature. At this stage, ice can be removed, if desired, by warming the sample (such as to –80° C), which allows the ice to sublimate in the vacuum. Sublimation only takes a short time (5–15 minutes) depending on the desired result. Such an approach can either remove obscuring surface ice or allow for the imaging of the superstructure within the topmost region of the sample.

A review of protocols would not be complete without considering maceration. Soft tissue can be removed by maceration in sodium or potassium hydroxide. After removal of the caustic chemicals by washing with water, the tissue is usually dried using CPD or HMDS. The dry sample can then be viewed with conventional SEM. This approach allows 3D cuticular complex of ducts and venom sac to be imaged without being obscured by the gland cells (Kheyri, Cribb, and Merritt 2013).

In summary, electron microscopy provides a range of techniques that enable imaging of venom systems at high resolution in 2D and 3D. While conventional approaches utilize bench chemical fixation and processing, alternatives exist such as cryofixation and cryoimaging, and these are becoming more widely adopted. For further information on SEM, TEM, and imaging of elements within samples, visit the online teaching site MyScope (http://ammrf.org.au/myscope/). The site includes sample preparation flow charts, videos, and virtual machines.

CHAPTER 8

THREE-FINGER TOXINS (3FTXS)

Y. UTKIN, K. SUNAGAR, T. N. W. JACKSON, T. REEKS, AND B. G. FRY

8.1 SUMMARY

Three-finger toxins (3FTxs) represent a peptide type that has undergone dynamic evolution to produce numerous apotypic structural and functional forms. The plesiotypic function was α-neurotoxicity, with this effect greatly potentiated by the deletion of two of the plesiotypic cysteine residues secondary to the evolution of advanced venom-delivery architecture in the elapid snakes. Subsequent to this deletion of the second and third plesiotypic cysteine residues, a plethora of new functions with differing pharmacology have emerged. This molecular diversity has been shown to be of significant therapeutic use, including as painkillers and anticancer compounds.

8.2 ENDOPHYSIOLOGICAL PLESIOTYPE

Three-finger toxins are so called because of their three-finger fold (3FF), three polypeptide loops that protrude from the molecule's compact hydrophobic core, giving them the characteristic fingerlike appearance. The 3FF was first described in snake toxins but was later found in numerous nonvenom proteins of different origin and function, which varied greatly in structure, some containing multiple 3FF domains (Greenwald et al. 1999; Huai et al. 2006; Galat 2008; Galat et al. 2008; Klein et al. 2008). The Ly-6/uPAR superfamily (Tsetlin 1999) includes both secreted molecules and membrane-bound receptors, including Ly-6 cell-surface glycoproteins (Gumley, McKenzie, and Sandrin 1995); membrane-bound proteins, such as urokinase-type plasminogen activator receptors (UPAR) (Llinas et al. 2005); GPI-anchored glycoproteins, such as lynx (Miwa et al. 1999; Ibanez-Tallon et al. 2002; Dessaud et al. 2006; Lyukmanova et al. 2011) and prostate stem-cell antigen (PSCA) (Reiter et al. 1998; Bahrenberg et al. 2000; Gu et al. 2000; de Nooij-van Dalen et al. 2003);

and soluble proteins lacking the GPI anchor, such as SLURP (Horie et al. 1998; Adermann et al. 1999; Miwa et al. 1999; Ibanez-Tallon et al. 2002; Chimienti et al. 2003; Mastrangeli et al. 2003; Tsuji et al. 2003; Arredondo et al. 2006; Dessaud et al. 2006; Grando 2008; Pettersson et al. 2008; Lyukmanova et al. 2011). While the precise 3FTx nontoxin ancestor has not yet been conclusively determined, central nervous system neuromodulators such as lynx and SLURP, which contain only one 3FF domain, are the most likely candidates. These 3FF peptides not only resemble 3FTxs structurally but also share the same biological targets with them; both bind nAChR in order to modify agonist affinity (Adermann et al. 1999; Miwa et al. 1999; Choo et al. 2008; Chernyavsky et al. 2009; Hruska et al. 2009; Tekinay et al. 2009).

8.3 PLESIOTYPIC STRUCTURAL AND FUNCTIONAL TOXIN FORMS

3FTxs constitute one of the largest superfamilies of snake-venom proteins. The plesiotypic 3FF peptide was recruited for use in envenoming at the base of the snake radiation (Fry et al. 2013). Relative secretion levels of this toxin type between snake lineages are extremely variable. Within the henophid snakes, not only have transcripts been sequenced from species such as *Aspidites melanocephalus* (pythonid snake) and *Cylindrophis ruffus* (cylindrophiid snake), but translation and secretion also have been confirmed (Fry et al. 2013). The nonconstricting *C. ruffus* expresses toxins in their seromucoidal glands (which stain strongly for protein secretion in histological examinations) in much higher amounts than constricting species such as *A. melanocephalus*, which secrete mucus as the primary glandular product and have glands that histologically stain accordingly (see chapter 1.2.3). Very low levels of 3FTx mRNA have been sequenced from viperid snake venoms (Junqueira de Azevedo et al. 2006; Pahari et al. 2007), but the corresponding proteins have not been isolated or characterized. In contrast, in the venoms of non-front-fanged advanced snakes, such as the fast-moving and gracile colubrid snakes, plesiotypic 3FTxs are the dominant toxin type (Fry et al. 2003a; Fry et al. 2003c; Fry et al. 2008). Similarly, most elapid snake venoms are dominated by 3FTxs. Most of the elapid snake venom 3FTxs, however, are extremely apotypic, while all other lineages secrete only the plesiotypic toxin type.

The 10-cysteine pattern of nonvenom neuroactive 3FF peptides such as lynx/SLURP is preserved in plesiotypic 3FTxs. These are the sole 3FTx type found in snake venoms, except for elapid snakes which typically secrete the plesiotypic 3FTxs in low levels and instead favor structurally and functionally apotypic forms (Fry et al. 2003a; Fry et al. 2003c; Fry et al. 2008). The plesiotypic 3FTx activity is binding to neuromuscular α1 nAChR, and plesiotypic 3FTxs are considerably more toxic to reptiles or birds than to mammals (Fry et al. 2003a; Pawlak et al. 2006; Pawlak et al. 2009; Heyborne and Mackessy 2013). This specificity of action led to plesiotypic α-neurotoxins being mistakenly referred to as "weak neurotoxins" (Utkin et al. 2001) and may have contributed to errors in experimental design in investigations of the toxicity of non-front-fanged venoms (Rochelle and Kardong 1993).

Evaluation of natural selection pressures on various elapid snake venom 3FTxs indicated that even forms containing the plesiotypic 10-cysteine arrangement continue to diversify structurally and functionally under positive selection (see color plate 26A). Neofunctionalized variants of this form target α7 and muscarinic receptors (Chang et al. 2002; Nirthanan et al. 2002; Fry et al. 2003b; Mordvintsev et al. 2009; Fry et al. 2013; Sunagar et al. 2013b). Similarly, in non-front-fanged advanced snake and viperid snake venoms, plesiotypic 3FTxs continue to mutate in an accelerated manner typical of predatory toxins (Sunagar et al. 2013b). The hypermutation of viperid venom 3FTxs, however, could be a result of the loss of selective constraints on this toxin type, which is secreted in extremely low levels by these snakes (Sunagar et al. 2013b).

8.4 APOTYPIC STRUCTURAL AND FUNCTIONAL TOXIN FORMS

8.4.1 APOTYPOSIS

Although the characteristic 3FF structural motif is preserved in all 3FTxs, there are several structural variations (see color plate 26A). Cysteine variations in the plesiotypic form have been noted for henophidian snake secretion 3FTxs and those from the venom of the colubrid genus *Boiga* (see color plate 26A). The henophidian snake secretion 3FTxs have an additional cysteine residue after the characteristic cysteine doublet near the C-terminus. The resultant cysteine triplet contains a "free cysteine" and therefore has the potential for dimerization. However, it is likely that one of the cysteine residues of this triplet is buried in the molecular core, since the toxin does not appear to form a dimer (Fry et al. 2013). The venom of snakes from the non-front-fanged genus *Boiga* contains heterodimers of plesiotypic-type α-neurotoxins (Fry et al. 2003c; Pawlak et al. 2009) which bind to the postsynaptic nicotinic acetylcholine receptors (see color plate 2B) resulting in flaccid paralysis (see color plates 11L–11N). Each monomer in the toxin possesses a different newly evolved cysteine residue, which together form the interchain disulfide bridge.

In elapid snake venoms, additional 3FTx forms have evolved following the deletion of the second and third plesiotypic cysteine residues. This removal of structural constraints led to an explosive radiation of the 3FTx toxin class, with the subsequent evolution of novel neurotoxic, coagulopathic, and cytotoxic pharmacological activities. The small hydrophobic core of the molecule is stabilized by the remaining four disulfide bridges. This structure is retained by the α-neurotoxins (type I [short-chain] and type III) and other apotypic forms such as cytotoxins, aminergic (adrenergic and muscarinic) toxins, L-type Ca^{2+} blockers, and platelet inhibitors. The apotypic type II α-neurotoxins and κ-neurotoxins have an additional pair of cysteine residues that stabilize loop 2 (see figure 8.1). Variations in the number of amino acid residues between cysteine residues (loop size), together with amino acid substitutions, result in considerable diversity of 3FTx biological properties. However, despite the extremely large scope and complexity of the snake 3FTx family, only certain neurotoxins and cytotoxins/cardiotoxins have been well characterized (Fry et al. 2003b; Fry et al. 2008) (see figure 8.2).

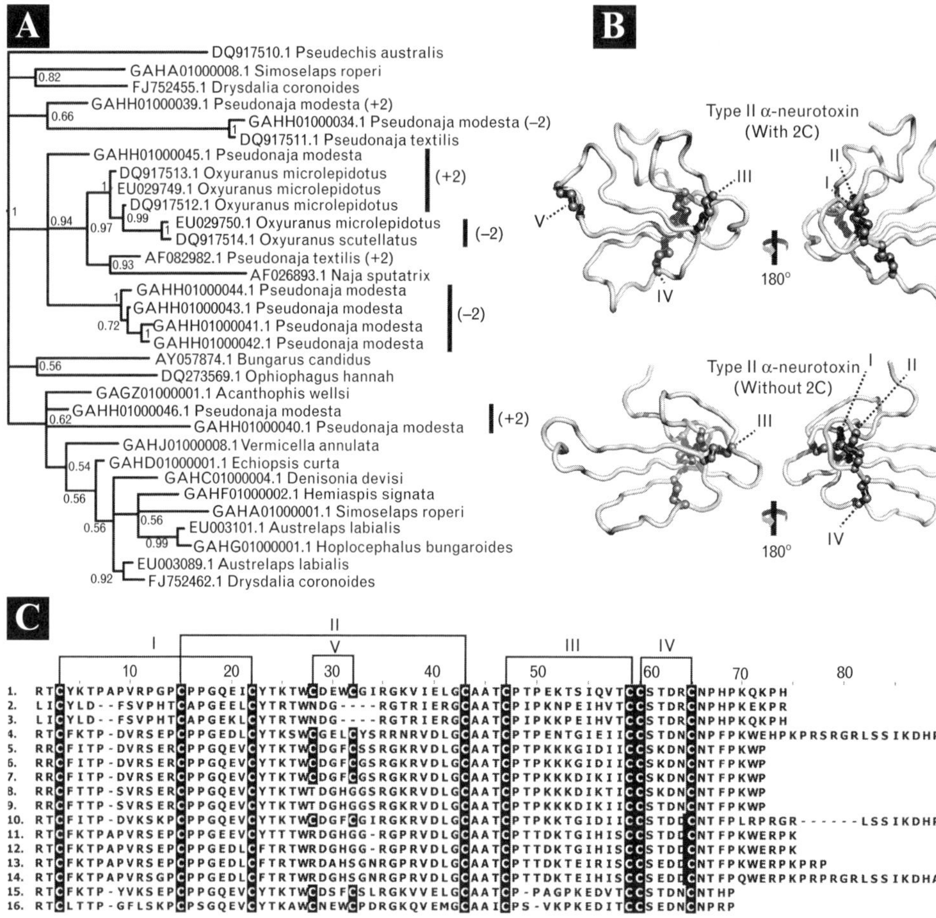

FIGURE 8.1: Structural evolution of type II 3FTxs (Jackson et al. 2013; Sunagar et al. 2013b). (A) Molecular phylogenetic tree showing that the cysteine pair characteristic of type II 3FTxs has been lost on multiple occasions in *Oxyuranus*/*Pseudonaja* species. (B) The structural variance between forms with and without this cysteine pair. (C) Sequence alignment and cysteine connectivity of representative type II 3FTxs.

8.4.2 NEUROTOXIC APOTYPIC 3FTXS

The deletion of the second and third plesiotypic cysteine residues in elapid snake venom 3FTxs resulted in a dramatic increase in the strength of binding to nAChRs and thus the potency of α-neurotoxicity (Fry et al. 2003b). These are the potent toxins responsible for the flaccid paralysis seen clinically (see color plates 11L–11N). Type I α-neurotoxins (also known as short-chain α-neurotoxins) and type III α-neurotoxins both bind to the neuromuscular α1 nAChR. Within the type I (short-chain) α-neurotoxins, a further modification to increase the cysteine number has occurred to isoforms from Australo-Papuan elapid snake venoms, which have this additional cysteine insertion after the first plesiotypic cysteine (Fry et al. 2003b). This cysteine, however, is buried in the molecule's core and thus does not have any clear structural or functional role. In contrast, the type II α-neurotoxins (also known as long-chain neurotoxins) have two new cysteine residues, which stabilize loop 2 (Fry et al. 2003b).

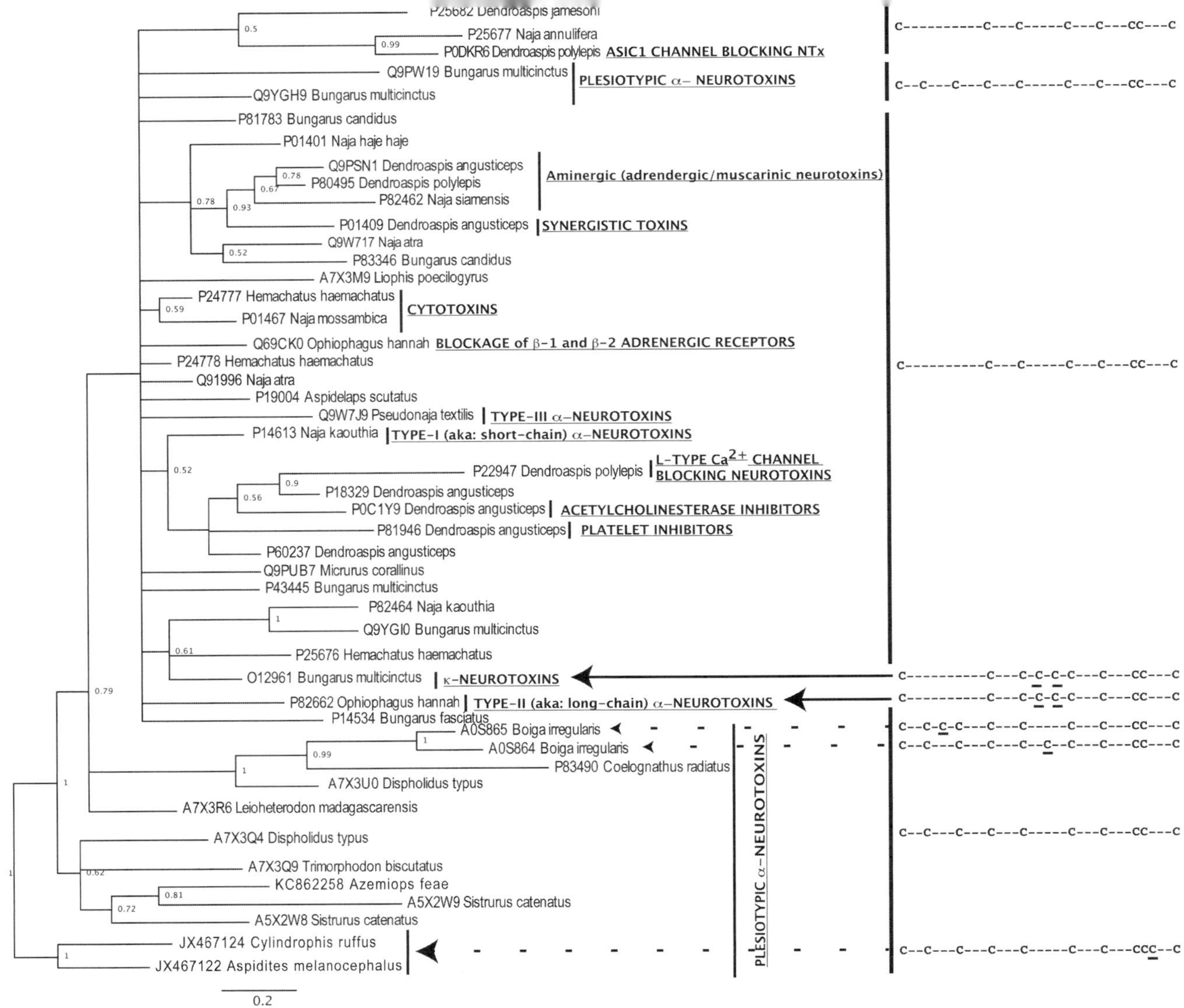

FIGURE 8.2: Molecular phylogeny, structural and functional evolution of 3FTxs (Jackson et al. 2013; Sunagar et al. 2013b). Cysteine patterns are depicted on the right with newly evolved cysteines underlined.

This structural change is accompanied by the novel function of targeting neuronal α7, α9, and α10 nAChR while retaining the plesiotypic activity of binding neuromuscular α1 nAChR (Servent et al. 1997). The presence of the fifth disulfide bond in the second loop is crucial for these interactions. Recently, an unusual type II (long-chain) α-neurotoxin (α-elapitoxin-Aa2a) was isolated from *Acanthophis antarcticus* (Blacklow et al. 2011). Despite containing the loop 2 stabilizing disulfide bond, this toxin lacks high affinity for neuronal α7-type nAChRs, while binding efficiently with muscle-type receptors. Intriguingly, the characteristic long-chain cysteine pair has been secondarily lost on at least two occasions within the venom of the *Oxyuranus/Pseudonaja* clade of Australian elapid snakes (Jackson et al. 2013; Sunagar et al. 2013b) (see figure 8.1). However, these toxins still retain the ability to bind to neuronal nAChRs (St Pierre et al. 2007b). Evaluation of evolutionary selection pressures on various α-neurotoxins (type I, II and III) revealed that they have been subject to a strong influence from positive selection. The rapid evolution of type II α-neurotoxins in the venom of *Oxyuranus* and *Pseudonaja* species probably resulted in the accumulation of mutations even in structurally important regions, which are typically constrained by negative selection, and led to the origination of the aforementioned "unusual" type II 3FTxs (Jackson et al. 2013; Sunagar et al. 2013b).

The deletion of the second and third plesiotypic cysteines not only potentiated α-neurotoxicity but also freed up the scaffold for the evolution of novel neurotoxic activities. Novel postsynaptic activities include targeting a wide array of adrenergic muscarinic nAChR subtypes (Jerusalinsky and Harvey 1994; Bradley 2000; Karlsson et al. 2000) and inhibition of neuronal α3β2 and α7 nAChR by κ-neurotoxins (which also have an apotypic pair of cysteines stabilizing loop 2) (Grant and Chiappinelli 1985) (see color plate 2B and figure 8.2). Synaptically, novel toxins inhibit acetylcholinesterase (Albrand et al. 1995), while presynaptically the L-type Ca^{2+} channels are targeted (de Weille et al. 1991).

Toxins within the aminergic (adrenergic/muscarinic) clade may be agonists or antagonists with exquisite receptor subtype specificity (Servent and Fruchart-Gaillard 2009). For example, muscarinic toxins MT1 and MT2 bind strongly to the M1 receptor and less effectively to the M3 (Kornisiuk et al. 1995) and M4 receptors (Harvey et al. 2002). MT3 is a selective ligand of the M4 receptor with significantly lower affinity for the M1 receptor (Olianas et al. 1999). MT1 is an agonist at M1 and antagonist at M4 receptors with the affinity for both receptors being similar (Sanchez et al. 2009). Similar examples may be found for other muscarinic toxins. Muscarinic toxins are known from the venoms of African snakes in the elapid genus *Dendroaspis*, and similar peptides have been found in cobra venoms (Kukhtina et al. 2000); thus, this toxin type is likely to be widespread within African/Asian elapid snake venoms. Within the aminergic (adrenergic and muscarinic) clade, additional 3FTxs acting on different subtypes of α-adrenoreceptors have been isolated from green mamba *Dendroaspis angusticeps* venom (Quinton et al. 2010; Rouget et al. 2010). One of the toxins (AdTx1) has subnanomolar affinity and high specificity for the human α_{1A}-adrenoceptor subtype (Quinton et al. 2010). AdTx1 forms an unusually stable complex with α_{1A}-adrenoceptor and has great potential for the development of new drugs targeting this receptor subtype. Another toxin, ρ-Da1b, inhibited rauwolscine by binding to the three α2-adrenoceptor subtypes with an affinity ranging from 14 to 73 nM (Rouget et al. 2010). It also demonstrated antagonism on α2A-adrenoceptor. Structurally, these two toxins are similar to muscarinic toxins, which were previously shown to interact weakly with adrenoreceptors (Harvey et al. 2002). The plesiotypic form from *Bungarus* venom (Chung et al. 2002) with muscarinic activity represents an additional neofunctionalization, reinforcing the importance of this activity in envenomation.

Novel neurotoxins from *Dendroaspis* venoms interact not only with nAChRs and mAChRs but also with sites involved in nerve impulse transmission. This pool of toxins includes those that bind the enzyme acetylcholinesterase (AChE) at picomolar concentrations and are highly specific for the enzyme peripheral site. When bound to AChE, the toxins block the access of acetylcholine to the active site of the enzyme and prevent neurotransmitter hydrolysis (Karlsson, Mbugua, and Rodriguez-Ithurralde 1984). Presynaptically, other unique *Dendroaspis* 3FTxs bind L-type Ca^{2+} channels, competing with 1,4-dihydropyridine for a binding site on the channel molecule and thus blocking the calcium current (de Weille et al. 1991). Another class of 3FTxs, also known only from *Dendroaspis* venoms, act as reversible gating modifiers by antagonistically binding to closed or inactivated ASIC1a-ASIC2a (ACCN2-ACCN1) channels in central neurons and ASIC1b-containing channels in nociceptors (Diochot et al. 2012).

Effects of α-neurotoxins have been assessed in animal/human tissue *in vitro*. It should be noted that the rich functional diversity of 3FTxs is yet to be fully explored, as several classes of functionally tested 3FTxs were sourced from only a few genera of advanced snakes. It was revealed that α-neurotoxins exhibit a range of specificity toward various *in vitro* tissue preparations, such as the chick biventer cervicis nerve-muscle preparation, frog sciatic nerve-sartorius-muscle preparation, and rat phrenic nerve diaphragm (Lee et al. 1972) (see table 8.1). Several functionally important residues in type I α-neurotoxins, retrieved from *Laticauda semifasciata* venom, were found to be important for binding to *Torpedo* nAChRs (Chiappinelli, Cohen, and Zigmond 1981; Stiles 1993). However, very few of these residues in type I α-neurotoxin from *Naja mossambica* were found to be important for binding to mouse muscle nAChRs (Ackermann and Taylor 1997; Ackermann et al. 1998; Osaka et al. 2000). Similarly, it was revealed that erabutoxin-a binds to the human nAChR with lesser affinity in comparison with α-bungarotoxin. However, both these neurotoxins were found to be capable of binding to the *Torpedo* nAChRs at greater affinity, highlighting the binding difference between 3FTxs in various species. Thus, it becomes important to test biochemical activity of 3FTxs on the natural prey to characterize residues that are actually involved in mediating binding to nAChRs. Several functionally important residues in α-cobratoxin from *Naja kaouthia* venom have also been identified (Antil, Servent, and Menez 1999). However, the sites were identified to be important for binding to *Torpedo* fish nAChRs and thus may not play an essential role in binding to reptilian or mammalian nAChRs. Moreover, many type I α-neurotoxins have not been completely sequenced (Barber, Isbister, and Hodgson 2013). Because amino acid sequencing is time-consuming and expensive, full sequencing is often not performed when the activity is not novel.

Interestingly, several residues were extremely well conserved in plesiotypic 3FTxs from *A. melanocephalus* and *C. ruffus* (Sunagar et al. 2013b). However, functional studies assessing the importance of these residues in receptor binding are still missing. Similarly, studies to determine the role of identified conserved residues in *Atractaspis* 3FTxs (Terrat et al. 2013) in receptor binding are yet to be undertaken. Thus, it becomes essential to test 3FTxs from various snake lineages on tissue preparations of natural prey in order to determine the actual role of residues in binding to their target receptors.

Taicatoxins are composed of three noncovalently linked subunits in a 1:1:4 stoichiometric ratio: (i) 3FTx, (ii) neurotoxic PLA_2 (see chapter 20), and (iii) kunitz peptides (see chapter 15). It has been demonstrated that this oligomeric complex is capable of blocking Ca^{2+} channels (Brown et al. 1987).

Table 8.1: Functional Resides of α-Neurotoxic 3FTxs

Residue	Tested on	3FTx from	Reference
Type I α-neurotoxin			
Lys-27, Arg-33, and Lys-47	*Torpedo* and mouse nAChR	*Naja mossambica* and *Laticauda semifasciata*	Pillet et al. 1993; Tremeau et al. 1995; Ackermann and Taylor 1997; Ackermann et al. 1998; Antil, Servent, and Menez 1999
Trp-29	*Torpedo* and mouse nAChR	*N. m. mossambica* and *L. semifasciata*	Pillet et al. 1993; Tremeau et al. 1995
Asp-31	*Torpedo* and muscle nAChR	*N. m. mossambica* and *L. semifasciata*	Tremeau et al. 1995; Antil, Servent, and Menez 1999
Arg-33	*Torpedo* and mouse nAChR	*N. m. mossambica* and *L. semifasciata*	Tremeau et al. 1995; Antil, Servent, and Menez 1999
Ile-36	*Torpedo* nAChR	*L. semifasciata*	Tremeau et al. 1995; Antil, Servent, and Menez 1999
Type II α-neurotoxin			
Lys-23	*Torpedo* nAChR	*N. kouthia*	Antil, Servent, and Menez 1999
Trp-25	*Torpedo* nAChR	*N. kouthia*	Antil, Servent, and Menez 1999
Asp-27	*Torpedo* nAChR	*N. kouthia*	Antil, Servent, and Menez 1999
Phe-29	*Torpedo* nAChR	*N. kouthia*	Antil, Servent, and Menez 1999
Arg-33	*Torpedo* nAChR	*N. kouthia*	Antil, Servent, and Menez 1999
Arg-36	*Torpedo* nAChR	*N. kouthia*	Antil, Servent, and Menez 1999
Lys-49	*Torpedo* nAChR	*N. kouthia*	Antil, Servent, and Menez 1999
Phe-65	*Torpedo* nAChR	*N. kouthia*	Antil, Servent, and Menez 1999

8.4.3 CYTOTOXIC APOTYPIC 3FTXS

These toxins are responsible for the severe localized tissue destruction that occurs in some elapid envenomations, particularly species such as *Naja nigricollis* (see color plates 13 and 14A–14F). Cytotoxins, which are characterized by a molecular surface that has flanking hydrophobic and cationic amino acids, exhibit cytolytic activity (Kini and Evans 1989b). In experimental manipulations of lysine and methionine residues in these regions, conformational change does not occur, but cytotoxic activity is lost (Kini and Evans 1989b; Kini and Evans 1989a; Kini and Doley 2010). Structure-function studies of CTX V from *Naja atra* revealed Met26 and Lys44 as particularly crucial (Jayaraman et al. 2000). Studies of the toxin form γ (gamma) from *N. nigricollis* revealed that Trp11 (loop I) and Tyr22 (loop II) were responsible in part for the cytotoxicity, with Tyr22 (loop II) being of structural importance (Gatineau et al. 1987). These peptides produce lysis by binding to cell membranes and disturbing their integrity, with some cell types (such as transformed cells) being particularly sensitive to these effects (Feofanov et al. 2004). Cell depolarization results in systolic heart arrest, smooth and skeletal muscle contracture, erythrocyte lysis, necrotic cell death, and apoptosis (Condrea 1976; Harvey 1985; Wang and Wu 2005; Yang et al. 2005). These toxins have also been referred to as cardiotoxins because of their ability to cause cardiac arrest at high concentrations. The mechanism of action remains unclear, and there are competing hypotheses, including interaction with cell-surface proteins (Wang et al. 2006a) and the primary target of lysis action being lysosomes (Feofanov et al. 2005). It is clear, however, that pore formation and increased membrane permeability are the result of interaction with anionic lipids in the cell membrane (Forouhar et al. 2003; Kao, Lin, and Chang 2009; Kao, Wu, and Chang 2009). A 3FTx type isolated from *Ophiophagus hannah* venom has been called β-cardiotoxin because of its dose-dependent action of blocking β-adrenergic receptors leading to a heart rate decrease without affecting contractility (Rajagopalan et al. 2007). However, these peptides are phylogenetically distinct from the "true" cytotoxins/cardiotoxins (Sunagar et al. 2013b) (see figure 8.2).

Interestingly, while the majority of 3FTx lineages exhibit considerable sequence diversity and are evolving rapidly under the influence of positive Darwinian selection, cytotoxins exhibit little sequence variation and are constrained by negative selection (Sunagar et al. 2013b). It has been theorized that a highly conserved hydrophobic patch on the cyotoxin's molecular surface interacts with hydrophobic regions of the cell's phospholipid bilayer (Condrea 1976; Dufton 1984). These hydrophobic residues, which facilitate such nonspecific interactions, are spread over the three loops of these cytotoxic 3FTxs and cover nearly 40% of the molecular surface (Bilwes et al. 1994). Moreover, cytotoxins are known to exhibit a diversity of other biological functions, such as lysis of various types of cells, including erythrocytes and epithelial cells; inhibition of enzymes, including protein kinase C and Na+/K+ ATPase; and depolarization and contraction of muscle cells and prevention of platelet aggregation (Harvey, Marshall, and Karlsson 1982; Bougis, Khelif, and Rochat 1989; Yu et al. 1993b; Kumar, Lee, and Yu 1996). The residues implicated in these activities are distributed all along their length. Thus, unlike α-neurotoxins that bind to specific molecular receptors, cytotoxins bind in a much more nonspecific manner and thus evade the predator-prey co-evolutionary arms race that influences the diversification of other predatory toxin types. Therefore, the lack of variation within this 3FTx type results from the negative selection pressure on these functionally/structurally important residues. In contrast, α-neurotoxins and the array of ion channels they target likely exert reciprocal evolutionary selection pressures on each other, which produce positive selection pressure for the diversification of the 3FTx gene.

8.4.4 COAGULOPATHIC APOTYPIC 3FTXS

Several 3FTxs interfering with blood-coagulation processes have been identified. The hemextin AB complex inhibits the enzymatic activity of factor VIIa and the factor VIIa/tissue factor complex (Banerjee et al. 2005). Hemextin was isolated from the venom of the elapid snake *Hemachatus haemachatus* and consists of hemextin A and hemextin B forming a tetramer

(Banerjee et al. 2007). In isolation, hemextin A exhibits only mild anticoagulant activity, while hemextin B is inactive. In the hemextin AB complex, hemextin B synergistically enhances the anticoagulant activity of hemextin A. Dendroaspin (or mambin), a type I neurotoxin analogue isolated from the venom of the elapid snake *Dendroaspis jamesoni* contains a disintegrin-like RGD motif on the tip of one of its polypeptide loops and is a potent inhibitor of platelet aggregation (McDowell et al. 1992). It is also able to inhibit platelet adhesion mediated by $\alpha_{II\beta}\beta_3$ integrin (Lu et al. 1994). ϒ-bungarotoxin is a plesiotypic 3FTx from the venom of the elapid snake *Bungarus multicinctus* that contains the RGD motif in the second polypeptide loop, where it is not easily accessible for interaction. As a result, this toxin is a weak antagonist of platelet aggregation (Shiu et al. 2004). The noncytotoxic cardiotoxin A5 from the venom of the elapid snake *Naja mossambica* interacts with integrin, although it does not influence blood coagulation. Cardiotoxin A5 does not exhibit the RGD motif in its amino acid sequence. Nevertheless, A5 binds to $\alpha_V\beta_3$ integrin and inhibits bone resorption (Wu et al. 2006).

8.4.5 MULTIMERIC APOTYPIC 3FTXS

Most 3FTxs are monomeric, although a few exist in multimeric forms, which may associate noncovalently or through disulfide bonds. The κ-bungarotoxins are a well-studied class of noncovalent dimeric 3FTxs (Dewan, Grant, and Sacchettini 1994). The hemextin-AB complex isolated from *H. haemachatus* venom is unique in being a noncovalently associated cytotoxin tetramer, consisting of two copies of hemextin A and two of hemextin B (Banerjee et al. 2005). A series of disulfide-bound dimers of 3FTxs lacking the second and third cysteines have been characterized from *N. kaouthia* venom (Osipov et al. 2008). α-Cobratoxin, a type II (long-chain) neurotoxin, may be a homodimer (αCT-αCT) or as one part of three heterodimer variants, each with a different cytotoxin. In all dimers, the disulfide bond in central loop II of α-cobratoxin is intact. The intermolecular disulfides are identified between Cys3 in one protomer and Cys20 of the second, and vice versa (Osipov et al. 2012), and disulfides from the core are taking part in monomer cross-linking. Both the homodimeric and heterodimeric forms bind to α1and α7 AChR, with the homodimer able also to bind to α3β2 nAChRs (Osipov et al. 2008). A type I α-neurotoxin disulfide-linked homodimer from *O. hannah* venom blocks both muscle-type and neuronal α7, α3β2, and α4β2 nAChRs, with the highest affinity for α7 nAChRs (Roy et al. 2010). The data obtained for the various dimeric toxins show that the dimerization is necessary for efficient toxin binding to heterooligomeric AChRs.

Although κ-neurotoxins evolve rapidly and accumulate variations, they are less diverse than the α-neurotoxins. The relative lack of variation in κ-neurotoxins could be attributed to their unique ability among 3FTxs to form noncovalent dimers (Grant and Chiappinelli 1985). Most residues in this class of 3FTx evolve under the constraint of negative selection (Sunagar et al. 2013b), thus preventing mutations that may affect the toxin's three-dimensional structure and ultimately its ability to form noncovalently linked dimers. The κ-neurotoxins are characterized by the presence of two stabilizing cysteines in loop II that maintain a rigid three-dimensional structure and are homologous to the same disulfide-bond in loop II of the type II α-neurotoxins. Besides the van der Waal's bond between Phe47 and Leu55, as many as six main chains and three side-chain hydrogen bonds are known to stabilize this toxin (Dewan, Grant, and Sacchettini 1994). Hence, the necessity to form such bonds exerts a strong structural constraint against the accumulation of variation in κ-neurotoxins.

Interestingly, besides disulfide-bridge formation, there has been only one other posttranslational modification found in 3FTxs. This modification is the glycosylation of asparagine 29 in some cytotoxins secreted by *N. kaouthia* (Osipov et al. 2004). Several 3FTxs contain consensus glycosylation sites (Osipov et al. 2004); however, no other modifications have been reported so far.

8.5 THERAPEUTIC POTENTIAL OF THE TOXIN TYPE

3FTxs are among the most toxic components of snake venoms. This feature strongly hampers their use *in vivo*. However, there are a number of examples that demonstrate the potential of 3FTxs in the development of therapeutics.

8.5.1 ANALGESIA

It has been demonstrated in several studies that both type I (short-chain) and type II (long-chain) α-neurotoxins possess analgesic properties (Zhang et al. 2006a; Zhang et al. 2006b; Chen et al. 2009; Cheng et al. 2009; Shi et al. 2011). Keluoqu, a painkilling drug with α-cobrotoxin as its primary ingredient, has been on the market in China since 1978. This drug is used to control severe pain in patients with advanced cancer and for postoperative pain (Xu et al. 2006). The biotechnology company Nutra Pharma Corporation and its subsidiary ReceptoPharm are involved in the development of drugs based on α-neurotoxins to treat pain, multiple sclerosis (MS), acquired immunodeficiency syndrome (AIDS) caused by the human immunodeficiency virus (HIV), and adrenomyeloneuropathy (AMN). Nutra Pharma has recently received approval from the US Patent and Trademark Office for its patent (US 7,902,152) titled "Use of cobratoxin as an analgesic," describing the use of α-cobratoxin as a treatment for chronic pain. An oxidized derivative of α-cobratoxin (Receptin preparation from ReceptoPharm) produced analgesic effects similar to those of the natural toxin but is far less toxic (Zhang et al. 2006a). The recently discovered acid-sensing ion channel (ASIC) targeting 3FTx, Malbalgin, from the venom of the elapid snake *Dendroaspis polylepis*, has significant therapeutic potential because of its painkilling activity (Diochot et al. 2012).

8.5.2 MULTIPLE SCLEROSIS

The α-cobratoxin and its derivatives are considered pro-drugs for the treatment of multiple sclerosis (MS) (Reid 2007). Modified forms of this toxin demonstrated neuromodulatory, antiviral, and analgesic activity. It also showed potent immunosuppressive activity in experimental allergic encephalomyelitis (Mohamed et al. 2006). Nutra Pharma is also actively developing this potential. It has received one more approval for its patent (US 8,034,777), entitled "Modified anticholinergic neurotoxins as modulators of the autoimmune reaction," describing the use of neurotoxins for the treatment of MS.

ReceptoPharm products RPI-MN and RPI-78M, developed from α-cobratoxin, have passed preclinical and phase 1 clinical trials for the prevention and/or treatment of MS, HIV, AMN, herpes, rheumatoid arthritis, and pain. However, it should be mentioned that use of modified α-cobratoxin in treatment of AMN has been strongly criticized (Mundy et al. 2003).

Esperanza Peptide Company produces oxidized α-cobratoxin under the trade name Immunokine. The company has developed a sublingual homeopathic spray containing Immunokine and uses it for treatment of MS and other diseases. However, the usefulness of this product has been thoroughly discredited http://www.dcscience.net/?p=3109.

8.5.3 CARDIOVASCULAR DISEASE

The recently discovered β-cardiotoxins show potential as tools for researching β-adrenergic receptors and as therapeutics for cardiovascular disease. Compared with β-agonists, the use of β-adrenergic receptor blockers as therapy for chronic heart failure resulted in a 30% to 35% reduction in mortality risk (Eichhorn and Bristow 2001). The β_1-receptors regulate pro-apoptotic signaling (cardiotoxicity), and β_2-receptors regulate anti-apoptotic signaling (cardioprotection). However, while these are the overarching activities guided by each receptor, the role in regulating cardiotoxicity and cardioprotection of these two β-adrenergic receptor subtypes is complex and variable and influenced by acute and chronic stressors (Bernstein, Fajardo, and Zhao 2011).

8.5.4 ANTICANCER

Antitumor activity (squamous-cell skin cancer and breast cancer) of a cardiotoxin and crotoxin (see chapter 21) combination called VRCTC-310 is currently in phase 1 clinical trials (Costa et al. 1998). Antimetastatic activity of the cytotoxin CTX 3 from the venom of the elapid snake *N. atra* has been shown to be a result of decreased expression and activity of MMP-9 (matrix metalloprotease 9) through the inactivation of p38 MAPK and PI3K/Akt signaling pathways and NF-κB activity in the human breast-cancer cell line MDA-MB-231 (Lin et al. 2012). In addition, the increase of membrane permeability caused by this toxin, along with cytotoxin γ from the venom of the elapid snake *Naja nigricollis* has been shown to have bactericidal effects (Chen et al. 2011a; Chen et al. 2011b). Apoptosis-induced cytotoxicity of human leukemic U937 and K562 cells has been shown by the cytotoxin NK-CT1 isolated from the venom of the elapid snake *N. kaouthia*, with this toxin also demonstrating the ability to produce auricular blockade (Debnath et al. 2010). The binding of $\alpha_V\beta_3$ integrin by cardiotoxin A5 may be of interest in the development of anticancer drugs, as this integrin is known to be important in tumor growth and is highly expressed on new blood vessels of tumors. The inhibition of this integrin may therefore inhibit tumor angiogenesis, leading to a reduction in tumor growth (Wu et al. 2006).

8.6 CONVERGENCE WITH OTHER TOXINS

Structurally, this is the only use of this peptide framework as a toxin. There are, however, areas of functional convergence (see color plate 2B). This is particularly the case for the *Dendroaspis* genus, which produces a complex, unique form of neurotoxicity (see chapter 1.3.5). Venoms from the *Dendroaspis* genus of the elapid snakes contain presynaptic-acting kunitz peptides (see chapter 15.4) that inhibit voltage-dependent potassium channels and L-type calcium channels in the same manner as the apotypic 3FTxs from the same venom. Voltage-dependent potassium channels have also been convergently targeted by scorpion venoms (Bergeron and Bingham 2012), while L-type calcium channels have also been convergently targeted by spider venoms (Kubista et al. 2007; Ono, Kimura, and Kubo 2011).

Another area of convergence within snake venoms is the muscarinic 3FTxs being functionally convergent with PLA_2s (type I) (see chapter 20) from *N. atra*, which also act on this receptor type (Huang et al. 2008). Muscarinic receptors have also been convergently targeted by spider venoms (Almeida et al. 2011). Nicotinic acetylcholine receptors have been convergently targeted by other snake-venom peptide types such as the unique SVMP propeptide apotypic forms from *Psammophis* venoms (see chapter 23.4.4), the neurotoxin azemiopsin from *Azemipos feae* (see chapter 19.4), and the waglerin peptides from *Tropidolaeumus* venoms (see chapter 24.25). Convergence extends to coagulopathic actions, such as the RGD functional motif of the antiplatelet 3FTxs (dendroaspins/mambins) (see chapter 8.4.4), which is a domain convergence with the disintegrin domains of the SVMP (see chapter 23) and with RGD antiplatelet toxins from other venomous animals (Fry et al. 2009a) (see color plate 2A).

CHAPTER 9

BETA-DEFENSINS

A. R. B. PRIETO DA SILVA, B. G. FRY, K. SUNAGAR, H. SCHEIB, T. N. W. JACKSON, G. RÁDIS-BAPTISTA, A. J. ZAHARENKO, P. L. DE SÁ JR., A. PEREIRA, N. OGUIURA, M. A. F. HAYASHI, A. KERKIS, T. YAMANE, AND I. KERKIS

9.1 SUMMARY

The β-defensins are a diverse group of small cationic antimicrobial peptides. They are not only an integral part of the vertebrate innate immune system, but they also serve as myotoxic venom proteins in advanced snakes. Despite extreme sequence variability and diverse biological activities, all β-defensins conserve the same three-dimensional structure, and their genes have the same organization. The most well-studied reptilian β-defensin venom homologues are from the genus *Crotalus*, of which the toxin crotamine from *Crotalus durissus terrificus* has received the most attention. Crotamine has a well-characterized ability to cause severe muscle necrosis through a nonenzymatic mechanism. Crotamine was long thought to affect sodium channels, but emerging evidences suggest that it blocks voltage-gated potassium channels (K_V). This peptide has also demonstrated biological versatility, not only acting as a toxin but also possessing cell-penetrating peptide (CPP) properties, targeting chromosomes, and acting as a DNA plasmid carrier. Additionally, it shows specificity for actively proliferating cells and has selective antitumor activity in preclinical melanoma models. All these biological properties of crotamines are of great pharmacological potential (Kerkis et al. 2010; Kerkis et al. 2014).

9.2 ENDOPHYSIOLOGICAL PLESIOTYPE

The β-defensins, being rich in lysine and arginine residues, are cationic peptides. They are found in both vertebrates and invertebrates. They have well-characterized antimicrobial activity (Lehrer and Ganz 2002). The high net-positive charge of β-defensins facilitates electrostatic interaction with negatively charged cell surfaces. Thus, they can cause membrane permeabilization, in which porelike structures form, leading to the efflux of cellular nutrients. They can also interact specifically with negatively charged targets or receptors. The compact structure of β-defensins consists of 18 to 45 amino acids, including a conserved cysteine pattern that results in a characteristic three-cysteine bridge linked 1-5, 2-4, 3-6. Despite extremely variable sequences and biological activities, β-defensins have well-conserved elements of secondary structure and also of residues responsible for the formation of the characteristic fold: all the cysteine residues; the glycine residues 8, 9, 26; and proline 20, including the γ-core motif G-$X_{(3)}$-C-$X_{(3\text{-}9)}$-C-K (Yount and Yeaman 2004). Cysteines involved in the formation of disulfide bonds, and also the α-helix and the γ-core motifs, are highly conserved. The β-defensins and their homologues exhibit impressive variation in their coding sequences, resulting in a considerable diversity of physicochemical properties (such as charge and hydropathicity) and potential oligomerization propensities. While a similar structural scaffold characterizes these peptides, each may individually exhibit numerous biochemical activities and solvent accessibilities. There is extreme variability in molecular-surface charge distribution. Physicochemical variability in similarly folded peptides might account for the observed differences in biological functions and specificities.

Because of the antimicrobial nature of β-defensins, they are an integral part of the host defense mechanism in multicellular organisms. Several avian β-defensins, which are found in various compartments of the egg (such as shell, egg white, and vitelline membranes), have been theorized to be involved in embryonic protection (Herve-Grepinet et al. 2010). Gallinacin-2, for instance, exhibits potent antibacterial activity not only against the Gram-negative bacterium *Eschericia coli* ML-35 but also against the Gram-positive bacterium *Listeria monocytogenes* EGD; however, it lacks antifungal activity against *Candida albicans* (Harwig et al. 1994). Gallinacins are expressed in circulating avian heterophil granulocytes, bone marrow, lungs, and testes. Moderate expression of gallinacin is observed in synovial bursa and intestine, while cloaca, gall bladder, brain, pancreas, trachea, air sacs, and spleen exhibit low-level expression of these peptides. Gallinacins are also expressed in vagina, ovarian stroma, and the theca layer of the ovarian follicle but not in the granulosa layer. Certain avian β-defensins, such as spheniscins, are known to play an important role in the preservation of food in the stomach of king penguins (*Aptenodytes patagonicus*) (Thouzeau et al. 2003; Landon et al. 2004). Similar to those of birds, the eggs of nonfeathered reptiles are also enriched with several defensin-like peptides, such as pelovaterin, which plays a dual role in the eggshell of the Chinese soft-shell turtle (*Pelodiscus sinensis*). It induces the nucleation and stabilization of vaterite, one of the crystalline polymorphs of calcium carbonate in eggshells, and also exhibits strong antimicrobial activity against *Pseudomonas aeruginosa* and *Proteus vulgaris* (Lakshminarayanan et al. 2008). Other cationic peptides with the β-defensin scaffold may replace lysozymes in reptilian eggs (Chattopadhyay et al. 2006). Certain reptilian β-defensins that are secreted by leukocytes in blood may participate in the innate immune response (Stegemann et al. 2009). In addition to being efficient in combating bacteria, β-defensins display antifungal and antiviral activities.

Crotasin (Cts-p2) is the closest endophysiological homologue of the venom-type β-defensins. This peptide was first identified in genomic libraries of crotamine-negative specimens of *C. d. terrificus* (Rádis-Baptista et al. 2004). Crotasin, whose physiological role is still unknown, is abundantly expressed in the pancreas, heart, liver, brain, and kidneys of snakes, while being scarcely secreted in the venom glands. The gene organization of crotasin and venom β-defensins indicates that both genes have likely evolved through the duplication of a single plesiotypic gene (Correa and Oguiura 2013). Indeed, crotasin is similar to the putative ancestor of the venom-gland forms (Rádis-Baptista et al. 1999; Rádis-Baptista et al. 2003). The signal peptide sequences of crotasin and venom β-defensin precursors are very similar, whereas the mature peptides share less than 30% identity. However, the conservation of disulfide-bond-forming cysteine residues suggests that they might adopt similar folds.

9.3 PLESIOTYPIC STRUCTURAL AND FUNCTIONAL TOXIN FORMS

Rattlesnake β-defensins, with similar chemical and biological properties to crotamines, have been referred to as small basic polypeptide myotoxins (SBPMs) (Ownby 1998). However, this nomenclature does not apply to all reptilian venom-gland β-defensins, since not all are cationic and myotoxic activity has only been characterized in *Crotalus* venom forms. Thus, the broader suite of venom-gland-expressed peptides are collectively referred to as β-defensins. While the putatively antimicrobial forms from iguanian lizards are diverse, and the anguimorpha lizard variants are also extensively mutated, the *Crotalus* venom forms are highly conserved (see figures 9.1 and 9.2 and color plate 27).

While the bioactivities of the lizard gland β-defensins remain to be characterized, the snake-venom homologues have been the subject of intense research. Crotamine from *C. d. terrificus* has been particularly well investigated. Crotamine, first isolated in 1947, is one of four major components isolated from the venom of the South American rattlesnake *C. d. terrificus* (Gonçalves and Polson 1947). Crotamines have a single chain with 42 amino acids (MW 4880 Da), six-cysteines pattern characteristic of β-defensins, which fold into the typical three disulfide-bond conformations (Nicastro et al. 2003; Fadel et al. 2005). Crotamine is rich in basic residues (nine lysines, two arginines, and two histidines) and has a higher net-positive surface charge (theoretical pI 9.51, experimental pI 10.3) than nonvenom β-defensin-like peptides (see color plate 27). It lacks alanine, asparagine, valine, and threonine residues. Crotamine is an amphipathic molecule that exposes a large positively charged patch on both sides of the molecular surface, with a narrow but well-delimited hydrophobic section (see color plate 27). Crotamine can be found in two isoforms with a minimal difference in primary sequence. In the crotamine-Ile-19 isoform, which was isolated from the venom of *Crotalus durissus ruruima*, the leucine residue at position 19 is replaced by isoleucine (Dos Santos et al. 1993). Previously the UniProt database entry P63327 having the isoform 1 crotamine sequence instead of crotamine Ile-19 (isoform 2), but this has now been rectified. It appears that the intrinsic chemical properties of leucine and isoleucine do not produce appreciable steric differences or influence the overall three-dimensional structures of these isoforms, as both peptides have the same biological activity and are immunologically indistinguishable. Moreover, neither form causes observable tissue damage (Dos Santos et al. 1993).

The electrostatic surface charge of the molecule is responsible for several of the biological activities of crotamine, particularly its selective cell-penetrating action toward certain cell types at particular phases of cell cycle (Kerkis et al. 2004; Kerkis et al. 2006). Crotamines possibly interact with specific targets such as ion channels and proteoglycans (Nascimento et al. 2007; Yount et al. 2009), resulting in the reduction of membrane potential, thus increasing the influx of ions through the membrane and the modification of conductance by a channel-mediated

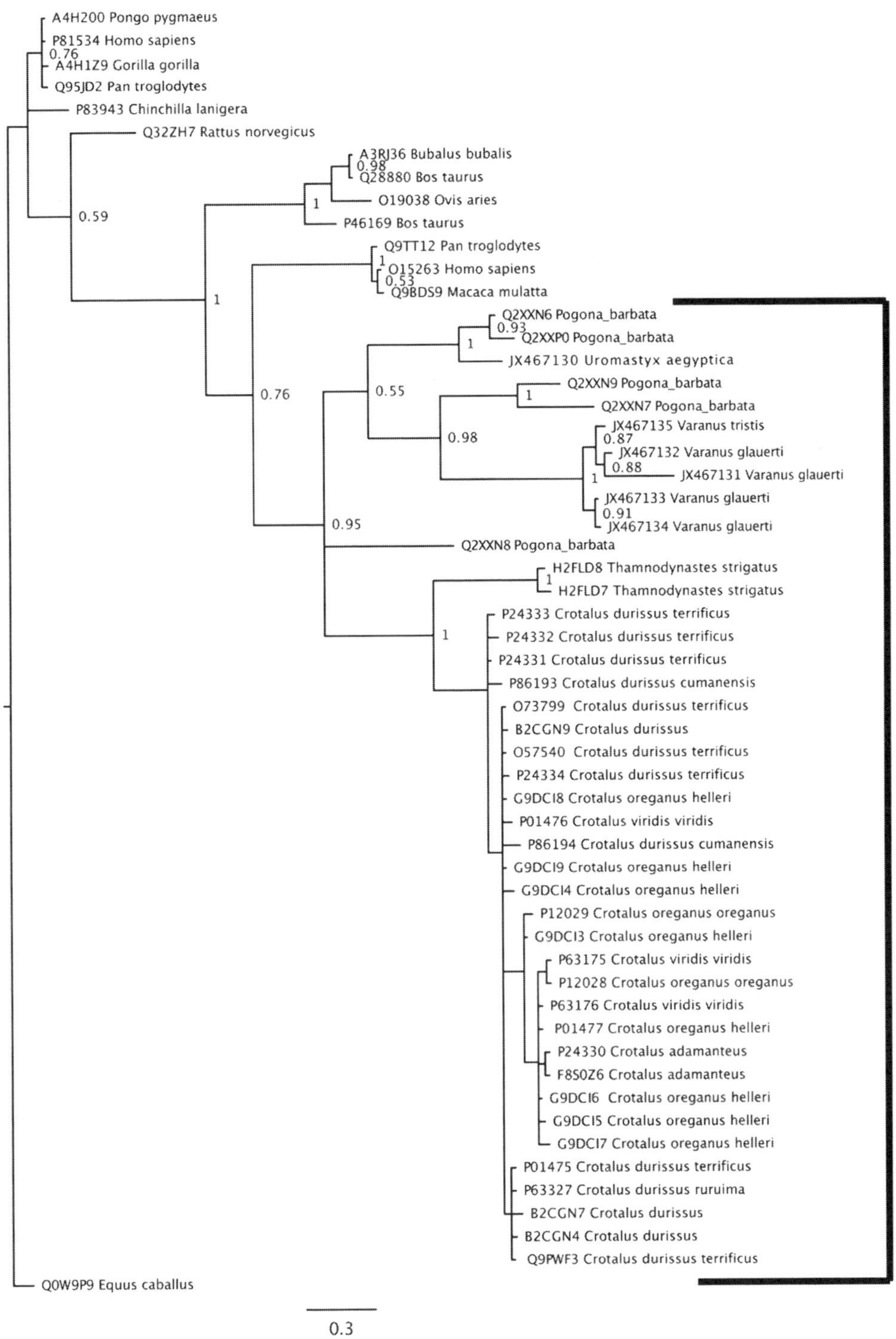

FIGURE 9.1: Molecular phylogenetic tree of β-defensin peptides and related nontoxin proteins (Fry et al. 2013).

```
1.  mkilyllfaflflaflsepgna------------------------------------------YKQ
2.  mkilyllfaflflaflsepgna------------------------------------------YKR
3.  mkilyllfaflflaflsepgna------------------------------------------YKR
4.  mkilyllfallflaflsepgna------------------------------------------QDL
5.  mkflyllcavlflgllqapgvtea-------------------------------------IKSYAE
6.  mkflyllcavlflgllqapgvtea-------------------------------------IKSYAK
7.  mrflyllfavaflfsvqaEDAELEQEQQGDPWEGLDEFQDQPPDDLGADDDAAEDRADIQFRSGFDG
8.  mkllailsavlffvlmaasvda---------------------------------------GPSSSG
9.  mkflyllcavlflgllqapgvtqa-------------------------------------VDSYEE
10. mktllllcllvlaaf-ea-sa-------------HPKPDGEPEPLQAADGGAPGMEGRTAVRDNSAW
11. mktllllcllvfaaf-qa-sa-------------HLKPDGEPEPLQVADGGAPGMEGRTSVRDDSAW

1.  C-HKKGGHCFPKEK-ICIPPSSDFGKMDCRW-RW---KCCKKGSGK
2.  C-HKKGGHCFPKTV-ICLPPSSDFGKMDCRW-KW---KCCKKGSVNNAISI
3.  C-HKKGGHCFPKEK-ICIPPSSDFGKMDCRW-KW---KCCKKGKWKIMPSPSRTMDIFKIWPRT
4.  C-HNLGGRCF-RNR--CSWSLRNHGGQDCPW-GS---VCCKP
5.  CERRRGTACY--SV-VCvRGTKTVGK--CNWYQ----KCCQSVIGK
6.  CERRRGTACY--SV-VCPRGTKTIGK--CNWYQKMLPKCHWQVIPP
7.  CRRS-RGKCY---RFMCPRGTATVGK--CS-FTF---FCCK
8.  C-HAARGYCYFMD---CPRGRRQIGN--CGSLRR---PCCK
9.  CSRTRASACY--S-FLCPRGTVTIGK--C-SWTQ---KCCQSVLGK
10. CG-SRGGKCY---LILCPRGTGRIGK--C-STMY---VCCK
11. CG-SRGGKCY---LILCPRGTGRIGK--C-STMY---VCCK
```

FIGURE 9.2: Sequence alignment of representative β-defensin toxins (Fry et al. 2013). (1) O57540 *Crotalus durissus terrificus*, (2) G9DCI6 *Crotalus oreganus helleri*, (3) G9DCI4 *C. o. helleri*, (4) H2FLD8 *Thamnodynastes strigatus*, (5) Q2XXN6 *Pogona barbata*, (6) Q2XXP0 *P. barbata*, (7) Q2XXN9 *P. barbata*, (8) Q2XXN8 *P. barbata*, (9) M9T5K8 *Uromastyx aegyptica*, (10) JX467132 *Varanus glauerti*, (11) JX467135 *Varanus tristis*. Cysteines are highlighted in black, prolines in gray; signal peptides are shown in lower-case.

mechanism, ultimately releasing Ca^{2+} ions from the sarcoplasmatic reticulum (Katagiri et al. 1998; Ownby 1998; Nascimento et al. 2012). Crotamine binds strongly to excitable membranes, leading to the contraction of skeletal muscles and spastic paralysis in the hind limbs of mice (Cheymol et al. 1971). Other effects include lachrymal hypersecretion, dyspnea, tachycardia, paralysis, and death (Brazil, Prado-Franceschi, and Laure 1979). The LD_{50} of crotamine is 1.5 mg/kg by intravenous injection (Allen, Tucker, and Green 1986).

Crotamine induces irreversible membrane depolarization and spontaneous repetitive firings of mammalian skeletal muscle (Chang and Tseng 1978; Brazil, Prado-Franceschi, and Laure 1979), apparently indicating that it interferes with Na^+ currents (Matavel et al. 1998). After several studies showing indirect evidence that crotamine might act on Na^+ currents, it was demonstrated that it actually did not directly affect mammalian voltage-dependent sodium channels (Rizzi et al. 2007). This was further supported by pharmacological comparisons between crotamine and toxins with proven sodium-channel activity, where the latter were unable to mimic the hind-limb paralysis caused by crotamine (Peigneur et al. 2012a). Computational docking models indicated that crotamine might act as a voltage-dependent potassium-channel blocker that specifically recognizes residues of the eukaryotic channel pore (Yount et al. 2009). This possible action on voltage-gated potassium K_V channels has been investigated in an effort to explain crotamine's *in vivo* effects (Peigneur et al. 2012a). Crotamine's effect on ion-channel function was investigated using the two-electrode voltage-clamp technique on 16 cloned ion channels (12 K_V channels and 4 Na_V channels) expressed in *Xenopus laevis* oocytes. Crotamine was shown to selectively inhibit $K_V1.1$, $K_V1.2$, and $K_V1.3$ channels with $IC_{(50)}$ values of 369.43 ± 55.51 nM, 386.14 ± 10.62 nM, and 286.53 ± 91.72 nM, respectively. The inhibition of $K_V1.3$ channels induced by crotamine occurred rapidly, was not voltage-dependent, and was reversible after wash, suggesting a classic channel-blockage effect without altering K_V kinetics.

Although there are several animal neurotoxins that can interact with potassium channels (Mouhat et al. 2008), snake toxins with this activity are uncommon. Previously known snake toxins with K_V-channel-inhibiting activity were restricted to apotypic CRiSP proteins from *Naja atra* venom (Wang et al. 2006a) (see chapter 10), and apotypic kunitz peptides from mamba venoms (see Harvey and Robertson 2004) (see chapter 15). Kunitz peptides with this activity have also been characterized from sea anemone venom (see Peigneur et al. 2011). However, none of these toxins possesses the β-defensin scaffold; thus, crotamine was the first β-defensin characterized with K_V-channel-inhibiting activity. However, some structural aspects of crotamine have a similar topology to type III K_V-channel toxins in anemones that potently and selectively modulate the hERG-related potassium channels (Diochot et al. 2003; Peigneur et al. 2012b).

The relative presence of β-defensin peptides in *Crotalus* venoms varies considerably between species and even within different populations of the same species (Boldrini-Franca et al. 2010). For example, two subpopulations of *C. d. terrificus*, "crotamine positive" and "crotamine negative", exist in Brazil (Schenberg 1959). Crotamine-positive rattlesnakes show variation in the amount of this toxin in the crude venom, although it makes up not less than 10% of the total venom weight. In crotamine-negative venom, even in high concentration of the venom, crotamine could not be detected by ELISA quantification assay (Oguiura et al. 2000). Northern blot analysis using crotamine cDNA as a probe with total RNA and poly(A^+) RNA samples extracted from proven crotamine-positive and crotamine-negative specimens revealed a band of approximately 0.4 kb that was present only in the total mRNA sample from crotamine-positive venom glands (Rádis-Baptista et al. 1999). This variation in venom composition is not restricted to South American subpopulations. The venom of *C. adamanteus* exhibits remarkable intraspecific variation, as do *C. horridus, C. oreganus helleri, C. scutulatus,* and *C. viridis* venoms (Straight et al. 1991).

While most venom-encoding genes accumulate tremendous variation, the β-defensins secreted in large quantities by various populations of *C. o. helleri* completely lack variations in their coding sequences and experience significant influence from negative selection (Sunagar et al. 2014). The few hypermutable sites that exist are located in nonsecreted regions of the toxin and likely do not contribute in the envenoming process (Sunagar et al. 2014). Nearly 76% of residues in these peptides are extremely well conserved (percent identity ≥ 90%). The lack of variation in the secreted regions of these toxins is likely an indicator of their nonspecific mode of action, binding to negatively charged membranes using their cationic amino acids, consequently resulting in membrane permeabilization. Therefore, it is expected that the evolutionary constraints of this toxin type favor the preservation of cationic residues required for toxicity. Of residues in these toxins, 29% are cationic (K, R, and H) and were extremely well conserved (percent identity ≥ 80%; Sunagar et al. 2014).

β-defensins isolated from *Crotalus* venoms include myotoxin-α from *C. viridis viridis* (Cameron and Tu 1977; Fox, Elzinga, and Tu 1979) peptide C from *C. o. helleri* (Maeda et al. 1978), myotoxin I and II from *C. viridis concolor* (Bieber, McParland, and Becker 1987), CAM-toxin from *C. adamanteus* (Samejima, Aoki, and Mebs 1991), and E toxin from *C. h. horridus* (Allen et al. 1996). There is a remarkably high degree of amino acid conservation among these members, with scores varying from 83% to 98%. For example, crotamine (UniProt entry P01475) shares 92.86% sequence identity with *C. v. viridis* myotoxin-α (UniProt entry P01476), differing only by two residues: at position 25, phenylalanine is substituted by leucine, and at position 33, conservative substitution has occurred with arginine substituted by lysine. Equivalent similarity exists between myotoxin-α and myotoxin 2 and 3 isoforms (UniProt P63175, P63176) from *C. v. viridis* venom. Myotoxin 1 and 2 (UniProt P12028, P12029) from *C. v. concolor* venom and crotamine-IV-2 and IV-3 (UniProt P86193, P86194) from *C. durissus cumanensis* venom share a similar percentage of residues. Individual variation occurs in myotoxin-α (Aird, Kruggel, and Kaiser 1991), and the number of isoforms suggests that myotoxin-α is the product of a duplicated locus producing natural variants in venom with substitutions of isoleucine by threonine at position 19, leucine by phenylalanine or lysine at position 25, and lysine by arginine at position 33. In a similar manner, myotoxin-2 from the venom of *C. v. concolor* occurs in several variants involving minor substitutions: glutamate by threonine at position 15, lysine by valine at position 16, threonine by leucine at position 19, and lysine by arginine at position 33 (Bieber and Nedelkov 1997). Interestingly, *C. v. viridis* myotoxin-2 is similar to myotoxin-1 from *C. v. concolor*, except that it possesses an additional C-terminal asparaginyl-alanine. With 45 residues, it is the longest known snake myotoxin-α homolog, and it is cleaved at position 43, producing the mature myotoxin-4 (Griffin and Aird 1990). Such divergence indicates active molecular evolution of this venom peptide type.

Crotamines were first isolated and characterized from *Crotalus* venoms, and they remain the most intensively investigated β-defensins; β-defensin was, however, recruited for use as a toxin at the base of the toxicoferan reptile chemical arsenal (Fry et al. 2006; Fry et al. 2010b; Fry et al. 2013). The β-defensin peptides are the most diverse (possessing hypervariable regions) and highly expressed peptide/protein type in the incipient venom glands of iguanian lizards. In this lineage, they are likely restricted to an antimicrobial role. Subsequently, they were weaponized for use in envenoming by other toxicoferan lineages.

9.4 APOTYPIC STRUCTURAL AND FUNCTIONAL TOXIN FORMS

None is documented.

9.5 THERAPEUTIC POTENTIAL OF THE TOXIN TYPE

The reported antinociceptive effect of crotamines (Mancin et al. 1998) is probably caused by the co-elution of these peptides with the 14 amino acid long crotalphine. Crotalphine is a degradation product of crotapotin γ-chain, a nonenzymatic and nontoxic acidic subunit of the major neurotoxic venom component crotoxin (Konno et al. 2008).

Unlike plesiotypic β-defensin peptides, crotamine inhibits microbes only at minimal concentrations, 25 μg/mL to 100 μg/mL against *E. coli* with little or no action against other species of Gram-negative and Gram-positive bacteria (Oguiura et al. 2011; Yamane et al. 2013). Regardless of its modest effect against bacteria and unlike mammalian β-defensins, crotamine acts against eukaryotic cells like those of the fungus *C. albicans*, altering membrane potential and inducing phosphatidylserine accessibility (Yount et al. 2009; Yamane et al. 2013). This supports the hypothesis that antimicrobial β-defensins and snake-venom myotoxins like crotamine differentially evolved to distinguish between microbial and mammalian ion channels. This selectivity is achieved through specific molecular surface interactions between toxin and target. Crotamine has antifungal activity against not only *Candida* but also other fungi such as *Trichosporon* and *Cryptococcus neoformans* in concentrations of 12 to 50 μg/mL (Yamane et al. 2013). Crotamine activity is clearly distinguishable from other AMPs because it is selective for proliferative eukaryotic cells (Kerkis et al. 2010; Radis-Baptista and Kerkis 2011; Yamane et al. 2013).

The inhibition of K_V channels by crotamine could be related to its selectivity for proliferating active cells and tumor cells. Potassium channels play an important role in membrane hyperpolarization, regulating cell-cycle progression from G_1/S (Wonderlin and Strobl 1996; Ouadid-Ahidouch and Ahidouch 2008; Becchetti 2011). The enhanced expression of $K_V1.3$ channels and their critical role in the proliferation of several types of carcinogenic cell types have been well established (Bielanska et al. 2012). The inhibition of $K_V1.3$ channels expressed in human Jurkat T cells may be involved in the antiproliferative and proapoptotic effects of drugs such as 8-prenylnaringenin, which act as immune suppressors (Gasiorowska et al. 2012). Recently described, mito$K_V1.3$ is the first potassium channel located in mitochondrial membranes and plays an important role in regulating cell apoptosis (Szewczyk, Jarmuszkiewicz, and Kunz 2009; Szabo, Zoratti, and Gulbins 2010). Mito$K_V1.3$ is inhibited by Bax, which induces hyperpolarization of the mitochondrial membrane. Interaction between Bax and the pore channel depends on a lysine residue similar to the functional binding dyad from K^+-channel-blocking toxins (Leanza et al. 2012). Crotamine might interact with K_V channels, including mitochondrial channels, in a similar way to neurotoxins from sea anemones, since a comparable "functional dyad" is hypothesized to occur on it (Peigneur et al. 2012a). Identification of this dyad might be helpful in comparing the effects and mechanisms of action of such toxins on organelles such as mitochondria. Recent studies have demonstrated that crotamine elicits mitochondrial depolarization and intracellular calcium release (Nascimento et al. 2012).

Crotamine in micromolar concentrations is nontoxic to any of the cell cultures tested thus far (including fibroblast 3T3, embryonic stem cells [ES cells], HEK293, PC12, CHO-K1, and primary culture astrocyte cells) and does not affect the pluripotency of ES cells or the development of mouse embryos (Kerkis et al. 2004). Despite its low cytotoxicity against nontumor cell lines, crotamine has potent and specific toxicity against aggressive mouse and human tumor cell lines such as murine melanoma cells (B16-F10), human skin-melanoma cells (SK-MEL-28), and a pancreatic carcinoma cell line (MIA PaCa-2). *In vivo* crotamine treatment significantly delayed implantation of tumor in preclinical melanoma model, inhibited tumor growth, and prolonged the life span of the mice (Pereira et al. 2011). Because of all the aforementioned properties, crotamine has great potential for use as an imaging agent for detecting dividing cells, for the intracellular delivery of hydrophilic biomolecules, and as an alternative chemotherapeutic compound against aggressive types of cancer.

Crotamine is potentially useful because of its selective biological action toward certain cell types at a given phase of cell cycle and for its ability to rapidly and efficiently translocate into actively proliferating cells (Radis-Baptista and Kerkis 2011). Crotamine has high selectivity for proliferating cells *in vitro* and *in vivo* and shows cell-penetrating ability in various cell lines and in mouse blastocysts. By accumulating in the cell nucleus, crotamine can strongly label cells from mouse bone marrow, spleen, and peritoneal liquid. Nuclear localization of it was observed in both unfixed and fixed cells. In the cytoplasm, crotamine specifically associates with centrosomes, facilitating the process of centriole duplication and separation. In the nucleus, it also binds to chromosomes at S/G2 phase, when centrioles start dividing (Kerkis et al. 2004). Actively proliferative cells uptake crotamine via endocytosis, an event that ensures crotamine binding to cell membranes via heparan sulfate (Nascimento et al. 2007). It is also useful for intracellular vesicle tracking and as a cell-cycle marker of proteoglycans (Nascimento et al. 2007; Hayashi et al. 2008). Crotamine interacts with DNA and acts as a carrier capable of delivering DNA into replicating mammalian cells with low cytotoxicity against normal proliferative cells (Nascimento et al. 2007; Hayashi et al. 2008; Kerkis et al. 2010; Hayashi et al. 2012).

Crotamine is capable of forming DNA-peptide complexes by electrostatically binding to negatively charged DNA molecules like plasmid vectors. Thus, it could be used as a vector for rapid and efficient translocation of coupled drug compounds into actively proliferating cells. It could also be used for labeling highly replicating cells (Nascimento et al. 2007; Radis-Baptista and Kerkis 2011). Recently, the binding of crotamine to single- and double-stranded DNA of different lengths and base compositions has been studied over a range of ionic conditions. Inspection of the three-dimensional structure of crotamine suggests that residues Arg31 to Lys35 could serve as a potential DNA-binding site. A derivate hex peptide containing this sequence behaves similarly to full-length crotamine (Chen et al. 2012a). Structure-guided deconstruction of crotamine in an attempt to define the minimal structural motif essential for membrane translocation and internalization has been used to identify and design therapeutically useful crotamine derivatives called nucleolar translocation peptides (NrTPs) (Radis-Baptista, de la Torre, and Andreu 2008; Radis-Baptista, de la Torre, and Andreu 2012; Rodrigues et al. 2012). NrTPs and cysteine-rich cell-penetrating peptide for cytosolic delivery (CyLop-1) are minimal structural motifs from crotamine essential for membrane translocation and internalization. CyLop-1 escapes from the endocytic pathway and exhibits cytosolic distribution. CyLoP enters the cytosol by endosomal uptake and delivers small cargos of pro-apoptotic molecules (Jha et al. 2011). Several distinct NrTPs-1 were created, which possess remarkable translocating properties, such as fast internalization and preferentially targeting the nucleolus of tumor cells (Radis-Baptista, de la Torre, and Andreu 2012). NrTPs conjugated with EGFP (enhanced green fluorescent protein) confirm the preferential nucleolar localization of the translocating and internalized peptide (Radis-Baptista, de la Torre, and Andreu 2008). The fluorescence vanished after 24 hours' incubation of HeLa cells with peptide-free culture medium, indicating relatively fast degradation of the internalized molecules. All peptides were found to be harmless to HeLa cells up to 100 μM. In order to demonstrate the capacity of NrTP to mediate the intracellular delivery of large molecules, it was conjugated with a large protein, β-Galactosidase (116.5 kDa), which far exceeds the thickness of cell membranes and is equivalent in size to many therapeutically relevant cargos (such as antibodies around 150 kDa). The NrTP-β-gal was efficiently translocated into HeLa cells, as monitored by stable β-galactosidase activity after cell lysate (Rodrigues et al. 2011). These results clearly demonstrate that spliced crotamine derivatives are very efficient in entering cells and accumulating in their cytoplasm or nucleoli. Crotamines and their derivatives, such as the CPPs, include a novel nucleolar localization signature, which confers on them a remarkable nucleolus-homing ability. This could be harnessed for a number of purposes, such as specific and precise delivery of diverse cargos or DNA vectors into tumor cells, as imaging agents and functional molecules. Thus, they have tremendous applications in the field of immunotherapy, cell cycle regulation, and transgenesis (Nascimento et al. 2007; Chen et al. 2012a; Radis-Baptista, de la Torre, and Andreu 2012).

9.6 CONVERGENCE WITH OTHER VENOMS

The β-defensin peptides have been convergently recruited into the venoms of various animals, including sea anemones, scorpions, spiders, platypus, and shrews (Fry et al. 2009a). Crotamine has evident structural similarities to the platypus venom defensin-like-peptide-2 (DLP-2) but shares only 25% of its sequence identity (Torres et al. 2000; Torres and Kuchel 2004; Whittington et al. 2008). DLP-2 is expressed by a precursor with a signal peptide of 21 residues and a propeptide consisting of only two residues. The mature chain comprises a peptide with 42 amino acids, including two variants at position 26, producing a stereo-inversion of L-Met-26 (in DLP-4) to D-Met-26 (in DLP-2). The most significant differences can be seen near the N- and C-termini compared with crotamine; the DLP-2 polypeptide is elongated at the N-terminus and truncated near the C-terminus. The activity of DLP-2 has not yet been well characterized, but it does not act on Na^+ ion channels or have any demonstrated antimicrobial, myotoxic, hemolytic, or cellular activities.

Recently, another animal toxin called maurocalcine (L-MCa), in its D-diastereomer conformation, was isolated and purified from the venom of a Tunisian scorpion (*Scorpio maurus palmatus*) and was also categorized as CPP. However, it has a different fold from crotamine and other β-defensins (Poillot et al. 2010). Despite the low sequence similar and the distinct sequential arrangement of the secondary structure elements of crotamine ($\alpha\beta_1\beta_2$) and toxin AaHII ($\beta_1\alpha\beta_2\beta_3$) from the venom of the scorpion *Leiurus quinquestratus*, the overall fold of the toxin is quite similar to some of the plant defensins and the scorpion potassium channel toxins, such as the charybdotoxin (Nicastro et al. 2003). Interestingly, this sodium-channel scorpion toxin (AaHII) from *L. quinquestriatus* is also a potent cytoskeletal agent *in vitro* and has pronounced effects on the *in vitro* polymerization and stability of neuronal microtubules purified by temperature-dependent cycles of assembly and disassembly (Brown, Johnston, and Tolbert 1983).

Some sea anemone neurotoxins secreted by nematocysts have a disulfide-bridge pattern similar to β-defensin-like peptides including crotamine and DLP-2 (Siqueira et al. 2002; Oliveira et al. 2006; Zaharenko et al. 2008). This group includes APETx-1, a neurotoxin from *Anthopleura elegantissima*, which potently and selectively modulates the voltage-dependence of the channel gating of human Ether-a-go-go-related potassium channels such as KCNH2, KCNH6, and KCNH7 (Chagot et al. 2005). This group of peptides delays inactivation during signal transduction and is much more potent against crustaceans than against vertebrates (Monks et al. 1994).

CHAPTER 10

CYSTEINE-RICH SECRETORY PROTEINS

K. SUNAGAR, T. N. W. JACKSON, T. REEKS, AND B. G. FRY

10.1 SUMMARY

Cysteine-rich secretory proteins (CRiSPs) are exclusively found in vertebrates and are associated with a broad range of functions. These proteins, however, are part of the CAP (CRiSP, antigen 5 [Ag5], and pathogenesis-related 1 [Pr-1]) superfamily of proteins which are distributed widely across the tree of life. Proteins from the CAP superfamily are present in the venoms of a range of venomous animal lineages, such as coleoids, cone snails, stinging insects, scorpions, spiders, lampreys, and even vampire bats. This diverse recruitment highlights the importance of this molecular scaffold in a range of envenomation strategies.

CRiSPs, which are characterized by sixteen universally conserved cysteine residues, were originally isolated from mammalian reproductive tracts. Venoms of toxicoferan reptiles, particularly certain non-front-fanged advanced snakes and anguimorph lizards, are also rich in CRiSP proteins. Despite their being characterized as a major component of venom in these lineages, there is considerable debate surrounding their biological function in venoms, which remains enigmatic in many cases. They are hypothesized to participate in envenoming by disrupting homeostasis in the prey through several mechanisms, including blockage of cyclic nucleotide-gated and voltage-gated ion channels and inhibition of smooth-muscle contraction. Similarly it is unclear whether CRiSPs have been developed into toxins on a single early occasion or on two separate occasions, but with phylogenetic analyses favoring the latter scenario.

10.2 ENDOPHYSIOLOGICAL PLESIOTYPE

The CAP superfamily consists of CRiSP, AG5 (from insects), and PR-1 (from plants). Although members of the CAP superfamily are distributed throughout the domains Archaea, Bacteria, and Eukaryota and are found virtually throughout the animal kingdom, CRiSPs are restricted to vertebrates alone. CRiSPs were originally identified as acidic epididymal glycoproteins from the reproductive tracts of rats (Cameo and Blaquier 1976; Kierszenbaum et al. 1981; Brooks et al. 1986). Since then, CRiSPs have been characterized from a wide variety of tissues, including the salivary glands, neutrophils, colon, thymus, ovary, pancreas, and blood (Nobile et al. 1994; Morrissette et al. 1995; Nobile et al. 1996; Olson et al. 2001; Ookuma, Fukuda, and Nishida 2003; Yamazaki and Morita 2004; Yamazaki and Morita 2007).

As a result of its high cysteine content, N-terminal signal peptide sequence, and extracellular localization (Haendler et al. 1993), the protein isolated from the rat epididymal fluid is now named CRiSP1, and several orthologues have been identified in humans, horses, and mice (Rankin et al. 1992; Giese et al. 2002). The androgen-dependent secretion of CRiSP1 by the proximal epididymis occurs on the sperm surface during maturation and the fusion of gametes. Initially, CRiSP1 secretion is localized on the dorsal region of the sperm head; however, it migrates to the equatorial segment during capacitation. CRiSP2, which is heavily secreted in the testes and likely participates in interactions involving Sertoli and spermatogenic cells, was initially retrieved from guinea pigs, mice, rats, humans, and horses (Maeda et al. 1998). CRiSP2 was previously known by several names, such as auto-antigen 1 and TPX1. CRiSP2 is also known to regulate RyR activity (Gibbs et al. 2006). It may play a role in either the initiation or regulation of sperm tail beating via ion channels controlling intracellular calcium stores (Harper, Barratt, and Publicover 2004; Marquez, Ignotz, and Suarez 2007). CRiSP3 was originally characterized from mice and soon after was retrieved from humans and horses. CRiSP3 are the most widely distributed of all CRiSPs and are secreted in the majority of vertebrate cells and tissues, including saliva, the pancreas, murine pre-B cells, human neutrophils, the thymus, the colon, and the ovaries. CriSP3 has also been hypothesized to play a vital role in the immune system. In contrast, CRiSP4 has only been identified in mice and rats to date (Jalkanen, Huhtaniemi, and Poutanen 2005; Nolan et al. 2006).

CRiSPs are secreted as single-chain proteins, and, depending on the kind of posttranslational modification (such as glycosylation) they undergo, their molecular mass ranges between around 20–30 kDa (Yamazaki and Morita 2004; Roberts et al. 2006). CRiSP isoforms share a high degree of amino acid sequence similarity and a highly conserved pattern of 16 cysteine residues that constitute eight disulfide bridges (Guo et al. 2005; Shikamoto et al. 2005; Wang et al. 2005a). The protein can be subdivided into two domains: a ~21-kDa PR-1 (or CAP) domain, which exhibits sequence similarity with the plant PR-1 (pathogenesis-related) proteins; and a ~6-kDa cysteine-rich domain (CRD). The PR-1 domain consists of all the structural motifs characteristic of a CAP domain and is known for its structural conservation. The CRD, in turn, is composed of two domains: a hinge region and an ion-channel regulator (ICR) domain. Of the eight disulfide bonds, 10 of the 16 cysteine residues involved are confined to the CRD region at the C-terminus. The presence of several unrelated proteins with similar domains that are enriched with cysteine residues adds to the already confusing nomenclature of CRiSPs. A new CRiSP nomenclature system was recently suggested (Gibbs et al. 2006). The cysteine spacing in the hinge region of CRiSPs is Cx2Cx3Cx4C. It should be noted that each member of the mammalian CAP subfamily differs in the spacing of cysteine residues in the hinge region. There seems to be a greater degree of rotational freedom between the hinge and the ICR domains, despite the fact that the former is rigidly attached to the PR-1 domain (Gibbs et al. 2006). CRiSPs are also characterized by a predicted signal peptide, which aids in localization to specific intracellular compartments. These structural domains of CRiSPs were initially identified by partial disulfide-bond mapping studies on CRiSP1. The domain organization of these proteins was finally confirmed when the crystal structures of several advanced snake CRiSPs were solved. Despite lacking transmembrane domains, CRiSPs are often found associated with membrane components either by glycosylation or through interaction with integral membrane proteins.

It has been suggested that these proteins may have a dual function, with both the N-terminal CAP domain and the C-terminal CRD possessing distinct roles. The biological activities of CRiSPs are mainly dependent on where they are expressed and in what quantity. Mammalian CRiSP1 and CRiSP2 primarily participate in reproductive pathways. CRiSP3 are secreted in large amounts not only in the reproductive tract but also in other places, such as immune tissues, salivary duct, and exocrine glands. Consistent with these differing roles, evolutionary assessment of CRiSP1 and CRiSP2 protein-encoding genes revealed them as being extremely well conserved and under the influence of negative selection, while CRiSP3 accumulates a greater number of variations (Sunagar et al. 2012). The lack of variation in reproductive homologues of CRiSP (CRiSP1 and CRiSP2) probably results from the fact that they play an important role in homeostasis, while the diversity in CRiSP3 is likely a result of their role in the immune system (Sunagar et al. 2012). Recently, however, it was revealed that CRiSP1, CRiSP2, and CRiSP4 might also have a wider distribution than previously recognized, including expression in nonreproductive tissues in the immune system and secretory epithelia, and therefore may have more diverse activities.

10.3 PLESIOTYPIC STRUCTURAL AND FUNCTIONAL TOXIN FORMS

Reflective of their wide salivary distribution, CRiSPs were among the first toxicoferan toxins to be recruited and amplified (Fry 2005). Along with kallikrein enzymes (see chapter 14) they remain the most highly expressed toxin class in the venom systems of anguimorph lizards and many non-front-fanged advanced snakes (Fry et al. 2003c; Jin et al. 2003; Osipov et al. 2005; Fry et al. 2006; Fry et al. 2008; Fry et al. 2010b; Fry et al. 2013). Although the precise role of these proteins in reptilian venoms is yet to be elucidated, as a class they are capable of a diverse array of biological functions: inhibition of neuronal Ca^{2+} and K^{+} (Nobile et al. 1994; Nobile et al. 1996); inhibition of vascular smooth-muscle contraction (Yamazaki et al. 2002), inhibition of BK (big potassium) Ca^{2+} channels (Wang et al. 2005a), inhibition of the release of Ca^{2+} from the sarcoplasmic reticulum of both cardiac and skeletal muscle (Morrissette et al. 1995), induction of hypothermia and lethality (Mochca-Morales, Martin, and Possani 1990), inhibition of K^{+} ion induced contraction of smooth muscles (Brown et al. 1999; Yamazaki et al. 2003), and blocking of ryanodine receptors, voltage-gated Ca^{2+}, and K^{+} ion channels (Morrissette et al. 1995; Nobile et al. 1996). Structures solved for forms from advanced snakes have allowed for detailed domain characterization of these proteins (Guo et al. 2005; Shikamoto et al. 2005; Wang et al. 2005a). However, the biological function of the majority of reptile venom CRiSPs, including plesiotypic forms, still remains completely unknown. Even the sites involved in biological activities remain unknown to date. It has been proposed recently that cooperation between the cysteine-rich and PR-1 domains facilitates the extreme specificity and affinity of reptile venom CRiSPs for ion channels (Suzuki et al. 2008). These authors also proposed the pore turret of CNGA2 as the potential target site for pseudechetoxin (PsTx) and pseudecin (Pdc), two CRiSP derivatives from *Pseudechis australis* and *Pseudechis porphyriacus* (respectively).

The molecular evolutionary history of these toxins also has proven to be intriguing. While they form close phylogenetic associations, the lizard and snake sequences do not form a monophyletic group (see figures 10.1 and 10.2). Thus, based on the principles utilized for determining single versus multiple evolutions of venom proteins (Fry and Wüster 2004), it must be currently hypothesized that these toxins are the result of convergent amplification, duplication, and diversification in lizards and snakes. The association of all the lizard toxins together, however, provides additional evidence for the organismal phylogeny of ([Anguimorpha + Iguania] + [Serpentes]) as the arrangement of the three toxicoferan lineages (see chapter 1.2.1). Each type has undergone differential molecular evolution (see color plate 26B) (Sunagar et al. 2012).

10.4 APOTYPIC STRUCTURAL AND FUNCTIONAL TOXIN FORMS

Assessment of natural selection pressures influencing the evolution of genes encoding these proteins in toxicoferan venoms reveals that they accumulate considerable variation under the influence of positive Darwinian selection (Sunagar et al. 2012). Moreover, reptile venom CRiSPs from caenophidian snake venoms accumulate significantly greater variation than those from anguimorph lizard venoms. It is probable that the evolution of reptile venom CRiSP homologues is influenced by the specific prey-subjugation strategy employed by the organism. Anguimorph lizards depend more on pace and powerful jaws for predation than venom, while advanced snakes are mainly dependent on venom, constriction, and/or a combination of the two. Likely as a result of these differing strategies, reptile venom CRiSPs experience a greater influence of positive selection in caenophidian snake venom than in anguimorph lizard venoms. Moreover, viperid and non-front-fanged advanced snake venom CRiSPs appear to have experienced a greater influence of positive selection than the elapid snake venom forms (Sunagar et al. 2012). This suggests that reptile venom CRiSPs may be more important functionally in the venoms of the aforementioned snakes. Normal body CRiSPs, on the other hand, experience negative selection and remain extremely well conserved. Nearly 46% of hypermutational sites are found in the CRD of advanced snake venom CRiSPs, the region suspected to be responsible for ion-channel-binding activity (Suzuki et al. 2008). The PR-1 domain accumulated the remaining positively selected sites, consistent with its suspected role in supplementing this ability.

For decades, it was assumed that cyclic nucleotides exert their diverse cellular effects primarily through the activation of the cyclic nucleotide-dependent protein kinases, PKA and PKG (Meinkoth et al. 1993). It was later revealed that they could also exert biological activities by binding directly to certain ion channels. Cyclic nucleotide-gated (CNG) channels were first characterized in the sensory epithelium of the visual and olfactory systems. They have subsequently been retrieved from other tissues, such as brain, skeletal muscle, heart, testes, liver, and kidneys. Unique reptile venom CRiSP homologues, called pseudechetoxin (PsTx)

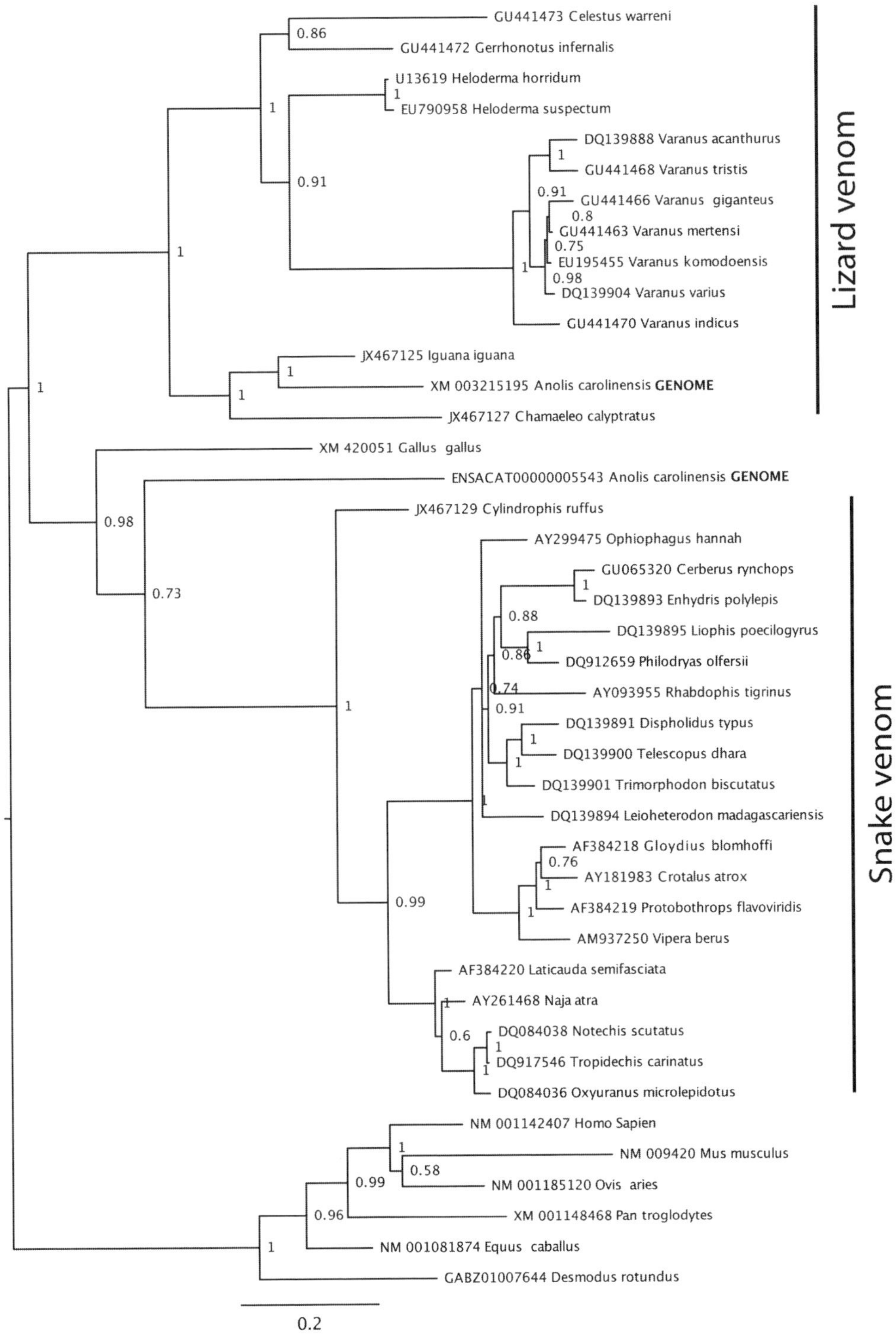

FIGURE 10.1: Bayesian molecular phylogenetic tree of lizard and snake CRiSP toxins and related nontoxin proteins.

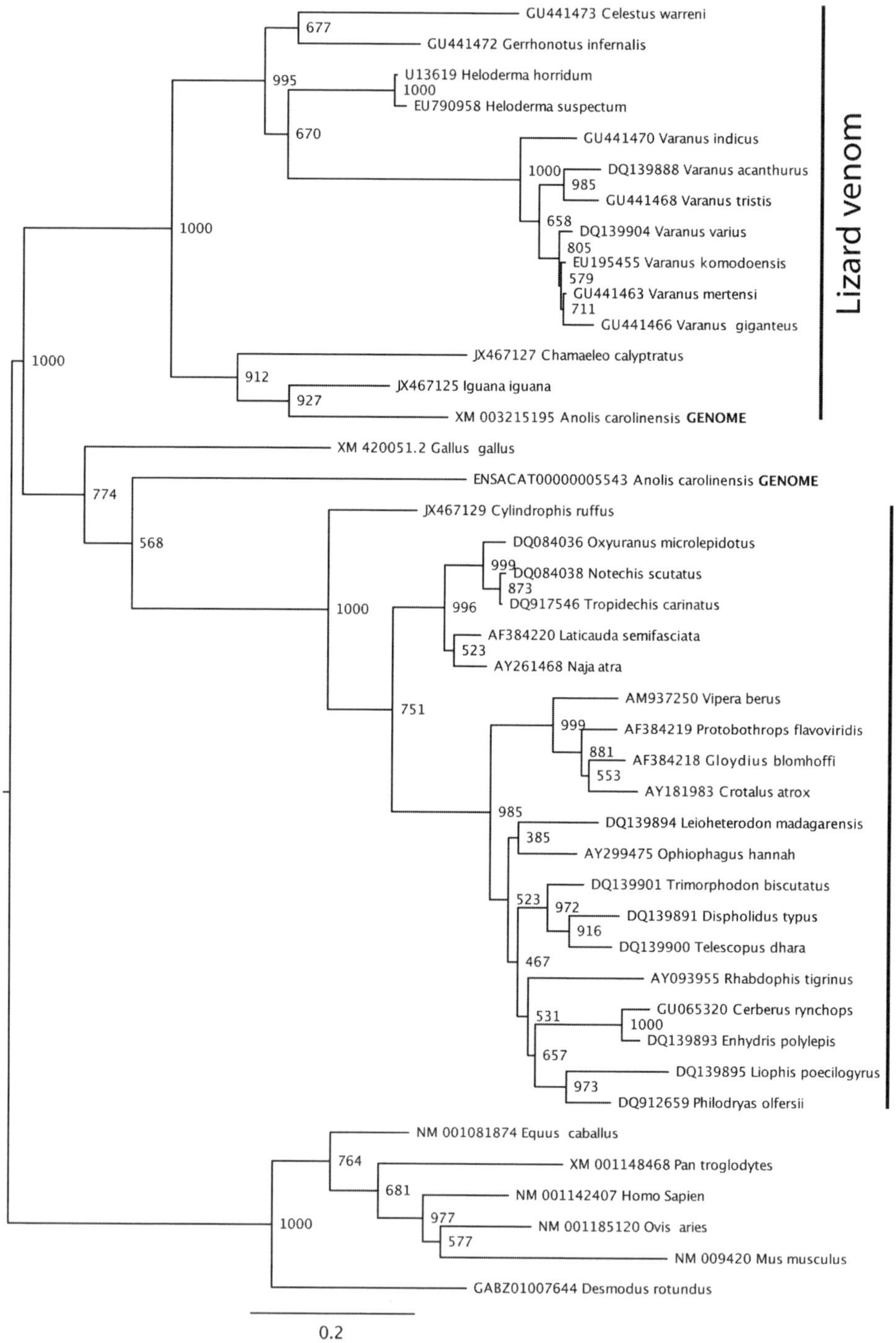

FIGURE 10.2: Maximum-likelihood molecular phylogenetic tree of lizard and snake CRiSP toxins and related nontoxin proteins.

and pseudecin (Pdc), from the venoms of Australian elapid snakes, are the only known protein blockers of CNG channels (Brown, Bishop, and Brooks 1999; Yamazaki, Brown, and Morita 2002; Brown et al. 2003). PsTx and Pdc are extremely similar in terms of their pI (highly basic, with a pI of 10) and primary (97% identity) and tertiary structures. Despite this, PsTx is more potent than Pdc and, unlike Pdc, is able to bind directly to the pore-turret region of CNG channels (Brown et al. 2003). Moreover, there is a large difference in their affinities for CNG channels (Yamazaki, Brown, and Morita 2002), with PsTx exhibiting a 15-to-30-fold greater affinity than Pdc for olfactory (CNGA2) and retinal CNG channels (CNGA1) (Yamazaki, Brown, and Morita 2002). Structural comparisons of the two homologues have revealed that the major difference between PsTx and Pdc is located around the concave surface formed between the PR-1 domain and the CRD (Suzuki et al. 2008). This highlights the importance of this region in binding and inhibition of CNG channels. It was also documented that Zn^{2+} ion binds to and facilitates the movement of the CRD, which, in turn, opens the concave surface (Suzuki et al. 2008).

10.5 THERAPEUTIC POTENTIAL OF THE TOXIN TYPE

None is currently documented.

10.6 CONVERGENCE WITH OTHER VENOMS

The CAP scaffold has been convergently recruited in the venoms of several animal lineages (see color plate 2B), such as cephalopods, cone snails, stinging insects, scorpions, spiders, and even vampire bats (Fang et al. 1988; Milne et al. 2003; Francischetti et al. 2013; Low et al. 2013; Ruder et al. 2013b). They exhibit a diversity of biological activities in the venoms of these animals. The first CAP homologue characterized from the venom of cone snails, known as Tex31, possesses a proteolytic activity (Milne et al. 2003). This function was attributed to the PR-1 domain of this protein. Like the reptile venom CRiSP homologues, cone snail venom CRiSPs have a CRD. However, this domain seems to be extremely divergent from that of the snake homologues and hence likely represents an independent derivation. Allergy-inducing hymenopteran venoms are enriched with CAP toxins. CAP homologues have also been recruited to the feeding secretion of various hematophagous insect taxa from the orders *Diptera* (mosquitoes [Arcà et al. 2005], sandflies [Kato et al. 2006], biting midges [Campbell et al. 2005], tsetse flies [Li, Kwon, and Aksoy 2001], and muscid flies [Ameri et al. 2008]), *Hemiptera* (triatomine bugs [Ribeiro et al. 2004a]), and *Siphanoptera* (fleas [Andersen et al. 2007]), with certain lineages expressing multiple homologues. The secretory glands of the mosquitoes *Aedes aegypti* (Ribeiro et al. 2007) and *Anopheles gambiae* (Arcà et al. 2005) contain three and four distinct CAP members, respectively. CAP have also been detected in the secretory concoctions of soft (*Argas*) and hard (*Ixodes*) ticks, each representing a distinct recruitment event. All these aforementioned arthropod CAP proteins differ from their vertebrate homologues in lacking the CRD, and their activities are yet to be characterized. CAP homologues in the infective nematode *Ancylostoma caninum*, present during the transition from larval to parasitic stages, lack the CRD (Jones 1996), suggesting that the loss of the CRD might have occurred in the early ecdysozoan ancestor. Although these ecdysozoan proteins fall outside the conventional definition of CAP proteins, in the phylogenetic tree they are found nested within the CAP protein family. Interestingly, unique monodomain CAP proteins have been retrieved from cnidarians, which lack the N-terminal CAP domain, indicating that the CAP-CRiSP combination is likely to be a bilaterian innovation. The buccal glands of the parasitic lamprey *Lethenteron japonicum* are also enriched with CRiSP proteins, which block L-type voltage-gated Ca^{2+} channels (CaV1) and act as potent vasodilators that facilitate hematophagy (Ito et al. 2007). CRiSP homologues have also been retrieved from the hematophagous secretions of vampire bats, *Desmodus rotundus* (Francischetti et al. 2013; Low et al. 2013). In addition, Pr-1 proteins are one of the most up-regulated proteins in plants after pathogen invasion (van Loon, Rep, and Pieterse 2006).

CHAPTER 11

EXENDIN PEPTIDES

B. G. FRY, K. ROELANTS, T. N. W. JACKSON, K. SUNAGAR, Z. TAKACS, T. REEKS, AND H. F. KWOK

11.1 SUMMARY

Exendin peptides are not only the most enigmatic of all toxins, but they are also among the most medically useful. Their genetic ancestry remains to be conclusively resolved, but the evidence and the principle of parsimony point toward a single evolutionary origin after duplication of an ancestral multidomain gene from a gene encoding hormones of the VIP/glucagon superfamily. The two glucagon-like toxins (exendins 3 and 4) are less cardiotoxic than the two toxins that are more similar to vasoactive intestinal peptides (VIP; exendins 1 and 2). Thus the evidence points towards a glucagon gene origin for this toxin type but further work remains to be undertaken. Exendin-4 has been developed into a drug to treat diabetes (exenatide, trade name Byetta) by enhancing insulin secretion in response to elevated plasma glucose levels. While this drug is effective and also is off-label useful for obesity treatment, it has a number of severe side effects that have limited its market penetration.

11.2 ENDOPHYSIOLOGICAL PLESIOTYPE

Exendin toxins in *Heloderma* venoms pose a fascinating molecular evolutionary riddle. They are cleaved from monodomain monoproduct precursor proteins, in contrast to the hormones of the glucagon/VIP superfamily which are cleaved from multidomain multiproduct precursor proteins. Sequence comparisons of exendin precursors and related glucagon and VIP precursors reveals a number of contrasting similarity patterns. The highly similar exendin-1/1b and exendin-2 (also known as helospectin-1/2 and helodermin, respectively) resemble the second peptide domain of the VIP precursor (corresponding to the VIP peptide itself). However, the highly similar exendin-3 and exendin-4 resemble either the first

```
              10        20        30        40        50        60        70        80
1.   mkg--thfsaglvlllvlvqssleLPLQDDTTSS--ETDASLLDERKDTMRARRHSEGTFSNDYSKYLETRRAQDFVQWLMNS
2.   mki-ilwlcvfglflatlfpiswqMPVESGLSSEDSASSES------------------------------------------
3.   mki-ilwlcvfglflatlfpiswqMPVESGLSSEDSASSES------------------------------------------
4.   mki-ilwlcvfglflatlfpvswqMPVESGLSSEDSASSES------------------------------------------
5.   mkmksiyfiaglflmivqgswqNPLQDTEEKSRSFKASQSEPLDESRQLNEVKRHSQGTFTSDYSKYLDSRRAQDFVQWLMST
6.   mtsmyfv--aglllmivqgswqSPLQETEEKSRSFKASQAEPLDDSRQLNEVKRHSQGTFTSDYSKYLDTRRAQDFVQWLMNT
7.   mesrskpqflailtlfsvlfsqslaWPLYGPP--------SSVRLDDRLQFEGAGDPDQVSLKADSDILQNALAENDTPYYDV
8.   mksilwlcvfglliatlfpvswqM-----------------------------------------------------------
9.   >
10.  >

          90       100       110       120       130       140       150       160
1.   KR-------------------------SDSPARRHADGTYTSDVSSYLQDQAAKDFITWLKSGQPKPD-AVETRSVDTLR-RR
2.   ---------------------------FASKIKRHGEGTFTSDLSKQMEEEAVRLFIEWLKNGGPSSG-APPPSG
3.   ---------------------------FASKIKRHGEGTFTSDLSKQMEEEAVRLFIEWLKNGGPSSG-APPPSG
4.   ---------------------------FASKIKRHSDGTFTSDLSKQMEEEAVRLFIEWLKNGGPSSG-APPPSG
5.   KRNGQQGQEDKENDEFPDQLSSNAISKRHSEFERHAEGTYTSDITSYLEGQAAKEFIAWLVNGRGRRD-FPEKALMAEEMGRR
6.   KRSGQQGVEEREKENLLDQLSSNGLARHHAEYERHADGRYTSDISSYLEGQAAKEFIAWLVNGRGRRD-FLEEAGTADDIGRR
7.   SRN-----------------------------ARHADGVFTSDYSRLLGQISAKKYLESLIGKRIS-----SSISEDPVPVKR
8.   -------------------------------------------------------------AIKSRLSSEDSETDQRLFESKR
9.   >
10.  >

      170       180       190       200       210
1.   HVDGSFTSDVNKVLDSLAAKEYLQWVMNSPSSASGKRH
2.
3.
4.
5.   HADGTFTSDINKILDDMAAKEFLKWLINTKVTQRDLLGEYQ
6.   HADGTFTSDYNQLLDDIATQEFLKWLINQKVTQRDLLGEYQ
7.   HSDAVFTDNYTRLRKQMAVKKYLNSILNGKRSSEGDSPDFLEELEK
8.   HSDAIFTEEYSKLLAKLALQKYLASILGSRTSPPPPSR
9.   HSDAIFTEEYSKLLAKLALQKYLASILGSRTSPPP
10.  HSDATFTAEYSKLLAKLALQKYLESILGSSTSPRPPSS
```

FIGURE 11.1: Sequence alignment of exendin toxins and representative related nontoxin peptides (Fry et al. 2010a). (1) The nontoxin, normal body glucagon peptide Q6RYB9 (*Ictalurus punctatus*). The glucagon-like venom peptides (2) EU790959 (*Heloderma suspectum cinctum*), (3) P26349 (*H. suspectum*), and (4) P20394 (*Heloderma horridum*). The nontoxin, normal body glucagon peptides (5) Q3HLJ1 (*Meleagris gallopavo*) and (6) O12956 (*H. suspectum*). (7) The nontoxin, normal body vasoactive intestinal peptide P01283 (*Rattus norvegicus*). The vasoactive intestinal peptide-like venom peptides (8) EU790960 (*H. suspectum*), (9) P04204 (*H. suspectum*), (10) P04203 (*H. suspectum*). Posttranslationally cleaved bioactive peptides are highlighted in black, and signal peptides are shown in lower-case. > designates incomplete N-terminal sequence.

or third domain of the glucagon precursor (see figure 11.1). Superficially, these similarities would suggest that some of the exendin precursors (exendin-1, -1b, and -2) are evolutionary derived from a VIP-like plesiotypic precursor, while others (exendin-3 and -4) are derived from a glucagon-like ancestor. All exendin precursors, however, share highly similar (nearly identical) N-terminal signal peptides and prepro-regions and form a monophyletic clade (see figure 11.2). This pattern instead suggests a close evolutionary relationship between all four exendin precursors.

As a result of these contrasting patterns, three alternative scenarios, each involving different mechanisms of molecular evolution, can be proposed for the origin of the four exendin toxins (Fry et al. 2010a; see color plate 28): (1) two independent origins of exendin precursor genes followed by loss of peptide domains and gene conversion; (2) a single origin of a multidomain exendin precursor gene followed by domain loss and convergent domain evolution; and (3) a single origin of a multidomain exendin precursor gene followed by gene fission and convergent domain evolution.

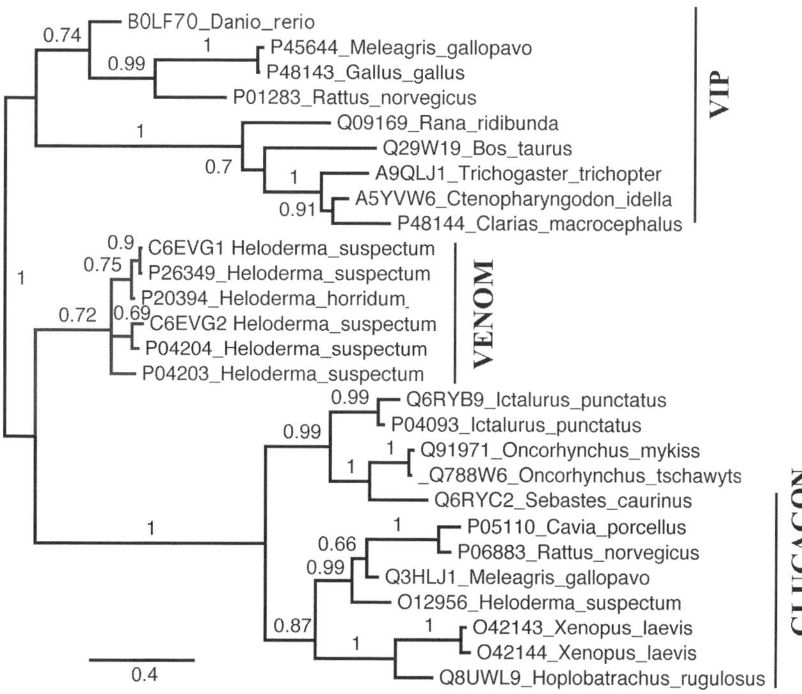

FIGURE 11.2: Phylogenetic tree of exendin toxins and related nontoxin proteins (Fry et al. 2010a).

Scenario 1 (color plate 28A) implicates two independent evolutionary origins of *Heloderma* exendins, once by recruitment of a VIP-like precursor (the ancestor of exendin-1, and -2), and another from a glucagon-like precursor (the ancestor of exendin-3 and -4). However, phylogenetic analyses (Fry et al. 2010a; see figure 11.2) recovered all exendin precursors in a single clade, hence suggesting a single toxin recruitment.

If the exendin pairs had indeed originated twice, the nearly identical signal peptides must be explained by transfer of the signal peptide-encoding exon from one ancestral exendin gene to that of the other. However, the likelihood that the two independently recruited ancestral toxin genes are located on the same chromosome would be very low. In the genomes of other vertebrates (chicken, human, rat, and zebrafish), the glucagon and VIP genes are situated on different chromosomes, indicating spatial separation of the related hormone genes early in vertebrate evolution. The same could be expected to hold true for the *Heloderma* hormone genes and thus for any toxin gene that would have duplicated from them. Consequently, unless the ancestral exendin genes were accidentally relocated to the same chromosome, the transfer of a signal peptide from one ancestral gene to the other would require an interchromosomal (ectopic) gene-conversion event or a retrotransposition event. Scenario 1 would therefore imply not only two independent toxin-recruitment events and subsequent peptide domain losses but also extremely unlikely interchromosomal exchange of one or several ancestral exendin gene segments between the independently evolved loci.

The two scenarios entailing a single evolutionary origin of exendin peptides (color plates 28B and 28C) are equally consistent with phylogenetic analyses (see figure 11.2; Fry et al. 2010a). Scenario 2 entails the loss of one or two domains (depending on whether the ancestral toxin gene evolved from the VIP or glucagon gene; compare color plates 28B with 28B′).

After an additional duplication events (yielding two monodomain exendin genes), convergent mutations in the peptide domains of the precursors yielded the presently observed differential similarity of exendins to the VIP and glucagon hormones. A second round of gene duplication events gave rise to the present-day exendin gene pairs.

Alternatively, scenario 3 explains the rise of distinct VIP-like and glucagon-like exendin precursors as a result of fission of a single multidomain ancestral gene through one or several gene conversion events, rather than gene duplication followed by domain losses. In this scenario, a single multidomain gene that either duplicated from the VIP-like or glucagon like hormone gene (compare color plates 28C and 28C′) first underwent convergent mutations to obtain both VIP-like and glucagon-like domains. Next, gene conversion (such as a result of unequal crossing-over) would insert the 5′-segment (including the signal peptide-encoding region and probably also the gene promoter and 5′-UTR region) in between two of the peptide-encoding regions, resulting in the single multidomain precursor gene being split into two tandem genes. Subsequent duplication of each of the two tandem genes would then give rise to the presently observed exendin precursor pairs.

Sequencing of the *Heloderma* genome (or at least the flanking regions of the VIP and glucagon hormone genes) could determine if the exendin genes lie in the neighborhood of one of the two hormone genes and thereby elucidate their origin by tandem duplication. Insight into the genomic organization of exendin genes with respect to the VIP and glucagon genes of *Heloderma* could determine which one of the two hormone genes gave rise to the exendins: adjacency of all exendin genes to one of the two hormone genes would be a strong indication of their origin by tandem duplication. Molecular evidence favors glucagon precursor gene as the plesiotypic gene (see color plate 28B′ or 28C′) with subsequent convergent mutations producing VIP-like exendins. Glucagon as the plesiotypic substrate is favored by PCR investigations that detected low levels of the VIP-like exendin-1 in *Heloderma horridum* and failed to detect VIP-like exendin-2 in the same species but with both showing up in high levels for *Heloderma suspectum* (Fry et al. 2010a). This is suggestive of cardiotoxic exendin isoforms having a larger role in *H. suspectum* venom than in *H. horridum.*

The current implication that at least some of the exendins evolved convergently with VIP/glucagon hormones is a strong indication of adaptative evolution to increase resemblance with either one of the two hormones. Therefore, in light of rapid prey immobilization, the evolution from a less cardiotoxic (glucagon-like) peptide toward a more cardiotoxic (VIP-like) one seems more plausible than the converse.

11.3 PLESIOTYPIC STRUCTURAL AND FUNCTIONAL TOXIN FORMS

The glucagon-like exendins-3 and exendin-4 are only moderately cardioactive (Fry et al. 2010a). Studies have shown that exendin-3 and exendin-4 are similar (50% and 53% identity, respectively) to human glucagon-like peptide-1 (GLP-1) and act as high-affinity agonists on the GLP-1 receptor (GLP1R) (Eng et al. 1990; Eng 1992). Several important functional residues, such as the N-terminal histidine, are conserved in both exendin and GLP-1. Exendin is much more resistant to cleavage by dipeptidyl peptidase-4 (DPP-4) than GLP-1 as a result of the substitution of the plesiotypic alanine with glycine at the second position. As a result, exendin-4 shares the bioactivities of GLP-1 but has a longer half-life (60 to 90 minutes) (Greig et al. 1999).

Interestingly, while exendin-4 acts as an agonist of GLP1R, the truncated version, exendin-4 (9–39), behaves as an antagonist and is able to block GLP-1-induced cAMP formation. In addition, truncated exendin-4 (9–39) inhibits insulin secretion triggered by GLP-1 (7–36), exendin-3, and full-length exendin-4 (Goke et al. 1993; Raufman 1996; Chen and Drucker 1997). Structurally, seven of eight N-terminal amino acids are identical between exendin-4 and GLP-1. On the remainder of the molecule, only nine of 22 amino acids are conserved between exendin-4, exendin-4 (9–39), and GLP-1. This suggests that some residues at the C-terminal end of these peptides may be sufficient for binding but that agonist activity requires the presence of the highly conserved eight N-terminal residues.

11.4 APOTYPIC STRUCTURAL AND FUNCTIONAL TOXIN FORMS

As argued above, molecular evolution from a glucagon-like (mildly cardiotoxic) form toward a VIP-like (strongly cardiotoxic) form is more plausible than the converse. As endogenous VIP hormones are more potent vasodilators than endogenous glucagon hormones, a peptide that mimics a generalized VIP hormone would indeed be a more effective toxin. This is consistent with exendin-1 and exendin-2 exhibiting greater cardioactivity than not only exendin-3 and exendin-4 but also the nontoxin peptide GLP-1 (O12956) (see figure 11.3) (Fry et al. 2010a). This is in agreement with other studies on exendin-1 and exendin-2 which demonstrated vasodilatory actions similar to VIP (Uddman et al. 1999; Tsueshita et al. 2004).

11.5 THERAPEUTIC POTENTIAL OF THE TOXIN TYPE

Pharmacological studies have reported that exendin-4 can act at GLP-1 receptors *in vitro* on certain insulin-secreting cells, at dispersed acinar cells from guinea pig pancreas, and at parietal cells from the stomach and the peptide is also reported to stimulate somatostatin release and inhibit gastrin release in isolated stomachs (Goke et al. 1993; Eissele et al. 1994; Schepp et al. 1994). Exendin-3 and exendin-4 were found to stimulate cAMP production in, and amylase release from, pancreatic acinar cells (Malhotra et al. 1992; Raufman et al. 1992; Singh, Eng, and Raufman 1994). Exendin-4 has a significantly longer duration of action than GLP-1, with dose-dependent lowering of blood-glucose levels by exendin-4 in diabetic mice persisting for up to 24 hours. As a result of their insulinotropic activities, the use of exendin-3 and exendin-4 for the treatment of diabetes mellitus and the prevention of hyperglycemia has been proposed.

Exendin-4 is one of the most well-known venom peptides because of its high-profile use as adjunctive treatment, first-line therapy, and/or monotherapy for type 2 diabetes mellitus. As a therapeutic agent, the generic name of exendin-4 is exenatide. Exenatide twice-daily (EBID)

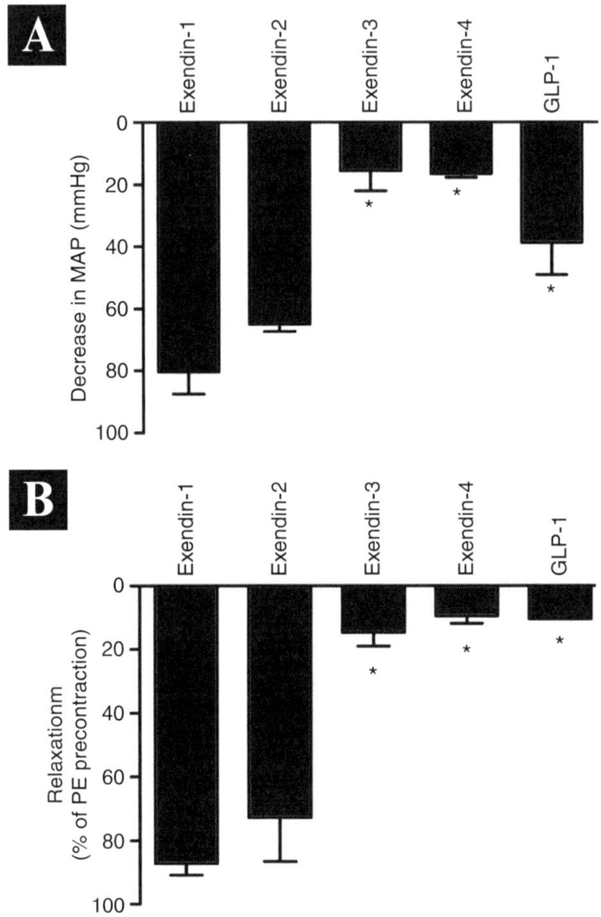

FIGURE 11.3: (A) Effect of exendins 1–4 or GLP-1 (100 μg/kg, i.v. n = 3) on the MAP of anesthetized rats. *P, 0.05, significantly different from exendin-1, one-way analysis of variance (ANOVA). (B) Effect of exendins 1–4 or GLP-1 (10 μg/ml, n = 3) on phenylephrine precontracted aorta. *P, 0.05, significantly different from exendin-1, one-way ANOVA (Fry et al. 2010a).

was approved in 2005 by the Food and Drug Administration (FDA) in the United States under the trade name Byetta. In 2013, global sales of Byetta and its newer formulation, Bydureon (to be injected only once a week; see below), grossed $698 million.

Exenatide and derived drugs act at GLP-1 receptors and enhance insulin secretion in response to elevated plasma glucose levels (Goke et al. 1993). By mimicking the insulinotropic activity of GLP-1, the drug enhances this pathway to improve glycemic control for the treatment of diabetes mellitus and the prevention of hyperglycemia (Singh, Eng, and Raufman 1994). Reports have found certain side effects of EBID, and weight loss is a common factor in a number of cases (Kendall et al. 2005; Buse et al. 2007). Therefore, in recent years, EBID has also been investigated for its antiobesity properties, since it slows stomach emptying, extending the period of satiation after eating.

Analysis of postmarketing results from 2004 to 2009 stored in the FDA's Adverse Events Reporting System database found a 6.7-fold increase in reported cases of pancreatitis and a 2.9-fold increase in reported cases of pancreatic cancer in patients taking EBID (Elashoff et al.

2011). These findings have been both refuted and corroborated in trials on animals, and there is general consensus among researchers that further studies on human subjects are necessary to either establish or rule out a link between pancreatic cancer and EBID. Although a definitive link between EBID and pancreatic cancer has not yet been clearly established, biomedical researchers have additional reasons to believe that the drug can increase a patient's risk of developing the disease. Diabetics typically lack a sufficient amount of hormones like GLP-1, which stimulate the pancreas to release insulin during and after a meal in order to lower blood glucose levels. By increasing GLP-1 receptor activity, EBID reduces the patient's blood glucose level. Unfortunately, many studies suggest that increased levels of GLP-1 receptor activity are strongly linked to a greater risk of pancreatic cancer (Giovannucci et al. 2010).

Aside from possible side effects, the foremost clinical disadvantage of venom-derived drugs such as EBID is that in the native peptide form, they generally cannot be administered orally. This is because either they are likely to be degraded by peptidase enzymes in the gastrointestinal tract, or they are too large to be absorbed. As a consequence, these drugs are typically delivered via the parenteral route, which makes the therapy intrusive for the patient and increases the costs for the healthcare provider. As such, it seems that the search for potential lead compounds in lizard venoms will need to proceed in tandem with research into formulation development to ensure that pharmacological discoveries are translated into commercially successful therapies. The drawbacks associated with EBID's parenteral route of administration and the twice-a-day frequency requirement are being addressed in different ways. In all of the following examples, the basic therapeutic indication tends to be the same as for EBID, while the route and frequency of administration differ. Efforts outlined next are also relevant for the development and formulation of other therapeutic agents based on peptide toxins.

Exenatide once-weekly (EQW, trade name Bydureon) is a long-acting formulation. Exenatide is encapsulated in 0.06-mm-diameter biodegradable microspheres of medical-grade poly-(D,L-lactide-co-glycolide) (PLG). Following injection, the microspheres degrade in situ by natural (such as noncatalyzed) hydrolysis into lactic acid and glycolic acid while providing an extended release of exenatide into the circulation (DeYoung et al. 2011). EQW is administered subcutaneously once a week, and it has been approved in 2011 in the EU and 2012 in the United States.

VRS 859 (exenatide-XTEN) is a fusion protein that contains exenatide and a long hydrophilic polypeptide segment. XTEN stands for half-life extension technology, which is an alternative to PEGylation, the covalent attachment of polyethylene glycol (PEG) to recombinant peptides (Payne, Murphy, and Manning 2011). XTEN is made of natural amino acids, is biodegradable, has low immunogenicity, and has no known toxic metabolites. VRS 859 has a long half-life, thus holding the potential for once-a-month administration intervals by the subcutaneous route. It is produced recombinantly and is undergoing clinical trials (Sivertsen et al. 2012). In experimental settings, site-specific PEGylation of exenatide has also been investigated as a comparable solution to address the short plasma half-life of the native peptide (Gong et al. 2011).

The noninvasive treatment option Exenatide Nasal is a nasal spray formulation of exenatide in clinical trials (Sivertsen et al. 2012; Millotti et al. 2013). In experimental animals, sublingual and intratracheal administration of exenatide have also been identified as potential means to reach therapeutic concentrations (Gedulin et al. 2008). Biotinylated exendin-4 analogues (Jin et al. 2009) and exenatide-loaded pH-responsive microspheres (Zhang et al. 2014) have been reported as potential solutions for oral administration. In the case of microspheres, tested in mice, the bioavailability of exenatide reached 10.2% relative to the subcutaneous injection (Zhang et al. 2014).

Other formulation attempts in various stages of preclinical and clinical trials are as follows (Madsbad et al. 2011; Sivertsen et al. 2012): exendin-4 with leucine substituted for methionine at position 14 (AC3174) (Hargrove et al. 2007), once-a-week exendin-4 conjugated to recombinant human albumin (CJC-1134-PC, PC-DAC exendin-4) (Baggio et al. 2008), continuous subcutaneous delivery exenatide (ITCA 650) (Henry et al. 2013), oral exenatide capsule (ORMD-0901) (Eldor et al. 2010), exenatide-based transdermal ViaDor-GLP1R agonist (formulated to be stable at body temperatures for up to one year), and exendin-4 fused to human transferrin (PF-04603629) (Gustavson et al. 2011).

GLP-1 based therapies have potential in cardiovascular diseases (Sivertsen et al. 2012). In experimental settings, exenatide has been shown to have cardioprotective properties, such as protection against reperfusion injury-induced cell death, thus reducing the infarct size in acute myocardial infarction (Sonne, Engstrom, and M. Treiman 2008; Timmers et al. 2009; Lonborg et al. 2012a; Lonborg et al. 2012b). The cardioprotection is believed to be mediated via GLP-1 receptors on the cardiomyocytes, which in turn activate the prosurvival signaling pathways of phosphatidylinositol 3-kinase (PI3K)–Akt and protein kinase A (PKA) (Hausenloy and Yellon 2012; Lonborg et al. 2012a). Suggested potential downstream mechanisms leading to cardioprotection are the inhibition of the mitochondrial permeability transition pore, activation of A-kinase anchoring proteins (AKAPs), increased myocardial glucose uptake, reduced apoptotic cell death, and transcription of various cardioprotective factors (Hausenloy and Yellon 2012).

11.6 CONVERGENCE WITH OTHER VENOMS

VIP/PACAP-like peptides have recently been characterized as components of vampire bat venom (Low et al. 2013), but their bioactivities remain to be elucidated.

CHAPTER 12

FACTOR VA PROTEINS

S. EARL, K. SUNAGAR, T. N. W. JACKSON, T. REEKS, AND B. G. FRY

12.1 SUMMARY

Among the most potent of all coagulopathic snake-venom components is the factor Xa-factor Va prothrombin activator complex found exclusively in the venoms of Australian elapid snakes from the genera *Oxyuranus* and *Pseudonaja*. The complex consists of a prothrombin-activating protease homologous to the mammalian coagulation factor Xa (snake venom factor Xa [SV-fXa]) and a cofactor (snake venom factor Va [SV-fVa]) homologous to the mammalian coagulation factor Va. By recruiting these two factors to their venoms, *Pseudonaja* and *Oxyuranus* have evolved a powerful and unique procoagulant toxin that can induce rapid venom-induced consumption coagulopathy in prey. Isolated SV-fVas have been found to share structural and functional similarities with endophysiological fVa but has also evolved additional features such as increased stability. These attributes have contributed to SV-fVas having applications as therapeutic agents to combat surgical and traumatic hemorrhage and is currently in preclinical development for these indications.

12.2 ENDOPHYSIOLOGICAL PLESIOTYPE

The plesiotypic precursor to SV-fVa is coagulation factor V (fV) (Fry 2005), the inactive precursor to fVa. These proteins function as part of the clotting cascade, aiding in the formation of a hemostatic plug to stop bleeding following vascular injury. Factor V is predominantly synthesized in the liver (Tracy, Eide, and Bowie 1982) and circulates in plasma as a large single-chain pro-cofactor. The human form has a molecular mass of 330 kDa (Kane and Majerus 1981; Suzuki, Dahlback, and Stenflo 1982; Nesheim et al. 1984). The total blood concentration of fV is 20 nM, and approximately 20% of this is found in the platelet α-granules, where it is critical for clot formation (Tracy et al. 1984; Nesheim et al. 1986; Janeway et al. 1996; Hayward et al. 1997; Guasch et al. 1998; Montefusco et al. 2000).

Factor Va functions as part of the prothrombinase complex, the central protein complex of the clotting cascade (see color plate 2A). This complex is composed of a serine protease (fXa), a protein cofactor (fVa), and Ca^{2+} assembled on an anionic membrane and functions by cleaving prothrombin to form thrombin (Orfeo et al. 2004). Thrombin is the key mediator of clotting, cleaving fibrinogen into fibrin to form a fibrin clot but also activating additional prothrombin, factor V, VIII, XIII molecules and platelets to propagate the clotting signal (Brummel et al. 2002). Factor Va plays a critical role in the output of the prothrombinase complex, since its function as an fXa receptor and catalytic effector increases the rate at which prothrombin is converted to thrombin by 300,000-fold relative to the rate of the reaction catalyzed by fXa alone (Nesheim, Taswell, and Mann 1979). In other words, the amount of thrombin formed in one minute by the prothrombinase complex would take six months to form if fXa acted alone (Mann and Kalafatis 2003). This slow rate of thrombin formation by fXa alone would be insufficient for clot formation, and evidence of the physiological importance of fVa in blood clotting can be observed from mice in which complete deficiency results in massive hemorrhage and death (Cui et al. 1996).

The inactive fV pro-cofactor is converted to the active form fVa by thrombin during the early stages of the blood-coagulation process (Butenas, vantVeer, and Mann 1997; Brummel et al. 2002; Butenas et al. 2002; Hockin et al. 2002). Thrombin cleaves fV sequentially at Arg709, Arg1018, and Arg1545 to liberate a 150-kDa fragment (B domain). Factor Va is a heterodimer composed of a 74-kDa light chain (A3, C1, and C2 domains) and a 105-kDa heavy chain (A1 and A2 domains) (see figure 12.1) held together with noncovalent interactions requiring divalent metal ions (Kane and Majerus 1981; Suzuki et al. 1982; Nesheim et al. 1984; Jenny et al. 1987; Krishnaswamy, Russell, and Mann 1989). The fVa light chain contains the hydrophobic binding site for platelet membranes, and both the heavy and light chains of fVa bind to fXa (Krishnaswamy and Mann 1988; Kalafatis, Jenny, and Mann 1990; Ortel et al. 1992; Kalafatis, Rand, and Mann 1994a; Kalafatis, Rand, and Mann 1994b; Lecompte, Bouix, and Mann 1994; Ortel et al. 1994). Thus, fVa anchors fXa to the membrane surface, and having both fXa and prothrombin bound to the membrane decreases the effective K_m for prothrombin activation by two orders of magnitude and increases the k_{cat} for the reaction by three orders of magnitude, resulting in a rate increase of five orders of magnitude in the presence of fVa (Mann and Kalafatis 2003).

While fVa plays a vital function in clot formation, it also prevents an overactive clotting process that would result in thromboembolism. First, the cofactor activity of fVa enables it to regulate fXa activity and thereby regulate clotting activity. The difference in the concentrations of fXa (present in blood at 100 nM) and fVa (effectively 4 nM) also makes fVa the limiting component in prothrombinase-complex formation and accordingly clot formation (Halkier 1991). Additionally, the coagulation process is down-regulated by proteolytic inactivation of fVa. Protein C is cleaved to form activated protein C (APC) by the action of thrombin. In the presence of the vascular cofactor thrombomodulin, APC subsequently competes with fXa for binding to fVa (Nesheim et al. 1982; Krishnaswamy, Mann, and Nesheim 1986; Krishnaswamy, Williams, and Mann 1986; Solymoss et al. 1988; Kalafatis and Mann 1993; Kalafatis, Rand, and Mann 1994a; Kalafatis, Rand, and Mann 1994b; Kalafatis et al. 1994; Hockin et al. 1999). APC then inactivates fVa by cleaving the heavy chain (of the human form) at Arg506, Arg306, and Arg679 (see figure 12.1), thus providing inhibition of clot formation by both competitive and inactivation mechanisms. Following single cleavage at Arg506, fVa shows reduced affinity for fXa, reducing the rate of thrombin formation. Subsequent cleavage

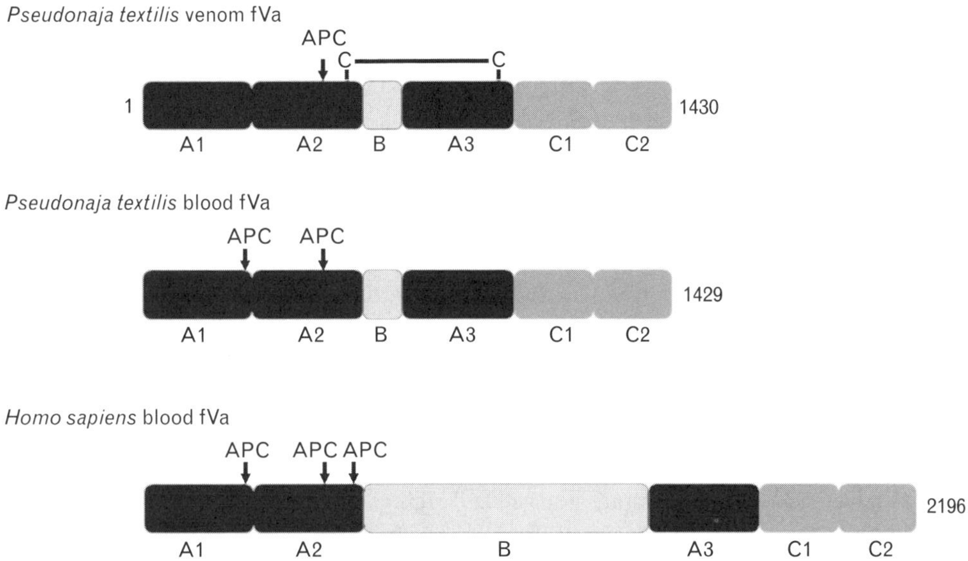

FIGURE 12.1: Schematic comparison of different factor V proteins. Relative sizes of the different domains, APC cleavage sites, and the predicted disulfide bond in *Pseudonaja textilis* SV-fVa connecting the A2 and A3 domains are shown.

at Arg306 leads to complete inactivation of fVa and dissociation of the prothrombinase complex (Kalafatis, Rand, and Mann 1994a; Kalafatis, Rand, and Mann 1994b; Hockin et al. 1997; Mann et al. 1997).

12.3 PLESIOTYPIC STRUCTURAL AND FUNCTIONAL TOXIN FORMS

The recruitment of endophysiological fV to form SV-fVa occurred in the common ancestor of the Australian elapid snake genera *Oxyuranus* and *Pseudonaja* (Fry et al. 2008; Fry et al. 2012a). SV-fVa mRNA transcripts have been sequenced from three species thus far: *Pseudonaja textilis, Oxyuranus scutellatus*, and *Oxyuranus microlepidotus* (Rao, Joseph, and Kini 2003; St Pierre et al. 2005; Welton and Burnell 2005). These share 95.7% sequence identity at the nucleotide level, suggesting that there is a highly conserved structure-function relationship within these forms (St Pierre et al. 2005). The cloning of the cDNA transcript for endophysiological (blood) fV from the liver of *P. textilis* and comparison between deduced amino acid sequences revealed that there is 96% sequence identity between the blood and venom homologues of fV in this species (Le et al. 2005). Comparison between SV-fVa and human fVa homologues showed that there is approximately 50% identity at the amino acid level and that the venom homologues maintain the same domain structure as their mammalian fVa counterparts. It is important that most residues in human fVa directly implicated in functional activity appear to be conserved in SV-fVas. For example, from a string of nine residues on the human fVa heavy chain identified as the binding site for fXa, eight are conserved in SV-fVas (St Pierre et al. 2005). The region of greatest difference between the mammalian and venom forms is the B domain, which is significantly smaller in SV-fVas (see figure 12.1). The SV-fVa mRNAs code for a 1429–1430 amino acid protein with a predicted mass of approximately 160 kDa. The observed molecular mass, however, is approximately 180 kDa, with contributions from glycosylation (Birrell et al. 2006; Birrell et al. 2007).

The limited distribution of SV-fVas within snake venoms has meant that functional characterization of the protein has been predominantly performed on only one form from *P. textilis*. When *P. textilis* SV-fVa is separated from *P. textilis* SV-fXa, prothrombin-activation activity and total clotting activity are reduced dramatically (Rao and Kini 2002). This confirms that SV-fVa is functioning as a SV-fXa cofactor in the SV-fXa/SV-fVa complex (see also chapter 13). Similarly, it has been demonstrated that recombinant *P. textilis* SV-fVa can bind to human fXa and form a hybrid prothrombinase complex (Bos et al. 2009). However, the affinity between *P. textilis* SV-fVa and human fXa is approximately 25 times weaker than those within the *P. textilis* SV-fXa/SV-fVa and the human fXa/fVa complexes. The relatively large size of both endophysiological fVa and SV-fVa coupled with its structural instability in the absence of SV-fXa has made crystallography studies challenging, and no crystal structures have successfully been solved for either form of the protein. However, molecular modeling has been performed on endophysiological fVa, giving clues to its three-dimensional structure (Delev et al. 2008). Despite many similarities between the forms of fVa, it is clear that the venom form has evolved several key features, not present in blood fVa, which give rise to unique structural and functional properties (Le et al. 2005; Bos et al. 2009).

12.4 APOTYPIC STRUCTURAL AND FUNCTIONAL TOXIN FORMS

The steric and conformational constraints of the large B domain maintain mammalian blood fV in an inactive state until thrombin cleavage of the B domain generates fVa (Toso and Camire 2004). The B domain in SV-fVas are only 127 residues in length (Le et al. 2005); therefore, the venom form is more similar in structure to endophysiological fVa than to fV. Indeed, the shortened B domain of SV-fVas means that they are constitutively active, unlike mammalian blood fV, possessing cofactor activity without thrombin cleavage. This was demonstrated with a recombinant form of the *P. textilis* SV-fVa with thrombin-cleavage sites in the B domain mutated. This mutant protein possessed similar cofactor activity to that of wild-type recombinant *P. textilis* SV-fVa activated with thrombin (Bos et al. 2009). Interestingly, *P. textilis* endophysiological fVa has a B domain that is one amino acid shorter than the venom homologue, but it is unclear whether this blood form is also constitutively active (Le et al. 2005). However, the B domain includes the greatest amount of sequence difference between *P. textilis* venom and blood fVa, suggesting that there may be tighter evolutionary constraints on the blood form. Another novel functional feature of *P. textilis* SV-fVa is its ability to act as a cofactor to SV-fXas in solution, without binding to anionic membranes (Bos et al. 2009). This is in stark contrast to mammalian fVa, which requires anchoring to a phospholipid membrane (usually the platelet surface) in order to function as a cofactor.

Two of the three APC cleavage-inactivation sites present in mammalian fVa are missing in SV-fVas (see figure 12.1) (Rao, Joseph, and Kini 2003). As a consequence of this, SV-fVas are functionally resistant to inactivation via cleavage by APC (Rao, Joseph, and Kini 2003; Bos et al. 2009), the major down-regulation mechanism for blood fVa utilized to attenuate the clotting process. APC cleavage of *P. textilis* SV-fVa does, however, occur at one site. There is a disulfide bond present in addition to the noncovalent interactions, which stabilize the light and heavy chains in SV-fVas (Bos et al. 2009). This disulfide bond confers increased stability to SV-fVas, as it enables the light and heavy chains to remain structurally intact, maintaining cofactor activity after cleavage by APC at position 507 (Bos et al. 2009). In contrast, APC cleavage at this site in blood fVa, which lacks the disulfide bond, leads to dissociation of the fVa subunits and a loss of activity (Mann et al. 1997). A schematic diagram of the comparative structural features of the different fVa molecules is presented in figure 12.1. From these examples, it is clearly evident that structural adaptations in SV-fVas have given rise to increased stability and unique functionality. These functional derivations demonstrate that SV-fVas have evolved unique attributes, which enable it to escape the normal hemostatic-regulation mechanisms present in blood. These adaptations generate a sustained and disseminated procoagulant stimulus, which is employed to great effect as a biological toxin to induce lethal thromboembolism. Indeed, venoms from the *Pseudonaja* and *Oxyuranus* species are among the most lethal of all snake venoms (Broad, Sutherland, and Coulter 1979), and at least some of this extreme toxicity can be attributed to the powerful procoagulant effects induced by the SV-fVa/SV-fXa complex (Fry et al. 2008; Fry et al. 2012a).

Clinically, the most relevant feature of the SV-fVa/SV-fXa complex from *Pseudonaja* venom is that the factors do not require an activation process and hence induce venom-induced consumption coagulopathy (VICC) rapidly in bite victims (Isbister, Duffull, and Brown 2009). Australian elapid snakes of the genera *Hoplocephalus, Notechis*, and *Tropidechis* secrete venoms that only contain SV-fXa prothrombin activators, while *P. textilis* venom contains both SV-fXa and SV-fVa; however, the coagulopathy associated with envenoming by these divergent snakes is extremely similar (Isbister, Duffull, and Brown 2009). A study examining coagulopathy in 167 patients observed that the symptoms associated with envenoming by both groups of snakes included: (i) complete consumption of fibrinogen, (ii) very high concentrations of D-dimers, (iii) nonrecordable international normalized ratio, and (iv) similar recovery time from VICC (Isbister, Duffull, and Brown 2009). The onset of coagulopathy is delayed, however, in *Hoplocephalus, Notechis*, and *Tropidechis* bites relative to *P. textilis* bites (Isbister, Duffull, and Brown 2009). Using serial clotting-factor assays, the study demonstrated that in *Pseudonaja* envenoming the concentrations of fibrinogen, factor V, and factor VIII dropped to very low levels almost instantly, whereas it took nearly one to two hours to detect a dip in concentration following bites from *Hoplocephalus, Notechis*, and *Tropidechis* (Isbister 2009).

12.5 THERAPEUTIC POTENTIAL OF THE TOXIN TYPE

Blood clotting is a tightly controlled process, but SV-fVas have evolved to bypass several control points. While these regulatory mechanisms prevent thromboembolism under endophysiological circumstances, they may also hinder rapid and effective clot formation and thereby compromise ability to reduce blood loss under circumstances of severe and prolonged hemorrhage. Furthermore, fVa is the component that is depleted to the greatest extent in surgical patients with persistent bleeding (Despotis, Joist, and Goodnough 1997). Given the critical role of fVa in clot formation and that it is limiting in concentration relative to fXa, it can be hypothesized that increasing the amount of fVa available and prolonging its activity could significantly improve clot formation and reduce blood loss in the case of severe and sustained hemorrhage. Therefore, SV-fVas may have therapeutic applications for the reduction of blood loss arising from surgery or trauma. There are multiple advantages of SV-fVas for this application. First, the combination of both constitutive activity and resistance to clot inactivation by APC suggests that SV-fVas could provide a transient prolonging of the clotting signal to assist clot formation in cases of severe bleeding. Second and most important, the mechanism of SV-fVas as cofactors is highly advantageous. Most hemostatic agents cannot be injected into the bloodstream because of risk of thromboembolism.

Intravenous administration is an ideal delivery route for some wounds, particularly in noncompressible hemorrhage, as it is often impossible to reach all bleeding sites and therefore unfeasible to apply topical hemostatic agents. Being cofactors, blood fVas are inactive until combined with fXas, which should only be present at the site of injury. Conceivably, therefore, SV-fVa could be administered intravenously without inducing thromboembolism and would be able to supplement endophysiological fVa at the site of bleeding and contribute to a transient clotting response to stop severe bleeding. Activity of SV-fVas would also eventually be down-regulated through inhibition and degradation of blood fXas. To evaluate SV-fVas as potential therapeutic agents, the protein from *P. textilis* venom was isolated and tested for its hemostatic properties *in vitro* and *in vivo* in a rodent bleeding model (Masci 2007). *P. textilis* SV-fVa was found to dramatically increase clotting *in vitro* and significantly reduce blood loss in a murine tail-vein excision model after intravenous administration. A patent was subsequently filed for this protein, and it is currently in preclinical development by Venomics and the University of Queensland under the candidate name CoVase and candidate code V0801. Studies to evaluate *P. textilis* SV-fVa in larger animal models of hemorrhage are currently in planning stages.

In summary, SV-fVas are unique cofactors that have maintained the plesiotypic activity but gained important new structural and functional properties that potentiate its toxic effects. However, under controlled circumstances, the novel structural and functional attributes of SV-fVas could potentially be harnessed to provide a next-generation therapeutic agent for intravenous administration to control severe surgical and traumatic bleeding. This is yet another example of exploiting the unique properties of venom proteins for the discovery and development of new human therapeutic agents.

12.6 CONVERGENCE WITH OTHER VENOMS

None is documented.

CHAPTER 13

FACTOR XA ENZYMES

M. TRABI, K. SUNAGAR, T. N. W. JACKSON, AND B. G. FRY

13.1 SUMMARY

Factor Xa is a serine protease that plays a crucial role in blood coagulation by cleaving prothrombin to activate thrombin, which in turn leads to the cleavage of fibrinogen and the formation of a clot. Snake venom homologues of factor Xa (SV-fXas) have retained this activity and upon envenomation initiate coagulation in the prey. Insertion of certain nucleotide segments in the promoter region of these venom homologues has elevated their level of expression in venom glands relative to physiological tissues. SV-fXas are found in two forms: as the enzymatic subunit of SV-fXa/SV-fVa complexes or as free proteins. The recruitment of SV-fXa occurred near the base of the Australian elapid snake radiation, and this toxin is exclusively found in the venoms of these snakes. SV-fXas have been invaluable in therapeutic research, particularly in the development of novel hemostatic drugs.

13.2 ENDOPHYSIOLOGICAL PLESIOTYPE

Hemostasis, the ability to stem bleeding in case of an injury, is one of the basic survival mechanisms of higher animals. Hemostasis has to be both highly efficient and tightly controlled in order to quickly form a blood clot at the site of an injury but prevent clotting under normal circumstances. This feat is accomplished by a complex series of activation steps and feedback mechanisms involving a cascade of enzymes and cofactors that ultimately leads to the formation of a fibrin clot (see color plate 2A). The endophysiological plesiotype of SV-fXa, coagulation factor Xa, is a vitamin K dependent glycoprotein synthesized in the liver and released into the bloodstream as an inactive zymogen, factor X. The human form of factor X consists of a 16-kDa light chain and a 42-kDa heavy chain held together by a disulfide bond. Upon initiation of the coagulation cascade, the zymogen is activated by either the intrinsic (IXa, VIIIa, phopholipids, and Ca^{2+}) or the extrinsic (VIIa, tissue factor, and Ca^{2+}) tenase complex. Activation is achieved through the excision of the activation peptide, a 52-residue segment at the N-terminus of the heavy chain. Factor Xa, a fully functional serine protease, then forms a complex with factor Va, phospholipids, and Ca^{2+} to activate prothrombin to thrombin. Like other serine proteases of the coagulation cascade, factor Xa shows characteristic domain architecture. The light chain consists of the Gla domain and two growth-factor-like domains, EGF-I and EGF-II, while the heavy chain contains the serine protease domain.

Prothrombin, a 72-kDa single-chain glycoprotein, circulates in plasma at a concentration of 100 to 200 mg/L (McDuffie et al. 1979; von Ahsen et al. 2000). Under physiological conditions, prothrombin is activated by the prothrombinase complex, a protein complex made up of the serine protease factor Xa, its cofactor factor Va, Ca^{2+}, and phospholipids. The activation involves cleavage of two peptide bonds (Arg274-Thr275 and Arg323-Ile324), which results in the excision of the activation peptide and the release of active α-thrombin (Joseph et al. 1999).

13.3 PLESIOTYPIC STRUCTURAL AND FUNCTIONAL TOXIN FORMS

13.3.1 STRUCTURE–FUNCTION RELATIONSHIPS

The most effective procoagulant venom proteins (or complexes) are those that activate prothrombin to thrombin, since this reaction is one of the central steps in blood coagulation. Factor Xa was recruited for use as a toxin near the base of the Australo-Papuan elapid radiation (see color plate 3). SV-fXas are found either as freestanding toxins or as the enzymatic subunits of SV-fXa/SV-fVa venom complexes (see chapter 12). Both stand-alone toxins and the enzymatic subunit of the fXa/fVa complex are the result of the same toxin recruitment event (Fry et al. 2005; Fry et al. 2012a; Fry et al. 2013) and will therefore be treated together in this chapter.

The stand-alone form is plesiotypic and is expressed in highly variable levels across the Australian radiation. It is abundant in the venoms of snakes from within the "tiger snake clade" (the genera *Hoplocephalus, Notechis*, and *Tropidechis*) in addition to that of *Pseudechis porphyriacus* and *Demansia vestigiata*. On the other hand, it is present in extremely low levels in *Acanthophis*, best detected by PCR or transcriptomics (see St Pierre et al. 2005; Birrell et al. 2007; St Pierre et al. 2007a; Jackson et al. 2013). SV-fXa is also part of the SV-fXa/SV-fVa toxin complex found only in the venoms of *Oxyuranus* and *Pseudonaja* (see chapter 12).

SV-fXas have retained the overall structure of their plesiotypic protein along with its maturation sequence from pro-protein to zymogen to active enzyme and its main activity, the cleavage of prothrombin to thrombin. However, a number of subtle changes have turned a moderately expressed coagulation protein into a powerful toxin. These include modifications of expression patterns, length of the activation peptide, a heavy-chain insert, β-hydroxylation, processing, storage, and function. Thus the liver and venom forms of factor Xa differ by ~20% sequence identity (Reza, Swarup, and Kini 2005; Reza et al. 2006).

13.3.2 GENE ORGANIZATION

Comparison of the complete trocarin D gene sequence with that of liver-derived factor X from *Tropidechis carinatus* revealed eight exons in each gene, identical exon-intron boundaries, and, with the exception of the first intron, 92% to 99% shared identity between introns (Reza, Swarup, and Kini 2006). However, the trocarin D promoter contains a 264-bp insertion with elements that induce high levels of expression. This segment, termed venom recruitment/switch element (VERSE), is responsible for approximately 30-fold higher expression of the trocarin D gene in the venom gland relative to the factor X gene in the liver (Reza, Swarup, and Kini 2005). A similar insertion (with approximately 95% shared identity) was found in the promoter region of the *Pseudonaja textilis* SV-fXa gene (Reza et al. 2006).

13.3.3 EXPRESSION

The recruitment of fXa as a toxin results in these snakes having two parallel prothrombin-activating systems, one produced in the liver and excreted into the bloodstream to achieve hemostasis in case of an injury and the other selectively expressed and stored in the venom gland to act as a toxin in bite victims. Both forms display extreme tissue specific expression, with expression in the corresponding tissue being up many times higher than the expression of the same gene in the reciprocal tissue (Reza, Swarup, and Kini 2005; Reza et al. 2006). For example, two genes for factor X have been documented as being expressed in the liver of *P. textilis*. One of these genes codes for a protein, PFX1, which is the endophysiological homologue and thus closely resembles liver-derived factor Xa from reptiles and mammals. The protein coded for by the second gene, PFX2, however, shows some features typical for venom SV-fXas, including a shorter activation peptide and an insertion in the heavy chain. The expression of PFX2 in the liver is approximately 56,000 times lower than that of PFX1.

13.3.4 ACTIVATION PEPTIDE

Maturation of the factor X zymogen to its active form requires excision of the activation peptide. In *T. carinatus*, the activation peptide of liver-derived factor X is 57 residues in length, whereas that of the venom form trocarin D is only 28 residues long, similar to that of other venom SV-fXas (Rao, Swarup, and Kini 2004). The amino acid sequence of this shortened activation peptide is highly conserved between most SV-fXas but differs considerably from liver-derived sequences of both mammalian and snake origin. Interestingly, the length of the activation peptide in vestarin D1 and D2 (from *D. vestigiata*) is somewhat in between that of liver-derived and other venom-derived SV-fXas, while *P. textilis* PFX2 has a short activation peptide despite its liver origin.

13.3.5 HEAVY-CHAIN INSERT

Most SV-fXas contain a 12-residue insert in the heavy chain, with the notable exceptions being porpharin D (*P. porphyriacus*) and vestarin D1 and D2 (*D. vestigiata*) (Birrell et al. 2007; St Pierre et al. 2007a). At this point in time, the function of this insert is unclear.

13.3.6 GLYCOSYLATION

While the activation peptide released during the processing of blood-coagulation factor X is heavily glycosylated, mature factor Xa does not contain any such posttranslational modifications. In contrast, studies of the venom forms trocarin D and the enzymatic subunit of pseutarin C are O-glycosylated at Ser52 in the EGF-1 domain of the light chain and N-glycosylated at Asn45 (Asn240 in pseutarin C numbering) in the heavy chain (Joseph et al. 1999; Joseph et al. 2003a; Joseph et al. 2003b; Rao, Swarup, and Kini 2004). O-glycosylation at Ser52 is also present in human and bovine coagulation factors VII and IX (Hase et al. 1988; Nishimura et al. 1989; Iwanaga et al. 1991). A Ser52Ala mutant of VIIa has reduced clotting activity but unaltered amidolytic or factor X-activating properties (Bjoern et al. 1991). The carbohydrate moiety at Ser52 may be required for the rapid association of factor VII/VIIa with its cellular receptor and its cofactor (tissue factor) (Iino, Foster, and Kisiel 1998). Human factor X contains two O-glycosylation and N-glycosylation moieties (Inoue and Morita 1993). However, they are located on the activation peptide that gets cleaved out during the activation process, and as a result, the activated factor X is not glycosylated.

The purpose of glycosylation in SV-fXas has not been investigated to date, but there is evidence that it enhances protein stability and even prolongs the circulatory time of serum proteins (Marshall 1972). Deglycosylation of trocarin D resulted in autolytic destruction, and the lack of glycosylation in mature mammalian factor Xa may be an evolutionary adaptation to make factor Xa short-lived in the bloodstream (Reza, Swarup, and Kini 2005). In contrast, glycosylation may prevent the inactivation of trocarin D through proteolysis and confer molecule thermal stability (Wang et al. 1996; Rao, Joseph, and Kini 2003).

13.3.7 Β-HYDROXYLATION

In human factor Xa, Asp63 in the EGF1 domain is β-hydroxylated to β-hydroxyaspartic acid (McMullen et al. 1983). Despite some sequence conservation around this hydroxylation site, the corresponding residue in trocarin D is not posttranslationally modified (Rao, Swarup, and Kini 2003).

13.3.8 PROCESSING AND STORAGE

Two enzymes are essential for the processing and release of fully functional factor Xa. First, a vitamin K dependent carboxylase is required to γ-carboxylate a series of Glu residues vital for the binding of factor Xa to phospholipid membranes. Second, the zymogen factor X needs to be converted to active factor Xa by excision of the activation peptide, an enzymatic cleavage that is performed by factor VIIa (in the extrinsic tenase complex) or factor IXa (in the intrinsic tenase complex). It is unclear how SV-fXas are processed in the venom gland, since no carboxylase has yet been identified. In contrast to coagulation factors, which circulate in blood in their zymogen form, SV-fXas are stored in the venom gland in their active form. It is conceivable that the proteolytic processing step (the excision of the activation peptide from the zymogen) is performed by mature SV-fXas present in the gland. Autoactivation of human factor X by factor Xa has been reported at very low levels (Colman 2006).

13.3.9 FUNCTION

SV-fXas have retained the procoagulant properties of liver-derived factor Xa, but it seems that some controlling elements have been altered or lost. The venoms of some Australian elapid snakes contain large quantities of prothrombin activators. Pseutarin C was reported to account for 20% to 30% of the total venom in *P. textilis* (Rao and Kini 2002), while as a stand-alone toxin, SV-fXas typically represent 3% to 6% of total venom protein (Joseph et al. 1999; Rao, Swarup, and Kini 2003), although it may be much higher in some genera such as *Hoplocephalus* or *Tropidechis*.

Considering the abundance of prothrombin activators in Australian snake venoms and their powerful properties, it is no surprise that venom-induced consumptive coagulopathy (VICC) is a common clinical sign of snake envenomation in Australia. VICC is seen after bites from *Pseudonaja, Oxyuranus, Notechis, Tropidechis*, and *Hoplocephalus* species (Isbister et al. 2008) and is generally diagnosed by a grossly prolonged activated partial thromboplastin time (APTT) and elevated D-dimer concentrations. In 112 cases of complete VICC, fibrinogen, factor V, and factor VIII were often totally depleted, but prothrombin levels never fell below 60% of normal (Isbister et al. 2010). While conversion of 10% to 15% of prothrombin is sufficient to completely activate the coagulation cascade and cause total fibrinogen depletion, this partial prothrombin activation can only be explained by the relatively rapid inactivation of the prothrombin activator. Indeed, a mathematical model of coagulopathy can only correctly predict the changes seen after an *Oxyuranus* bite if the half-life of the prothrombin activator is only 10 to 15 minutes (Tanos et al. 2008; Wajima, Isbister, and Duffull 2009).

Although onset of VICC is more rapid with venoms containing SV-fXa/SV-fVa complexes, resolution of VICC occurs within 24 to 36 hours irrespective of prothrombin activator type (Isbister et al. 2010), which points toward inactivation of the SV-fXa rather than of the SV-fVa cofactor. In addition to its procoagulant properties, trocarin D has been shown to cause inflammation and mitogenensis (Joseph et al. 2003a).

13.4 APOTYPIC STRUCTURAL AND FUNCTIONAL TOXIN FORMS

In contrast to a large number of venom protein families that have undergone rapid diversification, both in peptide sequence and in function, SV-fXas still closely resemble their plesiotypic endophysiological protein homologue, coagulation factor Xa. This is consistent with the slow molecular evolutionary rates typical of most large globular proteins (see chapter 1.3.2) (Fry 2005).

13.5 THERAPEUTIC POTENTIAL OF THE TOXIN TYPE

Haempatch (Q8009), developed from the *P. textilis* SV-fXa, has been patented as a hemostatic drug. In the venom, Haempatch is present in a complex with a factor-Va-like protein the properties and procoagulant activity of which were first described in 1988 (Masci, Whitaker, and de Jersey 1988). The efficacy of Haempatch has been assessed in a series of dermal injury and rat organ trauma models and its antibleeding properties compared to those of FloSeal (Baxter Healthcare), a thrombin and collagen preparation used in surgical settings (Warner et al. 2007; Warner et al. 2009). Standardized sections of the spleen, liver, or kidney were excised, the agent applied to the wound surface, and blood loss determined in one-minute intervals. Haempatch reduced the average overall blood loss by 56% to 90%, compared with 0% to 70% with FloSeal, and it shortened the average time to hemostasis (defined as a blood loss of less than 1 μL per minute) by 42% to 53%, compared with 0% to 22% for FloSeal. In the spleen model, total blood loss with FloSeal was actually higher than without any treatment, while in the liver model, time to hemostasis was longer than untreated. In both cases, Q8009 demonstrated greater hemostatic properties, with a reduction in total blood loss of as much as 56% (spleen model) and reduction in time to hemostasis by as much as 48% (liver model). Recombinant Haempatch with procoagulant activity comparable to that of the native SV-fXa has been successfully expressed in human kidney cells (H293F), and the highest expression levels were obtained with constructs in which the activation peptide was deleted or replaced with a peptide involved in insulin-receptor processing (Filippovich et al. 2002). The ability to produce a highly active recombinant molecule certainly increases the likelihood of a market application and opens the path for future clinical trials.

13.6 CONVERGENCE WITH OTHER VENOMS

Although recruitment of fXa as a procoagulant toxin has not been documented in other animal venoms, factor Xa inhibiting (anticoagulant) toxins have been widely exploited by hematophagous animals such as vampire bats (Fernandez et al. 1999), leeches (Tuszynski, Gasic, and Gasic 1987; Nutt et al. 1988), and ticks (Waxman et al. 1990) (see color plate 2A).

CHAPTER 14

KALLIKREIN ENZYMES

S. VAIYAPURI, K. SUNAGAR, J. M. GIBBINS, T. N. W. JACKSON, T. REEKS, AND B. G. FRY

14.1 SUMMARY

Toxicoferan venom kallikrein enzymes (TV-kallikreins) evolved from members of the kallikrein class of serine proteases. Although they have retained some of the plesiotypic activities, they have also acquired several apotypic biochemical functions. TV-kallikreins mainly affect blood pressure or coagulation. Despite sharing significant structural similarity, these enzymes vary widely in their functions. To date, more than 250 TV-kallikrein enzymes have been identified, and nearly half of them have been functionally characterized; however, very few of these have been recovered from the venoms of reptiles other than viperid snakes. The structural and functional mechanistic information regarding TV-kallikreins is somewhat limited, hindering a clear understanding of their pathology. These enzymes have, however, received a considerable amount of attention for their potential use in diagnostics and therapeutics, particularly in relation to hemostatic disorders in humans. The clinical and diagnostic applications of these proteins exemplify the need for further studies characterizing their sequences, structures, functions, and phylogeny and the molecular evolutionary regimes they are subject to.

14.2 ENDOPHYSIOLOGICAL PLESIOTYPE

TV-kallikreins are derived from glandular/tissue kallikrein-type serine proteases, which are a widely distributed family of cysteine cross-linked β-barrel enzymes with diverse activities (Page and Di Cera 2008). These hydrolytic enzymes contain a serine residue in their catalytic site and hydrolyze peptide bonds in target substrates. Serine proteases are present in all organisms and may be intracellular or extracellular in location. They play important roles in biological processes such as digestion, blood coagulation, neurotransmission, or posttranslational processing of secreted peptides or proteins. Chymotrypsin and chymotrypsin- (or trypsin-) like proteases belong to the SA clan and S1 family of serine proteases and have similar structural and mechanistic properties (Barrett and Rawlings 1995).

Kallikrein enzymes are a group of secreted serine proteases primarily found in glandular cells and biological fluids (Bhoola, Figueroa, and Worthy 1992). These enzymes release vasoactive substances called kinins from kininogens, which are their endophysiological substrates (Muller-Esterl 1989). There are two types of kallikrein enzymes: plasma and glandular/tissue. These two types differ in MW, substrate specificity, immunological characteristics, and the type of kinins released as a consequence of their activities (Bhoola, Figueroa, and Worthy 1992). Tissue kallikrein enzymes are secreted serine proteases that belong to the S1 family within the SA clan (Barrett and Rawlings 1995; Rawlings, Barrett, and Bateman 2012). These kallikrein enzymes are single-chain proteins synthesized as pre-proenzymes with a signal peptide 16 to 30 amino acids in length at the N-terminus, followed by a propeptide of 4 to 9 amino acids and a catalytic serine protease domain (Yousef and Diamandis 2001). The signal peptide and the propeptide are removed during the maturation process through enzymatic cleavage. The calculated MW of kallikrein enzymes is in the range of 23–26 kDa, although posttranslational modification such as glycosylation will increase the molecular mass further in some cases, such as hK1, hK3, hK5, and hK7 (Yousef and Diamandis 2001; Borgono, Michael, and Diamandis 2004). Tissue kallikrein enzymes possess a similar tertiary structure to chymotrypsin and trypsin, which consists of two juxtaposed six-stranded antiparallel β-barrels and two α-helices (Rawlings, Tolle, and Barrett 2004). The active site (H57, D102, and S195) lies between the β-barrels and brings the domains together. Structural variations in kallikrein enzymes have been observed mainly in the surface loops surrounding the substrate-binding domains, which dictate substrate and inhibitor specificities and control catalysis (Perona and Craik 1995; Bernett et al. 2002). In general, five or six covalently linked disulfide bonds are formed in kallikrein enzymes to provide structural rigidity. The presence of N-glycosylated moieties is known to influence the substrate and inhibitor specificity of kallikrein enzymes (Borgono, Michael, and Diamandis 2004). At the S1 substrate-binding specificity pocket of 12 kallikrein enzymes (at position 189 according to α-chymotrypsinogen numbering; this numbering system is followed throughout this chapter), aspartate or glutamate residues are present, and these can cleave the carboxyl side chains of basic amino acids (such as arginine or lysine) in their substrates in a similar manner to trypsin (Borgono, Michael, and Diamandis 2004). The remaining three kallikrein enzymes, hK3, hK7, and hK9, and chymotrypsin have nonpolar serine, asparagine, and glycine residues, respectively, which confer specificity toward large hydrophobic amino acids such as tyrosine and phenylalanine (Yousef and Diamandis 2001; Borgono, Michael, and Diamandis 2004).

Glandular/tissue kallikrein enzymes are expressed in a range of tissues and mostly as a group of several enzymes. For example, almost all are expressed in salivary glands, but certain subgroups of kallikrein enzymes are expressed in other organs, such as skin, prostate, breast, pancreas, and central nervous system. There are two major types of kininogens: low and high MW (Muller-Esterl 1989; Bhoola, Figueroa, and Worthy 1992). Plasma kallikrein converts high-MW kininogen into bradykinin (a nonapeptide), which mediates physiological processes such as blood coagulation, regulation of blood pressure, and vasodilation (Bhoola, Figueroa, and Worthy 1992; Sainz, Pixley, and Colman 2007). In contrast, tissue kallikrein converts low-MW kininogen into Lys-bradykinin (a decapeptide) to regulate blood pressure, vascular permeability, smooth-muscle contraction, and inflammatory processes (Bhoola, Figueroa, and Worthy 1992; Borgono, Michael, and Diamandis 2004).

Consistent with kallikrein enzymes being expressed in diverse tissues, they play a wide range of physiological roles (Bhoola, Figueroa, and Worthy 1992; Borgono, Michael, and Diamandis 2004). For example, hK1 cleaves low-MW kininogen and releases kallidin, which binds to bradykinin receptors and mediates biological processes such as the regulation of blood pressure, smooth-muscle contraction, pain induction, and vascular permeability (Bhoola, Figueroa, and Worthy 1992). In addition to kininogenase activity, hK1 is also involved in processing growth factors and peptide hormones, mainly in the pituitary gland and the pancreas. The hK1 enzyme has been suggested to convert inactive procollagenase and angiotensinogen into active collagenase and angiotensin (Borgono, Michael, and Diamandis 2004). The hK2 and hK3 enzymes possess lower kininogenase activity and are involved in the liquefaction of seminal fluids by hydrolyzing seminogelin I and II and fibronectin (Borgono, Michael, and Diamandis 2004). The hK3 enzyme (prostate-specific antigen) produces angiostatin-like fragments through the limited proteolysis of plasminogen and may inhibit proliferation, growth, and metastasis of cancer cells (Heidtmann et al. 1999; Sotiropoulou et al. 2003). The hK6 enzyme is also reported to cleave plasminogen (Borgono, Michael, and Diamandis 2004). It cleaves the Aα- and Bβ-chains of the clotting protein fibrinogen to release α- and β-fibrinopeptides and collagen type I and IV, and it may play a role in tissue remodeling and tumor invasion (Magklara et al. 2003). The hK5, hK6, hK13, and hK14 enzymes are able to cleave extracellular matrix proteins (Borgono, Michael, and Diamandis 2004). Several kallikrein enzymes (such as hK1) process peptide hormones such as insulin and glucagon in the pancreatic Islets of Langerhans (Bhoola, Figueroa, and Worthy 1992; Borgono, Michael, and Diamandis 2004). In addition, they have been reported to maintain epidermal hemostasis in skin. Some kallikrein enzymes, such as hK6, play a role in the central nervous system by regulating myelination and myelin turnover in mice (Yamanaka et al. 1999). Some kallikrein enzymes, such as hK8, activate protease-activated receptors (PAR) and G-protein-coupled receptors to trigger cellular signaling (Borgono, Michael, and Diamandis 2004), which highlights their possible function in hemostasis through the regulation of platelets and endothelial functions. The 15 kallikrein enzymes are differentially expressed in various cancer cells, and thus their expression pattern could be exploited in the diagnosis of cancer (Bhoola, Figueroa, and Worthy 1992; Borgono, Michael, and Diamandis 2004).

14.3 PLESIOTYPIC STRUCTURAL AND FUNCTIONAL TOXIN FORMS

14.3.1 PLESIOTYPIC STRUCTURAL FORMS

This toxin form contributes to the profound, rapidly developing swelling characteristic of most viperid snake and anguimorph lizard envenomations (see color plates 11D and 12C). Although TV-kallikrein is one of the major types of proteolytic enzymes found in the venoms of viperid snakes, its relative presence in the venoms of other toxicoferan reptiles is variable. While most elapid snakes secrete this toxin type in low levels, TV-kallikreins are a major component of the venom of some non-front-fanged caenophidian snakes (such as *Philodryas* species) and

```
                              10        20        30        40        50        60        70        80        90       100
1.   mevtkllatflllqlsf---------aLS-FRTRIIGGEICAEDGHPWLAWIY--GSNGLQCSGVLINQDWVLTAAHCYERSGMQLKLGVHNREVPREGE
2.   mepakllalllllpsl----------vsaYRKRIVGGQECNEAEHPWLVLLY--QSKKPFCSGILLDYNWVLTAAHCYVSGEIEIRLGLHNRNVPRGDE
3.   mepikllaflplll-af----------vsaKHKRIVGGQECSEDEHPWLVLFH--DSTGPFCSGVLLDHNWVLTAAHCYERGKMQMKLGVHNRNVLRGDE
4.   mepakllallllllpsl----------vsaHHKRILGGKECNEAEHPWLVLLY--ESKKPFCSGILLDRNWVVTAAHCYVSGEIEIRLGLHNRNMPRGDE
5.   mgpaklvvfvmlfhpsl----------vsaNPLRITGGEECNEDSHPWLVVIY--TEANTMCGATLLNQDWVLTAAHCYKRGKIWLNFGVHNREQARGDE
6.   mgpaklvafvmllhpsl----------vsaNPMRITGGEECNEDSHPWLVLLY--TEANTMCGATLLNQDWVLTAAHCYKRGKIWLSFGVHNREEARGDE
7.   mgpaklvafvlllqpsl----------vsaNPLRITGGQECNKDSHPWLVLLV--TEASTMCGATLLNQDWVLTAAHCYDREPIHLYFGIHNKKQPSGHE
8.   mgpaklvafvllmhpsl----------vsaNPLRITGGQECNEDSHPWLVLLY--TEAGTMCGATLLNQDWVLTAAHCYQRGKIWLNLGVYNREQAREDE
9.   mgsaklvalvlllqpsl----------vsaNPMRITGGQECNRDTHPWLVLLT--TKSGAMCGATLLNQDWVLTAAHCYAREPIHLKFGIHNDHLPSEHE
10.  mgpaklvvfvllmhpsl----------vsaNPMRITAGEECNEDSHPWLVLLT--TESGAVCGATLLNQDWVLTAAHCYEREKIWLSFGVHNREQRRGDE
11.  mgpaelmafvlllqssl----------vlaNPTRVIGGQECIEDEHPWLAAII--DHEFFICGATLLNQDWVLTAAHCYESTKLQVKFGVHHKGKPRGDE
12.  mgpaklvafvlllqpsl----------vsaNPMRITGGQECNQDTHPWLVLLI--TEAGTMCGATLLNQDWVLTAAHCYEREPIHLYFGIHNTKLPRGDE
13.  mgpaelttfvlllqlsl----------vlaNPTRIIGGQECIEEEHPWLAAII--DNENFICGATLLNQDWVLTAAHCYERGKLQVKFGVHHKGKPRGDE
14.  mgpaelmafvlllqssl----------vlaNPTRVIGGQECIEDEHPWLAAII--TSEFFVCAATLLNQDWVLTAAHCYESMKLQVKFGVHHKGKPRGDE
15.  mgpaklvafvlllqpsl----------vlaNPVRVVGGQECEKDGHRWLGLLQ--TEEGPYCGVALISQNWVLSAAHCYESMKLQVRLGVHHRGKPIADE
16.  mgpaklvvfvllmhpsl----------vsaNPMRITAGEECNEDSHPWLVLLI--TEAGSMCGATLLNQDWVLTAAHCYEREKTWLRFGVHNREQTRGDE
17.  mgpaklvtfvlllqpsl----------vsaNPTRIIGGQECNEDSHPWLVLLV--AEASFMCGATLLNQDWVLTAAHCYDSRPIHLYFGIHNKKQPRGHE
18.  mgpaklvafvlllqpsl----------vsaNPLRITGGQECNKDSHPWLVLLV--TEGSTMCGATLLNQNWVLTAAHCYDRKPIHLYFGIHNKKQPSGHE
19.  mgpgevmafvlllqpsl----------vlaNPTRIIGGQECFEDWHPWLAAIL--ANESFVCGATLLNQDWVLTAAHCYQSMNLQVKFGVHDKGKPRGDE
20.  mgpaklvafvllmhpsl----------vsaNPLRITGGEECNEDSHPWLVLLT--TEANTMCGATLLNEDWVITAAHCYERGKIWLSFGVHNREHARGDK
21.  mgpaklvafvlllqpsl----------vsaNPLRITGGQECNKDSHPWLVLLV--TEASTMCGATLLNQDWVLTAAHCYDREPIHLYFGIHNKKQPSGHE
22.  meppkllaflllvlpsf----------vsaYHKRIIGGQECPQLLHPWLGLLY--HSGGPYCSVVLINQDWVVTAAHCFYSGELRVGLGVHNRRVLTGNE
23.                            *IIGGQECDETGHPWLALLH--RSEGSTCSGVLLNQDWIVTAAHCFYLGELRIGLGVHNRRVLRGNE
24.                            *IIGGQECDETGHPWLALLH--RSEGSTWSGVLLNRDWILTAAHCEELGPMKICFGMKNRNVLRGDE
25.  maligvlanllilclsy---------aRT-APDRIIGGLECNQNEHRSLVLLY--NSGGFFCSGTLINHEWVLTAAHC-NRENIQIKLGVHNIHVPNEDE
26.  mplirvlasllilqlsygKSLDNGAKAIT-SLDRIIGGFECNPSEHRSLVYLY--NSAGFFCSGTLINHEWVLTAAHC-NREDIQIRLGVHNVHVHYEDE
27.  malirvlasllilqlsy---------aVT-PFDRIIGGFECNEYEHRSLVHLY--NSSGFFCSGTLINHEWVLTAAHC-NRDDIQIKLGVHNVSVNYEDE
28.  mvlirvlanllilqlsy---------aQKS-SELVIGGVECDINEHRFLAAFFKYQPWTFQCAGTLIHEQWVLGAAHCYKRGLNIYLGMHNQSIQFDDE
29.  mvlirvlanllvlqlsy---------aQK-SSELVTGGAECDINEHPFLVALYTASSSTFHCAGTLINQEWVLTAVHC-DRKKIRIKLGMHSKNMPNEDE
            110       120       130       140       150       160       170       180       190       200
1.   QLRTSAAEAKCYPDAASSDSNS-CAQN---TFINDI-------------MLVKLNAPVDYNDNVQPVPLSTAVPEEGTICKVMGWGSINASEDIYPLVPYCV
2.   QMRVSAATLCYPDTASTTQNS-CAD-----FTADI-------------MMIKLNSPVKYSKHIGPLLLPSGSVSVGTECRVMGWGTTTSKVTLPIVPHCA
3.   ESRVSAAARCFPDRPGATQNS-CEA-----FTADI-------------MMVKLNSPVKYNKHISSLLLPSASVFVGAKCRVMGWGSNTVVRNT--IVPHCV
4.   QMRVSAATLCYPDTASTTQNS-CAD-----FTADI-------------MMIKLNSPVKYSKHIGPLLLPTGSVFVGTKCRVMGWGTTTSEVTFPIVPQCA
5.   EVREAVGTFCYPDSPGTTTSS-CPCYTLDH-GDDI-------------MLIKLNASVTYNEHIAPMALPDHAVPLGTECDIIGWGETELIVDTASDVPLCA
6.   QVREAVGTFCYPDSPGTTNSS-CPSFTLDS-GDDI-------------MLIGLNASVTYNEHIAPMALPDHAAPLGTECNIIGWGETELIVGSPSEVPLCA
7.   EERVAVGTFCYPDSPGTTTNS-CAKFT-----HDI-------------MLIKLDTPVTYNEHIAPVALPDRAAPLGTKCNIIGWGETELTVGSPSHIPFCA
8.   QVREAVGTFCYPDSPGTTSSS-CPGYTLDR-GDDI-------------MLIKLNASVTYNEHIAPMALPDHAAPLGTECDIIGWGETELIVGSPSDVPLCA
9.   QVRDAVSTFCYPDSPSNTTNS-CADFTLDR-GHDI-------------MLIKLNAPVTYNEHIAPLPLPDHAAPLGTECNIIGWGETELTVGS--HVPLCT
10.  QVREAVGTFCYPDIPSTTNSS-CPSFTLDR-GDDI-------------MLIKLDTPVTYNEHIAPMALPDHAAPLGTECDIIGWGETELIVGSASDVPLCA
11.  QVRDAVSTFCFPDTPGTTNST-CPYEKNNT-KHDI-------------MLIKLNASVTYNEHIAPLALPDRAAPPGTECEIIGWGKTEVADEPFPKVPFCA
12.  QVRDAVSTFCYPDSPSTTTNS-CANFT-----HDI-------------MLIKLDTPVTYNEHIAPMALPDRAAPLGTECDIIGWGETELTVGSPSHIPLCA
13.  QVRDVVSTFCFPDTPGTTNST-CPYEKNNT-KHDI-------------MLMKLNASVTYNEHIAPLALPDRAAPMRTMCKIMGWGQTEVAEGPFPKVPFCA
14.  QVRDVVSTFCFPDTPGTTSST-CPYERNNT-KHDI-------------MLMKLNASVTYNEHIAPLALPDRAAPPGTECDIIGWGKTEVADEPFPKVPFCA
15.  QLRDVDSTFCYPDAPDTTNST-CPYLGFDAISNDI-------------MLMKLSSPVTYNEHIASIALPDHDAPMGTECTIMGWGLTGATNGSYPKVPLCA
16.  QVREAVGTFCYPDSPGTTNSS-CPSFTLDR-GDDI-------------MLIKLNASVTYNEHIAPMALPDHAAPLGTECDIMGWGETELIVGSPSDVPLCA
17.  QARDAVSTFCYPDSPGTTNSS-CPSFRLDR-GDDI-------------MLIKLNASVTYNEHIAPMALPDRAAPLGTECDIIGWGETELTIGSVSHIPFCA
18.  DERVAVGTFCYPDPPGTTTNS-CAKFT-----HDI-------------MLIKLDTPVTYNEHIAPMALPDRAAPLGTKCDIIGWGETELTVGSVSDIPFCA
19.  QLRDVVGTFCFPDTPGTTNST-CPYERNDT-KHDI-------------MLMKLNASVTYNKHIAPLALPDRAAPLGTYCQIMGWGKTEVADEPFPKVPFCA
20.  QVREAVGTFCYPDSPGTTTSS-CPCYTLDR-GDDI-------------MLIKLNASVTYNEHIAPMALPDRAPPVGTECDIIGWGETELIVGSASVVPLCA
21.  EERVAVGTFCYPDSPGTTSDS-CAKFT-----HDI-------------MLIKLDTPVTYNEHIAPMALPDRAAPLGTKCNIIGWGETELTVGSPSHIPFCA
22.  QFRVSERKICYPDTNSTTTNSSCSE-----YTD-I-------------MLIKLNSSVEYTRNVRPLRLPTSTVSEGTYCTVMGWGTTTSPNVTYPAVPHCV
23.  QVRVSARKKCYPATASIITNSSCSE-----YTDDI-------------MLIKLDSSVEYTERVRPLSLPTSPASEGAECTVMGWGTTTPPDVTYPAVPVCV
24.  QVKVAAVKKCYPATAGTIYN--CN------YVNTVLMNNDLLKRELFPMLIKLDSSVDYNERVAPLSLPTSPASLGAECSVLGWGTTTPDDVTLPDVPVCV
25.  QIRVPKEKVCCLGTMN------CTQ-----WNQDI-------------MLIRLNSSVNYSTHIAPLSLPSNPPSVGSVCRVMGWGTITSPEVTYPEVPHCV
26.  QIRVPKEKLCCLSTNN------CTQ-----FSQDI-------------MLIRLNSPVNYSEHIAPLSLPSNPPSMGSVCCVMGWGTITSPEVTYPEVPHCV
27.  QIRVPKEKLCCHSTNN------CTQ-----LGQDI-------------MLIRLNSSVNYSEHIAPLSLPSNRPSMGSVCRVMGWGLLTSPEVTFPKVPHCV
28.  QRRYAIEEHYYR----------CDEKLTK-WEKDV-------------VLLKLNKPVRNSTHIAPLSLPSSPPSIGSLCRVMGWGIMSSTKDILPDVPHCA
29.  QIRVPKKKVF-----------CHNSKFPNGMDKDI-------------MLIKLRRPVNNSAHIAPISLPSSPSSPRSRCRIMGWGKISSAEDTDSLVPHCA
            210       220       230       240       250       260       270       280       290       300
1.   DIEILNNGVCEAAYPWCTAYDKLLCAGVLEGGKDSCHGDSGGPLMCGDELQGILSFGGHPCAKPLEPGVYVKVDSYLDWIQSIIENNTPETS
2.   NINILENRVCAVAYPSWRMTNDMLCAGDQKGCKDSCQGDSGGPLICGGQLQGVVSMGISP-----GPGVYTNVNSYFNWILEFI
3.   DLTILENRVCTAALPVLDMSDDDLCAGDPKGCKDSCQGDSGGPLICGGQLQGIVSRGRSP-----GPGIYTKVDSYLNWILEFI
4.   NINILENRVCAVAYPSWRMTNDMLCAGDQKGCKDTCQGDSGGPLICGGQLQGIISMGKSP-----GPGVYTNVNSYFNWILEFI
5.   SISTRENHFCEDAYSA-NMTDDMLCAGDLKGGPDACRSDSGGPLLCGGQLQGLVSFGGIPCGQPKKPGVYTKIFNYRKWIRSHI
6.   SISTRENHFCEEAYSA-NMTDNMLCAGVLEGGPDACNGDSGGPLLCGGQLQGLVSFGGYPCGQPMMPGVYTKIFSYRKWIHSHI
7.   SINTMDNHFCQDAYSA-TITDNMLCAGVLEGGPDACKSDSGGPLLCGGQLQGLVSFGGYPCGQPMKPSVYTKIFNYRRWIRSHI
8.   SISTRENHFCEDAHSA-NMTDNMLCAGVLEGGPDACKGDSGGPLLCGEQLQGLVSFGGYPCGEPMMPGVYTKIFSYRKWIHSHI
9.   TINTMDNHFCQDVSTV-TITDNMICAGVLEGGPDACKGDSGGPLLCGGQLQGLVSFGSYPCGEPMVPGVYTKIFNYREWIQSHML
10.  SISIRENHFCEDVYLR-NITDDMICAGVLEGGPDACKGDSGGPLLCGGQLQGLVSFGGYPCGEPMVPGVYTKIVKYRKWIQSHML
11.  SIKILKNHFCQDAYPWWHMTDYQLCAGILGGGRDTCKGDSGGPLVCDRQIQGIVSFGSDPCAQLLEPGVYSSVYEYLNWIQGHTGQEQMRLVSKKKSLKSH
12.  SINTMENHFCQDVYSA-TITDNMLCAGVLEGGPDACKGDSGGPLLCDGQFQGLVSFGGYPCGEPMRPSVYTKIFNYRKWIRSHM
13.  SIKILENHFCQDAYPWWHMTDNQLCAGFLEGDRDTCKGDSGGPLVCGRKIQGIVSFGSEPCGEPLMPGVYTKVYNYNNWIQSHI
14.  SIKILKNHFCQDAYPWWHMTNNQLCAGILGGDRDTCKGDSGGPLVCDGRIQGVVSFGDDPCGQLLKPGVYTNVYQYLNWIHIHI
15.  SVNTFHNSVCDVVYPWWDLTDNQLCAGVLLGGIDTCKGDSGGPLVCGGRIQGVVSFGDDPCGQLLKPGVYTNVYKYVHWIHRYI
16.  SINTMENHFCQDVSSV-TITDDMICAGVLEGGPDACRGDSGGPLLCGGQLQGLVSFGGYPCGEPMVPGVYTKIFNYRKWIQSHML
17.  SINTMDNHFCQDVSSV-TITDDMICAGVLEGGPDACKGDSGGPLLCGGQLQGLVSFGGYPCGQPMMPGVYTKIFSYREWIYSHIR
18.  SINTMDNHFCQDAYSA-TITDDMICAGVLEGGPDACMRDSGGPLLCGGQLQGLVSFGGYPCGQPMKPSVYTKIFNYRKWIRSHI
19.  SIKILNNDFCQYAYPWWHMTNNQLCAGFLEGDRDTCEGDSGGPLVCGGRIQGVVSFGGQPCAEPLQPGVYTNVYEYLNWMRRYI
20.  SISTRENHFCEEAYSV-NMTDSMICAGVLEGGPDACKSDSGAPLLCGGQLQGLVSFGDYPCGKPRKPGVYTKVFNYREWIYSHI
21.  SINTMDNHFCQDAYSA-TITDNMLCAGVLEGGPDACKSDSGGPLLCGGQLQGLVSFGGYPCGQPMKPSVYTKIFNYRRWIRSHI
22.  RIRMLNNIDCELAYPWWNITDSVLCAGTWSGGKDSCKGDSGGPLICRGQLQGIVSFGGFPCAQPLEPGVYTKVNSYLNWIQSIIQ
23.  RIEMLNNAVCELARDLWNITDSVLCAGTWFGGKDSCKGDSGGPLICRGQLTGIVSWGGFPCEQPLEYGVYTKVISFLFWIQSI
24.  NIEIFNNAVCQVARDLWKFTNK-LCAGVDFGGKDSCKGDSGGPLVCDNQLTGNVSWG-FNCEQGEKYG-YIKLIKFNFWIQNIIQGGTT--CP
25.  NIQILHKELCEAAYPILLGNSNILCAGKLLGDKDSCKGDSGGPLICNGQIQGIVSWGGFPCAQILEPGVYTKVFDYIDWIQDIMAGNANVICPGDNF
26.  DINILHIPVCQAAYPTMSGK-NILCAGILEGGKDSCKGDSGGPLICNGQIQGIVSWGRFPCAQFLEPGIYTKVFDYKDWIEGIIAGNSNVICP
27.  DINILHIQVCQAAYPSMSEN-YLLCAGVLEGGKDSCKGDSGGPLICNREIQGIVSWGGFPCAQLLEPGVYTKVFDYIDWIEGIIAGNTSVTCPSDNF
28.  NIKLVNYTECVAPYPNIPVTTRLWCAGILEGGIDTCHQDSGGPLICDGQFQGIVAFGRYPCAQPRVPALYTKVSNYTDWIQNIIAGKTTTACPP
29.  NIFIVKHSWCEAIYPWVPADSRTLCAGILQGGKDTCHGDSGGPLICNGQIQGIVSGGSEPCGQRLKPAVYTKVFDYTDWIQSIIAGNTTVTCPP
```

anguimorph lizards (Fry et al. 2003c; Fry and Wüster 2004; Fry et al. 2006; Fry et al. 2008; Fry et al. 2009a; Fry et al. 2009b; Vaiyapuri et al. 2011; Vaiyapuri et al. 2012). Phylogenetic analysis shows that TV-kallikreins evolved from glandular/tissue kallikrein (Itoh et al. 1988; Fry and Wüster 2004; Fry 2005; Fry et al. 2006; Fry et al. 2013) and thus belongs to the SA clan and S1 family of serine proteases (Serrano and Maroun 2005). Similar to endophysiological kallikreins, TV-kallikreins are secreted as preproenzymes, with a signal peptide of approximately 18 amino acid residues in length and a 6-residue activation peptide followed by a 229–237 amino acid kallikrein domain (Fry 2005; Fry et al. 2006; Fry et al. 2008; Fry et al. 2010b; Vaiyapuri et al. 2011; Koludarov et al. 2012). As for endophysiological kallikreins, the MW of TV-kallikreins ranges from 26–28 kDa. However, in most cases, these enzymes are glycosylated posttranslationally, and thus the native MW may vary widely (Fry 2005; Fry et al. 2006; Fry et al. 2008; Fry et al. 2010b; Vaiyapuri et al. 2011).

Most of the TV-kallikreins sequenced to date from snake venoms are from viperid snakes, but more than 30 TV-kallikreins have also been recovered from anguimorph lizard venoms. All viperid-snake-venom TV-kallikreins contain a unique C-terminal extension of seven amino acids, including an additional cysteine (see figure 14.1). This C-terminal extension has been reported as present in two elapid-snake-venom kallikrein enzymes but is absent from all other non-viperid-snake sequences, including all those from lizards (see figure 14.1). It must be noted that this cysteine was present in the first reported sequence of anguimorph-lizard-venom-specific TV-kallikrein (Utaisincharoen et al. 1993). There are dramatic differences between this sequence and those obtained from *Heloderma* and other anguimorph lizards in a later study (Fry et al. 2010b). Sequence alignments show that the tertiary structure determined by disulfide bonds is likely to be affected if the previously reported sequence was correct, as it contained 13 cysteines (missing the second plesiotypic cysteine with two new cysteines in atypical positions) (see figure 14.1). Conspicuously, the last cysteine of the *Heloderma horridum* sequence is absent from all other lizard toxin sequences but is instead characteristic of snake-venom isoforms. Another anomaly in this sequence is an insertion at positions 126–132 that is not present in any other toxicoferan lizard sequence. RT-PCR investigations revealed that primers designed off of the reported *H. horridum* sequence do not react with *H. horridum* cDNA, while primers designed using the reported *Heloderma suspectum*

FIGURE 14.1: Kallikrein toxin sequence alignment (Fry et al. 2010b). (1) H9GAC6 (*Anolis carolinensis*), (2) K4I265 (*Abronia graminea*), (3) E2E4I5 (*Celestus warreni*), (4) E2E4H9 (*Gerrhonotus infernalis*), (5) Q2XXN1 (*Varanus acanthurus*), (6) E2E4G3 (*Varanus gilleni*), (7) E2E4H0 (*Varanus glauerti*), (8) E2E4G4 (*Varanus glauerti*), (9) E2E4H2 (*Varanus gouldii*), (10) E2E4G6 (*Varanus gouldii*), (11) E2E4G1 (*Varanus gouldii*), (12) E2E4G5 (*Varanus indicus*), (13) E2E4G2 (*Varanus indicus*), (14) B6CJU5 (*Varanus komodoensis*), (15) B6CJU6 (*Varanus komodoensis*), (16) E2E4G7 (*Varanus mertensi*), (17) Q2XXN0 (*Varanus mitchelli*), (18) E2E4H4 (*Varanus scalaris*), (19) E2E4G0 (*Varanus scalaris*), (20) E2E4G9 (*Varanus scalaris*), (21) E2E4H1 (*Varanus tristis*), (22) C6EVG4 (*Heloderma suspectum*), (23) P43685 (*Heloderma horridum*), (24) *H. horridum*) (Utaisincharoen et al. 1993), (25) Q09GK1 (*Philodryas olfersii*), (26) Q5MCS0 (*Hydrophis hardwickii*, formerly *Lapemis hardwickii*), (27) A8QL56 (*Ophiophagus hannah*), (28) Q512B6 (*Deinagkistrodon acutus*), (29) E9JG23 (*Echis coloratus*). The erroneous *H. horridum* sequence (24) (from Utaisincharoen et al. 1993) is highlighted in gray.

sequence reacted strongly with the *H. horridum* cDNA library (see figure 14.2). Resequencing of the TV-kallikrein encoded in the *H. horridum* cDNA library revealed it to be very similar to the *H. suspectum* sequence while contrasting starkly with that reported in the 1993 paper. Thus, it appears that the *H. horridum* sequence P43685 contains multiple errors. The aberrant cysteine pattern is particularly suggestive in this regard, as it is divergent from all other lizard-toxin isoforms, including those from the same species, but displays all the hallmarks of typical snake sequences.

While more than 250 TV-kallikreins have been characterized at sequence level (Fry et al. 2006; Fry et al. 2008; Fry et al. 2009b; Fry et al. 2010b; Vaiyapuri et al. 2012), only eight crystal structures have been solved to date and all from viperid snake venoms. Similar to normal body kallikrein enzymes and other members of the S1 family, TV-kallikreins possess two antiparallel β-barrel domains with an α-helical domain at the C-terminus. The catalytic site (H57, D102, and S195) is located between the two domains. All TV-kallikreins have been predicted to have a common trypsin-like kallikrein domain. In general, 12 cysteine residues are present in viperid snake venom TV-kallikreins (see figure 14.1), and they form 6 disulfide linkages, which stabilize the structure of these proteins. The disulfide linkage between C91-C245e appears to be unique to viperid-snake-venom TV-kallikreins, as this is formed by the viperid-venom-specific C-terminal extension (Parry et al. 1998). Since the C-terminal extension of 7 amino acids is absent in anguimorph-lizard-venom TV-kallikreins, it is not clear how many disulfide linkages stabilize their structure. The cysteine residue present in the internal extension at position 93G may compensate for the absence of the cysteine from the C-terminal extension, but further investigation is required to confirm this notion. Similarly, TV-kallikreins from the venom of sea snakes and *Ophiophagus hannah* contain the C-terminal extension with a cysteine residue at position C245e. However, those from other elapid snake venoms, and the venoms of non-front-fanged caenophidian snakes, lack this extension. In addition, a salt bridge formed between V/I16 and D194 (see figure 14.1) is involved in stabilizing the active site structure of venom kallikrein enzymes (Parry et al. 1998; Vaiyapuri et al. 2012). Similar to the case of physiological kallikreins, an aspartate residue is found at position 189 (S1 pocket) in most

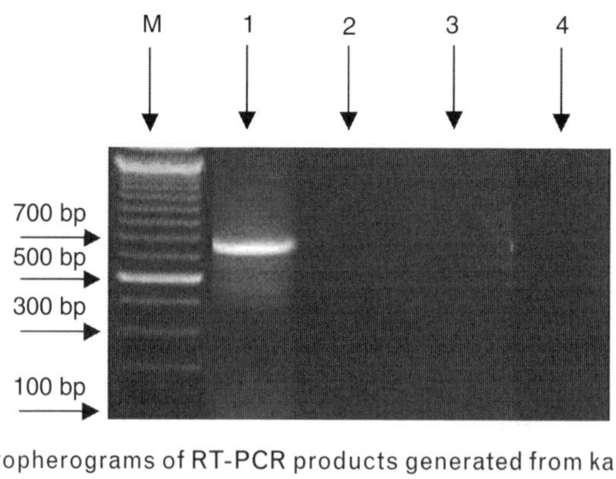

FIGURE 14.2: Gel electropherograms of RT-PCR products generated from kallikrein gene-specific primers. Lane M, standard DNA ladder, each band representing 100-base pair increments; lane 1, RT-PCR product from the venom library of *Heloderma horridum* with primers designed from *Heloderma suspectum* kallikrein sequence EU790962; lane 2, nontemplate control; lane 3, RT-PCR from venom libraries of *H. horridum* with primers designed from the *H. horridum* kallikrein sequence P43685 reported previously; lane 4, nontemplate control (Fry et al. 2010b).

venom kallikreins (see figure 14.1). This may confer affinity for basic amino acids such as arginine and lysine in substrates, a feature present in trypsin and thrombin (Vaiyapuri et al. 2012). Some venom enzymes have nonpolar amino acids such as glycine or serine at position 189, and this may confer affinity for large hydrophobic amino acids such as tyrosine or phenylalanine, similar to that of chymotrypsin (Vaiyapuri et al. 2012). Indeed, some of the TV-kallikreins, such as those from *Bothrops jararaca* (Kuniyoshi et al. 2012) and *Macrovipera lebetina* (Siigur et al. 2011), have specificity for tyrosine residues at the P1 position. The residues present at the sides of the substrate-binding pocket at positions 216 and 226 have been reported to influence the selection of substrates, and various combinations of these residues are present in different venom kallikrein enzymes. With the exception of a few TV-kallikreins, all these enzymes are predicted to have potential N-glycosylation sites, which upon glycosylation may influence the substrate and inhibitor specificities of these enzymes (Vaiyapuri et al. 2012). Glycosylation accounts for as much as 62% of the native MW of *Bothrops* protease A (BPA) from the venom of *B. jararaca* (Paes Leme et al. 2008).

Since very few TV-kallikreins have been recovered from caenophidian snakes other than viperid snakes, it is currently not possible to conclude whether the C-terminal extension is a common feature in snake-venom-specific kallikreins. However, it is consistently absent from all known toxicoferan lizard-venom TV-kallikreins, which instead contain a unique internal extension of 6 to 9 amino acids between positions 93 and 95 of chymotrypsinogen, with a conserved cysteine residue at position 93G (see figure 14.1). The length of the signal peptide and the activation peptide is greater (56 amino acids together) in two of the TV-kallikreins identified in the venoms of the elapid snakes *Bungarus multicinctus* and *Naja atra*, in comparison with other elapid-snake-venom sequences. Although for the majority of the TV-kallikreins the predicted isoelectric point is between 5 and 7 or 8 and 9, certain forms (such as one from the venom of the viperid snake *Gloydius ussuriensis*) have isoelectric points as low as 4.6 (Vaiyapuri et al. 2012).

14.3.2 PLESIOTYPIC FUNCTIONAL FORMS

14.3.2.1 Substrate Specificity

TV-kallikreins are synthesized and stored as inactive zymogens in the lumen of the venom gland. They are stored at low pH to prevent premature activation (Mackessy and Baxter 2006). During venom delivery, snake-venom kallikreins and other toxins flow through the rostral portion of the accessory gland, which is rich in serous secretions containing proteolytic enzymes that result in the removal of the propeptide and the activation of enzymes (Mackessy and Baxter 2006). Currently, the activation processes of toxicoferan lizard-venom-specific TV-kallikreins remain uncharacterized.

Many of the functionally characterized viperid-snake-venom TV-kallikreins have been shown to cleave more than one substrate, potentially indicating multiple targets and mechanisms of action in human bite victims or prey (Vaiyapuri et al. 2012). Some TV-kallikreins retain kallikrein-like activities by cleaving kininogen to release bradykinin. Plasma kallikrein is capable of converting high-MW kininogen into bradykinin, while plesiotypic TV-kallikreins can convert low-MW kininogen into kallidin (Bhoola, Figueroa, and Worthy 1992). In addition to this kininogenase activity, endophysiological tissue kallikrein enzymes have fibrinogenolytic and angiotensin-converting activities. Thus, these three activities are considered plesiotypic functional activities of TV-kallikrein. Although some tissue kallikrein enzymes cleave plasminogen to produce angiostatin (Heidtmann et al. 1999), this activity is not discussed in this chapter, as TV-kallikreins do not produce angiostatin and instead only liberate plasmins. Functional characterization of TV-kallikreins from toxicoferan reptiles other than viperid snakes has been limited to three *Heloderma* lizard-venom kallikreins to date (Hendon and Tu 1981; Utaisincharoen et al. 1993).

14.3.2.2 Kininogenase Activity

Enzymes that convert kininogen into biologically active peptides called kinins are known as kininogenases (Bhoola, Figueroa, and Worthy 1992). These include not only endophysiological kininogenases such as plasma and tissue kallikreins but also some TV-kallikreins. As a result of the plesiotypic nature of this activity, TV-kallikreins with kininogenase activities are present in the venoms of highly divergent toxicoferan lineages, viperid snakes and *Heloderma* lizards (Hendon and Tu 1981; Vaiyapuri et al. 2012). Numerous isoforms with this activity have been isolated and characterized from a wide array of viperid snake venoms spanning the full taxonomical range (see Hendon and Tu 1981; Markland et al. 1982; Bjarnason et al. 1983; Ohtani et al. 1988; Nikai et al. 1993; Matsui et al. 1998; Serrano et al. 1998; Oyama and Takahashi 2003; Oyama, Fukuda, and Takahashi 2008; Vaiyapuri et al. 2010). Documented activities include the cleavage of high- and low-MW kininogen to release kinins such as bradykinin and kallidin (Markland et al. 1982; Ohtani et al. 1988; Serrano et al. 1998; Oyama and Takahashi 2003; Oyama, Fukuda, and Takahashi 2008). Secondary to the liberation of kinins, these enzymes reduced blood pressure (Zaganelli et al. 1996). The limited testing of *Heloderma* lizard TV-kallikrein shows identical activities with those of viperid snake homologues, including the cleavage of kininogen to release bradykinin resulting in the lowering of blood pressure (Hendon and Tu 1981; Alagón et al. 1986; Utaisincharoen et al. 1993; Datta and Tu 1997), thus reinforcing this as a plesiotypic activity.

14.3.2.3 Angiotensin-Converting Activity

Angiotensinogen is a globular protein synthesized by the liver and released into circulation constitutively. It is a substrate for renin, which converts this protein into angiotensin I, a precursor for the active peptide angiotensin II. The hydrolysis of angiotensin I by angiotensin-converting enzymes liberates angiotensin II (hypertensive peptide), resulting in an increase of blood pressure (Peach 1977; Ondetti and Cushman 1982). Tissue kallikrein enzymes such as hK3 cleave angiotensin I and II to produce inactive peptides. Consistent with this, TV-kallikreins have been shown to destructively cleave angiotensin I or II (or both) and thus contribute to the induction of hypotension in envenomed victims or prey (Hung and Chiou 2001; Siigur et al. 2011; Kuniyoshi et al. 2012). Similarly, gilatoxin, a TV-kallikrein from *H. horridum* venom, degrades angiotensin I to II by removing dipeptides (D-R) and thus contributes to the prolonged hypotensive effects exhibited by envenomed victims (Hendon and Tu 1981).

14.3.2.4 Fibrinogenolytic Activities

This activity contributes to hemorrhage (see color plates 11E, 11I, 11J, 12C). Fibrinogen is a soluble protein produced in the liver and released into the bloodstream. This protein is made up of three chains: Aα, Bβ, and γ. During coagulation, thrombin, a kallikrein enzyme, digests fibrinogen to liberate fibrinopeptides and activate clotting factor XIII, which in turn forms an insoluble fibrin clot. Human kallikrein (hK6) digests the Aα and Bβ chains of fibrinogen (Magklara et al. 2003). Similarly, several TV-kallikreins cleave either or both Aα and Bβ fibrinogen chains (Vaiyapuri et al. 2012). The cleavage of fibrinogen by venom kallikrein results in the formation of unstable clots in circulation and in other cases reduces the level of functional fibrinogen in circulation, thus contributing to clotting inhibition (see color plates 11I–11K). As fibrinogenolysis is the major function of thrombin, structurally distinct TV-kallikreins that exhibit similar fibrinogenolytic activities are commonly known as thrombin-like enzymes. However, this nomenclature has led to some confusion, as they are thrombin-like through functional convergence but are not structurally similar to thrombin.

The nomenclature of TV-kallikreins is based on the fibrinogenolytic activity. TV-kallikreins with specific α-fibrinogenolytic activity have been classified as Venombin As. Such isoforms have been isolated and characterized widely from Crotalinae venoms but not from Viperinae venoms to date (Nolan, Hall, and Barlow 1976; Stocker and Barlow 1976; Markland and Pirkle 1977; Ascenzi et al. 1986; Kisiel et al. 1987; Orthner, Bhattacharya, and Strickland 1988; da Silva et al. 1989; Huang and Chiang 1994; Nishida et al. 1994; Serrano et al. 1998; Sanchez et al. 2000; Muanpasitporn and Rojnuckarin 2007; Costa et al. 2010). Venombin Bs, which are specifically β-fibrinogenolytic, are currently known exclusively from Crotalinae venoms (Nolan, Hall, and Barlow 1976; Stocker and Barlow 1976; Markland and Pirkle 1977; Ascenzi et al. 1986; Kisiel et al. 1987; Orthner, Bhattacharya, and Strickland 1988; da Silva et al. 1989; Huang and Chiang 1994; Nishida et al. 1994; Serrano et al. 1998; Sanchez et al. 2000; Muanpasitporn and Rojnuckarin 2007; Costa et al. 2010). The apparently unique occurrence of such enzymes in Crotalinae venoms may, however, be the result of a bias in toxinological research toward such species. In contrast, enzymes possessing both α- and β-fibrinogenolytic activities have been isolated from both Crotalinae and Viperinae (Komori et al. 1993; Gao et al. 1998; Lee et al. 1999; Zhang et al. 2001; Jiao et al. 2005; Vaiyapuri et al. 2010). Consistent with the fact that αβ-fibrinogenolytic activity is widespread in viperid snake venoms and α- or β-fibrinogenolytic specialization being apotypic, the TV-kallikrein isolated and characterized from the elapid snake *Ophiophagus hannah* exhibits αβ-fibrinogenolytic activity (Zhang et al. 1994b). While γ-specific activity has not been documented, generalized α-, β-, and γ-fibrinogenolytic activity are likely to be the plesiotypic state, since these activities have been documented for both caenophidian-snake- and anguimorph-lizard-venom TV-kallikreins (Hendon and Tu 1981; Marrakchi et al. 1995).

14.4 APOTYPIC STRUCTURAL AND FUNCTIONAL TOXIN FORMS

14.4.1 APOTYPOSIS

TV-kallikreins exhibit a wide range of apotypic functions. While most TV-kallikreins retain the aforementioned plesiotypic activities of endophysiological kallikreins, several perform additional apotypic coagulopathic functions, including clotting-factor activation, plasminogen activation, platelet aggregation, and protein C activation.

Several TV-kallikreins with substitutions in one or two of the catalytic triads have been recovered from viperid snake venoms (Vaiyapuri et al. 2011; Vaiyapuri, 2012b). These substitutions are mostly at position 57, and arginine replaces histidine in several cases. Some viperid-snake-venom TV-kallikreins have substitutions for S195, and one sequence has threonine at position 102 instead of the typical aspartate residue. Most TV-kallikreins have been identified only at transcript level, and only a couple have been functionally characterized; therefore, the functional significance of these substitutions is not always clear. Functional analysis of a TV-kallikrein from *Bothrops alternatus* with substitutions at position N57 and T102 has confirmed that this enzyme retains functional stability (Costa et al. 2010). Whereas a TV-kallikrein from the venom of *Protobothrops jerdonii* in which an arginine residue is substituted for H57 is apparently nonfunctional (Wu et al. 2008). The presence of kallikrein homologues has been widely observed in vertebrates and invertebrates. Some of these kallikreins no longer hydrolyze peptide bonds, although in most cases, they have acquired alternative functions (Pils and Schultz 2004). Thus, TV-kallikreins that have lost their plesiotypic functionality likely participate in envenoming through other means, such as by binding to substrates irreversibly and reducing their normal concentration.

14.4.2 ACTIVATION OF CLOTTING FACTORS

Blood coagulation is an essential process that prevents excessive loss of blood following injury. TV-kallikreins activate various blood-clotting factors (Kini 2005a). Factor V is a high-MW single-chain protein that circulates in plasma as a zymogen. Following activation by thrombin, factor Va forms a prothrombinase complex with activated factor X (Xa) to convert prothrombin into thrombin (see chapters 12 and 13). Viperid-snake-venom TV-kallikreins such as RVV-Vα and β from the venom of *Daboia siamensis* (Schiffman, Theodor, and Rapaport 1969; Tokunaga et al. 1988), factor V activator from the venom of *Macrovipera lebetina* (Siigur et al. 1998), thrombocytin from the venom of *Bothrops atrox* (Niewiarowski et al. 1979), and contortrixobin from the venom of *Agkistrodon contortix contortrix* have been reported to activate factor V (Amiconi et al. 2000). The Viperinae TV-kallikreins with these activities are phylogenetically distinct from the Crotalinae toxins possessing the same activity; this activity is either convergently apotypic in the two lineages or is present but undocumented in intervening lineages. TV-kallikreins from the venoms of *O. hannah, Bungarus fasciatus*, and *Cerastes vipera* possess diversity of other activities including factor X activation (reviewed in Kini 2005a), suggesting that this activity also either evolved convergently or was of plesiotypic origin.

TV-kallikreins with atypical coagulopathic activities include activators of factor X and prothrombin from *C. cerastes* (Marrakchi et al. 1995); protein C activators that also cleave human prothrombin, factor X, factor IX, and factor VII from *Agkistrodon contortix contortrix* (Kisiel et al. 1987), *Agkistrodon bilineatus* (Nakagaki, Kazim, and Kisiel 1990), *Gloydius halys* (Bakker et al. 1993), and *G. ussuriensis* (Kogan et al. 1993); activators of factor VIII from *B. jararaca* (Nishida et al. 1994); activators of factor XIII from *Agkistrodon contortix contortrix* (Amiconi et al. 2000); activators of factor XIII and prothrombin from *B. atrox* (Niewiarowski et al. 1979); plasminogen activators from *Viridovipera stejnegeri* (Zhang et al.

1995), *Lachesis muta muta* (Sanchez et al. 2000), *Gloydius blomhoffi brevicaudus* (Park et al. 1998), and *Cryptelytrops albolabris* (Muanpasitporn and Rojnuckarin 2007). Additional activities include lysis of fibrin clots directly in the absence of plasmin (Vaiyapuri et al. 2010), inactivation of plasminogen activator inhibitor-1 and α2-antiplasmin (Urano et al. 2000), and platelet aggregation (Niewiarowski et al. 1979; Nishida et al. 1994; Marrakchi et al. 1995; Serrano et al. 1995; Braud, Bon, and Wisner 2000). Although the mechanism of action by which most TV-kallikreins activate platelets is unknown, those from *B. jararaca* (Serrano et al. 1995) and *B. atrox* (Serrano et al. 1995) do so by cleaving protease-activated receptors 1 and 4 in a similar manner to thrombin. The phylogenetic positioning of these apotyposis events within the evolution of viperid snake venom is unclear, and further work will be necessary to resolve these timing issues.

As these TV-kallikreins also cleave fibrinogen, it is apparent that the activities of TV-kallikreins are much more complex than those of the endophysiological tissue kallikreins from which they were recruited. These additional apotypic activities may act synergistically with the activation of fibrinogen and clotting factors in victims or prey animals. Inappropriate activation of plasminogen leads to the cleavage of immature blood clots, resulting in increased bleeding. Protein C is a kallikrein enzyme that circulates in blood in an inactive form. This protein plays major role as an anticoagulant in the negative regulation of blood clotting. Upon activation by thrombin, the activated protein C proteolytically inactivates factors Va and VIIIa to control blood clotting. In contrast to normal activation, the inappropriate and excessive activation of protein C affected by TV-kallikreins leads to impaired blood clotting which results in a bleeding diathesis. Plasminogen is the inactive precursor to plasmin that circulates in the blood plasma. Upon activation by tissue or urokinase plasminogen activators, the fibrinolytic plasmin dissolves fibrin clots during the wound-healing process. Additional plasminogen activation by TV-kallikreins will therefore result in the rapid breakdown of any clots that form. Platelets are small circulating blood cells that play critical roles in hemostasis. Following blood-vessel wall damage, platelets are activated by exposed collagen, and thrombin is generated at the site of injury to form a platelet plug (or aggregate) to prevent loss of blood. Inappropriate activation of platelets under pathological conditions such as rupture or atherosclerosis leads to the formation of platelet aggregates within circulation which results in a life-threatening blockade of blood flow to organs such as the heart and the brain (Gibbins 2004). Activation of platelet aggregation by TV-kallikreins may therefore result in the rapid death of prey animals.

14.4.3 OTHER APOTYPIC FUNCTIONS OF TV-KALLIKREIN

Other potentially pathological noncoagulopathic activities by TV-kallikreins include cleavage of type I collagen and gelatin (Reichl et al. 1993); increase of blood-brain barrier permeability, causing barrel rotation syndrome (known as gyroxin syndrome when caused by TV-kallikrein) in mice (Alves et al. 2011); cleavage of C3 to activate the complement cascade (Yamamoto et al. 2002); increase of capillary permeability cause by TV-kallikreins from the venoms of *G. ussuriensis*, *Crotalus ruber*, and *Crotalus viridis viridis* (Shimokawa and Takahashi 1993a; Shimokawa and Takahashi 1993b); and albuminolytic, myonecrotic, and hemolytic activities caused by *Bothrops alternatus* TV-kallikrein. In addition, this enzyme can induce hyperemia, lymphocytic interstitial pneumonitis in the lung, necrosis, inflammatory infiltration in the liver, and glomerular congestion in the kidneys (Costa et al. 2010); and release of endothelium-dependent relaxing factor from the vascular endothelium leading to the relaxation of coronary arteries is caused by TV-kallikrein from *B. atrox* (Kirby et al. 1979; Niewiarowski et al. 1979).

14.5 THERAPEUTIC POTENTIAL OF THE TOXIN TYPE

Snake-venom components that affect hemostasis have received broad attention for their potential in the development of diagnostics and therapeutics for hemostatic disorders.

14.5.1 DIAGNOSTIC USES

RVV-V, a TV-kallikrein from the venom of *D. siamensis*, cleaves coagulation factor V, converting it into active factor Va (Schiffman, Theodor, and Rapaport 1969). Because of its selectivity for factor V, RVV-V has been prepared as a reagent (Pentapharm, Switzerland) for the routine assay of factor V in plasma (Marsh 2001; Marsh and Williams 2005). A fast-acting protein C activating TV-kallikrein (Protac, Pentapharm, Switzerland) from the venom of *Agkistrodon contortrix contortrix* has been used for the functional assay of protein C and S in plasma (Stocker et al. 1987). As this TV-kallikrein does not require thrombomodulin to cleave protein C, it simplifies and reduces the cost of assaying protein C in plasma. Thus, Protac has been used to identify patients with defects in protein C activation pathways (Robert et al. 1996; Kirschbaum et al. 1999). It has also helped in identifying dysfunctional protein C (Girolami et al. 1993).

TV-kallikreins primarily cleave either α or β (sometimes both) chains of fibrinogen and generally do not activate (with one known exception) factor XIII, which is important for stabilizing fibrin clots (Vaiyapuri et al. 2012). Thus, TV-kallikreins either form easily dispersible clots or do not form clots but just degrade fibrinogen. These fibrinogen-degradation products are eliminated by fibrinolysis. This reduction of functional fibrinogen levels makes TV-kallikreins useful in assaying fibrinogen and its breakdown products in plasma (Marsh 2001; Marsh and Williams 2005). TV-kallikreins are not inhibited by heparin and thus can be used for assaying antithrombin III and other hemostatic variables in heparin-containing samples. The most widely used TV-kallikreins in fibrinogen functional assays are reptilase (commercial name Bothrombin from Pentapharm, Switzerland) from the venom of *B. atrox* and Ancrod (Knoll, Germany) from the venom of *Calloselasma rhodostoma* (Marsh and Williams 2005). Since conventional thrombin time is used to measure the speed of coagulation in patients, the use of this assay in plasma is limited by the presence of thrombin inhibitors, such as antithrombin and heparin. Thus, "reptilase time" is a convenient functional clotting assay in heparin-containing samples (Funk et al. 1971). Reptilase is also used in assaying antithrombin III in fibrinogen-free plasma samples (Howie, Prentice, and

McNicol 1973). Reptilase is a very useful tool for the identification of hypofibrinogenemia, fibrin-degradation products in plasma, and defects in fibrin polymerization (dysfibrinogenemia) (Latallo and Teisseyre 1971). TV-kallikreins also have potential for use in studies of fibrin assembly (Torbet 1987), thrombin-dependent feedback activation of the clotting system (Kumar, Beguin, and Hemker 1994), and platelet-force development (Carr, Carr, and Greilich 1996). In addition, some TV-kallikreins activate prothrombin and thus could be used in measuring prothrombin time (Marsh and Williams 2005). These plasminogen activators may also be useful in assaying functional plasminogen and plasmin in place of tissue plasminogen activator.

14.5.2 THERAPEUTIC USES

Because of its fibrinogen-degrading activity, the TV-kallikrein Ancrod (from *C. rhodostoma*) has been introduced as an anticoagulant under the commercial name of Arvin for the treatment of stroke, heart attack, and deep-vein thrombosis (Marsh and Williams 2005). Use of Ancrod/Arvin in ischemic-stroke patients may be better and more cost-effective than use of other existing drugs (Samsa et al. 2002). Ancrod/Arvin may also be of therapeutic benefit to patients with heparin-associated thrombocytopenia and thrombosis syndrome (Illig and Ouriel 1996). Ancrod/Arvin has been used as an alternative anticoagulant to heparin in cardiopulmonary bypass surgeries (von Segesser et al. 2001). Another TV-kallikrein, defibrase (commercial name of Batroxobin, Pentapharm) from the venom of *B. atrox*, is used to reduce the fibrinogen content in plasma in order to prevent inappropriate blood clotting (Marsh and Williams 2005). Both of these TV-kallikreins have been used to treat deep-vein thrombosis and peripheral occlusive diseases. They also reduce the incidence of postoperative deep-vein thrombosis and the recurrence of thrombosis after vascular surgeries (Marsh and Williams 2005).

14.6 CONVERGENCE WITH OTHER VENOMS

The kallikrein toxins found in *Blarina* species shrews liberate kinins (Kita et al. 2004; Kita et al. 2005). The errors in the sequence of a published *Heloderma* TV-kallikrein (Utaisincharoen et al. 1993) addressed earlier in this chapter led astray a study that examined convergent evolution between lizard and *Blarina* kallikrein toxins. The demonstration that this *Heloderma* TV-kallikrein sequence was not authentic (Fry et al. 2010b) unfortunately invalidated the central conclusion of convergence that relied on the questionable insert at 126–132 as a key element (Aminetzach et al. 2009). This inauthentic relationship has caused considerable confusion, spreading through the literature and being cited on a number of occasions as an example of convergence (Brodie 2010; Janes et al. 2010; Lawrence, Lai, and Clements 2010; Pavlopoulou et al. 2010; Zhu et al. 2010; Boutemy et al. 2011; Losos 2011; Simmer et al. 2011; Yennamalli et al. 2011; Davies et al. 2012; Green and Extavour 2012; Ligabue-Braun, Verli, and Carlini 2012; Song et al. 2012; Wong and Belov 2012; Zhu et al. 2012; Garb and Hayashi 2013; Martin and Orgogozo 2013; Meyer et al. 2013).

Kaillikrein toxins are also found in *Lonomia* caterpillar venoms, and these have been demonstrated to possess fibrinolytic activity (Amarant et al. 1991). The venom of the parasitic wasp *Cotesia rubecula* contains a protein derived from the serine protease scaffold (Vn50), which, like the insect homologues from which it evolved, lacks a functional catalytic triad (His-Asp-Ser), as the plesiotypic serine residue is replaced by a glycine (Asgari et al. 2003). The insect endophysiological enzymes from which Vn50 is derived are involved in a proteolytic cascade that causes hemolymph melanization, a primary defensive reaction upon infection or invasion by parasites. Vn50 instead inhibits hemolymph melanization and, given its strong structural similarity, is postulated to be a competitive inhibitor of endogenous serine protease homologues in the host.

Serine proteases in the venom of coleoids (cuttlefish, octopuses, and squid) exhibit a molecular diversity comparable to those observed in venomous reptiles (Fry, Roelants, and Norman 2009; Ruder et al. 2013b). Their variation in intraloop (functional) residues appears to constitute an adaptive neofunctionalization pattern that seems analogous to those observed in other multigene toxin families. However, knowledge of the activity, potency, and differentiation of these enzymes awaits detailed functional analyses.

In addition to being widely present in predatory venoms, the venoms of hematophagous organisms are rich in serine proteases (Fry et al. 2009a). For example, serine proteases from the venom of vampire bats are powerful plasminogen activators and display structural variation in relation to the diets of the bats, some of which specialize in feeding on the blood of either birds or mammals (Kratzschmar et al. 1991; Schleuning et al. 1992; Low et al. 2013). Serine proteases have also been independently recruited into the feeding secretions of diverse lineages of hematophagous arthropods, including hard ticks (Xu, Bruno, and Luft 2005), nematoceran flies such as mosquitoes (Ribeiro et al. 2007) and biting midges (Campbell, Wilson, and Manninen 2005), and triatomine bugs (Santos et al. 2007). While the specific function of any of these proteins is not known, the presence of CUB domains suggests substrate-specific activity (Ribeiro et al. 2007). Possibly, these serine proteases target specific host proteins at the bite site or are involved in activating other proteins following secretion into the host.

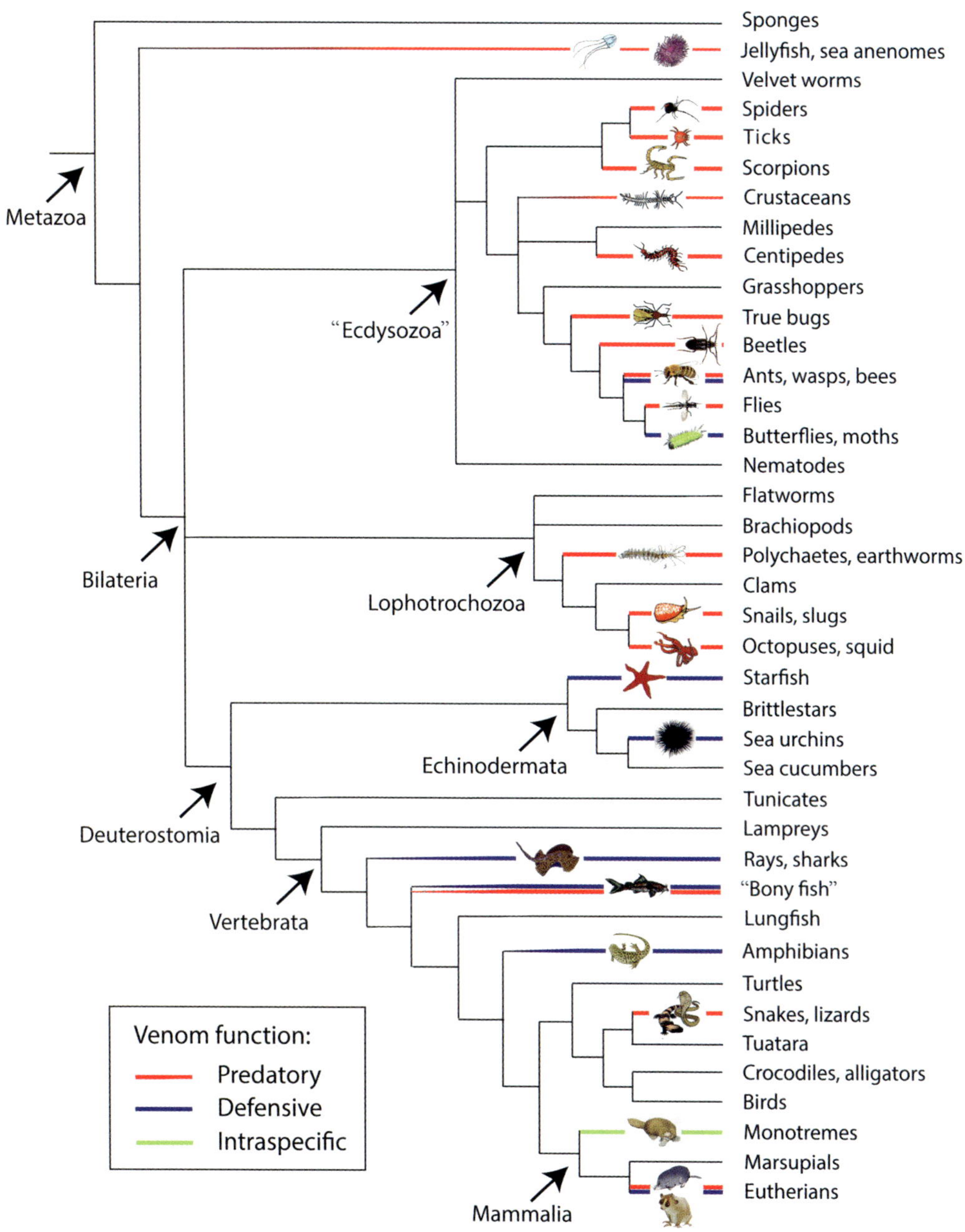

PLATE 1: Schematic tree of venomous life in the animal kingdom (see chapter 1.1) (Casewell et al. 2013). The tree depicts the evolutionary relationships between animal lineages and highlights the frequency with which venom systems are found in the animal kingdom. Colored branches highlight major animal lineages that include members with venom systems. Red branches indicate a predatory role for venom, blue a defensive role, and green a role in intraspecific competition. The phylogeny is based on the tree of life presented by Pennisi (2003). Note that several animal lineages lacking venom have been pruned from the tree to facilitate presentation.

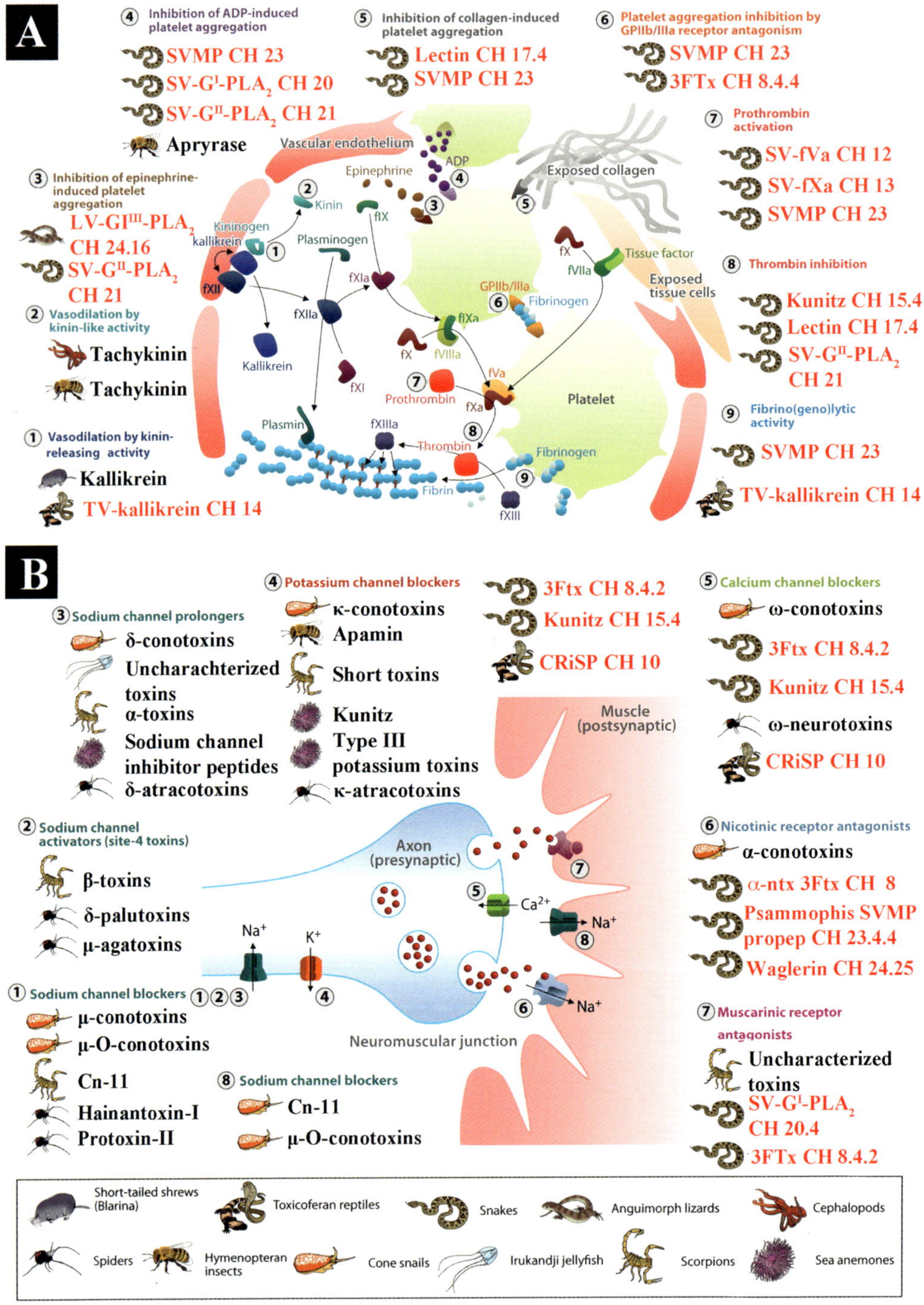

PLATE 2: Convergence of toxin action in the animal kingdom (see chapter 1.1) (Fry et al. 2009a; Casewell et al. 2013). (A) Sites of convergent coagulopathic activity are displayed and are represented by numbers 1-9. (B) Sites of convergent neurotoxic activity are displayed and are represented by numbers 1-8. Each number represents a different physiological target that toxins in various venomous lineages have convergently evolved to attack. Names of toxins and the lineages from which they originate are listed below each numbered legend, with pictures of the venomous lineages relating to the key at the bottom of the figure. For snake toxins, the relevant chapter is cross-referenced.

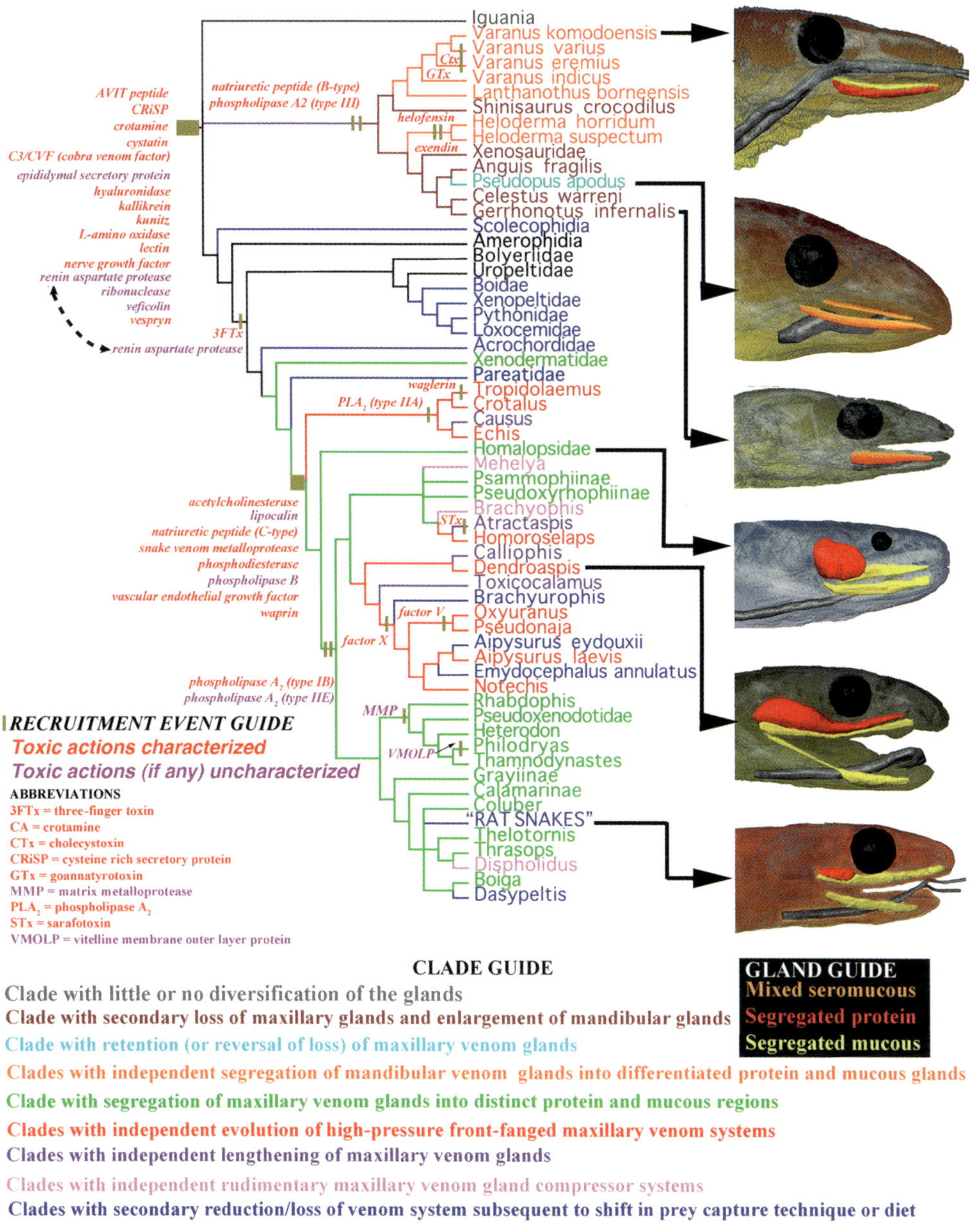

PLATE 3: Toxicofera evolutionary tree showing the timing of toxin-recruitment events and gland derivations (see chapter 1.2.1 and 1.3) (Fry et al. 2006; Fry et al. 2008; Fry et al. 2010a; Fry et al. 2010b; Fry et al. 2012b; Fry et al. 2013). *RAP has been sequenced from the *Boa constrictor* glands, but a homologous sequence has been recovered from the *Anolis carolinensis* genome (Fry et al. 2013). Thus, this toxin type was recruited at least at the level of the henophidian snakes but perhaps all the way to the base of the toxicoferan reptiles. (Phylogeny based on Vidal 2002; Vidal and David 2004; Vidal and Hedges 2005; Fry et al. 2006; Vidal and Hedges 2009; Vidal et al. 2009.)

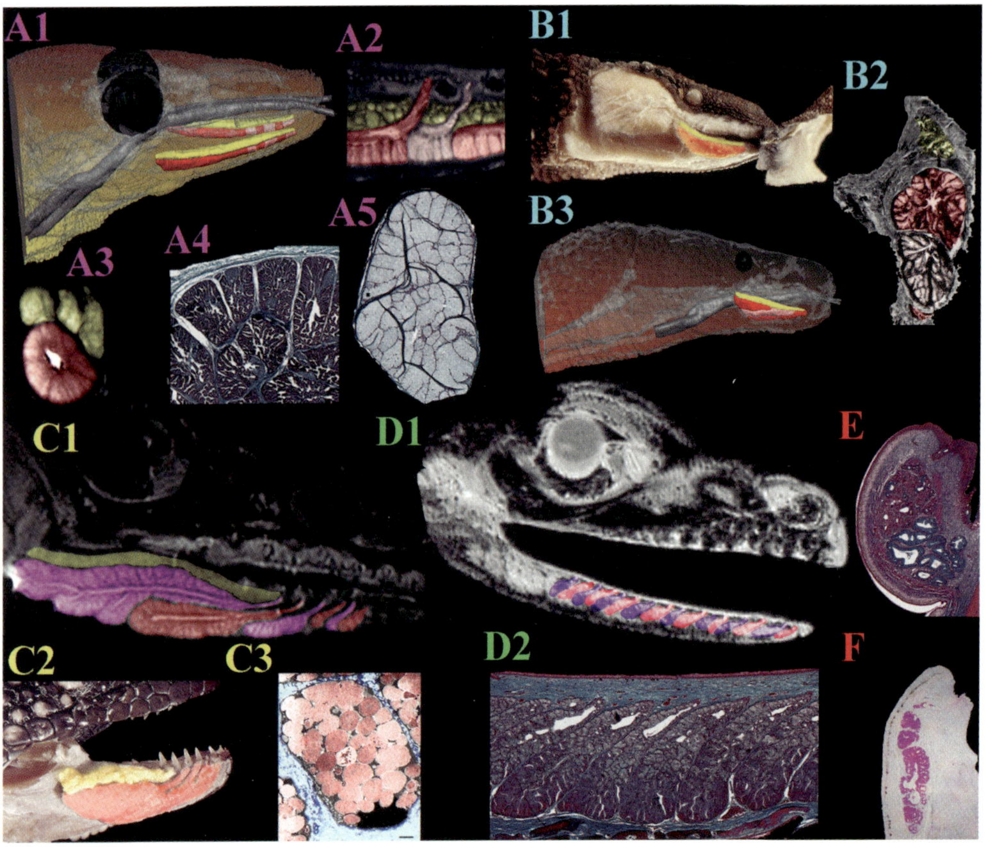

PLATE 4: Anatomical variations in the anguimorph lizard venom systems (see chapter 1.2.2). (A) *Varanus komodoensis*, magnetic resonance imaging: (A1) showing the six compartments (pink/red) of the mandibular venom gland and the infralabial mucus gland (yellow); (A2) longitudinal MRI section showing large ducts emerging separately from each mandibular venom gland compartment; and (A3) transverse MRI section showing the mandibular venom gland large central lumen (red) and labial gland individual lobes (yellow). Transverse histology of Masson's Trichrome stained sections: (A4) the intratubular lumina of the mandibular venom gland that feed into the large central lumen; (A5) a mucus infralabial lobule (note that the six large dark folds are histology artifacts). (B) *Lanthanotus borneensis*: (B1) dissection of the mandibular venom gland (computer highlighted red) and the mucus gland (computer highlighted yellow); (B2) black and white photo of transverse histology section of the mandibular venom gland compartments (computer colored to show protein [red/pink] and mucus [yellow] as partitioned glands) (Kochva 1974); (B3) longitudinal MRI of the mandibular venom gland and the infralabial mucus gland. (C) *Heloderma horridum*: (C1) MRI of the mandibular venom gland; (C2) dissection of the mandibular venom gland (computer highlighted red) and the mucus gland (computer highlighted yellow); and (C3) black and white photo of histology of venom gland intracellular storage granules (computer highlighted red) and the nucleus (computer highlighted black). (D) *Gerrhonotus infernalis*: (D1) longitudinal MRI of the multiple compartments (one per tooth); and (D2) Masson's Trichrome stained longitudinal histology revealing the seromucous mixed gland arrangement with the protein-secreting regions (dark reddish-purple) ventral to the mucus-secreting region (grayish-green). Mandibular gland histology sections: (E) the Iguania species *Anolis equestris* stained with Masson's Trichrome histology showing the mixed seromucous gland; or (F) the Xantusiidae species *Lepidophyma flavimaculatum* stained with Periodic Acid Schiff's showing the purely mucus-secreting gland. (Photos B1, B2, C2, C3: Elazar Kochva.)

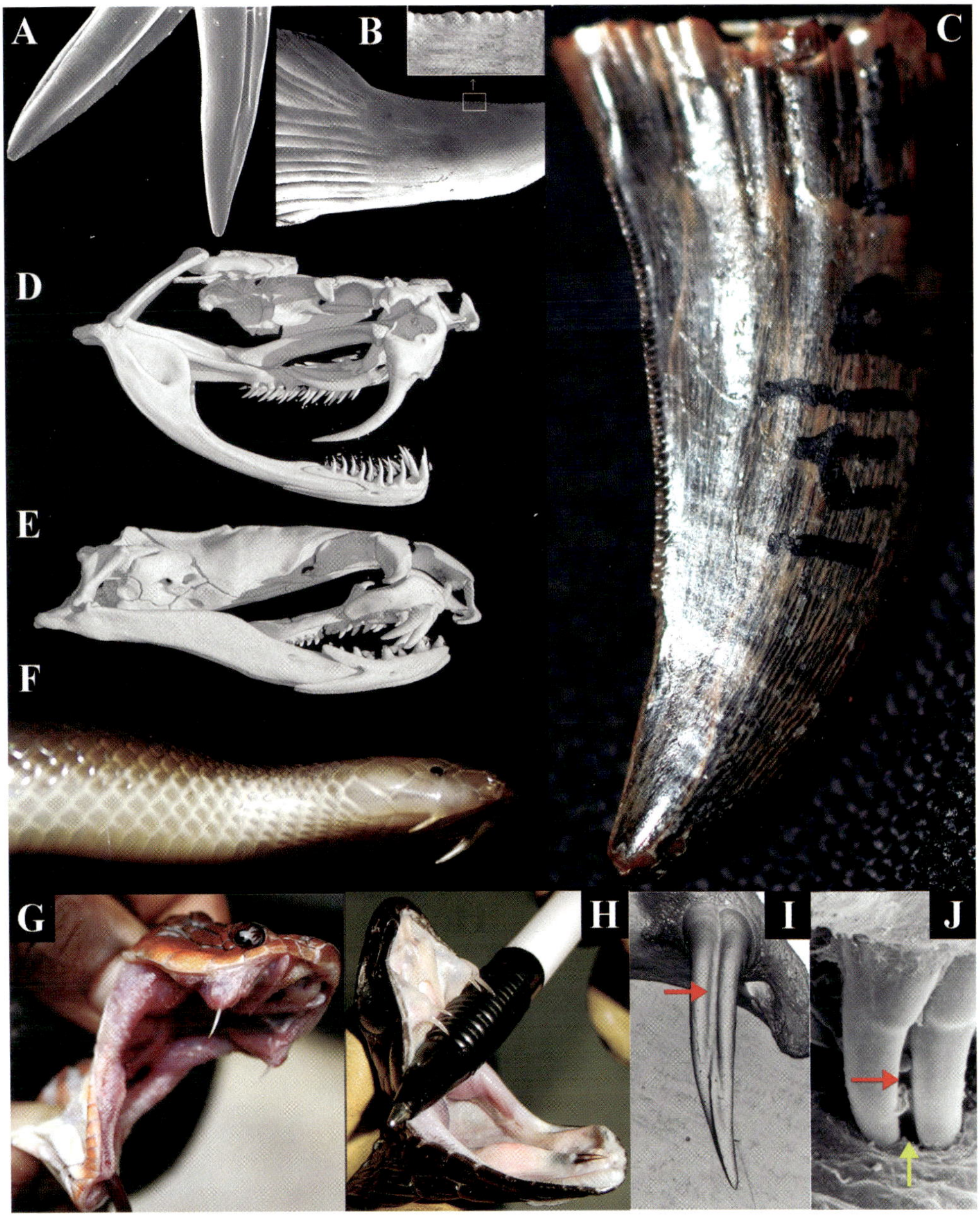

PLATE 5: Representatives of the diversity of specialised venom-delivery teeth (see chapter 1.2.2 and 1.2.3). Scanning electron microscopy of (A) *Heloderma suspectum* showing the deep venom-delivering grooves and (B) *Varanus komodoensis* showing the posterior cutting edge; (C) a fossilized tooth from the extinct seven-meter giant lizard *Varanus prisca* (formerly Megalania priscus) showing the serrated posterior cutting edge; CT scanning of the skulls of (D) *Lachesis muta* and (E) *Micrurus fulvius* showing their independently evolved front fangs; (F) the unique side-stabbing fangs of *Atractaspis bibroni*; the tremendously enlarged rear fangs of (G) *Macropisthodon rudus* and (H) *Dispholidus typus*; scanning electron micrographs showing the prominent grooves (horizontal arrows) on the fangs of (I) *Rhinobothryum lentiginosum* and (J) *Boiga dendrophila* (note: the Boiga specimen was prepared with the fang embedded in prey tissue, so only the base of the fang is visible; the prey tissue has separated slightly from the fang, forming a clear venom tube [vertical arrow]). (Photos A, B, C: Bryan G. Fry; D, E: Digimorph/Michael Kearney/Oliver Rieppel; F: Devon Massyn; G: Duncan McRae; H: Timothy N. W. Jackson; I, J: Bruce Young.)

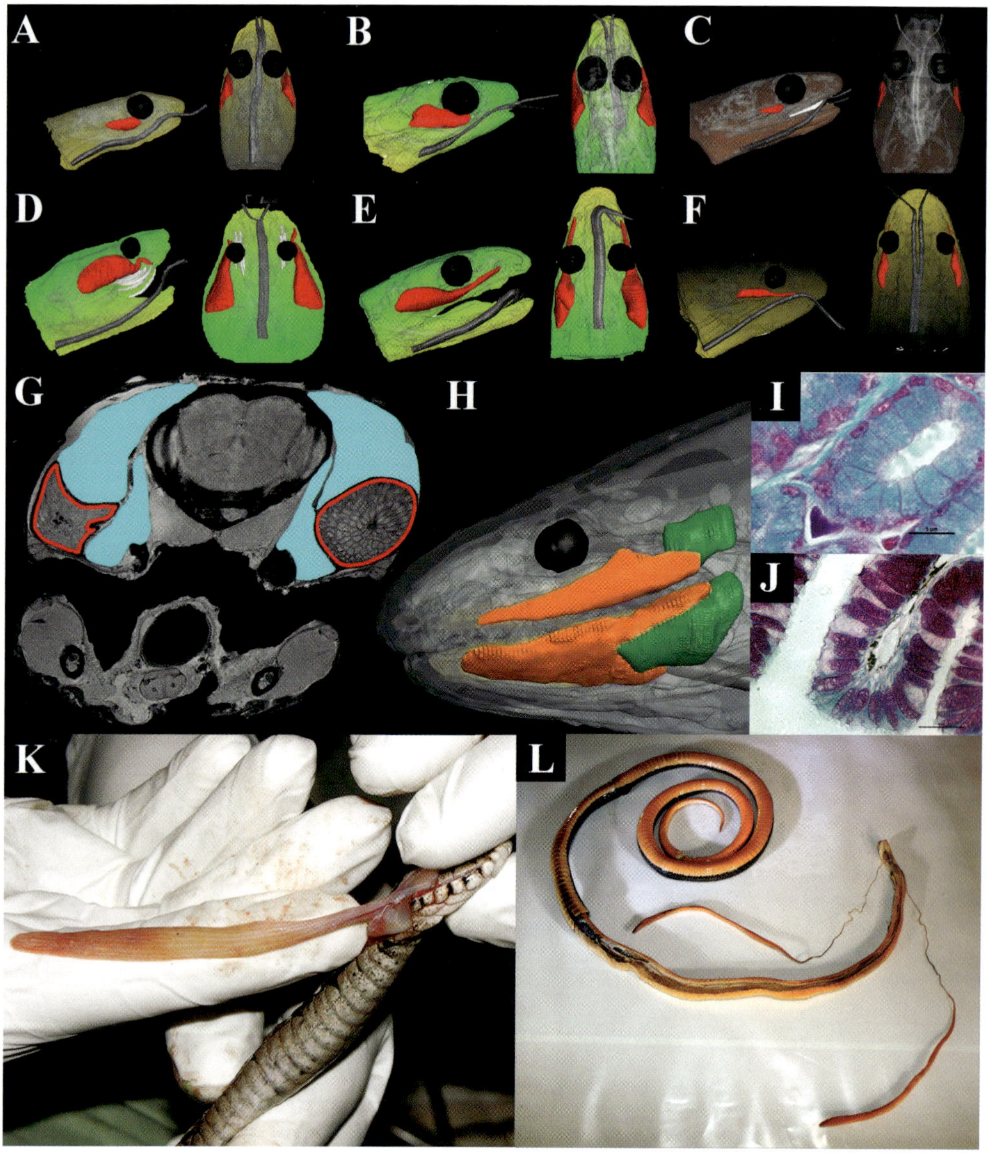

PLATE 6: Diversity of caenophidian snake oral glands (see chapter 1.2.3). MRI showing the size and structure of the venom glands (red) of (A) *Coluber constrictor*, (B) *Dispholidus typus*, (C) *Pantherophis obselata*, (D) *Cryptelytrops albolabris*, (E) *Dendroaspis jamesoni*, (F) *Aipysurus mosaicus*. (G) MRI cross-section of *Dendroaspis jamesoni* showing one gland compressed. (H) *Cylindrophis ruffus* MRI showing the maxillary and mandibular glands (orange) and the rictal glands (green). Histochemically stained glands: (I) *Cylindrophis ruffus* rictal gland stained with Masson's Trichrome and (J) *Aspidites melanocephalus* mandibular gland stained with Periodic Acid Schiff's. (K) and (L) are dissections showing the spectacularly elongated glands of a freshly euthanized *Causus rhombeatus* and a preserved *Calliophis bivirgata* (respectively). Dissected *C. bivirgata* (ZRC 2.6682) deposited at the Lee Kong Chian Natural History Museum (formerly known as the Raffles Museum), National University of Singapore. (Photos: Bryan G. Fry.)

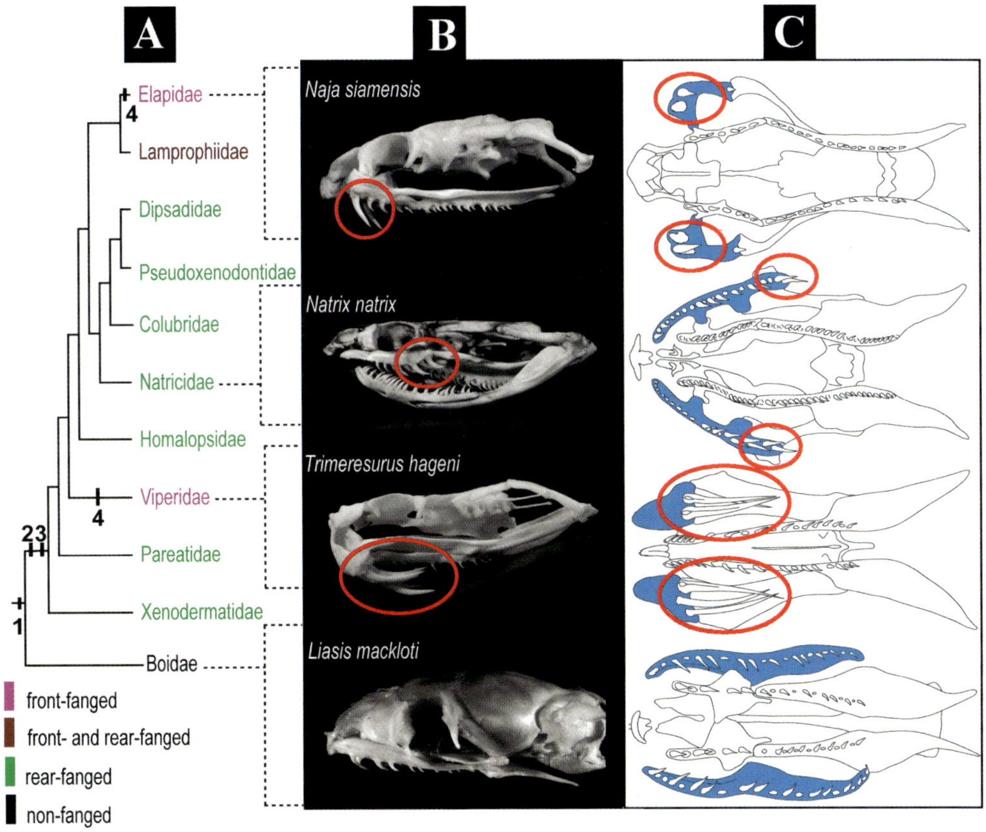

PLATE 7: Fang embryological development (see chapter 1.2.3) (Vonk et al. 2008). Adult maxillary dentition, mapped onto a molecular snake phylogeny to show relative positions of the various fang types. (A) The evolutionary changes leading from an unmodified maxillary dentition to the different fang types in advanced snakes are indicated at the nodes: (1) the plesiomorphic condition of continuous maxillary dental lamina without specialized subregions; (2) evolution of posterior maxillary dental lamina, developmental uncoupling of posterior from anterior teeth, (3) starting differentiation of the posterior teeth with a distinct venom gland; (4) secondary loss of anterior dental lamina and the development of hollow front fangs. (B) Lateral views of dentition and palate of dissected skulls (fangs circled). (C) Ventral view of palate schematic (fangs circled). Phylogeny is based on Vidal 2002; Vidal and David 2004; Vidal and Hedges 2005; Vidal et al. 2009). (Photos: Freek Vonk.)

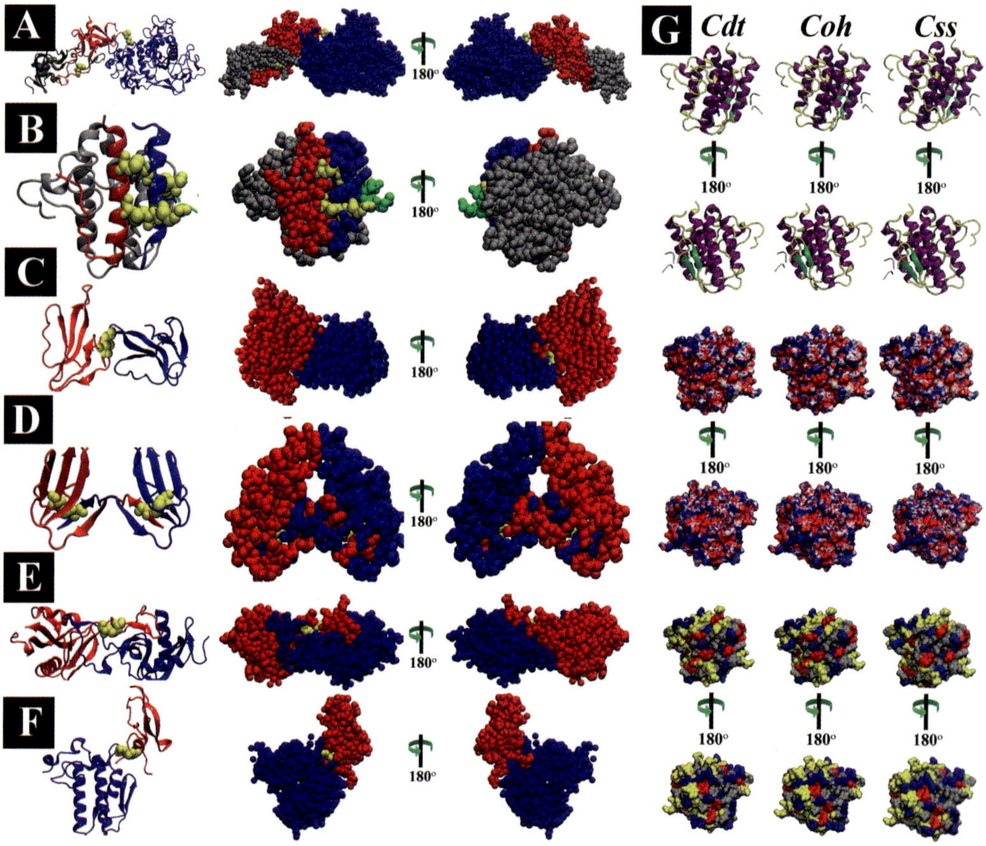

PLATE 8: Disulfide-linked multimeric toxins (Protein Data Bank codes for each shown in brackets; www.rcsb.org) (see chapter 1.3.4). (A) *Daboia russellii siamensis* [2E3X] SVMP with linkages of the cysteine-rich domain (gray), disintegrin (red), and metalloprotease (blue) (see chapter 23); (B) *Crotalus durissus terrificus* [3R0L] presynaptic neurotoxic complex consisting of two non-covalently-interacting SV-G^{II}-PLA$_2$ chains, acidic (red, blue, green) and basic (gray), with the acidic chain cleaved into three subchains held together by internal disulfide bonds (see chapter 21); (C) *Boiga irregularis* [2H7Z] dimer of heterologous plesiotypic 3FTx (see chapter 8); (D) *Naja kaouthia* [4AEA] dimer of homologous Type I 3FTx (see chapter 8); (E) *Protobothrops flavoviridis* [1BJ3] dimer of heterologous lectins (see chapter 17); (F) *Bungarus multicinctus* [1BUN] heterodimeric complex of a kunitz peptide (red) and SV-G^{I}-PLA$_2$ (blue) (see chapters 15 and 20); (G) The extraordinary conservation of the presynaptic neurotoxic SV-G^{II}-PLA$_2$ complex found in *Crotalus* venoms (Cdt = *Crotalus durissus terrificus*, Coh = *Crotalus oreganus helleri*, Css = *Crotalus scutulatus scutulatus*) (see chapter 21). Surface charge potentials were mapped on surfaces allowing for color-scale data range values of −10.00 (red) to +10.00 (blue). Surface residue hydrophobicity mapping of residue-type surfaces depicts acidic residues in red, basic residues in blue, polar residues in yellow, and nonpolar residues in silver.

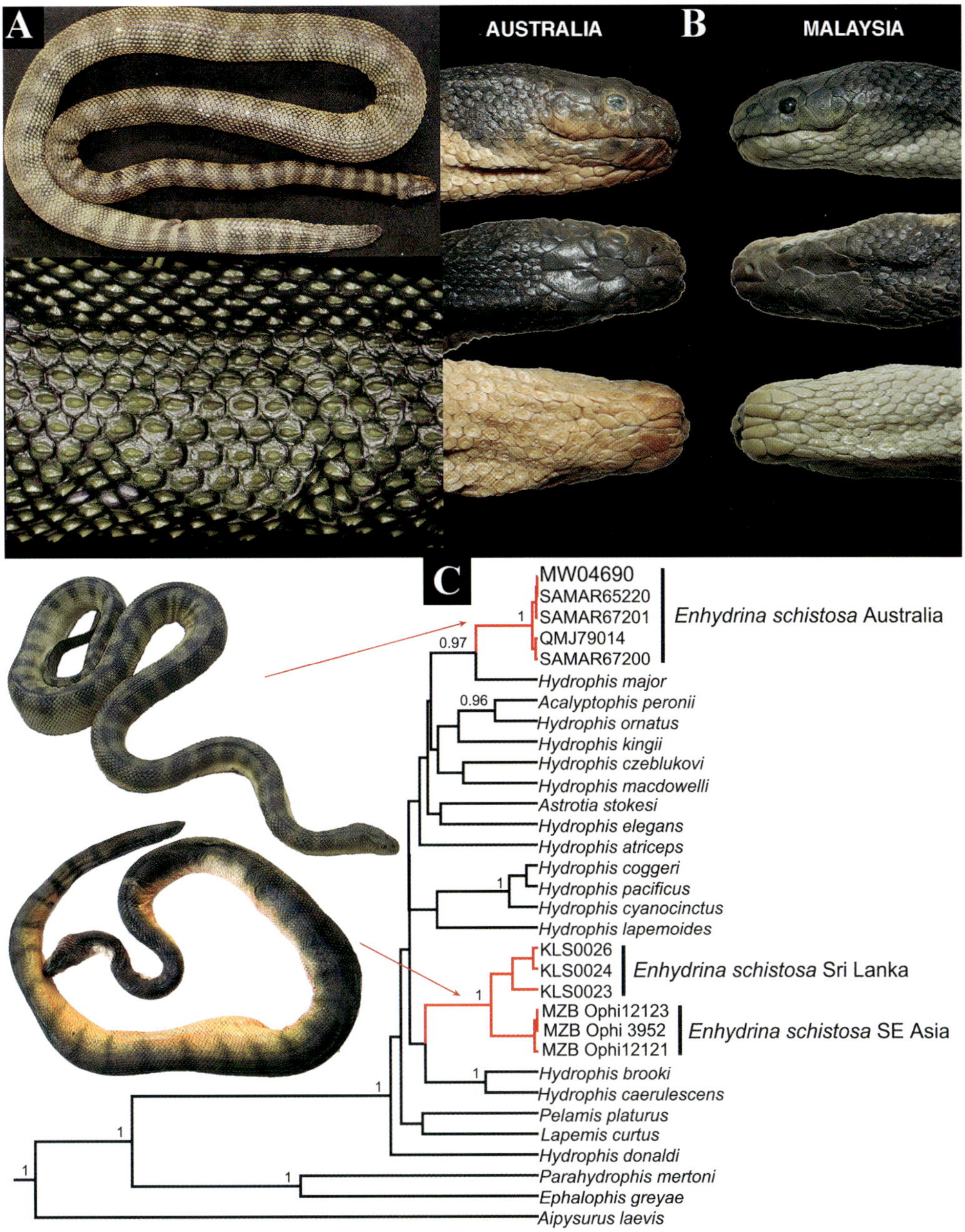

PLATE 9: Notable recently discovered sea snake species (see chapter 1.4. (A) *Hydrophis donaldi* (rough-scaled sea snake), with the uniquely enlarged scales putatively used to protect it from lacerations from the sharp-rock microhabitat it occupies. (B) and (C) The remarkable convergence between two unrelated sea snakes formerly considered to be simply different populations of *Enhydrina shistosa*. The Southeast Asian species retains the species name shistosa but is now in the enlarged genus *Hydrophis*, while the Australo-Papuan species is named *Hydrophis zweifeli*. (Photos A, B, and lower snake image in C: Kanishka Dimithra Bandara Ukuwela; upper snake image in C: Mahree-Dee White.)

PLATE 10: Notable recently discovered terrestrial species (see chapter 1.4). (A) *Oxyuranus temporalis*, a new species of taipan from the Australian western desert. (B) *Crotalus ericsmithii*, a long-tailed rattlesnake from northern Mexico. (C) *Naja ashei*, a giant spitting cobra from Kenya. (D) *Protobothrops mangshanensis*, an extremely large viper from a very restricted montane habitat in China. (E) *Pseudocerastes arachnoides*, a viper from Iran with an exceptional novel caudal lure that resembles a spider. (Photos A: Jordan Voss; B: Eric Smith; C: Wolfgang Wüster; D: Johan Pennanen; E: Behzad Fathinia.)

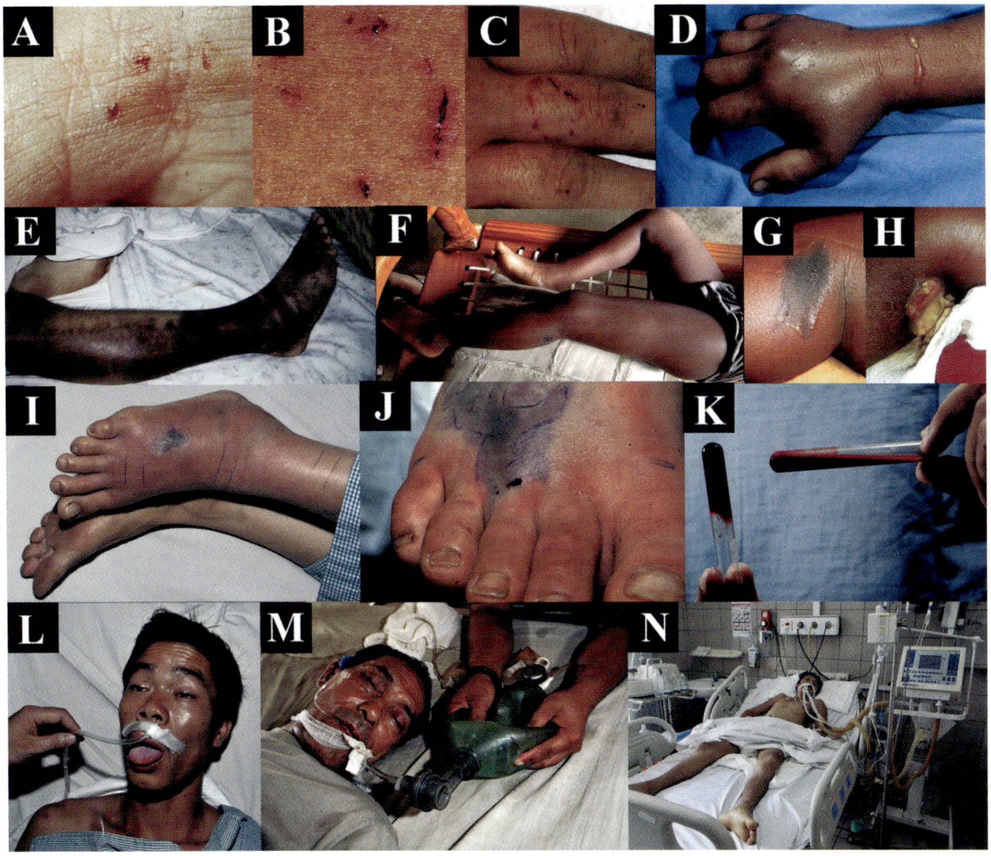

PLATE 11: Clinical signs of snakebite (see chapter 2). (A) The two characteristic puncture wounds, taken three minutes postbite by a *Vipera ursinii* in Hungary. (B) A series of scratches taken 36 hours postbite by an unidentified *Bungarus* species in Nepal. (C) The characteristic crescent-shaped wounds following a *Heloderma suspectum* bite in Arizona. (D) Edema and blistering six hours postbite by an unidentified viper in Nepal. (E) Subcutaneous hemorrhage caused by a *Crotalus atrox* in Arizona that was not treated with antivenom. "Black" discoloration of this sort is generally *not* an indication of tissue necrosis, although it often becomes more pronounced over several days' time owing to reduction in overlying edema and conversion of hemoglobin to methemoglobin. Edema and necrosis taken (F) one day, (G) three days, and (H) seven days postbite by a *Naja mossambica* in Swaziland. (I, J) Edema and hematoma postbite by an unidentified viper species in Vietnam. (K) A 20-minute whole-blood clotting test conducted on an adult patient displaying signs of consumption coagulopathy postbite by an *Ovophis monticola* in Nepal (control sample is on the left). (L) Partial ptosis in a patient instructed to "look up" and "protrude the tongue"; photo taken postbite by an unidentified *Bungarus* species in Vietnam. (M) A patient in Nepal requiring artificial ventilation because of diaphragm paralysis six hours postbite by an unidentified *Naja* species. (N) A patient in Vietnam requiring artificial ventilation because of diaphragm paralysis postbite by an unidentified *Bungarus* species. Effects of D, E, I, J, and K were contributed to by TV-kallikrein (see chapter 14), TV-kunitz (see chapter 15), TV-lectin (see chapter 17), TV-G^{II}-PLA$_2$ (see chapter 21), and SVMP (see chapter 23). Effects of F, G, and H were contributed to by cytotoxic 3FTx (see chapter 8.4.3). Effects of L, M, and N were contributed to by neurotoxic 3FTx (see chapter 8) and TV-GI-PLA2 (see chapter 20). (Photos A, B, D, I, J, K, L, M, N: Zoltan Takacs; C, E: University of Arizona archives; F, G, H: Thea Litschka-Koen.)

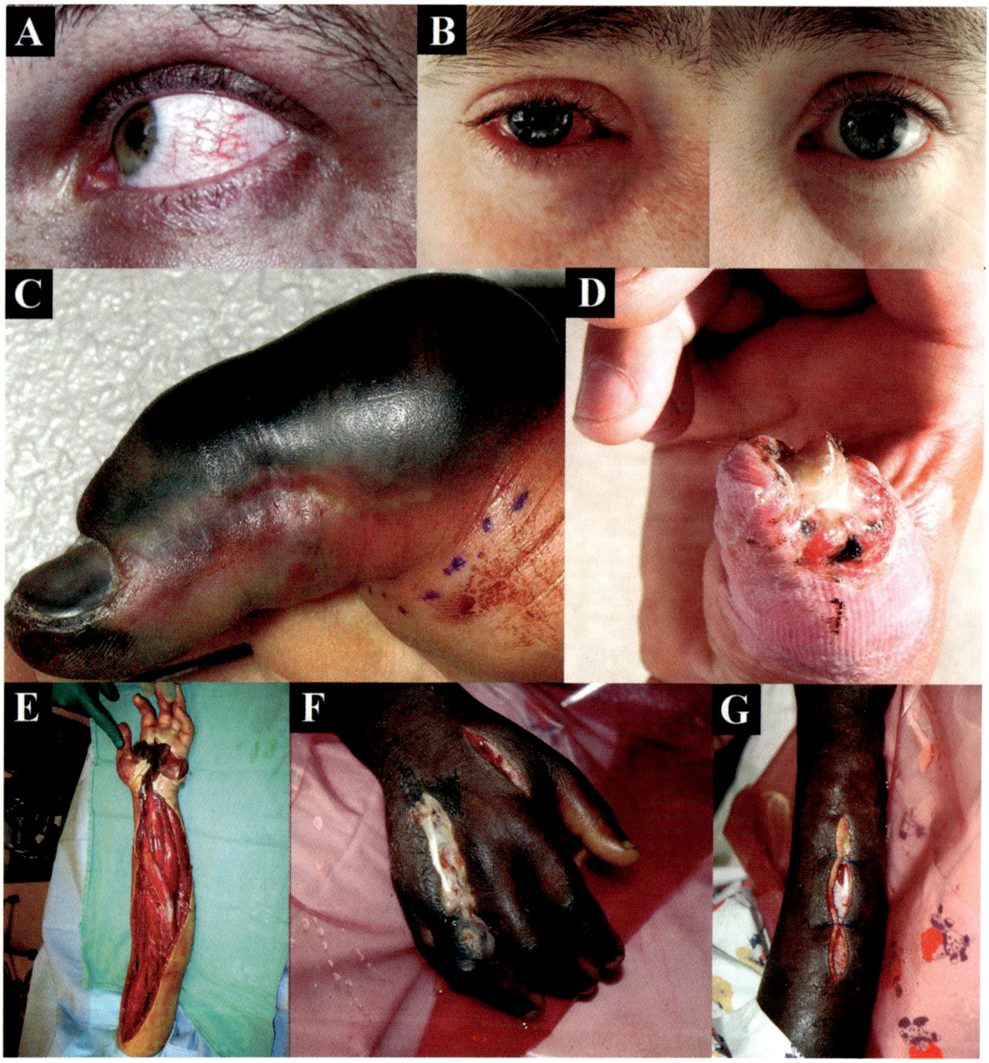

PLATE 12: The local effects of snake envenomation requiring emergency intervention (see chapter 2). (A) Early conjunctival inflammation following spitting in the eye by a *Naja mossambica*. (B) Author C. Cochran with toxic conjunctivitis following accidental inoculation of *Crotalus mitchelli* venom in the right eye. Note the exaggerated curvature of the bulbar conjunctiva on the affected side, which shows a pattern of vascular prominence and edema over the white scleral surface, with sparing of the central cornea. In contrast, the smooth curvature of the normal eye's surface is clear from the ambient light reflex. Note: mono-brow is a photoediting artifact. *Atractaspis bibroni* envenomation to author M. Clarkson (note photos are flipped horizontal for photo composition reasons, bite was actually to the left hand): (C) Ninety-six hours postbite, showing the rapid development of extensive local tissue death. (D) Surgical amputation of the affected thumb 14 days postbite. (D) Intervention required surgical amputation of the affected thumb 14 days postbite. (E) Surgical resection of subcutaneous tissue performed 12 hours postbite from a *Crotalus adamanteus* in an attempt to relieve perceived (but untested) intramuscular pressure. (F, G) Inappropriate deep self-inflicted wounds following an envenomation by a *Naja nigricollis*; intervention required extensive suturing repair of the affected areas. Effects of A were contributed to by cytotoxic 3FTx (see chapter 8.4.3). Effects of B and C were contributed to by SVMP (see chapter 23). Effects of E were contributed to by TV-kallikrein (see chapter 14) and SVMP (see chapter 23). (Photo A: Alejandro Alagón, B: Chip Cochran; C, D: Myke Clarkson; E: Justin Schwartz; F, G: Thea Litschka-Koen.)

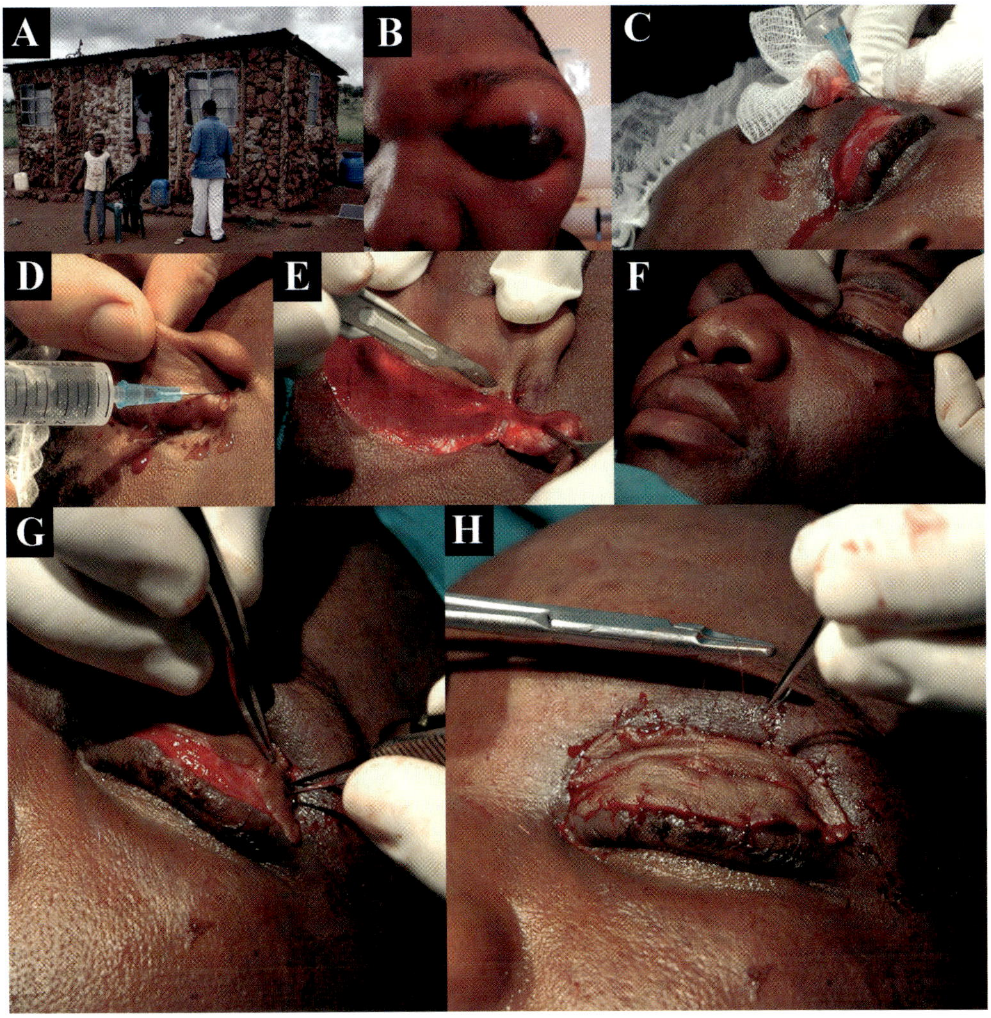

PLATE 13: Skin-graft procedure following necrosis resultant from a bite to the eye by a *Naja nigricollis* (see chapter 2). Effects were contributed to by cytotoxic 3FTx (see chapter 8.4.3). (Photo: Thea Litschka-Koen.)

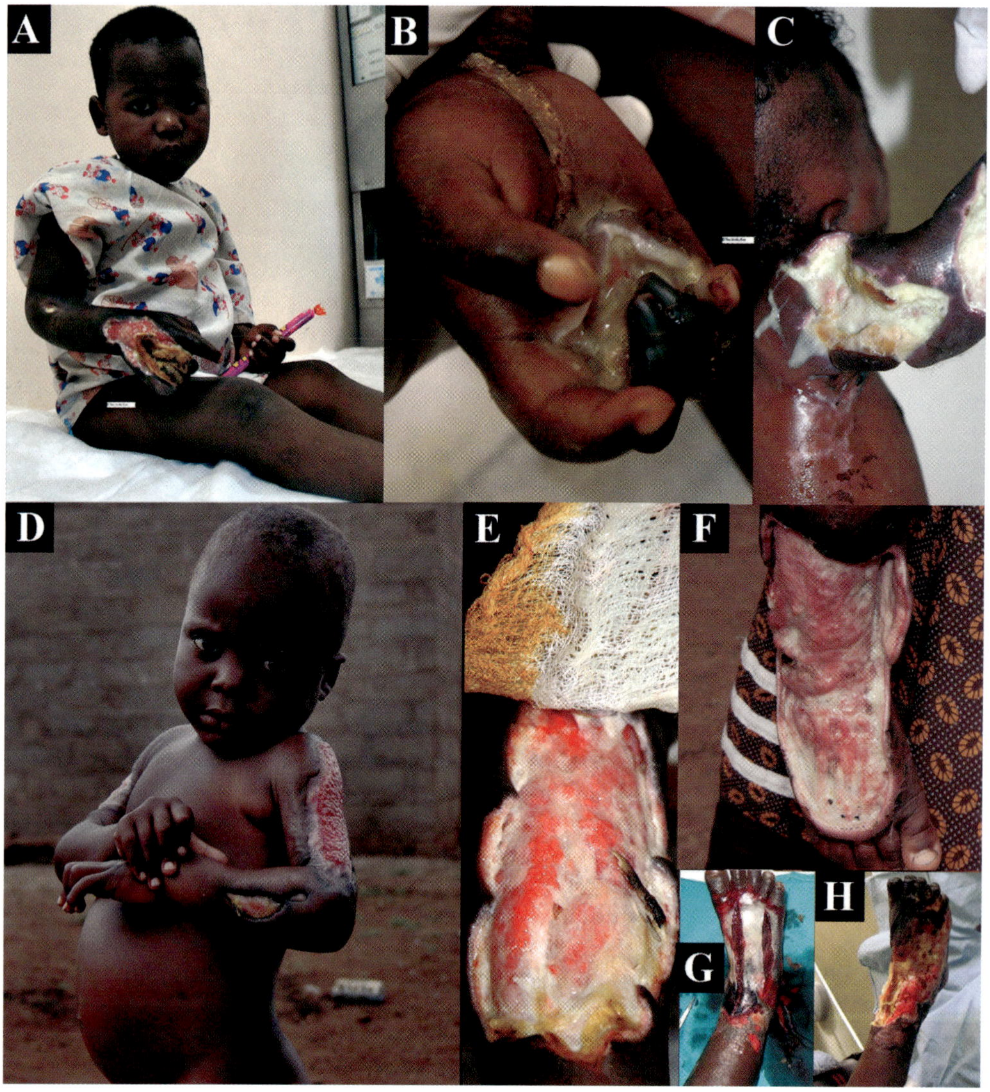

PLATE 14: The catastrophic effects that occur when tourniquets are used and indigenous remedies attempted, instead of promptly seeking out proper medical care and antivenom (see chapters 2 and 4). Bites from *Naja nigricollis* (A-G) and *Bitis arietans* (H). In all cases, a tourniquet was used, but in none was antivenom given. Intervention in all cases was many days postbite and included extensive skin debridement and grafts followed by treatment for secondary infection. Number of days postenvenomation: A and B: 15; C: 14; D: 60; E: 60; F: 17; G: 4; H: 11.. A-G effects were contributed to by cytotoxic 3FTxs (see chapter 8.4.3). (Photos: Thea Litschka-Koen.)

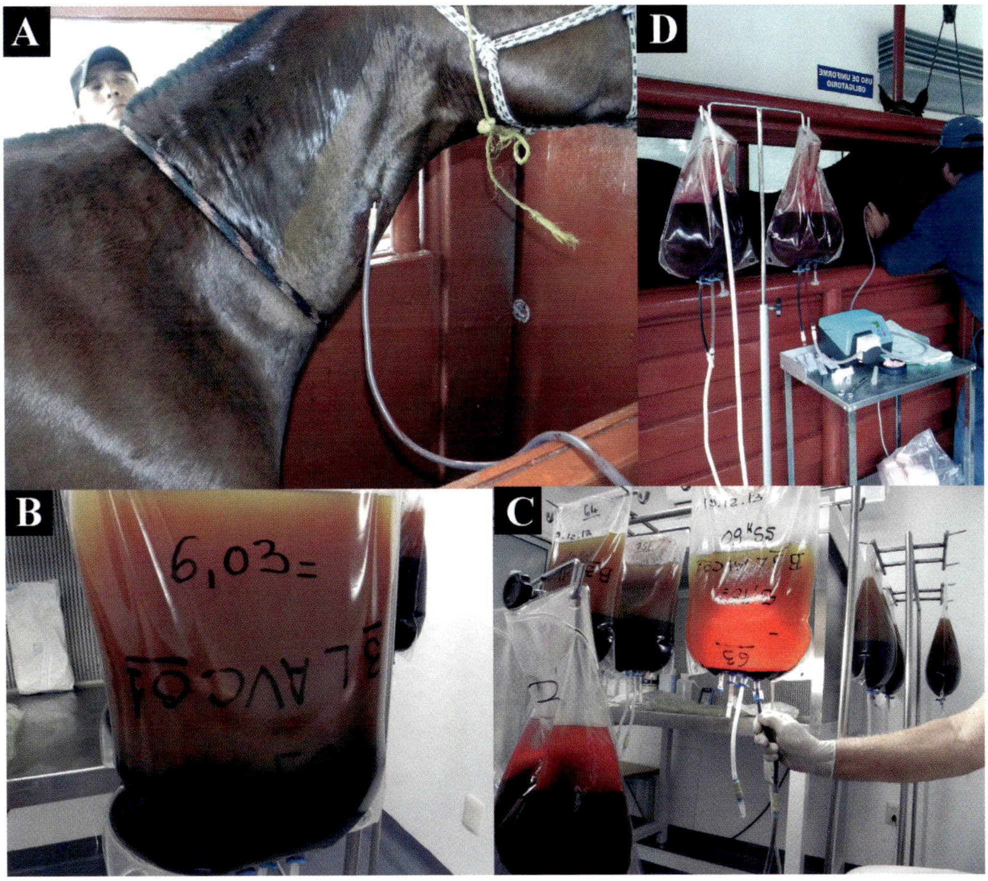

PLATE 15: Antivenom production (see chapter 3). (A) Blood extraction from the right jugular vein. (B) Collection of blood into bags with anticoagulant in a sterile room and sedimentation of erythrocytes. (C) Separation of the cellular fraction of the blood into a new bag with injectable isotonic-glucose solution. (D) Return of the blood cellular fraction through the left jugular vein using a peristaltic pump. Photos are by Alejandro Alagón.

PLATE 16: Techniques for safely working with venomous reptiles (see chapter 5.4). (A) Author B. G. Fry hooking and tailing a *Pseudechis australis*. (B) Tubing a *Crotalus mollosus*. (C) Double-pinning a *Crotalus lepidus klauberi*. (D) Using a pinner and a padded brace on a *Bitis rhinoceros*. Bite-resistant gloves are used to work with (E) Author B. G. Fry using 7 mm neoprene gloves to capture an *Aipysurus laevis*, (F) Author B. G. Fry using kevlar reinforced heavy leather gloves to capture a *Naja kaouthia*, and (G) *Heloderma exasperatum* biting a hexarmor glove. (Photo A: Iwan Hendrikx; B: Marco Sassoe; C: Bryan G. Fry; D: Nick Casewell; E: David Wachenfeld; F: Timothy N. W. Jackson; G: Bryan G. Fry.

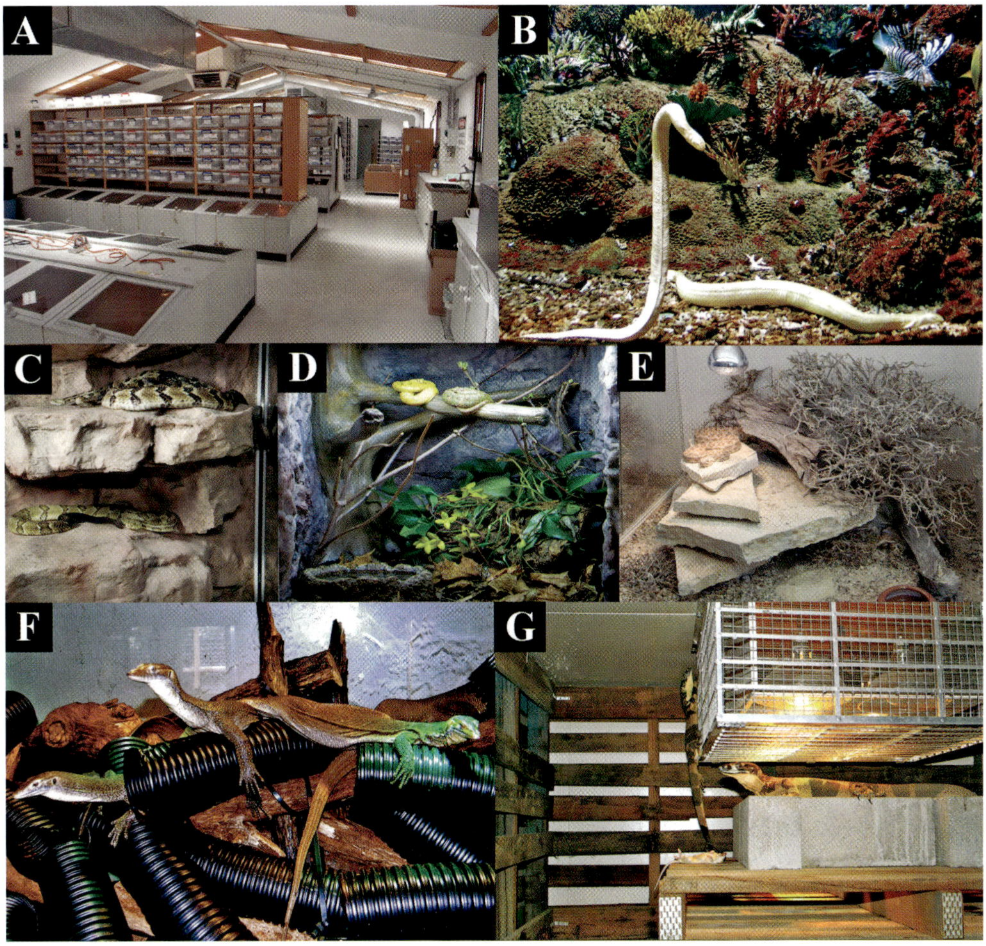

PLATE 17: Enclosures (see chapter 5.5.3). (A) A high-density approach to keeping snakes in a commercial venom-extraction facility with large, top-opening, horizontal cages for large, fast-moving elapid snakes and rack systems for small specimens. Naturalistic enclosures that not only meet the basic husbandry needs of the specimens but are aesthetically pleasing: (B) a specialized 70,000-L enclosure for *Hydrophis curtus* (formerly *Lapemis curtus*); (C) *Crotalus horridus*; (D) *Bothriechis schlegeli*; and (E) *Crotalus oreganus concolor*. Large, powerful varanid lizards are best served with a more industrial approach that, while less aesthetically pleasing, prevents catastrophic events such as fires resulting from torn lighting wire: (F) *Varanus mertensi*; (G) *Varanus varius*. (Photo A: Jeffrey A. Abraham; B: Rob Jones; C, D: Rob Carmichael; E: André Weima; F, G: Bryan G. Fry.

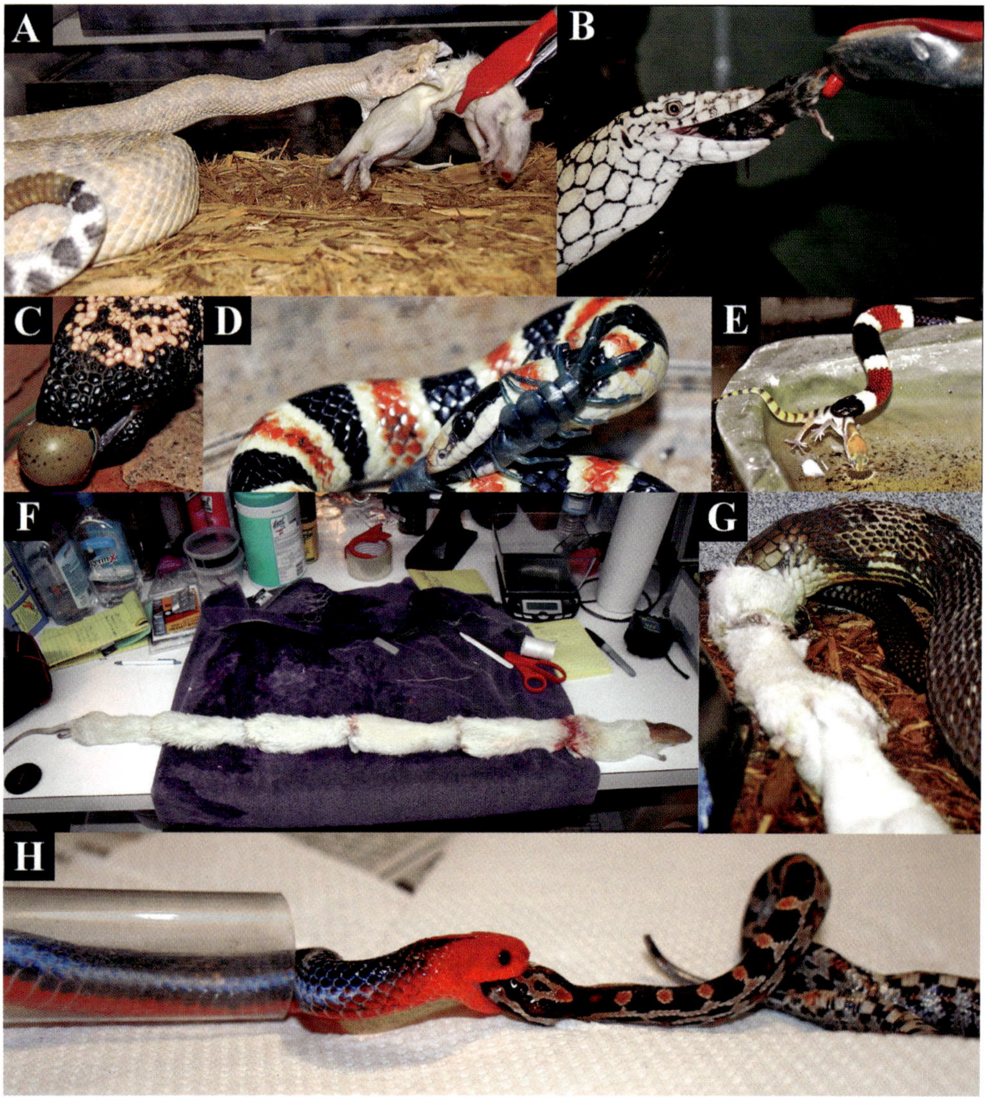

PLATE 18: Feeding techniques (see chapter 5.7). Using tongs to safely present a dead prey to (A) an aberrantly patterned *Crotalus atrox* and (B) a juvenile *Varanus giganteus*. Specialist food may be required, such as (C) eggs as part of a varied diet for a *Heloderma suspectum*; (D) *a Scolopendra species for a Chionactis occipitalis*; (E) *Coleonyx variegatus* for a *Micruroides euryxanthus*. (F, G) In the absence of suitable feeder snakes, *Ophiophagus hannah* may be induced to eat a series of rats sewn together using dissolving sutures with the additional suturing on of a snake head. (H) The ideal situation for ophiophagus snakes such as *Calliophis bivirgatus* is to feed dead captive-bred specimens, as they are likely to be parasite-free. (Photos A: Clint Guadiana, C, D, E, F, G: Cale Morris; B: Bryan G. Fry; H: Iwan Hendrikx.

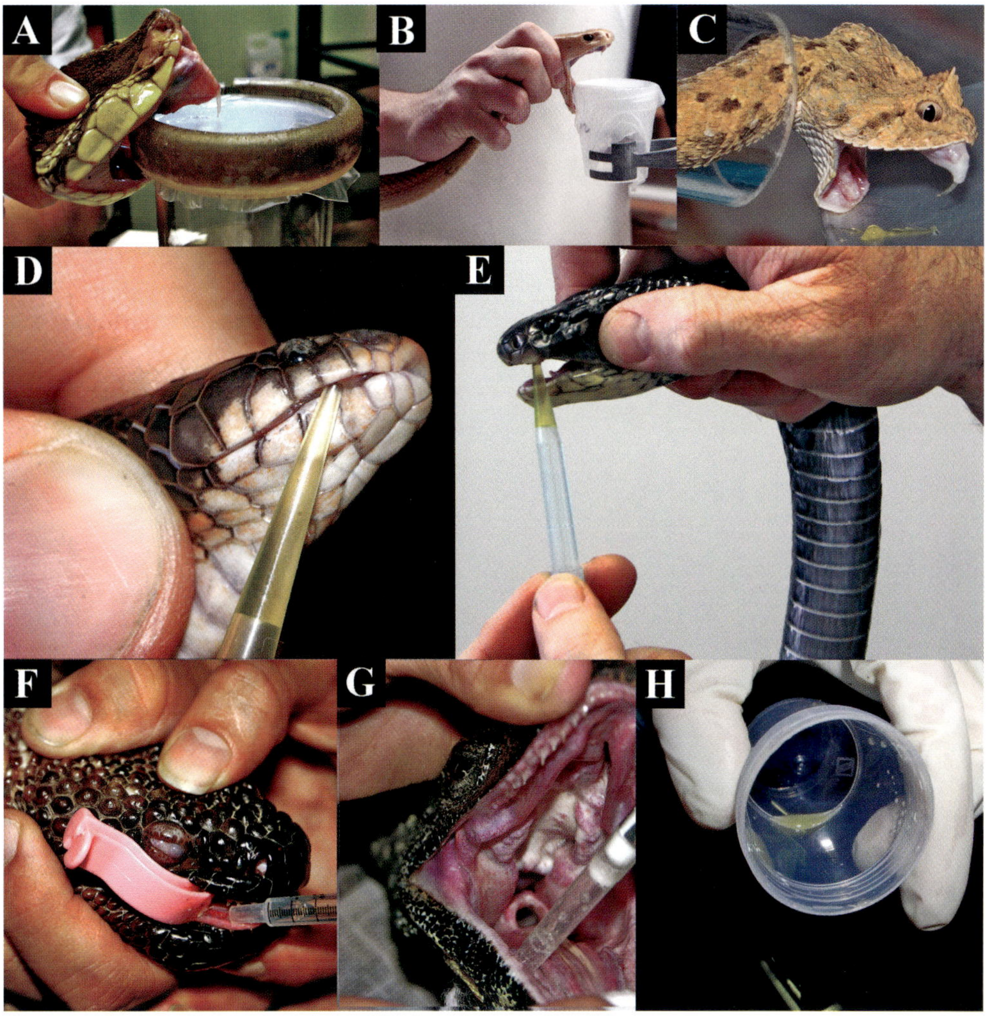

PLATE 19: Venom extraction (see chapter 5.9). Vipers or large elapids may be milked into sterile, parafilm-covered plastic containers (glass should be avoided because of its high protein-binding capacity), such as (A) *Bothrops asper*, (B) *Oxyuranus scutellatus*, or (C) *Bitis caudalis*. Small to medium-sized elapids are more efficiently milked by sliding a pipette over the fang and gently wiggling it to trigger the nerve, such as (D) *Elapsoidea boulengeri* or (E) *Pseudechis papuanus*. (F) Helodermatid lizards such as *Heloderma exasperatum* are milked by having the animal chew on soft plastic or rubber tubing. (D) Varanid lizards such as *Varanus varius* are milked by anesthetizing with general anesthesia and then stimulating rapid and profuse gland secretion through the periglandular injection of pilocarpine, with the venom collected as it emerges between the teeth. This method is also used for other nonhelodermatid anguimorph lizards in addition to non-front-fanged snakes. (H) Extracted venom must be immediately processed (see chapter 7.1). (Photo A: Carlos Rivera-Fernández; B, E: Jeffry Abraham; C: Daniel Liepack; D: Harold van der Ploeg; F-H: Bryan G. Fry.

PLATE 20: Field collecting (see chapter 5.10). (A) A wide range of handling equipment must be brought, enough for the entire team, for a variety of uses and situations and to accommodate breakage or loss. (B) For transport of live venomous reptiles, use a well-ventilated rigid outer box, with compartments inside for the placement of bags or containers. (C) A dry shipper is an efficient piece of equipment for the cryogenic storage of venom and tissues. (D) Funnel traps with drift barriers are an efficient method of collecting snakes and small lizards. (E) As the field locations are often remote, proper camping gear must be brought. (F) Flooded road crossings must only be attempted with a suitably equipped 4WD vehicle. (Photos: Bryan G. Fry.)

PLATE 21: Occupational snake training (see chapter 5.11) such as in Pakistan by authors B. G. Fry and S. McCarthy (A) Training areas must be clearly marked as off-limits to nonparticipants. Site evaluation must include (B) identification of snake-prevalent areas requiring permanent warning signage; (C) assessment of buildings for routes of entry by snakes; (D) proper storage of flat materials on blocks or rollers to elevate them off the ground, making them less favorable refuge habitat for snakes and also making it easier to visually ascertain if a snake is present. Training must include representative real-world scenarios such as (E) this *Naja naja* in a garbage dump in garbage dumps and (F) difficult capture situations as per this *Daboia russelii* intertwined in a glass bottle storage container. (Photos: Bryan G. Fry.)

PLATE 22: Snake rescuing (see chapter 5.11). (A) Author G. Shankar removing an *Ophiophagus hannah* from the roof of a house. (B) *O. hannah* under a set of clothes drawers. (C) Author G. Shankar removing an *O. hannah* from a village water well. (D) *Pseudonaja textilis* caught in garden netting. (E) *Dendroaspis polylepis* in an office drawer. (Photo A: Sharmila Gowri Shankar; B: Nazzareno Miele; C: Sharmila Gowri Shankar; D: Barry Goldsmith; E: Thea Litschka-Koen.

PLATE 23: (A) The Sweetwater Rattlesnake Roundup at the Nolan Convention Center in Sweetwater, Texas (see chapter 5.11.3). (B) An enclosed area containing large numbers of rattlesnakes at the Sweetwater Rattlesnake Roundup. Most of the snakes were collected in areas distant from the festival weeks or months ahead of time and kept without food or water before being killed at the event. (C) *Crotalus atrox* being skinned after being crudely beheaded at the Sweetwater Rattlesnake Roundup. (D) Children being encouraged to skin decapitated *Crotalus atrox* and then place their bloody handprints on the wall at the Sweetwater Rattlesnake Roundup. (E) Live *Crotalus atrox* placed in freezers to slow them down enough that the fangs can be torn out with pliers and the mouths sewn shut so that tourists can pay five dollars to pose for photos with them at the Apache Rattlesnake Festival in Oklahoma. (F) *Crotalus atrox* body parts for sale as tourist novelties at the Big Spring Rattlesnake Roundup in Texas. (G) *Crotalus atrox* venom collected at the Sweetwater Rattlesnake Roundup in unsterile conditions and with venom from snakes originating from different localities being mixed together, resulting in it being of no use for venom research or antivenom production. (Photos A, B, C, G: Michael Smith; D, F: Sky Stevens; E: Kim LaForest.

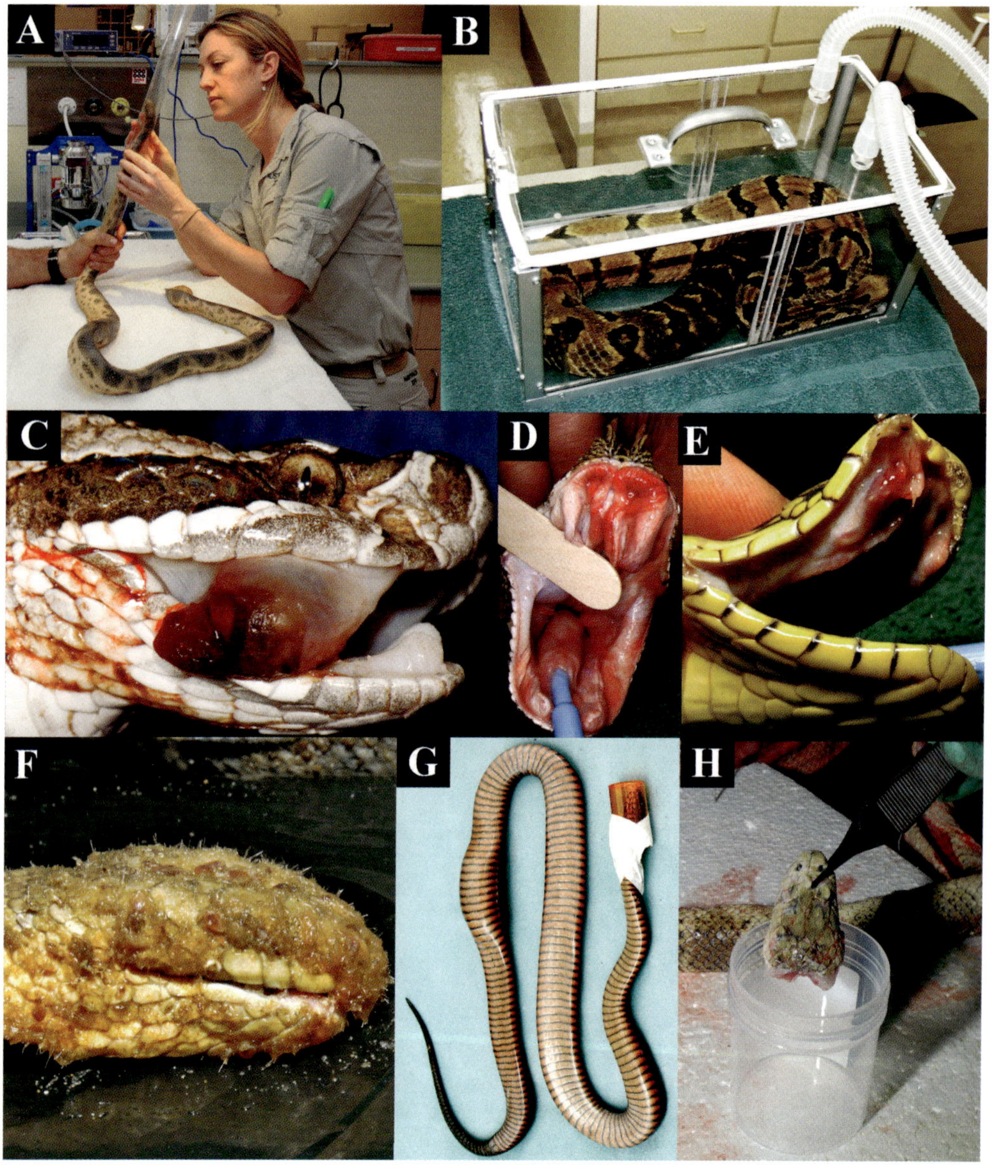

PLATE 24: Veterinary techniques (see chapter 6). Safe restraint for examination, minor procedures, and anesthesia of venomous snakes may be accomplished by the utilization of (A) clear acrylic tubes (seashore-stranded *Hydrophis elegans* patient being examined by author A. Gillett) or (B) sedation or anesthesia in an induction chamber (*Crotalus horridus* patient). Infection of the oral cavity of a venomous snake poses special challenges and dangers, such as (C) common appearance of a fang-sheath lesion (the etiology of the majority of these lesions is bacterial, but viral, fungal, and traumatic causes have been reported) (*Sistrurus catenatus tergeminus* patient); (D) atypical stomatitis caused by larval nematode worms (this case illustrates the importance of diagnostic testing in order to achieve a definitive diagnosis for effective treatment) (*Bitis nasicornis* patient); (E) bilateral granulomas of the fang sheaths secondary to a poorly performed venomoid procedure by an individual lacking any veterinary qualifications (*Dendroaspis jamesoni* patient); (F) heavy epibiotic infestation (stranded *Hydrophis major* patient). Venomous snakes can pose a postmortem envenomation threat to personnel, so the head must be isolated in some manner, such as by (G) taping a container securely over the head (*Pseudechis porphyriacus* patient) or (H) severing the head and placing it in a container (*Hydrophis elegans* patient). (Photo A: Ben Beaden; B, C, D, E: Stephen Barten; F, H: Amber Gillett; G: Robert Johnson.

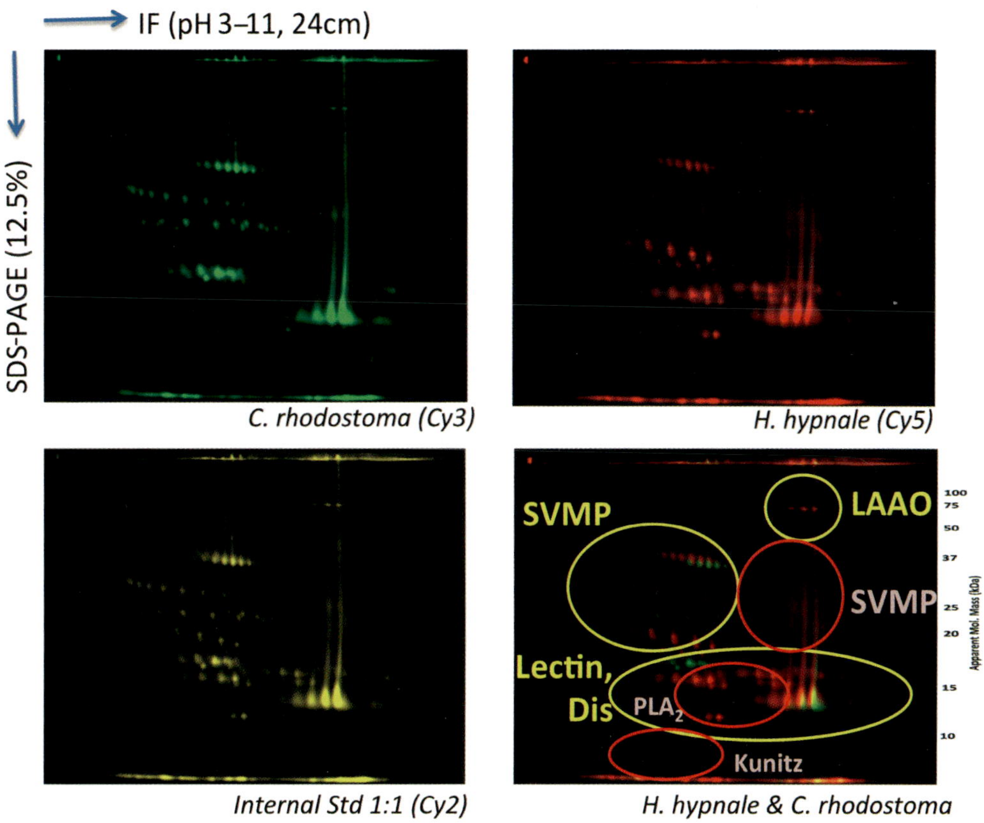

PLATE 25: 2D-DIGE comparison of *Calloselasma rhodostoma and Hypnale hypnale* venoms. In the bottom right panel, yellow indicates shared toxin types, red those unique to *H. hypnale* in this particular venom set comparison (Ali et al., 2013a).

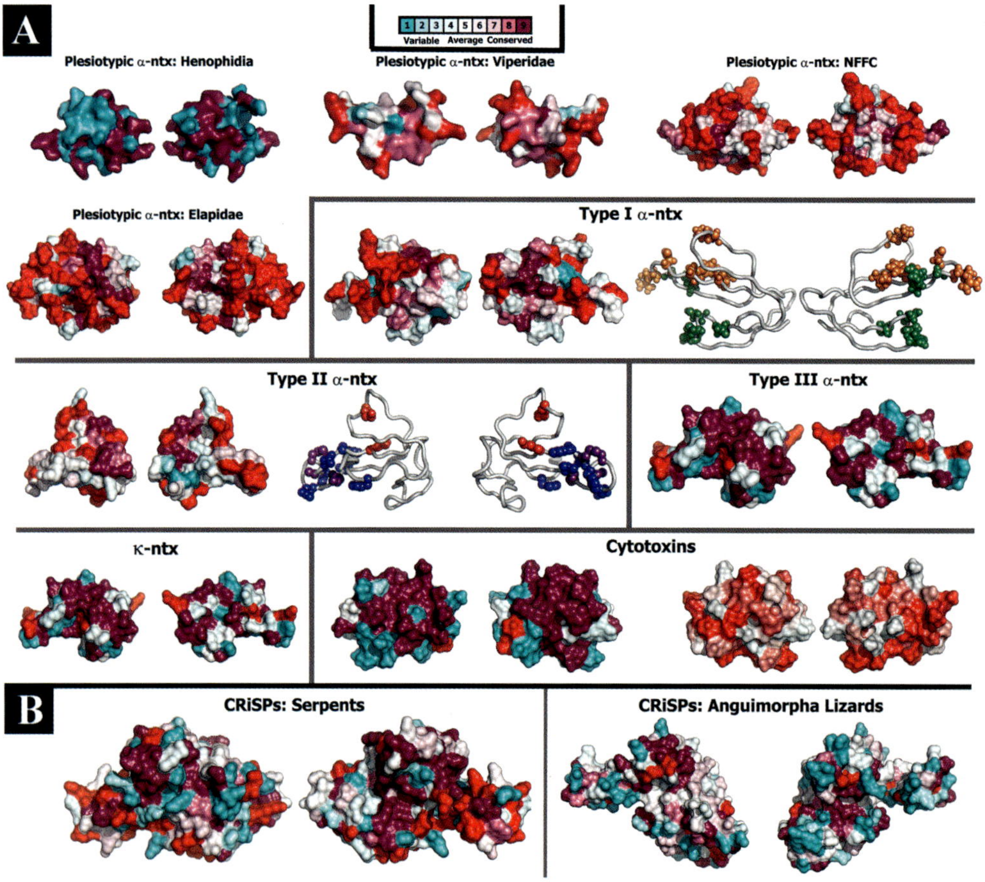

PLATE 26: Molecular evolution of (A) 3FTx (see chapter 8) and (B) CRiSP (see chapter 10) toxins. The evolutionary variations on the molecular surfaces of toxins are colored to depict relative variability (see chart), while the hypermutational sites are represented in red. Cardiotoxin hydrophobicity is additionally depicted through a red gradient. In the second set of figures for the type I α-neurotoxins, residues that bind muscular nAChRs (α-1) are colored by the model species utilized to establish binding affinity: fish and mammal (orange) and fish only (green). In the second set of figures for the type II α-neurotoxins, residues known to bind muscular nAChRs are depicted in red, while residues that interact with neuronal nAChRs are depicted in blue, and those that target both neuronal and muscular nAChRs are depicted in purple.

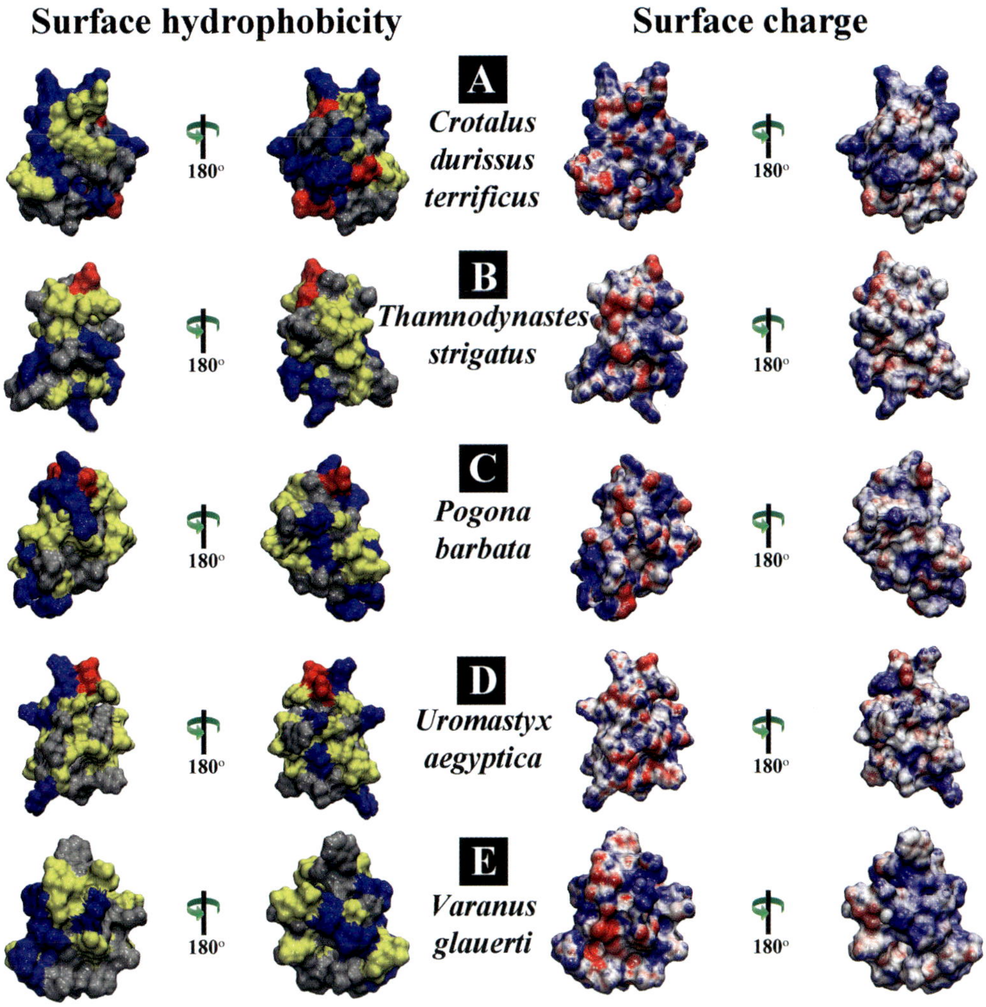

PLATE 27: Molecular variation of β-defensin peptides (see chapter 9).

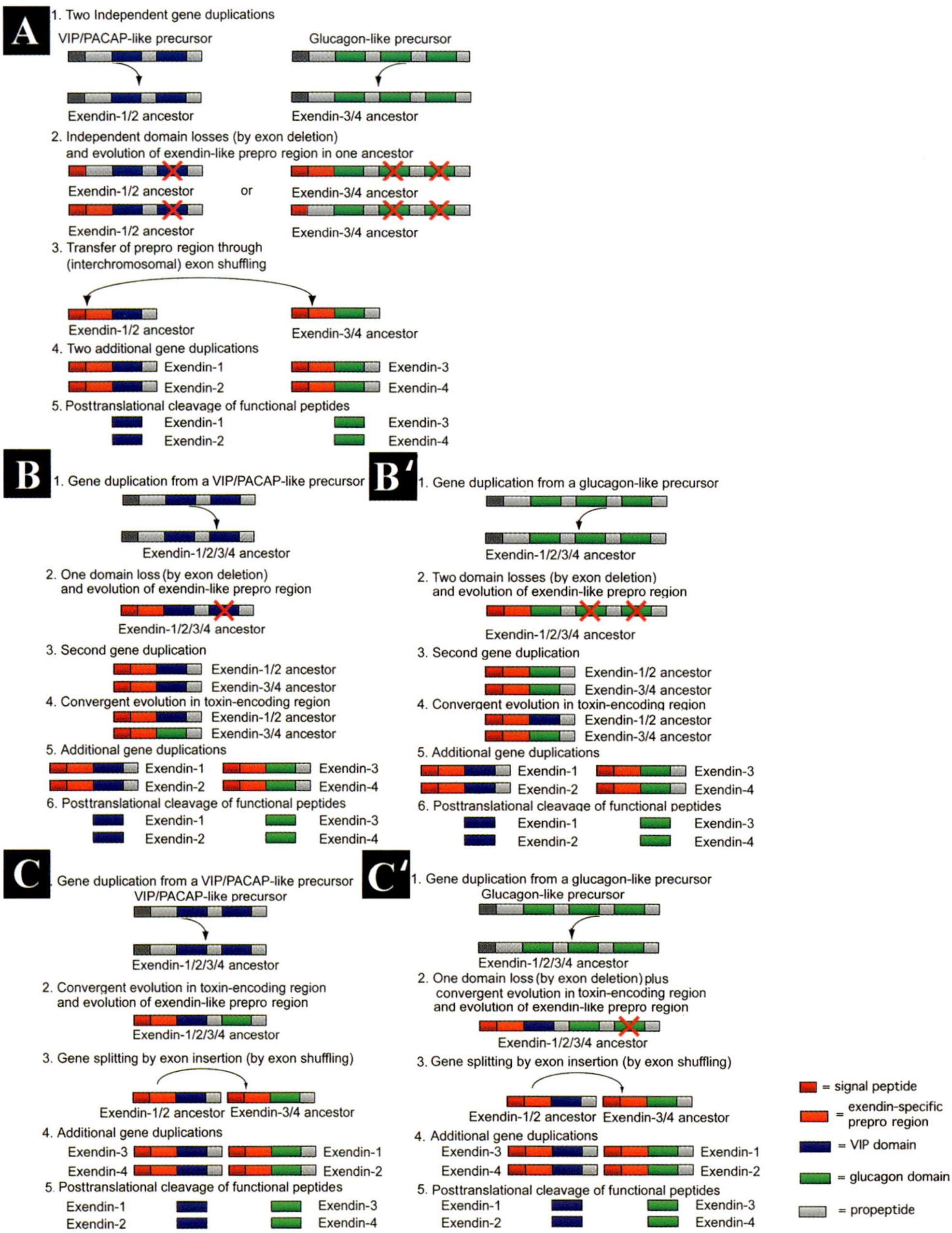

PLATE 28: Three evolutionary scenarios explaining the origin of the two monodomain exendin toxin pairs (exendins 1 and 2 and exendins 3 and 4) in *Heloderma* venoms, starting from either a VIP or a glucagon gene (see chapter 11). (A) Ancestors of the two exendin pairs may have independently duplicated from separate ancestral multidomain hormone precursor genes. This scenario implies independent domain losses and an exon-shuffling event to explain the closely related signal peptides among the two pairs and is unlikely in light of our phylogenetic results favoring a mono origin of exendin peptides. (B) A second scenario implies a monoexendin ancestor that duplicated from an unknown multidomain hormone precursor gene. The monoexendin ancestor subsequently underwent domain loss and multiple successive gene duplication events to obtain four monodomain exendin genes. (C) Alternatively, an unknown multidomain ancestor was split into two tandem genes by interdomain insertion of a signal peptide through exon shuffling, and each tandem gene subsequently underwent an additional gene duplication to obtain the four exendin genes.

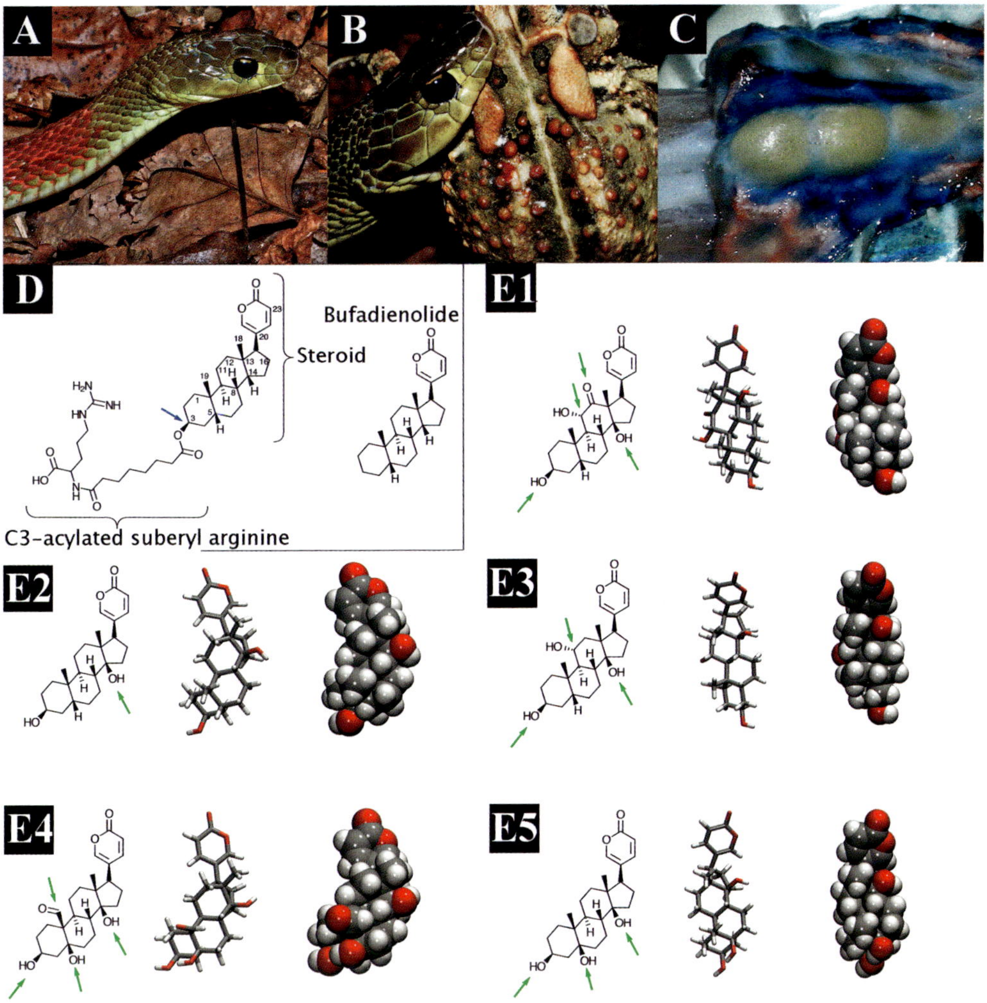

PLATE 29: The snake poison system (see chapter 25). (A) The aposematic warning coloration of *Rhabdophis subminiatus*, with the red coloration on the neck corresponding to the location of the nuchal glands. (B) The toxins are sourced from the dermal glands of toads, such as the large parotid glands. (C) Dissection of *Rhabdophis tigrinus* to reveal the nuchal glands. (D) Left: Bufotoxins are modified bufadienolides with a suberyl arginine moiety acylated to C3 (blue arrow). Suberyl arginine is the product of a condensation reaction of one of the carboxyl groups of suberic acid and the amino group of arginine. Right: General structure of a bufadienolide. Bufadienolides possess a steroid scaffold with an unsaturated lactone ring attached to C17. (E) The 2D (left) and 3D structures (center and right) of selected bufadienolides. Green arrows in the 2D models highlight positions with chemical changes relative to the basic bufadienolide. In the 3D structures C atoms are colored gray, O atoms are red, and H atoms are white. At the center, the molecule is shown in licorice, at right with its atoms depicted as van der Waals radii. E1 is Arenobufagin from the Argentine toad *Rhinella arenarum*. E2 is Bufalin found in Chan Su, a traditional Chinese medicine from the Asian toads *Duttaphrynus melanostictus* and *Bufo gargarizans*. E3 is Gamabufagin from the Japanese toad *Bufo japonicus*. E4 is Hellebrigenin from the North American toads *Anaxyrus terrestris* and *Anaxyrus fowleri*. E5 is Telocinobufagin found in Chan Su, a traditional Chinese medicine from the Asian toads D. *melanostictus* and B. *gargarizans*. (Photos A-C: Kevin Messenger.)

PLATE 30: The (un)natural history of *Varanus komodoensis* and *Bubalus bubalis* (see chapter 25). These two animals share a habitat only because of the recent (< 300 years) introduction of *B. bubalis* onto the islands by Dutch settlers. Consequently, (A) *B. bubalis* does not recognize *V. komodoensis* as a natural predator, while *V. komodoensis* will experimentally attack any animal. (B) *B. bubalis* always escape but sometimes with deep wounds to the hindquarters such as this specimen with an artery bleeding prolapse. (C) Upon seeking refuge in the tiny stagnant rocky water holes that are the only water source on the small islands, the wounds become infected because of the high concentrations of *B. bubalis* fecal matter. (D, E) If death occurs, multiple *V. komodoensis* will feed on the fresh carcass, using their powerful necks and forelegs to remove large chunks of meat. The feeding group rarely if ever includes the original specimen who inflicted the initial wound days earlier. (F) Only once the adults have finished will the smaller *V. komodoensis* feed on the remains. (Photos A, D, E: Chris Kugelman; B, C: Bryan G. Fry; F: Gembong Riyadi Nurrasa.)

CHAPTER 15

KUNITZ PEPTIDES

W. S. ENG, B. G. FRY, K. SUNAGAR, Z. TAKACS, T. N. W. JACKSON, AND L. W. GUDDAT

15.1 SUMMARY

Reptile venom kunitz peptides have undergone extraordinary structural and functional molecular adaptation, from multiproduct precursors to single-product precursors and from enzyme inhibitors to channel-blocking neurotoxins. These small peptides are useful not only for probing neurological activity but also as potential therapeutics in preventing blood loss during surgery.

15.2 ENDOPHYSIOLOGICAL PLESIOTYPE

Kunitz-type protease inhibitors are ubiquitous among plants and animals. The MEROPS database (http://merops.sanger.ac.uk/) classifies these molecules into two subclasses: kunitz-A (family I2, isolated from animal) and kunitz-P (family I3, isolated from plants) (Rawlings, Barrett, and Bateman 2010). The most typical function of kunitz-type inhibitors is inhibition of serine proteases, a class of enzyme characterized by the presence of a catalytic triad consisting of histidine, aspartate, and serine amino acid residues (Rawlings, Barrett, and Bateman 2010). Important members of the serine protease family include trypsin, chymotrypsin, elastase, plasma kallikreins, tissue kallikreins, and plasmin. Precursors for plesiotypic kunitz-type inhibitors encode two kunitz peptides that are posttranslationally cleaved from each other. However, derivatives in which the precursor encodes a single peptide are common among apotypic versions (such as BPTI P00974).

Bovine pancreatic trypsin inhibitor (BPTI, also known as aprotinin or by its commercial name, Trasylol) was first characterized in the early 1930s (Kunitz and Northrop 1936). The three-dimensional structure was determined using X-ray crystallography (Huber et al. 1970), and its solution structure was subsequently solved by nuclear magnetic resonance (NMR) spectroscopy (Wüthrich 1989). The structure of BPTI may be used to describe the general features of this class of peptide.

Both crystal and NMR structures show that BPTI, which consists of 58 amino acid residues, folds into an α/β/α motif, with the tertiary structure stabilized by three disulfide bonds (Cys5-Cys55, Cys14-Cys38, Cys30-Cys51). An important feature of BPTI is a loop located between the first α-helix (α1) and the first β-sheet (β1). This region, known as the canonical loop, binds in a complementary manner (in terms of shape, charge, and hydrogen bonding) to the active site of a serine protease but is not cleaved by the enzyme (Rühlmann et al. 1973). There are six amino acid residues critical to serine protease inhibition by BPTI: Pro13, Cys14, Lys15, Ala16, Arg17, and Ile18. The nomenclature describing the locations of these amino acids identifies them as P_3, P_2, P_1, P_1', P_2', and P_3', respectively. Proteolytic cleavage ordinarily occurs at the peptide bond between the P_1 and P_1' residues (Schechte and Berger 1967). However, cleavage does not occur when BPTI forms a complex with a protease (such as trypsin). Instead, BPTI becomes irreversibly bound to the enzyme, thereby acting as an inhibitor. An additional region of kunitz peptides that contributes to the binding of proteases is referred to as the secondary binding loop. This is the part of the polypeptide that is directly after the second β-sheet (β2) and includes amino acid residues from positions 36–39 in the sequence. When BPTI and trypsin bind, all four of these residues form contacts (mostly through van der Waals attractions) and may contribute significantly to the inhibition of serine proteases.

15.3 PLESIOTYPIC STRUCTURAL AND FUNCTIONAL TOXIN FORMS

Like the precursors of endophysiological homologues, in the plesiotypic state of Toxicofera kunitz peptides (TV-kunitz), the precursor encodes two peptides, which are liberated from each other through posttranslational cleavage and secreted separately (Doley et al. 2008; Koludarov et al. 2012; Fry et al. 2013; Jackson et al. 2013). Phylogenetically, the dual TV-kunitz-domain encoding precursors are located at the base of the toxin trees (Fry et al. 2013) (see figure 15.1). Only this plesiotypic form has been recovered from anguimorph lizards and iguanian lizards (Fry et al. 2013). In contrast, the dominant form present in caenophidian snakes is an apotypic precursor that encodes for only a single TV-kunitz peptide and has a distinct signal peptide relative to the two-domain plesiotypic precursor (Fry et al. 2008; Koludarov et al. 2012; Fry et al. 2013; Jackson et al. 2013) (see figure 15.2). Despite the isolation and characterization of TV-kunitz peptides from a wide array of venoms over a long period of time, they remain poorly characterized. For some, the inhibitory activity against a range of proteases has been assayed, while for most others, very little information other than the primary sequence data is available. For example, an extensive genomic investigation of the Australian snake venoms has identified 44 distinct TV-kunitz peptide sequences (St Pierre et al. 2008). Filippovich and colleagues identified six TV-kunitz-type inhibitors, referred to as textilinins, in the venom of *Pseudonaja textilis* (Filippovich et al. 2002). However, among these to date, only textilinin-1 has been extensively characterized in terms of structure and activity (Masci et al. 2000; Filippovich et al. 2002; Flight et al. 2005; Millers et al. 2006; Flight et al. 2009) (see section 15.4).

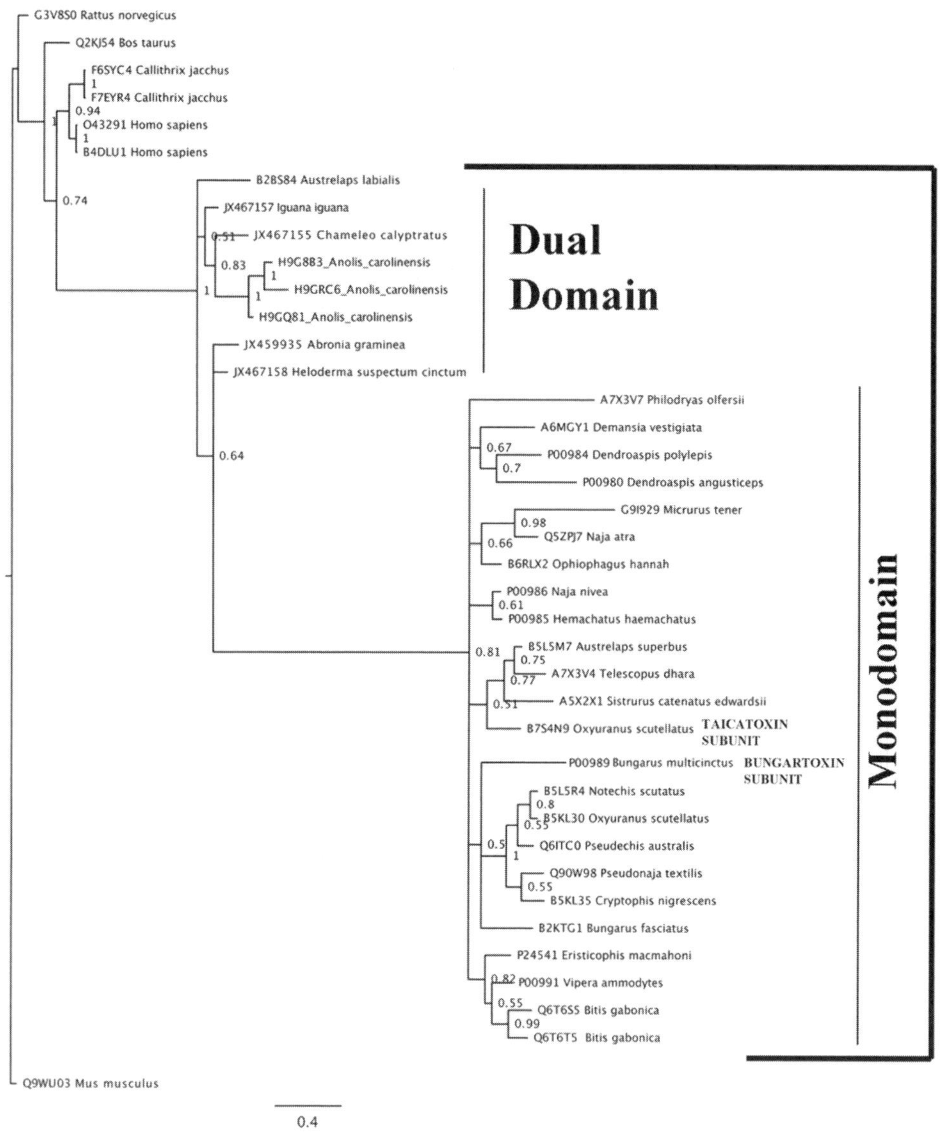

FIGURE 15.1: Molecular phylogeny of kunitz toxins and related nontoxin peptides (Fry et al. 2013; Jackson et al. 2013).

The plesiotypic function is inhibition of trypsin or chymotrypsin, and some variants exhibit specific affinity for trypsin or chymotrypsin. Potent trypsin-inhibitory action has been attributed to the presence of a charged lysine residue at the P_3' position, as opposed to the hydrophobic isoleucine that is found in the same position in BPTI or the aromatic phenylalanine found in the majority of Australian elapid snake venom forms. OH-TCI from the elapid snake *Ophiophagus hannah* is an inhibitor of both α-chymotrypsin and trypsin with K_i values of 174 and 172 nM, respectively (He et al. 2008). The P_1 residue for this inhibitor is lysine, emphasizing the importance of this amino acid in trypsin inhibition but also indicating that α-chymotrypsin-inhibiting TV-kunitz peptides do not always possess hydrophobic residues at

```
             10        20        30        40        50        60        70        80        90
1.                                                                 >FGGCGGNANKFRTEAQCLQNCSTEGDEAPMETP----
2.                                                            >CQRFIFGGCKGNPNNFLTQDNCVQRCTTGGDEKAAETP----
3.        >gatallvvlllvlgteaGALS--ETCHLPKVVGRCRAAFQRWWYNATSQTCQKFIFGGCRGNLNNFLTEQECHQACVLDGQEEATAAPEEPG
4.   >
5.                                  >CHLPKVVGRCRASFPRWWYNATSQTCQRFIFGGCRGNSNNFLSEEECKKGCAT-GE-EVSIVELRLR
6.        >vkirtlplqirvlsanEILFQSPKNCHLPKVVGRCRASFPRWWYNATSQTCQRFIFGGCRGNSNNFLSEEECKKGCATDGDIEAPVTHDGPA
7.                                >ENCHLPKVVGRCRASFPRWWYNATSQTCQRFIFGGCRGNSNNFLSEEECKKGCAT-GE-EVSIVE--LS
8.   mtr-------ekslallitlaaalaaaESPPGRCHSPKTVGPCRASFHRWRYNATSQMCQEFIFGGCKGNANNFVSKQDCFQTCIRGGAAEATVVPSGP-
9.   m-ssgglllllglltlwaeltpvsgq--------------------------------------------------------------------------
10.  m-ssgglllllglltlwaeltpvsgk--------------------------------------------------------------------------
11.  m-ssgglllllglltlwaeltpvsgq--------------------------------------------------------------------------

    100       110       120       130       140       150       160       170       180       190
1.   -----------------------GWLFMKEYCA-EPQDVGPCRASFPRWYFDMESQTCKKFTYGGCGGNKNNYVFEEQCLKQCSG--EIVEEPAEFH--P
2.   -----------------------GWLSTREYCA-APQAVGPCRASFSRWYFDPESQTCKMFIYGGCGGNKNNYLFEEQCLRRCSGAGEILEEPEEPHSPP
3.   EDTVMPVTTKDVSKSGRRLEPGDNRVGFREFCV-VPQAVGPCRASFPRWFFDIETRTCKMFIYGGCGGNKNNYIFEDHCLSQCTGNGEITEEHLEPDTLS
4.                                                                               >RLLNQCTGSGEITDESLEPETPA
5.   IYS-NSAVPTAISRSGRRHEPGDDRVGFEEFCA-APKEVGHCRASFPRWYFDTETRTCKMFIYGGCGGNKNNYIFEEHCLNQCSGNGGLDSLP-------
6.   EEI-MTSTTKAISRSGRRHEPGDDRVGFEEFCA-APKEVGHCRASFPRWYFDTETRTCKMFIYGGCGGNKNNYIFEEHCLNQCSGNGDIINEPLGPDSSS
7.   NNR-KQACLHLISI--RMSQSAEVTFNTKEFCA-APKEVGHCRASFPRWYFDTETRTCKMFIYGGCGGNKNNYIFEEHCLNQCSGNGDFHEGPMPV----
8.   ------ATEVATPRAGHLPEAYENRPGFREFCA-APRVVGPCRASFLRWYFDLESRMCKMFIYGGCRGNKNNYLFEEHCWSQCTGDGEITEEPGDAGAQP
9.   --------------------------DRPKFC-HLPANPGPCRATITRFYYNSDSKQCEKFTYGGCHGNENNFETKDKCHYTCVGK
10.  --------------------------KRPDFC-YLPADTGPCMANFPRFYYDSASKKCKKFTYGGCHGNANNFETREECRKKCFASAARRPT
11.  --------------------------DRPKYC-DLPADSGPCRARKRHFYYNSRSNQCEEFTYGGCRGNENNFETKDKCHYTCVAPPRRRPL

    200       210       220       230       240       250       260
1.   HLFSDPLIHSTRAVVLAVLLAVMAAILLGS<
2.   YIFSDPLIHSTRAAVLAVLLAIMAAVLLASLVVFFVKVCRKNQQLSLGTVWSTLDDKECLMSNAHTL
3.   HHFTDTLIHSTRAVVLTVLLAIMAAVLLGSLVVFFVKMCRKDQGVTLGTVWNSLDDKEYLMNNAYTL
4.   HHFPEPLIHSTRAVVLAVLLALMVAILLGSMVVFFVKVCRKNQELSLGTVWSTLDDKEYLMSNXYTL
5.   ----------SAAVVLAILLAVMAAILLGSMVFFFIKLCRRNQDGTLERVWSTMDDKESLMNNAYTL
6.   HYFPGPLIHSTRAVVLAILLAVMAAILLGSMVFFFIKLCRRNQDGTLERVWSTMDDKESLMNNAYTL
7.   ----------TPTVVLAILLAVMAAILLGSMVFFFIKLCRRNQDGTLERVWSTMDDKESLMNNAYTL
8.   PLPSEPFSFSTRAVVLAVLPAILVTILLGSMGVFFVKICRKNPELSVGTVWSTLDDKEYLMSNAYTL
9.
10.
11.
```

FIGURE 15.2: Sequence alignment of kunitz toxin peptides (Fry et al. 2013; Jackson et al. 2013). (1) JX459935 (*Abronia graminea*), (2) JX467158 (*Heloderma suspectum cinctum*), (3) JX467155 (*Chameleo calyptratus*), (4) JX467157 (*Iguana iguana*), (5) H9G8B3 (*Anolis carolinensis*), (6) H9GQ81 (*Anolis carolinensis*), (7) H9GRC6 (*Anolis carolinensis*), (8) B2BS84 (*Austrelaps labialis*), (9) B5L5M7 (*Austrelaps superbus*), (10) Q6T6S5 (*Bitis gabonica*), (11) A7X3V4 (*Telescopus dhara*). Posttranslationally cleaved bioactive peptides are shaded in gray. > and < indicate incomplete sequences.

the P1 site. *Bungarus fasciatus* fraction IX (BF9) is an example of a variant that exhibits greater affinity for chymotrypsin than for trypsin (Liu, Wu, and Lo 1983). Surface plasmon resonance and isothermal titration calorimetry (ITC) reveal that it binds tightly to α-chymotrypsin, with dissociation constants of 58 and 230 nM, respectively (Chen et al. 2001). ITC studies demonstrate that this molecule does not inhibit trypsin. BF9 is unable to bind trypsin because the P_1 site is occupied by an asparagine. Another TV-kunitz peptide that has asparagine in this position is Oh11-1 from *O. hannah*, although its K_i value is significantly higher, 3.52 μM (Chang et al. 2002), than that of the inhibitors previously described. However, when a nonpolar amino acid is found in this position (such as leucine in a TV-kunitz peptide from the venom of the viperid snake species *Vipera ammodytes ammodytes* [Ritonja, Meloun, and Gubensek 1983]), the inhibitory constant is significantly lower (13 nM), which is in keeping with chymotrypsin's general substrate-specificity profile.

Based on the similarities and differences of these sequences with those of TV-kunitz peptides with characterized activities, it is possible to speculate on the activities of proteins for which only the sequence data are available. About half of all TV-kunitz peptides for which sequence data are available have either an arginine or a lysine at P_1, suggesting that these molecules are protease inhibitors with specificity for trypsin-like proteases. Textilinin-3 from *P. textilis* venom, for example, possesses an asparagine residue in the P_1 position, which is also observed in TV-kunitz peptides from the venoms of *B. fasciatus* (P25660) and *O. hannah*

(P82966), which are known to exhibit not only chymotrypsin-like inhibitory activity but also limited trypsin-like inhibitory activity. Thus, it is likely that textilinin-3 would exhibit a similar specificity profile. Textilinin-4 has a lysine residue at P_1, suggesting that it is likely to be a potent and specific inhibitor of trypsin-like proteases but may also be a weak inhibitor of chymotrypsin-like proteases. Textilinin-5 and textilinin-6 both have negatively charged amino acids at P_1 and are therefore unlikely to be inhibitors of either chymotrypsin or trypsin. The MEROPS peptidase database indicates that enzymes that have preference for substrates with negatively charged amino acids at P_1 include aminopeptidase A, streptogrisin E, glutamyl peptidases I BI and BL, blotched snakehead birnavirus Vp4 peptidase, caspases 1–3 and 6–10, and granzyme B, suggesting that textilinin-5 and textilinin-6 are potentially inhibitors of these enzymes.

15.4 APOTYPIC STRUCTURAL AND FUNCTIONAL TOXIN FORMS

It is interesting to observe that about half of the TV-kunitz peptides sequenced to date do not have an arginine or a lysine in the P_1 site; therefore, the role these molecules might play in the snake's venom is unknown. Some of these, such as those from *Demansia vestigiata* (St Pierre et al. 2007a), have a lysine and a leucine at positions 5 and 9 in the polypeptide chain, an arrangement that suggests the molecule is likely to have potassium-channel blocking activity. As with the TV-kunitz peptides that inhibit proteases, our understanding of the mode of action of these molecules would be greatly enhanced if structural data of the complexes with the target ion channel become available. Nonetheless, there are enough activity data to be confident that the major binding site for these molecules is associated with the N-terminus of the polypeptide.

Several structurally and functionally apotypic forms of venom TV-kunitz peptides have been isolated and characterized. Precursor derivations include those that encode a single TV-kunitz peptide and a splice variant that encodes a TV-kunitz peptide and a waprin peptide (Jackson et al. 2013). As the TV-kunitz-waprin dual domain has only been recovered in transcriptome studies, it remains to be established whether the TV-kunitz-waprin fused toxin is posttranslationally processed into two separate toxins or if it exerts its toxic function as a protein complex.

Functional derivatives include plasmin inhibitors (Flight et al. 2005; Flight et al. 2009), potassium-channel blockers (Skarzynski 1992; Danse et al. 1994; Harvey et al. 1997; Gasparini et al. 1998; Harvey 2001) and calcium-channel blockers (Schweitz et al. 1994; Gilquin et al. 1999). Structural derivatives have roles that include disulfide-linked subunits of toxin complexes such as β-bungarotoxin from *Bungarus* venom (Kondo, Narita, and Lee 1978a; Kondo, Narita, and Lee 1978b) and taicatoxin from *Oxyuranus* venom (Possani et al. 1992) (see chapter 20).

Textilinin-1 and textilinin-2 from the venom of *P. textilis* are uniquely effective at preventing blood loss in an animal model (Masci et al. 2000). Textilinin-1 is a potent inhibitor of plasmin with a K_i^* value of 3.5 nM and is also effective at inhibiting thrombin (Flight et al. 2005; Flight et al. 2009). The reason for textilinin-1's potency as a plasmin inhibitor lies in the fact that its P_1 residue is positively charged (arginine) and therefore fits neatly into the S1 specificity pocket of trypsin-like proteases. Textilinin-1 has been crystallized and its three-dimensional structure determined (Millers et al. 2006). It has a fold that strongly resembles that of BPTI, but it possesses the unusual feature of a canonical loop that displays a great deal of structural flexibility. In the crystal structure, there are two conformations for this loop: one with the standard canonical structure where the P_1 and P_1' residues are exposed to the solvent and a second where the fold of the loop is inverted around the P_1–P_1' bond. As a result of this, the side chain of the P_1' residue, valine, becomes partially buried and forms contacts with residues located in the mid-section of the protein (such as Pro15, Thr 13, and Ile36). An important difference between the inhibitory activity of textilinin-1 and aprotinin is their specificity for different proteases. Textilinin-1 is at least a 3-fold to 1.3 x 10^4–fold weaker inhibitor of tissue plasminogen activator, urokinase, activated protein C, elastase, plasma and tissue kallikrein, and trypsin than aprotinin. This tighter binding specificity and more reversible binding may give textilinin-1 an advantage as a therapeutic antibleeding agent compared with aprotinin (Filippovich et al. 2002). Many of the enzymes associated with the blood-clotting cascade are trypsin-like proteases; thus, it is easy to see how these peptides may interfere with hemostasis in victims of snake envenomation. However, what is unclear is how small changes in the amino acid sequences of these molecules can modulate specificity for particular proteases. One of the reasons for this lack of understanding is that structural data for complexes of these inhibitors with their target enzymes are presently of limited availability. Once these data have been expanded, it will be possible to more precisely define the roles of these molecules and their direct effects on enzymes of the blood-coagulation pathways.

Dendroaspis angusticeps venom uniquely potentiates and prolongs the action of the neurotransmitter acetylcholine (Osman, Ismail, and El-Asmar 1973). The molecule responsible for this activity is a TV-kunitz peptide (α-dendrotoxin) that acts by blocking potassium channels, thereby facilitating the release of acetylcholine (Harvey and Karlsson 1980; Harvey and Anderson 1985). In contrast with BPTI, it is only a very weak inhibitor of trypsin, and BPTI does not interact with potassium channels. These differences suggest that different parts of the molecules are responsible for the observed activities (Harvey 2001). A sequence comparison of α-dendrotoxin with other *Dendroaspis* venom TV-kunitz peptides (δ-dendrotoxin, dendrotoxin-I, and dendrotoxin K) and BPTI shows that Lys28, Lys29, and Lys30 are conserved in dendrotoxins but absent in BPTI, suggesting that these residues may be responsible for the binding of the former to potassium channels. Surprisingly, site-directed mutagenesis experiments demonstrated that these residues are in fact not essential for binding (Danse et al. 1994). It was subsequently observed that acetylation of Lys5 results in the loss of potassium-channel binding activity, suggesting that this is the functional site for α-dendrotoxin (Harvey et al. 1997). Site-directed mutagenesis and chemical synthesis experiments on α-dendrotoxin also suggest that the functional site of α-dendrotoxin is at the N-terminal region and that the most important residues in this interaction are Lys5 and Leu9 (Gasparini et al. 1998).

Dendroaspis venoms not only include TV-kunitz peptides with potassium-channel inhibiting activity but also uniquely contain additional apotypic forms that block voltage-gated calcium channels (Schweitz et al. 1994), exemplifying the functional plasticity of this peptide type. Calcicludin (P81658) is a TV-kunitz peptide isolated from *D. angusticeps* venom that blocks L-, N-, and P-type Ca^{2+} channels (Schweitz et al. 1994). It has greatest affinity for L-type channels, with reported IC_{50} values as low as 0.2 nM (depending on the cell type), and has no affinity for dendrotoxin-sensitive K^+ channels. Calcicludin was the first peptide with a TV-kunitz-type protease inhibitor domain determined to block

Ca^{2+} channels. It was previously believed that only proteins with either a three-stranded β-sheet and cysteine-knot motif or the three-fingered scaffold were capable of blocking Ca^{2+} channels. The three-dimensional structure of calcicludin was determined by NMR spectroscopy (Gilquin et al. 1999) and is structurally similar to all other TV-kunitz-BPTI-type protease inhibitors. However, a difference in structure at the N-terminus of the protein is evident between calcicludin and the dendrotoxins. A comparison of N-terminal amino acid sequences confirms that the major differences between these two forms occur in this region. Of most significance is that Lys5 and Leu9, which appear to be important for dendrotoxin binding to potassium channels, are replaced by tryptophan and glutamate, respectively, in calcicludin. Based on these differences, it is hypothesized that the N-terminal region of calcicludin is responsible for binding to Ca^{2+} channels (Nishio et al. 1999). Calcicludin binding to the outer vestibule of L-type Ca^{2+} channels may promote structural changes in the pore that stabilize the channel, bracing it in a nonconducting blocked state (Wang, Du, and Peterson 2007).

Snake-venom TV-kunitz peptides can also occur as components of larger multi-subunit complexes. Two such complexes that have been widely studied are β-bungarotoxins and taicatoxins (see chapter 20) with other complexes also known but lesser studied.

The β-bungarotoxin was first isolated from the venom of the elapid snake *Bungarus multicinctus* and has presynaptic neurotoxic properties (Chang and Lee 1963). It consists of two dissimilar polypeptide chains (referred to as A- and B-chains) that are cross-linked by a disulfide bond (Kondo, Narita, and Lee 1978a; Kondo, Narita, and Lee 1978b). The A-chain is structurally homologous to phospholipase A_2 (PLA_2), while the B-chain is a TV-kunitz peptide consisting of 61 amino acid residues (Wu et al. 1998). The B-chain has an extra cysteine residue (Cys55) not normally observed in TV-kunitz peptides. It is this residue that cross-links with Cys15 of the A-chain forming the interchain disulfide bridge and thus the heterodimer (Wu and Chang 2001). It is hypothesized that the A-chain acts as a presynaptic neurotoxin, while the B-chain is involved in the binding of specific targets on the cell membrane, likely blocking voltage-dependent potassium channels (Kondo, Narita, and Lee 1978a; Kondo, Narita, and Lee 1978b; Kini and Iwanaga 1986; Benishin 1990; Wu et al. 1998). Although more than 16 β-bungarotoxins isoforms have been isolated from *B. multicinctus* venom alone (Kondo et al. 1982), to date only eight of them have been studied in detail: β_1- to β_5-bungarotoxin and SPI-III (Kondo, Narita, and Lee 1978a; Kondo, Narita, and Lee 1978b; Kondo et al. 1982; Chu et al. 1994). The X-ray crystal structure of β_2-bungarotoxin has been determined and is composed of chain A_1 (P00617) and chain B_2 (P00989) (Kwong et al. 1995). Chain B_2 possesses a TV-kunitz-BPTI domain but does not exhibit protease-inhibitory activity. The amino acid sequence contains a glutamine in the P_1 position of the canonical loop, making it unlikely to be an inhibitor of trypsin-like proteases. None of the common amino acids at the N-terminus that are generally associated with the blocking of ion channels is present.

Lesser known complexes have also been discovered. Taicatoxins are composed of three noncovalently linked subunits in a 1:1:4 stoichiometric ratio: (i) 3FTx (see chapter 8), (ii) neurotoxic PLA_2 (see chapter 20), and (iii) kunitz peptides. It has been demonstrated that this oligomeric complex is capable of blocking Ca^{2+} channels (Brown et al. 1987). MitTx, a heteromeric TV-PLA_2-GI complex made up of a kunitz peptide and a TV-PLA_2-GI (see chapter 20). This complex acts as a powerful agonist on acid-sensing ion channels (ASICs), inducing excruciating pain (Bohlen et al. 2011).

15.5 THERAPEUTIC POTENTIAL OF THE TOXIN TYPE

Based on its specific inhibition of and binding affinity for plasmin, textilinin-1 may be suitable as an alternative antihemorrhagic treatment to BPTI (Trasylol), which had until recently been the medication of choice to control blood loss during cardiac surgery. Data collected from the "Blood Conservation Using Antifibrinolytics in a Randomized Trial" (BART) study demonstrated an increased rate of death in patients receiving Trasylol compared with those receiving lysine analogues (tranexamic acid or aminocaproic acid) during high-risk cardiac surgery (Fergusson et al. 2008). As a result of these findings, the Food and Drug Administration (FDA) has suspended the use of this drug in such surgeries in the United States and in many other countries.

Some nonvenom kunitz peptides have been shown to prevent inflammation and tissue injury and therefore may promote tissue remodeling (Shigetomi et al. 2010). Trypsin-dependent activities of epithelial Na^+ ion channels in the kidney, gut, and airway, which are associated with embryonic development, organogenesis, and maintaining blood pressure and volume, are influenced by BPTI and bikunin, and thus TV-kunitz could be therapeutically useful in these contexts (Planes and Caughey 2007). With an enhanced understanding of the structure-function relationships of the active components of the TV-kunitz peptides in snake venoms, there is great potential to develop new therapeutics to treat a broad range of diseases.

The development of ecallantide (KALBITOR) (Williams and Baird 2003) is a prime example that highlights the therapeutic potential of the molecular scaffolds on which animal toxins are built. As noted above and similar to many other venom toxin superfamilies, TV-kunitz scaffolds can tolerate a wide variety of amino acid substitutions, which may confer novel biological activities to the molecule. The template for the development of ecallantide was a TV-kunitz scaffold present in the human lipoprotein-associated coagulation inhibitor (LACI), also known as tissue-factor pathway inhibitor (TFPI) (Ley, Markland, and Ladner 1996). Iterative amino acid variations were introduced in the region that corresponds to the BPTI-trypsin recognition site, while the parts of the molecule responsible for scaffold integrity (such as disulfide patterns) were left intact. Employing a phage-display system, the molecules with the greatest affinity for human plasma kallikrein were selected from the resulting libraries. The ultimate result was DX-88 (the investigational name for ecallantide), a selective and reversible inhibitor of human kallikrein. Ecallantide is a 60 amino acid residue TV-kunitz scaffold differing in only seven residues from the template LACI domain. Upon binding to kallikrein, ecallantide inhibits the generation of bradykinin from kininogen. The US FDA approved it in 2009 for the treatment of acute attacks of hereditary angioedema, a rare but life-threatening genetic disorder linked to the kallikrein-kinin cascade (Zuraw, Yasothan, and Kirkpatrick 2010). Ecallantide is produced in the yeast *Pichia pastori*.

15.6 CONVERGENCE WITH OTHER VENOMS

Kunitz peptides are a particularly fascinating toxin type. They represent an extraordinary example of dual convergence: not only have they been independently recruited into the venom arsenal of numerous organisms (see color plate 2), but they are also convergently apotypic for the same neurotoxic and coagulopathic activities in multiple lineages (Beress 1982; Antuch et al. 1993; Minagawa et al. 1997; Harvey and Robertson 2004; Bayrhuber et al. 2005; Hisada et al. 2005; Honma and Shiomi 2006; Fry et al. 2008; Koludarov et al. 2012). Kunitz-type toxins belong to a superfamily of protease inhibitors and are short α/β proteins constrained by three disulfide bridges. The three-dimensional fold has been conserved during the derivation of novel activities. Two apotypic sites of action (potassium channels and coagulation factor X) have been convergently targeted on several occasions (see color plate 2). The functional convergence of *Dendroaspis* kunitz peptides on calcium and potassium channels with 3FTx (see chapter 8.4.2) also from *Dendroaspis* venoms is particularly notable (see chapter 1.3.5).

Apotypic kunitz peptides from the venom of cone snails (Bayrhuber et al. 2005), sea anemones (Schweitz et al. 1995), snakes (Harvey and Robertson 2004), and spiders (Yuan et al. 2008) convergently block potassium channels. This is not a plesiotypic activity but is convergently apotypic. Conspicuously, despite significant overall sequence variation, the toxins share a motif in which the minimum ion-channel pharmacophore is defined by a critical basic residue (usually a lysine) located 7 Å from a hydrophobic residue. This functional site, assisted by various constellations of "secondary" residues for K^+ channel binding, is quite distinct from the protease inhibition site. The same fold is therefore being used for two distinct functions, but different side chains of the molecule are involved in each. Sea anemones contain a range of kunitz peptides (Beress 1982; Antuch et al. 1993; Minagawa et al. 1997; Honma and Shiomi 2006), some of which are also dual-action polypeptides (Schweitz et al. 1995). For example, kalicludines 1-3, also known as AsKC1-3, not only inhibit trypsin but also block K^+ ion channels. Compared with the dendrotoxins (described above), they are weaker inhibitors of $K_v1.2$, and compared with BPTI, they are less potent trypsin inhibitors. *Conus striatus* venom contains a kunitz-type neurotoxin that shares 33% sequence identity with BPTI and 35% with dendrotoxin I. This toxin, named Cn-kunitzin-S1, blocks Shaker potassium channels with an IC_{50} of about 60 nM. The kunitz-type toxins in mygalomorph spider venoms also have a dual activity, acting as both trypsin inhibitors and K_v ion channel blockers (Yuan et al. 2008). For example, huwentoxin-XI, a kunitz-type toxin from the Chinese tarantula *Ornithoctonus huwena*, competitively inhibits trypsin with a K_i value of 68 nM (Liang 2004) and is also a weak blocker of $K_v1.1$ ion channels. The epitopes responsible for protease inhibition and blockage of K_v ion channels are spatially distinct and located at opposite ends of the cone-shaped molecule (Yuan et al. 2008).

Kunitz peptides are also a major constituent of the secretions of ticks and some insects, arthropod groups that have independently evolved hematophagy. The kunitz peptides in the secretions of these discrete lineages inhibit coagulation factor Xa and are convergent in this activity both with those of each other and with those of snakes (Wei et al. 1998; Campbell et al. 2005). A diversity of apotypic kunitz scaffolds occur within tick secretions, including representatives with one, two, or five kunitz domains, which have been named the monolaris (which includes three distinct subgroups), bilaris (with four distinct clades), and penthalaris groups (Francischetti et al. 2002; Ribeiro et al. 2006). The bilaris kunitz peptides also have 15 to 23 putative O-galactosylation sites that may aid in targeting and protection from degradation by carboxypeptidases. The bilaris protein ixolaris is a TFPI mimic (Francischetti et al. 2002). This protein uses fX as a scaffold, binding to the active site through one kunitz domain. This prevents diffusional loss, and when factor X docks on the factor VIIa/TF complex, the second kunitz domain binds to the active site of factor VIIa. As a result, ixolaris is active at picomolar concentrations. A different protein, ornithodorin, from the argasid tick *Ornithodoros moubata*, also has two kunitz domains, but, in contrast to ixolaris, it functions as a slow, tight-binding inhibitor of thrombin (van de Locht et al. 1996). The crystal structure of the ornithodorin-thrombin complex reveals, surprisingly, that neither kunitz fold is involved in binding to thrombin. Rather, the C-terminal region interacts with the fibrinogen recognition exosite, and a series of residues near the N-terminus interact with the catalytic cleft. Neither kunitz "reactive site loop" contacts the substrate. A similar situation occurs with tick anticoagulant peptide (TAP) binding factor Xa (Wei et al. 1998). A number of other double-kunitz domain proteins have been identified in the secretory-gland transcriptome of a hematophagous nematoceran fly, *Culicoides sonorensis* (Campbell et al. 2005). Although the function of these proteins has not been experimentally demonstrated, three of them (TFPI1–TFPI3) are presumed to be TFPIs, based on the pronounced anti-factor-Xa activity of the secretion of this species and their apparent orthology (32% identity and 38% similarity) with known anti-factor-Xa proteins of tick feeding secretion (Francischetti, Andersen, and Ribeiro 2002).

Kunitz peptides have also been recovered from the venom of the vampire bat *Desmodus rotundus* (Francischetti et al. 2013; Low et al. 2013). Through transcriptome analysis, the precursors of these peptides were found to be multidomain. Nucleotide and amino acid level selection assessments evaluating the regime of natural selection influencing the evolution of these vampire bat kunitz homologues indicated that they have rapidly accumulated mutations under the influence of positive Darwinian selection, and a majority of these hypermutable sites were located on the molecular surface (Low et al. 2013). It was thus revealed that these toxins evolve rapidly under the influence of focal mutagenesis and accumulate a large proportion of mutations on the molecular surface and in structurally and/or functionally unimportant regions. It has been suggested that focal mutagenesis not only aids in the conservation of the structural and functional integrity of these toxins but may also prevent or delay the mounting of immunological resistance against these molecular scaffolds in prey animals. Prey animals of *D. rotundus* have been shown to develop immunity to venom components such as draculin if targeted and fed upon over prolonged periods (Delpietro and Russo 2009), and therefore, mechanisms for evading the immune system of prey animals are likely to be selected for.

CHAPTER 16

L-AMINO ACID OXIDASE ENZYMES

N. H. TAN, B. G. FRY, K. SUNAGAR, T. N. W. JACKSON, T. REEKS, AND S. Y. FUNG

16.1 SUMMARY

L-amino acid oxidases (LAAOs) catalyze the oxidative deamination of an L-amino acid to produce the corresponding α-keto acid. In the process, hydrogen peroxide (H_2O_2) and ammonia are generated. Snake venoms are a rich source of LAAOs. The enzyme is a flavoprotein with two identical subunits, and MW ranging from 110 kDa to 140 kDa. LAAOs from different snake venoms are highly similar in terms of their amino acid sequence, and hence they may possess similar three-dimensional structures. X-ray diffraction studies have indicated that each subunit of these enzymes consists of three well-defined domains: the FAD-binding domain, the substrate-binding domain, and a helical domain that provides access to the active site of the enzyme. The roles of snake venom LAAOs (SV-LAAOs) are not yet completely understood. However, a cytotoxic action, caused mainly as a result of H_2O_2 generated by the oxidative action of the enzyme, may contribute to hemorrhagic and necrotic effects of the venom. In addition, H_2O_2 can trigger inflammation and also result in blood clots. Recent studies have revealed that SV-LAAOs exhibit several interesting pharmacological activities with potential therapeutic applications. The pharmacological effects are mainly mediated by the H_2O_2 generated, and the binding of the enzyme to the target cells appears to play an important role in its cytotoxicity. Effects include platelet aggregation induction, platelet aggregation inhibition, and antibacterial, antiproliferative, antiparasitic, and antiviral (HIV) activities. Although the practicality of therapeutic applications of this enzyme is yet to be fully demonstrated, its potential usefulness as a molecular probe for investigating cellular processes and the utilization of its cytotoxic action against certain cells should not be underestimated.

16.2 ENDOPHYSIOLOGICAL PLESIOTYPE

L-amino acid oxidase (L-amino acid: O_2 oxidoreductase, EC 1.4.3.2) is a dimeric protein containing FAD as a prosthetic group and belonging to the flavin monoamine oxidase family. The N-terminal sequence contains a common βαβ-fold, which is the binding site of FAD. LAAO is a glycoprotein, with nearly 3% to 4% of its mass made up of carbohydrate. Glycosylation appears to be important for the secretion and solubility of the protein (Geyer et al. 2001). Deglycosylation does not, however, appear to affect enzymatic activity (Stabeli et al. 2004; Izidoro et al. 2006; Rodrigues et al. 2009). The role of the glycans in the pharmacological activities of the enzyme is, therefore, still not fully understood. This enzyme catalyzes the oxidative deamination of an L-amino acid to produce the corresponding α-keto acid, ammonia, and hydrogen peroxide. The oxidation of L-amino acids by the enzyme proceeds in two steps, resulting in the formation of an α-imino acid as the intermediate product (Tan and Fung 2009). A mechanism of direct hydride transfer from the amino acid to FAD during the reaction has been supported by a recent study (Chen et al. 2012b). LAAOs from different sources differ substantially in their specific activity; however, they generally possess similar substrate specificity. Typically, the preferred substrates are hydrophobic amino acids. Substrate-specificity studies indicated the presence of an alkyl side-chain binding site which includes at least four subsites, each accommodating a methyl/methylene carbon:

$$RCH(NH_3^+)COO^- + O_2 + H_2O \rightarrow RCOCOO^- + NH_4^+ + H_2O$$

16.3 PLESIOTYPIC STRUCTURAL AND FUNCTIONAL TOXIN FORMS

SV-LAAOs are flavoproteins with one flavin coenzyme per subunit; the flavins are responsible for the yellowish color of the enzyme (Tan 1998). The enzymes are generally stable at room temperature and at 4° C but are unstable in alkaline conditions and generally reversibly inactivated by freezing (Tan and Fung 2009). SV-LAAOs generally have a MW ranging from 110 kDa to 140 kDa. The enzyme is a dimer with two identical subunits. The complete amino acid sequences of several snake-venom L-amino acid oxidases have been elucidated by deduction from cDNA sequences (Raibekas and Massey 1998; Macheroux et al. 2001; Takatsuka et al. 2001; Zhang et al. 2003; Franca et al. 2007; Chen et al. 2012b). All sequences elucidated so far are highly similar, displaying a pattern of sequence conservation typical of large globular venom proteins. High-resolution three-dimensional structures show that each subunit of the enzyme consists of 15 α-helices and 22 β-strands that fold into three well-defined domains: the FAD-binding, substrate-binding, and helical domains (Zhang et al. 2004a; Moustafa et al. 2006; Alves et al. 2011; Georgieva et al. 2011). The motif of the FAD-binding domain constitutes the classic nucleotide-binding fold seen in many FAD-binding enzymes. The interface between the substrate-binding and helical domains forms a 25-Å-long funnel, which provides access to the active site.

Snake venoms are rich sources of SV-LAAOs, which may constitute as much as 30%, or as little as 1%, of the total venom protein. The roles of SV-LAAOs in the pathophysiological actions of snake venoms are not clearly understood, but the enzyme has recently attracted attention because it exhibits a wide range of pharmacological activities that have therapeutic potential. The enzymatic and pharmacological activities of SV-LAAOs have been extensively reviewed (Du and Clemetson 2002; Tan and Fung 2009; Zuliani et al. 2009). It has been suggested that SV-LAAO-induced apoptosis is facilitated by promoting the generation of intracellular reactive oxygen species and then p53 protein expression (Sun et al. 2003). It has also been suggested that apoptosis could occur as a result of oxidative stress induced by H_2O_2 that is generated by SV-LAAOs (Tempone et al. 2001). The oxidative stress could then activate heat-shock proteins and initiate cell-membrane disorganization and then apoptosis. It is likely that the mitochondrial pathway plays an important role in the apoptosis induced by a SV-LAAO from *Deinagkistrodon acutus* venom (Zhang and Wei 2007), with the apoptosis mediated by caspase (Alves et al. 2008).

More recently, the antiproliferative effect of a SV-LAAO from *Bothrops atrox* venom was demonstrated to be accompanied by an accumulation of cells in the G_0/G_1 phase boundary in JL-60 cells, indicating that this SV-LAAO blocked G_0-G_1 transition, resulting in G_0/G_1 phase cell cycle arrest (Alves et al. 2011). Recently, it has been reported that injection of a SV-LAAO from *Gloydius ussuriensis* venom induced lung lesions (pneumorrhagia) and also liver and cardiac damage (cardiac interstitial edema and bleeding) in mice, at a dose of 0.25 mg/kg (Wei et al. 2007). A previous report demonstrated that a SV-LAAO isolated from *Bothrops insularis* venom caused acute tubular necrosis in rats (Braga et al. 2008). These effects may be related to the cytotoxic action of the enzyme. SV-LAAOs also exhibit potent rat paw edema-inducing activity (Tan and Fung 2009), indicating that the enzyme contributes to the inflammatory activity of the venom. The hemorrhagic effect of SV-LAAOs may result from the cytotoxic activity of the enzyme on endothelial and other vascular cells. Both edema-inducing and hemorrhagic actions of SV-LAAOs are likely related to the generation of H_2O_2 by the action of the enzyme on L-amino acid substrates (Du and Clemetson 2002).

The results of one study indicated that deglycosylation of a SV-LAAO from *Ophiophagus hannah* by endo F enzyme does not affect the cytotoxicity of the enzyme (Ahn, Lee, and Kim 1997). On the other hand, a separate study reported that while the secreted recombinant apoxin 1 (a SV-LAAO from *Crotalus atrox* venom) exhibited strong cytotoxicity, the recombinant protein in cells (such as without glycosylation) did not possess cytotoxic activity (Torii et al. 2000). In addition, a separate study demonstrated that desialylation of the enzyme reduced its cytotoxicity by 20% (Ande et al. 2006). Clearly, more investigations are needed to firmly establish the role of glycans in the various pharmacological activities of SV-LAAOs.

The intravenous LD_{50} of SV-LAAOs is approximately 5 to 9 μg/g, which is higher than the intravenous LD_{50} of most snake venoms, and so the enzyme is unlikely to play a major role in the toxic action of the venom. It may, however, potentiate the action of other toxins by contributing to the hemorrhagic and inflammatory effects, and the blood-clotting defects, caused by snake venoms.

16.4 APOTYPIC STRUCTURAL AND FUNCTIONAL TOXIN FORMS

It has been reported that a SV-LAAO purified from *Gloydius blomhoffi* venom possessed anticoagulant activity (Sakurai et al. 2003). However, further investigation of the anticoagulant effect of the enzyme suggested that the anticoagulant activity was actually caused by a metalloprotease, which was present as a minor contaminant in the SV-LAAO preparation (Fujisawa, Yamazaki, and Morita 2009).

SV-LAAOs from various sources (such as *Bothrops alternatus, Bothrops pirajai, Crotalus durissus cascavella*, and *Eristocophis macmahoni*) exhibited platelet aggregation-inducing activity with ED_{50} values ranging from 30–50 μg/mL (Ali et al. 2000; Stabeli et al. 2004; Izidoro et al. 2006; Toyama et al. 2006). Platelet aggregation-inducing activity was inhibited by catalase, a H_2O_2 scavenger, indicating that H_2O_2 is involved in the platelet-aggregating action of the enzyme. It has been suggested that the H_2O_2 generated by the enzyme promotes elevated thromboxane A_2 synthesis that in turn induces platelet aggregation (Li, Yu, and Lian 1994).

On the other hand, the SV-LAAOs from many venoms (such as the viperid snake species *Bothrops leucurus, Daboia russelii, G. blomhoffi, G. ussuriensis, Macrovipera lebetina*, and the elapid snake species *Naja kaouthia*) dose-dependently inhibits both agonist-induced (using ADP) platelet aggregation and shear-induced platelet aggregation (Sakurai et al. 2001; Takatsuka et al. 2001; Sun et al. 2003; Tonismagi et al. 2006; Naumann et al. 2011; Chen et al. 2012b). For example, a SV-LAAO from the venom of *D. russelii* venom inhibited ADP- and collagen-induced platelet aggregation with an IC_{50} of 0.27 (34 μg/mL) and 0.82 μM (103 μg/mL), respectively (Chen et al. 2012b). The platelet-aggregation inhibitory activity of the enzyme was also inhibited by catalase, indicating that H_2O_2 is also involved in the platelet-aggregation inhibitory activity. This activity may be related to the reduced binding sites for ADP in platelets exposed to H_2O_2, or interference in the interaction between activated platelet integrin GPIIb/IIIa and fibrinogen (Raibekas and Massey 1998; Samel et al. 2006).

Thus, investigations of the effects of SV-LAAOs on platelet aggregation seem to yield contradictory results: while some authors report that the enzyme induces platelet aggregation, others report that it inhibits platelet aggregation. The discrepancy might be a result of differences in specific activity of the enzyme. However, this is not likely, as different laboratories working on the effects on platelet aggregation of the same SV-LAAO isolated from *O. hannah* venom also yielded contradictory results. Initially, it was reported that *O. hannah* SV-LAAO induced platelet aggregation in platelet-rich plasma (PRP), with an ED_{50} of 50 μg/mL (Li et al. 1994). In a more recent study, however, SV-LAAO isolated from the same species and from the same geographical region of China, did not induce platelet aggregation in PRP even at a concentration of 100 μg/mL (Jin et al. 2007). On the contrary, the enzyme inhibited platelet aggregation induced by ADP dose-dependently, at venom concentrations ranging from 4.8 μg/mL to 48 μg/mL, and the IC_{50} was determined to be 0.15 μM (approximately 18 μg/mL).

What, then, is the reason for this discrepancy? There are various possibilities: differences in experimental procedures (including the preparation of blood samples), or the involvement of mechanisms other than H_2O_2 in the pharmacological actions of certain SV-LAAOs (Takatsuka et al. 2001), or, as has already been proven in one case, contamination by other toxin types. To answer this question, collaborative investigations in separate laboratories using various SV-LAAOs under standardized experimental conditions are required.

16.5 THERAPEUTIC POTENTIAL OF THE TOXIN TYPE

16.5.1 ANTIBACTERIAL ACTIVITY OF SV-LAAOS

The antibacterial activities of SV-LAAOs are well known (see Tan and Fung 2009). Table 16.1 summarizes the antibacterial activity of some SV-LAAOs. The enzymes inhibit a wide spectrum of bacteria, both Gram-positive [G(+)] and Gram-negative [G(–)] with MIC (minimum inhibitory concentration) ranges from 0.78 μg/mL to 288 μg/mL. *O. hannah* SV-LAAO possesses the most potent antibacterial activity. It has MIC values against the G(+) bacteria *Staphylococcus aureus* and *S. epidermidis* (0.78 μg/mL and 1.56 μg/mL, respectively) comparable (by weight) to those of common antibiotics such as cefotaxime and vancomycin, and more potent (by molarity) than penicillin (Lee et al. 2011). Table 16.1 also shows that SV-LAAOs from different venom sources exhibit different selectivity in their antibacterial action. For example, SV-LAAOs from *Bothrops marajoensis, Daboia siamensis, Crotalus durissus cascavella*, and *O. hannah* venom

Table 16.1: Antibacterial Actions of Some Snake-Venom L-Amino Acid Oxidases

Venom source	Antibacterial actions	Reference
Bothrops marajoensis	Active against G(+) *Staphylococcus aureus* (most active, MIC = 50 μg/mL), G(+) *Candida albicans*, and G(–) *Pseudomonas aeruginosa*	Costa Torres et al. 2010
Bothrops moojeni	Almost equally active against G(–) *Escherichia coli*, G(–) *Salmonella typhimurium*, G(–) *Pseudomonas aeruginosa*, and G(+) *Staphylococcus aureus*	Stabeli et al. 2007
Bothrops pauloensis	More active against G(–) *Escherichia coli* than against G(+) *Staphylococcus aureus*	Rodrigues et al. 2009
Crotalus durissus cascavella	More active against G(+) *Staphylococcus mutans* than against G(–) *Xanthomonas axonopodis pv passiflorae*	Toyama et al. 2006
Daboia siamensis	Active against G(+)*Staphylococcus aureus* (MIC = 9.0 μg/mL), G(–) *Pseudomonas aeruginosa* (MIC = 144 μg/mL), and G(–) *Escherichia coli* (MIC = 288 μg/mL)	Zhong et al. 2009
Gloydius halys	Almost equally active against G(–) *Escherichia coli* and G(–) *Bacillus subtilis*	Zhang et al. 2004b
Macrovipera lebetina	More active against G(–) *Escherichia coli* than against G(+) *Bacillus subtilis*	Tonismagi et al. 2006
Naja oxiana	Slightly more active against G(+) *Bacillus subtilis* than against G(–) *Escherichia coli*	Samel et al. 2008
Ophiophagus hannah	More active against G(+) *Staphylococcus aureus* (MIC = 0.78 μg/mL) and *Staphylococcus epidermidis* (MIC = 1.56 μg/mL) than against G(–) *Escherichia coli* (MIC = 50 μg/mL) and *Pseudomonas aeruginosa* (MIC = 25 μg/mL)	Lee et al. 2011

are more potent against G(+) bacteria than against G(–) bacteria, whereas SV-LAAOs from the venoms of *Bothrops pauloensis, M. lebetina*, and *Naja oxiana* are more potent against G(–). On the other hand, SV-LAAOs from *Bothrops moojeni* and *Gloydius halys* inhibit G(+) and G(–) bacteria almost equally. These differences in selectivity of the antibacterial action of SV-LAAOs are probably related to differences in bacteria-binding affinity (Lee et al. 2011).

The antibacterial mechanism of SV-LAAOs involves the H_2O_2 generated as a result of the oxidation of L-amino acid substrates (Tan and Fung 2009). Treatment with SV-LAAOs results in the rupture of bacterial membranes and the extravasation of plasma contents (Toyama et al. 2006). SV-LAAOs also cause cytolysis of common reptilian and amphibian pathogens such as *Aeromonas hydrophilia* (Stiles, Sexton, and Weinstein 1991).

It is likely that binding of SV-LAAOs to bacteria is important for the antibacterial action of the enzyme, as this results in a localized concentration of H_2O_2 that is sufficiently potent to kill the bacterium (see Ehara et al. 2002; Kitani et al. 2008). In a recent study, it was estimated that 1,800 molecules of *O. hannah* SV-LAAO were bound per G(+) bacterial cell, but only 1,100 molecules of the enzyme were bound per G(–) *Escherichia coli* or *Pseudomonas aeruginosa* cell. Thus, the greater affinity of *O. hannah* SV-LAAO for G(+) bacteria than for G(–) bacteria could explain the greater potency of the enzyme against the former (Lee et al. 2011).

Some of the SV-LAAOs with highly potent antibacterial action may have the potential to be developed into therapeutic antibacterial agents. Despite the fact that the enzyme is cytotoxic, animal experiments indicate that its lethal toxicity is generally very low (Tan and Fung 2009), and therefore, this should not present a barrier to its therapeutic use. Microorganisms are generally several-fold more sensitive to reactive oxygen species than are human cells (Sorg 2004). Indeed, it has been argued that there is a "bactericidal window" in which the concentration of reactive oxygen species generated by the enzyme is sufficient to stop bacterial growth without harming host (human) cells (Ciscotto et al. 2009). However, the therapeutic application of SV-LAAOs as bactericides will always be limited because of the proteinaceous nature and relatively high cost of the enzyme (purified from venoms or as cloned products). Nevertheless, an understanding of the antibacterial action of SV-LAAOs via H_2O_2 generation may enable the designing of new drugs and/or therapeutic approaches.

16.5.2 ANTIPARASITIC ACTIVITY OF SV-LAAOS

A number of authors have reported the antiparasitic activity of SV-LAAOs, including its inhibitory actions against the promastigote (but not amastigote) of various *Leishmania* species (Tempone et al. 2001; Toyama et al. 2006; Ciscotto et al. 2009; Alves et al. 2011; Naumann et al. 2011) and *Trypanosoma cruzi* (Deolindo et al. 2010). *Leishmania* species are the cause of Leishmaniasis, a potentially lethal human infectious disease that affects as many as 12 million people a year and is prevalent in perhaps half of the world's nations. *T. cruzi* is the cause of Chagas disease, which affects 16 million to 18 million people a year in Latin America. The H_2O_2 generated by the enzyme induces apoptosis in promastigotes of *Leishmania* species cells, with an IC_{50} of approximately 1 μg/mL against *Leishmania amazonensis, L. panamensis*, and *L. chagasi*. The likely mode of action is that H_2O_2 induces oxidative stress in the promastigote, which leads to activation of heat-shock proteins and eventually cell death (Tempone et al. 2001). On the other hand, amastigotes, which contain much more catalase and glutathione oxidase, are more resistant to the action of the enzyme.

The H_2O_2 generated may also be responsible for the induction of programmed cell death in *T. cruzi* (Deolindo et al. 2010). At present, there are not many drugs that are effective in the treatment of Leishmaniasis. Gaining an understanding of the mode of action of SV-LAAOs on parasites may aid in the design of new drugs or therapeutic approaches for the treatment of parasite-induced diseases.

16.5.3 ANTIPROLIFERATIVE ACTIVITY OF SV-LAAOS

Many SV-LAAOs exhibit antiproliferative activity against cancer cells (Souza et al. 1999; Takatsuka et al. 2001; Sun et al. 2003; Ande et al. 2006; Izidoro et al. 2006; Samel et al. 2006; Naumann et al. 2011). Table 16.2 lists the antiproliferative potency (IC_{50}) of some SV-LAAOs. The IC_{50} after 24-hour incubation ranges from 0.3 to 25 μg/mL. The general consensus has been that the antiproliferative activity of the enzyme results from its ability to induce

Table 16.2: Antiproliferative Activity of Snake Venom L-Amino Acid Oxidase

Cell line	IC_{50} (incubation time)	Source	Reference
Human lung cancer cells A549	10 μg/mL (24 hr)	*Deinagkistrodon acutus*	Zhang and Cui 2007
Human cervical cancer cells (Hela)	10 μg/mL (48 hr)	*Deinagkistrodon acutus*	Zhang and Cui 2007
Malignant glioma cells	1.9 μg/mL (48 hr)	*Protobothrops flavoviridis*	Sun et al. 2003
C6 cells	2.5 μg/mL (48 hr)		
RBR 17T cells	2.1 μg/mL (48 hr)		
U251 cells			
Ehrlich ascites tumor cells	5 μg/mL (1 hr)	*Bothrops moojeni*	Stabeli et al. 2007
HL60 cells	~50 μg/mL (24 hr)	*Bothrops atrox*	Alves et al. 2008
Jurkat cells	25 μg/mL (24 hr)		
B16F10 cells	25 μg/mL (24 hr)		
PC12 cells	25 μg/mL (24 hr)		
Human breast cancer cells, SKBR-3 cells, Jurkat cells, Erlich ascitic tumor (EAC) cells	~0.5 μg/mL (24 hr)	*Bothrops pauloensis*	Rodrigues et al. 2009
Human lung cancer cells A549	1.24 μg/mL (12 hr)	*Bungarus fasciatus*	Wei et al. 2009
Murine melanoma B16/F10 cells	0.17 μg/mL (24 hr)	*Ophiophagus hannah*	Ahn, Lee, and Kim 1997
Human fibrosarcoma HT-1080 cells	0.6 μg/mL (24 hr)		
Chinese hamster ovary (CHO) cells	0.3 μg/mL (24 hr)		
Murine epithelial Balb/3T3 cells	0.45 μg/mL (24 hr)		

apoptosis. Apoptosis-inducing activity is abolished by catalase and other H_2O_2 scavengers, indicating that H_2O_2 (generated by SV-LAAO action) plays an important role in this activity. One report, however, suggested that SV-LAAOs caused cell death by apoptosis only when the enzyme was at low concentrations (0.0015 μM, or 0.18 μg/mL) and that at higher concentrations (0.003–0.03 μM, or approximately 0.36–3.6 μg/mL), necrosis is the main cause of cell death (though apoptosis continues to occur) (Ande et al. 2006). It also suggested that SV-LAAO-induced apoptosis appears to be mediated by the glycan moiety of the enzyme, as desialylation reduces cytotoxicity by 20%. Moreover, D-amino acid oxidase, which does not have glycan moiety, could trigger necrosis (which is mainly caused by the H_2O_2 generated) but not apoptosis. While the antiproliferative effect of SV-LAAOs on cell lines has been demonstrated, the therapeutic benefit of SV-LAAOs in cancer treatment is yet to be established, as the *in vivo* antiproliferative effect of the enzyme on solid tumor growth has yet to be investigated. However, SV-LAAOs have been proposed to have practical applications in the clinical treatment of glioma (Sun et al. 2003).

16.5.4 ANTI-HIV ACTIVITY OF SV-LAAOS

A SV-LAAO isolated from the venom of *Viridovipera stejnegeri* exhibits dose-dependent inhibition of HIV-1 infection and replication, with EC_{50} values of 1.5 nM (0.19 μg/mL) and 10.6 nM (1.3 μg/mL), respectively (Zhang et al. 2003). This EC_{50} is 16 and 6 times lower, respectively, than the 50% cytotoxic concentrations of the enzyme against C8166 cells. Antiviral activity is inhibited by catalase, indicating that H_2O_2 generated by SV-LAAO action on amino acid substrates plays an important role. However, under the same conditions, exogenous addition of H_2O_2 does not exhibit anti-HIV-1 activity, indicating that binding of SV-LAAOs to the cell membrane is necessary for antiviral action. The mechanism of the antiviral action of SV-LAAOs remains to be clearly elucidated; however, H_2O_2 may act as a secondary messenger, triggering signal reactions and activation of host cells to inhibit HIV infection and/or replication. The anti-HIV activity of SV-LAAOs may contribute to development of new generations of anti-HIV drugs. Unfortunately, there has been little progress in this area since 2003 (Zhang et al. 2003).

16.6 CONVERGENCE WITH OTHER VENOMS

None is documented.

CHAPTER 17

LECTIN PROTEINS

F. T. ARLINGHAUS, B. G. FRY, K. SUNAGAR, T. N. W. JACKSON, J. A. EBLE, T. REEKS, AND K. J. CLEMETSON

17.1 SUMMARY

C-type lectins are among the largest protein families in mammals and reptiles and are involved in defense mechanisms and a variety of other functions. Lectin scaffold (TV-lectin) toxins enrich toxicoferan venoms, particularly those from viperid snakes. The lectin scaffold was recruited into the venom arsenal at the base of the toxicoferan reptile tree, and the homomeric, carbohydrate-binding form is the plesiotypic state. After the viperid snakes split off from the remaining caenophidian snakes, a lectin subtype was mutated in their venoms to remove the carbohydrate-binding ability. These heterodimeric apotypic lectin toxins are called snake venom C-type lectin-like proteins (snaclecs) to recognize their structural and functional derivations. Since they attack targets that are often otherwise difficult to inhibit, they have considerable potential not only in drug design and development and biochemical assays but also in the identification of potential therapeutic targets. Besides in snakes, toxic lectins have also been convergently recruited into the venoms of fish and caterpillars.

17.2 ENDOPHYSIOLOGICAL PLESIOTYPE

Lectins are a large family of carbohydrate-binding proteins and glycoproteins, which exist in many structural and functional variants within the body. The Ca^{2+}-dependent (C-type) lectins, which include the snake-venom C-type lectins, are nonenzymatic proteins that can affect cell adhesion, endocytosis, or pathogen neutralization (Ogawa et al. 2005; Crocker, Paulson, and Varki 2007; van den Berg, Gringhuis, and Geijtenbeek 2012). The C-type lectin domain (CTLD) includes the carbohydrate-recognition domain (CRD), which binds sugar residues. A C-type lectin is usually a multidomain protein containing one or more CRDs, which consist of 110 to 130 amino acid residues each (Zelensky and Gready 2005). The N- and C-termini join to form an antiparallel β sheet and encompass two α helices and a second antiparallel β sheet. The CRD loops are involved in Ca^{2+} binding, and a single CRD can bind up to four Ca^{2+} ions. Furthermore, the second loop within the CRD structure is involved in carbohydrate binding and, in snake venom C-type lectins, domain-swapping dimerization (Mizuno et al. 1997). A number of amino acid residues characteristic of the CRD are highly conserved, notably four cysteine residues that form two disulfide bonds. Additionally, CRDs contain conserved motifs such as the WIGL motif, which is utilized in the identification of CRDs, and motifs for sugar binding such as EPN, QPD, and WND (Drickamer 1999).

The endophysiological C-type lectins are divided into two categories: intracellular C-type lectins, which are responsible for protein trafficking and sorting, and extracellular (such as secreted) C-type lectins. The venom (exophysiological) homologues of C-type lectins are also divided into several groups depending on their structural characteristics. However, all belong to one of the two aforementioned categories (Zelensky and Gready 2005). C-type lectin group I (lecticans) consists of four subtypes: versican, aggrecan, neurocan, and brevican lectins. Versican lectins have a wide tissue distribution, whereas aggrecan lectins are mainly found in cartilage. Neurocan and brevican lectins are both expressed in the central nervous system (Yamaguchi 2000). They consist of a proteoglycan core, an immunoglobulin (Ig) domain, an epidermal growth factor (EGF)-like domain, a complement control protein (CCP) domain, and a CTLD. C-type lectins belonging to group II are type 2 transmembrane proteins. The first lectin ever discovered, the asialoglycoprotein receptor, belongs to this group of lectins. The asialoglycoprotein receptors are mainly found on liver cells and remove serum glycoproteins from circulation (Rigopoulou et al. 2012). The subterminal galactose of serum glycoproteins must be exposed in order for them to interact with asialoglycoprotein receptors, and deglycosylation is induced upon the removal of a sialic acid residue. Group III C-type lectins play a major role in the immune system by complement activation. The mannose-binding protein (MBP) and surfactant proteins SP-A and SP-D are major constituents of this group, as are the collectins, which have a collagen-like domain and associate to form homotrimers, thus increasing their specificity for glycans. All proteins in this group bind mannose-type sugars (Kishore et al. 2006). Group IV C-type lectins (selectins) are type 1 transmembrane proteins with an EGF-like domain, several CCP domains, and a CTLD. There are three subtypes of selectins: P-selectins, which are expressed on endothelial cells and platelets; E-selectins, expressed on endothelial cells; and L-selectins, expressed on leukocytes. L-selectins are leukocyte adhesion receptors and mediate tethering and rolling of leukocytes on vascular endothelial cells. With the subsequent help of integrins, they can extravasate via interaction with the P- and E-selectins, thus recruiting leukocytes to sites of inflammation (McEver 2002). Group V lectins are type 2 transmembrane proteins and have cysteine-rich extracellular domains that enable homo- or heterodimerization to generate tandem CTLDs. Natural killer cell receptors and structurally similar receptors are group V C-type lectins that regulate leukocyte function, among other things (Pyz et al. 2006). The group VI C-type lectins are type 1 transmembrane proteins with large, cysteine-rich extracellular domains. These consist of a fibronectin type II domain and 8 to 10 CTLDs. The members of this group are all mannose receptors and are chiefly expressed on macrophages. They bind and internalize foreign and endogenous glycoconjugates (Cambi and Figdor 2003). DEC-205, a group VI C-type lectin, internalizes antigens for processing and presentation to T-cells (Erbacher et al. 2009). Group VII C-type lectins are quite different from those of the first six groups, as they consist solely of a CTLD. Proteins belonging to this group are mostly located in the pancreas and serve as growth factors with antiapoptotic and antiinflammatory properties (Zelensky and Gready 2005). However, there are also members of this group that are expressed in other tissues and have functions in adhesion and migration. Many more groups of C-type lectins have been characterized over the years; however, reviewing details on all these groups is beyond the scope of this chapter and is not directly relevant to the discussion of venom C-type lectins.

17.3 PLESIOTYPIC STRUCTURAL AND FUNCTIONAL TOXIN FORMS

Toxicofera reptile venom lectins (TV-lectin) are group VII C-type lectins. Most of the plesiotypic TV-lectin are homodimers, with each subunit consisting of approximately 135 amino acid residues, including seven cysteines. An interchain disulfide bond between the Cys86 residues of each subunit links the homodimers (Clemetson 2010). The plesiotypic TV-lectins retain the EPN motif that confers specificity for Ca^{2+}-dependent agglutination of erythrocytes via the mannose pathway (Clemetson and Clemetson 2001; Clemetson, Lu, and Clemetson 2005). They can also activate leukocytes and platelets and may even activate endothelial cells. The mannose-binding function is also related to antibacterial and antifungal activities (van Asbeck et al. 2008), which are important roles for venom during the relatively slow digestive process. The plesiotypic form has been sequenced widely from toxicoferan lizards and snakes (see figure 17.1).

17.4 APOTYPIC STRUCTURAL AND FUNCTIONAL TOXIN FORMS

17.4.1 CHANGES OF THE PLESIOTYPIC FUNCTIONAL MOTIFS AND CYSTEINE PATTERNS

The first major evolutionary modification in snake-venom TV-lectin was the mutation of the EPN motif in the loop to QPD, which led to a change in specificity from mannose to galactose and thus to mediation of the alternative erythrocyte-agglutination pathway. Forms containing the QPD motif form a monophyletic clade within the EPN forms (see figure 17.1). QPG, EPG, KPG, KRN, EPD, and other motif variants are known but have not been functionally characterized (see figure 17.1). However, sialic acid and N-acetyl glucosamine are likely targets of these variants. Within the snake-venom apotypic TV-lectin forms, additional cysteines have been inserted, with some clades possessing a single additional cysteine and others possessing two new cysteines. These new cysteines may promote alternative connectivities, including monomeric forms (with eight cysteines) or novel dimers (with nine cysteines).

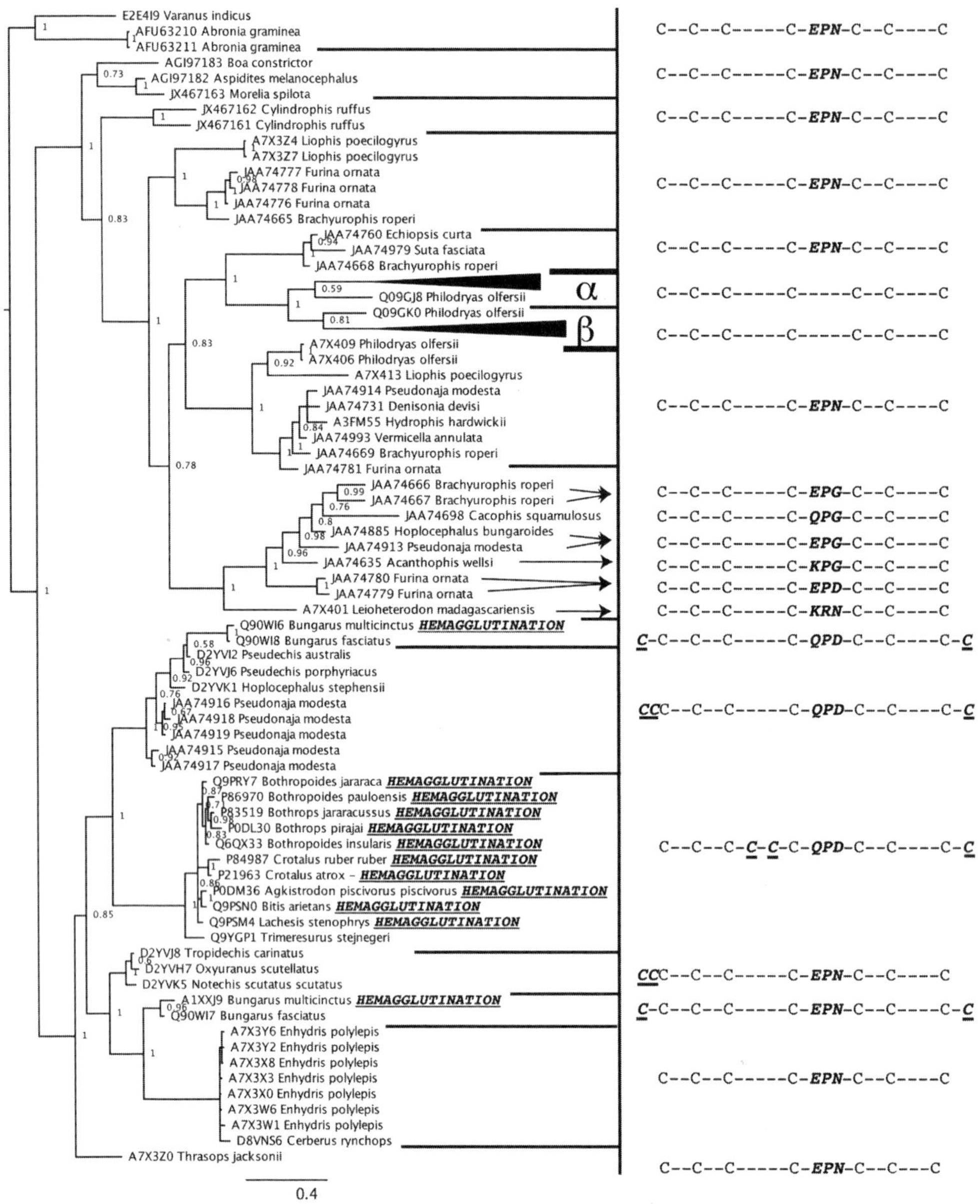

FIGURE 17.1: Molecular phylogeny of lectin toxins with known activities annotated. Cysteine patterns, with newly evolved cysteines underlined and functional tripeptide motifs, are also shown.

17.4.2 SNACLECS: STRUCTURE–FUNCTION RELATIONSHIPS OF SNAKE-VENOM-SPECIFIC APOTYPIC HETERODIMERS LACKING THE SUGAR-BINDING FUNCTIONAL MOTIFS

Snaclec is the name used for apotypic heterodimeric TV-lectin forms from snake venom that lack carbohydrate-binding ability but instead possess numerous neofunctionalized activities. The timing of these structural and functional derivations remains unclear. While viper venoms are extremely rich in heterodimeric forms, it remains unclear if they are present in other snake lineages. Two forms that sit basal to the alpha and beta viperid venom sequences in the phylogenetic tree have been sequenced from the transcriptome of the non-front-fanged caenophidian snake *Philodryas olfersii* (Ching et al. 2006). These may represent the plesiotypic state of the heterodimers. As these sequences are known only from transcriptome sequencing, both their bioactivity and their heterodimer structure remain to be elucidated. Conversely, while a heterodimer has in fact been isolated from the venom of *Ophiophagus hannah* (ophioluxin) (Du et al. 2002), it is known only from very small sequence fragments, and thus the phylogenetic affinity of each chain remains enigmatic. Forms from *Bungarus* that were erroneously said to be of the apotypic heterodimeric form (Du and Clemetson 2010) were actually of the plesiotypic homomeric type.

The apotypic TV-lectin heterodimeric forms are covalently linked by a single disulfide bond, and each subunit contains three internal disulfide bonds (Andrews et al. 1989; Usami et al. 1993). Mutations in the basic CRD unit, including the deletion of approximately six amino acid residues in the central loop, led to destabilization of the sugar-binding structural loop and thus the loss of Ca^{2+}-mediated sugar-binding activity (Koh et al. 2011). Forming loop-swapping dimers, which create a structure with a concave and a convex face, may restore stability to the CRD unit. The concave face interacts with its targets either sterically or by shape fitting (Morita 2005b). Mutations then led to the two subunits drifting apart, resulting in a heterodimeric structure and simultaneously, via modification of residues on the concave face, providing binding specificity for various target molecules. The homologous α and β subunits of the heterodimers have molecular masses of 14–15 and 13–14 kDa, respectively. These "simple" heterodimers as such can directly target essential hemostatic components, or they can multimerize either noncovalently or covalently, to form homodimers $(\alpha\beta)_2$ or heterodimers $(\alpha\beta)(\gamma\delta)$ of the basic heterodimer unit, or they can form homotetramers $(\alpha\beta)_4$ or heterotetramers $(\alpha\beta)_2(\gamma\delta)_2$. These larger multimers can dimerize to form octomeric structures.

These structurally apotypic forms have a wide variety of targets, including membrane receptors, coagulation factors, and proteins essential to hemostasis (see figures 17.2 and 7.3). They can induce or inhibit aggregation or agglutination of platelets by interacting with different platelet receptors. Among others, these include the von Willebrand factor (vWF)-binding GPIb receptor, the collagen-binding receptors GPVI and α2β1, and CLEC2, an innate immune receptor involved in vascular development. Other snaclecs cause hemorrhaging by inhibiting coagulation factors with or without the help of cofactors (Peng, Lu, and Kirby 1991). Procoagulant snaclecs affect blood clotting by consuming coagulation factors, leading to a loss of hemostasis and resulting in hemorrhage. It is intriguing that phylogenetic analyses of the alpha and beta subunits do not produce identical trees (see figures 17.2 and 17.3). In addition, the alpha tree (see figure 17.2) is more structured than the beta tree (see figure 17.3), which suggests that the alpha subunits guide activity.

17.4.2.1 Anticoagulant Snaclecs

Several Ca^{2+}-dependent snaclecs targeting coagulation factors have been described. The Ca^{2+}-binding sites differ from those of the classic C-type lectins (Suzuki et al. 2005), because the Ca^{2+} ions are utilized to bind to gamma-carboxyglutamic acid (Gla) domains in the coagulation factors and not to sugars. Thus, the Gla domain is blocked and unable to bind to

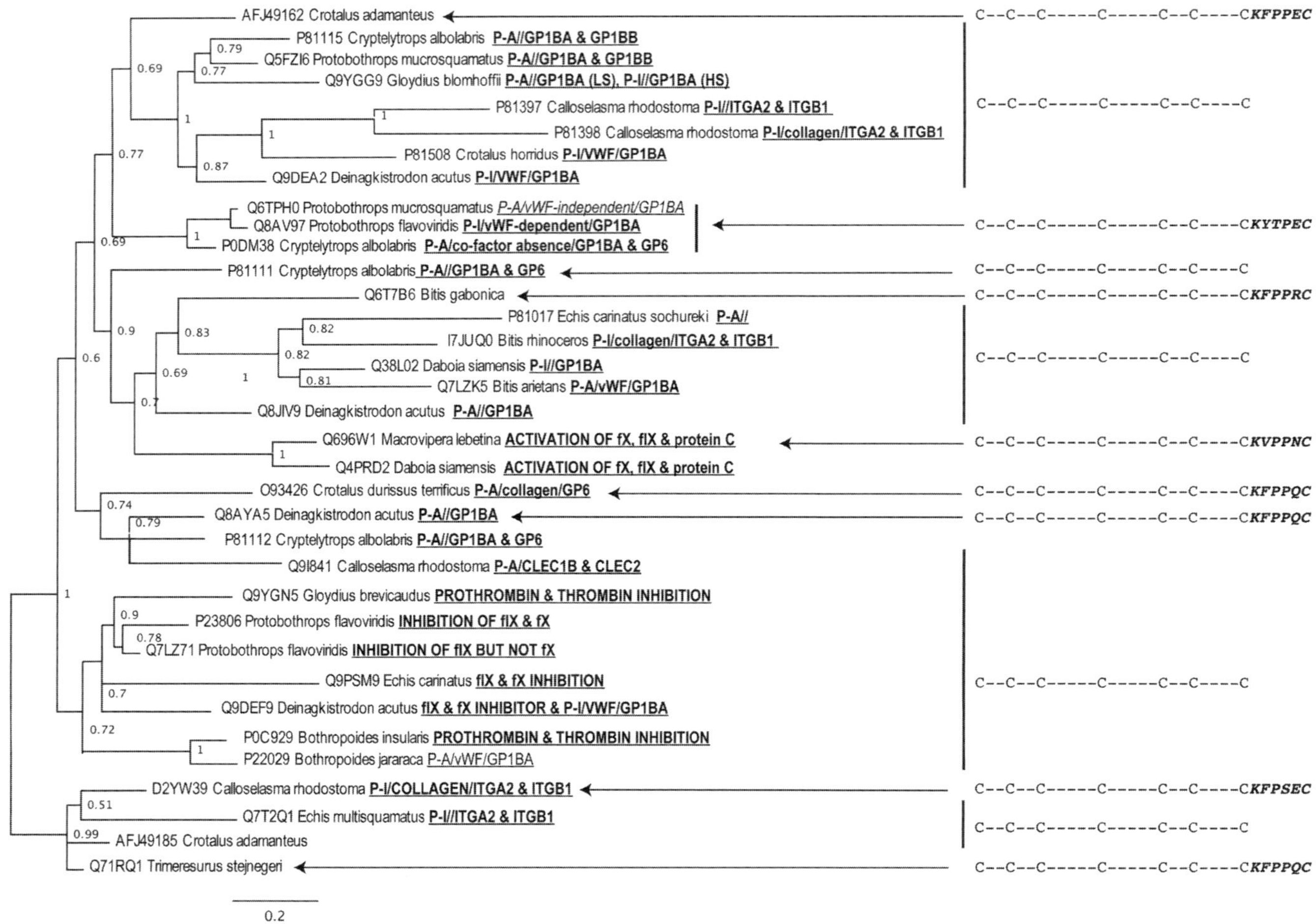

FIGURE 17.2: Molecular phylogeny of the α-subunit of the viperid venom specific apotypic disulfide-linked dimers. Known receptor specificities and bioactivities are annotated. Cysteine patterns are shown, with newly evolved cysteines underlined.

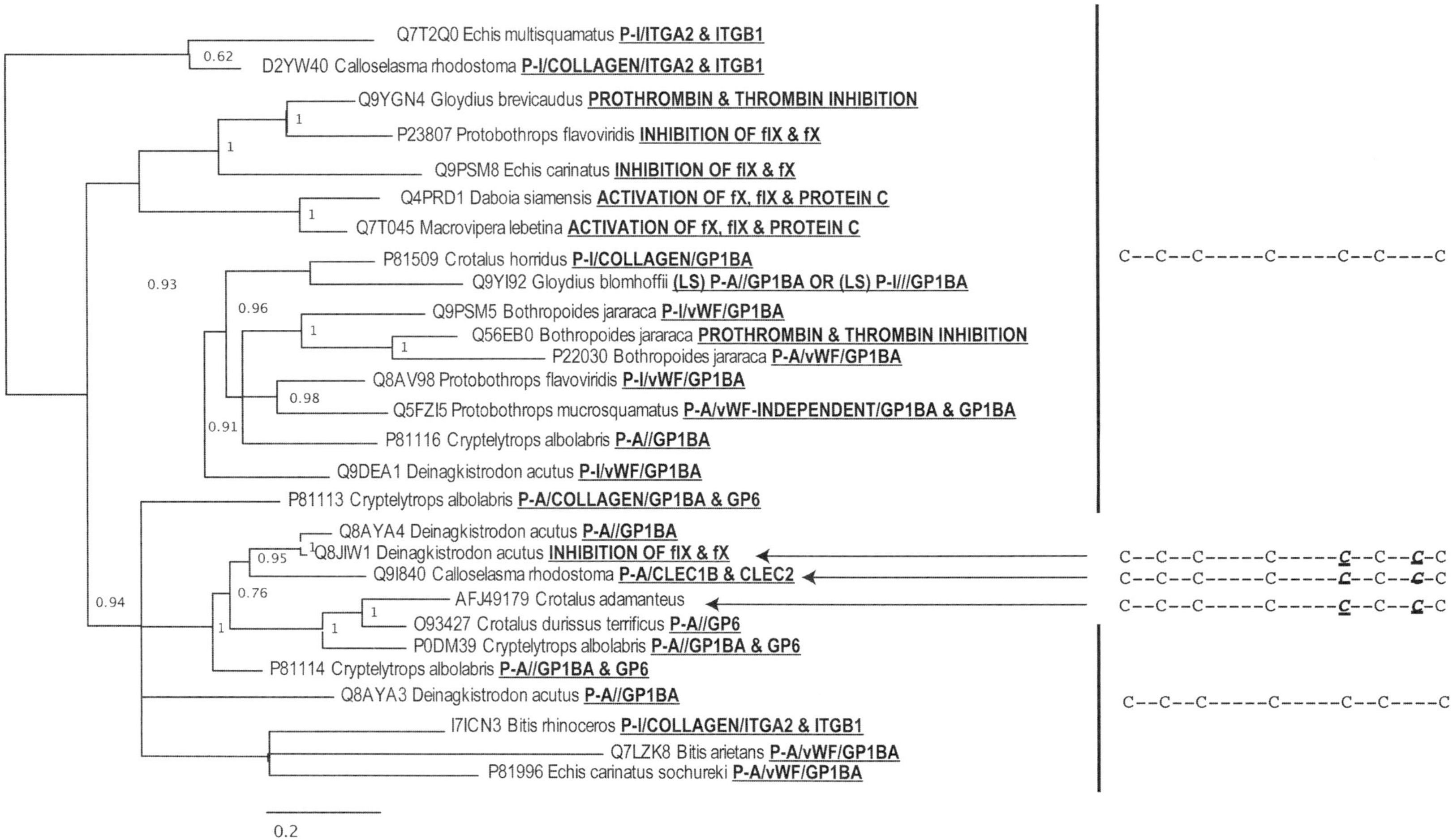

FIGURE 17.3: Molecular phylogeny of the β-subunit of the viperid venom specific apotypic disulfide-linked dimers. Known receptor specificities and bioactivities are annotated. Cysteine patterns are shown, with newly evolved cysteines underlined.

phospholipid membranes (Mizuno et al. 2001). The first snake-venom protein found to bind to a coagulation factor was one of these factor-IX/X-binding snaclecs (Atoda, Hyuga, and Morita 1991). Subsequently, such factor-IX/X-binding snaclecs were isolated and characterized from species across the phylogenetic diversity of the viperid snake venoms, including Crotalinae such as *Bothrops jararaca* and *Deinagkistrodon halys* and Viperinae such as *Echis carinatus leucogaster* (Sekiya, Atoda, and Morita 1993; Chen and Tsai 1996; Ishikawa et al. 2009). Isolated and characterized from *Protobothrops flavoviridis* is a cofactor-independent snaclec that binds only fIX and efficiently inhibits (0.4 nM) blood coagulation (Atoda et al. 1995). While the β subunit of the fIX- and fIX/fX-binding snaclecs from *P. flavoviridis* are identical, the α subunits differ at 19 sites (Atoda et al. 1995). In contrast, the α and β subunits of the fX-binding snaclec from *D. acutus* differ from the corresponding subunits of the fIX/fX-binding snaclec from *P. flavoviridis,* particularly in the region of the concave face (Atoda et al. 1998; Mizuno et al. 2001). The fIX and fX are in the middle of the platelet-activation cascade, and during endophysiological responses, they amplify the coagulation cascade and facilitate the formation of a hemostatic plug after injury. Snake-venom proteins that target this step in the activation cascade are only some of the numerous interventions in regular physiological processes. The coagulation cascade exists mainly to produce thrombin, which is a major contributor to thrombus formation. Not only is thrombin a major activator of platelets, but it also converts fibrinogen to fibrin, an essential step in stable clot formation. Thus, mutations conferring toxic interference in the production of thrombin would be evolutionarily selected for in predatory venoms. Many snake-venom thrombin inhibitors exist. They are found within several different molecular classes and use various inhibitory mechanisms. The first snaclec thrombin inhibitor to be discovered and studied was bothrojaracin (Zingali, Bianconi, and Monteiro 2001; Monteiro 2005; Zingali et al. 2005), which acts by binding to and blocking exosite II on thrombin, preventing thrombin from binding to fibrinogen and to GPIb on platelets. It was later found that bothrojaracin also binds to prothrombin, blocking the activation-peptide cleavage site and therefore blocking the conversion of prothrombin to thrombin. Clearly, this mechanism has a major effect on coagulation and hemostasis. A related molecule was later discovered in *Bothrops insularis* venom (Oliveira-Carvalho et al. 2008).

17.4.2.2 vWF-Binding Snaclecs

As mentioned earlier, snake-venom proteins target not only blood-coagulation factors but also membrane receptors and other proteins essential to hemostasis, such as vWF (Clemetson et al. 2001). The vWF monomers consist of four different types of domains (A, B, C, and D) and multimerize in a head-to-head and tail-to-tail fashion into fibrillar polymers which are covalently linked by disulfide bonds. There are four different A domains (A1–A4), which are essential in vWF function, as they interact with collagen and the platelet receptor GPIb and also with ADAMTS13, which regulates vWF multimer size in circulation. When injury occurs, subendothelial collagen is exposed, and vWF adheres to it via the A1 and A3 domains. These two domains make up the collagen-binding site (Matsui and Hamako 2005). The A1 domain contains the GPIb-binding site, which is exposed by conformational changes of vWF that are induced by binding to collagen (Nuyttens et al. 2011) and by shear stress from blood flow. This subsequently leads to platelet binding and arterial thrombus formation. Initially, it was believed that snaclecs such as botrocetin induced conformational changes in the vWF-A1 domain, activating it and initiating binding to GPIb. This mechanism was supported by the ability of botrocetin, vWF, and GPIb antibodies to individually prevent platelet agglutination (Read et al. 1989). However, botrocetin was eventually shown to bind to the low-affinity conformation of the vWF-A1 domain, thus extending the GPIb-binding surface on the A1 domain when preferentially binding to resting vWF (Fukuda et al. 2002).

A similar vWF-binding snaclec, bitiscetin, which binds both vWF and GPIb, has been purified from the viperid snake species *Bitis arietans* (Hamako et al. 1996). With the help of antibodies, the bitiscetin binding site has been identified as the α4 and α5 helices of the vWF-A1 domain (Matsui et al. 2002). Crystal-structure analysis has confirmed that the α5 loop of the A1 domain lies within the concave binding surface of the bitiscetin heterodimers (Hirotsu et al. 2001). Further binding sites between bitiscetin and vWF are present at each end of the concave binding surface (Maita et al. 2003). Bitiscetin still binds to vWF even when a specific antibody against the α subunit blocks platelet agglutination, indicating that the α subunit is positioned close to the GPIb-binding site (Matsui et al. 2002). Thus, both botrocetin and bitiscetin function by clamping the A1 domain of vWF to GPIb, but each uses a completely different binding site. Presumably, these mechanisms evolved convergently in South American and African snakes, subsequent to the breakup of Gondwana.

17.4.2.3 Antithrombotic Snaclecs

The vWF receptor GPIb and the collagen receptors GPVI and α2β1 integrin on platelets are of special interest for the development of antithrombotic drugs (Clemetson and Clemetson 2008). GPIb is one of the more important platelet receptors because of its critical role in hemostatic and thrombotic pathways. Alboaggregin-B, isolated from the venom of *Cryptelytrops albolabris*, was among the first snaclecs found to bind to GPIb (Peng, Lu, and Kirby 1991; Usami et al. 1996). Alboaggregin-B binds to the 45-kDa N-terminal domain of GPIb and induces platelet agglutination. A monoclonal antibody against the GPIb N-terminal domain blocks this activity of alboaggregin-B (Peng, Lu, and Kirby 1991; Yoshida et al. 1993). Several years after the characterization of alboaggregin-B, a tetrameric analogue of this toxin, alboaggregin-A, was isolated from the same snake's venom and shown to induce platelet aggregation via GPIb (Kowalska et al. 1998). As platelet aggregation requires platelet activation, an increase in phospholipase-C and tyrosine kinase activity can be detected following alboaggregin-A-induced aggregation (Andrews et al. 1996).

Analysis of the venom of *E. carinatus* led to the isolation of the GPIb-binding snaclec, echicetin. Echicetin is different from alboaggregin-B, as it inhibits, rather than induces, platelet agglutination by vWF (Peng et al. 1993; Peng, Holt, and Niewiarowski 1994). However, when echicetin is cross-linked or clustered, it induces platelet aggregation in a fashion similar to alboaggregin-A (Navdaev et al. 2001). As with alboaggregin-B, a monoclonal antibody against the 45-kDa N-terminal domain of the GPIb receptor inhibits echicetin binding to platelets; a K_d of 30 nM was measured for the binding of echicetin to platelets (Peng et al. 1993). An overlap of the echicetin-binding site with the thrombin-binding site on GPIb was demonstrated (Peng et al. 1995). Using echicetin, another GPIb-binding snaclec, purpureotin, was identified from *Cryptelytrops purpureomaculatus* venom. In contrast to other snaclecs described, the α and β subunits of purpureotin are associated noncovalently by hydrophobic and electrostatic interactions despite the usual seven cysteines being present for each chain (Li et al. 2004).

Two further antithrombotic snaclecs, which induce small platelet aggregates, have been purified from the venom of *P. flavoviridis* (Taniuchi et al. 1995). Flavocetin-A is a heterodimeric CLP, and in contrast to most other snaclecs (such as alboaggregin-A), both subunits contain an additional cysteine residue (Shin et al. 2000). Using these additional residues, flavocetin-A is able to form interchain disulfide bonds in a head-to-tail fashion (Fukuda et al. 1999). The resulting tetrameric structure initiates a platelet response by cross-linking platelets, possibly utilizing cooperative binding activity (Fukuda et al. 2000; Taniuchi, Kawasaki, and Fujimura 2000). Flavocetin-B has the same effect on platelets and is able to prevent flavocetin-A binding. Flavocetin-B is composed of three different subunits: α and β, which are identical to the α and β subunits of flavocetin-A, and γ. Flavocetin-B has not been analyzed further; however, it was discovered recently that flavocetin-A also binds to the collagen receptor α2β1. In this study a fourth subunit was discovered, and this led to a new structure, $(\alpha\beta)_2(\gamma\delta)_2$, being proposed. The study demonstrated that this tetramer containing two different heterodimers binds GPIb and α2β1 simultaneously, with the αβ heterodimer proposed as the GPIb-binding moiety and the γδ heterodimer proposed as the integrin-binding moiety (Arlinghaus and Eble 2013). This possible structure of flavocetin-A is comparable to the structure of bilinexin, a snaclec isolated from *Agkistrodon bilineatus*, which also binds GPIb and α2β1 (Du et al. 2001).

The snaclec convulxin, from *Crotalus durissus terrificus*, is another tetrameric protein. Convulxin, which was initially described as a trimer, induces strong platelet aggregation (Francischetti et al. 1997; Leduc and Bon 1998; Batuwangala et al. 2004). Antibodies against α2β1 (6F1) and GPVI were described to inhibit convulxin-induced aggregation (Jandrot-Perrus et al. 1997). Platelet aggregation is induced by clustering of the GPVI receptor, facilitated by the tetrameric structure of convulxin. Clustered GPVI receptors become activated, leading to an increase in intracellular Ca^{2+} levels (Batuwangala et al. 2004). However, convulxin forms dimeric, noncovalently linked octomers in solution, and these octomers may be the basic active form of this snaclec. Convulxin can also bind GPIb weakly but only at high concentrations. It is not clear if this indicates that the GPVI-binding function is still evolving from a plesiotypic snaclec GPIb-binding activity. Similarly, a more active GPVI-binding snaclec from the venoms of *Ophiophagus hannah* does not show any binding to GPIb (Du et al. 2002). GPVI-binding snaclecs are not yet known from South American snakes other than the *Crotalus durissus* group.

Aggretin, also known as rhodocytin, is a snaclec that has generated considerable controversy. It was isolated from the venom of *Calloselasma rhodostoma* and has a typical snaclec cysteine pattern and structure (Huang, Liu, and Yang 1995; Chung, Au, and Huang 1999). As 6F1, an anti-α2β1 monoclonal antibody, inhibited aggretin-induced platelet aggregation, this activity was believed to be facilitated by binding to the α2β1 integrin (Huang, Liu, and Yang 1995). Another study showed that aggretin-induced platelet aggregation, effected via the collagen receptor α2β1, could only be blocked when the α2 antibody A2IIE10 was used in combination with agkistin, a GPIb antagonist. Each antibody only partially inhibited the aggregation response when used separately (Chung, Peng, and Huang 2001). Other cell-based studies demonstrated that aggretin-induced proliferation and migration of human umbilical-vein endothelial cells (HUVEC) could be inhibited by an antibody against α2β1 (Chung, Wu, and Huang 2004). Furthermore, aggretin induces α2β1 integrin-dependent proliferation of vascular smooth-muscle cells (VSMC) (Chung et al. 2009). In contrast to the conclusions of these studies of aggretin-induced platelet aggregation (via the collagen receptor α2β1) and cell proliferation and migration, aggretin did not bind α2β1 integrin in a cell-free protein interaction assay (Eble et al. 2001). Further controversy was elicited by a study that showed that aggretin induced platelet aggregation despite neither α2β1 integrin or GPIb/GPVI being present on platelets (Bergmeier et al. 2001). Following these studies, another interaction partner of aggretin was revealed: the C-type lectin receptor CLEC-2. On platelets, the binding of aggretin to CLEC-2 results in platelet aggregation by activating a signaling pathway via the single tyrosine phosphorylation site on the cytosolic tail of CLEC-2 (Suzuki-Inoue et al. 2006). Aggretin binding to CLEC-2 on macrophages leads to activation of mitogen-activated protein kinase (MAPK) and nuclear factor (NF) κB signaling pathways and ultimately to the release of tumor necrosis factor (TNF) α and interleukin (IL) 6 (Chang et al. 2010). Crystallization studies reveal that two aggretin heterodimers align antiparallel in a tetrameric structure. Using this structure for modeling led to two different mechanisms of interaction in which aggretin interacts either with two CLEC-2 monomers or with one (or two) CLEC-2 dimers (Hooley et al. 2008; Watson and O'Callaghan 2011).

Integrin α2β1 is another receptor involved in platelet activation. It is the sole collagen-binding integrin on platelets, and it is required for the binding of platelets to subendothelium (Sixma et al. 1997). When α2β1 integrin expression is reduced, platelet responses are dysfunctional (Nuyttens et al. 2011). Although not lethal, the absence of the integrin α2 subunit in knockout mice results in defects in hemostasis (Chen et al. 2002). Likewise, an increased α2β1 expression leads to an increased affinity of platelets for collagen, thus increasing the risk of thrombosis (Kritzik et al. 1998). The α2β1 receptor is a major integrin inhibited by snaclecs, although disintegrins are typically employed for inhibiting other integrins (Calvete et al. 2005). However, there are other reports of snaclecs inhibiting additional integrins, for example, lebectin and lebecetin, which inhibit α5β1 and αV-containing integrins (Sarray et al. 2007).

To date, there are five known snaclecs that target the α2β1 integrin and inhibit its binding to collagen. Most of these bind to the α2A domain, a domain present in only some integrin α subunits and homologous to the vWF-A domain. The α2A domain, which changes conformation from an inactive to an active form, is the domain primarily responsible for collagen binding (Emsley et al. 2000). This conformational change takes place not only within the α2A domain but also throughout the inhibited integrin, ultimately resulting in integrin signaling (Arnaout, Mahalingam, and Xiong 2005). The first α2β1 antagonist, EMS16, was isolated from *Echis multisquamatus* and inhibits collagen-induced platelet aggregation and cell migration (Marcinkiewicz et al. 2000; Okuda et al. 2003). This heterodimer has the typical domain-swapping interface of snaclecs, and its crystal structure, resolved at 1.9 Å, reveals a unique positively charged electrostatic potential patch on the concave surface (Horii et al. 2003). Another crystallization study of EMS16 in complex with the α2A domain reveals that EMS16 sterically blocks access to the collagen-binding site. The binding is independent from divalent cations, and this suggests that EMS16 stabilizes inactive α2A domain conformation (Horii et al. 2003).

The second α2β1 antagonist discovered is rhodocetin, and studies indicate that there is great potential for this snaclec in drug design and medical research. Rhodocetin was isolated from *C. rhodostoma* and is an inhibitor of collagen-induced platelet aggregation. Initially, it was believed to be a heterodimer with noncovalently bound subunits (Wang, Kini, and Chung 1999; Eble et al. 2001). The cysteine residues typically involved in the formation of an intersubunit disulfide bonds are replaced in rhodocetin by Ser79 and Arg75 in the α and β subunits, respectively (Wang, Kini, and Chung 1999). The dimeric structure of rhodocetin is stabilized by an additional β sheet in each subunit and the hydrogen bonds that are formed between the backbones of the two β sheets (Paaventhan et al. 2005). The α (15.9 kDa) and β (15.1 kDa) subunits only inhibit collagen-induced platelet aggregation when paired and not when used separately (Wang, Kini, and Chung 1999). Additionally, rhodocetin not only inhibits platelet aggregation but also inhibits cell adhesion and migration (Eble et al. 2001); however, rhodocetin is unable to cluster α2β1 integrin to generate focal adhesions (Eble et al. 2002). Rhodocetin also blocks extravasation of tumor cells into the liver stroma, thus inhibiting the generation of micrometastases (Rosenow et al. 2008). Similar to EMS16, rhodocetin preferentially binds to less active conformations of integrin α2β1. This is supported by the fact that rhodocetin binds more readily to α2β1 in the presence of Ca^{2+} ions than does collagen. Furthermore, rhodocetin interacts with the α3 and α4 loops and adjacent helices of the collagen-binding α2A domain (Eble and Tuckwell 2003). It is now known that rhodocetin is a heterotetramer of four homologous subunits: α, β, γ, and δ. As mentioned previously, the α and β subunits form a noncovalently bound heterodimer by domain swapping, whereas a disulfide bond between Cγ81 and Cδ74 stabilizes the γ and δ subunits. The αβ and γδ heterodimers associate with one another and form a cruciform molecule (Eble et al. 2009). Compared with subunits of other snaclecs, the rhodocetin α subunit contains the additional peptide turn N^{74}KGQR, and the rhodocetin γ subunit contains the additional peptide turn K^{77}EQQC. These additional turns make up the interface of rhodocetin's cruciform structure: gly76 of the α subunit fits into a groove formed by the peptide turn of the γ subunit (Eble et al. 2009). Once the heterotetrameric rhodocetin has bound to the α2A domain, it dissociates into its two heterodimers: the γδ heterodimer remains firmly attached to the integrin, while the αβ heterodimer is released (Bracht et al. 2011). The αβ heterodimer subsequently binds to the platelet receptor GPIb and inhibits vWF/ristocetin-induced platelet aggregation. In contrast, the cross-linked αβ heterodimer results in small platelet agglutinates, eventually leading to phosphorylation of the p72SYK kinase in platelets (Navdaev, Lochnit, and Eble 2011). Furthermore, rhodocetin αβ binds avidly to neuropilin-1 (Nrp1), a receptor that usually interacts with vascular endothelial growth factor (VEGF)-A. Upon αβ binding, Nrp1 associates with cMet, which is subsequently phosphorylated and activated, ultimately leading to redistribution of focal adhesions to focal contacts (Niland et al. 2013).

The α2A domain-binding CLP vixapatin—first published as VP12—was purified from *Daboia palaestinae* venom (Staniszewska et al. 2009). After the isolation of vixapatin, an additional purification method was applied to *D. palaestinae* venom, which led to the isolation of a second α2A domain-binding toxin, VP-i (Arlinghaus et al. 2013). A comparison of vixapatin and VP-i reveals them to be homologous proteins with similar, but unique, functions. Both of these snaclecs bind the integrin α2A domain in cell-free adhesion assays and also inhibit integrin-dependent cell adhesion to, and migration on, collagen (Arlinghaus et al. 2013). Vixapatin also has an inhibitory effect during angiogenesis (Momic et al. 2012). Furthermore, VP-i inhibits collagen-induced platelet aggregation (Arlinghaus et al. 2013). The venom of *Bitis rhinoceros* also contains a snaclec, rhinocetin. Rhinocetin inhibits α2β1 integrin-dependent endothelial cell functions and collagen-induced platelet aggregation (Vaiyapuri et al. 2012a).

17.4.2.4 Heterodimeric Lectins That Form Complexes with PIII SVMP

Heterodimeric snaclecs represent an extreme structural derivation and form a complex with PIIId type SVMP. Examples of this apotypic form are RVV-X and carinactivase-I, which activate factor X and prothrombin, respectively (Morita 2005a). The linkage between the snaclec domain and the metalloprotease/cysteine-rich and disintegrin domains involves an additional disulfide bond in the chain homologous with the alpha chain of other snaclecs. The confusing nomenclature of these complexes must be noted. The snaclec alpha chain is referred to as the beta chain, while the chain homologous with the beta chain of other snaclecs is called the gamma chain. The SVMP chain is known as the alpha chain.

17.5 THERAPEUTIC POTENTIAL OF THE TOXIN TYPE

The vWF receptor GPIb, and the collagen receptors GPVI and α2β1 integrin on platelets, are of special interest for the development of antithrombotic drugs (Clemetson and Clemetson 2008). Thus, TV-lectins that act on these sites are immediately useful as investigational ligands and also have long-term potential as lead structures in drug design and development.

17.6 CONVERGENCE WITH OTHER VENOMS

Toxic lectins have been found in fish and caterpillars in addition to snakes. The fish toxins are galactose-specific lectins that exhibit hemagglutination activity (minimum hemagglutination concentration 2.5 μg/well) in a calcium-independent fashion and also possess remarkable pro-inflammatory activity, which they induce via neutrophil mobilization (Lopes-Ferreira et al. 2011). The caterpillar types have been sequenced from the lethal caterpillar stage of the moth species *Lonomia obliqua* and may play a role in the hemorrhagic syndrome that results from human envenomation by this species (Veiga et al. 2005).

CHAPTER 18

B-TYPE NATRIURETIC PEPTIDES

B. G. FRY, K. SUNAGAR, T. N. W. JACKSON, T. REEKS, AND H. F. KWOK

18.1 SUMMARY

Natriuretic peptides are normally secreted as propeptides that require posttranslational modification to release the functional domains. They represent a unique class of proteins that perform a plethora of activities in anguimorph lizard venoms. Endophysiological forms are broadly classified into two groups, depending on whether they possess the C-terminal tail domain (A-, B-, V-type natriuretic peptide) or not (C-type natriuretic peptide). Anguimorph lizard venom forms are of the B-type and thus are called LV-BNPs. Like endophysiological BNPs, LV-BNPs target guanylyl cyclase receptor A. Apotyposis in the anguiid and helodermatid LV-BNPs resulted in the *de novo* evolution of proline-rich bioactive peptides in the propep region of the precursor. LV-BNPs and the apotypic proline-rich peptides from anguimorph lizard venoms have significant potential as therapeutics, especially in the treatment of congestive heart failure, chronic respiratory disease, and renal dysfunction. In addition, the propeptide region novel proline peptides (helokinestatins) have been shown to have tremendous potential as anticancer drugs.

18.2 ENDOPHYSIOLOGICAL PLESIOTYPE

Natriuretic peptides have diverse physiological roles such as diuresis, natriuresis, and cardiovascular homeostasis through the regulation of vascular smooth-muscle tone, fat metabolism, and ventricular hypertrophy. Natriuretic peptides are secreted as precursors containing a signal peptide, a propeptide, and the natriuretic peptide domain. This precursor is posttranslationally processed to liberate the natriuretic peptide (Wu et al. 2003). All natriuretic peptides are characterized by a 17-amino-acid disulfide loop (Sudoh et al. 1990). Two broad groups exist, one containing ANPs, BNPs, and VNPs with the other containing CNPs. These groups differ primarily in the relative presence of a C-terminal tail, with all forms containing an N-terminus of variable length. To date, three receptors have been identified as potential molecular targets of natriuretic peptides: NPR-A (GC-A), NPR-B (GC-B), and NPR-C. GC-A and GC-B are the primary targets (binding ANPs/BNPs/VNPs and CNPs, respectively) and mediate activity through the triggering of intracellular guanylyl cyclase secondary pathways (Chang et al. 1989; Lowe et al. 1989). While the truncated receptor NPR-C (which can bind all natriuretic peptide forms) was initially thought to have a simple clearance role, additional studies have indicated that these receptors can in fact trigger some biological effects through G-protein-dependent pathways (Bennett and Rose 1991; Koh et al. 1992; Murthy and Makhlouf 1999; He, Dukkipati, and Garcia 2006). Unlike ANPs, which are stored in atrial granules, CNPs are not stored in granules and are heavily secreted in the brain (Ogawa et al. 1992), the vascular epithelium (Suga et al. 1992a; Suga et al. 1992b), cartilage, and other peripheral tissues (Hagiwara et al. 1994a; Hagiwara et al. 1994b). BNPs are similar to ANPs in their action, and small amounts of BNP are stored in atrial granules, with larger amounts secreted in the ventricle as a result of ventricle-wall stretch (Mukoyama et al. 1991).

18.3 PLESIOTYPIC STRUCTURAL AND FUNCTIONAL TOXIN FORMS

In contrast to the complicated and enigmatic molecular evolutionary history of the TV-CNP found in snake venoms (see chapter 19), the evolutionary history of anguimorph lizard venom LV-BNP is relatively simple and easily elucidated. LV-BNPs are clearly nested within the BNP phylogenetic clade of the natriuretic peptide superfamily, and all forms retain the N- and C-terminal tails (Fry et al. 2006; Fry et al. 2009b; Fry et al. 2010a; Fry et al. 2010b; Koludarov et al. 2012). As with BNP, there is a correlation between LV-BNP GC-A receptor binding and the ability to relax precontracted endothelium-denuded aortic rings. However, there exists a more complex interaction between LV-BNPs and the GC-A receptor than simple binding correlation. Residue replacement studies (site-directed mutagenesis) of a novel natriuretic peptide from *Gerrhonotus infernalis* revealed that both aspartate at ring position 7 and isoleucine at ring position 9 (see figure 18.1) are required for full activity in endothelium-denuded aortic rings (Fry et al. 2010b). However, the mutated peptides were not equipotent in inducing a drop in the mean arterial blood pressure when administered intravenously to anesthetized rats. The isoleucine 9 mutant produced a much greater effect than the aspartate 7 mutant (see figure 18.2). Conversely, a difference of activity in the two assays was evident in isoforms from *Celestus warreni* and *Varanus glauerti* (see figures 18.1 and 18.2). Both contained the functional residues Asp-7 and Iso-9, but both had only a slight effect on mean arterial blood pressure in anesthetized rats. However, the *V. glauerti* isoform was more potent than *C. warreni* in its ability to relax precontracted endothelium-denuded aortic rings.

Such variation in the ability to affect blood pressure and the ability to relax the endothelium aortic ring indicates that there is a difference in the receptors present on the surface of the endothelium-denuded aortic ring and those encountered by peptides in anesthetized rats. Bioactivity testing revealed that in endothelium-intact aortic rings, the mutant *G. infernalis* natriuretic peptide containing isoleucine at position 9 was indeed more potent than the mutant with aspartate at position 7 (see figures 18.1 and 18.2). However, the degree of

```
              10        20        30        40        50        60        70        80        90
1.   mnfkltcsgslllllvlftydqqraNPLKRGQ---SLSKRFEDDSRELRQKRAAEENVPPLGPDVQKPRAEENLPPPYMPTVPRDVDENAPQLD
2.   mnpkltcsswllvllvlcshdqaraNPLKRRQ---PLSTRDEGDSRELRPKRGTEAHLPLQGPDVGRPAAQ-NRPPPFVP---RDTKDGAPQLV
3.   mnfkltcsgsfllllvlftydqqraNPLKRGQ---SLSKRFEDDSRELRQKRAAEENVPPMGPDVQKPRAEENLPPPYMPTVPRNVDENAPQLD
4.   mnprlacstwlplllvlftldqgraNPVERGQEYRSLSKRFDDDSTEL----------------ILEPRASEENGPPYQPLVPR----------
5.   mnprlacstwlplllvlftldqgraNPVERGQEYRSLSKRFDDDSRKL----------------ILEPRASEENGPPYQPLVPR----------
6.   mnprlacstwlplllvlftldqgraNPVERGQEYRSLSKRFDDDSTEL----------------ILEPRASEENGPPYQPLVPR----------
7.   mdprlvragclvlllvllvqdrgaAHPARAGQKYKPLIRRSEEDSQALGQEVHVAARAADE---------------------------------
8.   mdprlvragclvlllallvqdqggAHPARTGQKYKPLIRRSEEDSQALGQEGDMAVRAADE---------------------------------
9.   mdprlvragslvlllallvqdqgaAHPARPGQKYKPLIRRSEEDSQALGQEGDVAARAADE---------------------------------
10.  mdprlvragslvlllallvqdqgaAHPARAGQKYKPLIRRSEEDSQALGQEGDVAARAADE---------------------------------
11.  mdprlvragclvlllvllvqdrgaAHPARAGQEFPLRSKLSADGSEVPRRDASLAARAPGE---------------------------------

             100       110       120       130       140       150       160       170       180
1.   LSLATASEENVPSPIFPFVPSAEGENVPPYQLLVPRSSIAKPPPFMPFIPKSS---VAKPPPFLPLVPRSSIAKPPPFVPLVPRSSDENPPQFV
2.   ----DVGRPAAQNRPPPFVPRDTKDGAPQLVSYLARSSREKPPPFLPLLPRA EDKISPYQPRVPRPSG---ETVPPIVGHEPRSKDGGGFRK
3.   LSLARASEENVPSPIFPFVPRAEGENAPPYHPLVPRSSIAKPPPFLPFIPR------------------SSVAKPPPFVPLVPRSSDEKPPQFV
4.   -----ASDENVPPAYVPLVPRASDENVPP------------PPLQMPLIPRAS------------------EQKGPPFNPP-P---------FV
5.   -----ASNENVPPAFVPLVPRASDENVPP------------PPLQMPLIPRASDENVPPPPLQMPLIPRASEQKGPPFNPP-P---------FV
6.   -----ASDENVPPAYVPLVPRASDENVPP------------PPLQMPLIPRAS------------------EQKGPPFNPP-P---------FV
7.   --------------------------------------------------------------------------------------------E-
8.   --------------------------------------------------------------------------------------------E-
9.   --------------------------------------------------------------------------------------------E-
10.  --------------------------------------------------------------------------------------------EE
11.  --------------------------------------------------------------------------------------------ED

             190       200       210       220       230       240       250
1.   NYE-PRSFDENALRKLIHKVQRSFWRSPRRNKLLYPGDGCFGQKIERIGTVSGMGCG----SSNTGYSGKK
2.   --------------------VGRLRPVWGAPRKHYPGNGCFGHKLDKIGAVSGMGCRGSIRRPQPGRDA
3.   NYE-PRSFDENALRKLIHKVQRSFWRSPRRDKRLYPGDGCFGQKIERVGTVSGMGCR----SSNTGYSGKK
4.   DYE-PRAANENALRKLIK---RSFERSPGRNKRLSPGDGCFGQKIDRIGAVSGMGCN----SVSS--QGKK
5.   DYE-PRAANENALRKLIK---RSFERSPGRNKRLSPGDGCFGQKIDRIGAVSGMGCN----SVSS--QGKK
6.   DYE-PRAANENALRKLIK---RSFERSPGRNKRLSPGDGCFGQKIDRIGAVSGMGCN----SVSS--QGKK
7.   DAAGP-GDALQQPALK--------TLLASREKRLQPGESCFGQKIDRIGHVSGMGCNTFDPNKGSSSTEKK
8.   DAAGP-GDALQQPALK--------TLLASREKRLQPGDSCFGQKIDRIGHASGMGCNTFDPNKESSSAEKK
9.   DAAGP-GDALRQPAFK--------MLLASREKRIQPEGSCFGQKIDRIGHVSGMGCNKFDPNKESSSTGKK
10.  DAAGP-GDALRQPAFK--------TLLASREKRLQPEGSCFGQKMDRIGHVSGMGCNKFDPNKGSSSTGKK
11.  DAHSTFQEIIRDRRTP--------PGPPRKNKRLSPGSGCFGNRLDRIGSTSGLGCGSGSSSSGTGTAGKK
                                           12345678901234567
```

FIGURE 18.1: Alignment of LV-BNPs (Fry et al. 2010b; Koludarov et al. 2012): (1) K4I268 (*Abronia graminea*), (2) E2E4J4 (*Celestus warreni*), (3) E2E4J3 (*Gerrhonotus infernalis*), (4) C6EVG7 (*Heloderma suspectum*), (5) E8ZCG5 (*Heloderma horridum*), (6) C6EVG7 (*Heloderma suspectum*), (7) E2E4J0 (*Varanus glauerti*), (8) E2E4J1 (*Varanus scalaris*), (9) B6CJV0 (*Varanus komodoensis*), (10) Q2XXL8 (*Varanus varius*), and (11) E2E4J2 (*Varanus glauerti*). Posttranslationally cleaved bioactive peptides are shaded in gray. Signal peptides are shown in lower-case.

difference was less than the very pronounced difference displayed in anesthetized rats. Thus, it appears that there is indeed a difference in the type of natriuretic peptide receptors exposed in endothelium-intact versus endothelium-denuded aortic ring and that natriuretic peptides present in other vascular tissues may further contribute to induce a drop in the blood pressure.

18.4 APOTYPIC STRUCTURAL AND FUNCTIONAL TOXIN FORMS

As with snake-venom natriuretic genes (see chapter 19), the LV-BNP plesiotypic monodomain BNP gene was medially extended through further recruitment of multiple tandem-repeated proline-rich peptides (helokinestatins in this case) to become a multidomain gene encoding multiple products. The discrete peptides are liberated from one another by posttranslational proteolysis (see figure 18.3). The helokinestatin derivation evidently occurred basal to the anguid/helodermatid lizard clade as these domains are absent from varanid lizard venom forms (see figure 18.1).

The [ASDEN] sequence, which links the posttranslationally cleaved *Heloderma* helokinestatin peptide isoforms, was partially conserved between variants in *G. infernalis* ([AAEEN],

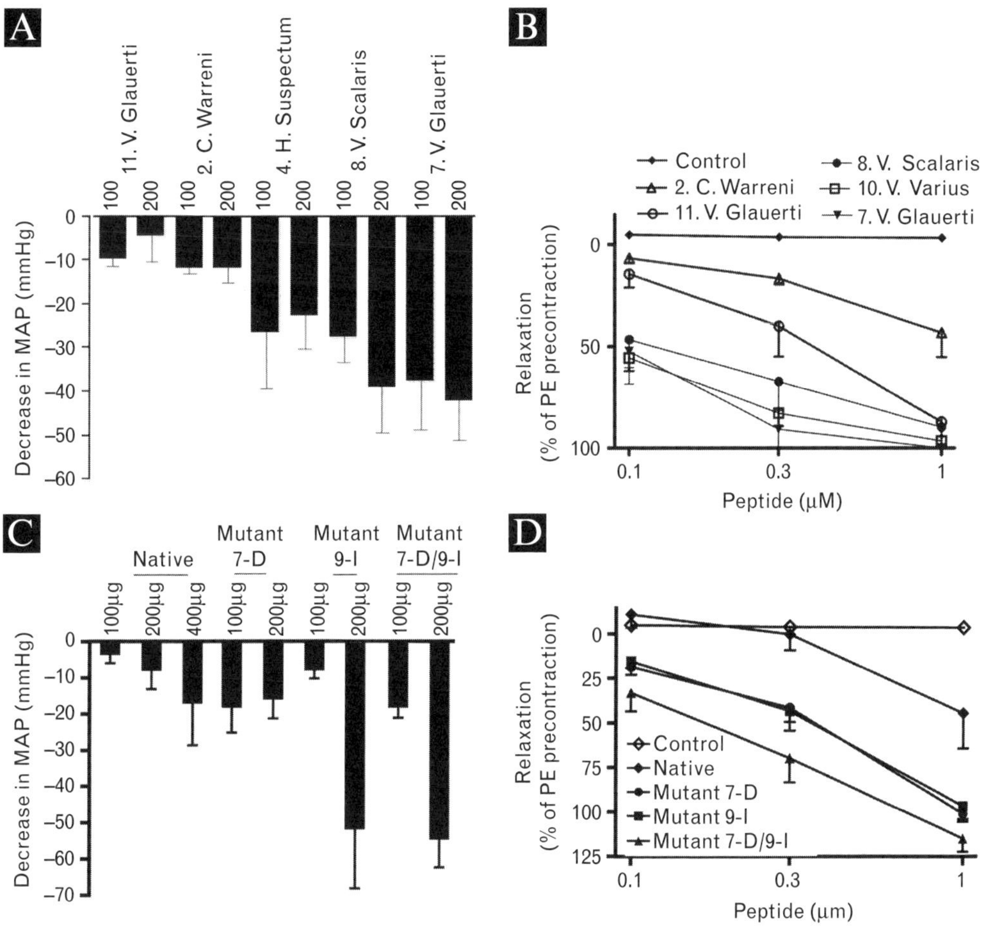

FIGURE 18.2: LV-BNP. (A, B) Relative decrease in mean arterial pressure by different isoforms. (C, D) Structure-function relationships of a variant from *Gerrhonotus infernalis* that lacks the two invariant amino acids 7-D and 9-I. Numbers before each taxon in B and C indicate relative alignment row in figure 18.1 (Fry et al. 2010b).

[AEEN], and [AEGEEN], respectively) but not in the same proline-rich region of the *C. warreni* transcript (see figure 18.1). Proline brackets also guide the activities of helokinestatin. Bioactivity testing of a representative taxonomical sampling of helokinestatin variants encoded by the natriuretic gene demonstrated variable contractile-inhibiting responses to bradykinin in the guinea pig ileum (see figure 18.3).

Helokinestatin represents an entirely novel structural class of lizard-venom peptide displaying a unique noncompetitive bradykinin-receptor antagonist activity mediated via B2 receptors. The discovery of helokinestatin reveals an additional and novel target within the kinin/kininogen system at the level of selective bradykinin receptor antagonism (Kwok et al. 2008). These peptides are proline-rich, bradykinin-antagonist peptides (Kwok et al. 2008; Fry et al. 2010a; Fry et al. 2010b; Koludarov et al. 2012; Ma et al. 2012). Being enriched in proline residues confers great conformational flexibility in addition to a measure of resistance to proteolysis. They all possess a C-terminal "Pro-Arg motif," which is distinct from the C-terminal "Ile/Val-Pro-Pro motif" of most bradykinin-potentiating peptides (Menin et al. 2008). The

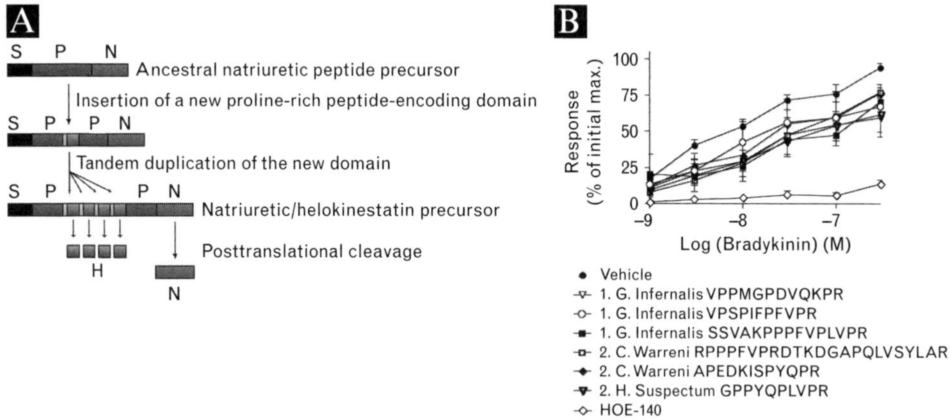

FIGURE 18.3: (A) Domain mutation producing the tandemly repeated helokinestatin peptides medial to the LV-BNP encoding domain and the subsequent posttranslational proteolysis of the multidomain product to liberate the discrete peptides (Fry et al. 2010a). H = helokinestatin, N = LV-BNP, P = propeptide, S = signal peptide. (B) Effect of different helokinestatin peptides (0.1 mM; n = 5) on cumulative concentration-response curves to bradykinin (1 nM–0.3 mM). Numbers before each taxon in B indicate relative alignment row in figure 18.1 (Fry et al. 2010b).

high degree of conservation of these core structural features across taxa suggests an important role for this group of peptides in the venoms of anguirmoph lizards. They are encoded in tandem within two virtually identical biosynthetic precursors of 177 and 178 amino acid residues, differing by only a single Pro residue. Synthetic replicates of all helokinestatins were devoid of any direct action on the smooth muscle of rat-tail artery but were found to be potent inhibitors of bradykinin-induced relaxation in this preparation, in a manner suggestive of a noncompetitive mechanism. Helokinestatin-3 (VPPPPLQMPLIPR) and helokinestatin-5 (VPPPLQMPLIPR) were found to be the most potent in this respect, causing almost complete inhibition of bradykinin-induced relaxation (Fry et al. 2010b; Ma et al. 2012). Helokinestatins and bradykinin-potentiating peptides (BPPs) may have a shared evolutionary history, but the former do not inhibit ACE. A study investigating the molecular evolutionary history of helokinestatin and natriuretic peptides in *Abronia graminea* venom revealed differential evolutionary rates for the two molecular scaffolds, despite the fact that they are tandem-encoded by the same precursor (Koludarov et al. 2012). While helokinestatin was found to accumulate variations, the coding sequences of natriuretic peptides were evolving under a negative-selection regime and lacked variations. Examination of other anguid and helodermatid lizard venom precursors shows this two-speed evolution to be a consistent feature.

18.5 THERAPEUTIC POTENTIAL OF THE TOXIN TYPE

Because of their variable activity in different assays and thus the suggestion of differential receptor interactions, the unique lizard LV-BNP peptides may prove to be of therapeutic use, just as snake venom forms have (see chapter 19). In addition, they may be useful as investigational ligands in discovering new receptor subtypes, just as the mamba-snake-venom muscarinic toxins were instrumental in the discovering of novel subtypes of muscarinic receptors (see chapter 8). Finally, the bradykinin-receptor antagonist helokinestatin has the potential to act synergistically with existing anticancer drugs and opens a new avenue for anticancer therapy, since bradykinin evokes prostaglandin production (Kwok et al. 2008). The potential of helokinestatins as bradykinin inhibitors may also be exploited in the local control of chronic inflammation.

18.6 CONVERGENCE WITH OTHER VENOMS

Natriuretic peptides have been sequenced from the venoms of *Ornithorhynchus anatinus* (de Plater, Martin, and Milburn 1998; Kourie 1999c; Kourie 1999a; Kourie 1999b; Torres et al. 2002) and *Desmodus rotundus* (Francischetti et al. 2013). Both are similar to snake forms (see chapter 19) in being the result of a recruitment of the C-type natriuretic peptide, rather than the B-type recruited in to lizard venoms. Only the *O. anatinus* form has been tested, and it displayed vasorelaxation activity (de Plater, Martin, and Milburn 1998).

CHAPTER 19

C-TYPE NATRIURETIC PEPTIDES

B. G. FRY, T. N. W. JACKSON, Z. TAKACS, T. REEKS, AND K. SUNAGAR

19.1 SUMMARY

Natriuretic peptides are normally secreted as propeptides that require posttranslational maturation to release the functional domains. Endophysiological forms are broadly classified into two groups, depending on whether they possess the C-terminal tail domain (A-, B-, V-type) or not (C-type). Caenophidian snake venom natriuretic peptide forms are of the C-type (tailless) and are known as SV-CNPs. Like endophysiological CNPs, SV-CNPs target guanylyl cyclase receptor B. Apotyposis has resulted in some SV-CNPs acquiring a C-terminal tail and thus targeting guanylyl cyclase receptor A typically targeted by the tailed forms ANP/BNP/VNP. In addition, in some forms, *de novo* evolution of proline-rich bioactive peptides in the propep region of the SV-CNP precursor has occurred. SV-CNPs have had tremendous success as therapeutics, especially in the treatment of high blood pressure by captopril. This drug, based on one of the propeptide region proline-rich peptides, has saved millions of lives and has been rightly called a wonder drug. Captopril has been viewed as a flagship compound for the isolation and development of clinically significant drugs from animal venoms.

19.2 ENDOPHYSIOLOGICAL PLESIOTYPE

See chapter 18.

19.3 PLESIOTYPIC STRUCTURAL AND FUNCTIONAL TOXIN FORMS

To date, the molecular evolutionary history of SV-CNPs has been a riddle. Two generalized forms of SV-CNPs exist in snake venoms, one with and one without a C-terminal tail (see figure 19.1).

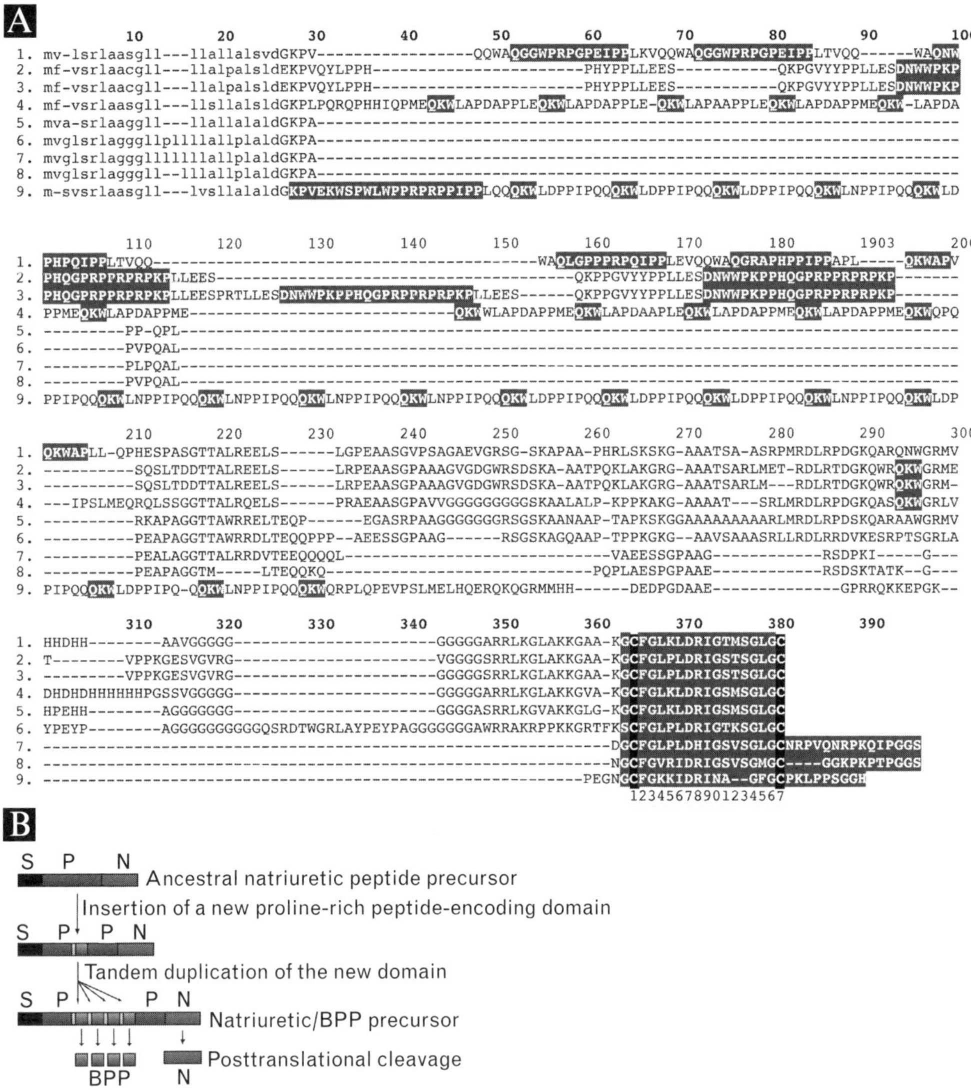

FIGURE 19.1: (A) Alignment of SV-CNPs (Jackson et al. 2013). **(1)** P68515 (*Bothrops insularis*), (2) K4J3K2 (*Azemiops feae*), (3) K4IT20 (*Azemiops feae*), (4) A8YPR6 (*Echis ocellatus*), (5) Q09GK2 (*Philodryas olfersii*), (6) GAHI01000013 (*Suta fasciata*), (7) P83228 (*Oxyuranus scutellatus*), (8) GAHI01000016 (*Suta fasciata*), and (9) A8YPR9 (*Cerastes cerastes*). Posttranslationally cleaved peptides are shaded in gray. Signal peptides are shown in lower-case. (B) Domain mutation producing the tandemly repeated bradykinin-potentiating peptides medial to the SV-CNP encoding domain and the subsequent posttranslational proteolysis of the multidomain product to liberate the discrete peptides (Fry et al. 2010a). BPP = bradykinin-potentiating peptides, N = SV-CNP, P = propeptide, S = signal peptide.

These two forms have been proposed to originate from either the same or two different toxin-recruitment events (Fry and Wüster 2004; Ching et al. 2006; Fry et al. 2009a). The scarcity of full-length sequences has impeded the successful reconstruction of the phylogenetic history of this molecular scaffold, in which most variation accumulates in the region separating the signal peptide and the natriuretic encoding domain, while the short mature peptides are highly conserved at most sites. As new homologues are being retrieved from the caenophidian snake families Dipsadidae, Elapidae, Lamprophiidae, Natricidae, and Viperidae, the phylogenetic tree of this intriguing toxin class is beginning to be resolved. Recent molecular phylogenetic assessments revealed that the advanced snake venom natriuretic peptides are monophyletic (Ching et al. 2006; Fry et al. 2010b; Jackson et al. 2013). Consistent with their placement within the CNP clade, SV-CNPs without the C-terminal tail occupies basal position in the phylogenetic tree. SV-CNP precursors recovered from the venoms of the elapid snakes *Suta fasciata* and *Cacophis squamulosus* lack the apotypic C-terminal tail typical of other sequences recovered from elapid snake venoms (Jackson et al. 2013) (see figure 19.2). The discovery of tailless elapid SV-CNPs supports the aforementioned hypothesis, particularly as these sequences also possess the glycine-rich region of the propeptide that is characteristic of tailless viperid snake venom forms (see figure 19.1). These results add further strength to the hypothesis that the C-terminal extension represents a derivation of this toxin type, and thus the action of the tailed forms on GC-A (Fry et al. 2005) is not indicative of descent from the atrial natriuretic peptide (ANP) but instead represents a remarkable case of convergence for receptor-specific targeting (Fry et al. 2010b). Recently, a study was published in which a SV-CNP possessing the C-terminal extension was sequenced from the venom of *Crotalus oreganus abyssus* (da Silva et al. 2012b). The published sequence represents only the final processed amino acid sequence and does not include the precursor domains.

The aforementioned molecular phylogenetic studies further elucidated the SV-CNP plesiotypic action by shedding light on whether they initially targeted GC-A or GC-B receptors. As intraring residues of body forms are either highly conserved or hypervariable, specificity for GC-A or GC-B is largely conferred by the relative presence or absence of the C-terminal tail. The snake-venom homologues follow this same pattern. Testing of tailless SV-CNPs, such as that from the venom of the viperid snake *Protobothrops flavoviridis*, show that they produce hypotension by effecting relaxation of aortic smooth muscle, mediated by GC-B receptor binding (Michel et al. 2000). Conversely, the forms that have gained the tail at the C-terminus target GC-A. PNP from the venom of the viperid snake *Pseudocerastes persicus* induced a similar cGMP response to that induced by ANP, suggesting that PNP acts on the GC-A receptor (Amininasab et al. 2004). Similarly, DNP from the venom of the elapid snake *Dendroaspis angusticeps* was functionally analogous to ANP in stimulating the production of guanylate cyclase via GC-A (Schweitz et al. 1992). DNP showed pM (picomolar) affinity for the GC-A receptor and no affinity for GC-B (Ki > 1000 nM). DNP was nearly 10-fold more potent than ANP at stimulating cGMP production in GC-A-expressing cells (Johns et al. 2007). DNP was shown to produce relaxation in canine arterial smooth muscle by acting on both GC-A (endothelium-denuded aortic rings) and NPR-C (endothelium-intact aortic rings) receptors (Collins et al. 2000). In a comparison of three tailed SV-CNPs found in *Oxyuranus* venoms, which differed most significantly in the relative presence at ring position 8 of the typical R residue (TNP-c) or a unique substation of H (TNP-a and TNP-b), 0.1 μM TNP-c was equipotent to ANP and DNP in binding GC-A and relaxing both endothelium-denuded and endothelium-intact aortic rings (Fry et al. 2005) (see figure 19.3). Conversely, neither TNP-a nor TNP-b bound GC-A, and both were inactive in the endothelium-denuded aortic ring assay. Both produced a 20% relaxation of the endothelium-intact aortic rings, however, suggesting that their activity is mediated by NPR-C. In a separate study that compared two more tailed SV-CNPs, the homologue from *Pseudonaja textilis* with R at ring position 8 bound GC-A while the *Pseudechis australis* homologue with H at this position was unable to bind to GC-A (St Pierre et al. 2006). Thus, the plesiotypic activity of this group of toxins is targeting

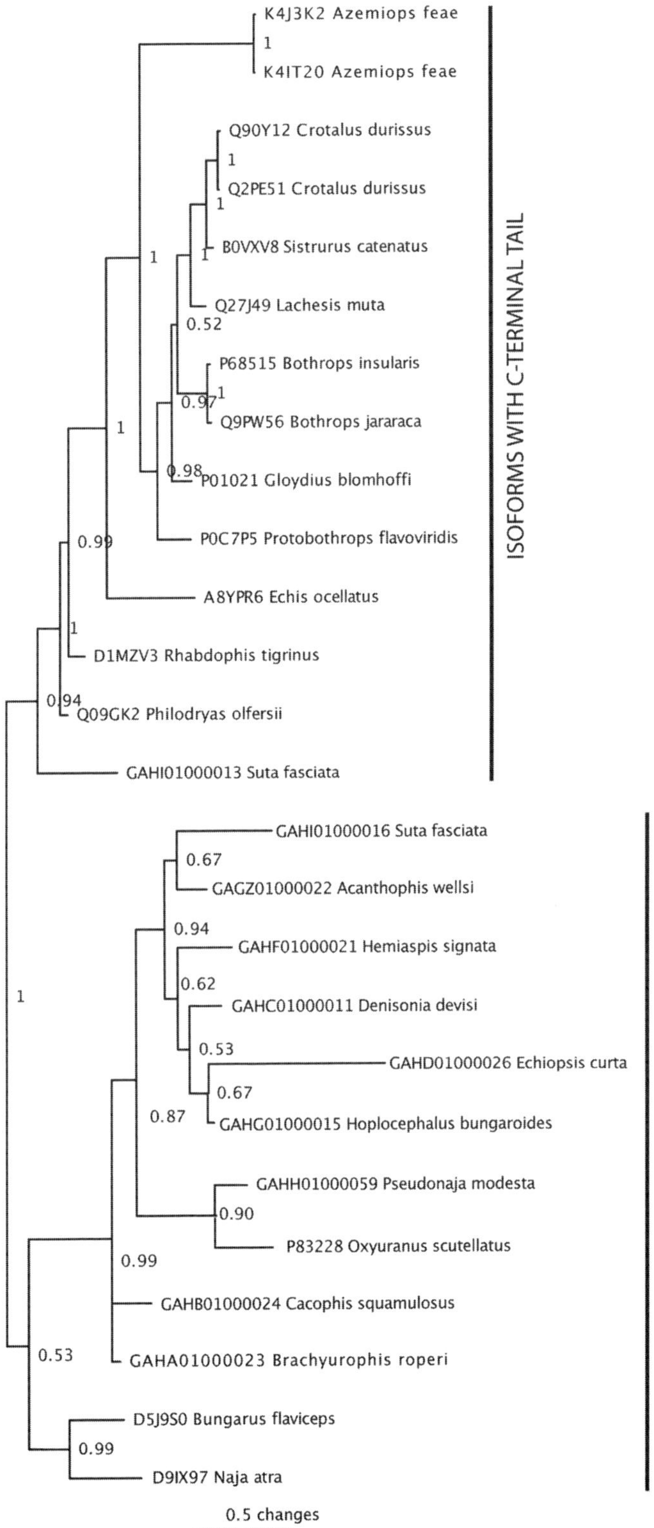

FIGURE 19.2: Molecular phylogeny of SV-CNP tailed and tailless forms (Jackson et al. 2013).

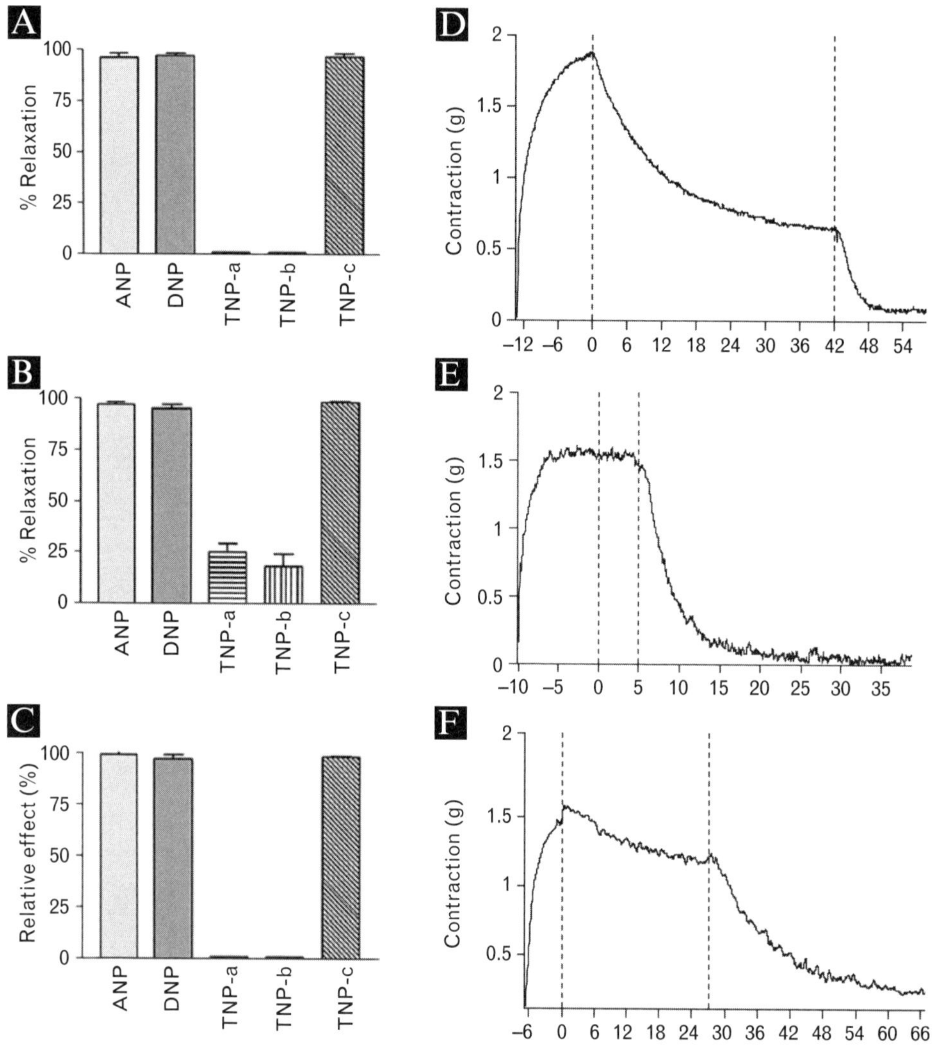


```
ANP   =      SLRRSSCFGGRMDRIGAQSGLGCNSFRY
DNP   =      EVKYDPCFGHKIDRINHVSNLGCPSLRDPRPNAPSTSA
TNP-a = SDPKIGDGCFGLPLDHIGSVSGLGCNRPVQNRPKK
TNP-b = SDSKIGDGCFGLPLDHIGSVSGLGCNRPVQNRPKK
TNP-c = SDSKIGNGCFGFPLDRIGSVSGLGCNRIMQNPPKKFSGE
```

FIGURE 19.3: Bioactivity of SV-CNPs (Fry et al. 2005). Comparative effects of *Oxyuranus* SV-CNPs (0.1 μM) on precontracted with 40 mM KCl (N = 5) rat (A) endothelium-intact aortic rings and (B) endothelium denuded aortic rings. (C) 0.1 μM comparative cGMP production in 293 cells overexpressing GC-A (N = 3). TNP-c (10 nM; t = 0) followed by TNP-c (0.1 μM; t = 42) on (D) endothelium-denuded aortic rings precontracted with 40 mM KCl. TNP-b (0.1 μM; t = 0) followed by TNP-c (0.1 M) on (E) endothelium-denuded (t = 5) and (F) endothelium-intact aortic rings precontracted with 40 mM KCl (t = 27) (N = 3). Cysteines are shaded in black; prolines are shaded in gray.

the GC-B receptor, while the mere presence of a C-terminal tail in a residue nonspecific manner sterically guides GC-A-specific activity in the toxin forms, which is consistent with our understanding of the nontoxin physiological homologues (Fry et al. 2005).

Recently, a proline-rich peptide was isolated from the venom of *Dendroaspis angusticeps* (Quinton et al. 2011) that was described as being of an unknown peptide or protein type. It was subsequently revealed that this proline-rich peptide is part of the propetide region of the natriuretic peptide precursor. This fact was apparently missed by the aforementioned *Dendroaspis* study despite an *Oxyuranus* precursor being available (Uniprot: P83228) that showed significant similarity to this peptide in an early stretch of the propeptide region. Additional sequencing showed these peptides to be ubiquitous in elapid venom natriuretic precursors (Jackson et al. 2013).

19.4 APOTYPIC STRUCTURAL AND FUNCTIONAL TOXIN FORMS

Gaining a C-terminal tail in SV-CNPs switched receptor specificity of the plesiotypic CNP (targeting of GC-B) to targeting the GC-A receptor in a manner convergent with ANP and BNP. In addition, numerous other structural and functional derivations have occurred within this venom-peptide class in both the natriuretic peptide-encoding domain and the propeptide region in which *de novo* toxin formation has occurred.

Within the SV-CNP natriuretic peptide domain, structural derivations include truncations. For example, a SV-CNP isolated from *Protobothrops flavoviridis* has lost the entire N-terminal tail and consists of the ring residues alone (CFGHKLDRIGSTSGLGC). Despite its truncation, this SV-CNP retained vasorelaxant activity in rat aortic strips and potent diuretic activity in anesthetized rats (Michel et al. 2000). SV-CNP functional derivations include inhibitions of the angiotensin-converting enzyme (St Pierre et al. 2006), although the specific residues that confer this apotypic activity have not been elucidated to date. The ability of the SV-CNP from *Macrovipera lebetina* to inhibit collagen-induced platelet aggregation is mediated by the N-terminal sequence (GDNKPPKKGPPNG) of the natriuretic peptide domain (Barbouche et al. 1996; Barbouche et al. 1998).

Within the viperid snake venoms, the propeptide region of the SV-CNP precursor has mutated to encode additional, posttranslationally cleaved bioactive proline-rich peptides: BPP (brakykinin-potentiating peptides) (see Ondetti et al. 1971) and BIP (bradykinin-inhibiting peptides) (see Graham et al. 2005) from pit-viper venoms and the α-neurotoxin azemiopsin from *Azemiopis feae* (Utkin et al. 2012; Brust et al. 2013) (see figure 19.1). Understudied apotypic peptides within this precursor include the *Atheris* pHpG (poly-histidine, poly-glycine peptide) and potential QPW metalloprotease inhibitors from *Echis*. BPPs, also known as angiotensin-converting enzyme inhibitors, retard the degradation of bradykinin and also inhibit the activity of the angiotensin-converting enzyme (ACE). This activity results in the lowering of blood pressure (Ferreira 1965). Curiously, some precursors that encode multiple BPPs in New World Crotalinae also encode BIPs (Graham et al. 2005; Soares et al. 2005).

19.5 THERAPEUTIC POTENTIAL OF THE TOXIN TYPE

Because of their endogenous protective role in the cardiovascular and renal systems (Scriven and Burnett 1985; Yoshimura et al. 1991; Bando et al. 1999; Bargnoux et al. 2008), there is significant interest in the therapeutic development of SV-CNPs. As it was the first SV-CNP to be isolated, the therapeutic value of DNP has been investigated in the greatest detail. The proline-rich C-terminal tail has been shown to confer greater stability and potency than endophysiological mammalian forms, a characteristic likely shared with other SV-CNP possessing the C-terminal tail. A particularly promising technique has been the fusion of the tail from DNP with the ring structure of mammalian CNP to create a chimera called CD-NP (Lisy et al. 2008). CD-NP has increased natriuretic and diuretic potency with lower hypotensive side effects than human BNP. The molecular mechanism by which this chimera operates is binding GC-A while also partially binding GC-B (Dickey, Burnett, and Potter 2008). CD-NP is currently in clinical trials for the treatment of heart failure because of its ability to confer renal protection with minimal side effects following myocardial infarction (Lee et al. 2009).

The BPPs encoded in the propeptide region of crotaline SV-CNP precursors have yielded the most successful venom-derived drug to date: captopril (Ondetti and Cushman 1981). This drug utilizes the antihypertensive effect of these small, proline-rich peptides produced by selective inhibition of the angiotensin-I-converting enzyme (Krieger, Salgado, and Assan 1971; Brunner, Gavras, and Laragh 1974; Gavras et al. 1974). Captopril is an orally active peptidomimetic synthetic analogue of the active tripeptide motif Phe-Ala-Pro contained within the BPPs (see figure 19.4). The medical and economic importance of this drug cannot be overstated; it has been a multi-billion-dollar "wonder drug" that has saved countless lives.

Physiological homologues of SV-CNP have also been investigated as candidates for treatment of congestive heart failure resulting from myocardial infarction and/or hypertension (Yoshimura et al. 1991; Colucci et al. 2000). A drug called carperitide, a synthetic form of ANP (Saito 2010), was approved in Japan in 1995. Although 82% of patients with acute heart failure showed improvement after treatment with carperitide, adverse side effects such as low blood pressure and renal function disturbance discouraged its use (Suwa et al. 2005; Nomura et al. 2008). Similarly, a recombinant form of normal body BNP called nesiritide was demonstrated to improve left-ventricular function by vasodilation (Yoshimura et al. 1991), in addition to improving dyspnea and fatigue (Colucci et al. 2000) in patients suffering from congestive heart failure. Unfortunately, the drug seems to have an effect on renal function and causes high mortality rate in patients (Sackner-Bernstein and Aaronson 2005; Sackner-Bernstein et al. 2005; Sackner-Bernstein, Skopicki, and Aaronson 2005). The CD-NP chimera was created by the Mayo Clinic to induce the cardiovascular and renal effects of these peptides without causing harmful hypotensive side effects (Lisy et al. 2008). In canine models of experimental heart failure, these chimeric peptides were natriuretic and diuretic while increasing the cardiac unloading and glomerular filtration rate and without the hypotensive effects of BNP (Lisy et al. 2008). Natriuretic peptides with platelet-inhibiting action, called lebetin peptides, have been isolated from *Macrovipera lebetina*. These peptides possess proline- and lysine-rich domains as part of the N-terminal tail (Barbouche et al. 1996; Barbouche et al. 1998).

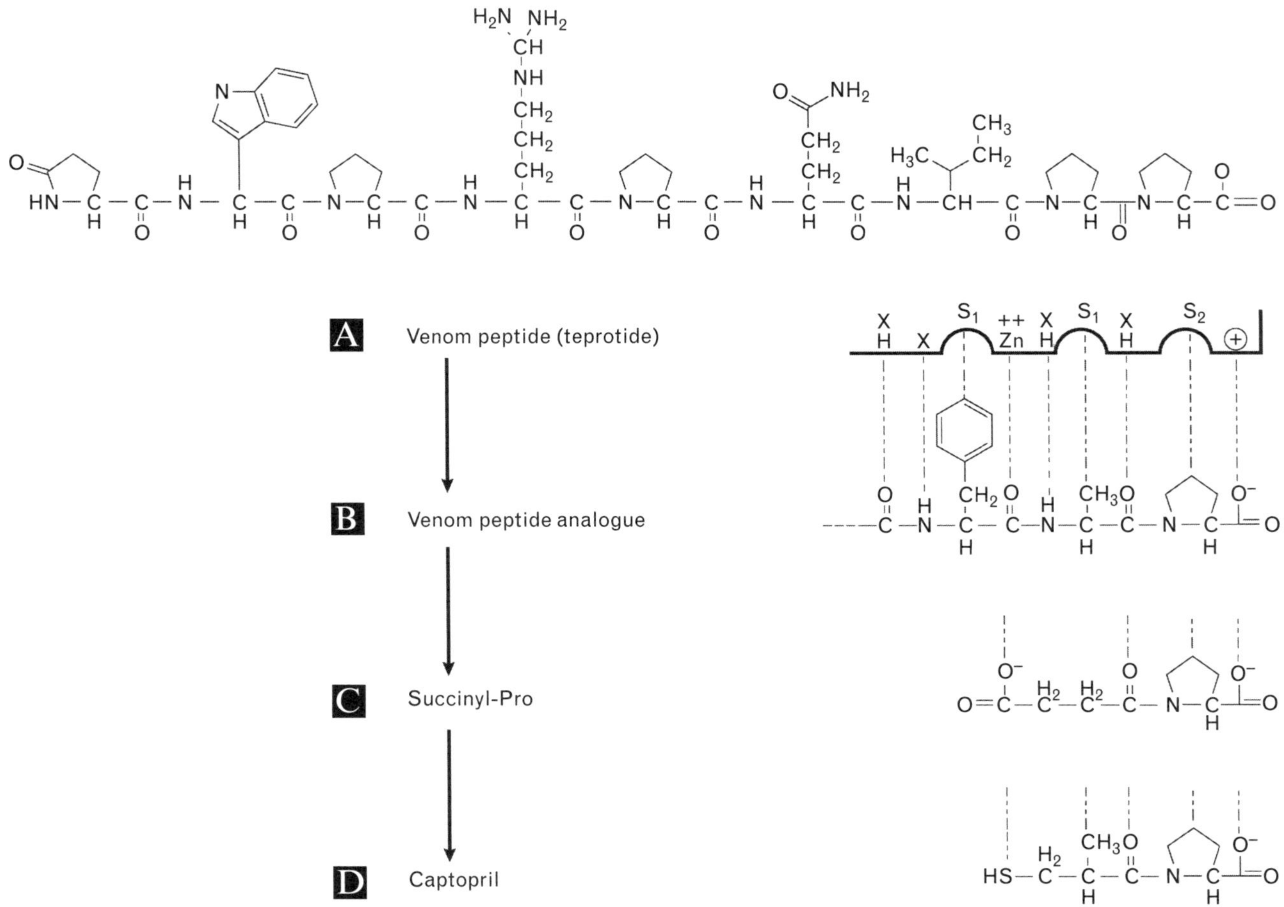

FIGURE 19.4: The chemical steps in the development of captopril from a SV-CNP (Lewis and Garcia 2003). (A) The long-lasting tetropride was isolated from *Bothrops jararaca* venom that had the crucial limitation in not being orally available (Ondetti et al. 1971; Ondetti 1988). (B) The minimum pharmacophore for the binding to angiotensin-converting enzyme (ACE) was determined to be Phe-Ala-Pro sequence (Cushman and Ondetti 1999). (C) While ACE-selective, succinyl-pro displayed only low affinity (Cushman and Ondetti 1999). (D) The orally available peptidomimetic captopril was designed by the addition of a sulfhydryl group to succinyl-Pro (Cushman et al. 1977; Natesh et al. 2003).

19.6 CONVERGENCE WITH OTHER VENOMS

See chapter 18.

CHAPTER 20

GROUP I PHOSPHOLIPASE A_2 ENZYMES

K. SUNAGAR, T. N. W. JACKSON, T. REEKS, AND B. G. FRY

20.1 SUMMARY

Phospholipase A_2 are esterolytic enzymes that catalyze the hydrolysis of 1,2-diacyl-3 -sn-glycerophospholipids at the C2 position, resulting in the formation of lysophospholipids and a free fatty acid. They constitute one of the largest families of lipid-hydrolyzing enzymes. PLA_2s are known to be among the most potent components found in snake venoms, where they exhibit a diverse array of pharmacological activities. Pancreatic-type group I PLA_2 was recruited into the common ancestor of elapid snakes and some non-front-fanged caenophidian snakes (see color plate 3) to form the snake venom group I toxin type (SV-G^{I}-PLA_2). In contrast, synovial-type group II PLA_2 was recruited into the viperid snakes to form the snake venom group II toxin type (SV-G^{II}-PLA_2-) (see chapter 21). SV-G^{I}-PLA_2s are plesiotypically neurotoxic, and within elapid snake venoms, they have been shown to evolve rapidly under the influence of positive Darwinian selection. This has led to the neofunctionalization of a diversity of apotypic activities, including coagulopathy and myotoxicity. As a result, SV-G^{I}-PLA_2s have tremendous potential as pharmacological exploratory probes and possibly as therapeutics.

20.2 ENDOPHYSIOLOGICAL PLESIOTYPE

Phospholipase A_2 (PLA_2; EC 3.1.1.4) belongs to the family of esterolytic enzymes that catalyze the hydrolysis of the sn-2 fatty acyl bond of glycerophospholipids, releasing lysophospholipids and a free fatty acid. Both of these products act as primary messengers for numerous physiological pathways. For example, hydrolysis of the arachidonic acid sn-2 bond and further processing by various enzymes result in the formation of eicosanoids

(Funk 2001), which in turn play a role in a diversity of physiological functions. The PLA_2 family is one of the largest families of lipid-hydrolyzing enzymes, larger than those of phospholipase C and phospholipase D (Valentin and Lambeau 2000). PLA_2s are involved in numerous endophysiological activities, including apoptosis, lipid digestion, signal transduction, hypersensitization, host defense, smooth-muscle contraction, reactive oxygen species generation, fertilization, maintenance of cellular phospholipid concentration, membrane repair through deacylation/reacylation, production of prostaglandins and leukotrienes that participate in signal transduction and smooth-muscle contraction, production of eicosanoids and other lysophospholipid derivatives with potent biological activities, and cell migration, proliferation, and differentiation (Arita et al. 1991; Dennis et al. 1991; Fry et al. 1992; Nakajima et al. 1992; Sommers et al. 1992; Kudo et al. 1993; Vadas et al. 1993; Dennis 1994; Roshak et al. 2000; Balsinde, Perez, and Balboa 2006; Herbert and Walker 2006; Zhang et al. 2006c; Bao et al. 2007; Song et al. 2007). The normal functioning of PLA_2 enzymes is therefore essential for homeostasis. Malfunctioning of these enzymes has been associated with numerous diseases, such as adult respiratory distress syndrome, asthma, chronic inflammation, osteoarthritis, psoriasis, and rheumatism (Balsinde et al. 1999; Touqui and Alaoui-El-Azher 2001). They are also associated with autoimmune uveitis, myocardial infarction, and septic and endotoxic shock (Mukherjee et al. 2014).

PLA_2 enzymes are characterized by the presence of a Ca^{2+} binding loop (residues 25–33; motif Y25-G-C-Y/F-C-G-X-G-G33) that acts as a binding site for Ca^{2+} ions, which are required for catalytic functioning. They also require histidine, tyrosine, and aspartic acid at 48, 52, and 99, respectively, for catalytic activity. An active region around 15 Å deep is found in all PLA_2s, which acts as a binding site for phospholipid substrates (Lambeau and Gelb 2008). Five major classes of PLA_2 enzymes are recognized, based on amino acid sequence, three-dimensional structure, site of expression, and catalytic specificity: (i) secreted PLA_2 ($sPLA_2$), (ii) cytosolic PLA_2 ($cPLA_2$), (iii) Ca^{2+}-independent $iPLA_2$, (iv) platelet-activating factor (PAF) acetylhydrolases, and (v) lysosomal PLA_2.

Secreted PLA_2s, which require histidine at their active site for functioning, are abundantly secreted in a range of cell types. They have a low MW and are also dependent on Ca^{2+} ions for enzymatic catalysis. They are characterized by the presence of six conserved disulfide bonds, and an additional one or two disulfide bonds may also be present (Six and Dennis 2000; Schaloske and Dennis 2006). SV-G^{I}-PLA_2 was recruited from within the secreted PLA_2, specifically from pancreatic group I PLA_2. Pancreatic group I PLA_2s aid in the digestion of phospholipids, are expressed as zymogens or proenzymes, and require the liberation of the propeptide domain (around eight residues) as part of posttranslational modification in presence of trypsin.

Genes encoding group I PLA_2s have four exons and three introns (Seilhamer et al. 1986; Kerfelec et al. 1990). These PLA_2s are stabilized by seven disulfide bonds, require Ca^{2+} for enzymatic catalysis, and are characterized by a low MW of around 13–14 kDa (Heinrikson, Krueger, and Keim 1977). They possess two antiparallel beta sheets and three alpha helices and are typically 114–120 residues long. The highly conserved N-terminal domain, which also contributes significantly to the hydrophobic channel, is believed to stabilize the adjacently located β-wing (Scott 1997). The absence of this domain has been correlated with the loss of catalytic function in certain PLA_2 enzymes (Scott et al. 1991). Characteristic of the pancreatic group I PLA_2s is a loop known as the pancreatic loop, which connects the catalytic α-helix to the β-wing.

20.3 PLESIOTYPIC STRUCTURAL AND FUNCTIONAL TOXIN FORMS

Snake venom group I PLA_2s (SV-G^I-PLA_2s) have been isolated mostly from elapid snake venoms but have also been isolated from other snake families, excluding viperid snakes. The PLA_2 are regarded as one of the most toxic enzymatic venom components in the secretions of advanced snakes and are well known for the diversity of pharmacological activities they exhibit. PLA_2 enzymes were isolated for the first time from the venoms of *Naja naja* and *N. tripudians* (now *N. kaouthia*) (De 1944). Because of their ability to induce hemolysis of red blood cells (RBC), they were initially known as hemolysins Numerous additional PLA_2s have been characterized since then. Despite sharing a high degree of sequence similarity, they exhibit a plethora of biological and physiological activities (Braganca and Sambray 1967; Salach et al. 1971; Vishwanath, Kini, and Gowda 1987; Takasaki, Suzuki, and Tamiya 1990; Kini and Chan 1999; Kini 2003).

While most group I PLA_2s excise the propeptide domain during posttranslational modification (Francis et al. 1997; Huang et al. 1997), some SV-G^I-PLA_2s retain them in the mature state (Pearson et al. 1993; Jackson et al. 2013). Based upon the similarity in structure and function of the form isolated from the colubrid snake *Trimorphodon biscutatus* and that of elapid snakes, it is hypothesized that the plesiotypic form of SV-G^I-PLA_2 is a neurotoxic monomer that retains the pancreatic loop (Fry et al. 2003c; Huang and Mackessy 2004; Lumsden et al. 2004b; Fry et al. 2008). Neurotoxicity results from physical damage to nerve terminals (Harris et al. 2000; Fry et al. 2003c; Huang and Mackessy 2004; Lumsden et al. 2004a; Rigoni et al. 2004; Rigoni et al. 2005; Fry et al. 2008). Such SV-G^I-PLA_2s have been isolated and characterized widely from elapid snake venoms, including *Micrurus frontalis frontalis* (Francis et al. 1997), *Notechis scutatus* (Francis, Bdolah, and Kaiser 1995), *Ophiophagus hannah* (Huang et al. 1997), *Oxyuranus scutellatus* (Fohlman, Lind, and Eaker 1977), and *Pseudonaja textilis* (Pearson et al. 1993). They have also been recovered from the venom of the colubrid snake *T. b. lambda* (Fry et al. 2003c; Huang and Mackessy 2004; Lumsden et al. 2004b; Fry et al. 2008). These neurotoxins bind irreversibly to motor nerve terminals (Cull-Candy et al. 1976; Prasarnpun et al. 2005), leading to depletion and/or impaired release of acetylcholine vesicles and resulting in the degeneration of the terminal (Cull-Candy et al. 1976; Dixon and Harris 1999; Harris and Goonetilleke 2004; Prasarnpun et al. 2005). The neuromuscular blockade induced by these toxins occurs in the following steps: (i) a sudden decline in the release of acetylcholine, (ii) an increase in the release of acetylcholine, and (iii) the complete inhibition of the neuromuscular junction (Su and Chang 1984; Dixon and Harris 1999; Aird 2002; Hodgson and Wickramaratna 2002; Lewis and Gutmann 2004; Doley and Kini 2009). Because of damage caused to nerve terminals, clinical recovery is slow and depends on the rate of regeneration and formation of new neuromuscular junctions (Harris and Goonetilleke 2004; Prasarnpun et al. 2005).

Venom composition and activity not only varies across lineages but may also vary between closely related species within the same clade (Mackessy 2010). The evolution of prey-specific toxins in species with specialized diets also induces intraspecific variation in venom composition (Jorge da Silva and Aird 2001; Barlow et al. 2009). A correlation between diet and variation in PLA_2 has been demonstrated in a number of snake lineages. *Aipysurus eydouxii*, *Aipysurus mosaicus*, and *Emydocephalus annulatus* have evolved to feed exclusively on fish eggs (Voris and Voris 1983; Gopalakrishnakone and Kochva 1990). Consequently, these species of sea snake have atrophied venom glands and have even lost effective fangs (McCarthy 1987; Gopalakrishnakone and Kochva 1990). As a result of this remarkable dietary shift, preexisting negative selection pressures constraining the diversification of PLA_2s in their venoms have been alleviated, and the genes encoding PLA_2 enzymes have accumulated several deleterious mutations (Li, Fry, and Kini 2005a). Not surprisingly, there is a 50- to 100-fold decrease in the toxicity of venom of these species relative to that of other species in the same genus.

20.4 APOTYPIC STRUCTURAL AND FUNCTIONAL TOXIN FORMS

A wide diversity of structurally and functionally apotypic forms of SV-G^I-PLA_2s have been isolated and characterized. The most obvious structural variation is the loss of the pancreatic loop. Depending on whether they possess a pancreatic or an "elapid" loop, SV-G^I-PLA_2s have been divided into group IA and group IB enzymes, respectively. However, as the loop has been lost on several independent occasions (see figures 20.1 and 20.2), this simplistic division is artificial, and these groups are not reciprocally monophyletic.

Presynaptic neurotoxic SV-G^I-PLA_2s may be structurally apotypic because of the aggregation of two or more chains, which may be held together by covalent or noncovalent bonds or a combination thereof (Yang, Dignam, and Gentry 1997). Such complexes include β-bungarotoxins, taipoxins, textilotoxins, MitTx, and taicatoxin.

The β-bungarotoxins (β-Btx), found in the venom of *Bungarus* species, are a well-studied presynaptically active complex (Strong et al. 1976). They are composed of two chains linked by a disulfide bond; chain A is a pancreatic-loop containing a SV-G^I-PLA_2, while chain B is a kunitz peptide (see chapter 15) (Kondo, Narita, and Lee 1978a). The two chains are stabilized by a disulfide bond (Bon 1997). Like plesiotypic SV-G^I-PLA_2 toxins, β-bungarotoxins induce presynaptic neurotoxicity characterized by the depletion of synaptic vesicles, destruction of motor nerve terminals, and reinnervation following axonal degeneration (Dixon and Harris 1999). However, β-bungarotoxins are also known to cause calcium influx through voltage-gated Ca^{2+} ion channels. They also induce an increased release of acetylcholine via the SNARE (SNAP [soluble N-ethylmaleimide-sensitive factor attachment protein] receptor) complex, resulting in the depletion of synaptic vesicles (Prasarnpun et al. 2005).

Taipoxins are extremely potent neurotoxic heterodimeric SV-G^I-PLA_2 complexes with an LD_{50} value of 2 μg/kg in mice (Fohlman et al. 1976). They were isolated for the first time from *O. scutellatus* venom but were later also retrieved from the venom of *O. microlepidotus* (Fohlman 1979; Kuruppu et al. 2005a; Hodgson, Dal Belo, and Rowan 2007). Subsequently, this complex was revealed to have evolved at the base of the Australian snake radiation (Jackson et al. 2013). This complex is composed of an extremely toxic α subunit, a nontoxic/neutral β subunit (with two interchangeable β1 and β2 isoforms), and a moderately toxic γ subunit (Fohlman et al. 1976; Lind 1982). The α and β subunits lack the pancreatic loop, while the slightly larger (and glycosylated) γ subunit retains the pancreatic loop and has an eight-residue N-terminal extension resulting from upstream cleavage of the propeptide region. While the neurotoxic activity is driven by the α subunit, the toxicity of the α subunit by itself is much lower than that of the full complex (Fohlman et al. 1976; Francis et al. 1993; Kuruppu et al.

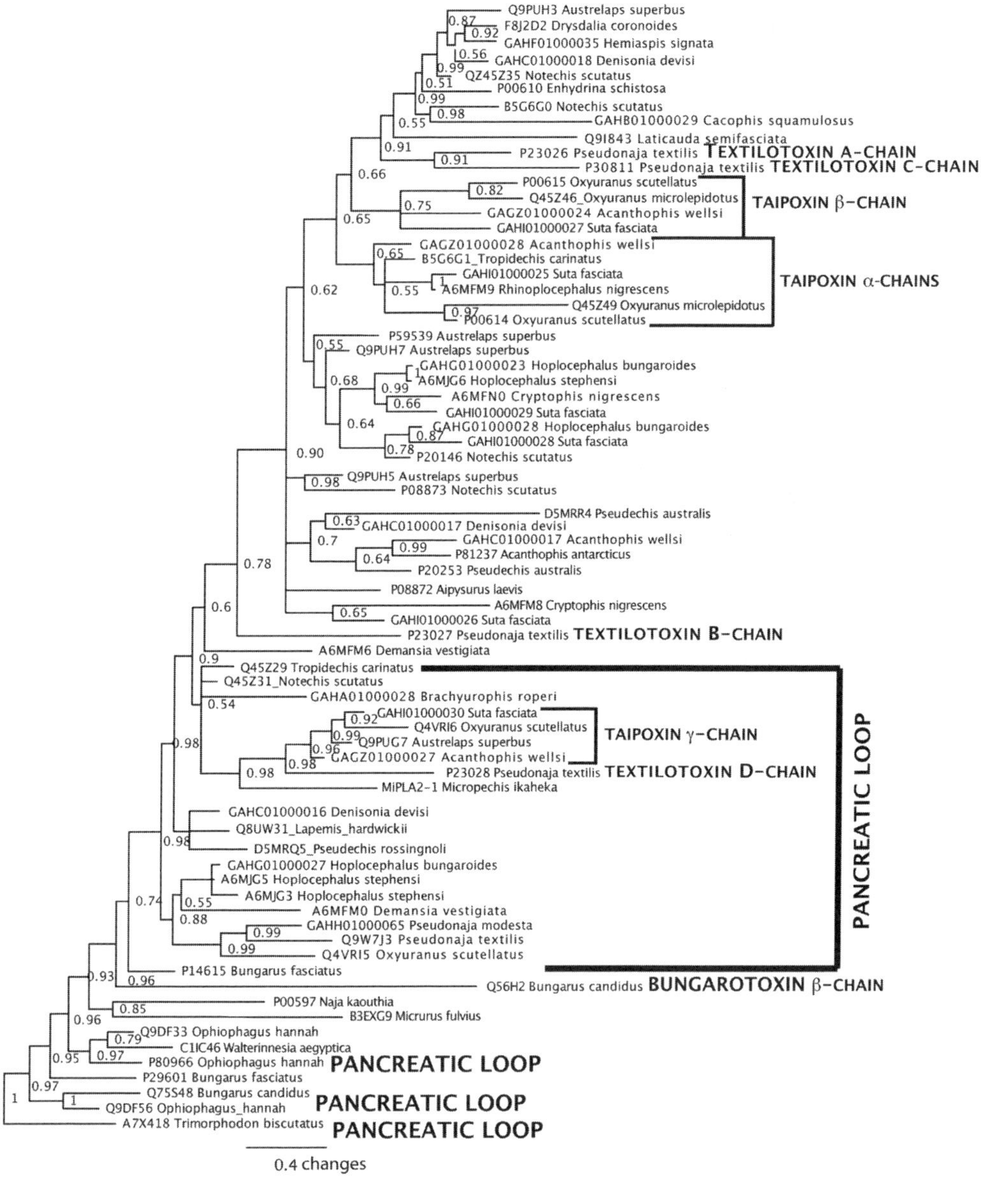

FIGURE 20.1: Molecular phylogeny of SV-G^{I}-PLA$_2$s showing the multiple losses of the pancreatic loop and the multiple derivations for use in disulfide-linked complexes (Jackson et al. 2013).

2005a). The SV-G^{I}-PLA$_2$s isolated from *Austrelaps superbus* venom that exhibit weak neurotoxic activity (van der Weyden et al. 1997; van der Weyden et al. 2001) might in fact be γ subunits that have been separated from the rest of the complex.

Textilotoxins are extremely potent (murine LD_{50}: 1 μg/kg) SV-G^{I}-PLA$_2$ complexes that cause presynaptic blockade of neuromuscular transmission by disrupting the regulatory mechanism that controls acetylcholine release (Su et al. 1983; Wilson et al. 1995b). This toxin has the most complex structure and highest toxicity of any identified snake-venom neurotoxin (Tyler et al. 1987). The structure was initially reported as being composed of five noncovalently

```
                10        20        30        40        50        60        70        80        90       100
1.   mhpahllvllavCvsllgasdIPPLP-LNLAQFGFMIKCANRGSRPVSHYMDYGCYCGKGGSGTPVDELDRCCQVHDECYGEAEKRFKCV-----PYMTL
2.   mhpahllvllavCvsllgasdIPPLP-LNLLQFGYMIRCANRRTRPVWHYMDYGCYCGKGGSGTPVDDLDRCCQVHDECYGEAKRRFRCA-----PYWTL
3.   mypahllvllavCvsllgasdIRPQP-LNLLQFGFMIRCANKRRRPVWHYMDYGCYCGKGGSGTPVDALDRCCKIHDDCYAEAEKRYGCY-----PYWTL
4.   mypahllvllavCvsllgasdIPPQP-LNLIQFGFMIKCGNKRRQPLFRYMDYGCYCGKGGSGTPVDDLDMCCKIHDDCYAKATKSYSCA-----PYWTL
5.   mylahllvllavCvsllgasdNSPQP-LNLIQFGFMIKCGNKRRQPLFRYLDYGCYCGKGGRGTPVDDLDMCCKIHDDCYAKATKSYSCS-----PYWTL
6.   mypahllvllavCvsllgasdIRQQP-LNLLQFAFMIRCANNRRRPVWHYMNYGCYCGKGGSGTPVDELDRCCKIHDDCYARATKSYSCA-----PYWTL
7.   mhpahllvllavCvsllgasdIPPLP-LNLLQFGFMIECAIRNRQPALDFMNYGCYCGTVGHGTPVDDLDRCCKTRNECYAEAEKH-GCY-----PSLTT
8.   mhpahllvlsavCvsllgasdIPPLP-LNLVQFGKMIECAIRNRRPALDFMNYGCYCGKGGSGTPVDDLDRCCQVHDECYAEAEKH-GCY-----PSLTT
9.   mypahllvllavCvsllgaarIPPLP-LNLAQFGKMIECANKGSRPSLDYMNYGCYCGTEDRGTPVDELDRCCKIHDDCYAEAEK-HGCF-----PKMTS
10.  mypahllvllavCvsllgatgIPPRP-LTLLQFDTMIECANNRSRSSNDYMDYGCYCGEGGSGTPVDDLDRCCKIHDDCYGEAKK-HECD-----PLSLK
11.  mhpahllvllavCvsllgsseIPQPS-LDFEQFSNMIQCTIPCGESCLAYMDYGCYCGPGGSGTPIDDLDRCCKTHDECYAEAGKLSACKSVLSEPNNDT
12.  mypahllvllavCvsllgaasIPVPS-LNFEQFGNMIQCTIPCEESCLGYMDYGCYCGPGGSGTPMDELDRCCQTHDNCYAEAGKLPACKAMLSELYNDT
13.  mypahllvllavCvsllgasNIPLPS-LDFEQFGKMIQCTIPCEESCLAYMDYGCYCGPGGSGTPSDELDRCCQTHDNCYAEAGKLPACKAMLSEPYNDT
14.  mypahllvllavCvsllgasdI-PQPSFDFEQFSNMIQCTIPCEESCSGYMDYGCYCGPGGSGTPVDDLDRCCQTHDNCYAEAGKLPACKAMLSESDNDS

                110       120       130       140       150
1.   YSWKCYGTAPSC-NTKTDCQRFVCNCDAKAAECFARSPYQNKNWNINTKARCK
2.   YSWKCYGTAPSC-NTKTQCEGFVCNCDAKAAECFARSPYQPSNRNINTKERCK
3.   YSWKCYGSGPYC-NSKTQCQRFVCACDSRAAICFAQNPYNNANWNINTKTRCR
4.   YSWKCYASGPYC-NSKTRCQRFVCACDAAASRCFARARYNPANYNINTKKRCR
5.   YSWKCYASGPYC-SSKTRCQRFVCACDAAASRCFARAHYNPANYNINTKKRCQ
6.   YSWKCYGNGPTC-NSKTRCQRFVCDCDSKAAMCFARAPYNNANWNIDTKKRCR
7.   YRWQCGRVGLHC-NSKTQCEVFVCACDLAAAKCLAQEDYNPAHFNINTKARCR
8.   YTWECRQVGPYC-NSKTQCEVFVCACDFAAAKCFAQEDYNPAHSNINTGERCK
9.   YSWKCANDVPVC-NSKTQCQHFVCYCDLWATKCFSQETYNKNNFGIDTKTRCK
10.  YRWRCGRQGPYCR-SKTQCQRFVCLCDLLAAKCIAQETYNRENYNIDTKKRCQ
11.  YSYECNEGQLTCNDDNDECKAFICNCDRTAVTCFAGAPYNDLNYNIGMIEHCK
12.  YSYSCIERQLTCNDNNDECKAFICNCDRTAANCFAGAPYNDLNYNIDTTEHCQ
13.  YSYSCIERQLTCNDDNDECKAFICNCDRAAVICFSGAPYNDSNYDIGTIEHCK
14.  YSYSCIEGQLTCNDDNDECKAFICNCDRTAVNCFARAPYNDSNYDIGTTEHCK
```

FIGURE 20.2: Sequence alignment of "taipoxin/paradoxin"-like SV-G^{I}-PLA$_2$ presynaptic complex subunits (Jackson et al. 2013). α-subunit (1) Q45Z43 (*Oxyuranus microlepidotus*), (2) Q45Z48 (*Oxyuranus scutellatus*), (3) GAGZ01000028 (*Acanthophis wellsi*), (4) A6MFM9 (*Rhinoplocephalus nigrescens*), (5) GAHI01000025 (*Suta fasciata*), (6) B5G6G1 (*Tropidechis carinatus*). β-subunit (7) Q45Z46 (*O. microlepidotus*), (8) Q45Z53 (*O. scutellatus*), (9) GAGZ01000024 (*A. wellsi*), (10) GAHI01000027 (*S. fasciata*). γ-subunit (11) Q4VRI6 (*O. scutellatus*), (12) GAGZ01000027 (*A. wellsi*), (13) Q9PUG7 (*Austrelaps superbus*), (14) GAHI01000030 (*S. fasciata*).

linked subunits (A, B, C, and D, with D existing as a covalently linked dimer) (Tyler et al. 1987; Pearson et al. 1993), but more recent evidence points instead to two alternative hexameric structures of $(A/B)_2C_2D_{2a}$ or $(A/B)CD_{2a}D_{2b}$, where D_{2a} and D_{2b} refer to differentially glycosylated dimers of the dimeric subunit D (Aquilina 2009). The subunits differ from one another structurally and functionally: (i) subunits A, B, and C lack the pancreatic loop, while subunit D retains it; (ii) subunit A exhibits toxic activity, whereas subunits B and C are nontoxic; (iii) subunit D exhibits weak phospholipase activity (Pearson et al. 1993). To date, only studies concerning isolation and characterization from *P. textilis* have been published, and therefore, the relative presence and diversity in other species of *Pseudonaja* remain to be elucidated. It is anticipated that textilotoxin will be recovered from the venoms of other species of *Pseudonaja*, although it appears that *P. modesta* lacks textilotoxin (Jackson et al. 2013).

Lesser studied complexes include MitTx and taicatoxins. MitTx is a heteromeric SV-G^{I}-PLA$_2$ complex made up of a SV-G^{I}-PLA$_2$ and a kunitz peptide (see chapter 15). This complex acts as a powerful agonist on acid-sensing ion channels (ASICs), inducing excruciating pain (Bohlen et al. 2011).Taicatoxins are composed of three noncovalently linked subunits in a 1:1:4 stoichiometric ratio: (i) 3FTx (see chapter 8), (ii) neurotoxic PLA$_2$, and (iii) kunitz peptides (see chapter 15). It has been demonstrated that this oligomeric complex is capable of blocking Ca^{2+} channels (Brown et al. 1987).

As outlined in the sections above, the plesiotypic action of SV-G^{I}-PLA$_2$s is presynaptic neurotoxicity (Bon 1997). However, numerous SV-G^{I}-PLA$_2$s with a diversity of apotypic toxic activities have also been isolated and characterized, including cardiotoxins, myotoxins, platelet-aggregation inhibitors, and activation of the muscarinic receptor. Examples include the acidic OHVA-PLA_2 isolated from the venom of *O. hannah*, which exhibits myotoxicity, cardiotoxicity, and antiplatelet effects (Huang, Wang, and Liu 1993; Huang and Gopalakrishnakone 1996; Huang et al. 1997). The cardiotoxic action of OHVA-PLA_2 is a result of it causing an increase in the level of intracellular Ca^{2+} (Huang, Wang, and Liu 1993). In addition, potent myonecrotic SV-G^I-PLA_2s have been retrieved from the venoms of Australasian elapids (Mebs, Ehrenfeld, and Samejima 1983), including *Acanthophis, Micropechis, Notechis, Pseudechis*, and various sea snakes (Ducancel et al. 1988; Kuruppu et al. 2005b). Sea snakes and *Pseudechis* species are known to be particularly rich in such PLA_2s (Lind and Eaker 1981; Kamiguti et al. 1994; Laing et al. 1995; Fry et al. 2003c). MiPLA1, retrieved from the venom of *Micropechis ikaheka*, was the PLA_2 known to cause hemoglobinuria (Gao and Xie 1999). This activity was attributed to leakage in the kidneys rather than lysis of red blood cells. PLA_2 toxins in the venoms of other Australian elapid snakes, such as *A. superbus*, are capable of causing platelet-aggregation inhibition (Subburaju and Kini 1997; Chow et al. 1998). Three PLA_2 isoenzymes (praelongins 2bIII, 2cII, and 2cIV) that exhibited antiplatelet activity were isolated from the venom of *Acanthophis praelongus* (Sim 1998). Certain PLA_2 enzymes, such as CM-I, CM-II, and CM-IV from the venom of *Naja nigricollis* (Evans et al. 1980), exhibit an anticoagulant activity that appears to be independent of their hydrolytic activity (Boffa et al. 1976; Evans et al. 1980; Verheij et al. 1980a). These PLA_2s are categorized into strongly (inhibits coagulation of blood at concentrations of less than 2 μg/ml, such as CM-IV) and weakly (anticoagulant at concentrations ranging from 3 to 10 μg/ml, such as CM-I, CMII) anticoagulant enzymes (Boffa and Boffa 1976; Verheij et al. 1980a). Their molecular targets are soluble proteins of the coagulation pathway. Activation of the muscarinic receptor causes a dose-dependent contraction of smooth muscle (Huang et al. 2008).

20.5 THERAPEUTIC POTENTIAL OF THE TOXIN TYPE

SV-G^I-PLA_2s exhibit numerous pharmacological activities: anticoagulation, antiparasitic, bactericidal, cytotoxic, cardiotoxic, edema-inducing, hypotension-inducing, myotoxic, platelet-aggregation inducing and inhibiting, and pre- or postsynaptically neurotoxic, which may or may not be dependent on their phospholipid hydrolytic activity (Kini 2003). As a result, SV-G^I-PLA_2s have tremendous potential as pharmacological exploratory probes and as direct therapeutics. For example, N-type and M-type receptors, a group of PLA_2-binding proteins, were discovered using the neurotoxic SV-G^I-PLA_2 OS_2 from the venom of *O. s. scutellatus* (Lambeau and Lazdunski 1999). In addition, ASIC channels, which are known to participate in the sensation of pain, were characterized using MitTx (Bohlen et al. 2011; Olivera and Teichert 2011). A SV-G^I-PLA_2 called NN-XI_a-PLA_2, which was isolated from the venom of *N. naja*, was found to be cytotoxic toward Ehrlich ascites tumor cells (Basavarajappa and Gowda 1992). The antineoplasmic activity of SV-G^I-PLA_2s has significant therapeutic potential. A diversity of pharmacological properties of bee venoms, such as neurite outgrowth induction, caused by PLA_2 have recently been reported (Nakashima, Kitamoto, and Arioka 2004). Group IB PLA_2s (pancreatic PLA_2) exhibits antiparasitic activity and is capable of preventing *Plasmodium falciparum* infection of erythrocytes (Deregnaucourt and Schrevel 2000). SV-G^I-PLA_2s isolated from snake venoms also have demonstrated bactericidal activity (Forst et al. 1986).

20.6 CONVERGENCE WITH OTHER VENOMS

The PLA_2s found in viperid snake venoms are structurally similar to the PLA_2s found in mammalian inflammatory secretions and are an independent recruitment, this time of a group II PLA_2 (SV-G^{II}-PLA_2) (see chapter 21) (Fry and Wüster 2004). The PLA_2s in the venom of Anguimorpha lizards are another independent recruitment, this time of group III PLA_2 (see chapter 24). PLA_2s have also been independently recruited in the venoms of multiple arthropod orders (such as bees, centipedes, scorpions, spiders, ticks) and of coleoids and jellyfish (Kini 1997; Valdez-Cruz, Batista, and Possani 2004; Hariprasad et al. 2007; Fry, Roelants, and Norman 2009; Ruder et al. 2013b; Zamudio et al. 1997). In the venoms of these animals, they possess numerous toxin activities, such as myotoxicity, neurotoxicity, cardiotoxicity, cytotoxicity, hemolytic activity, muscarinic inhibition, edema induction, platelet inhibition, hypotension, and inflammatory and anticoagulant action (Harris 1985; Rosenberg 1990; Hawgood and Bon 1991; Yang 1994; Gutiérrez and Lomonte 1995; Kini and Evans 1997; Ownby 1998; Teixeira et al. 2009). Group III PLA_2s constitute the major component of honeybee (*Apis mellifera*) venom (up to 12% of the venom) (Habermann 1972; Sobotka et al. 1976; King and Spangfort 2000). Group III PLA_2s have also been isolated from the venoms of coleoids and scorpions (Kini 1997; Zamudio et al. 1997; Valdez-Cruz, Batista, and Possani 2004; Hariprasad et al. 2007; Fry, Roelants, and Norman 2009; Ruder et al. 2013b). Similar to the other Group III PLA_2s, the coleoid PLA_2 sequences have a lengthy propeptide region, followed by a 10-cysteine arrangement in the mature toxin. However, PLA_2 homologues in squid lack the ninth plesiotypic cysteine, resulting in an odd number of cysteines, which could facilitate dimerization (Ruder et al. 2013b). The genes encoding these PLA_2s have been shown to be evolving under the significant influence of negative selection (Ruder et al. 2013b). Functionally, the muscarinic receptor action of apotypic SV-G^{I}-PLA_2s is convergent with that of apotypic 3FTxs (see chapter 8.4.1) but with the difference that the SV-G^{I}-PLA_2s are agonists while the muscarinic 3FTxs are antagonists.

CHAPTER 21

GROUP II PHOSPHOLIPASE A_2 ENZYMES

K. SUNAGAR, I. H. TSAI, B. LOMONTE, T. N. W. JACKSON, AND B. G. FRY

21.1 SUMMARY

Phospholipase A_2 are esterolytic enzymes that catalyze the hydrolysis of 1,2-diacyl-3-sn-glycerophospholipids at the C2 position, resulting in the formation of lysophospholipids and a free fatty acid. They constitute one of the largest families of lipid hydrolyzing enzymes. PLA_2s are known to be one of the most potent components found in snake venoms, where they exhibit a diverse array of pharmacological activities. Group II PLA_2 was recruited into the viperid snakes to form the snake venom group II toxin type (SV-G^{II}-PLA_2). In contrast, pancreatic-type group I PLA_2 was recruited into the venom of the common ancestor of elapid snakes/non-front-fanged caenophidian snakes (see color plate 3) to form the snake venom group I toxin type (SV-G^{I}-PLA_2) (see chapter 20). In addition, SV-G^{II}-PLA_2s are further divided into two major subcategories, D^{49}-SV-G^{II}-PLA_2 and K^{49}-SV-G^{II}-PLA_2, depending on whether the enzyme contains an aspartic acid or a lysine residue at the 49th position. Given the essential role of the active site Ca^{+2} in the esterolytic mechanism, the replacement of the calcium-binding Asp^{49} with a Lys (or Asn, Arg, and Ser) in PLA_2s results in the loss of catalytic activity. Thus the ability to induce toxicity is a result of some noncatalytic mechanism by the enzyme homologue. Snake-venom PLA_2 enzymes have undergone neofunctionalization under the significant influence of positive selection, resulting in a diversity of apotypic activities, such as antiplatelet, anticoagulant, neurotoxic, or antibacterial activities. Most, if not all, SV-G^{II}-PLA_2s appear to have special pharmacological sites besides the catalytic center, thus facilitating apotypic effects. Interestingly, many SV-G^{II}-PLA_2s show multiple activities with varied potencies. The acidic PLA_2s usually are inhibitors of platelet aggregation, while the basic PLA_2s are diversified to play myotoxic, neurotoxic, or anticoagulant roles. Most of the basic SV-G^{II}-PLA_2s can cause hypotension, myonecrosis, and edema in experimental animals but with great variation in the duration and potency of their effects. Because of this diversity, these toxins possess tremendous potential as investigational ligands, if not as therapeutics.

21.2 ENDOPHYSIOLOGICAL PLESIOTYPE

The 14-kDa $sPLA_2$ (PLA_2; EC 3.1.1.4) found in the body fluid of vertebrates are membrane- or interfacial-active esterases that catalyze the hydrolysis of the sn-2 fatty acyl bond of glycerophospholipids and release lysophospholipids and free fatty acids; these products may further participate in a diversity of physiological pathways. The $sPLA_2$ have been implicated in an array of physiological activities, such as lipid digestion, signal transduction, antimicrobial or host defense, apoptosis, prey capture, smooth-muscle contraction, reactive oxygen species generation, fertilization, production of eicosanoids, and generation of lysophospholipid derivates with special biological activities, such as cell migration and proliferation, and differentiation (Roshak et al. 2000; Balsinde et al. 2006; Herbert and Walker 2006; Zhang et al. 2006c; Bao et al. 2007; Song et al. 2007).

Based on the amino acid sequence, catalytic specificity, site of expression, and three-dimensional structure, vertebrate PLA_2 have been classified into five major classes: (i) secreted PLA_2 ($sPLA_2$), (ii) cytosolic PLA_2 ($cPLA_2$), (iii) Ca^{2+}-independent $iPLA_2$, (iv) platelet-activating factor (PAF) acetylhydrolases, and (v) lysosomal PLA_2. The $sPLA_2$ members are further classified into 10 categories: group IB, IIA, IIC (pseudogene in humans), IID, IIE, IIF, III, V, X, and XII. Group IIA PLA_2 is one of the $sPLA_2$ types that have been extensively studied. Ca^{2+} ions are extremely important for the catalytic activity of these enzymes (Yu, Berg, and Jain 1993). In addition to the catalytic site ($D^{42}XCCXXHD^{49}$), group II PLA_2s are characterized by a Ca^{2+} ion-binding loop domain ($X^{28}CGXGG^{33}$), which experiences a significant influence of negative selection pressure and as a result remains extremely well conserved (Murakami and Kudo 2002). This domain has been demonstrated to be a result of interaction between the ß-carboxyl group of Asp^{49} and the backbone carbonyl groups of Tyr/X^{28}, Gly^{30}, and Gly^{32} (Fleer, Verheij, and de Haas 1981). These enzymes are known to be acute-phase proteins that are expressed in various cells and tissue types, in response to proinflammatory chemical messengers (cytokines). Thus, they play a role in amplifying the inflammatory response.

21.3 PLESIOTYPIC VENOM TOXIN FORMS

Snake venom group II PLA_2 ($SV\text{-}G^{II}\text{-}PLA_2$) appear to be restricted to the viperid snake lineage. Population-level variations in the composition of $SV\text{-}G^{II}\text{-}PLA_2$s have been widely noted in pit viper venoms (see Creer et al. 2003; Tsai et al. 2004). The potently myotoxic forms (see Kihara et al. 1992) that contain aspartic acid at the 49th alignment position are the plesiotypic form and are referred to as $D^{49}\text{-}SV\text{-}G^{II}\text{-}PLA_2$. It has been hypothesized that binding of a water molecule at 3 Å from N-terminus is required for esterolysis, as the Ca^{2+}-conjugated water molecule acts as a nucleophile (Verheij et al. 1980b). Since the $D^{49}\text{-}SV\text{-}G^{II}\text{-}PLA_2$ plesiotypic function (enzymatic activity) is dependent on the active site Ca^{2+} ion, the PLA_2 structure shows slight variation in the Ca^{2+} binding loop ($X_{28}CGXGG_{33}$), and mutations at this region result in a significant reduction in catalytic action (Fukagawa et al. 1993). Additionally, the His residue at the 48th position plays the role of a general base in the reaction mechanism. Chemical modifications of His^{48} resulted in the loss of both enzymatic and toxic activities (Rodrigues et al. 1998; Soares et al. 2001; Soares and Giglio 2003). Site-directed mutagenesis of these residues has found concordant results (van den Bergh et al. 1988; Ward et al. 2002; Petan, Krizaj, and Pungercar 2007).

21.4 APOTYPIC VENOM TOXIN FORMS

In addition to the plesiotypic myotoxic activity, apotypic monomeric D^{49}-SV-G^{II}-PLA_2s have been characterized. Pharmacological studies reveal certain functional types of the D^{49}-SV-G^{II}-PLA_2s, including platelet-interacting hemotoxins that are commonly or abundantly expressed in pit viper venoms (see Fuly et al. 1997; Fuly et al. 2000; de Faria et al. 2001; Xu et al. 2002; Fuly et al. 2003; de Albuquerque Modesto et al. 2006; Wang et al. 2008; Vargas et al. 2012) and presynaptic neurotoxins that block acetylcholine release, so far identified in neotropic rattlesnake venom and venoms of certain *Daboia, Gloydius*, and *Vipera* (Ritonja and Gubensek 1985; Mancheva et al. 1987; Jan et al. 2002; Chen et al. 2004; Gao et al. 2009; Yang et al. 2015). Some D^{49}-SV-G^{II}-PLA_2s also show weak presynaptic neurotoxicity in addition to medium anticoagulant effects by binding to coagulation factor FXa and inhibiting the prothrombinase activity (Faure, Gowda, and Maroun 2007). Interestingly, other D^{49}-SV-G^{II}-PLA_2s with more potent presynaptic neurotoxic activities usually show lower anticoagulant effects (see Petan et al. 2005). The variation between closely related forms that differ in apotypic activity (coagulopathy versus neurotoxicity) allows the elucidation of the structure-function relationships (Saul et al. 2010; Tsai, Wang, and Hseu 2011b).

In addition to existing as monomers, neurotoxic apotypic D^{49}-SV-G^{II}-PLA_2s also exist as covalently linked neurotoxin complexes that may be homodimers (Tsai et al. 1995) or heterodimers characterized by a basic toxic subunit and an acidic nontoxic subunit that acts as a chaperone (Mancheva et al. 1984; Perbandt et al. 1997; Banumathi et al. 2001; Gao et al. 2009; Faure et al. 2011; Petrova, Atanasov, and Balashev 2012). The stoichiometry of both subunits is one-to-one in the venom (of some *Crotalus, Daboia*, and *Vipera* species), and no free subunits have been detected. Interestingly, a point mutation (His^{48} to Gln^{48}) in the acidic subunit impairs the catalytic activity of the acidic subunit of some of the heterodimeric PLA_2 toxins from Viperinae venom, resulting in a unique evolutionary scenario, where the catalytic and toxic functions of one subunit are being modulated by an inactive and nontoxic subunit (Banumathi et al. 2001).

The presynaptically active, noncovalently linked D^{49}-SV-G^{II}-PLA_2 complexes from *Crotalus* venoms are a particularly intricate toxin complex. They are characterized by subunit A (designated as CA) being posttranslationally cleaved into three polypeptide chains stabilized by the seven intrachain disulfide bonds and bind to chain B (designated as CB) (Aird et al. 1985; Bouchier et al. 1991; Faure et al. 1991). Unlike CA, which is noncatalytic and nontoxic, CB by itself is capable of causing a weak neuromuscular blockade, while the CA/CB complex is extremely potently neurotoxic. When the complex binds to the synaptic membranes through CB, there is a dissociation and release of the CA subunit (Bon et al. 1979). This extremely potent complex has been retrieved from numerous *Crotalus* venoms, with isoforms from different populations of *Crotalus durissus terrificus* alone (Faure et al. 1991), each differing by only a few residues (Bouchier et al. 1988; Faure and Bon 1988). Other species for which this complex has been isolated include *C. horridus, C. oreganus helleri, C. scutulatus scutulatus, C. vegrandis, C. viridis concolor, C. tigris*, and *Sistrurus catenatus tergeminus* (Bieber, Tu, and Tu 1975; Ho and Lee 1981; Pool and Bieber 1981; Straight and Glenn 1988; Chen et al. 2004; Calvete et al. 2012; Sunagar et al. 2014). While this indicates that this complex is widely distributed among rattlesnake venoms, there may be extreme expression differences within a population, such as has been demonstrated for *C. d. terrificus, C. o. helleri*, and *C. s. scutulatus* (Glenn and Straight 1978; Lourenco et al. 2013; Sunagar et al. 2014). Despite the differences in expression levels, it is quite notable that this complex evolves at a very slow rate, displaying extreme conservation between these three divergent species (see color plate 8G).

A different form of neurotoxic PLA_2 has been characterized in *Bitis* species (Vulfius et al. 2011). The isolation and characterization of this toxin type from the giant species *Bitis arietans* in addition to the small species *B. atropos* suggest that it is ubiquitous in the venoms of species within this genus. However, the large species *B. arietans, B. gabonica, B. nasicornis*, and *B. rhinoceros* have venoms that are dominated by hemorrhagic SVMP among other destructive toxins. In contrast, the venoms of the small species *B. atropos, B. caudalis, B. cornuta, B. heraldica, B. inornata, B. peringueyi, B. rubida, B. schneideri*, and *B. worthingtoni* appear to be rich in this toxin type, particularly *B. atropos*. It is for this reason that as part of first aid for envenomations by these small viper species, pressure-immobilization bandages are not contraindicated, as these venoms do not produce massive local tissue swelling or necrosis as is characteristic of large *Bitis* species (see chapter 2).

Based on whether they have an Asp (D) or Lys (K) residue at position 49 in their amino acid sequence, SV-G^{II}-PLA_2s in pit viper venoms are further divided into two subtypes: D^{49}-SV-G^{II}-PLA_2 and K^{49}-SV-G^{II}-PLA_2 (Ownby et al. 1999; Six and Dennis 2000; Lomonte and Rangel 2012).

Until the discovery of myotoxic K^{49}-SV-G^{II}-PLA_2 in *Agkistrodon piscivorus piscivorus* snake venoms (Maraganore et al. 1984), only D^{49}-SV-G^{II}-PLA_2s were known (Marcussi et al. 2007), because biochemists usually assay the lipolytic or hydrolytic activity during PLA_2 purification from the crude venoms. Later, by HPLC-Mass and/or cloning, other noncatalytic PLA_2 variants including N^{49}-SV-G^{II}-PLA_2s and R^{49}-SV-G^{II}-PLA_2s have been found in some Asian pit viper venoms (Tsai et al. 2004; Chijiwa et al. 2006; Mebs et al. 2006; Wei et al. 2010). S^{49}-SV-G^{II}-PLA_2s, where Asp^{49} is replaced by a Ser residue, have been found in the venoms of certain true vipers, such as *Echis carinatus sochureki* (Polgar et al. 1996), *Vipera ammodytes ammodytes* (Krizaj et al. 1991), and *V. renardi* (Tsai et al. 2011a).

The K^{49} substitution results in loss of the hydrolytic activity on substrates (artificial or natural) (Fernandez et al. 2013). However, they are toxic despite the loss of enzymatic activity, with toxic modes of action including cytolysis, edema induction, liposome disruption, and myotoxicity (Lomonte, Angulo, and Calderon 2003). The toxic activities of these K^{49}-SV-G^{II}-PLA_2s are attributed to residues 115–129 (Lomonte et al. 1994), numbering at the C-terminal region (Renetseder et al. 1985; Scott et al. 1992; Chioato et al. 2002; Chioato and Ward 2003; Chioato et al. 2007). Some have proposed that the low catalytic activity of K^{49}-SV-G^{II}-PLA_2 enzymes is a result of contamination with D^{49}-SV-G^{II}-PLA_2s (van den Bergh et al. 1988; Krizaj et al. 1991; Scott et al. 1992), while others consider it an intrinsic property (Liu et al. 1990; Shimohigashi et al. 1995). However, more recent studies based on the generation of recombinant Lys^{49} and Ser^{49} toxins support the former possibility in solving this controversy, by demonstrating that such recombinants proteins lack PLA_2 activity (Ward et al. 2002; Petan, Krizaj, and Pungercar 2007).

The effects of K^{49}-SV-G^{II}-PLA_2s on skeletal muscles are histopathologically indistinguishable from their D^{49}-SV-G^{II}-PLA_2 counterparts (Fletcher and Rosenberg 1997; Gutiérrez and Lomonte 1997). While both myotoxin types induce Ca^{2+} entry and release of ATP, their mechanisms for sarcolemmal damage differ, since the former lack phospholipolytic activity (Fernandez et al. 2013). Numerous basic myotoxic PLA_2 isoforms have been isolated and characterized from diverse pit viper venoms (Gutiérrez, Ownby, and Odell 1984; Homsi-Brandeburgo et al. 1988; Soares et al. 1998; Andriao-Escarso et al. 2000; Angulo et al. 2005; Lomonte et al. 2009). They have been shown to exhibit a diversity of toxic activities, such as edema induction, cell lysis, and myonecrosis (Lomonte et al. 2009). Myotoxic PLA_2s are known to act more rapidly, relative to their nonenzymatic counterparts (Fletcher, Selistre de Araujo, and Ownby 1997). A noncovalently linked homodimeric K^{49}-SV-G^{II}-PLA_2 known as bothropstoxin I has been isolated from the venom of *Bothrops jararacussu*. At pH 5.0, bothropstoxin I dissociates into monomers and loses the ability to disrupt liposomes (de Oliveira et al. 2001). Later studies indicated that the cell-damaging effects of these enzymes are certainly reduced at lower pH, although not completely abolished, and that the dimeric nature of these enzymes enhances their toxic activity (Angulo et al. 2005).

21.5 THERAPEUTIC POTENTIAL OF THE TOXIN TYPE

Besides playing a significant role in the digestion of dietary phospholipids, cationic or basic SV-G^{IIA}-PLA_2s, as well as some acidic PLA_2s, are known to exhibit potent antimicrobial activity against both Gram-positive and Gram-negative bacteria (Paramo et al. 1998; Koduri et al. 2002; Nevalainen, Graham, and Scott 2008; Vargas et al. 2012). They are characterized by high positive surface charges that bind and disrupt the lipopolysaccharide layer of Gram-negative bacteria and sometimes in association with the bactericidal or permeability-increasing protein or the membrane attack complex (Gronroos et al. 2005; Tsai et al. 2007; Nevalainen, Graham, and Scott 2008). A study that examined eight myotoxic $sPLA_2$s suggested that this effect could be common to most SV-G^{IIA}-PLA_2s (Santamaria et al. 2005).

Although the substitution of Asp with a Lys at the 49th position results in loss of catalytic activity, K^{49}-SV-G^{IIA}-PLA_2s are known to exert several toxic effects, such as myotoxicity, lethality, edema, and pain (Lomonte and Gutiérrez 1989; Chaves et al. 1998; Landucci et al. 2000; Soares et al. 2000; Chacur et al. 2003). They are also known to be cytotoxic to a large number of cell types in culture (Lomonte, Tarkowski, and Hanson 1994; Lomonte et al. 1999), and to disrupt the liposomal bilayers (Diaz et al. 1991; Rufini et al. 1992) and degranulate mast cells (Landucci et al. 1998). Such toxic activities are not dependent on the hydrolysis of phospholipids. Instead, a cationic and hydrophobic C-terminal region is implicated in their toxic actions (Lomonte et al. 1994; Nuñez, Angulo, and Lomonte 2001; Chioato et al. 2002). As a result, K^{49}-SV-G^{II}-PLA_2s are excellent models for investigating the diversity of aforementioned cellular effects that are not dependent on the PLA_2 activity. Moreover, K^{49}-SV-G^{II}-PLA_2s have been demonstrated to exhibit a spectrum of activities in a lymphoblastoid cell line, including apoptosis, cell proliferation, and necrosis (Mora et al. 2005).

The diversity of biological activities exhibited by venom PLA_2s, including anticoagulant (Lakshminarayanan et al. 2003; Karbovs'kyi et al. 2007; Zouari-Kessentini et al. 2009), anti-HIV (Fenard et al. 1999), antimalarial, and antiparasitic (Deregnaucourt and Schrevel 2000), antitumor (Jerusalinsky et al. 1995; Jebali et al. 2009; Zouari-Kessentini et al. 2009), antiangiogenic effects (Fujisawa et al. 2008; Bazaa et al. 2010; Kessentini-Zouari et al. 2010), and bactericidal activities (Páramo et al. 1998; Perumal Samy et al. 2007), makes them ideal candidates for drug design and development. The recently demonstrated antiangiogenic and antitumor activity of certain viper venom PLA_2s, such as that of CC-PLA_2-1, CC-PLA_2-2, and MVL-PLA_2 from the venoms of *Cerastes cerastes*, has been promising in anticancer therapeutics (Zouari-Kessentini et al. 2009; Bazaa et al. 2010; Kessentini-Zouari et al. 2010).

21.6 CONVERGENCE WITH OTHER VENOMS

PLA_2s have convergently recruited in the venoms of non-front-fanged caenophidian snakes and elapid snakes (see chapter 20) and anguimorph lizards (see chapter 24). PLA_2s were also independently recruited in the venoms of multiple arthropod orders including centipedes, insects, scorpions, spiders, ticks and have also been documented in the venoms of cnidaria and coleoids.

CHAPTER 22

SARAFOTOXIN PEPTIDES

A. BDOLAH, F. DUCANCEL, K. SUNAGAR, T. N. W. JACKSON, AND B. G. FRY

22.1 SUMMARY

Sarafotoxins (SRTxs) are only found in the venoms of the enigmatic side-stabbing snake genus *Atractaspis*. These peptides belong to the vasoconstrictive endothelin (ET) family. The plesiotypic ETs modulate the contraction of cardiac and smooth muscles in physiological tissues of vertebrates. ETs, which are physiological homologues of toxic sarafotoxins, seem to be only present in vertebrate animals and are capable of causing envenoming actions upon overdosage, suggesting that these proteins share not only similar structures but also biochemical functions. The chief envenoming action of sarafotoxins is from its effect on the vascular smooth-muscle system.

22.2 ENDOPHYSIOLOGICAL PLESIOTYPE

ETs are considered the most potent vasoconstrictors (Yanagisawa et al. 1988; Barton and Yanagisawa 2008). The vasoactive properties of endothelin-1 have made this toxin peptide best known for its role in inducing hypertension. However, studies in the past twenty years have revealed that the ET system plays a much wider role in mammalian biology (see Khimji and Rockey 2010) and probably has an important role as a vertebrate-specific signaling pathway (Braasch, Volff, and Schartl 2009). ETs were shown to have effects on the cardiac system. They contribute to myocardial contractility, chronotropy, and arrhythmogenesis (Barton and Yanagisawa 2008).

Reconstruction of the evolutionary history of the endothelin system, which consists in ET ligands and their G-protein-coupled receptors, indicated that this system is a vertebrate-specific innovation (Braasch, Volff, and Schartl 2009). It was shown that ET ligands and their receptor have been duplicated during whole-genome duplication events. Generally, each ET receptor can bind all ET peptides but with different affinities. Mammalian ET_A has higher affinity to ET-1 and ET-2 (Arai et al. 1990), whereas ET_B has similar affinities to all ET ligands (Sakurai et al. 1990). Frog ET_C preferentially binds ET-3 (Karne, Jayawickreme, and Lerner 1993).

22.3 PLESIOTYPIC STRUCTURAL AND FUNCTIONAL TOXIN FORMS

Sarafotoxins are unique toxins found only in the venom of *Atractaspis* species. So far, SRTx-like peptides have been isolated from the venoms of four *Atractaspis* species: *A. engaddensis* (SRTx-a, b, c, and d/e), *A. bibroni* (bibrotoxin); *A. microlepidota* (from east Africa, SRTxm1-5), and *A. irregularis* (SRTxi1-3) (Takasaki et al. 1988; Becker et al. 1993; Hayashi et al. 2004; Quinton et al. 2005). The SRTxs from *A. engaddensis* and from *A. bibroni* are 21 amino acid peptides (similar to the ETs), whereas SRTxs (called "long SRTx) isoforms that were isolated from the other two *Atractaspis* species are 24 or 25 amino acid peptides.

The evolutionary history of the SRTx/ET genes indicates that the SRTx lineage diverged from a common plesiotypic gene prior to the first gene duplication event that took place in the ETs' lineage (Landan et al. 1991a; Landan et al. 1991b). This notion is further supported by a comparison of the ET and the SRTx gene structures. The genes encoding the different ET isopeptides contain a single stretch of a mature ET peptide and are located on separate chromosomes (Arinami et al. 1991), whereas a complete cDNA encoding the SRTx from *A. engaddensis* shows a polycistronic gene organization (Ducancel et al. 1993). This cDNA encodes for a polypeptide precursor that consists in 12 successive stretches of 40 residues, each contain a "spacer" sequence followed by a sequence of one of six SRTx isoforms. A SRTx precursor possessing a polycistronic organization was also identified in the venom gland of *A. microlepidota* (Hayashi et al. 2004). The longest cDNA fragment that was isolated from this gland corresponds to the C-terminal of the precursor. It consists of seven tandem stretches of 48 amino acids corresponding to five different SRTx isoforms (see figure 22.1). Each contains a "spacer" sequence followed by a sequence of one of five 24 amino acid long SRTx isoforms. It should be noted that even though the lengths of the tandem repeat stretches in the two latter described SRTx precursors are different, their mature and spacer domains are highly similar

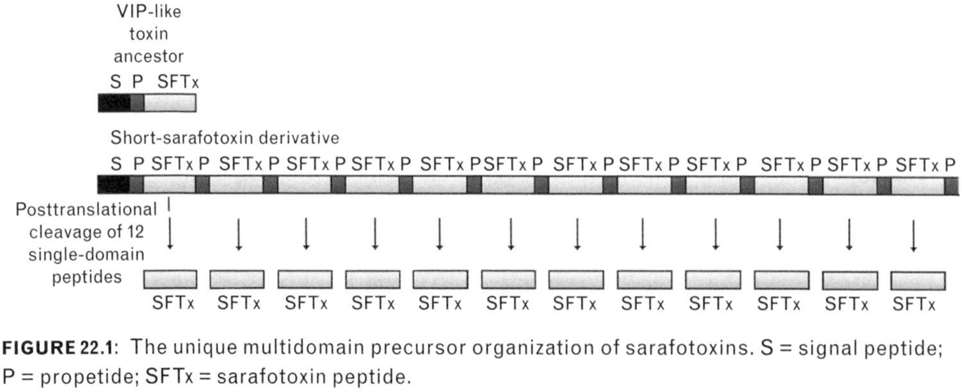

FIGURE 22.1: The unique multidomain precursor organization of sarafotoxins. S = signal peptide; P = propetide; SFTx = sarafotoxin peptide.

(see Bdolah 2010). Two clones that encode for two precursors of SRTx isoforms were identified in the venom gland of *A. irregularis* (Ducancel 2005). Only one form of mature SRTx emerges from each precursor.

Comparison of three-dimensional structures of ET-1 and SRTx-b revealed that they adopt (i) an α-helical motif stabilized by a disulfide bond and characterized by an extended structure of the first three or four amino acids, (ii) a β-turn in +5 to +8 residues, (iii) an α-helical conformation from Lys9-Cys15, and (iv) a linear C-terminus (Tamaoki et al. 1991; Atkins, Martin, and Smith 1995). Site-directed mutagenesis of SRTx-b revealed that serine at position +2 was vital for vasoconstriction (Lamthanh et al. 1994). Similarly, site-directed mutagenesis of ETs has highlighted the importance of carboxy and amino acid side chains of terminal residues, such as Asp8, Glu10, and Phe14, in influencing the binding affinities of ETs to their native receptors (Nakajima et al. 1989). Iodine-labeled SRTxs revealed that they have a high affinity for rat atrial and brain membranes, and their binding is quite rapid yet reversible.

The resemblance of the structure of the SRTx precursors of *A. engaddensis* and *A. microlepidota* may suggest that these two species have evolved from a common ancestor. It should be noted that *Atractaspis* has been proposed as being of two distinct lineages (Underwood and Kochva 1993). The *microlepidota* group was considered to include species found in western, central, and eastern Africa, the Sinai Peninsula, and much of Arabia and Israel; however, this was before the distribution was revised, with the distribution of *A. microlepidota* confined to western Africa, while the populations of Saudi Arabia are defined as *A. andersoni* and *A. engaddensis* (Trape, Mané, and Ineich 2006; Dobiey and Vogel 2007; David and Vogel 2010). The *bibroni* group extends across western, central, eastern, and southern Africa (including *A. bibroni* and *A. irregularis*). Members of the *bibroni* group have venom glands that are confined to the head region, whereas members of the *microlepidota* group have venom glands that continue down from the neck to reach from one-tenth to one-fifth of the body length. Among the *microlepidota* group, both *A. engaddensis* and *A. microlepidota* have a polycistronic SRTx gene organization, but whereas the former is the most toxic among the species of *Atractaspis*, the venom of *A. microlepidota* (from east Africa) shows a very low toxicity, probably because of a much lower SRTx-like content (see Bdolah 2010).

Having a common molecular origin and sharing homology in three-dimensional structures, it is not surprising that the ETs and the SRTxs also have similar pharmacological characteristics. Overdose of ET-1 was shown to have lethal capacity in mice similar to that of SRTx-b (Bdolah et al. 1989). The prominent cardiotoxic effects of the crude venom of *A. engaddensis* and of its isolated SRTx have been clearly established (Weiser et al. 1984; Wollberg et al. 1988). Experiments with anesthetized mice and with isolated cardiac preparations have shown that the SRTx cause coronary vasoconstriction in addition to exerting direct effects on the conduction system of the mammalian heart. The toxicity of a 24 amino acid long SRTx-isoform (a synthetic SRTx-m of *Atractaspis microlepidota*) was also analyzed. In comparison to SRTx-b, this peptide has a lower lethality (two- to three-fold higher LD_{50}); it has similar cardiovascular effects but of lower potency (Hayashi et al. 2004). The cardiotoxic effects of SRTxs have also been demonstrated in other vertebrates (see Kochva, Bdolah, and Wollberg 1993).

The main toxic effect of the SRTxs derives from their effect on the vascular smooth-muscle system. Using rabbit aortic rings, it was demonstrated that SRTx-b, which is the most toxic peptide of *A. engaddensis*, induced the highest maximal tension as compared with the other SRTxs of this species. SRTx-a induced only 60% maximal response, but its potency in this system was similar to that of SRTx-b. SRTx-c, which has a very low toxicity in mice, also induced a negligible contractile response in the rabbit aortic system (Wollberg, Bdolah, and Kochva 1989). SRTx-m induced similar maximal contractile response in this system; however, the potency of SRTx-m was more than a magnitude lower than that of SRTx-b (Hayashi et al. 2004). Bibrotoxin and SRTx-b showed similar contractile responses when tested with rat thoracic preparations (Becker et al. 1993).

The contractile effects of SRTxs were also demonstrated in nonvascular smooth-muscle systems (Kochva, Bdolah, and Wollberg 1993). In the guinea pig ileum, the most toxic isopeptide, SRTx-b, once again induced the most effective contractile response, whereas SRTx-c induced a very weak response, and SRTx-d/e produced practically no response. In the rat uterus, at the proestrus, estrus, and metestrus stages, SRTx-b, SRTx-c, and ET-1 showed essentially similar effects in isolated uteri preparations but to different degrees (Wollberg et al. 1992). These included (i) a slight increase in maximal tension, (ii) suppression of the spontaneous rhythmic contraction, and (iii) an increase in the rate of rhythmic contractions.

The use of radio-labeled SRTx-b made it possible to demonstrate high-affinity binding sites (K_d 3–4 nM, B_{max} ~100 fmol/mg protein) in rat heart and brain preparations. These earlier experiments also showed that the SRTxs are potent activators of the phosphoinositide-signaling pathway (Kloog et al. 1988). Competition-binding experiments between the labeled SRTx-b and other SRTxs, and also with ETs, revealed that peptides of the ET/SRTx superfamily compete for similar receptors and activate similar signaling pathways (Kloog and Sokolovsky 1989). However, some differences in the response to ETs or SRTxs were reported. In the Egyptian mongoose, which is resistant to the venom of *Atractaspis*, intravenous administration of SRTx-b at a high dose of 13 times the LD_{100} for mice resulted in reversible electrocardiographic disturbances. Unlike ET-1, SRTx-b failed to induce contraction of the isolated mongoose aorta (Bdolah et al. 1997). Binding studies revealed SRTx- and ET-specific binding sites in brain and cardiovascular preparations of the mongoose. It was thus suggested that SRTx/ET receptors in the mongoose contain features that enable them to differentiate between the two peptides. SRTxs and ETs also activate different signal-transduction cascades (including cyclic AMP and cyclic GMP production and others) in different tissues. These diversified effects have been attributed to the existence of different SRTx/ET receptor subtypes and/or activation of different G-protein pathways in different tissues (see Sokolovsky and Shraga-Levine 2001).

Recently, a study was carried out to explore the structural effect and functional significance of the C-terminal extensions of three or four amino acids that are observed in the case of long SRTxs in the venoms of *A. microlepidota* and *A. irregularis* (Mourier et al. 2012). They are characterized by different C-terminus extensions that follow the invariant Trp21, which plays a crucial role in endothelin-receptor binding: namely, SRTx-m (24 amino acids including the extension D-E-P) and SRTx-i3 (amino acids including the extension V-N-R-N). Both peptides were shown to be highly toxic in mice and displayed the cysteine-stabilized α-helical motif that characterizes ETs and short SRTxs, to which a longer C-terminus with variable flexibility is added. To discern the functional and pharmacological consequences of the supplementary amino acids, different chimerical in addition to truncated forms of SRTxs were designed and synthesized: (i) removing the extra C-terminal residues of SRTx-m or i3 or (ii) grafting the latter onto the C-terminal extremity of the short SRTx-b from *A. engaddensis*. Competitive-binding assays toward iodinated endothelin-1 binding to cloned ETAR and ETBR receptor subtypes overexpressed in CHO cells revealed the essential role of the C-terminus extensions for ET-receptor recognition. Indeed, l-SRTx display an affinity three to four orders of magnitude lower as compared with SRTx-b for the two receptor subtypes. Moreover, grafting the C-terminus extension to SRTx-b induced a drastic decrease in affinity, while its removal (truncated l-SRTx) yielded an affinity for ET receptors similar to that of s-SRTx. Furthermore, it has been shown by intracellular Ca^{2+} measurements that l-SRTx, along with s-SRTx, display agonistic activities. Thus, this new family of SRTx/ET displays a major difference in potency and also the crucial role of the C-terminus extension in their various pharmacological profiles. Finally, one of the chimeric toxins chemically synthesized in this study appears to be one of the most potent and selective ligand of the ET_B receptor known to date. These results emphasize the interest in systematically exploring *Atractaspis* venoms and venom-gland transcriptomes to tentatively identify and categorize novel SRTx isoforms capable of displaying diverse functional activities. Finally, it appears from this last study that natural long SRTxs display very low affinities for the two main abundant ET receptors and a high toxicity in mice. That contradictory result suggests that some SRTx isoforms may interact and recognize different targets from the short SRTxs. Thus, these new SRTxs may constitute attractive tools to enhance our understanding of the ET axis functioning and complexity. Also, these new isoforms may be the starting point in designing new SRTx-derived drugs.

22.4 APOTYPIC STRUCTURAL AND FUNCTIONAL TOXIN FORMS

It is currently unclear whether the long or short form is the plesiotypic state.

22.5 THERAPEUTIC POTENTIAL OF THE TOXIN TYPE

In humans, ETs are produced by various organs following a complex biosynthetic pathway requiring the cleavage of propeptides by endothelin-converting enzymes to obtain the mature and physiologically active endothelin. Among the three ETs, ET-1 is the most abundant isoform and is mainly generated within the vascular wall. Once secreted, the ETs exert their biological action in a paracrine or autocrine fashion and intervene in a wide range of physiological functions, such as vascular-tone homeostasis, neural-crest development, ovarian cycle, cell proliferation, angiogenesis, and inflammation. In addition to their pleiotropic physiological properties, ETs (and particularly ET-1) have also been involved in a variety of diseases, such as pulmonary hypertension, congestive heart failure, atherosclerosis, chronic kidney disorders, neoplasia, asthma, myocardial arrhythmias, and ischemia. Moreover, their overexpression is also associated with numerous types of cancers, such as prostate, ovarian, colorectal, bladder, breast, and lung carcinomas (Kawanabe and Nauli 2011). To mediate their numerous physiopathological effects, ETs activate two distinct G-protein-coupled receptors: ETA receptor (ETAR) and ETB receptor (ETBR). ETBR equally binds all three endothelin isoforms, whereas ETAR shows a higher affinity for ET-1 and ET-2 than for ET-3. Both receptors present a quasi-ubiquitous expression pattern, but ETAR predominates on vascular smooth-muscle cells and cardiomyocytes, while ETBR is particularly abundant on vascular endothelial cells. Of the two receptors, ETBR is often considered the regulating and beneficial subtype, as it is able to counteract the prohypertensive, prohypertrophic, and proliferative signaling pathways triggered by ETAR. Moreover, ETBR is also involved in the clearance of endothelins from bloodstream and tissues, which prevents harmful endothelin accumulation. Consequently, a decrease in ETBR expression (often coupled to an augmented ETAR expression) has been associated with several pathological situations involving the endothelin axis. Nevertheless, the apparent dichotomy between the "good" ETBR and the "damaging" ETAR is not so straightforward. Indeed, ETBR overexpression and/or overactivation (most of the time paralleled by an ETAR up-regulation) have been correlated to the progression of diseases such as atherosclerosis, liver cirrhosis, or systemic sclerosis. Furthermore, the pathological role of ET receptors has also been particularly documented in the field of oncology (Bagnato and Rosano 2008; Bagnato, Spinella, and Rosano 2008). In addition to its role as a potent endogenous vasoconstrictor and mediator of cardiovascular and renal disorders, ET-1 is also a candidate progression factor in many tumor types. ET-receptor activation seems to promote tumor progression by means of several mechanisms, including cell proliferation, inhibition of apoptosis, matrix remodeling, and bone deposition in skeletal metastases through activation of osteoblasts.

Following the discovery of the ETs and their receptors, antagonists for ET receptors became available (Aubert and Juillerat-Jeanneret 2009). It took several more years of basic research, studies with experimental animals, and clinical trials before the first ET-receptor antagonist, bosentan (Tracleer), was approved in the United States and Europe for treatment of arterial hypertension (Rubin and Roux 2002). This is a life-threatening condition that can severely affect the function of the lung and the heart. Bosentan was shown to improve functional status of patients with this disease and slow clinical deterioration. Ongoing clinical trials are evaluating ET-receptor antagonists for treatment of cardiovascular disorders, cancers, and other diseases. Clearly, ET antagonists are promising new agents in the treatment of the different pathologies and diseases in which the ET axis (ETs and their receptors) is involved. Several ET-receptor antagonists are in clinical trials (Tamkus et al. 2009).

Among other potential ET-antagonists, "weak" SRTxs, such as SRTx-d/e, were considered as potential peptides that may interfere with the binding of ETs to their receptors. Preliminary experiments have indicated that SRTx-d/e inhibited smooth-muscle contraction by SRTx-b (Wollberg et al. 1998). Thus, an exploration of the hitherto unknown SRTx-like peptide in the different species of *Atractaspis* may provide a useful contribution to the discovery of novel antagonists. The recent discovery of the so-called long SRTxs (Ducancel 2005) illustrates the interest of further exploring the natural molecular, structural, and functional biodiversity that seems to exist within SRTxs in *Atractaspis* species. From a pharmacological point of view, it has been shown that natural long SRTxs or engineered ones are able to display variable affinities for ETAR and/or ETBR and, most interestingly, different specificity of recognition that may be particularly useful in designing new clinical drugs (Mourier et al. 2012). Thus, one of the chimeric SRTxs synthesized in this study appears to be one of the most potent and selective ligand of the ET_B receptor known to date.

22.6 CONVERGENCE WITH OTHER VENOMS

There is none.

CHAPTER 23

SNAKE VENOM METALLOPROTEASE ENZYMES

N. R. CASEWELL, K. SUNAGAR, Z. TAKACS, J. J. CALVETE, T. N. W. JACKSON, AND B. G. FRY

23.1 SUMMARY

Snake venom metalloprotease enzymes (SVMPs) are members of a large multilocus gene family that encodes functionally and structurally diverse venom proteins responsible for impairing hemostasis in bite victims. SVMPs are related to ADAM-TS and ADAM proteins; the SVMP scaffold recruited for a role in venom at the base of the caenophidian radiation is related to ADAM 7, ADAM 28, and ADAM decysin 1 proteins. Following its recruitment, the plesiotypic SVMP lost protein domains associated with the ADAMs, resulting in the formation of P-III SVMP, which contain just metalloprotease, disintegrin-like, and cysteine-rich domains. Following the divergence of viperid snakes from the remaining caenophidian snakes, a duplicated SVMP P-III gene evolved by positive selection into a P-II SVMP via loss of the cysteine-rich domain. Subsequently, P-I SVMP have evolved independently in many viperid snake lineages by convergent loss of the P-II disintegrin domain and adaptive evolution of the metalloprotease domain. The consequence of this diverse evolutionary history is the expression of multiple, structurally distinct SVMP isoforms in the venom of viperid snakes. Members of all three major classes of SVMP (P-I, P-II, and P-III) have been characterized as hemorrhagic, but as the loss of nonprotease domains has occurred over time, the hemorrhagic potency of the toxins appears to have decreased. Consequently, the plesiotypic P-III class of SVMP exhibits the lowest minimum hemorrhagic doses observed to date, and they are more hemorrhagic than the apotypic P-I and P-II forms. These proteins exhibit an array of functional activities, some of which are pertinent to the development of novel human therapeutic and diagnostic compounds. In nature, the various forms likely act synergistically to impair the cardiovascular system of their prey and, in the case of snakebite accidents, humans.

23.2 ENDOPHYSIOLOGICAL PLESIOTYPE

SVMPs are classified as adamalysins, which, alongside the matrixins, astacins, and serralysins, are members of the metizincin gene superfamily (Bode, Gomis-Ruth, and Stockler 1993; Huxley-Jones et al. 2007). Within the adamalysin subfamily, the SVMPs are grouped with ADAM (a disintegrin and metalloprotease) and ADAM-TS (ADAM with thrombospondin motifs) proteins, because they share the disintegrin-like and metalloprotease domains that characterize the latter groups (Hite et al. 1994; Jia et al. 1996; Fox and Serrano 2005; Huxley-Jones et al. 2007). ADAMs, ADAM-TSs, and SVMPs exhibit complex structural arrangements that consist of a number of conserved domains, alongside some that are unique to each group (see figure 23.1). The prodomain and metalloprotease domain are typically conserved throughout the adamalysins, which is indicative of their shared evolutionary history (Andreini et al. 2005; Fry 2005; Casewell 2012). However, the domains that sequentially extend these regions are highly variable: ADAM-TS proteins have diverse C-terminal regions which typically contain a disintegrin-like and cysteine-rich domain followed by thrombospondin domain repeats; while the ADAMs contain disintegrin-like, cysteine-rich, epidermal-growth-factor- (EGF-)like, transmembrane, and cytoplasmic domains (see figure 23.1). The domain variability observed in the C-terminal region of SVMP has been used for toxin classification: those that do not extend past the metalloprotease domain are classed as P-Is, while P-IIs contain a disintegrin domain and P-IIIs the disintegrin-like and cysteine-rich domains also present in their ancestors (see figure 23.1).

ADAMs are widely recognized as highly diverse peptidases and are often referred to as "sheddases," because a number of family members are responsible for catalyzing the shedding of membrane-bound protein domains, effectively acting as the molecular switch that activates biological receptors (Black et al. 1997; Srour et al. 2003; White 2003; Blobel 2005). Proteolysis by ADAMs influences a variety of physiological receptors, growth factors, and cytokines, and this activity is associated with key roles in angiogenesis, neurogenesis, and, potentially, fertilization (Blobel 2005). Some ADAMs appear to function through proteolytically independent mechanisms, for example, by acting as receptors themselves (Fukata et al. 2006). The ADAM-TS proteins can be broadly defined as substrate-specific secreted proteases (Apte 2004). The ancillary domains of these adamalysins may be responsible for the substrate specificity they exhibit, as the protease domain alone is unable to process native substrates (Cal et al. 2002). ADAM-TS may influence angiogenesis and coagulation and regulate cell-matrix

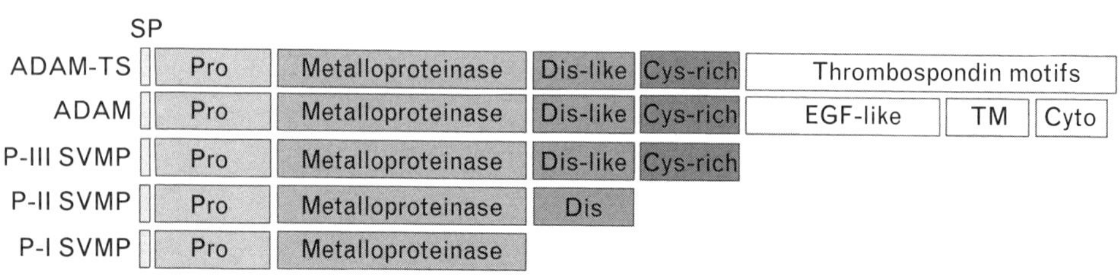

FIGURE 23.1: Structural schematic of adamalysins. SP = signal peptide, Pro = prodomain, Dis = disintegrin, Cys-rich = cysteine-rich domain, EGF-like = epidermal-growth-factor-like domain, TM = transmembrane domain, Cyto = cytoplasmic domain.

interactions by processing extracellular molecules such as von Willebrand factor, procollagens and aggrecan (Hall et al. 2003; Apte 2004).

The adamalysins originate from a common ancestor (Andreini et al. 2005), with the ADAM-TS proteins diverging at the base of the adamalysin gene radiation (see figure 23.2) (Casewell 2012). Within this metizincin subfamily, the ADAMs and SVMPs form a clade, and the SVMPs are monophyletic, indicating that the recruitment of this toxin type into the venom of the advanced snakes is likely the result of a single event (see figure 23.2) (Moura-da-Silva et al. 1996; Fry and Wüster 2004; Fry 2005; Fry et al. 2008; Casewell 2012). The ADAM scaffold recruited into snakes was likely also the ancestor of ADAM 7, ADAM 28, and ADAM decysin-1 (ADAM DEC1) proteins, as these three proteins form a sister clade to the SVMPs (see figure 23.2) (Casewell 2012). These three ADAMs cluster on the same chromosome in humans, and they may have originally shared a novel function distinct from other ADAMs (Bates, Fridman, and Mueller 2002). However, their functional activities have subsequently diverged: (i) ADAM 7 is expressed in the epididymis and may enable sperm motility and fertilization (Oh, Han, and Cho 2009); (ii) ADAM 28 is found in a variety of tissues and is associated with integrin binding, facilitating enzymatic cleavage of the extracellular matrix, and the transendothelial migration of lymphocytes (Fry 2005; McGinn et al. 2011); and (iii) ADAM-DEC1 is expressed in macrophages and dendritic cells and associated with pathogenesis in the lungs,

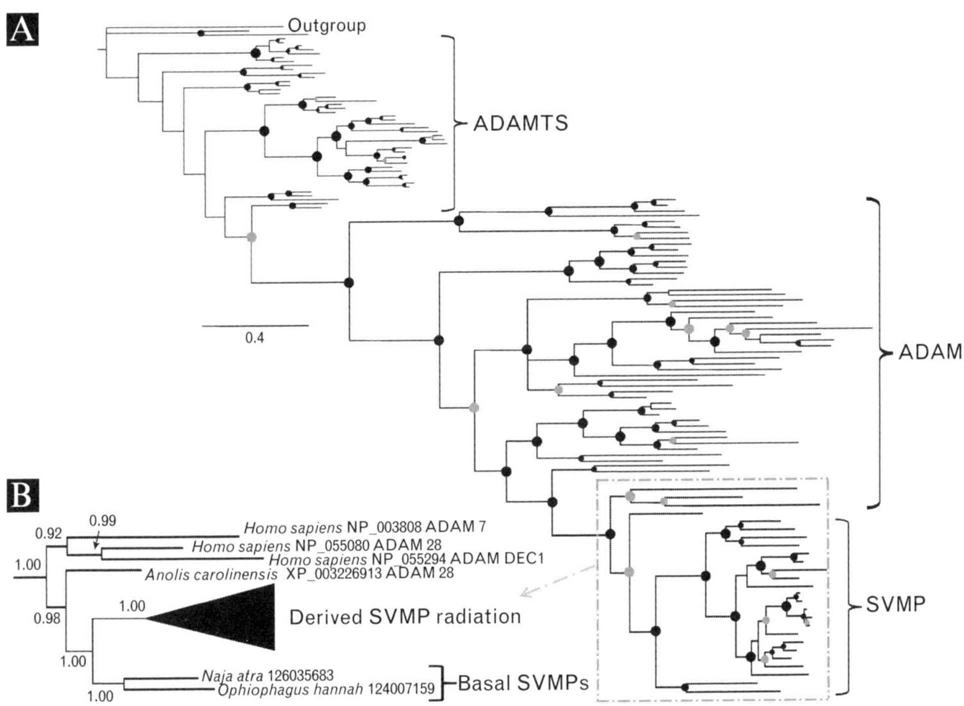

FIGURE 23.2: Evolutionary history of the adamalysins (Casewell 2012). (A) ADAMs and SVMPs form a sister group to the ADAM-TS proteins. Nodes with black and gray circles indicate significant support (Bayesian posterior probabilities of 1.00 and > 0.95, respectively). (B) Mammalian ADAM 7, 28 and DEC1 and reptilian ADAM 28 form a sister group to the SVMPs. Note that the derived SVMP radiation has been collapsed (indicated by black triangles) to facilitate the display of the evolutionary relationship.

although its specific proteolytic function remains unknown (Shapiro 2003; Crouser et al. 2009). Perhaps relevant to the subsequent evolution of SVMPs in snake venom is the fact that ADAM DEC1 has lost a number of C-terminal domains, including the cysteine-rich domain and part of the disintegrin domain (Bates, Fridman, and Mueller 2002). The ADAM scaffold recruited into caenophidian snake venom lost the EGF-like, transmembrane, and cytoplasmic domains over evolutionary time (Gutiérrez et al. 2009), resulting in the P-III structure observed at the base of the SVMP radiation. This is notable because the evolutionary history of the SVMP toxin type is punctuated by domain loss (Casewell et al. 2011a), suggesting that the ancestor of these two sister groups (SVMPs and ADAM 7/ADAM 28/DEC1) may have been predisposed to gene truncation. These metalloproteases retain the zinc-binding motif (HEXXHXXGXXH).

23.3 PLESIOTYPIC STRUCTURAL AND FUNCTIONAL TOXIN FORMS

The plesiotypic P-III SVMP consist of metalloprotease, disintegrin-like, and cysteine-rich domains that exhibit close similarity to apotypic SVMPs. Notably, patterns of cysteine residues observed in the plesiotypic SVMP are also conserved in their nonvenom ADAM ancestors, mammalian ADAM 7, ADAM 28, and ADAM DEC1 and reptilian ADAM 28 (Casewell 2012). The SVMP toxin family is associated with a wide range of functional activities, but the plesiotypic activity is the induction of hemorrhage. This hemorrhage is the result of proteolytic cleavage of basement membrane components present in capillary vessels, which likely results in the distention of the capillary wall, the disruption of endothelial cell integrity, and, ultimately, extravasation (Gutiérrez et al. 2005; Gutiérrez et al. 2006; Gutiérrez, Rucavado, and Escalante 2009).

P-III SVMPs have been isolated from a wide variety of taxa, including members of all three front-fanged caenophidian snake lineages, in addition to non-front-fanged lineages (Ovadia 1987; Tan and Saifuddin 1990; Assakura et al. 1992; Assakura, Reichl, and Mandelbaum 1994; Kamiguti et al. 2000; Fox and Serrano 2005; Fry et al. 2008; Casewell et al. 2011a). To date, all SVMPs isolated from snakes other than viperids are of the P-III class and form two clades at or near the base of the toxin radiation (see figure 23.2) (although some have been described as atypically truncated; see section 23.4.4). The amount of SVMPs in elapid snake venoms is typically low (Leao, Ho, and Junqueira de Azevedo 2009; Correa-Netto et al. 2011; Fernandez et al. 2011; Jiang et al. 2011; Petras et al. 2011), implying that these toxins may only play a minor or perhaps conserved role in prey envenoming. In contrast, the amount in the venoms of non-front-fanged and viperid snakes is extremely variable. For example, the non-front-fanged caenophidian snakes *Thamnodynastes strigatus* and *Philodryas olfersii* exhibit contrasting SVMP venom-gland expression levels of around 5% and around 52% of all toxins, respectively (Ching et al. 2006; Ching et al. 2012), and some other non-front-fanged lineages also express very large amounts (Fry et al. 2003c; Fry et al. 2008). Nonetheless, the diversity of SVMPs in nonviperid snakes is typically lower than that observed in viperid snake venoms. The plesiotypic P-III SVMPs from nonviperid venoms have been largely uncharacterized. Those that have been characterized include the form from *Ophiophagus hannah*, which inhibits platelet aggregation and produces fibrinolysis characterized by the degradation of the α-chain of human fibrinogen (Guo et al. 2007), activities similar to those of plesiotypic P-III SVMPs from viperid snake venoms. While it has yet to be demonstrated in the laboratory, the extremely large amounts of SVMPs in the venoms of *Atractaspis* species are likely responsible for the extreme tissue destruction observed following envenomations by species from this genus (see color plates 12C–12D).

23.4 APOTYPIC STRUCTURAL AND FUNCTIONAL TOXIN FORMS

The P-III-like structure that was recruited into the venom of snakes at the base of the caenophidian snake radiation has undergone considerable structural and functional alteration over evolutionary time. In light of the conservation of cysteines in the P-III plesiotypic SVMPs, the gain and loss of additional cysteine residues important for facilitating structural changes and posttranslational modifications of apotypic SVMPs are the direct result of amino acid substitutions following the recruitment of the SVMP scaffold into venom.

23.4.1 APOTYPIC P-III SVMPS

While P-III SVMPs are structurally plesiotypic, a number of them possess apotypic activities. For example, while the P-III SVMP from the elapid snake *O. hannah* exhibits fibrinolytic activity, this function is atypical for P-III SVMPs and is more commonly associated with the P-I SVMP class (see section 23.4.3). In addition, the apotypic capability of prothrombin activation appears to have evolved convergently at least three times in the P-III SVMPs, once near the base of the SVMP radiation in elapid snakes, once in the viperid snake genus *Echis*, and once in the viperid snake genus *Bothrops* (Nishida et al. 1995; Yamada, Sekiya, and Morita 1996; Gao, Kini, and Gopalakrishnakone 2002; Loria et al. 2003; Silva et al. 2003). An elapid snake venom P-III SVMP that activates prothrombin is the form from the venom of *Micropechis ikaheka* (Gao, Kini, and Gopalakrishnakone 2002). While it has yet to be demonstrated in the laboratory, the large amounts of P-III SVMPs in *Dispholidus typus* venom are likely responsible via prothrombin activation for the profound disruption of blood chemistry characteristic of envenomations by this species.

The SVMP diversification observed in viperid snake venoms is indicative of the importance of these toxins. Proteomic and transcriptomic analyses of venom from numerous members of the viperid snakes have revealed SVMPs to be the most highly expressed protein family, making up as much as 74% of the venom (Junqueira de Azevedo and Ho, 2002; Bazaa et al. 2005; Wagstaff and Harrison 2006; Gutiérrez et al. 2008; Casewell et al. 2009; Wagstaff et al. 2009). It is worth noting that P-I and P-II SVMPs, absent from nonviperid snakes, often contribute substantially to this representation. Although the general trend in the viperid snakes is venom with large quantities of SVMPs, the venom of some species contains surprisingly low levels (see Junqueira de Azevedo et al. 2006). Following the split of the viperid snakes from the remaining caenophidian snakes, gene duplication resulted in considerable diversification of P-III SVMPs within the former, with multiple P-III isoforms typically retained in the venom of any one species (Junqueira de Azevedo and Ho 2002; Bazaa et al. 2005; Wagstaff and Harrison 2006; Gutiérrez et al. 2008; Casewell et al. 2009; Wagstaff et al. 2009).

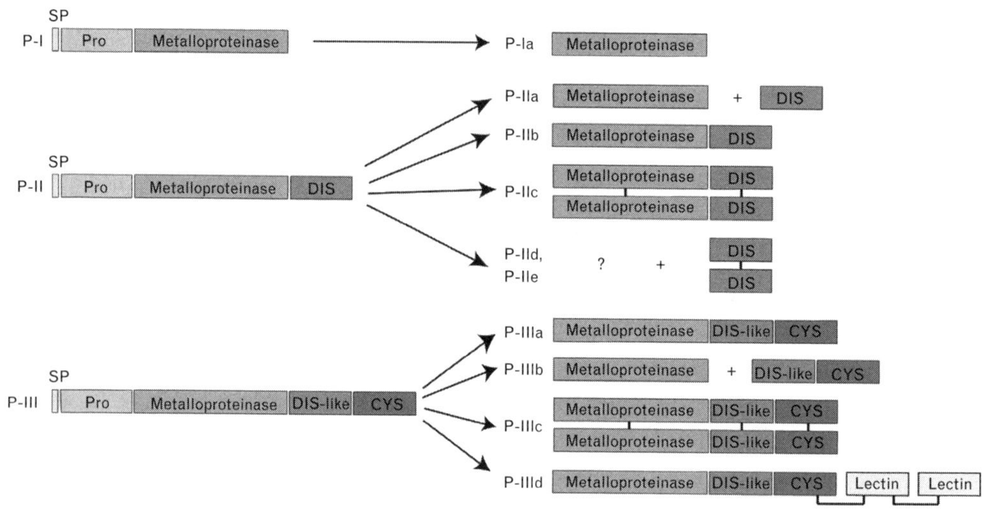

FIGURE 23.3: Structural schematic of the posttranslational modifications of SVMP subclasses (Fox and Serrano 2008). Subclasses of P-I, P-II, and P-III classes are indicated by letters a–e. SP = signal peptide, Pro = prodomain, DIS = disintegrin domain, ? = not yet identified, CYS = cysteine-rich domain, LECTIN = lectin non-SVMP toxin component.

Apotypic P-III SVMP subclasses include those that remain intact (P-IIIa), those that proteolytically process the disintegrin-like and cysteine-rich domains (P-IIIb), those that form intact dimeric structures (P-IIIc), and those that bond covalently with C-type lectin venom components (P-IIId) (see figure 23.3) (Fox and Serrano 2008). The P-IIIb and P-IIIc SVMP subclasses are independently monophyletic (see figure 23.4), indicating that the evolution of their posttranslational modifications have evolved only once each. The P-IIIc SVMP subclass evolved early in the diversification of viperid snake venom SVMPs, as evidenced by their position near the base of the phylogenetic tree (see figure 23.4). Representatives of the P-IIIc SVMP subclass have been isolated from members of both the Crotalinae and the Viperinae. Contrastingly, the P-IIIb SVMP subclass has only been recovered from Crotalinae, although proteolytically processed P-III SVMPs have been isolated proteomically from Viperinae (Bazaa et al. 2005; Wagstaff et al. 2009). This posttranslational modification may therefore have evolved in parallel, though the mechanism remains uncharacterized. Two independent origins of the P-IIId SVMP subclass exist in separate Viperinae lineages, one prior to the divergence of *Macrovipera* and *Daboia* (P-IIId-I) and another restricted to members of the genus *Echis* (P-IIId-II) (see figure 23.4). The majority of the P-III SVMP isoforms sequenced to date remain structurally uncharacterized. While the hemorrhagic activity of P-III SVMPs is predominately dependent on the metalloprotease domain, the potency of this activity is seemingly associated with the presence of the additional domains that are absent from the P-I and P-II SVMP class. These domains presumably augment the proteolytic activity (Fox and Serrano 2005).

High levels of hemorrhagic activity are not a prerequisite for P-III SVMP functionality, with minimally active and even entirely nonhemorrhagic P-III SVMPs isolated from members of both the Crotalinae and Viperinae (Loria et al. 2003; Silva et al. 2003; Leonardi et al. 2007). These particular SVMPs exhibit evidence of additional functional activities; the minimally hemorrhagic "ammodytase" isolated from *Vipera ammodytes* exhibits fibrinolytic activity (Leonardi et al. 2007), while "basparin A" and "berythractivase" from the genus *Bothrops* have activate prothrombin (Loria et al. 2003; Silva et al. 2003).

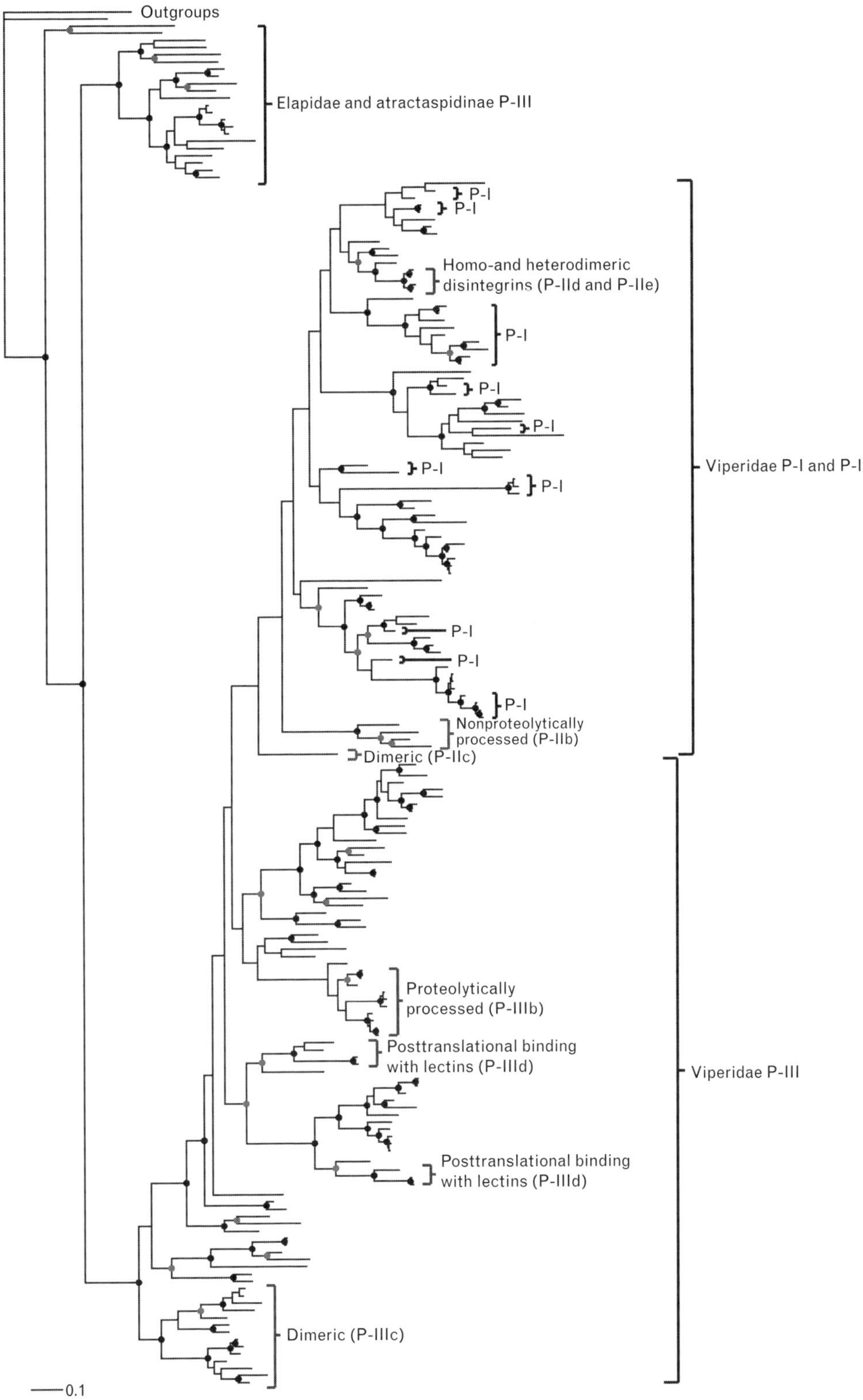

FIGURE 23.4: Evolution of SVMPs (Casewell et al. 2011a). The P-III scaffold was recruited into venom at the base of the caenophidian snakes. Viperid P-IIIs have diversified and subsequently evolved into the P-II structure. Convergent evolution of the P-I structure has occurred following the loss of the disintegrin domain from P-II SVMPs. Subclasses of P-II and P-III SVMPs have been mapped to the tree. Nodes with black and gray circles indicate significant support (Bayesian posterior probabilities of 1.00 and > 0.95, respectively).

Some P-III SVMPs have evolved additional procoagulant functions by activating other components of the clotting cascade, such as factor X (Kisiel, Hermodson, and Davie 1976; Hofmann and Bon 1987; Takeya et al. 1992; Siigur et al. 2004); the molecular characterization of these components implies a shared evolutionary origin of this functionality. The pathophysiology resulting from procoagulant SVMPs is the consumption of fibrinogen, which results in defibrination, where the depletion of coagulant factors renders an individual susceptible to hemorrhage (see color plates 11E, 11I–11K, 12C). The combined action of hemorrhagic and procoagulant SVMPs can therefore dramatically affect the severity of a snakebite incident. Other apotypic activities of P-III SVMPs include the induction of apoptosis and cleavage of the blood glycoprotein von Willebrand factor (vWF). Apoptotic P-III SVMPs have been characterized from a number of viperid snake venoms, and the capability to selectively induce endothelial cell apoptosis also appears to arise from a single evolutionary origin (Masuda, Hayashi, and Araki 1998; Masuda et al. 2000; Masuda et al. 2001). Toxin interference with integrin binding to extracellular matrix components may be the mechanism of apoptotic action, as *in vitro* analyses suggest apoptosis occurs following the loss of contact between endothelial cells and the extracellular matrix (You et al. 2003; Diaz et al. 2005; Tanjoni et al. 2005). Some P-III SVMPs, such as jararhagin, contain cysteine-rich domains that are capable of binding to proteins that contain an "A domain," such as vWF, FACIT collagens XII and XIV, and matrillins 1–3. Binding to the "A domain" of these proteins blocks their function and appears to target the protein for proteolytic cleavage (Serrano et al. 2000; Serrano et al. 2005; Serrano et al. 2006). This activity can result in the disruption of the structure of the extracellular matrix and may also hinder vWF-mediated platelet aggregation (Serrano et al. 2007). A summary of P-III SVMP functions characterized to data can be observed in table 23.1, and a more thorough review of the combinatory roles of SVMPs acting to cause pathology has been described (Gutiérrez et al. 2009).

23.4.2 P-II SVMP

Following the divergence of viperid snakes from the remaining caenophidian snakes and prior to the divergence of Crotalinae from Viperinae, a P-III SVMP gene was duplicated and subsequently lost its cysteine-rich domain (Moura-da-Silva, Theakston, and Crampton 1996). This process may have occurred by a stop-codon-mediated deletion of the P-III-specific cysteine-rich domain followed by the mutation of a cysteine residue present within the P-III-specific ECD motif (Sanz, Harrison, and Calvete 2012). This modification removed the structural constraint imposed by a disulfide bond, thereby enabling the emergence of the loop observed protruding from the newly formed disintegrin domain, which is largely responsible for the functional activities of P-II SVMPs (Sanz, Harrison, and Calvete 2012). Accelerated evolution of the P-II SVMP precursor is responsible for causing additional changes to the disintegrin-like domain (Juarez et al. 2008; Casewell et al. 2011a), and in combination, these two processes ultimately resulted in the formation of the disintegrin domain. Positive selection acting on surface-exposed amino acid residues present in both the disintegrin-like and metalloprotease domains led to the formation of a novel SVMP class (Juarez et al. 2008; Casewell et al. 2011a).

P-II SVMPs have been isolated from a wide taxonomic variety of viperid snake venoms, and the majority characterized are capable of posttranslational proteolytic liberation of the disintegrin domain from the metalloprotease domain. Subclasses of P-II SVMPs include those

Table 23.1: Biological Activities of Selected P-III SVMPs (Fox and Serrano 2005)

Species	P-III SVMP	Activity	References
Bothrops erythromelas	Berythractivase	Activation of prothrombin	Silva et al. 2003
Bothrops jararaca	Jararhagin	Hemorrhagic; inhibition of platelet aggregation	Paine et al. 1992b; Kamiguti, Hay, and Zuzel 1996
Bothrops jararaca	HF3	Hemorrhagic; activation of macrophage phagocytosis	Assakura, Reichl, and Mandelbaum 1986; Silva et al. 2004
Crotalus atrox	Atrolysin A	Hemorrhagic; inhibition of platelet aggregation	Fox and Bjarnason 1995; Jia et al. 1997
Crotalus atrox	Catrocollastatin	Inhibition of platelet aggregation	Zhou, Smith, and Grossman 1995; Zhou, Dangelmaier, and Smith 1996
Crotalus atrox	VAP1	Apoptotic	Masuda, Hayashi, and Araki 1998; Masuda et al. 2000
Daboia russelii	RVV-X	Activation of factor X	Takeya et al. 1992; Gowda et al. 1994a
Deinagkistrodon acutus	Acurhagin	Hemorrhagic; inhibition of platelet aggregation	Wang and Huang 2002
Echis carinatus	Ecarin	Activation of prothrombin	Kornalik and Blomback 1975; Nishida et al. 1995
Macrovipera lebetina	VLFXA	Activation of factor X	Siigur et al. 2001; Siigur et al. 2004
Naja kaouthia	Kaouthiagin	Cleavage of vWF; inhibition of platelet aggregation	Hamako et al. 1998; Ito et al. 2001
Protobothrops flavoviridis	HR1a, HR1b	Hemorrhagic	Omori-Satoh and Sadahiro 1979; Kishimoto and Takahashi 2002
Protobothrops flavoviridis	HV1	Apoptotic	Masuda et al. 2001

that proteolytically process the disintegrin domain (P-IIa), remain intact (P-IIb), form intact dimeric structures (P-IIc), and exhibit homodimeric (P-IId) or heterodimeric (P-IIe) processed disintegrins (see figure 23.3) (Fox and Serrano 2008). The two subclasses that do not process the disintegrin domain posttranslationally (P-IIb and P-IIc) have only been recovered so far from Crotalinae and are positioned phylogenetically at or near the base of the P-II SVMP radiation (see figure 23.4), implying that (i) the plesiotypic P-II SVMP did not exhibit a processed disintegrin domain and (ii) these subclasses may have been lost in Viperinae snakes. The presence of dimeric clades of SVMPs near the base of both the P-III and P-II radiations (see figure 23.4) may be coincidental but suggests that the formation of a dimeric structure may facilitate subsequent diversification and modification of SVMP domain structures. Subsequent evolutionary loss of dimerism could provide a functional advantage by reducing toxin size and enabling more rapid physiological diffusion in prey (Doley and Kini 2009). It is notable that the apotypic forms of P-II SVMPs (such as excluding the P-IIb and P-IIc subclasses) appear to have diversified more extensively in the Crotalinae than the Viperinae. A single clade in the SVMP tree (see figure 23.4) contains all representatives isolated to date from members of the Viperinae, while Crotalinae P-II SVMPs are represented in multiple clades. Surprisingly, this apparent increase in diversification is not reflected by increased abundance of SVMP in the venom of these taxa; although processed disintegrins are readily identified in the majority of viperid snake species (typically as venom components with representation of < 10%), in general, representation is higher in the Viperinae (2%–18% of toxin components, mean = 7.6%, n = 11) than in the Crotalinae (0%–8%, mean = 2.9%, n = 14) (Bazaa et al. 2005; Sanz et al. 2006; Calvete et al. 2007a; Calvete et al. 2007c; Alape-Giron et al. 2008; Angulo et al. 2008; Gutiérrez et al. 2008; Sanz, Ayvazyan, and Calvete 2008; Sanz et al. 2008; Tashima et al. 2008; Calvete et al. 2009a; Calvete et al. 2009b; Nunez et al. 2009; Wagstaff et al. 2009).

The majority of P-II SVMPs undergo posttranslational proteolytic processing, resulting in the release of the disintegrin domain, which is typically responsible for the functional activity of these SVMPs (Calvete 2011a). However, some P-IIs are not proteolytically processed, including bilitoxin from *Agkistrodon bilineatus*, which forms an intact dimeric structure and exhibits potent hemorrhagic activity (Imai et al. 1989; Nikai et al. 2000); and jerdonitin from *Protobothrops jerdonii*, which is intact and monomeric and dose-dependently inhibits ADP-induced platelet aggregation (Chen et al. 2003a) (see table 23.2). Inhibition of platelet aggregation is a characteristic function of the disintegrin domains, despite their variable size and disulfide-bond arrangements (Calvete et al. 2005). They function by competitively binding to integrin receptors, with platelet-aggregation inhibition typically mediated by blocking β_1 and β_3 integrins and guided by a tripeptide motif (usually RGD) found in the integrin binding loop (Trikha, De Clerck, and Markland 1994). The absence of an RGD motif in bilitoxin (replaced by MGD) is likely responsible for the absence of platelet-aggregation inhibition activity in this protein (Nikai et al. 2000; Chen et al. 2003a). The inhibitory tripeptide sequences found in the integrin-binding loops of disintegrins seemingly modulate the selectivity and binding affinity for different integrin ligands. For example, the RGD tripeptide targets $\alpha_5\beta_1$, $\alpha_8\beta_1$, $\alpha_V\beta_1$, $\alpha_V\beta_3$, and $\alpha_{IIb}\beta_3$ integrins, whereas MLD targets $\alpha_4\beta_1$, $\alpha_4\beta_7$, $\alpha_3\beta_1$, $\alpha_6\beta_1$, $\alpha_7\beta_1$, and $\alpha_9\beta_1$ integrins; VGD $\alpha_5\beta_1$ integrin; KGD $\alpha_{IIb}\beta_3$ integrin; and KTS $\alpha_1\beta_1$ integrin (Scarborough et al. 1991; Marcinkiewicz et al. 1997; Marcinkiewicz et al. 1999a; Marcinkiewicz et al. 1999b; Oshikawa and Terada 1999; Wierzbicka-Patynowski et al. 1999; Marcinkiewicz et al. 2003). Of these targets, the $\alpha_{IIb}\beta_3$ integrin plays a central role in the aggregation of platelets; following vascular damage, this receptor is converted into an active form enabling the binding of ligands such as fibrinogen and von Willebrand factor, ultimately facilitating platelet aggregation. Because of this, venom disintegrins that interfere with this process have received attention as antiplatelet agents. Both tirofiban (Aggrastat) and eptifibatide (Integrillin) were designed based on snake-venom disintegrins isolated from viperid snake venoms and have been used in the treatment of acute coronary syndromes (see section 23.4).

Table 23.2: Biological Activities of Selected P-II SVMPs (Fox and Serrano 2005)

Species	P-II SVMP	Activity	References
Agkistrodon bilineatus	Bilitoxin-I	Hemorrhagic	(Imai et al. 1989; Nikai et al. 2000)
Crotalus atrox	Atrolysin E	Hemorrhagic; inhibition of platelet aggregation	(Hite et al. 1992; Shimokawa et al. 1996; Shimokawa et al. 1998)
Gloydius halys brevicaudus	MT-d	Proteolytic; inhibition of platelet aggregation	(Jeon and Kim 1999b; Jeon and Kim 1999a)
Protobothrops jerdonii	Jerdonitin	Inhibition of platelet aggregation	(Chen et al. 2003a)

23.4.3 P-I SVMPS

Following the loss of the cysteine-rich domain to form the P-II SVMP subtype, the disintegrin domain has been lost from the P-II structure on multiple occasions (see figure 23.4), resulting in the formation of the P-I SVMP class (Casewell et al. 2011a). Consequently, P-I SVMPs do not contain additional domains that extend the C-terminal end of the metalloprotease domain (see figure 23.1). As with the P-II class, P-I SVMPs have only been identified from the venom of viperid snakes. Convergent evolutionary origins of the P-I scaffold have occurred on at least nine occasions in a variety of members of the viperid snakes, including genus-specific losses of the disintegrin domain in *Echis* and *Macrovipera* in the Viperinae and on multiple occasions in the Crotalinae lineages of *Agkistrodon, Bothrops, Crotalus,* and *Deinagkistrodon* (Casewell et al. 2011b). As sampling is incomplete, disintegrin domain loss events have likely occurred on a number of additional occasions throughout the evolutionary history of the viperid snakes (Casewell et al. 2011a). In accordance with current models of gene evolution (Hughes 1994; Bergthorsson, Andersson, and Roth 2007), it is likely that the precursor of each P-I lineage (such as a P-II SVMP) was capable of encoding a primary (disintegrin domain) and secondary function (metalloprotease domain). Consequently, selective pressures acting to promote the secondary function may have induced gene amplification and facilitated the loss of the primary function in the duplicate gene through loss of the disintegrin domain. Subsequently, adaptive evolution of the metalloprotease domain, with positive selection directed against surface-exposed amino acid residues (in a similar way to the evolution of the P-II structure, although in different residues), has presumably directly resulted in the neofunctionalization of this SVMP class (Casewell et al. 2011a). Abundance of the P-I class in venom varies considerably, with proportional representation in the venom glands and venoms of viperid snakes ranging from complete absence to as high as 32% of all toxins (Alape-Giron et al. 2008; Casewell et al. 2009).

Unlike many P-II and P-III SVMPs, which undergo posttranslational cleavage of domains found downstream of the metalloprotease domain, the P-I SVMPs are produced as intact molecules containing only the metalloprotease domain (see figure 23.3). These proteins do not have a disintegrin or cysteine-rich domain, as these domains have been lost from the genes encoding such proteins over evolutionary time. Nonetheless, P-I SVMPs are capable of inducing a variety of functional activities, including hemorrhage, proteolysis, fibrinolysis, myonecrosis, and apoptosis (see table 23.3). These toxins contribute to the profound local tissue destruction (see color plates 12C–12D) and swelling (see color plates 11D and 12C) characteristic of envenomations by *Atractaspis* and most viperids. Notably, some of these SVMPs possess multiple activities, such as BaPI from the viperid *Bothrops asper*, which digests the alpha and beta chains of fibrinogen and also exhibits hemorrhagic activity via the proteolytic degradation of basement membrane components found in microvessels (Gutiérrez et al. 1995; Rucavado et al. 1995). These activities ultimately combine to induce hemorrhage, myonecrosis, blistering, inflammation, and edema *in vivo* (Gutiérrez et al. 1995; Rucavado et al. 1995; Gutiérrez et al. 2009). Similarly, BpirMP, isolated from the venom of a related species, *Bothrops pirajai*, exhibits fibrin(ogen)olytic, thrombolytic, and weak hemorrhagic activities through the hydrolysis of basement membrane components, the cleavage of both α and β chains of fibrinogen and the degradation of fibrin and blot clots (Bernardes et al. 2013). Batx-I, from the venom of *Bothrops atrox*, also exhibits both hemorrhagic and fibrinogenolytic activities (Patino et al. 2010), while BJ-PI2 from *Bothrops jararaca* induces fibrinolysis by degrading the α and β chains of fibrinogen (da Silva et al. 2012a). BJ-PI2, in contrast to the others, exhibits no hemorrhagic, coagulant or myonecrotic activities (da Silva et al. 2012a). P-I SVMPs isolated from other viperid genera have been predominantly characterized as hemorrhagic proteins. The *Crotalus atrox* venom protein Ht-d degrades type IV collagen and other basement membrane proteins, contributing to pathology by disrupting capillary basement membranes and causing hemorrhage (Shannon et al. 1989). Finally, LHF-II, isolated from the venom of *Lachesis muta*, is a hemorrhagic SVMP that also selectively degrades the α chain of fibrinogen (Sanchez et al. 1991).

Table 23.3: Biological Activities of Selected P-I SVMPs (Fox and Serrano 2005)

Species	P-I SVMP	Activity	References
Agkistrodon contortrix	Fibrolase	Fibrinolytic	Markland 1996
Bothrops asper	BaP1	Hemorrhagic; myonecrotic; inflammatory	(Gutiérrez et al. 1995; Rucavado et al. 1995)
Bothrops pirajai	BpirMP	Fibrinolytic, weakly hemorrhagic	(Bernardes et al. 2013)
Crotalus atrox	Atrolysin C	Hemorrhagic	(Shannon et al. 1989; Zhang et al. 1994a)
Crotalus atrox	Atroxase	Fibrinolytic	(Willis and Tu 1988)
Crotalus ruber	HT-2	Hemorrhagic	(Mori et al. 1987; Takeya et al. 1990)
Deinagkistrodon acutus	Acutolysin A	Hemorrhagic	(Gong et al. 1998; Liu et al. 1999)
Lachesis muta	LHF-II	Hemorrhagic	(Sanchez et al. 1991)
Protobothrops flavoviridis	H2-protease	Proteolytic; nonhemorrhagic	(Takeya et al. 1989)
Trimeresurus gramineus	Graminelysin I	Apoptotic	(Wu et al. 2001)

23.4.4 OTHER APOTYPIC SVMP PRECURSORS

Mapping the evolution of the well-characterized SVMP classes (P-I, P-II, and P-III) and their respective subclasses (such as P-IIIb-d and P-IIb-e) onto a species tree of the advanced snakes serves to elucidate the likely timing of the origin of these different structural and posttranslational modifications in this diverse toxin family (see figure 23.5). In addition to these structural arrangements, a number of atypical SVMPs have been described that exhibit different domain loss events (figure 23.6).

There are two known examples of P-III SVMP genes that are extensively truncated, resulting in the expression of proteins containing only parts of the prodomain (Fry et al. 2008; Casewell et al. 2011a; Brust et al. 2013). These atypical SVMPs have been identified from the lamprophiid genus *Psammophis* and the viperid snake genus *Echis*, and they lack the metalloprotease domain (and additional C-terminal domains) and zinc-binding motif that characterize the adamalysins. Although these genes have yet to be identified as translated and secreted in *Echis* venom, the resultant toxins are present in *Psammophis* venom (Fry et al. 2003c;

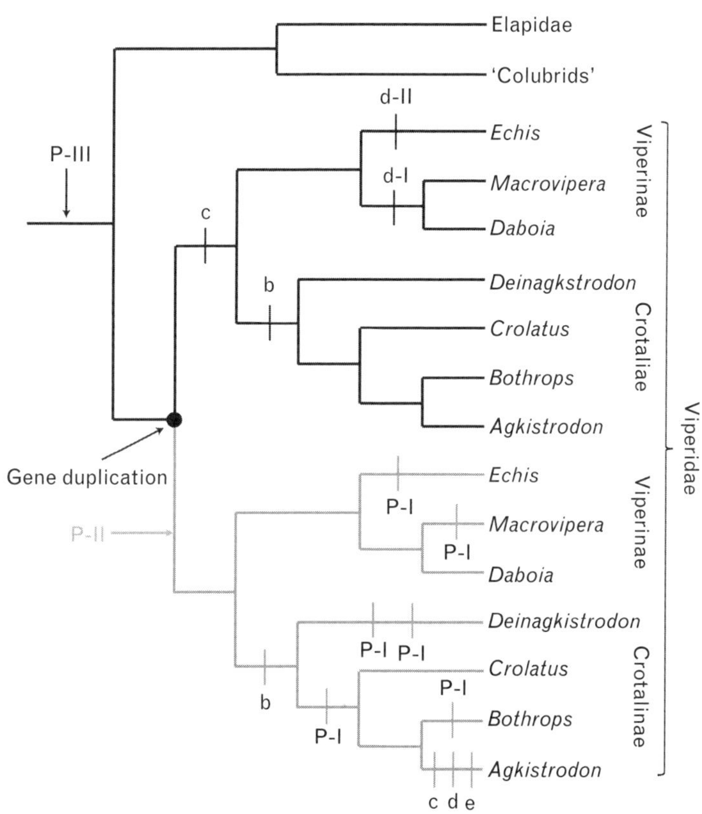

FIGURE 23.5: The inferred taxonomic timing of SVMP class and subclass diversification events. The species tree for viperids is duplicated to enable mapping of P-III diversification events (black lines) and P-II events (gray lines) following a gene-duplication event. Vertical lines represent the most parsimonious taxonomic timing of the formation of SVMP subclasses (indicated by letters) and the convergent evolution of P-I SVMPs identified to date. Note that only representative genera from the viperid snakes are displayed.

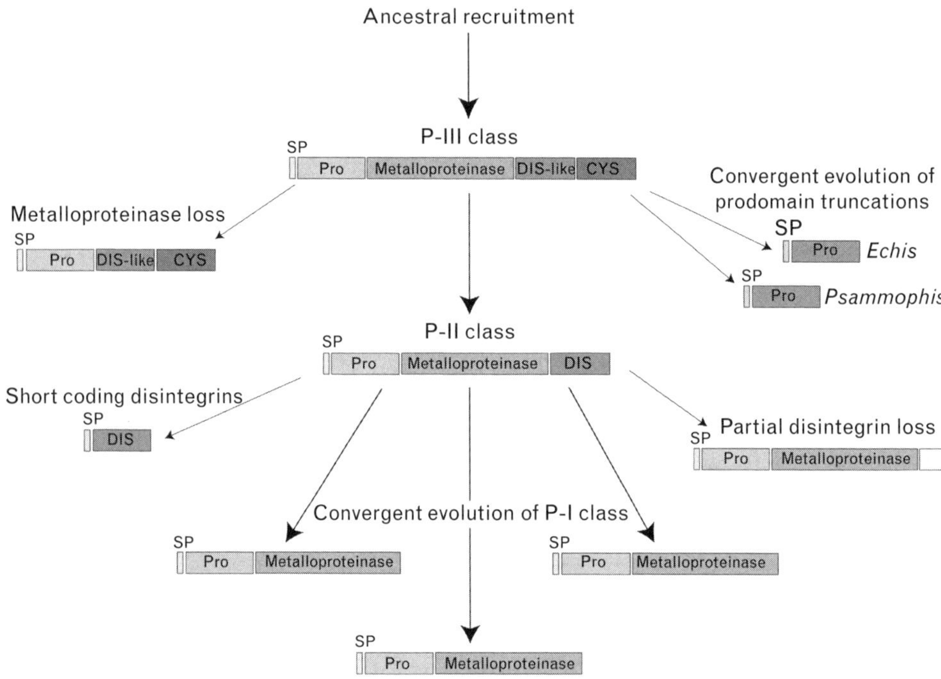

FIGURE 23.6: Schematic of the evolutionary history of domain loss in the SVMPs (Casewell et al. 2011a). SP = signal peptide, Pro = prodomain, DIS = disintegrin domain, CYS = cysteine-rich domain.

Brust et al. 2013), where they exhibit an entirely novel neurotoxic activity: inhibition of postsynaptic α7 nicotinic acetylcholine receptors (Brust et al. 2013). Notably, domain loss in *Psammophis* has resulted in increased selection pressure, which has driven a rapid rate of mutations in the prodomain and resulted in protein neofunctionalization, analogous to the processes observed in the evolution of P-I and P-II SVMPs (Casewell et al. 2011a; Brust et al. 2013).

Another P-III-related structure that lacks the metalloprotease domain has been identified from the genus *Echis* (Casewell et al. 2009; Casewell et al. 2011a); these transcriptomically identified components have yet to be identified directly in venom, but the transcripts contain a deletion of the entire metalloprotease domain, with the prodomain being directly followed by the disintegrin-like and cysteine-rich domains present in typical P-III SVMPs. Also identified from the genus *Echis* is a P-II SVMP component that exhibits partial loss of the disintegrin domain; although originally discovered in the venom-gland transcriptome (Wagstaff and Harrison 2006), this component has subsequently been identified proteomically (Wagstaff et al. 2009).

Short-coding disintegrins have been identified proteomically in venom; these are snake-venom disintegrins that are encoded by disintegrin-only genes (signal peptide and disintegrin domain only) and in some cases are identical to those obtained from proteolytic processing of P-II SVMP genes (Okuda, Koike, and Morita 2002; Francischetti et al. 2004; Juarez et al. 2006; Wagstaff et al. 2009). Short-coding disintegrins have presumably evolved from P-II SVMP genes that have lost the prodomain and the metalloprotease domain. However, this process appears to have occurred in parallel, with different short-coding disintegrin molecules evolving convergently (Sanz-Soler et al. 2012). A summary schematic of the evolutionary history of domain loss in SVMPs can be found in figure 23.6, which outlines both the loss of domains resulting in the formation of the P-I and P-II classes and the "atypical" domain losses described above.

23.5 THERAPEUTIC POTENTIAL OF THE TOXIN TYPE

The disintegrin domain in particular exhibits pharmaceutical activities relevant for human drug development. The most well-known and successful products, tirofiban (AGGRASTAT) and eptifibatide (INTEGRILIN), were designed based on snake-venom disintegrins and have been approved for use as antiplatelet agents. Tirofiban is an $\alpha_{IIb}\beta_3$, $\alpha_v\beta_3$, and $\alpha_v\beta_1$ integrin antagonist based on the RGD containing disintegrin echistatin, which was isolated from the venom of *Echis carinatus* (Chen et al. 1991; Barrett et al. 1994; Topol, Byzova, and Plow 1999). The antithrombotic properties of this protein have great potential for preventing coronary thrombosis in animal models (Shebuski et al. 1990), which prompted pharmaceutical modification for human use. Eptifibatide was designed based on the disintegrin barbourin from the venom of *Sistrurus miliarius barbouri* (Scarborough et al. 1991). Barbourin exhibits desirable selectivity for $\alpha_{IIb}\beta_3$ integrin binding via its KGD motif, and this characteristic was exploited during the commercial development and synthesis of eptifibatide (Scarborough et al. 1993). The consequence of both drugs binding to $\alpha_{IIb}\beta_3$ integrins is the blockage of this site for fibrinogen binding, ultimately resulting in the inhibition of platelet aggregation (Huang et al. 1987). In terms of chemistry, tirofiban is nonpeptide, while eptifibatide is a peptidomimetic of the respective lead toxin molecule. The FDA approved both of these drugs for human use at the end of the last century; they are indicated for parenteral delivery for the treatment of acute coronary syndromes (O'Shea and Tcheng 2002; Menozzi, Merlini, and Ardissino 2005; King 2011).

Other disintegrins also show pharmaceutical promise. Rhodostomin, isolated from the venom of *Calloselasma rhodostoma*, exhibits antiinflammatory and tumor-suppressing activities (Yeh et al. 2001; Tseng, Peng, and Huang 2004). This RGD-containing disintegrin is capable of binding to activated neutrophils, thereby blocking their adhesion to fibrinogen and decreasing superoxide production, which is relevant to inflammation pathways (Tseng, Peng, and Huang 2004). The specific binding of rhodostomin to $\alpha_v\beta_3$ integrins can also result in the inhibition of basic fibroblast growth factor, thereby reducing cell proliferation, as demonstrated in an *in vivo* murine melanoma tumor model (Yeh et al. 2001). Another disintegrin, salmosin from the venom of *Gloydius halys brevicaudus*, also exhibits inhibitory activity on cell proliferation by activating integrin-mediating signaling molecules. This results in the detachment of cells by interfering with adhesion pathways, ultimately leading to cell apoptosis (Hong et al. 2003). Some disintegrins, therefore, offer potential as novel anticancer agents. Another disintegrin, crotavirin from the venom of *Crotalus viridis*, exhibits potential for use in treating endocarditis, as it interferes with the virulent mechanism by which *Staphylococcus aureus* bacterial infections cause pathology. Crotavirin is a glycoprotein antagonist and is capable of interacting with circulating platelets by occupying $\alpha_{IIb}\beta_3$ integrins, thereby preventing the activation and aggregation platelets by the bacteria (Liu et al. 2005). Finally, in addition to the R/KGD-containing disintegrins described above, a number of disintegrins containing a KTS motif exhibit functional activities pertinent to pharmaceutical development. For example, the disintegrins obtustatin (from *Macrovipera lebetina obtusa* venom) and lebestatin (*Macrovipera lebetina*) exhibit specificity for $\alpha_1\beta_1$ integrins, resulting in the inhibition of integrin-mediated cell-adhesion migration and tumor growth *in vitro*, and also show antiangiogenic activity *in vivo* (Marcinkiewicz et al. 2003; Olfa et al. 2005; Brown et al. 2008). Obtustatin in particular has great potential as a novel therapeutic compound for the treatment of melanomas that contain a high degree of vascularization (Brown et al. 2008).

In addition to development of disintegrin-derived pharmaceuticals, the enzymatic domains also have potential therapeutic use. One of these, fibrolase, is a fibrinolytic P-I SVMP isolated from the venom of *Agkistrodon contortrix contortrix* (Bajwa et al. 1982; Randolph et al. 1992). A recombinant form of this protein, termed alfimeprase, was developed as a novel pharmaceutical for the treatment of stroke and catheter occlusion (King 2011). Alfimeprase preferentially cleaves the Aα-chain of fibrin(ogen) and is effective at lysing clots in animal thrombosis models (Swenson et al. 2004). Despite promising results in phase I and II clinical trials (Ouwens 1912; Moll et al. 2006), phase III trials have been discontinued because of insufficient efficacy (King 2011; Koh and Kini 2012).

While the therapeutic use of the enzyme domain is very much in its infancy, the enzymes have been tremendously useful as diagnostic agents. *Daboia russelii* venom was among the very first snake venoms utilized in modern medicine. In 1934, trials with locally applied *D. russelii* venom diluted at 1:10,000 resulted in "success to stop hemorrhage" following dental extraction in patients with hemorrhagic diathesis (MacFarlane and Barnett 1934). In 1938, *D. russelii* venom, under the name Rusven, was offered as a "powerful blood coagulant for local application in the control of external bleeding" (anonymous 1938). Today, this venom (and pure toxins) are used in a wide array of diagnostic assays. The P-III metalloprotease ecarin, the venom of *Echis carinatus*, is a potent prothrombin-activator. Ecarin specifically catalyzes prothrombin to a-thrombin via the formation of a meizothrombin intermediate (Morita, Iwanaga, and Suzuki 1976b; Morita, Iwanaga, and Suzuki 1976a) and is also able to generate α-thrombin from abnormal prothrombin produced by treating mammals with vitamin K antagonists (Nelsestuen and Suttie 1972c; Nelsestuen and Suttie 1972b; Nelsestuen and Suttie 1972a). Consequently, ecarin has been used clinically for many years in the "ecarin clotting time" test, where it acts as a standard to quantify the use of direct thrombin inhibitors in patients, particularly in anticoagulant therapy utilizing r-hirudin (Nowak 2003). Another ecarin-based clinical diagnostic assay is a confirmatory test for lupus anticoagulants, utilizing the clotting time ratio between ecarin and TV-fXa (see chapter 13) from either *Pseudonaja textilis* or *Oxyuranus scutellatus* venoms (Triplett et al. 1993; Moore, Smith, and Savidge 2003; Moore 2007).

D. russelii venom is also used for the determination of factor X activity in human plasma. Russell's viper venom time (RVVT, Stypven time) is the plasma clotting time induced by *D. russelii* venom (Stocker 1998; Kini 2005b; Siigur and Siigur 2010). The factor-X-activating P-IIId from *D. russelii* hydrolyzes human factor X heavy-chain Arg52-Ile53 bond, converting factor X into factor Xa; this is the same cleavage site as used by the endogenous physiological factor X activators, factor IXa and VIIa (Fujikawa, Legaz, and Davie 1972; Di Scipio, Hermodson, and Davie 1977; Takeya et al. 1992; Gowda et al. 1994a; Chen et al. 2008a). RVVT can be used in differential diagnosis of abnormalities in the extrinsic systems of coagulation.

Purified RVV-X is used, to a varying extent, as test for factor X assay, differentiating factor VII and factor X deficiency, and in lupus antigen assay (Perchuc and Wilmer 2010). The functional diagnostic test (ProC Ac R) (Pejkic et al. 2014) for the identification of resistance to activated protein C (APC) is also dependent, among others agents, on RVV-X activity. APC resistance is associated with increased risk for venous thrombotic events and is often linked to factor V Leiden mutation. RVV-X is also part of a diagnostic kit (CRYOCHECK CLOT S) for the quantitative determination of protein S activity (Mulder et al. 2010). Protein S plays a role in the inactivation of factors Va and VIIIa. Individuals with protein S deficiency are predisposed to recurrent venous thromboembolism and to an increased risk of perinatal mortality (ten Kate and van der Meer 2008).

Today *D. russelii* venom is still widely used in clinical testing for the presence of lupus antigen (Thiagarajan, Pengo, and Shapiro 1986), a major risk factor for arterial and venous thrombosis, accounting for about 15% of patients with thromboembolic events. Lupus antigen has also been linked to recurrent spontaneous abortions, transverse myelopathies, and nonbacterial thrombotic endocarditis. Lupus antigens are immunoglobulins of autoimmune origin that bind to proteins such as β2-glycoprotein I, prothrombin, or others in complex with negatively charged phospholipids (Tripodi 2007). As a result, lupus antigens prolong the clotting time in phospholipid-dependent *in vitro* coagulation tests. Dilute Russell's viper venom time (dRVVT, such as DVVtest, DVVconfirm) is a test of choice to screen and confirm the presence of lupus antigen.

Remarkably, although not surprisingly, diagnostic assays performed with *D. russelii* venoms have been reported to exhibit a significant variability when compared with other tests systems. The variability was attributed, among other factors, to the intrinsic heterogeneity of venom samples used for the assay process. Such venom variation could arise, among other factors, as a result of the geographical origin of the snake, ontogeny, and venom extraction and preparation methods (Triplett 2000). Standardization of venoms and recombinant toxin production could address such variability issues. The availability of direct diagnosis (such as by sequencing) is expected to overtake the utilization of a subset of toxin-based clinical laboratory tests.

23.6 CONVERGENCE WITH OTHER VENOMS

There is no current evidence that other animals have recruited ADAM-related proteins for use in venom. This is somewhat surprising, considering the prevalence of toxin convergence (Fry et al. 2009a; Casewell et al. 2013) and the ubiquity of ADAM genes (Huxley-Jones et al. 2007) in the animal kingdom. However, there are areas of functional convergence, such as nicotinic acetylcholine receptors being convergently targeted by other snake-venom peptide types such 3FTx (see chapter 8) and the waglerin peptides from *Tropidolaeumus* venoms (see chapter 24.25). Similarly, the disintegrin RGD functional motif has been used in apotypic 3FTx (see chapter 8.4.4) in a wide array of antiplatelet toxins from other venomous animals (Fry et al. 2009a) (see color plate 2A).

CHAPTER 24

LESSER-KNOWN OR PUTATIVE REPTILE TOXINS

B. G. FRY, R. RICHARDS, S. EARL, X. COUSIN, T. N. W. JACKSON, C. WEISE, AND K. SUNAGAR

24.1 SUMMARY

The historical body of reptile venom research has concentrated on the venoms of a restricted number of species, the large, readily obtainable snake species typically involved in medically significant envenomations. In addition, research has also concentrated on the dominant (most abundant) toxins from these venoms, as these are the most easily isolated. This trend has been particularly pronounced in cases where synthesis or expression of toxins was not possible and therefore obtaining the raw material for bioactivity testing required purification of toxins from crude venom. In such cases, large amounts of venom are required, and consequently, the toxins that were the most abundant in the venom were quite understandably the first studied. The combination of a restricted number of species being venom extracted and only select toxins within each venom being examined has meant that the literature is heavily biased toward only a small percentage of the total toxinological biodiversity. Thus, there are more papers published on some single toxin types (such as 3FTxs) than all the papers combined for the neglected toxin types discussed in this composite chapter.

These toxin types represent some of the first known and also some of the most recently discovered. For example, despite the early discovery of the presence of acetylcholinesterase (AChE) in snake venom, the potent activity of the enzyme making it easy to detect and measure, molecular characterizations of snake AChE are scarce. To date, only a few AChE sequences have been reported, and only one purified form (from *Bungarus* venoms) has had any significant work undertaken to characterize its bioactivity. Some lesser-known proteins are present in venoms subjected to vast amounts of previous research but had been long overlooked by research that has been target-driven or relying on relatively crude purification techniques favoring only toxins of certain sizes and relative abundance. Others were discovered as the result of studying novel snake taxa, such as the veficolins found in the venom of the non-front-fanged caenophidian snake *Cerberus rynchops*, others from investigations of anguimorph lizards (such as celestoxin, cholecystoxin, and goannatyrotoxin). Many of these toxins were discovered not by conventional protein purification but through the sequencing of transcriptome libraries. Of those that have been at least partially characterized, most have been studied on synthetic peptides made using the sequences from translated transcriptome libraries. Many, however, have remained entirely uncharacterized and are known only from the transcriptome sequence. This pool of underinvestigated proteins thus not only represents an exciting opportunity for furthering our understanding of the evolution of venom but is also a treasure trove of source material for drug design and development.

24.2 ACETYLCHOLINESTERASE

24.2.1 ENDOPHYSIOLOGICAL PLESIOTYPE

Acetylcholinesterase (AChE) is an enzyme involved in the regulation of nervous signal transmission at cholinergic synapses (see figure 24.1). Depending on the synapse, AChE can be located on pre- and/or postsynaptic membranes and/or in the synaptic cleft. In all cases, its function is to hydrolyze the neurotransmitter acetylcholine (ACh) and to avoid cholinergic receptor desensitization (Massoulié et al. 1993).

One prototypic situation is the neuromuscular junction, where AChE is involved in the clearance of ACh released by the motor neuron in order to activate postsynaptic nicotinic receptors. In this particular context, AChE is primarily located within the synaptic cleft and associated with ColQ, a trimeric partner protein containing a collagenic domain and a proline-rich domain at its N-terminus. In mammals, AChE is also present in the brain in cholinergic synapses. In this case, AChE is associated with a 20-kDa hydrophobic anchor called PRiMA (proline-rich membrane anchor) (Massoulié et al. 1999). In both cases, AChE is associated with the accessory proteins through their C-terminal extremity (Massoulié and Bon 2006). In vertebrates, a single gene encodes AChE, but the short C-terminal domain is encoded by two 3′ exons that can be alternatively used. In most cases, the T (tailed) exon is used. The C-terminal peptide is organized as a hydrophobic helix, which interacts with the proline-rich domains of accessory proteins. The alternative H (hydrophobic) exon encodes a peptide that is cleaved and associated with a hydrophobic glycolipid anchor (GPI). H peptides are found as monomers or dimers and are directly associated with cell membranes through the GPI anchor (Massoulié et al. 1998). In invertebrates, the situation is more complex and varies among species; in *Drosophila*, there is one gene with a single 3′ exon encoding an H-type C-terminal peptide, which is cleaved. In nematodes, there are four genes; each of the two primary genes (producing more than 95% of overall AChE activity) possesses a single 3′ exon encoding, respectively, a T-type or an H-type terminal peptide (Selkirk, Lazari, and Matthews 2005).

One peculiarity of AChE is that it is an extraordinarily efficient enzyme with an unusually high turnover rate (close to the substrate diffusion limit). This is particularly surprising since, as a member of the α/ß-hydrolase-fold protein family, the active site of AChE is buried in the middle of the protein (Marchot and Chatonnet 2012). This puzzling issue, coupled with the fact that AChE is a target of many chemicals used as pesticides but also as chemical weapons,

```
            10        20        30        40        50        60
1.  mpscqpgkmpapwpwwlqlllcipscvavLPGRAGELKVSTQTGSVRGLSLPVLDGHVSA
2.  mpscqpgkmpapwpwclqlvlcilssvavLPGRASELKVSTQMGWVRGLSLPVLDGHVSA

            70        80        90       100       110       120
1.  FLGIPFAEPPLGRMRFLRPEPVKPWQHVLDATSYKPACYQMVDTSYPGFQGTEMWNPNRG
2.  FLGVPFAEPPVGRMRFLRPEPVKPWQHVLDATSYQRACYQMVDTSYPGFQGTEMWNPNWG

           130       140       150       160       170       180
1.  MSEDCLYLNIWVPSPRPKDAPVLVWIYGGGFYSGAASLDVYDGRFLTYTQNVILVSLSYR
2.  MSEDCLYLNIWVPSPRPRDVPVLVWIYGGGFYSGAASLDVYDGRFLTYTQNVILVSLSYR

           190       200       210       220       230       240
1.  VGAFGFLGLPGSPEAPGNMGLLDQRLALQWIQNNIHPFGGNPRAVTVFGESAGAASVGMH
2.  VGALGFLGLPGSPEAPGNMGLLDQRLALQWIQNNINYFGGNPNAVTIFGESAGAASVGMH

           250       260       270       280       290       300
1.  LLSTQSRTLFQRAILQSGGPNAPWATVTPAESRGRAALLGKQLGCHFNNDSELVSCLRSK
2.  LLSTHSRGLFQRAILQSGGPNAPWATVTPAESRRRAALLGKQLGCQFNNDSELVSCLRSK

           310       320       330       340       350       360
1.  NPQELIDEEWSVLPYKSIFRFPFVPVIDGDFFPDTPEAMLSSGNFKETQVLLGVVKDEGS
2.  TPQELIDEEWSVLPYKSIFRFPFVPVIDGNFFPDTPEAMLSSGNFKETQVLLGVVKDEGS

           370       380       390       400       410       420
1.  YFLIYGLPGFSKDNESLISRADFLEGVRMSVPHANDIATDAVVLQYTDWQDQDNREKNRE
2.  YFLIYGLPGFSKDNESLINQADFLEGVRMSVPHANDIATEAVVLQYTDWQDQDNGEKNRE

           430       440       450       460       470       480
1.  ALDDIVGDHNVICPVVQFANDYAKRNSKVYAYLFDHRASNLLWPPWMGVPHGYEIEFVFG
2.  AMDDIVGDHNVICPVVQFANDYAKRNNKVYAYLFDHRASNLLWPPWMGVPHGYEIEFVFG

           490       500       510       520       530       540
1.  LPLNDSLNYTPQEKELSRRMMRYWANFARTGNPTDPADKSGAWPTYTASQPQYVQLNTQP
2.  LPLNDSLNYTPQEKELSRRMMRYWVNFARTGNPIDSTEKNGAWPTYTASRPQYVQLNTQP

           550       560       570       580
1.  LATQPSLRAQICAFWNHFLPKLLNATVDPPRADRRRRSARA
2.  LATQPSLRAQICAFWNQFLPKLLNATVRPPRADRRRRSARG
```

FIGURE 24.1: Alignment of the venom-derived AChE from (1) Q92035 (*Bungarus fasciatus*) and (2) R4FKE6 (*Suta suta*). Elapid-venom-specific substitutions and C-terminal exon are highlighted in gray. Signal peptides are shown in lower-case.

has promoted the attempt to identify mechanisms underlying this very high enzymatic activity and particular structure. The two main findings are that a large permanent dipole, resulting from uneven distribution of charges, drives ACh to the active site and that access to the active site is facilitated by the presence of thirteen aromatic residues lining the catalytic entry gorge (Porschke et al. 1996). One tool that has been used in order to analyze the catalytic properties of AChE is a unique type of 3FTx found in *Dendroaspis* venom: fasciculin (Duran, Cervenansky, and Karlsson 1996) (see chapter 8). AChE is the natural target of this toxin, and fasciculin has a high affinity for it, with a K_i in the picomolar range for mammalian AChE. Fasciculin, like some other inhibitors of AChE, binds to the peripheral anionic site, located at the entrance of the aromatic gorge, which is defined by a set of charged and aromatic residues. Joint mutation of AChE and fasciculin has allowed the identification of the residues involved in the interaction of both proteins (Marchot et al. 1997; Sharabi et al. 2009).

24.2.2 PLESIOTYPIC STRUCTURAL AND FUNCTIONAL TOXIN FORMS

AChE was one of the first toxin types discovered (Iyengar et al. 1938), and AChE activity has been detected widely in the venoms of advanced snakes (Frobert et al. 1997; Mackessy et al. 2006; Weldon and Mackessy 2010). However, despite this, and the potent activity of the enzyme making it easy to measure, molecular characterizations of snake AChE are scarce. Consistent with the globular nature of the protein (see chapter 1.3.2), AChE sequences are highly conserved (see figure 24.1). While the presence of this enzyme in viperid snake venoms has not been conclusively proven, activity has been detected in at least one species (Tan and Tan 1988b; Tan and Tan 1988a). Currently, sequences are only available from elapid snake venoms (Jackson et al. 2013). There is a conspicuous absence of AChE activity in *Dendroaspis* venoms, and that lack of activity is a result of the absence of the protein and not its inhibition by the AChE inhibitors (see chapter 8.4.2) that are secreted in the venom as part of the chemical arsenal (Frobert et al. 1997) (see chapter 15). This is consistent with the unique two-stage envenomation profile of *Dendroaspis* venoms, in which step one facilitates the massive release of neurotransmitter but step two blocks neurotransmitters (see chapter 1.3.5). Inhibition of AChE (which regulates neurotransmitters) is a key component of step one, as inhibition of AChE results in more neurotransmitter being available for muscle stimulation. Consequently, there would be a strong selection pressure against the presence of AChE in *Dendroaspis* venom.

The overall catalytic properties of AChE in elapid snake venoms do not differ appreciably from those of endophysiological (body) versions characterized from other vertebrates such as mammals. They are similar in terms of k_{cat}, K_m, inhibition in the presence of an excess of substrate and preference for ACh as substrate over bulkier substrates such as propionylthiocholine (Frobert et al. 1997; Cousin et al. 1998). There are differences, however, in the response of purified venom AChE to certain inhibitors. In the case of active site ligands (such as edrophonium), venom AChE is inhibited with a similar K_i to the AChE of other vertebrates. The situation is completely different, however, with ligands binding to the peripheral anionic site. For small compounds such as propidium, elapid snake venom AChE is less sensitive to these ligands than endophysiological AChE from other vertebrates, with, for example, a K_i 10- to 20-fold higher than that of canonical *Torpedo marmorata* AChE. In the case of fasciculin (described above), the situation is much more complex, with a K_i ranging from 5.10^{-11} M for *Ophiophagus hannah* venom AChE to no inhibition at concentrations as high as 10^{-6} M for *Naja* AchE (Frobert et al. 1997). *Bungarus fasciatus* AChE has an intermediate K_i. Sequence analysis revealed two residues lying in the peripheral anionic site that differ from those of the endophysiological vertebrate AChE that are more sensitive to inhibition. These changes result in the suppression of an aromatic amino acid in the set of aromatic residues lining the active site gorge (Y70 residue, according to *T. marmorata* numbering, changed to M) and in the exchange of an anionic residue for a cationic residue (D285 changed to K). In both cases, the amino acid substitutions also introduce a significant modification of side-chain size or length. Site-directed mutagenesis of these residues in recombinant *Bungarus fasciatus* AChE confirmed that these residues are responsible for the lower sensitivity of the enzyme toward fasciculin (Cousin et al. 1996a). The residues in position 276–277 (A276 and K277 in place of D and E in *B. fasciatus*) are responsible for the insensitivity of *Naja* AChE to fasciculin, lowering K_i from 5000 nM to 18 nM (AK➔DV) or 6.7 nM (AK➔DE). An additional change of charge in residue 285 (K➔D) results in a four- to 10-fold decrease of K_i (Weise and Cousin, unpublished). Interestingly, AChE from *Suta fasciata*, an Australian elapid snake, has the same substitutions as that of *B. fasciatus*, though the two species are only distantly related (Jackson et al. 2013).

As stated above, molecular forms of AChE in snake venom differ from those classically found in vertebrates, which are GPI-anchored monomers or dimers or, in most cases, tetramers associated with accessory proteins. In the body of *B. fasciatus*, endophysiological AChE exists as mixtures of tetramers or groups of tetramers associated with a ColQ subunit. In *B. fasciatus* venom, however, it is exclusively present as a hydrophilic and soluble monomer (Cousin et al. 1996b). The same is true for the AChE of other elapid snake venoms (Frobert et al. 1997). Sequencing of cDNA encoding AChE in venom-gland and genomic DNA indicates that this particular form results from an alternative exon, which encodes a short C-terminal peptide (Cousin et al. 1998). The exon is called the S (soluble) exon. This exon/peptide is unique to snakes and has never been identified in other taxa. Since the first identification of a majority monomeric form, the gene AChE B has been identified in the parasitic nematode *Nippostrongylus brasiliensis* (Hussein et al. 1999; Hussein, Grigg, and Selkirk 1999b). It also encodes a monomeric soluble form of AChE, which is abundantly secreted from specialized amphidial and secretory glands. From an evolutionary perspective, it is important to note that the S exon/peptide is unrelated to other 5′ exon/C-terminus peptides of either vertebrates or invertebrates. Since this first identification of a specific exon encoding the short C-terminus in *B. fasciatus*, additional sequences have been identified, all of which are highly conserved (Jackson et al. 2013), which is consistent with the slow evolutionary rates of globular proteins (see chapter 1.3.2). Considering the phylogenetic distance between *S. fasciata* and *B. fasciatus*, it can be inferred that production of soluble monomers of AChE by the venom gland is a common feature of most elapid snakes and that this is facilitated through the use of a common C-terminus. This innovation appears to be unique to snakes and is evidently a favorable character state that has been retained during evolution.

AChE's role in snake venom remains to be elucidated as testing to-date has not ascertain a toxic activity (Cousin et al. 1996b). This may reflect assay choice as the specific effect of venom AChE on the snakes that are the primary prey of an ophiophagus specialist such as *B. fasciatus* has not been investigated. One may hypothesize that since the venom of elapid snakes contains α-neurotoxins, which block the nicotinic receptor at neuromuscular junctions, AChE may clear the venom and the synaptic cleft of the target animal of ACh that would compete with the action of these toxins. The reduction of available ACh, combined with the blockage of any that remains, would result in potentiated neurotoxicity.

24.2.3 APOTYPIC STRUCTURAL AND FUNCTIONAL TOXIN FORMS

As is consistent for most globular proteins, very little molecular diversification has occurred.

24.2.4 THERAPEUTIC POTENTIAL OF THE TOXIN TYPE

None is documented.

24.2.5 CONVERGENCE WITH OTHER VENOMS

None is documented.

24.3 AVIT

24.3.1 ENDOPHYSIOLOGICAL PLESIOTYPE

The plesiotypic endophysiological AVIT/prokineticin (PK) peptides are expressed at high levels in testis and at lower levels in brain, lung, ovary, spleen, thymus, and uterus and have similarly diverse functions, including constriction of intestinal smooth muscle.

24.3.2 PLESIOTYPIC STRUCTURAL AND FUNCTIONAL TOXIN FORMS

AVIT peptides were recruited as toxins at the base of the toxicoferan radiation, having been isolated from both snakes and lizards (Boisbouvier et al. 1998; Schweitz et al. 1999; Fry et al. 2006) (see figure 24.2). The plesiotypic toxic activity is high-affinity binding to mammal PKR1 and PKR2 (prokineticin receptors 1 and 2), with an affinity that exceeds that of the mammalian endophysiological peptide PK2 by one order of magnitude and that of PK1 by two orders of magnitude, mimicking an overdose (Negri et al. 2007). This results in a rapid constriction of intestinal smooth muscle, leading to painful cramping and induction of hyperalgesia.

```
           10        20        30        40        50        60
1. mrsllcvpllllllsagesAVITGACDKDLQCGEGMCCAVSLWIRSIRICTPLGSSGEDCHPL
2. mrsllcapllllllsagesAVITGACDKDLQCGEGMCCAVSLWIRSIRICTPLGSSGEDCHPL
3. mrsllcapllllllsagesAVITGACDKDLQCGEGMCCAVSLWIRSIRICTPLGSSGEDCHPL
4.                   >AVITGACERDLQCGKGTCCAVSLWIKSVRVCTPVGTSGEDCHPA
5. mkcfaqivvlllviafshgAVITGVCDRDAQCGSGTCCAASAFSRNVRFCVPLGNNGEECHPA

           70        80        90       100
1. SHKVPFDGQRKHHTCPCLPNSACVQTSPGKYKCSPEFKNVFF
2. SHKVPFDGQRKHHTCPCLPNLVCGQTSPGKYKCLPEFKNVF
3. SHKVPFDGQRKHHTCPCLPNLVCGQTSPGKHKCLPEFKNVF
4. SHKIPFSGQRMHHTCPCAPNLACVQTSPKKFKCLSKS
5. SHKVPYNGKRLSSLCPCNTGLTCSKSG-EKFQCS
```

FIGURE 24.2: Sequence alignment of AVIT peptides from reptile venom (1) B6CJU8 (*Varanus komodoensis*, (2) Q2XXR8 (*Varanus varius*), (3) Q2XXR7 (*Varanus varius*), and (4) P25687 (*Dendroaspis polylepis*) and the colipase-derived peptide from spider venom (5) P81803 (*Hadronyche versuta*). Signal peptides are shown in lower-case.

24.3.3 APOTYPIC STRUCTURAL AND FUNCTIONAL TOXIN FORMS

None is documented.

24.3.4 THERAPEUTIC POTENTIAL OF THE TOXIN TYPE

None is documented.

24.3.5 CONVERGENCE WITH OTHER VENOMS

Spider venoms contain peptides (MIT-like ACTX) that are distantly related members of the same peptide superfamily, containing the characteristic conserved 10-cysteine pattern of the vertebrate AVIT peptides (Szeto et al. 2000; Wen et al. 2005). However, these peptides lack the N-terminal AVIT sequence, are much more diverse in amino acid composition and sequence length, and test negative in assays of PK1/PK2 activity. Consistent with the likelihood that they possess a distinct function, comparative three-dimensional modeling of MIT-like ACTX peptides and AVITs reveals dissimilar tertiary structures and markedly different surface chemistries. These results are perhaps not surprising, since insects, which constitute the majority of spider prey, appear to lack the PK/PKR system. So far, insecticidal assays have not revealed a lethal effect of MIT-like ACTX peptides, even at high doses. Nevertheless, the diversity of these peptides within individual species, their presence in a broad phylogenetic range of spiders (suggesting their ancient recruitment and long-term existence in spider venom), and their resistance to proteolytic breakdown suggest that they have an active role in spider venom. Recently, peptides similar to these spider venom peptides have been found in the hematophagous secretion glands of the tick *Ixodes scapularis* and may play a role in its feeding behavior (Ribeiro et al. 2006).

Similar bioactive peptides are found in the defensive skin secretions of *Bombina* fire-bellied toads (Bm8, Bo8, and Bv8) (Mollay et al. 1999; Chen et al. 2003b; Chen et al. 2005). As with the reptile venom forms, the *Bombina* skin-secretion defensive toxins are potent agonists of mammalian PK receptors, and their action mimics an overdose of endophysiological PK. As with the reptile venom forms, this results in gastric smooth-muscle contraction and hyperalgesia. However, over a longer time frame, the AVIT peptides in frog skin defensive secretions additionally produce an anorexogenic effect (Negri et al. 2007), inhibiting the feeding response of the affected predator even when food is presented after a period of starvation.

24.4 CELESTOXIN

24.4.1 ENDOPHYSIOLOGICAL PLESIOTYPE

This toxin type displays no homology with any known protein type.

24.4.2 PLESIOTYPIC STRUCTURAL AND FUNCTIONAL TOXIN FORMS

Celestoxin is novel, compared with other venom proteins, in having the secondary structures determined entirely by significant numbers of prolines but with cysteines absent (Fry et al. 2010b) (see figure 24.3). Two isoforms (GU441474 and GU441475) differ only in a 12-amino-acid insertion at alignment positions 135–147. Both peptides have an arginine-rich cleavage motif to liberate the bioactive domain from the propeptide. Contained within the bioactive domain is a sequence repeat (ALGVVGGLPVPK and LPGVAGGLPVPK) that differs in the formation of a proline bracket in the second repeat that is missing in the first repeat, which has an L residue in place of the P residue. Celestoxin analogue testing revealed this internal proline-bracketed motif LPGVAGGLPVPK to be responsible for full activity, with this fragment being equipotent in activity to the full-length peptide in producing hypotension. The ALGVVGGLPVPK fragment is inactive. Neither adrenoceptor antagonists nor 5HT antagonists can prevent or reverse the response.

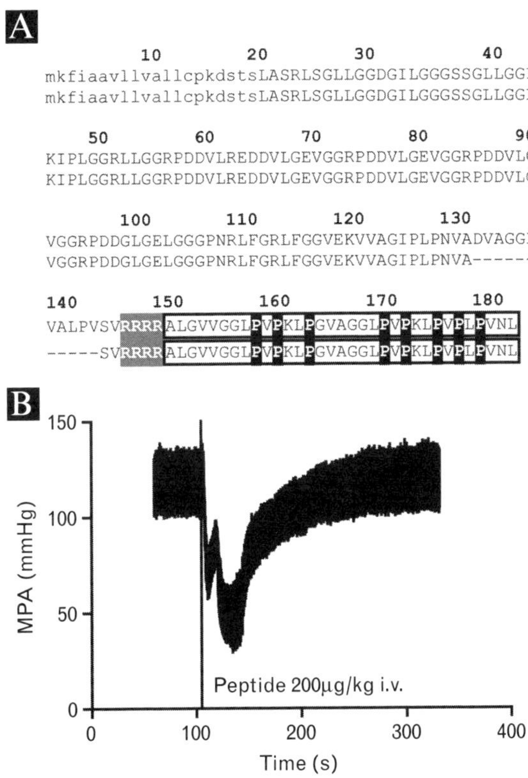

FIGURE 24.3: (A) Sequence alignment of *Celestus warreni* celestoxin precursors from the mandibular venom gland: GU441474 and GU441475. Gray highlights cleavage motif, and bioactivity important prolines are highlighted in black (Fry et al. 2010b). (B) Effect of intravenous injection of posttranslationally cleaved peptide (Fry et al. 2010b). Vertical line marks time of injection. Signal peptides are shown in lower-case.

24.4.3 APOTYPIC STRUCTURAL AND FUNCTIONAL TOXIN FORMS

None is documented.

24.4.4 THERAPEUTIC POTENTIAL OF THE TOXIN TYPE

None is documented.

24.4.5 CONVERGENCE WITH OTHER VENOMS

None is documented.

24.5 CHOLECYSTOXIN

24.5.1 ENDOPHYSIOLOGICAL PLESIOTYPE

The endocrine effects of very low concentrations (subnanomolar) of circulating cholecystokinin (CCK) are critical for nutrient homeostasis in mammals, stimulating exocrine pancreatic secretion, gallbladder emptying, and bowel motility and also inducing satiety.

24.5.2 PLESIOTYPIC STRUCTURAL AND FUNCTIONAL TOXIN FORMS

Cholecystoxin was sequenced from *Varanus varius* (Fry et al. 2010b) (see figure 24.4). Analogue testing of variable lengths of cholecystoxin determined that the smallest active peptide with full cardiovascular activity and potent CCK-A receptor binding activity was an octapeptide domain with a posttranslationally modified sulfatyrosine crucial to activity: DY^{S}LGWMDF (see figure 24.4).

24.5.3 APOTYPIC STRUCTURAL AND FUNCTIONAL TOXIN FORMS

None is documented.

24.5.4 THERAPEUTIC POTENTIAL OF THE TOXIN TYPE

None is documented.

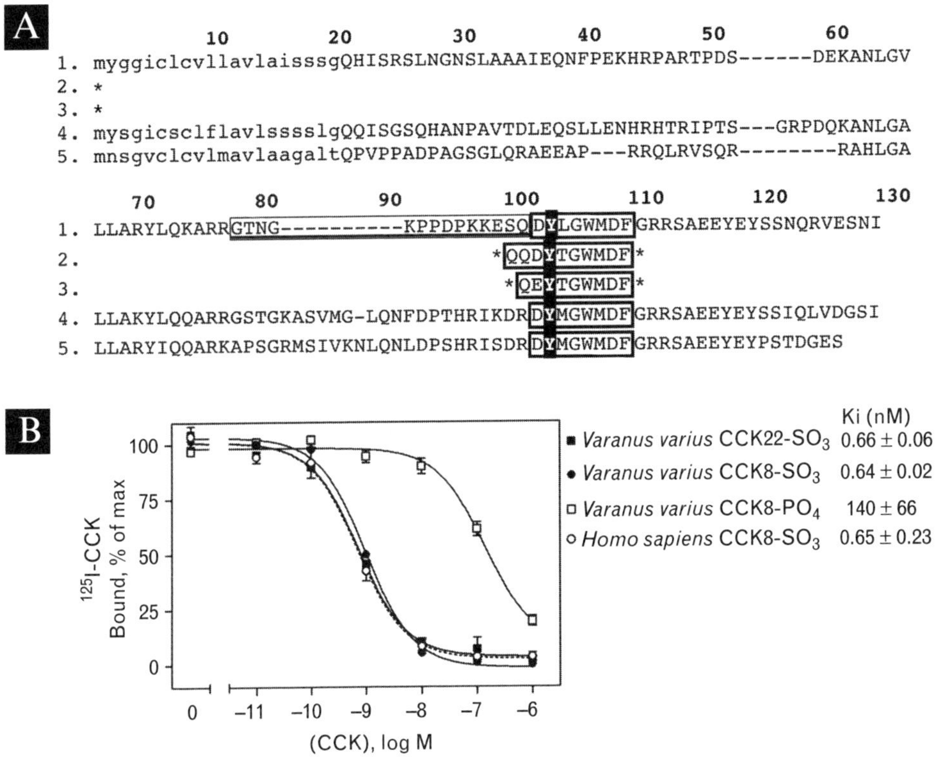

FIGURE 24.4: (A) Sequence alignment of precursors of (1) *Varanus varius* cholecystoxin (GU441461) from the mandibular venom gland and the frog skin defensive skin toxins from (2) *Litoria xanthomera* (P56264) and (3) *Phyllomedusa sauvagei* (Q7LZC4), in addition to the body peptides from (4) *Python molurus* (Q8JHB9) and (5) *Homo sapiens* (P06307). Box indicates minimal bioactive processed peptide, with both the long (22AA) and short (8AA) forms of (1) shown); Y represents the key bioactive posttranslationally modified sulfatyrosine; * designates the only bioactive peptide domain sequenced to date. (B) Relative binding of cholecylstokinin receptor A of the two variants highlighted in boxes in (A) above (one 22 amino acids, one 8 amino acids), a variant of CCK8 with a phosphotyrosine instead of the sulfatyrosine, and human CCK8 with a sulfatyrosine (Fry et al. 2010b). Signal peptides are shown in lower-case.

24.5.5 CONVERGENCE WITH OTHER VENOMS

Cholecystoxin represents an additional case of convergence of the selection of a gene type for use as a toxin, having been previously characterized as a bioactive component in frog skin chemical arsenals. Derived cholecystokinin frog peptides such as caerulein are potent ligands of CCK receptors. They interfere with the physiological pathways of endogenous CCK and gastrin and cause acute pancreatitis, vomiting, diarrhea, decreased blood pressure, changes in thermoregulation, and inhibition of exploratory and feeding behavior (Bowie and Tyler 2006). Higher concentrations of frog skin CCK-like peptides (caerulein) stimulate pancreatic acinar-cell basolateral exocytosis of zymogens, central to the development of the rodent model of pancreatitis. It is interesting that CCK-like peptides secreted via the exocrine route can exhibit toxic effects previously only for venom-derived forms.

24.6 COMPLEMENT C3/COBRA VENOM FACTOR

24.6.1 SUMMARY

The complement system participates in both innate and adaptive immune responses and is a major component of the immune system. It is characterized by approximately 30 distinct plasma proteins, serosal proteins, cell membrane receptors, and regulatory proteins. A unique complement-activating protein called cobra venom factor (CVF) is found in the venoms of the elapid snakes *Naja, Ophiophagus*, and *Hemachatus*. CVF is a structural and functional analogue of complement component C3 and functionally resembles C3b, the activated form of C3 (see figure 24.5). A humanized derivative of CVF has tremendous potential in therapeutics. Its use as a drug in the treatment of dangerous diseases such as paroxysmal nocturnal hemoglobinuria (PNH) and myocardial ischemia and reperfusion (MI/R) injuries has been explored recently.

24.6.2 ENDOPHYSIOLOGICAL PLESIOTYPE

The complement system was first discovered in the 1890s, when it was found to "complement" or assist the clearing of bacterial cells by heat-stable antibodies (Walport 2001). It was initially presumed to only participate in the innate immune response. Recently, its role in the adaptive immune response has been established, and its interaction with T and B cells is becoming clear (Molina et al. 1996; Janeway et al. 2001; Dunkelberger and Song 2010). It participates in homeostasis by complementing the adaptive immune response in clearing pathogens. The system is characterized by more than 30 proteins that are present as either soluble plasma proteins or membrane-associated receptors. Most plasma proteins are synthesized in the liver and circulate in the blood as inactive serum propeptides. The complement system can be activated through three pathways: (i) the classical pathway, (ii) the lectin pathway, and (iii) the alternative pathway. All three pathways involve a recognition

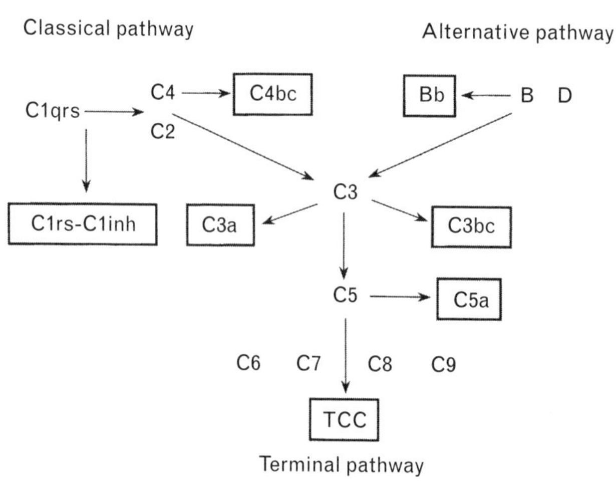

FIGURE 24.5: Activation of the complement cascade.

event, which is activated by the antigen-antibody complex, followed by an amplifying cascade, which sequentially activates numerous proteolytic enzymes. These enzymes in turn result in the generation of numerous biologically active complement activation products, such as the macromolecular membrane attack complex (MAC). This protein impairs the normal functioning of target cells, such as bacteria, viruses, tumor cells, and parasites, by inserting itself into target cell membranes. Interestingly, the complement system is not only involved in the innate and adaptive responses, but its role in tissue regeneration, tumor growth (Qu, Ricklin, and Lambris 2009), and human pathology, such as atypical hemolytic uremic syndrome and age-related macular degeneration, is also well established (Wagner and Frank 2010).

C3 appears to be the most essential component of the complement system, as it is activated by all the three pathways and essentially directly or indirectly mediates all the biological activities of the complement system. It is found in echinoderms and arthropods in addition to vertebrates, indicating that its evolution in the animal lineage occurred about 600 million to 700 million years ago. It is secreted as a single-chain prepropeptide, which matures into a two-chain peptide in the presence of a furin-type protease, which removes a stretch of four arginine residues that separate the N-terminal β-chain from the C-terminal α-chain in the propeptide (de Bruijn and Fey 1985; Vik et al. 1991). Mature C3 has a molecular mass of 190 kDa and consists of two chains (α and β) linked by a disulfide bond (Janatova 1986). It is the most abundant plasma protein in all vertebrates and occurs at a concentration of 1.2 mg/mL in human blood plasma. Like most complement proteins in the blood plasma, C3 is mostly synthesized in the liver; however, cells of the immune system are capable of secreting C3 (Botto 2000). C3b is another vital member of the complement system and is a functional derivative of C3. This protein initiates opsonization, which is a process of tagging infected cells to be cleared by macrophages. Proteolytic degradation of C3b results in the generation of smaller peptides, such as iC3b, C3dg, and C3d, which participate in the adaptive immune response. The sequential cascade of enzymatic reactions triggered by the activation of the complement system also results in the formation of potent anaphylatoxins such as C3a and C5a that elicit an array of physiological and biochemical responses. Among other functions, these proteins are capable of triggering chemoattraction, which recruits other components of the immune system and/or elicits apoptosis of infected cells.

24.6.3 PLESIOTYPIC STRUCTURAL AND FUNCTIONAL TOXIN FORMS

Structural homologues of C3 complement proteins known as CVF are found in the venoms of Asiatic elapid snakes of the genera *Naja, Ophiophagus*, and *Hemachatus* (Vogel et al. 1996). CVF has a MW of around 150 kDa and structurally resembles C3c, a derivative of C3 that is generated by the proteolytic removal of the C3a and C3d domains. The structural homology of CVF and C3 proteins was established through a number of biochemical (Alper and Balavitch 1976; Eggertsen, Lind, and Sjoquist 1981; Lundwall et al. 1984) and immunological studies (Alper and Balavitch 1976; Eggertsen et al. 1983; Vogel, Smith, and Muller-Eberhard 1984). Molecular cloning (Fritzinger et al. 1992; Fritzinger, Bredehorst, and Vogel 1994) and the determination of the structure of CVF (Janssen and Gros 2006; Janssen et al. 2009; Krishnan et al. 2009) supported the conclusions of these studies. Functionally, CVF resembles C3b, a derivative of C3. In the presence of Mg^{2+} ions, CVF forms a complex with factor B (Hensley et al. 1986), which is further processed into the activation peptide Ba and the bimolecular complex CVF, Bb by factor D (Vogt et al. 1978; Vogel and Muller-Eberhard 1982).

The structural homology of CVF and C3 proteins is further confirmed by crystallographic studies of C3, C3b, C3c (Janssen et al. 2005; Janssen and Gros 2006; Wiesmann et al. 2006), and CVF (Janssen et al. 2009; Krishnan et al. 2009). After maturation, CVF is made up of 11 domains, which include a linker domain, a C-terminal C345C domain with C3c, eight macroglobulin domains, and a complement C1r/C1s, Uegf, Bmp1 (CUB) domain, which is only present in C3b. Since C3 proteins and CVF share great sequence similarity, they are also functionally very similar. Despite the fact that CVF resembles C3c more than C3b structurally, CVF is a functional homologue of C3b, binding factor B in the presence of Mg^{2+} ions. The resultant CVF,B proconvertase is then cleaved by factor D into the bioactive peptide Ba and a CVF,Bb biomolecular complex. Although this complex exhibits a weak C3-convertase activity (Xu et al. 2001), when factor B is removed by factor D, CVFBb is formed, which is an efficient C3-convertase that targets the same scissile bond of C3 as typical convertases.

The precise role of CVF in the venoms of elapid snakes is unknown to date. In contrast to more potent neurotoxins such as three-finger toxins (3FTx), which are present in large quantities in the venoms of these snakes, the action of CVF is not lethal. When injected intravenously, CVF causes massive complement activation and leads to the activation of neutrophils and their sequestration to the lungs, damaging the lung tissue. Such effects are temporary (Till et al. 1982; Till et al. 1987; Mulligan et al. 1996) but may still play a contributing role in prey subjugation, perhaps working synergistically with slower-acting lethal toxins. Potentially lethal massive intravascular activation of complement by CVF is avoided by the presence of serum protein carboxypeptidase N, which removes the C-terminal arginine residue and rapidly inactivates C3a and most proinflammatory activities of C5a (Plummer and Hurwitz 1978; Huey et al. 1983). Certain components in snake venoms, such as nerve growth factors (see section 24.14) (Sunagar et al. 2013a) and hyaluronidases (see section 24.11) (Girish et al. 2002), have been hypothesized to indirectly participate in prey envenoming; CVF could play a similar indirect envenoming role by inducing massive complement activation locally at the bite site, eliciting the release of anaphylatoxins C3a and C5a, which are capable of increasing the vascular permeability and blood. Thus, CVF could act as a spreading factor.

24.6.4 APOTYPIC STRUCTURAL AND FUNCTIONAL TOXIN FORMS

None is documented so far.

24.6.5 THERAPEUTIC POTENTIAL OF THE TOXIN TYPE

For decades, CVF has been used for decomplementation of serum in evaluating the function of complement activation in numerous disease models (Vogel 1991; Vogel and Fritzinger 2007). Its therapeutic potential in the setting of myocardial ischemia and reperfusion (MI/R) is very well documented (Hill and Ward 1971; Maclean et al. 1978; Maroko et al. 1978; Pinckard et al. 1980). However, its inherent capability to cleave C5 and result in the generation of C5a anaphylatoxin, a potent inflammatory component, along with its immunogenicity, has kept it away from therapeutics (Gowda et al. 1994b; Gowda et al. 2001). Recently, a human C3 derivative with CVF-like functions, known as humanized CVF, was synthesized by replacing the short stretch of amino acids from the C-terminus of the C3 α-chain with the homologous sequences from CVF. This component was capable of activating human C3 but not C5 (Vogel and Fritzinger 2007). Recently, the use of recombinant humanized CVF (HC3-1496) in therapeutics was evaluated using a rodent model, which revealed that these animals were protected from myocardial ischemia and reperfusion (MI/R) injuries with resultant preservation of cardiac function. These results indicated that humanized CVF could protect the ischemic myocardium from reperfusion injuries induced by complement activation. Thus, recombinant versions of CVF exemplify a novel anticomplement therapy with potential for clinical use.

The usefulness of humanized CVF has also been explored in the treatment of paroxysmal nocturnal hemoglobinuria (PNH), a rare life-threatening disease in which the patient's red blood cells (RBC) undergo lysis as they fail to express MCP (CD55) and CD59 complement regulatory proteins on their molecular surface. Incubation of RBC from these patients with a recombinant truncated form of factor H (rH19-20) renders these cells highly susceptible to complement lysis in normal serum by blocking their regulatory functions. However, when these cells are incubated with partially or completely complement-depleted serum, in the presence of the humanized CVF protein HC3-1496, lysis is completely averted.

24.6.6 CONVERGENCE WITH OTHER VENOMS

None is documented.

24.7 CYSTATIN

24.7.1 ENDOPHYSIOLOGICAL PLESIOTYPE

The endophysiological plesiotype of all members of the cystatin superfamily is the family 1 stefin, a single-chain, 100-residue polypeptide cysteine-protease (CP) inhibitor (Kordis and Gubenek 2000). Data-mining analyses indicate that family 1 and family 2 cystatins are the intracellular and secreted endophysiological plesiotypes, respectively, for a whole range of multicystatin domain proteins, which do not necessarily function as CP inhibitors (Kordis and Turk 2009).

Prototypical type 2 cystatins, such as chicken ovocystatin and human cystatin C, form 1:1 stoichiometric complexes with C1 family CPs in competition with a substrate (Abrahamson, Alvarez-Fernandez, and Nathanson 2003). Crystal structure analyses indicate that the N-terminal region (Gly-11), loop 1 (Gln-55-Gly-59), and loop 2 (Pro-105-Trp-106) form a wedge-shaped enzyme-binding region required for inhibition of C1 family CPs (cystatin C numbering) (Bode et al. 1988). A number of cystatins also inhibit the C13 family CP legumain, and this is mediated by a separate region known as the backside loop (BSL) (Alvarez-Fernandez et al. 1999). Asn-39 is an essential component of the BSL, which is absent in snake venom cystatins (Richards et al. 2011).

Type 2 cystatins are implicated in a variety of cellular processes beyond the simple inhibition of exogenous and endogenous CPs for protection against tissue injury. Lysosomal CPs (cathepsins) facilitate tissue remodeling and tumor-cell invasion through degradation of ECM proteins, and these processes are elegantly regulated by cystatins (Obermajer et al. 2008). CP regulation by cystatins also plays an essential role in the modulation of multiple proteolytic cascades, including those of the immune system (Hartmann and Lucius 2003), the inflammatory response (Meyer-Hoffert 2009) and apoptosis (Repnik and Turk 2011). Moreover, some cystatins exhibit CP inhibition-dependent and -independent antimicrobial and antiviral activities (Bjorck, Grubb, and Kjellen 1990; Blankenvoorde et al. 1998; Abrahamson, Alvarez-Fernandez, and Nathanson 2003).

Type 2 cystatins have been found in various tissues and fluids from diverse organisms, including horseshoe crabs (*Tachypleus tridentatus*) (Agarwala et al. 1996), ticks (*I. scapularis*) (Kotsyfakis et al. 2006), jellyfish (*Cyanea capillata*) (Yang et al. 2003), various nematode species (Hartmann and Lucius 2003), and spiders (*Chilobrachys jingzhao*) (Chen et al. 2008b), in addition to mammals (Kordis and Turk 2009). Despite the fact that the venom gland is a modified salivary gland, multiple phylogenetic analyses of type 2 cystatins have placed snake venom cystatins on the same branch as human cystatin M, rather than with mammalian salivary cystatins S/D (Brillard-Bourdet et al. 1998; Dickinson 2002). Cystatin M was first discovered in a differential display screen, which showed it to be down-regulated in metastatic human breast cancer (Sotiropoulou, Anisowicz, and Sager 1997). Cystatin M inhibits breast cancer cell migration and matrix invasion (Shridhar et al. 2004) and is being touted as a novel tumor-suppressor gene (Zhang et al. 2004c). The precise mechanism of action remains unclear, but legumain inhibition appears to be important (Briggs et al. 2010). Cystatin M/E null mice present with disturbed skin barrier formation and die from dehydration at 6 to 12 days of age (Zeeuwen et al. 2002; Zeeuwen et al. 2004). The mouse phenotype is strongly associated with uncontrolled legumain activity.

24.7.2 PLESIOTYPIC STRUCTURAL AND FUNCTIONAL TOXIN FORMS

Unlike most other venom components, which have undergone rapid mutation to obtain novel activities and increased potency, venom cystatins vary little from their endophysiological plesiotype, despite having been recruited into the venom arsenal in the earliest evolutionary stages. The level of conservation with other family 2 cystatins is very high, and all species examined thus far have retained the plesiotypic active site. It would thus appear that the cysteine protease inhibitory activity, rather than some novel activity of cystatins, serves an indispensable but as yet uncertain purpose in snake venoms. Venom cystatins belong to the class 2 cystatin family, with closest identity to cystatin M/E in mammals (Brillard-Bourdet et al. 1998; Fry et al. 2006; Richards et al. 2011). Toxicoferan cystatins are highly conserved, with strong conservation of all essential structural elements, including the four cysteine residues involved in intramolecular disulfide-bond formation. Venom cystatins show remarkably high identity across genera. The 12 cloned elapid snake venom cystatins, for example, are all 141 amino acids in length, with 75% identity across the entire deduced preprotein sequences (Richards et al. 2011) (see figure 24.6). Recombinant AsCystatin from *Austrelaps superbus* (Australian lowland copperhead snake) exhibits similar potency and selectivity as cystatin C and ovocystatin (Richards et al. 2011), and no major functional difference between venom and prototypical type 2 cystatins has yet been shown. There has clearly been selection pressure to maintain similar structure and function to the nonvenom type 2 cystatins, with, in contrast to many other venom toxins, minimal changes from the plesiotypic protein observed.

Nevertheless, postulating the function of venom cystatin based on the activities of its most closely related nonvenom counterparts is dubious. Knowledge of cystatin M functions is of limited use in understanding the potential roles of venom cystatins because of the absence of Asn-39 (alignment position 70 in figure 24.6), crucial for legumain inhibition, in the majority of the toxicoferans (Richards et al. 2011). The best evidence of a role for cystatins in a venomlike context comes from invertebrate secretions (Fry et al. 2009a). Nematode parasite cystatins assist in evasion of the host defense system and modulation of the immune response (Hartmann and Lucius 2003), while secreted tick cystatin suppresses the immune response of the host (Kotsyfakis et al. 2006). These functions are imperative if the parasite is to achieve resident status. However, modulation of the prey immune response would be of no obvious benefit to a venomous snake that needs to rapidly immobilize and kill its prey. A toxic or toxic support function for venom cystatins might be ruled out altogether, given that the venom gland is a modified salivary gland, and salivary cystatins play various roles in the maintenance

```
              10        20        30        40        50        60        70        80        90       100
1.   m------asggspvwalvallsllplvargERAVG-------APEEVPVSNPGVQEAARFALEAYNRASNSMFCYKPVRVLKAETQVVSGMQYYLTVEIV
2.   mdasrllslgcrlswallvllpppaapalgHDHDHDQPGIVGGWRARSPSDPDVVEATAFCVGAYNEASNSAFLFKAARVVEAESQVVAGTKYSLTIELV
3.   m---------------mllpaltalsvpmpgAPMD-----------RPVTDPGVQQAAAFAVQAFNKASNSAFYYKAMRIVSARSQVVEGVKYYLTVELV
4.   m------vhsqlpvaaslrllcallllpsatmIPG-------GLSPRSVTDPDVREAAEFAVQEYNALSANAYYYKQLRIVEAQSQVVAGAKYYLTMELM
5.   m------vhsqlpvagplrllcallllpsatmIPG-------GLSPRSVTDPDVQEAAEFAVQEYNALSANAYYYKQLRIVEAQSQVVSGAKYYLTMELM
6.   m------vhsqlpvagplrllcallllpsatmIPG-------GLSPRSVTDPDVQEAAEFAVQEYNALSANAYYYKQLRIVEAQSQVVSGAKYYLTMELM
7.   m------vhsqlpvvallrllcallllpsatmIPG-------GLSPRSVTDPDVQEAAEFAVQEYNALSANAYYYKQLRIVEAQSQVVAGAKYYLTMELM
8.   m------lhsqlpvaaplrllcallllpsvtmIPG-------GLSPRSVTDPDVQEAAEFAVQEYNALSANAYYYKQLRIVEAQSQVVTGAKYYLTMELM
9.   m------vhsqlpvaaplrllcallllpsatmIPG-------GLSPRSVTDPDVQKAAEFAVQEYNALSTNAYYYKQLRIVEAQSQVVAGAKYYLTMELM
10.  m------vhsqlpvaaplrllcallllpsatmIPG-------GLSPRSVTDPDVQKAAEFAVQEYNSLSTNAYYYKQLRIVEAQSQVVAGAKYYLTMELM
11.  m------vhsqlpvaaplrllcallllpsatmIPG-------GISPRSVTDPDVQKAAEFAVQEYNALSTNAYYYKQLRIVEAQSQVVAGAKYYLTMELM
12.  m------vhfqlpvaaplcllcallllpsatmIPG-------GLSPRSVSDPDVQKAAAFAVQEYNARSANAHYYKELRVVEAQSQVVAGEKYYLMMELV
13.  m------vhsqlpvaaplrllcallllplatmIPG-------GLSPRSVTDPDVQEAAEFAVQEYNALSANAYYYKQLRIVEAQSQVVTGAKYYLTMELM
14.  m------vhsqlpvaaplrllcallllpsatmIPG-------GLYPRSVTDPDVQEAAECAVQEYNALSANAYYYKQLRIVEAQSQVVTGAKYYLTMELM
15.  m------vrsqlpvaaplrllcallllpsatmIPG-------GLSPRSVTDPDVQEAAAFSVQEYNALSANAYYYKELRIVEAQSQVVSGAKYYLTMELM
16.  m-------hsrlpvpaslclllllpsvlpasm-PG-------GLSPRDVTDPEVQEAAVFAVEEYNARSTNSNYFKALRLVQAESQVVSGAKYYLTMELV

             110       120       130       140       150       160
1.   NTRCEKKVGCGLKNMGSENCAVPSEAEQ-KQICEFVVWSRPWMQDTRLSSISCTPSSS
2.   RTVCEKKGGSGLKDVDLERCEVASDAEQEKEICTFQVWSRPWLNDTQLLQMSCKAASS
3.   NTLCEKKGGSGLNHVDLEHCAVASVADQRKQTCEFQVWFLPRLNDTRLEYMSCKAASS
4.   KTKCAKTTGKPKVYKEIQNCELPPKAQQEKLTCRFQVWSRPWLKKTELTKMSCN
5.   KTKCAKTTGKPKVYKEIQNCELPPKAQQEKLTCHFQVWSRPWLGKIELTKMSCN
6.   KTKCAKTTGKPKVYKEIQNCELPPKAQQEKLNCRFQVWSRPWLEKMELTKMSCN
7.   KTKCAKTTGKPKVYKEIQNCELPPKAQQEKLTCRFQVWSRPWLQKIELTKMSCN
8.   KTKCAKTTGKPKVYKEIQNCELPPKAQQEKLTCRFQVWSRPWLEKIELTKMSCN
9.   KTKCAKTTGKPKVYKEIQNCELPPKAQQEKLTCHFQVWSRPWLEKVELTKMSCN
10.  KTKCAKATGKPKVYKEIQNCELPPKAQQEKLTCHFQVWSRPWLEKMELTKMSCN
11.  KTKCAKTTGKPKVYKEIQNCELPPKAQQEKLTCHFQVWSRPWLEKMELTKMSCN
12.  KTKCAKTAGKPKVYKEIQNCELPPKAQQEKLTCHFQVWSRPWLDKTELTKMSCN
13.  KTKCAKTTGKPKVYKEIQNCELPPKAQQEKLTCRFQVWSCPWLQKIELTKMSCN
14.  KTKCAKTTGKPKVYKEIQNCELPPKAQQEKLTCHFQVWSRPWLQKIELTKMSCN
15.  KTKCAKATSKPKVYKEIQNCELPPKAQQEKLTCHFQVWSRPWLEKMELTKMSCN
16.  KTSCRKTNGNPKGYQEIQNCRLPPRNQQEKLTCHFEVWSRPWLNKTLLTKVTCN
```

FIGURE 24.6: Alignment of the cystatin sequences (1) Q2XXN5 (*Pogona barbata*), (2) M9T289 (*Iguana iguana*), (3) K4I4J3 (*Abronia graminea*), (4) E3P6P1 (*Rhinoplocephalus nigrescens*), (5) E3P6N8 (*Pseudechis australis*), (6) E3P6N7 (*Pseudechis porphyriacus*), (7) E3P6P2 (*Notechis scutatus*), (8) E3P6P0 (*Hoplocephalus stephensii*), (9) E3P6N6 (*Oxyuranus scutellatus*), (10) E3P6N4 (*Pseudonaja textilis*), (11) E3P6N5 (*Oxyuranus microlepidotus*), (12) E3P6P4 (*Naja kaouthia*), (13) E3P6N3 (*Austrelaps superbus*), (14) E3P6N9 (*Tropidechis carinatus*), (15) E3P6P3 (*Micropechis ikaheka*), and (16) J3RYX9 (*Crotalus adamanteus*). Signal peptides are shown in lower-case.

of oral health (Dickinson 2002). However, venom cystatins do not group phylogenetically with the salivary cystatins S/D (Brillard-Bourdet et al. 1998; Dickinson 2002; Fry 2005), and the cystatin S/D family exhibits much weaker inhibition of cathepsins in comparison with classical type 2 cystatins. This suggests a role for cystatin S/D family members in the inhibition of unidentified exogenous proteases, such as those from microbial organisms (Dickinson 2002). Antimicrobial activities for snake venom cystatins are certainly possible, but potent cathepsin inhibitory properties implicate them in more classical pathways (Richards et al. 2011). The current evidence favors inhibition of prey extracellular matrix proteases by snake venom cystatins as a means to protect the integrity of other venom components (Brillard-Bourdet et al. 1998), but this is yet to be proven.

Reptile venom cystatins differ from prototypical human cystatin C in three distinct ways. All contain a relatively truncated N-terminus (except for some lizard venom forms), while lacking Asn-39 (alignment position 70 in figure 24.6), essential for legumain inhibition, and possessing a small insertion between the cysteine residues that form the first disulfide bond (Richards et al. 2011). No functional significance has been assigned to this small insertion of amino acids, which is also present in cystatin M. The addition is unlikely to affect conformation, as it is in a largely unstructured region of the protein, on the opposite side to the C1 CP inhibitory wedge (Sotiropoulou, Anisowicz, and Sager 1997). Despite the truncated N-terminus, *A. superbus* venom cystatin retains the Gly-11 structural element (alignment position 35 in figure 24.6) and displays an enzyme selectivity profile and level of potency consistent with cystatin C and ovocystatin (Richards et al. 2011). The BSL structural element is responsible for legumain inhibition in type 2 cystatins and consists of a sequence of at least four amino acids located on the opposite side to the C1 CP binding surface and which fits the following consensus: nucleophilic-38, Asn-39, X-40, hydrophobic-41 (alignment positions 69, 70, 71, 72 in figure 24.6) (Alvarez-Fernandez et al. 1999). The BSL is absent in some type 2 cystatins (such as cystatin D), suggesting a second recruitment or a deletion event subsequent to the original recruitment of the plesiotypic protein. The absence of Asn-39 (alignment position 70 in figure 24.6) in elapid snake venoms precludes the inhibition of legumain (Richards et al. 2011). However, it is interesting that Asn is present at position 70 in *Pogona barbata* (see figure 24.6) cystatin and also convergently in a spider venom peptide (from *C. jingzhao*; Chen et al. 2008b). It is not certain if the surrounding sequences amount to functional BSLs that impart legumain inhibition, but both conform to the BSL consensus. If the lizard and spider venom cystatins were shown to be functional legumain inhibitors, it would suggest that the ability to inhibit legumain is unnecessary for snake venom cystatin and that the BSL was deleted at an early stage in snake evolution.

24.7.3 APOTYPIC STRUCTURAL AND FUNCTIONAL TOXIN FORMS

None is documented.

24.7.4 THERAPEUTIC POTENTIAL OF THE TOXIN TYPE

There is not always a clear delineation between the roles of C1 versus C13 CP inhibition by type 2 cystatins in normal cellular processes and pathology. However, there are several disorders in which cathepsins have been particularly noted as potential therapeutic targets and for which snake venom cystatins may be examined as drug leads. Although cathepsins are lysosomal proteases, they are often up-regulated and present both at the cell surface and in the secretions of cancer cells. Cathepsins play roles at various levels in cancer metastasis, including proteolysis of ECM components to allow for cancer cell migration, promotion of proteolytic cascades that promote invasion, and cleavage of E-cadherin which disrupts cell adhesion (Gocheva and Joyce 2007). Moreover, there is a tight association between cathepsin and the apoptotic caspase proteolytic cascade (Castino et al. 2002). Specific cathepsin B inhibitors induce apoptosis in a caspase-dependent fashion in neuroblastomas (Castino et al. 2002), and cathepsin L has shown promise as a target for cancer therapy, especially in combination with cytotoxic drugs (Lah, Duran Alonso, and Van Noorden 2006). A clearer understanding of the proteolytic mechanisms that underpin cancer progression will advance the development of protease inhibitors as effective cancer therapeutics. Cystatins may also be potential drug leads for Alzheimer's disease. β-Secretases induce cleavage of amyloid precursor protein (APP) to produce A-β, the main constituent of the brain amyloid plaques implicated in Alzheimer's disease pathology. Specific cathepsin B inhibitors have been shown to inhibit the conversion of APP to Aβ. Cathepsin B is thus implicated as a β-secretase, a potential target for cystatins in the treatment of Alzheimer's disease (Hook et al. 2005).

24.7.5 CONVERGENCE WITH OTHER VENOMS

Cystatin forms found in *Lonomia* spicule venom (Veiga et al. 2005) and tick hematophagous secretions (Grunclova et al. 2006) both exhibit mutations within the protease inhibiting the reactive site. While the secondary contact site is poorly conserved in *Lonomia*, it is largely conserved in ticks, such as within the group 2 cystatin (sialostatin L) identified in the hematophagous secretion gland transcriptome of the deer tick *I. scapularis* (Ribeiro et al. 2006) that specifically inhibits cathepsin L (Kotsyfakis et al. 2006). RNAi knockdown experiments demonstrate that this protein inhibits T-cell proliferation and cytokine secretion at the feeding site, in part by inhibiting processing and presentation of hematophagous secretion peptides by antigen-presenting cells. Cystatins are also secreted by the bush tick *Haemaphysalis longicornis* (Zhou et al. 2006), and a similar function may be inferred. The feeding secretion of the soft tick *Ornithodoros moubata* contains two cystatins that may have a similar effect, although one of them is also expressed in the gut, where it has a digestive function (Grunclova et al. 2006). The transcriptome of the hematophagous secretion gland of the mosquito *Aedes aegypti* recovered a truncated cDNA encoding a cystatin (Ribeiro et al. 2007), suggesting that this protein family may also have been recruited to the mosquito hematophagous secretion.

It is also interesting to note that there is a Pro-Ser substitution present in loop 2 of the spider venom form from *C. jingzhao* (Chen et al. 2008b), which would affect formation of the inhibitory wedge. At present, it is not clear if this substitution is genuine or if the spider cystatin is an active CP inhibitor, but a single nucleotide error could cause this shift in predicted proteins.

24.8 EPIDIDYMAL SECRETORY PROTEIN

24.8.1 ENDOPHYSIOLOGICAL PLESIOTYPE

This protein type is expressed in high amounts in the epididymis, but the endogenous function remains uncharacterized.

24.8.2 PLESIOTYPIC STRUCTURAL AND FUNCTIONAL TOXIN FORMS

This is known only from transcriptome sequencing (Fry et al. 2010b) (see figure 24.7).

```
            10        20        30        40        50        60        70        80
1. masfiaflvyswvllslaag---------------------------QVTYPESPACKFPFIYEGKAFTTCTEYGSTDKTPWCA
2. masfiaflvyswvllslvag---------------------------QVTYPESPACKFPFIYEGKAFTTCTEYGSTDKTPWCA
3.                                                <GGNSNGASCTFPFIYLGRTYYTCTNKMEHKGRYWCA
4. masfiaflvyswvllslaag---------------------------QVTYPESPACKFPFIYEGKAFTTCTEYGSTDKTPWCA
5. masfiaflvsswvllslva---------------------------GQVPFPESPACKFPFIYEGKAFTTCTEYGSTDKSPWCA
6. mnp-----wssy-llgwttfllyfyetsgKIPNLSSLGKHEFTKPWISIKEDQKDSCVFPFVYKGSSYFSCIKTNSFSP--WCA
7. mtr-----wssy-llgwttfllysyessgg----------------------MHEECVFPFTYKGSVYFTCTHIHSLSP--WCA

          90       100       110       120       130       140       150       160
1. TTSNYDRDRKWKPCAVKEYG----------------------------------------------------ELSVNYPEDPCR
2. TTSNYDRDRKWKPCAVKEYGGNSNGASCTFPFIYLGRTYYTCTNKLEHKGRYWCATTGSYDKDRKWSFCADIKLSVNYPKDPCR
3. TTGSYDKDRKWSLCADIKLS-------------------------------------------------------VDYPENPCR
4. TTSNYDGDRKWKPCAVKEYGGNSNGAPCTFPFIYLGRTYYTCTNKLEHKGRYWCATTGSYDKDRKWSFCADIKLSVNYPEDPCR
5. TTSNYDRDRKWKPCAIKEYGGNSNGASCTFPFIYLGRTYYTCTNKMEHKGRYWCATTGSYDKDRKWSLCADIKLSVDYPENPCF
6. TRAVYNG--QWKFCMADDYP------RCIFPFIFRGKSHNSCITEGSFLRRLWCSVTSSFDENQQWKYCETNEYGGNSFSKPCI
7. TRAVYNG--QWKYCQSEDYP------RCIFPFIYRGKAYNSCISQGSFLGSLWCSVTSVFDEKQQWKFCETNEYGGNSLRKPCI

      170       180       190       200       220       230       240       250      260
1. FPFTYKGKTYSGCTGAGREDGKLWCSISKNHDDNSQLVFCEPS------DPAPCYFPFKYKQKSYSDCTMNGSFDGHWWCATTA
2. FPFTYKDKTYSGCTGAGREDGKLWCSLSKNHDENSQLVYCEPS------DPAPCHFPFKYKKKSYSDCTMNGSIDGQLWCATTA
3. FPFTYKGKTYSGCTGDGREDGKLWCSLSKNHDENPQLVYCEPS------DPAPCHFPFKYKKKSYSDCTKNGSFDGQLWCATTA
4. FPFTYKGKTYSGCTGAGREDGKLWCSISKNHDDNSQLVFCEPS------DPAPCYFPFKYKQKSYSDCTMNGSFDGHLWCATTA
5. FPFIYHGKSYHSCTKVDSEKG-PWCATTANYDKSPQWKYCTSS------DIHPCVFPFIYHGTSYLSCTEVDN-QNRPWCATT>
6. FPSIFRNSTIFECMED--ENNKLWCPTTENMDEDGKWSLCADTRISSLVPGFPCHFPFSYKNKNYYNCIGKGTKENLTWCATSY
7. FPSIYRNNVVSDCMED--ESNKLWCPTTENMDKDGKWSFCADTRISALVPGFPCHFPFNYKNKNYFNCTNKGSKENLVWCATSY

           270       280       290       300       310       320       330       340
1. DYDKDSKWKACITEEYGGNSNGGQCVFPFTYMDKEYTTCTNEDTDGRFWCATTANYDKDKKWSFCADTRMSNGEIVKPGNVGSS
2. DYDKDSKWKACITEEYGGNSNGGQCVFPFTYMDKEYTTCTNEDADGRFWCATTANYDKDKKWSFCADTRLSNGEIVKPGNVGSS
3. DYDKDSKWKACVTEEYGGNSNGGQCVFPFTYMDKEYTTCTNEDTDGIFWCATTANYDKDKKWSFCADTRLSNGEIVNSGNVGSS
4. DYDKDSKWKAC>
5. >
6. NYDRDHTWVYC
7. NYDQDHTWVYC

       350       360       370       380       390       400       410       420
1. KAPCSFPFIYKGKTYTECTSKGTLTWKLWCSLTSNYDRDKRWRYCQPSDLETQNQLQSNEEKQTINISKE
2. KAPCFFPFIYKGKTYTECTSKGTLTWKLWCSLTSNYDRDRRWRYCQPSDLETQNQLQSNEEKQTINISKIGSGHL
3. KEPCSFPFIYKGKTYTACTSKGTVTWKLWCSLTSNYDRDKRWRYCQPSDLQTQNQLQSNEEKQTINISKE
4. >
5. >
6.
7.
```

FIGURE 24.7: Sequence alignment of epididymal secretory protein precursors from the mandibular venom glands of (1) *Varanus komodoensis* (GU441521), (2) *Varanus gouldii* (GU441519), (3) *Varanus indicus* (E2E4K3), (4) *Varanus komodoensis* (GU441522), and (5) *Varanus indicus* (GU441523) and the related nontoxin body forms from (6) *Canis familiaris* (Q9GL25) and (7) *Homo sapiens* (Q96BH3). < and > designate incomplete N- or C-terminal of precursor. ^ designates protein sequencing fragment (Fry et al. 2010b). Signal peptides are shown in lower-case.

24.8.3 APOTYPIC STRUCTURAL AND FUNCTIONAL TOXIN FORMS

The venom epididymal secretory protein sequences recovered from *Varanus* venom-gland transcriptome libraries (Fry et al. 2010b) have an additional eight cysteines in relation to the endophysiological plesiotype, which significantly changes their emergent three-dimensional structures (see figure 24.7). An additional mutation deleted four of the 20 plesiotypic cysteines, with potential consequences for structure and function. This variation was present in two species (*V. indicus* and *V. komodoensis*), both of which also possessed an isoform that did not have this deletion, indicating that this mutation occurred in one isoform prior to their divergence from a common ancestor.

24.8.4 THERAPEUTIC POTENTIAL OF THE TOXIN TYPE

None is documented.

24.8.5 CONVERGENCE WITH OTHER VENOMS

None is documented.

24.9 GOANNATYROTOXIN

24.9.1 ENDOPHYSIOLOGICAL PLESIOTYPE

YY-peptides are typically expressed in the intestine. Documented activities include inhibition of exocrine pancreatic secretion, vasoconstriction, and inhibition of jejunal and colonic mobility.

24.9.2 PLESIOTYPIC STRUCTURAL AND FUNCTIONAL TOXIN FORMS

Unlike most other venom proteins, goannatyrotoxin is novel in having its secondary structures determined by a lack of significant numbers of prolines and cysteines present in the endophysiological plesiotype (Fry et al. 2010b). Goannatyrotoxin produces a unique potent triphasic action: rapid biphasic hypertension followed by prolonged hypotension in anesthetized rats (see figure 24.8).

24.9.3 APOTYPIC STRUCTURAL AND FUNCTIONAL TOXIN FORMS

None is documented.

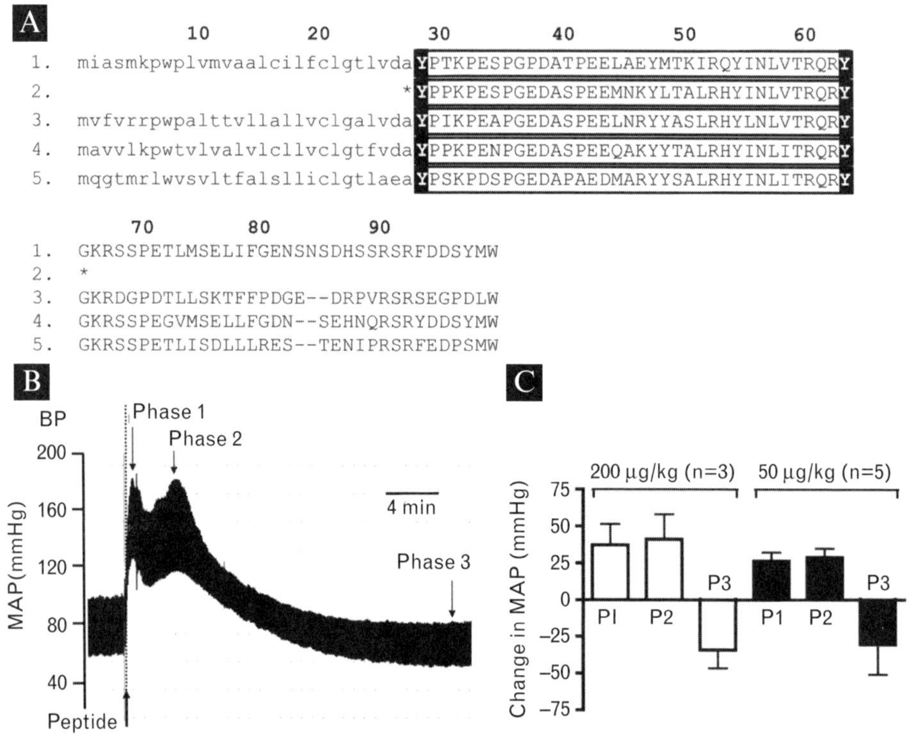

FIGURE 24.8: (A) Sequence alignment of goannatyrotoxin and related peptides (Fry et al. 2010b). (1) *Varanus eremius* goannatyrotoxin (GU441529), (2) the frog skin antimicrobial peptide P80952 from *Phyllomedusa bicolour*, and the related nontoxin, ordinary body peptides from (3) *Homo sapiens* (P10082), (4) *Anguilla japonica* (Q76CL2), (5) *Gallus gallus* (P28673). Box indicates posttranslationally processed bioactive peptide. Second **Y** is in amide form. * designates N-terminal and C-terminal of posttranslationally processed peptide sequence. (B, C) Effect of intravenous injection of *V. eremius* goannatyrotoxin (GU441529) (Fry et al. 2010b). P1 = phase 1; P2 = phase 2; P3 = phase 3. MAP = mean arterial pressure. Signal peptides are shown in lower-case.

24.9.4 THERAPEUTIC POTENTIAL OF THE TOXIN TYPE

None is documented.

24.9.5 CONVERGENCE WITH OTHER VENOMS

Goannatyrotoxin represents an additional case of convergence in the selection of a gene type for use as a toxin, with a YY-peptide having been previously characterized as a bioactive component in frog skin chemical arsenals. While goannatyrotoxin has a triphasic effect on the cardiovascular system, the frog skin YY-peptides have broad-spectrum antimicrobial activity (Mor et al. 1994; Vouldoukis et al. 1996).

24.10 HELOFENSIN

24.10.1 ENDOPHYSIOLOGICAL PLESIOTYPE

See chapter 9.

24.10.2 PLESIOTYPIC STRUCTURAL AND FUNCTIONAL TOXIN FORMS

This unique toxin type is constructed by four tandemly repeated β-defensin motifs, resulting in a single protein with a novel multidomain topology (Genbank: EU790964) (see figure 24.9) (Fry et al. 2010a). This demonstrates that high toxin diversity in venoms is attained not only by the expression of multiple paralogs but also by domain variation in a multidomain single-product gene formed by tandem repeats within a plesiotypic monodomain gene. All precursors sequenced have the same cysteine arrangement, and there is high overall sequence conservation, indicative of conservation of activity (see figure 24.9).

A

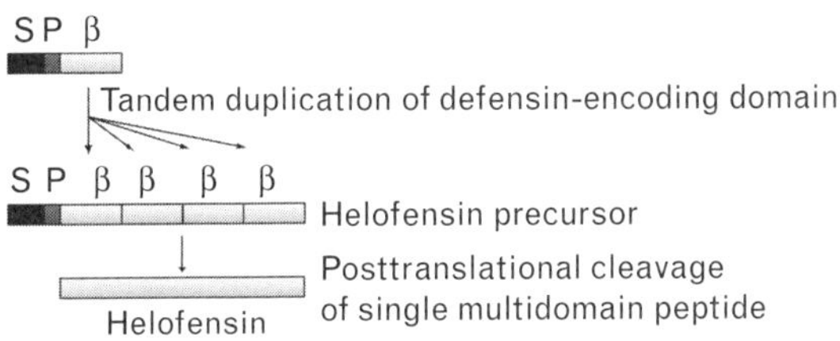

B

```
             10        20        30        40        50        60        70
1.   mqmdwlfiavisgigllssgvp-gtqg---AYTTEQCRALNG-TCNFYA---CFPKNVVIG--KCDWLGWGCCR-
2.   mqmdwlfiavisgigllssgvp-gtqg---AYTTEQCRALNG-TCNFYA---CFPKNVVIG--KCDWLGWGCCR-
3.   mqmdwlfiavvsaigllssgvp-gtqg---AYTTEQCRALNG-TCRFYA---CFPKNVVIG--KCDWLGWGCCR-
4.                                >*AYTTEQxRALxG-TxRFYA---RFPKNVVIG--KRDWLG*
5.   milyllfaflflaflsep------------gnaYKQCHKKGG-HCFPKEKI-CIPPSSDFGKMDCRWRWK-CC--
6.                               >FVQHRPRDCESING-VCRHKDTVNCREIFLAD----CYNDGQKCCRK
7.   mrihyllfallflflvpvpghg----GIINTLQKYYCRVRGG-RCAVLS---CLPKEEQIG--KCSTRGRKCC--
8.   mklhslisvlllfvtlipkgkt------GVIPGQKQCIALKG-VCRDKL---CSTLDDTIG--ICNEGKK-CCE-
9.   mkilyllsaflflaflsesgna-----------QPQCRWLDG-FCHSSP---CPSGTTSIGQQDCLWYES-CC--
10.  mrtlcsllligcllfsyatpvagILGPLRIQTDYHRCLREKG-FCLNAV---CPRSTLFVG--TCFPYKFYCC--

        80        90        100       110       120       130       140       150
1.   ARTPLECTAKKGTCTKTGCTKTDTDHGPCDGGAQCCQRDPVKYCKFHGNVCGRGKCPMDHIPIGEQCMPGYPCCKR
2.   ARTPLECTAKKGTCTKTGCTKTDTDHGPCDGGAQCCQRDPVKYCKFHGNVCGRGKCPMDHIPIGE-CTPGYPCCKR
3.   ARTPLECTAKKGTCTASGCTETDTDHGPCDGGAQCCQRDPVKYCKFHGNVCGRGKCPMDHIPIGEQCMPGYPCCKR
4.   *
5.   KKGSGK
6.
7.   RRKK
8.   RRWWILPYPTPVPKGKSP
9.   IPRYEK
10.  KFKR

          160       170       180       190
1.   DGPAYCKSKGGKCLRRCSQIVPTDIIGVCADGVPCCKSRQSTG
2.   DGPAYCKSKGGKCLNRCPQIVPTNVIGVCATGVPCCKSRQSTG
3.   DGPAYCKSKGGKCLRRCSQIVPTDIIGVCADGVPCCKSRQSTG
4.   *
5.
6.
7.
8.
9.
10.
```

FIGURE 24.9: (A) Domain mutation producing the helofensin tandem β-defensin domains. β = β-defensin domain, P = propeptide, S = signal peptide. (B) Sequence alignment of the helofensin isoforms and related proteins (Fry et al. 2010a). The representative *Heloderma* venom proteins (1) GQ918270, (2) GQ918271, and (3) EU790964 from *Heloderma suspectum* and (4) the protein fragment Q7LZ31 from *Heloderma horridum*. The representative β-defensin convergently found in other venoms: (5) O73799 (*Crotalus durissus terrificus*) and (6) P82172 (*Ornithorhynchus anatinus*). The representative nonvenom β-defensins: (7) P81534 (*Homo sapiens*), (8) Q30KJ9 (*Pan troglodytes*), (9) Q6HAA2 (*Crotallus durissus terrificus*), and (10) Q32ZI5 (*Rattus norvegicus*). * designates N-terminal fragment. Signal peptides are shown in lower-case.

Very little structural or functional work has been done on this toxin type. Indeed, over two decades have passed since the first study on this toxin type (Komori, Nikai, and Sugihara 1988), in which only a very small fragment was sequenced, and thus homology with any endophysiological protein was unable to be established. Later work identified the unique tetradomain, monoproduct precursor (Fry et al. 2010a). However, more work needs to be done, as the initial report gave an MW (as determined by SDS-PAGE) of around 28 kDa (Komori, Nikai, and Sugihara 1988), while the predicted proteolytically cleaved product in the recent study was 17 kDa (Fry et al. 2010a). As the SDS-PAGE gave the same MW under both nonreducing and reducing conditions, this precludes dimerization. Bioactivity testing revealed extremely potent activity, with an LD_{50} of approximately 0.135 mg/kg (based on a test range of 0.057–0.311) (Komori, Nikai, and Sugihara 1988). The inhibition of direct twitches is strongly suggestive of myotoxicity, but this needs to be confirmed.

24.10.3 APOTYPIC STRUCTURAL AND FUNCTIONAL TOXIN FORMS

None is documented.

24.10.4 THERAPEUTIC POTENTIAL OF THE TOXIN TYPE

None is documented.

24.10.5 CONVERGENCE WITH OTHER VENOMS

The β-defensin domain has been convergently recruited for use in a number of other venoms (see chapter 9).

24.11 HYALURONIDASE

24.11.1 ENDOPHYSIOLOGICAL PLESIOTYPE

Hyaluronidase is an enzyme that catalyzes the hydrolysis of hyaluronic acid, a major constituent of interstitial barriers.

24.11.2 PLESIOTYPIC STRUCTURAL AND FUNCTIONAL TOXIN FORMS

Like the normal body form, the venom form of this globular enzyme contains only five disulfide bridges in more than 300 amino acid residues, a very low proportion of disulfide bridges for a venom protein, and six of the cysteine residues are concentrated in the last 74 amino acids (see figure 24.10). Hyaluronidase is found in the venom of toxicoferan reptiles, with isoforms sequenced from snakes (Harrison et al. 2007; Fry et al. 2008) and helodermatid lizards (Fry et al. 2010b). It is speculated that it acts as a "diffusion factor," enhancing tissue permeability in order to allow a more efficient spreading of toxins or hemostatic factors (see Tu and Hendon 1983). While this has not been proven, evidence such as its degradation of extracellular matrix is certainly suggestive of such a role (Girish et al. 2002).

24.11.3 APOTYPIC STRUCTURAL AND FUNCTIONAL TOXIN FORMS

Hyaluronidase enzymes across animal taxa show lower levels of sequence diversity than other toxins, and no new activities have been reported for either venom or body forms (see figure 24.10). This is consistent with the basic molecular evolutionary principles of globular proteins (see chapter 1.3.2).

24.11.4 THERAPEUTIC POTENTIAL OF THE TOXIN TYPE

None is documented.

```
          10        20        30        40        50        60        70        80        90       100
1. >
2. myhlwikclaawiflkrfngVHVMHAKAPMYPNEPFLVFWNAPTTQCRLRYKVDLDLKTFHIVANANDTLSGSAVTIFYPTHLGIYPHIDDRGHFFHGII
3. myhiwikflaawiflkrfngVHVMQAKAPMYRNEPFLVFWNAPTTQCRLRYKVDLDLKTFHIVSNANDSLSGSAVTIFYPNHLGVYPHIDDRGHFFHGII
4. myhlwikclaawiflkrcngVHAMPAKAPMYPNEPFIVLWNAPTTQCPLRYKVDLDLKTFHIVANANDSLSGSVVAIFYPNHLGVYPHIDERGHFFHGII
5. myhlwikclaawiflkrfngVHVMQAKAPMYPNEPFLVFWNAPTTQCRLRYKVDLDLNTFHIVTNANDSLSGSAVTIFYPTHLGFYPHIDGRGHFFNGII
6. mchlwikclatwillkrfngVHVMQTRTPMYPNEPFLVFWNAPTTQCQLRYKVALDLKTFHIVTNANESLSGSAVTIFYPTQLGIYPHIDDHGNFVNGII
7. myhiwikclaawiflkrfngVHVMQAKAPMYPNEPFLVFWNAPTTQCLLRYKVDLDLKTFHIVTNANESLSGSAVTIFYPNHLGVYPYIDDRGHFFNGII

         110       120       130       140       150       160       170       180       190       200
1. >
2. PQNESLTKHLDKSKSDINRIIPLKTFHGLGVIDWENWRPQWDRNWGNKNVYRNRSIQFARDLHPELSENKIRRLAKAEYEKAAKSFMRDTLLLAEEMRPD
3. PQNESLTKHLNKSKSDINRIIPLKAFHGLGVIDWENWRPQWDRNWGSKNVYRNRSIQFARDLHPELSEDKIRRLAKKEYEKAAKSFMRDTLLLAEEMRPD
4. PQNESLTKHLNKSKSDINRMIPLKTFHGLGVIDWENWRPQWDRNWGSKNVYRNRSIQFAKKLHPELSEDKIKRLAKKEYEKAAKSFMRDTLLLAEEMRPN
5. PQNESLAKHLNKSKSDINRMIPLRTFHGLGVIDWENWRPQWDRNWGSKNVYRNRSIQFARDLHPELSEDEIKRLAKQEYEKAAKSFMRDTLLLAEEMRPY
6. PQNESITKHLNKTKSNINHMIPLKTFHGLGVIDWENWRPQWDRNWGSKNVYRNRSIQFAKELHPDLSEHAIKRLAKEEFEKAGKSFMRDTLLLAENMRPD
7. PQNESLAKHLNKSKSDINRMIPLRTFHGLGVIDWENWRPQWDRNWGSKNVYRNRSIQFARDLHPELSEDKIKRLAKQEFEKAAKSFMRDTLLLAEEMRPD

         210       220       230       240       250       260       270       280       290       300
1. >
2. GYWGYYLYPDCQNYDYKTKGDQYTGKCPDIEMSRNDQLLWLWRESTALFPNVYLEIILRSSDNALKFVHHRLKESMRIASMAREDYALPVFVYARPFYAY
3. GYWGYYLYPDCQNYDYKTKGDQYTGKCPEIEMSRNDQLLWLWRDSTALFPNVYLEIILRSSDNALKFVHHRLKEAMRIASMAREDYALPVFAYARPFYAY
4. GYWGYYLYPDCQNYDYKTKGDQYTGKCPDIEMSRNDQLLWLWRDSTALFPNVYLEIILRSSDNALKFVHHRLKESMRIASMAREDYALPVFVYARPFYAY
5. GYWGYYLYPDCQNYNYKTKPDQYTGECPDIEITRNNQLLWLWRDSTALFPNIYLETVLRSSDNALKFVHHRLKESMRIASMARKDYALPVFPYARPFYAY
6. GYWGYYLYPDCYNYNYKKKPEQYTGKCPDIEMSRNDQLIWLWRDSTALFPSIYLETILKSSANALKFVHHRLKESMRISSMARKDYALPVFVYARPFYAY
7. GYWGYYLYPDCHNYNYKTKPDQYTGECPDIEMSRNNQLLWLWRDSTALFPNIYLETILRSSDNALKFVHHRLKESMRIASMARKDYALPVFVYARPFYAY

         310       320       330       340       350       360       370       380       390       400
1.       >EDLVNTIGESAALGVAGVALWGSMQYASTKDSCEAVKQYINGAFGHYIVNVTYAAKLCSEELCKKNGRCIRKNSDSYDLHLPPQSFQISVDK
2. TFEPLTQEDLVTTVGETAAMGAAGIVFWGSMQYASTVDSCQKVKKYMNGPLGRYIINVTTAAKICSHALCRKNGRCVRKHSDSNAFLHLFPESFRIMVHA
3. TFEPLTQEDLVTTVGETAAMGAAGIVFWGSMQYASTVDSCQKVKKYMNGPLGRYIVNVTTAAKICSRVLCRKNGRCVRKHSDSNAFLHLFPESFRIMVYA
4. TFEPLTQEDLVTTVGETAAMGAAGIVFWGSMQYASTVDSCQKVKTYMNGPLGRYIVNVTTAAKICSHALCRKNGRCVRKHSDSNAFLHLFPESFRIMVHA
5. TFEPLTEEDLVNTVGETAAMGAAGIVFWGSMQYASTVDSCRKVKDYIDGPLGRYIVNVTTAAKICSHFLCKKHGRCVRKHSDSNAFLHLFPDSFRILVHG
6. TFEPLTEEDWVTTIGETAAMGAAGIVFWGSMQYASTIESCQKVKDYIKGPFGHYIINVTSAAKICSHFLCKGKGRCVRKHSDSNAFLHLFPESFRIMVQA
7. TFEPLTQEDLVTTVGEAAAMGAAGIVFWGSMQYASTVDSCGKVKDYMDGPLGRYIINVTTASKICSHFLCKKHGRCVRKHSDSNAFLHLFPDSYRIMVHG

         410       420       430       440       450
1. SKTDKKVIVQGNLKQEDIDAMKDKFVCQCYQGWEGIFCEIPSSTDGCPSN
2. NATEKKVIVKGKLELENLIYLRENFMCQCYQGWQGLYCEEYSIKDIRKI
3. NATEKKVIVKGKLELENLIYLRENFMCQCYQGWKGLYCEEYSIKDIRKI
4. NATEKKAIVKGKLELKDLIYLRKNFMCQCYQGWKGLYCEEYSIKDIRKI
5. NATEKKVIVKGKLELENLIFLINNFMCQCYQGWKGLYCEKHSIKDIRKI
6. NATHKKAIVKGKLELENLKYLRKNFICQCYQGWKGLYCEEHHNKEGN
7. NATEKKVIVKGKLELENLIFLRNNFMCQCYQGWKGLYCEKHSIQDIRKI
```

FIGURE 24.10: Sequence alignment of the hyaluronidase sequences from the venoms of (1) *Heloderma suspectum* (C6EVG3), (2) *Echis ocellatus* (A3QVN2), (3) *Cerastes cerastes* (A3QVN3), (4) *Bitis arietans* (A3QVP0), (5) *Crotalus adamanteus* (J3S820), (6) *Micrurus fulvius* (U3FYQ4), and (7) *Ovophis okinavensis* (U3TBU1).

24.11.5 CONVERGENCE WITH OTHER VENOMS

Hyaluronidase has been sequenced from the venoms of coleoids (Fry, Roelants, and Norman 2009; Ruder et al. 2013b), stonefish (Ng et al. 2005), and hymenopteran insects (Gmachl and Kreil 1993; King et al. 1996) and also from the hematophagous secretions of dipteran insects (Charlab et al. 1999; Ribeiro et al. 2004b). In addition to venoms for which protein sequences have been obtained, hyaluronidase activity has been reported for spider (Nagaraju, Devaraja, and Kemparaju 2007) and scorpion venoms (Batista et al. 2007), but although the corresponding enzymes have been isolated, they have not yet been sequenced. The masses of the enzymes isolated from spiders and scorpions (41–45 kDa) are similar to those of the shorter forms of hyaluronidase found in hymenopteran and dipteran venoms; the longer forms found in the venoms (and bodies) of reptiles, stonefish, and other vertebrates are typically 52–55 kDa. Thus, the short form appears to be restricted to arthropods.

24.12 LIPOCALIN

24.12.1 ENDOPHYSIOLOGICAL PLESIOTYPE

Lipocalins are a functionally diverse family of proteins that generally bind small hydrophobic ligands (Montfort, Weichsel, and Andersen 2000; Andersen et al. 2005).

24.12.2 PLESIOTYPIC STRUCTURAL AND FUNCTIONAL TOXIN FORMS

This protein type is known only from transcriptome sequences obtained from snakes: *Azemiops feae*, *Disphoids typus*, *Liophis miliaris*, *R. tigrinus*, and *Trimorphodon biscutatus* (Fry et al. 2012b).

24.12.3 APOTYPIC STRUCTURAL AND FUNCTIONAL TOXIN FORMS

None is documented.

24.12.4 THERAPEUTIC POTENTIAL OF THE TOXIN TYPE

None is documented.

24.12.5 CONVERGENCE WITH OTHER VENOMS

Lonomia obliqua venom contains a number of lipocalins (Veiga et al. 2005; Ricci-Silva et al. 2008), and one of these, Lopap, has a unique serine protease-like activity that activates prothrombin (Reis et al. 2006); there are currently no reports of other lipocalins with protease activity.

Several groups of hematophagous arthropods have independently recruited this protein family to serve a variety of antihemostatic functions. Arthropod lipocalins display pronounced molecular biodiversity, such as in the Reduviidae family of predatory insects, which includes the triatomids or kissing bugs (Andersen et al. 2005). For example, 341 isoforms were recovered from a hematophagous secretion-gland cDNA library of *Triatoma brasiliensis* (Santos et al. 2007). In *Rhodnius prolixus*, the nitrophorins, lipocalins with a unique heme-binding domain, function to store nitric oxide and deliver it to the skin during feeding (Champagne, Nussenzveig, and Ribeiro 1995; Andersen et al. 1997). These proteins also sequester histamine, preventing the host response to histamine released from mast cells and platelets (Ribeiro and Walker 1994), and inhibit coagulation by interfering with factor IXa (Ribeiro, Schneider, and Guimaraes 1995; Gudderra, Ribeiro, and Andersen 2005). Another lipocalin (nitrophorin 7, without a heme group) inhibits coagulation by binding to anionic membrane phospholipids, interfering with the assembly of procoagulant complexes (Andersen et al. 2004). Yet other lipocalins disrupt hemostasis by sequestering ADP, thus preventing platelet activation and aggregation (Francischetti et al. 2000; Francischetti, Andersen, and Ribeiro 2002), or by sequestering amines such as serotonin, thereby preventing vasoconstriction and possibly coagulation (Andersen et al. 2003). The venoms of species of the hemipteran genus *Triatoma* lack nitrophorins, but diverse lipocalins present within these secretions nevertheless fulfill a variety of antihemostatic roles. Characterized *Triatoma* lipocalins include pallidipin and triplatin, which have anticollagen activity, and the antithrombin molecule triabin (Noeske-Jungblut et al. 1994; Noeske-Jungblut et al. 1995; Morita et al. 2006).

Lipocalins also account for several of the pharmacological activities in the blood-meal feeding secretion of ticks, into which they have been independently recruited (Mans, Louw, and Neitz 2003). Moubatin from *Ornithodoros moubata* inhibits collagen-mediated platelet aggregation (Waxman and Connolly 1993) via a mechanism that includes binding thromboxane A2 and leukotriene B(4) (Mans and Ribeiro 2008). A second lipocalin from the same tick shares 46% amino acid identity with moubatin but specifically inhibits the complement pathway by antagonizing activation of complement component C5. Two related lipocalins, TGSP4 (from *Ornithodoros savignyi*) and AM-33 (from *Argas monolakensis*), bind cysteinyl leukotrienes with high affinity (Mans et al. 2008). Cysteinyl leukotrienes, produced by neutrophils, macrophages, basophils, and mast cells in response to injury, increase endothelial permeability leading to edema. This would be expected to lead to occlusion of blood vessels and accumulation of a nutritionally deficient, serous-rich, and erythrocyte-poor fluid at the tick feeding site. Sequestration of these molecules helps ensure a nutritionally adequate meal, rich in erythrocytes. A lipocalin (FS-HBP) specifically expressed by female hard ticks (*Rhipicephalus*) binds histamine in a manner completely distinct from the *Rhodnius* nitrophorins: rather than binding through a heme moiety, this protein has two histamine-binding pockets, one low-affinity and one high-affinity, in the interior of the β-barrel structure (Paesen et al. 1999; Paesen et al. 2000; Sangamnatdej et al. 2002). Additionally, the lipocalins monomine and monotonin, from *O. monolakensis*, inhibit vasoconstriction by binding serotonin (5-hydroxytryptamine) in addition to histamine (Mans et al. 2008). These molecules have a single amine-binding pocket and appear to have acquired their amine-binding properties as a separate evolutionary event distinct from the evolution of FS-HBP. Tick hematophagous secretion-gland cDNA libraries reveal the presence of several other lipocalins of unknown function (Ribeiro et al. 2006). Lipocalins are also found in the feeding secretion of blood-meal dipterans, such as the mosquito *A. aegypti*, and these forms are presumably anticoagulant (Ribeiro et al. 2007).

24.13 MATRIX METALLOPROTEASE

24.13.1 ENDOPHYSIOLOGICAL PLESIOTYPE

Found in macrophages and granulocytes, these enzymes cause proteolysis of the extracellular matrix.

24.13.2 PLESIOTYPIC STRUCTURAL AND FUNCTIONAL TOXIN FORMS

Bioactivities remain to be determined (Ching et al. 2012).

24.13.3 APOTYPIC STRUCTURAL AND FUNCTIONAL TOXIN FORMS

None is documented.

24.13.4 THERAPEUTIC POTENTIAL OF THE TOXIN TYPE

None is documented.

24.13.5 CONVERGENCE WITH OTHER VENOMS

None is documented.

24.14 NERVE GROWTH FACTOR

24.14.1 SUMMARY

In animals, NGF plays an essential housekeeping role, regulating the health and maintenance of the peripheral and central nervous systems. NGF is expressed ubiquitously in toxicoferan venoms. NGFs isolated from caenophidian snake venom glands share similar structures and functions with their housekeeping counterparts, including the ability to induce outgrowth of neurite cells. Although the toxic role of these proteins in envenomation remains to be elucidated, several direct and indirect mechanisms through which venom homologues of NGF could participate in envenoming have been postulated. Endophysiological NGF has been of great therapeutic interest as a potential treatment for a number of neurodegenerative diseases, most notably Alzheimer's disease. At least one study has suggested that venom NGF may have additional beneficial activities not exhibited by mammalian NGF.

24.14.2 ENDOPHYSIOLOGICAL PLESIOTYPE

The endophysiological plesiotype of venom NGF has been identified as β-NGF, which functions in the development and maintenance of the nervous system. Mammalian NGFs exist as multimeric complexes of three subunits, α, β, and γ, with only the β subunit possessing nerve growth activity (Bax et al. 1997). As would be expected, the predominant tissue localization of NGF is within the nervous system. Use of ELISA and mRNA assays have demonstrated that NGF is synthesized and secreted by sympathetic and sensory target organs and target tissues in the brain (Korsching 1993). In the adult brain, the highest concentrations of NGF are localized in the hippocampus, cortex, and olfactory regions (Korsching et al. 1985; Large et al. 1986). Apart from the nervous system, NGF is also found in various other tissue compartments. The salivary glands of adult male mice are particularly rich in NGF, while lower concentrations can also be found in guinea pig prostate, seminal fluid, and human skin (Levi-Montalcini and Angeletti 1968). NGF is also synthesized and secreted by Schwann cells, fibroblasts, and mast cells (Korsching 1993; Levi-Montalcini et al. 1996).

As its name implies, NGF functions to stimulate the growth of nerve cells, but it also has various other activities within the nervous system. NGF controls cell survival and differentiation, cessation of cell growth, and apoptosis of sensory neurons. These activities are mediated by binding to two distinct classes of cell-surface receptors, the p75 neurotrophin receptor ($p75^{NTR}$) and the TrkA receptor. Binding of NGF to TrkA leads to dimerization of the receptor and subsequent autophosphorylation of several tyrosine residues in the cytoplasmic domain. This triggers the intracellular signal-transduction cascade and induces a trophic effect (Meakin and Shooter 1992). Conversely, binding of NGF to $p75^{NTR}$ initiates apoptosis in the context of TrkA-negative neurons. However, if TrkA is co-expressed on the cell surface, NGF binding to $p75^{NTR}$ leads to increased survival, neurite outgrowth, and synaptic plasticity (Thoenen et al. 1987; Barrett 2000; Miller and Kaplan 2001; Chao and Bothwell 2002). In addition to functions in the nervous system, there is emerging evidence of other roles for NGF. For example, NGF is present in CNS limbic areas involved in mood and cognition, in the orchestration of neuroendocrine responses and circadian activities, and also in cells of the immune and endocrine system. This has led to the hypothesis that it may function as an intracellular messenger regulating endocrine responses to stress (Smith et al. 1995; Hristova and Aloe 2006). In support of this hypothesis, it has been found that the plasma concentration of NGF is increased in fighting male mice (Aloe et al. 1986; Lakshmanan 1987) and in soldiers experiencing parachute jumping for the first time (Aloe et al. 1994).

24.14.3 PLESIOTYPIC STRUCTURAL AND FUNCTIONAL TOXIN FORMS

NGF constitutes a very small proportion (1–5 mg/g of venom) of the total venom in most snakes (Siigur et al. 1985; Siigur et al. 1986; Koyama et al. 1992; Sanz et al. 2008). However, in snake species of the genus *Oxyuranus*, NGF makes up nearly 0.5% (weight weight) of the total venom injected (Morrison, Pearn, and Coulter 1982; Morrison et al. 1984; Earl et al. 2006). Such elevated amounts of NGF could result in toxicity to the prey.

Venom NGFs in elapid snake venoms have been shown to have undergo rapid mutation under the influence of positive Darwinian selection (Sunagar et al. 2013a). By mapping detected hypermutational sites onto a three-dimensional homology model of elapid snake venom NGFs, it was revealed that a large number of mutations were localized on the molecular surface of these proteins. In fact, nearly 85% of the hypermutational sites identified were either located on the molecular surface or had their side chains exposed to the surrounding medium. Moreover, nearly 74% of these mutations were found in the secreted β-chain of elapid NGF, highlighting that certain regions of these proteins are more favored than others in the accumulation of evolutionary variation.

Since NGF is capable of inhibiting venom metalloprotease-dependent proteolysis, secretion of NGF in the venom of certain vipers has been viewed as indicative of their role in prevention of venom autolysis (Cohen and Levi-Montalcini 1956; Wijeyewickrema et al. 2007a; Wijeyewickrema et al. 2007b; Wijeyewickrema et al. 2010). There are several other direct and indirect mechanisms through which NGFs could perform a role in envenomation, potentially resulting in an increase in the potency of the toxicoferan venom arsenal. For example, injection of a large dose of NGF into the lymphatic system of the prey would result in NGF overdose, which would trigger a variety of atypical reactions. Caenophidian NGFs, which rapidly accumulate variations under the influence of positive Darwinian selection, could induce a plethora of pharmacological effects by nonspecifically interacting with novel receptors of the prey (Sunagar et al. 2013a). NGF is also known to cause apoptosis of cells that lack TrkA receptors (Frade, Rodriguez-Tebar, and Barde 1996; Frade and Barde 1998; Frade and Barde 1999), and thus the caenophidian venom NGF homologues could function as proapoptotic factors and contribute to prey cell cytotoxicity, particularly after being introduced into tissues where they are not typically found.

As discussed above, venom hyaluronidase, which is capable of degrading hyaluronic acid in the extracellular matrix, may act as a spreading factor and facilitate diffusion of other venom components (Girish et al. 2002). This may be a role fulfilled by NGF also. Endophysiological NGFs are capable of triggering the release of granules containing histamines, serotonins, and other chemical mediators from mastocytes (Lindholm et al. 1987; Woolf 1996). Degranulation of mast cells located primarily in perivascular spaces, often close to neurons and blood vessels, increases vascular permeability and neurogenic inflammation (Theoharides 1990). Although controlled degranulation can aid in orchestrating and mounting an acute inflammatory reaction, massive releases are associated with anaphylaxis, bronchoconstriction (suffocation), and vasodilation (Payne and Kam 2004). Therefore, venom NGF homologues could perform an ancillary envenoming role, similar to that of hyaluronidases, by facilitating the efficient absorption of other venom components. Both glycosylated and nonglycosylated NGFs have been recovered from snake venoms (Pearce et al. 1972; Pearce and Thompson 1986; Katzir et al. 2003; Earl et al. 2006). Serum proteins undergo glycosylation as part of posttranslational modification to prolong their circulation time. Interestingly, venom NGFs in nearly all squamate lineages (except in a few elapid snakes, such as genus *Naja*, *Notechis scutatus*, *Pseudechis australis*, and *Pseudechis porphyriacus*) have conserved aspargine 23 (Asn 23), a putative glycosylation site in mature venom NGF (Kornfeld and Kornfeld 1985; Gavel and von Heijne 1990). This site is completely unconserved in mammalian and turtle NGF. Glycosylation of snake venom NGF homologues could prolong their circulation time in the bloodstream of prey animals and allow them to spread to various tissues and exert their toxic action. Hence several direct and indirect mechanisms can be postulated through which caenophidian NGFs might participate in prey envenoming.

24.14.4 APOTYPIC STRUCTURAL AND FUNCTIONAL TOXIN FORMS

There are two identified examples of covalently bound NGF homodimers, from the venoms of *Bitis arietans* (Smith et al. 1992b) and *Bungarus multicinctus* (Furukawa and Hayashi 1978). Attempts to solve the X-ray crystal structure of a venom NGF have so far proved unsuccessful, although crystals diffracting at 3.2 Å have been analyzed from *Naja kaouthia* venom NGF (Gu et al. 2002). A molecular modeling study of *Bothrops jararacussu* venom NGF, employing the coordinates of the human NGF structure, showed that the core of the monomer is formed by a pair of twisted β-sheets and that this core is highly conserved between human and snake NGF (Kashima et al. 2002). A high percentage of shared sequence identity between snake venom NGFs and human NGF suggests that the overall three-dimensional structure is conserved across these proteins. The only structural difference within venom NGFs appears to be in glycosylation. Some venom NGFs, such as those from *Bothrops atrox, Agkistrodon rhodostoma, Macrovipera lebetina*, and *O. scutellatus*, contain up to 20% carbohydrate, while NGFs from *N. kaouthia* and *P. australis* are nonglycosylated (Glass and Banthorpe 1975; Siigur et al. 1985; Kostiza and Meier 1996; Earl et al. 2006). The glycosylation site has been identified as Asn 23, with this residue usually not conserved in those NGFs lacking sugar groups (Earl et al. 2006). Interestingly, both lizard venom NGF sequences determined to date contain Asn at position 23, suggesting the presence of N-linked glycans on the mature proteins. An apotypic form has been isolated from *Crotalus adamanteus* venom that is unique in containing a γ subunit with arginyl esterase activity that is noncovalently bound to the β subunit (Perez-Polo et al. 1978). The bioactivity of this apotypic form has not yet been determined, so it is unclear whether this structural variation translates to unique function.

24.14.5 THERAPEUTIC POTENTIAL OF THE TOXIN TYPE

NGF is an important trophic factor for basal forebrain cholinergic neurons, stimulating their functioning and preventing their death. Given that these cholinergic neurons are essential for brain functioning and are degenerated by aging and in neurodegenerative diseases, NGF has been suggested as a potential therapeutic drug for the treatment of these conditions. Several animal studies, including primate studies, have shown that NGF is effective in preventing cholinergic basal forebrain neuronal death and in stimulating cell function (Rosenberg et al. 1988; Emerich et al. 1994; Chen and Gage 1995; Conner et al. 2001). However, delivering NGF to the brain in a therapeutic context has proved challenging, as it cannot cross the blood-brain barrier and cannot be directly infused because of the occurrence of side effects (Tuszynski 2002). To overcome this, a phase 1 clinical trial of NGF gene therapy for Alzheimer's disease was undertaken. The results were promising and a follow-up phase 2 clinical trial is currently under way (Tuszynski et al. 2005). While only human NGF has been evaluated in these studies, there may be potential for the application of venom NGFs in neurodegenerative disorders. *In vitro* work has shown that NGF from *Naja sputatrix* venom exhibits comparable potency to mouse NGF in stimulating neurite differentiation. Surprisingly, this form of NGF also has other beneficial activities not evident in the mammalian protein. Unlike mouse NGF, *N. sputatrix* NGF up-regulates endogenous expression of β-NGF in PC12 cells and also up-regulates expression of prosurvival cell-surface receptors and ion channels (Koh, Armugam, and Jeyaseelan 2004). No further data have been reported for *N. sputatrix* NGF, but it is hoped that supportive studies in animal models of neurodegeneration can be performed, potentially paving the way for the eventual testing of venom NGFs in human clinical trials.

24.14.6 CONVERGENCE WITH OTHER VENOMS

None is documented.

24.15 PHOSPHODIESTERASE

24.15.1 ENDOPHYSIOLOGICAL PLESIOTYPE

Secreted in the kidney, this enzyme cleaves a variety of phosphodiester and phosphosulfate bonds, including those of deoxynucleotides, nucleotide sugars, and NAD.

24.15.2 PLESIOTYPIC STRUCTURAL AND FUNCTIONAL TOXIN FORMS

Phosphodiesterase (PDE) activity is known from the venoms of many snakes. It has also been recovered from a number of venom-gland transcriptome libraries, including those of elapid snakes (such as *Micrurus fulvius*; Margres et al. 2013) and viperid snakes (such as *Sistrurus catenatus edwardsii*; Pahari et al. 2007). A protein sequence exists for PDE from *Bothrops jararaca* venom (Santoro et al. 2009). Activity testing of the *B. jararaca* form revealed it to be an effective inhibitor of platelet aggregation.

24.15.3 APOTYPIC STRUCTURAL AND FUNCTIONAL TOXIN FORMS

None is documented.

24.15.4 THERAPEUTIC POTENTIAL OF THE TOXIN TYPE

None is documented.

24.15.5 CONVERGENCE WITH OTHER VENOMS

The spingomyelinase toxin from sicarid spiders is a modified form of PDE.

24.16 PHOSPHOLIPASE A_2 (GROUP III)

24.16.1 ENDOPHYSIOLOGICAL PLESIOTYPE

Lizard venom group III PLA_2 (LV-G^{III}-PLA_2) were recruited from within the type III group of the phospholipase A_2 superfamily. Their recruitment into reptile venom is thus convergent with their recruitment into bee venom, rather than being closely related to either of the more common forms found in snake venom, SV-G^{I}-PLA_2 or SV-G^{II}-PLA_2 (see chapters 20 and 21).

24.16.2 PLESIOTYPIC STRUCTURAL AND FUNCTIONAL TOXIN FORMS

Initial characterization of this toxin type was from *Heloderma suspectum* (Gomez et al. 1989). The first characterization of its toxicity was from *H. horridum* venom, with anticoagulation activity demonstrated via the ability to block epinephrine-induced platelet aggregation (Huang and Chiang 1994). Full sequencing of *Heloderma* isoforms revealed significant variation in primary structure (Vandermeers et al. 1991). This toxin type then received little attention for 15 years, until it was revealed to be part of the plesiotypic toxin arsenal of anguimorph lizards and that the antiplatelet activity was also plesiotypic (Fry et al. 2006; Fry et al. 2009b; Fry et al. 2010b; Koludarov et al. 2012). Notably, a long C-terminal tail was revealed in the protein translations of the *H. suspectum* PLA_2 mRNA sequences that had not been seen in either of the reported *H. suspectum* sequences previously determined by protein sequencing (Swiss-Prot P80003 and P16354) (see figure 24.11). The regions of the full-length mRNA sequences corresponding to the points at which the protein sequences end (G for P80003 and GR for P16354) are rich in dibasic proteolytic sites (GRKRSQKKRK for EU790967 and GRKRSRKKRK for EU790968) (see figure 24.11). These are common recognition sites for the proteolytic removal of segments from both vertebrate (Hook et al. 2008) and invertebrate (Veenstra 2000) protein precursors. Therefore, P16354 may represent an incompletely proteolyzed product relative to P80003. The C-terminal tail of EU790968 is much longer than that of EU790967, which

```
                 10        20        30        40        50        60        70        80
1.      mlasmasvgalllamnalvmplasgSWSKDTTSSHTIAKGGRGRDRGQVLSFLWRRPRGLPPVLVQSIWDPQGHLLACALRDEPAHTN
2.      >
3.      >
4.      >
5.      >
6.      >
7.      >
8.      >
9.      >
        90        100       110       120       130       140       150       160       170
1.      WYLEAYKQQQALNFSGTELQHYLDTLEVQNDTCWGAEYTGEDLDLGVRRTLALEEQEAQETNRIQRGLTMPGTLWCGAGNIASNFSHL
2.                                                                      >GAFIMPGTLWCGAGNAASDYSQL
3.      >
4.                                                                      >GAFIMPGTLWCGAGNAASDYSQL
5.      >
6.      >
7.                                                                                        >DFSDL
8.                                                                    >MMRGLTMNGTLWCGAGNSAESYSDL
9.                                                                            >MPGTLWCGAGNAASDYSQL

        180       190       200       210       220       230       240       250       260
1.      GTFKGPDMCCRDHDHCDIQISGLKYNYGVFNFRPHTISHCDCDTRFRNCLMSLSDSIADFIGKTYFN---VMFVPCFELKEG-NACVE
2.      GTEKDTDMCCRDHDHCENWISALEYKHGMRNYYPSTISHCDCDNQFRSCLMKLKDGTADYVGQTYFN---VLKIPCFELEEG-EGCVD
3.             >SCCREHDQCSVQITALQRKHGIFNLRPYTISHCDCDTRFRTCLMDLNDTIADFIGTTYFS---VLQIPCFYLEESDEACLE
4.      GTEKDTDMCCRDHDHCSDTMAALEYKHGMRNYRPHTVSHCDCDNQFRSCLMNVKDRTADLVGMTYFT---VLKISCFELEEG-EGCVD
5.             >MCCRDHDHCENWISALEYKHGMRNYYPSTISHCDCDNQFRSCLMKLKDGTADYVGQAYFN---VLKIPCFELEEG-EGCVD
6.                                                                             >EIPCFYLEESDEACLE
7.      GAFQGPDSCCQQHDQCSVQITALQRKHGIFNLRPYTISHCDCDTRFRTCLMDLNDTIADFIGTTYFS---VLQIPCFYLEESDEACLE
8.      GTFKETDMCCRDHDHCDVSITGLTKRYSMFNYRLYTVSHCDCDTQFRNCLMGHEDSISDFVGQVYFN---LMDIPCIVLEEE-DVCVS
9.      GTEKDTDMCCRDHDHCSDTMAALEYKHGMRNYRPHTVSHCDCDNQFRSCLMNVKDRTADLVGMTYFT---VLKISCFELEEG-EGCVD

        270       280       290       300       310       320       330       340       350
1.      WNWMYMCTNYSQMPVAHLVDPSPYVPIGLPTPTLQPGTQSSQDRKRSQKKWK---QLRHAHPNGAGQRRRLGR
2.      WNFWLECTESKIMPVAKLVSAAPY--------QAQAETQSGEGR<
3.      WAWWGGCNKRGRMLMAHTVPPSAYGSVPSVTTLPQPGTETTKDRKHRKRKLQGPEHPGVPWQRGRPGSRRKLLL<
4.      NNFSQQCTKSEIMPVAKLVSAAPY--------QAQAETQSGEG<
5.      WNFWLECTESKIMPVAKLVSAAPY--------QAQAETQSGEGRKRSRKKRK---WLGQTHLDAAGQRGRLGTQGGLHYPPPLHFILT
6.      WAWWGGCSKQGRMLMAHTVPPRAYGSGPTVTPLPEPGTETNQDRKLRKRKLQGPEQPGGPRQRGRPGSRRKLLL<
7.      WSW<
8.      STWWFKCTEYSRMPVAHTVDQPRYENAAPPMPARGSGAQGGQARKRKELHRPTGMELAEAPPRGAERNIRYEESRHARSK<
9.      WNFWQQCTKSEIMPVAKLVSAAPY--------QAQAETQSGEGRKRSQKKRK<

        360       370       380
1.      VVNRYIQILPVLKEYYSSHI
2.      <
3.      <
4.      <
5.      TTTL
6.      <
7.      <
8.      <
9.      <
```

FIGURE 24.11: Sequence alignment of LV-G^{III}-PLA_2 (Fry et al. 2010b; Koludarov et al. 2012): (1) K4I293 (*Abronia graminea*), (2) P16354 (*Heloderma suspectum*), (3) B6CJU9 (*Varanus komodoensis*), (4) P80003 (*Heloderma suspectum*), (5) C6EVH0 (*Heloderma suspectum*), (6) E2E4K7 (*Varanus gilleni*), (7) Q2XXL5 (*Varanus varius*), (8) E2E4K8 (*Celestus warreni*), and (9) C6EVG9 (*Heloderma suspectum*). Dibasic cleavage sites are highlighted in gray, and propep is underlined. Signal peptides are shown in lower-case.

terminates with the proteolysis motif. Full-length sequences from anguiid and varanid lizards reveal that this (RK) proteolytic cleavage motif is ubiquitously present in anguimorph isoforms and thus represents a plesiotypic condition (see figure 24.11).

24.16.3 APOTYPIC STRUCTURAL AND FUNCTIONAL TOXIN FORMS

None is documented.

24.16.4 THERAPEUTIC POTENTIAL OF THE TOXIN TYPE

None is documented.

24.16.5 CONVERGENCE WITH OTHER VENOMS

See chapters 20 and 21.

24.17 PHOSPHOLIPASE A_2 IIE

24.17.1 ENDOPHYSIOLOGICAL PLESIOTYPE

Expression is limited to the brain, heart, lung, and placenta and mediates progression of inflammatory processes.

24.17.2 PLESIOTYPIC STRUCTURAL AND FUNCTIONAL TOXIN FORMS

This is known only from transcriptome sequencing of *Atractaspis aterrima, Leioheterodon madagascarensis, and D. typus* (Fry et al. 2012b).

24.17.3 APOTYPIC STRUCTURAL AND FUNCTIONAL TOXIN FORMS

None is documented.

24.17.4 THERAPEUTIC POTENTIAL OF THE TOXIN TYPE

None is documented.

24.17.5 CONVERGENCE WITH OTHER VENOMS

See section 24.16 and chapters 20 and 21.

24.18 PHOSPHOLIPASE B

24.18.1 ENDOPHYSIOLOGICAL PLESIOTYPE

Detected in white blood cells (granulocytes and neutrophils), intestinal enterocytes, and the epidermis, it facilitates the removal of fatty acids from both the sn-1 and sn-2 positions of phospholipids.

24.18.2 PLESIOTYPIC STRUCTURAL AND FUNCTIONAL TOXIN FORMS

Only one form has been tested, from the venom of *Pseudechis colletti*, and it was strongly hemolytic (Bernheimer et al. 1987). Other forms are known only from transcriptome sequencing (Chatrath et al. 2011; Rokyta et al. 2011).

24.18.3 APOTYPIC STRUCTURAL AND FUNCTIONAL TOXIN FORMS

None is documented.

24.18.4 THERAPEUTIC POTENTIAL OF THE TOXIN TYPE

None is documented.

24.18.5 CONVERGENCE WITH OTHER VENOMS

None is documented.

24.19 RENIN ASPARTATE PROTEASE

24.19.1 ENDOPHYSIOLOGICAL PLESIOTYPE

Expression is restricted to the kidney. The primary activity is generation of angiotensin I from angiotensinogen, thus acting as a mediator of extracellular volume and vasoconstriction.

24.19.2 PLESIOTYPIC STRUCTURAL AND FUNCTIONAL TOXIN FORMS

It is known only from transcriptome sequencing (Wagstaff et al. 2006; Fry et al. 2013).

24.19.3 APOTYPIC STRUCTURAL AND FUNCTIONAL TOXIN FORMS

None is documented.

24.19.4 THERAPEUTIC POTENTIAL OF THE TOXIN TYPE

None is documented.

24.19.5 CONVERGENCE WITH OTHER VENOMS

None is documented.

24.20 RIBONUCLEASE

24.20.1 ENDOPHYSIOLOGICAL PLESIOTYPE

The nearest nonvenom protein is a pyrimidine-specific C-preferring nuclease expressed predominantly in the pancreas.

24.20.2 PLESIOTYPIC STRUCTURAL AND FUNCTIONAL TOXIN FORMS

This is known only from transcriptome sequencing (Fry et al. 2010b; Fry et al. 2012b). The plesiotypic ribonuclease toxin sequences have deletions of two plesiotypic cysteines and thus may have novel conformational folding and differential surface chemistry (see figure 24.12).

24.20.3 APOTYPIC STRUCTURAL AND FUNCTIONAL TOXIN FORMS

The two isoforms of an apotypic variant recovered from the *Gerrhonotus infernalis* library have a frame-shift mutation near the C-terminal, not only resulting in slightly truncated sequences and the loss of two of the plesiotypic cysteines but also producing two new cysteines (see figure 24.12) (Fry et al. 2010b). The resultant novel disulfide-bridging pattern may produce significant changes in three-dimensional structure and consequently biotargeting.

24.20.4 THERAPEUTIC POTENTIAL OF THE TOXIN TYPE

None is documented.

```
                10        20        30        40        50        60        70
1. maakgscvfwfllasawivmks--daADTFESFKELHVDYPKTEAPNDNE-YCKKVMGGRGQTK-----LKANTYIH
2. maakgscvfwfllasawivmks--daADTFESFKELHVDYPKTGAPNDDE-YCKKVMGGRGRTK-----LKANTYIH
3. maakgscvfwfllasacivmks--daADTFESFKELRVDYPKTEAPNDNE-YCKKVMRGRGQTK-----LKANTYIH
4. maakgscvfwfllasawivmks--daADTFESFKELHVDYPKTEAPNDNE-YCKKVMGGRGQTK-----LKANTYIH
5. maakgscvfwfllasawivmks--daADTFESFKELHVDYPKTEAPNDNE-YCKKVMGGRGQTK-----LKANTYIH
6. maakgpcllwlfllftcmtvrt--eaDHDFEYFQKHHVDFPKPGT-WNDDVYCRMILYKRCIGFG----LEWTYIIH
7. mvmglgvlllvfvlglgltpptlaqDNSRYTHFLTQHYDAKPQGR---DDRYCESIMRRRGLTSP---CKDINTFIH
8.                           *QDWSSFQNKHIDYPETSAS-NPNAYCDLMMQRRNLNPT--KCKTRNTFVH
9.                          *ETRYEKFLRQHVDHPRTLGL-MGHRYCAVMLARRQVTAPGRPCKPSNTFVH

      80        90        100       110       120       130       140       150
1. APDSELLAACNRKKYKL---NHEYGRTSRLPTTLCTYGDRV------FLGSSLPGTIKVLCVNGKPVAFRGFNA
2. APDSELLAACNRKKYKL---NHEYGRTSRLPTTLCTYGDRV------FLGS>
3. APDSELLAACNRKKYKL---NHEYGRTSRLPTTLCTHDDGR------FFGSSLPGTIKVLCVNGKPVAFRGFTA
4. APDSELLAACNRKKYKL---NHEYGRTSRLPTTLYGCKGILCVLVPFGIRLHSDEE
5. APDSELLAACNRKKYKL---NHEYGRTSRLPTTLYGCKGILCVLVPFGIRLHSDEE
6. ATEEQLKAICTVDGKRL---TGAYRNLTPLPLASCYDTNNDY----AFHGESLTGIFEVQCKDGLPVTFLKFIK
7. GNKRSIKAICENKNGNPHRENLRISKSSFQVTT-CKLHGGSPWPPCQYRATAGFRNVVVACENGLPVHLDQSIFRRP
8. ASPSEIQQVCGS-GGTHYEDNLYDSNESFDLTD-CKNVGGTAPSSCKYNGTPGTKRIRIACENNQPVHFELVLS
9. APAEDLVATCTR---PADATGFHSTSTPMDITA-CRLRGGDTRPPCNYRARQLHHHVRVSCLDGLPVHLAGTHAS
```

FIGURE 24.12: Sequence alignment of ribonuclease precursors (Fry et al. 2010b). Venom-gland forms: *Gerrhonotus infernalis* (1) GU441512, (2) GU441513, (3) GU441514, (4) GU441516, and (5) GU441517), and (6) *Celestus warreni* (GU441518). The related nontoxin body forms from (7) *Homo sapiens* (P03950), (8) *Iguana iguana* (P80287), and (9) *Gallus gallus* (P81476). > designates incomplete C-terminal of precursor. * designates N-terminal posttranslationally processed protein sequence. Signal peptides are shown in lower-case.

24.20.5 CONVERGENCE WITH OTHER VENOMS

None is documented.

24.21 VASCULAR ENDOTHELIAL GROWTH FACTOR

24.21.1 ENDOPHYSIOLOGICAL PLESIOTYPE

Expressed in various tissues ranging from the brain to the heart, vascular endothelial growth factor (VEGF) promotes angiogenesis, vasculogenesis, endothelial cell growth, and endothelial cell proliferation; it promotes cell migration and induces permeabilization of blood vessels.

24.21.2 PLESIOTYPIC STRUCTURAL AND FUNCTIONAL TOXIN FORMS

VEGF causes an increase in the permeability of the vascular bed, resulting in hypotension and shock (see Yamazaki et al. 2003; Yamazaki et al. 2009).

24.21.3 APOTYPIC STRUCTURAL AND FUNCTIONAL TOXIN FORMS

None is documented.

24.21.4 THERAPEUTIC POTENTIAL OF THE TOXIN TYPE

None is documented.

24.21.5 CONVERGENCE WITH OTHER VENOMS

None is documented.

24.22 VEFICOLIN

24.22.1 ENDOPHYSIOLOGICAL PLESIOTYPE

These oligomeric proteins are found in peripheral blood leukocytes. Detection in the lung, spleen, and thymus may be a result of the presence of tissue macrophages or trapped blood in these tissues. They are not detected in lymphocytes and granulocytes. Characteristic domains include an N-terminal region collagen-like domain and a fibrinogen-like domain located in the C-terminal region (Kakinuma et al. 2003). Complex interchain disulfide formations produce dodecamers from sets of trimers. In terms of bioactivity, these proteins are involved in serum-exerting lectin activity via the binding of N-acetylglucosamine.

24.22.2 PLESIOTYPIC STRUCTURAL AND FUNCTIONAL TOXIN FORMS

This toxin class has only recently been discovered (Fry et al. 2010b; OmPraba et al. 2010; Fry et al. 2012b) (see figure 24.13A). Consequently, little is known about its molecular evolution or functional variation. The full-length sequences from *C. rynchops* are longer than the endophysiological mammalian sequences and contain an additional cysteine. They have either seven or eight cysteines, depending on whether the first cysteine is single or part of a doublet (see figure 24.13B). The seven-cysteine form could potentially have an exposed cysteine, which could facilitate dimerization. While they have not yet been tested, it has been speculated that veficolins may interfere with platelet aggregation and/or blood coagulation (OmPraba et al. 2010). However, bioactivity testing of venom forms remains to be undertaken.

24.22.3 APOTYPIC STRUCTURAL AND FUNCTIONAL TOXIN FORMS

The form from *V. komodoensis* lacks the early cysteine(s) of the snake venom forms and also has an internal deletion (but one that does not result in loss of any cysteines) (Fry et al. 2010b) (see figure 24.13B).

24.22.4 THERAPEUTIC POTENTIAL OF THE TOXIN TYPE

None is documented.

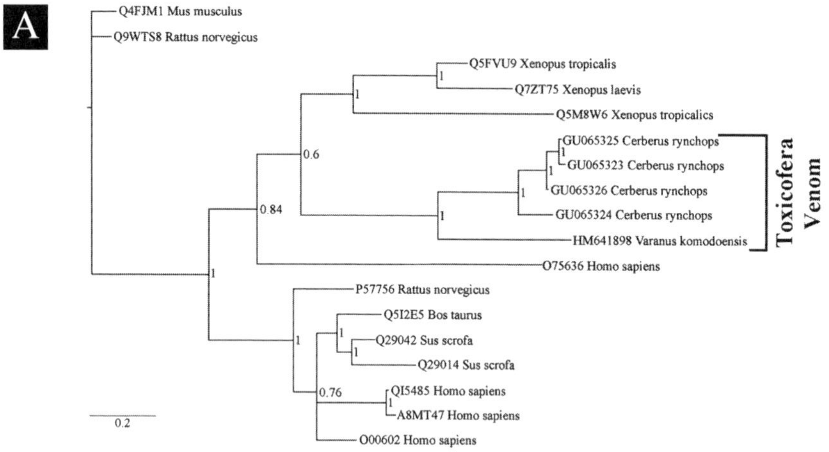

B

```
              10        20        30        40        50        60        70        80        90
1.  mrpwagfhlif--lvalsverevlg------QDTEANCCAQLKGLTQCGGDKIILQGQAGIPGIPGVPGTNGSPGLKGDLGPQGPPGER
2.  mtawldfplalsplvvvsmkg--gSFGQGSEANGSP----QLTGLSECGADRIFLQGQAGIPGIPGVPGTNGLPGAKGDLGPQGPPGER
3.  mkpwaafhlif--lvasslegEVSNYGTRGAQDTEPTCRTE----HQCTRDKIILQSQPGIPGIPGVPGTNGSEGLKGDPGPQGPPGIR
4.  mkswaafhlif--lvassvegEVSNYGTGGAQYTEPTYRTE----HQCISDKIILQGQPGIPGIPGVPGINGSEGLNGDPGPQGLSGEI

             100       110       120       130       140       150       160       170       180
1.  GSAGMTGKAGPKGDKGD------------KGDACSLDNVANCQQKGTEARNCKDLLERGETLSGWHTIYPTKGNPMMVFCDMETDGGGW
2.  GSTGIPGKAGPKGDKGDQG------------EACSL---ASCQQQEAGAKDCKELLDRGETLTGWYMIYPTTGRGMRAYCDMETDGGGW
3.  GPDGIRGEAGPKGDKGDQGDKGDKGDKGDKGEDCNLDD---CLPT--EVRNCQDLLERGETLNGWYKIYPTEEVSMLVFCDMSTDGGGW
4.  GFDGIPGVAGPKGDKGDQGDKGGQGDKGDKGDACILDD---CLPTDVEVRNCQDLLERGEILNGWYTIYPTKENPMVVFCDMETDGGGW

             190       200       210       220       230       240       250       260       270
1.  LVFQRRRDGSVDFYRDWESYKKGFGSFESEFWLGNDKIHLLTSNETQQLWIDLEDFNSSRTFAKYSSFRIANENQKYRLMVGSYLDGNM
2.  LVFQRRLDGSVDFYRDWEAYKKGFGRQVSEFWLGNDKIHLLTSSGIQQLRIDVEDFNNSKTFAK<
3.  LVFQRRQDGSVNFYREWESYKKGFGRQGSEFWLGNDKIHLLTNSGTQQLRIDLADFENIHTFATYSSFSIGNETEKYKLTLGSHLDGNM
4.  LVFQRRQDGSVDFYLDWQSYKKGFGKQESEFWLGNDKIHLLTSSGTQQLRIDLEDFEGSPTFATYSSFSIGDEKKK<

             280       290       300       310       320       330       340       350       360
1.  GDSFSGHNNQAFSTKDRDNDIYDGNCAVSFKGAWWYSKCHSSNLNGLYHKGEHTSYADGINWSAGKGQYYSYKYADMKIRPQ
2.  <
3.  GDSLTLQNGRSFSTKDRDNDVSHIHCAVNFKGAWWYRKCHESNLNGLYHKGKHATYANGINWLTGKGYHYSYKYADMKIRPQKSETTVS
4.  <
```

FIGURE 24.13: Veficolin (A) phylogenetics and alignment of (B) the sequences (1) M9T1P5 (*Cylindrophis ruffus*), (2) E2IYB3 (*Varanus komodoensis*), (3) D8VNT0 (*Cerberus rynchops*), and (4) I0BWT0 (*Enhydris polylepis*). Signal peptides are shown in lower-case.

24.22.5 CONVERGENCE WITH OTHER VENOMS

None is documented.

24.23 VESPRYN

24.23.1 ENDOPHYSIOLOGICAL PLESIOTYPE

It is secreted by hemopoietic lineages, and its activity remains uncharacterized.

24.23.2 PLESIOTYPIC STRUCTURAL AND FUNCTIONAL TOXIN FORMS

Vespryn induces hypolocomotion and hyperalgesia. It is unknown which of these, if either, is the plesiotypic activity and which is apotypic.

24.23.3 APOTYPIC STRUCTURAL AND FUNCTIONAL TOXIN FORMS

None is documented.

24.23.4 THERAPEUTIC POTENTIAL OF THE TOXIN TYPE

None is documented.

24.23.5 CONVERGENCE WITH OTHER VENOMS

In stonefish venoms, this toxin type produces lethal hemolysis through pore formation in cell membranes while inducing potent endothelium-dependent hypotension and also irreversibly interfering with neuromuscular function (Low et al. 1994; Chen et al. 1997; Sung, Low, and Khoo 2002). In addition, stonustoxins display edema-inducing activity, increase of vascular permeability, induction of endothelium-dependent vasorelaxation, platelet aggregation, and myotoxicity.

24.24 VITELLINE MEMBRANE OUTER LAYER PROTEIN

24.24.1 ENDOPHYSIOLOGICAL PLESIOTYPE

This is secreted by the oviduct and is a component of the outer membrane of the vitelline layer of the egg. Its function remains unknown.

24.24.2 PLESIOTYPIC STRUCTURAL AND FUNCTIONAL TOXIN FORMS

This is known only from transcriptome sequencing (Fry et al. 2012b).

24.24.3 APOTYPIC STRUCTURAL AND FUNCTIONAL TOXIN FORMS

None is documented.

24.24.4 THERAPEUTIC POTENTIAL OF THE TOXIN TYPE

None is documented.

24.24.5 CONVERGENCE WITH OTHER VENOMS

None is documented.

24.25 WAGLERIN

24.25.1 ENDOPHYSIOLOGICAL PLESIOTYPE

This is unknown, as waglerins display no homology to any known peptide.

24.25.2 PLESIOTYPIC STRUCTURAL AND FUNCTIONAL TOXIN FORMS

Selectively blocks the epsilon subunit of muscle nicotinic acetylcholine receptors, resulting in tachypnea, ocular proctosis, rapid collapse, and spasms in mice; the primary cause of death is respiratory failure (Schmidt, Weinstein, and Smith 1992; Lin, Smith, and Lee 1995; McArdle et al. 1999).

24.25.3 APOTYPIC STRUCTURAL AND FUNCTIONAL TOXIN FORMS

None is documented.

24.25.4 THERAPEUTIC POTENTIAL OF THE TOXIN TYPE

Waglerin is used in an antiwrinkle face cream.

24.25.5 CONVERGENCE WITH OTHER VENOMS

None is documented. However, nicotinic acetylcholine receptors have been convergently targeted by other snake venom peptide types (see color plate 2B), such as 3FTx (see chapter 8) and the unique neurotoxic propeptide forms of SVMP from *Psammophis* venoms (see chapter 23.4.4).

24.26 WAPRIN

24.26.1 ENDOPHYSIOLOGICAL PLESIOTYPE

Secretory leukocyte protease inhibitor (SLPI) is an acid-stable proteinase inhibitor with strong affinities for trypsin, chymotrypsin, elastase, and cathepsin G. It may prevent elastase-mediated damage to oral and possibly other mucosal tissues.

24.26.2 PLESIOTYPIC STRUCTURAL AND FUNCTIONAL TOXIN FORMS

The plesiotypic condition is a bidomain form, with the two peptides encoded liberated by posttranslational processing (see figure 24.14). Only one form has been tested, and it displayed an antimicrobial activity (Nair et al. 2007).

24.26.3 APOTYPIC STRUCTURAL AND FUNCTIONAL TOXIN FORMS

The apotypic precursor state is a monodomain precursor with a divergent signal peptide (see figure 24.26.1). The extensive sequence variation displayed by this peptide type suggests that apotypic activities are likely to be documented on further testing. Derivations of precursors include those that encode a waprin-kunitz fusion (Jackson et al. 2013). As the TV-kunitz-waprin dual domain has only been recovered in transcriptome studies, it remains to be established whether the TV-kunitz-waprin fused toxin is posttranslationally processed into two separate toxins or if it exerts its toxic function as a protein complex.

```
                10        20        30        40        50        60        70        80        90       100
1.  mtrekslallitlaaalaaae--------SPPGRCHSPKTVGPCRASFHRWRYNATSQMCQEFIFGGCKGNANNFVSKQDCFQTCIRGGAAEATVVPSGP
2.  mtcekslallitlaaalaaae--------SPPGRCHSPKTVGPCRASFHRWWYNATSQMCQEFIFGGCKGNANNFVSKQDCFQTCIRGGAAEATVVSSGP
3.  mssggvllllglltlwegltpvss----KDRPHFCHLPADTGPCKANFLAFYYHPVHRKCLEFTYGGCEGNANNFKTIDECKRTCAV*
4.  mssgglllllglltlwegltpvss----KDRPHFCHLPHDTGPCKGMFPAFYYHPGHRTCLQFIYGGCRGNANNFKTIDECKRICAA*
5.  mssgglllllglltlwegltpvss----KDRPKFCELPPDKGRCYGWFLAFYYNPVQHQCLQFIYGGCPGNANYFKTIEECKSTCAA*
6.  mssgglllllglltlwevltpvss----KDRPHFCHLPYDTGPCKARFIAFYYNSVQRKCLEFVYGGCEGNANNFKTIDACKRTCAA*
7.  mssgglllllglltlwevltpvss----KDRPEFCELPDDSGPCRGIFHAFYYNPDQRQCLEFIYGGCYGNANNFKTIDECKRICAA*
8.  mssgglllllglltlwevltpvss----KDRPNFCELPAETGLCNAHNTRFYYNPRQHQCIKFLYGGCGGNANNFKTIEECKSTCIA*
9.  mssgglllllglltlwegltlvss----TDRPDFCELPDDSGPCKGSFEAFYYNADQHKCLEFIYGGCDGNANNFKTIEECEHTCAA*
10. mssgglllllgfltlwaeltpvsg----QDPPRFCSLPAETGECRGRIPRFYYNSASKQCEQFFYSGCGGNANNFETKDQCHYTCVEKRGVCP-QGPQQQ
11. mssgglllllgfltlwaeltpvsg----QDPPRFCSLPAETGECRGRIPRFYYNSASKQCEQFFYSGCGGNANNFETKDQCHYTCVEKRGVCP-QGPQQQ
12. mvsgglllllglltlwaeltpvsg----QDRPRFCHLPADPGPCRAKITRFYYNSGSKQCEKFIYGGCQGNANNFETKDRCHYTCVEKPGVCP-QAPQQQ
13. mssgglllllglltlwxemxxisg----QDRPKYCDLPANPGPCRARITRFYYNSDSKQCEEFIYGGCHGNANNFETKDKCHYTCVEKPGVCPKPGPDVV
14. mtlrrgscplllfslvglltcaQEPDAAGQNTTA-VAEKAGTCPQAELEMPDRNCTEACQSDAGCEENKKCCRTGCGTSCQIP-DEKPGSCPNV-DTPI
15. mqar-flllftvillgmmgp------------------------------------------------------------VISTGIEKAGSCPDV-NMPI
16. mqarvfllllgvillgmmgpmvsa------------------------------------------------------------QDGKAGSCPDV-NQPI
17. mkatllllllfavil---pgtis------------------------------------------------------------AEQEKPGSCPNV-DMPI
18. mkarllllsv--vil---vgmvsa------------------------------------------------------------ENEKAGSCPDV-NQPI
19. mssgglllllglltlwaeltpvss----------------------------------------------------------LDRPKKPGLCP-PRPQK-
20. mssgglllllglltlwaeltpvsg----------------------------------------------------------QDRPKKPGLCP-PRPQK-

            110       120       130       140       150       160       170       180       1903      200
1.  ATEVATPRAGHLPEAYENRPGFREFCAAPRVVGPCRASFLRWYFDLESRMCKMFIYGGCRGNKNNYLFEEHCWSQCTGDGEITEEPGDAGAQPPLPSEPF
2.  ATEVATPRAGHLPEAYENRPGFREFCAAPRVVGPCRASFLRWYFDLESRMCKMFVYGGCRGNKNNYLFEEHCWSQCTGDGEITEEPGDAGAQPPLPSEPF
3.  *
4.  *
5.  *
6.  *
7.  *
8.  *
9.  *
10. GQQGKN-----CENDWKCPGQQKCCRYGG-KTECKDALFEKFASTTRTPTSVGSTPVGSSVPT*
11. GQQGKN-----CENDWKCPGQQKCCRYGG-KTECKDALFEKFASTTRTPTSVGSTPVGSSVPT*
12. QQQKLPC-LERCQHDWNCPGEQKCCRY-----GCLSE-CKDIVYSIHASATKRPSRIGTTVRT*
13. T----ICLVR-CESDWKCPGKEKCCSYAC-LRDCMEPV*
14. PPLGL-CKTT-CSKDSDCSETKKCCKNGCGFMTCTTARP*
15. PPLGI-CRTT-CQTDSNCPDIKKCCKNGCGFMTCSTPKA*
16. PPLGV-CKTT-CATDSNCPDIQKCCKNGCGHMSCTRPST*
17. PPLGL-CKTT-CSKDSDCSETKKCCKNGCGFMTCTTARP*
18. PPLGL-CRNM-CESDSGCPNNEKCCKNGCGFMTCSRPR*
19. PP----CVRE-CKNDWRCPGEQKCCRYGC-IYECRDPIFVK*
20. PP----CVKE-CKNDWSCPGQQKCCSYGC-IDECRDPIFVN*

          210       220       230       240       250       260
1. SFSTRAVVLAVLPAILVTILLGSMGVFFVKICRKNPELSVGTVWSTLDDKEYLMSNAYTL*
2. SFSTRAVVLAVLLAILVTILLGSMGVFFVKICRKNPELSVGTSVEHAG*
3. *
4. *
5. *
6. *
7. *
8. *
9. *
10. *
11. *
12. *
13. *
14. *
15. *
16. *
17. *
18. *
19. *
20. *
```

FIGURE 24.14: Domain combinations of waprin and kunitz peptides: Waprin-waprin dual domain (1) B2BS84 (*Austrelaps labialis*), (2) GAHG01000009 (*Hoplocephalus bungaroides*). Waprin monodomain (3) GAGZ01000019 (*Acanthophis wellsi*), (4) GAGZ01000017 (*Acanthophis wellsi*), (5) GAHB01000016 (*Cacophis squamulosus*), (6) GAHD01000011 (*Echiopsis curta*), (7) GAHG01000008 (*Hoplocephalus bungaroides*), (8) GAHH01000051 (*Pseudonaja modesta*), (9) GAHI01000010 (*Suta fasciata*). Waprin-kunitz dual domain (10) D3U2B9 (*Sistrurus catenatus edwardsii*), (11) D3U0D3 (*Sistrurus catenatus tergeminus*), (12) GAHB01000034 (*Cacophis squamulosus*), (13) GAHI01000009 (*Suta fasciata*). Kunitz-kunitz dual domain (14) A7X4K1 (). Kunitz monodomain (15) GAHC01000021 (*Denisonia devisi*), (16) A7X4J4 (*Rhabodophis tigrinus*), (17) A7X4K7 (*Philodryas olfersii*), (18) A7X4I7 (*Thrasops jacksonii*), (19) B5G6H4 (*Notechis scutatus*), (20) B5G6G8 (*Oxyuranus scutellatus*). Signal peptides are shown in lower-case. Posttranslationally cleaved bioactive domains are highlighted in gray.

24.26.4 THERAPEUTIC POTENTIAL OF THE TOXIN TYPE

None is documented.

24.26.5 CONVERGENCE WITH OTHER VENOMS

None is documented.

CHAPTER 25

POISONOUS SNAKES AND KOMODO DRAGON WEAPONIZED BACTERIA: WHICH IS MYTH AND WHICH IS REALITY?

B. G. FRY, H. SCHEIB, K. MESSENGER, K. L. TYRRELL, S. HOCKNULL, S. WROE, K. SUNAGAR, E. J. C. GOLDSTEIN, D. M. CITRON, AND T. N. W. JACKSON

25.1 POISONOUS SNAKES

The potently venomous genus *Rhabdophis* is one of three genera of Asian snakes in the Natricidae family that can also be legitimately called "poisonous." Most members of this genus possess a unique gland positioned at the base of the neck called a nuchal gland. In *Rhabdophis*, the nuchal gland is paired on either side of the spine. The gland is very superficial, lying between the muscle and the epidermis (see color plate 29). Trauma to the neck causes the gland to rupture, spilling the toxic contents of the organ. In some members of the genus, the gland also extends dorsolaterally along the snake's body. In these snakes, the nuchal poison is used for defense, whereas the oral venom is used for predation.

The nuchal glands are unique among vertebrates in several regards. Most animals that store defensive chemicals usually store them in preexisting organs such as the skin, muscles, or viscera (Mori et al. 2012). However, the nuchal glands appear to have evolved *de novo* as a protective structure for storing sequestered toxins. Ontogenetically, the glands are of mesodermal origin, whereas the primary secretory tissue of most (if not all) other skin glands of terrestrial vertebrates is ectodermal (Mori et al. 2012). Another fascinating aspect of the structure of the glands is that they lack lumina, ducts, and secretory epithelium (Hutchinson et al. 2007). Instead the defensive compounds are delivered via the circumferential blood vessels and central capillary beds of the circulatory system (Hutchinson et al. 2007; Mori et al. 2012).

Unlike the production of its venom, the snake cannot generate the poisonous substance found in the nuchal glands. Instead, it must sequester the poisons from toxic prey. Females also pass poisons they have stored on to their unborn offspring through the eggshell. To date, the toxic substances are only known to come from toads of the family Bufonidae. These defensive steroids are called bufadienolides. Bufadienolides are steroids originally secreted from the skin of both American and Japanese toads (Hutchinson et al. 2007), which led to the prefix "bufo-." They possess a six-member lactone ring attached to C17 of the steroid scaffold (leading to the ending "-olide"). In bufodienolides, the lactone ring contains two double bonds ("-dien-"). Bufadienolides lack the suberylarginine side chain of bufotoxins, which are C3-acylated bufadienolides (see color plate 29). Slight variations to this pattern are also toxic, such as bufanolides (fully saturated lactone ring) and bufenolides (one double bond in lactone ring).

The chemical makeup of the glands is completely dependent on the toad species the snakes are preying upon. In this regard, only *Rhabdophis tigrinus* from Japan have been examined. Specimens were taken from Japan and brought to the United States, where they were fed the three North American toad species *Anaxyrus fowleri, A. terrestris*, and *A. quercicus*. In their native habitat, these specimens would have fed on *Bufo japonicus*. Seventeen bufadienolides from the nuchal glands of Japanese *R. tigrinus* have been found (based on experiments using North American toads); a few examples include gamabufagin, hellebrigenin, telocinobufagin, and 11α-hydroxytelocinobufagin, in addition to several new natural products (Hutchinson et al. 2007). Some of the bufadienolides, such as gamabufotalin, appear to be sequestered unaltered or are possibly derived from side-chain hydrolysis of bufotoxins. Most of the remaining bufadienolides undergo additional hydroxylation (Hutchinson et al. 2007). In dendrobatid frogs, which also sequester toxins through diet, in one species endogenous hydroxylation has been shown to dramatically increase the toxicity of a sequestered alkaloid (Daly et al. 2003). Color plate 29 reveals that arenobufagin from the Argentinian toad *Bufo arenarum* and bufalin from *B. vulgaris* show only minor differences to the bufadienolides found in *Rhabdophis* snakes. Bufalin is a traditional Chinese medicine called Chan Su. It is a potent biocide that has been shown to induce apoptosis of human tumor cells (Yamada et al. 1998; Takai et al. 2012).

The possession of these defensive poison glands results in some very interesting behaviors. Upon confrontation, and when escape seems unlikely, most members of the genus will perform a set number of behaviors that are somewhat unique to this group of snakes. The most striking behavior is called neck arching, in which the snake lowers its chin and raises its neck to the would-be predator. If the snake is touched during this performance, it will often swing its arched neck into the stimulus touching it. This is called "neck-butting." Although these snakes possess extremely potent venom, most individuals do not bite but instead rely on their nuchal-gland behaviors. Most members also have strikingly colorful necks, which is emphasized during the neck-arching behavior.

In most cases, the poison is not easily excreted; in fact, because the glands themselves lack ducts, in order for the gland to rupture, a significant amount of pressure has to be placed on the neck. One apparent exception is the red-necked keelback (*Rhabdophis subminiatus*) of southern Asia. Members from Hong Kong are known to readily and easily exude their poisonous secretions from their necks without even being touched (Messenger et al. 2012). Some specimens have been known to seemingly "spray" their nuchal fluid into the air when approached by a predator. The mechanisms behind this ability have not yet been studied.

Of the 21 species of *Rhabdophis*, three to four are suspected to have secondarily lost the nuchal glands, and for another six species, it is still unknown whether the nuchal gland is present. The other two genera known to possess similar glands (though not identical) are *Balanophis*, a monotypic genus from Sri Lanka, and *Macropisthodon*. Within *Macropisthodon*, one member, *Macropisthodon rudis*, has secondarily lost its nuchal gland, although retaining potent venom and very large fangs (see color plate 5G).

Conversely, in areas without bufonid prey, therefore rendering the animals nonpoisonous (but still venomous), individuals do not commit to such nuchal-gland-associated behaviors and instead just rely on fleeing as their only means of defense (Mori and Burghardt 2008). Similarly in areas that lack toads, the nuchal-gland fluid is filled primarily with water and lipids, with trace amounts of nonbufadienolide steroids, mainly cholesterol and carbohydrates (Hutchinson et al. 2007). If the females are deprived of toxic prey prior to becoming gravid, the offspring will hatch out nonpoisonous. However, experiments have shown that even after the female has laid such non-toxic eggs, the embryo can still absorb bufadienolides through the eggshell via topical application (Hutchinson et al. 2012), thus reinforcing the readiness by which these remarkable snakes bioaccumulate and sequester toad toxins.

25.2 BACTERIA AS A WEAPON IN *VARANUS KOMODOENSIS* (KOMODO DRAGON)?

The previous lack of recognition of the venom system of anguimorph lizards has hindered the complete understanding of their predatory ecology. Thus the role of the toxin-secreting oral glands in feeding and/or defense requires further investigation. The only anguimorph lizard species besides helodermatid lizards that has been directly investigated in this regard is *Varanus komodoensis*. This giant evolved in Australia during the Neogene, with fossils dating from some 3 million to 4 million years ago. Three extinct giant species of *Varanus* are known from the Neogene: *V. sivalensis* from the Pliocene Siwalik Hills of India; *V. priscus* from the late Pleistocene of eastern Australia; and a yet-to-be-named species from the middle Pleistocene of central Australia and the island of Timor. *V. sivalensis* was similar in body size to *V. komodoensis* but left no post-Pliocene fossil record. Both of the remaining two species evolved a much larger body size than *V. komodoensis* (Hocknull et al. 2009). Competition with mammalian carnivores on the subcontinent may have led to this species' short-lived record. Fossils of all three species were found with contemporaneous vertebrate fauna that includes gigantic forms of birds, snakes, crocodilians, and marsupials. As the largest terrestrial carnivores, they probably preyed on megafauna, but direct evidence of their feeding behavior has not been clearly ascertained.

It is likely that these giant species of *Varanus* occupied predatory niches in Australia that were filled elsewhere by placental mammals (*Eutheria*). The absence of eutherian predators in Australia may account for the success of small-bodied varanid species there (typically members of the subgenus *Odatria*), and predation rather than competitive exclusion by placental mammal carnivores (Order Carnivora) has been suggested as the likely reason for the absence of small varanids elsewhere in the world (Sweet and Pianka 2007). It seems plausible, therefore, that a lack of competition from eutherian predators may also have facilitated the evolution of a clade of giant, predatory varanids, unique to Australopapua and Wallacea. *Varanus komodoensis*, famous today as the largest living lizard, is one of the largest members of this clade. Like other members of this varanid clade characterized by large and serrated teeth, it has an Australian origin. Other living members of this clade include the moderately large *V. varius* and the exceptionally long-tailed *V. salvadorii*. The largest of this serrated teeth clade included the two extinct species, *Varanus* (formerly *Megalania*) *priscus*, the world's largest terrestrial lizard, and a nearly as large extinct species that radiated out to Timor. It has been proposed that *V. komodoensis* and the Timor species dispersed westward to the Lesser Sunda Islands of Indonesia and that *V. komodoensis* further dispersed as far as the island of Java (Hocknull et al. 2009). The lack of extensive fossil records indicates that *V. komodoensis* failed to establish on Java, and, like *V. sivalensis* from India, it was probably outcompeted by mammalian carnivores.

On the island of Flores, three major faunal turnovers, including the extinction of its megafauna 12,000 years ago, have occurred during coexistence with *V. komodoensis*. After the extinction event, only relatively small prey items were available to *V. komodoensis*, until the introduction of pigs by man 5,000 years later. Thus during the intervening period, *V. komodoensis* was restricted to feeding upon smaller prey items. The fossil record illustrates their evolutionary adaptive flexibility, a feature common for many other species of varanid.

The currently available prey options for adult *V. komodoensis* include introduced megafauna (deer and water buffalo) that were brought in by Dutch settlers over the last few centuries. *V. komodoensis* therefore is living in a novel manmade ecosystem. Encounters between *V. komodoensis* and these larger potential prey animals are recent, having been facilitated by the anthropogenic dispersal of the large mammals, and have taken place over far too short a period of time to have influenced the evolution of *V. komodoensis*. Having said this, the *V. komodoensis* lineage is millions of years old and has coexisted with megafauna for most of its prehistory, making it potentially preadapted to preying on megaherbivores.

Compared with other large carnivores, such as crocodilians, *V. komodoensis* have lightweight skulls and a relatively weak bite force in relation to their mass and the size of the large mammals on which they currently feed (Moreno et al. 2008; Fry et al. 2009b; D'Amore et al. 2011). Instead of a powerful bite, *V. komodoensis* employs many recurved and serrated teeth (see color plate 5B) as its primary weapon, using a bite-and-slice strategy where the teeth act in concert to produce a near-singular cutting blade and motion. Mechanical damage (see color plate 30) alone is the probable cause of death in many cases, as it results in rapid massive blood loss (such as caused by slicing the femoral or carotid arteries). Moreover, *V. komodoensis* venom contains anticoagulant toxins that increase blood loss and other toxins that induce hypotension and shock (Fry et al. 2009b).

The body size of *V. komodoensis* has remained stable over the past 3–4 million years and in particular during the last 900,000 years on Flores. Although pigs and deer are of manageable size (40–50 kg) and may be killed rapidly as described above, water buffalo are unlike any past megafaunal species *V. komodoensis* may have evolved to prey on. Defensive strategies such as horns and hooves are not present in the extinct Australian megaherbivores. This fact is reflected in the observed success rate of predation attempts by *V. komodoensis*. Attacks on pigs and deer are typically successful (Bull, Jessop, and Whiteley 2010; Fry, personal observations), with the majority of such predation attempts resulting in rapid death as previously described. Prey animals that escape the initial attack with less severe injuries succumb in a few hours due to persistent bleeding, likely resulting from the anticoagulant effects of the venom. This extra fitness advantage results in significant selection pressure for the maintenance of venom, which, because of the physiological cost of its production, would otherwise be lost.

Although *V. komodoensis* are unlikely to be capable of bringing down an adult water buffalo, they often inflict deep wounds to the lower hindquarters of these animals (see color plate 30). The evolutionary reflex of the water buffalo is to seek refuge in swamps in their native habitat, which have slow-moving, renewable water. However, on the *V. komodoensis* islands where buffalos have been introduced, stagnant water holes replenished only by rain are typically the only water resource available (see color plate 30). The standing water of these wallows is contaminated with bacteria accumulated from feces of the buffalo and other fauna, rotting vegetation, and soil. Many of these bacteria are virulent pathogens, such as *Clostridium perfringens*, which can cause gangrene, sepsis, or septic shock, or other opportunistic pathogens such as fecal flora, can eventually cause life-threatening sepsis in the wounded water buffalo. Ecologist Walter Auffenberg noticed that wounded water buffalo were sometimes overcome with infection and thus susceptible to predation (Auffenberg 1981). He therefore postulated that "induction of wound sepsis and bacteremia through the bite of the *V. komodoensis* may be a mechanism for prey debilitation and mortality". Central to his "bacteria as venom" theory was that fatal infections were caused by virulent, pathogenic oral flora. However, this theory suffered from a conspicuous absence of evidence. Expansions proposed later, such as lizard-lizard swapping of "weaponized bacteria" during mutual feeding of prey (Bull, Jessop, and Whiteley 2010) are evolutionarily complex and thus highly unlikely. Extraordinary claims require extraordinary evidence. Bacteria used as a weapon would be an unprecedented form of predation strategy, for which none of the above studies provided such a standard of evidence.

A later study of the aerobic oral flora in 39 wild and captive *V. komodoensis* isolated a variety of fecal, skin, and environmental (vegetation, soil, and water) bacteria (Montgomery et al. 2002). In this study, 54 of the 57 isolated species were considered to be "potentially pathogenic." *Pasteurella multocida* was reported in only 5% (2 out of 39) of the examined *V. komodoensis* and was interpreted by the authors as supporting the bacteria-as-venom theory. No bacterium was found common to all *V. komodoensis*. Crucially, potential transient sources were not cultured, such as drinking water, stagnant wallows or the normal skin and fecal flora of prey. Greater clinical scrutiny reveals that although some of the species can be opportunistic pathogens, they are actually of low virulence and rarely reported in the literature to cause sepsis, thus they would not be considered to have a primary pathogenic role causing rapid, fatal infection of prey. In particular, although infection by *Pasteurella multocida* in bite wounds from mammals such as dogs, cats, lions, and Tasmanian devils is common and can be pathogenic, it is associated with sepsis infrequently, especially in healthy individuals (Weber et al. 1984; Khan et al 2012). Moreover, *P. multocida* has not been reported in oral flora studies of many other reptiles including various snakes, alligators, and turtles (Jackson and Jackson 1971; Goldstein et al. 1981; Flandry et al. 1989; Theakston et al. 1990; Blaylock 2001; Hejnar et al. 2007; Shek et al. 2009; Draper et al. 1981).

A more recent study screened the saliva and gingival flora of 10 adult and 6 hatching captive *V. komodoensis* from three zoos in the United States for aerobic and anaerobic bacteria (Goldstein et al. 2013). Saliva and gingival samples were taken at least one day after feeding. Adult *V. komodoensis* diets varied by zoo and consisted largely of whole prey, such as poultry, fish, and small mammals. Bacterial strains were identified biochemically as in the previous study and by molecular methods. The study identified 128 strains (87 aerobic and 41 anaerobic), with a median of 10 isolates per culture, including 18 different Gram-negative and 21

Gram-positive aerobic species and 6 different Gram-negative and 15 Gram-positive anaerobic species.

Although the *V. komodoensis* food and environment were not cultured, the bacteria isolated correlated with normal skin and fecal flora of prey mammals and the environment. There was also correlation between different *V. komodoensis* flora by age, sample source (saliva or gingiva), and diet between zoos. All adult *V. komodoensis* had oral flora that largely reflected their respective zoo food sources. *V. komodoensis* that were fed fish, for example, had fish-related bacteria not found in other *V. komodoensis*. Hatchlings showed only aerobic bacteria that were less diverse than in adult *V. komodoensis*. Gingival and saliva samples contained higher proportions of bacteria consistent with normal skin flora and normal fecal flora of prey, respectively. Marked interzoo variation was observed, with the majority of the *V. komodoensis* tested at each zoo having unique strains not isolated at the other zoos. Interestingly, the bacteria recovered from the captive *V. komodoensis* in this study more closely resembled those of the wild *V. komodoensis* in the Montgomery study (Montgomery et al. 2002), which could be a reflection of diet; the Montgomery study did not state whether the captive *V. komodoensis* were fed cut meats or whole prey as were those in this later study.

Rather than *V. komodoensis* oral flora consisting of virulent bacteria that are the likely cause of sepsis and rapid death, this recent investigation found that captive *V. komodoensis* oral flora is reflective of the skin and gut flora of their recent meals and of environmental flora, a finding similar to that for many other venomous and nonvenomous reptiles and other carnivores. In fact, numerous studies detailing the oral aerobic and/or anaerobic flora of both captive and wild reptiles, including rattlesnakes, garter snakes, cobras, and vipers, in addition to alligators and turtles, have all isolated varieties of organisms that largely overlap with the findings of this more recent study (Jackson and Jackson 1971; Goldstein et al. 1981; Flandry et al. 1989; Theakston et al. 1990; Blaylock 2001; Hejnar et al. 2007; Shek et al. 2009).

The potentially fatal infections contracted by water buffalo postbite are thus the result of a novel encounter between predator and potential prey that has only taken place for a small fraction of the evolutionary history of either species. In summary, the above-mentioned findings demonstrate that the bacterial flora associated with *V. komodoensis* saliva does not stand out from its environment or from flora found in the mouths of other predators and instead is associated with the prey record of an individual *V. komodoensis*.

The absence of a specialized pathogenic bacteria challenges the "weaponized bacteria" theory proposed three decades ago. This theory is also evolutionarily implausible, because infections would act too slowly to be under any sort of evolutionary selection pressure. Further, because of their size, water buffalo always escape the initial attack and flee a considerable distance. In all the extensive tracking and filming of *V. komodoensis*, there has never been a single documented case of a *V. komodoensis* biting a water buffalo and following it until it dies; therefore, the original dragon is not going to be the beneficiary. There is no fitness advantage to be gained if the biting animal does not obtain the prey.

GLOSSARY

ACETYLCHOLINE A neurotransmitter in the autonomic nervous system. An organic molecule that transmits nerve signals from one neuron to the next, or from a neuron to a muscle cell by being released in the synapse and binding one of two receptor types

ACTION POTENTIAL A momentary switch in the electrical potential across a cell's membrane that moves along the cell's length (e.g. from dendrites to synapse in a neuron) to transmit a signal or initiate a contraction

AGONIST A chemical that activates a receptor upon binding

ALPHA-NEUROTOXIN A postsynaptic neurotoxin that antagonizes nicotinic acetylcholine receptors in neuromuscular junctions, leading to paralysis

ALTERNATIVE SPLICING A genetic process during gene expression that involves the differential exclusion or inclusion of exons in the production of mRNA, allowing a single gene to produce multiple protein isoforms

ANGUIMORPHA A clade of toxicoferan reptiles, including the extant families Anguidae, Helodermatidae, Lanthanotidae, and Varanidae and several extinct ones (e.g. Mosasauridae)

ANION A negatively charged ion

ANTAGONIST A chemical that blocks a receptor upon binding

ANTICOAGULANT Prevents the coagulation (clotting) of blood

ANTIVENOM Antibodies used in the treatment of envenomation

APOTYPIC Representing a newly evolved or derived form, as opposed to an older ancestral one

AXON A nerve fiber, the typically long, slender part of a neuron that guides the action potential from cell body to the next synapse

BIOINFORMATICS Development and usage of software to computationally investigate and interpret biological data

BRUMATION A period of time for captive reptiles and amphibians where temperatures are lowered during the winter months

CALLOUT When a trained professional is called to remove a venomous snake from a business or residence where they often relocate them to a suitable nearby habitat

CARDIOTOXIN A toxin that damages and/or functionally impairs the heart

CATALYTIC Increasing the rate of a chemical reaction

CATION A positively charged ion

CDNA Abbreviation of complementary DNA, DNA obtained by reverse-transcription of mRNA

CLADE A monophyletic group of organisms

COAGULOPATHIC Affecting the coagulation (clotting) of blood

CODING DNA SEQUENCE (CDS) The portion of a gene that codes for protein

COFACTOR A molecule that is required for a protein's biological activity

COMMON ANCESTOR The most recent individual from which all organisms in a group are directly descended

CONTIGUOUS SEQUENCES (CONTIGS) A set of overlapping DNA segments that together represent a consensus region of DNA

CONVERGENT EVOLUTION Independent evolution of similar structures or features, typically from different starting conditions

C-TERMINAL The side of a peptide chain at which translation is ended, normally characterized by a carboxyl (–COOH) end group

CYSTEINE MOTIF A specific positioning of cysteine residues in the amino-acid sequence of a protein that sustains its folding by formation of disulfide bridges

CYTOTOXIN A molecule showing toxicity towards individual cells, often leading to cell death (e.g. through lysis or necrosis)

EDEMA Swelling caused by accumulation of body fluids in tissue

EFFECTIVE DOSE 50 (MEDIAN EFFECTIVE DOSE) Amount of a substance required to produce a specific effect in half of an animal population that makes up a test sample

ENDOPHYSIOLOGICAL Physiologically active within the body of the producing organism under normal circumstances

ENDOPHYSIOLOGICAL PLESIOTYPE The endophysiological protein form which a specialized exophysiological protein (e.g., a venom toxin) was recruited and subsequently diversified

ENVENOMATION The delivery of venom through a wound, resulting in toxicity

ENZYME A macromolecule (in most cases a protein) that catalyzes a biochemical reaction. Enzymes typically receive a name ending with the suffix '-ase' (e.g. phospholipase)

EVOLUTIONARY RADIATION An pattern in which a single evolutionary lineage rapidly (in a short geological time span) diversifies into multiple distinct lineages, e.g. through the rapid succession of speciation events. Although typically referring at species lineages, radiation may also be used to describe the diversification of a protein family (through the rapid succession of gene duplication events)

EXON Part of a gene that may contribute to its mRNA, as opposed to introns (which are always cleaved out form the pre-mRNA by a spliceosome)

EXON SHUFFLING An evolutionary process that involves the duplication of part of a gene (one or more exons) into another gene, often as a result of unequal crossover or transposon-mediated

EXOPHYSIOLOGICAL Physiologically active outside the body of the producing organism under normal circumstances (e.g., venom and poison toxins)

FIBRINOLYSIS An anticoagulant process during which fibrin is degraded

FOCAL MUTAGENESIS (SITE-DIRECTED MUTAGENESIS) Point mutations

FOSSORIAL Relating to reptiles or amphibians that inhabit a subterranean niche

FUNCTIONAL PLASTICITY Showing the capacity to serve alternative functions depending on the biological context

GENE AMPLIFICATION An evolutionary increase in gene expression, e.g. by mutation of its gene-regulatory elements, or by alteration of its transcription factors

GENE CONVERSION An evolutionary process by which one part of a gene is duplicated and replaces a homologous part (e.g. of another paralogous gene), such that the sequences become identical. As a result, two anciently duplicated genes may share more recently evolved similarity

GENE DUPLICATION An evolutionary process by which a stretch of DNA sequence containing a gene is duplicated in the genome through a mutational mechanism (unequal crossover or transposon-mediated), yielding two gene copies (paralogues)

GENE EXPRESSION The genetic process through which a gene is being activated and used to produce a biologically functional gene product (an RNA molecule or a protein)

GENOME (GENOMICS) The collective of genetic material (DNA) of an organism. Genomics is the biological research discipline dealing with the study of genomes

GRANULE A small cellular particle, often used to refer at vesicles containing secretory molecules

HEMATOPHAGOUS Feeding on blood

HEMORRHAGE Uncontrollable bleeding

HEMOSTASIS A process involving the interaction of multiple cell types and proteins that causes bleeding to stop, as a first step in wound healing

HEMOTOXIN A toxin that damages blood cells and/or functionally impairs normal processes in the blood, like hemostasis

HOMOLOGY/COMPARATIVE MODELING Constructing a protein structure model in the absence of experimental data

HOOP BAGS Large bags that to attach to a frame and handle designed to hold venomous snakes

HYDROLYSIS Cleavage of chemical bonds by the addition of water

HYDROPHILIC Having a strong affinity for water

HYDROPHOBIC Strongly repelled by water

HYPERALGESIA Increased pain sensitivity

HYPERMUTATION Frequent mutation under strong positive selection mutation that has an omega value greater than 1

HYPOTENSION Low blood pressure

INFLAMMATION An innate immune response of vascular tissue and immune cells induced by injury, or exposure to pathogens, irritants or toxins

ISCHEMIA A restriction of blood supply to tissues, causing a lack of oxygen and potentially leading to tissue damage

JIGGER STICK A pole with a U-shaped or Y-shaped end with a thick piece of rubber stretched across and are used to safely restrain venomous snakes

LEAD COMPOUND A molecule with promising bioactivity used as a design template during the development of new drugs

LIGAND A molecule capable to bind a biomolecule (often a receptor protein), thereby inducing a biological reaction

LUMEN The interior space of an anatomical structure (e.g. a vein, intestine or gland)

LYOPHILIZATION A dehydration technique (also known as "freeze-drying") during which a liquid solution (e.g. a venom) is brought to low pressure and freezing temperatures to sublimate aqueous components (e.g. water) and retain solid components (e.g. toxin proteins)

MANDIBLE (MANDIBULAR) Lower jaws

MAXILLA (MAXILLARY) Upper jaws

MONOPHYLETIC Representing all evolutionary descendants of a single ancestral lineage

MRNA Abbreviation of messenger RNA, the type of RNA molecule that passes on sequence information of a gene to the ribosomes for protein synthesis. After transcription of a protein-coding gene, primary transcript mRNA (known as pre-mRNA) is spliced and the resulting mRNA molecule is translated into a protein by ribosomes

NECROSIS Injury that results in the premature death of the affected cell or tissue

NEGATIVE SELECTION The selective elimination of newly arisen alleles from a population as a result of their fitness-reducing effect, preserving low allele variation and constraining change in proteins or phenotypes over a macroevolutionary scale

NEOFUNCTIONALIZATION Process whereby one paralogous copy derives a new function after gene duplication

NEUROMUSCULAR JUNCTION The interface between a neuron and the muscle cell it triggers

NEURON Nerve cell

NEUROTOXIN A toxin that targets the nervous system and disrupts the signaling that allows neurons to communicate effectively

NEUROTRANSMITTER Substance that transmits a signal from one neuron to another target neuron or muscle cell

NEUTRAL MUTATION A mutation that does not affect the fitness of an organism (i.e. is neither beneficial, nor maladaptive)

NONSYNONYMOUS SUBSTITUTION A nucleotide substitution in a protein-coding sequence that yields an amino-acid substitution

N-TERMINAL The side of a peptide chain from which translation is initiated, normally characterized by an amino (NH2-) end group

OMEGA (Ω, DN/DS) The ratio of nonsynonymous nucleotide substitutions to synonymous nucleotide substitutions

ORTHOLOGUES Genes with a common ancestry that have split as a result of a speciation event. Orthologues always represent genes in different organisms

PARALOGUES Genes with a common ancestry that have split as a result of a duplication event. Paralogues may reside in a single organism or in different ones

PARALYSIS Loss of muscle function, may be either excitatory (spastic) or inhibitory (flaccid)

PARAPHYLETIC Representing all evolutionary descendants (taxa, genes or proteins) of a single ancestral lineage, except one of its daughter lineages

PHYLOGENY (PHYLOGENETICS) An evolutionary hypothesis representing relationships among species or sequences as a dichotomous tree. Phylogenetics is the discipline in evolutionary biology that deals with the reconstruction of phylogenies

PLATELET AGGREGATION The clumping together of platelets

PLESIOTYPIC Representing an evolutionarily conserved or ancestral form, as opposed to a newly derived one

POSITIVE SELECTION The selective increase of the frequency of alleles in a population as a result of their fitness-increasing effect. At a macroevolutionary scale, positive selection may be reflected by directional evolutionary change in a protein sequence or phenotype

POSTSYNAPTIC Part of a neuron or muscle cell situated posterior to (at the 'receiver side' of) a synapse

POSTTRANSLATIONAL MODIFICATION Any biochemical modification of a peptide or protein following its translation, including cleavage, folding and modification of amino-acid side groups

PRESYNAPTIC Part of a neuron situated anterior to (at the 'donor side' of) a synapse

PROCOAGULANT Promoting the coagulation (clotting) of blood

PROPIECE Part of a precursor protein that is posttranslationally cleaved to release a functional protein or peptide

PROTEASE An enzyme that catalyzes a proteolytic reaction, i.e. cleavage of a peptide or protein by hydrolysis of a peptide bond

PROTEIN DOMAIN An amino-acid sequence with specific structural and functional features (e.g. a cysteine motif, a metal-binding site or an enzymatic activity) that typically represents an evolutionarily conserved building unit in multiple proteins, either alone, or in combination with other domains

PROTEIN FAMILY A group of proteins presumed to be evolutionarily related based on their structural similarity

PROTEOME (PROTEOMICS) The collective of proteins synthesized by an organism or one of its tissues. Proteomics is the biological research discipline dealing with the study of proteomes

PURIFYING SELECTION See negative selection

RECEPTOR A protein typically anchored into a cell membrane that accepts binding of another biomolecule (a ligand) in order to induce a cellular response

RESTRAINING TUBES Clear plastic tubes of different diameters that are used to safely hold venomous snakes in

SHOCK A life-threatening medical condition that occurs due to inadequate substrate for aerobic cellular respiration, often associated with a decrease in blood pressure and related symptoms (e.g., increased heart rate, fainting)

SIGNAL PEPTIDE An N-terminal amino-acid sequence in a precursor protein that is recognized by a protein complex (a signal recognition particle) upon translation, and then targeted to the endoplasmic reticulum for further processing in preparation of secretion

SNAKE HOOK A pole with an L-shaped or U-shaped piece of metal at the end used to pick up or restrain venomous snakes

SNAKE TONGS A tool used to safely pick up venomous reptiles which consists of a gripping handle and mouth that gently grasps the reptiles

SQUAMATA A clade of reptiles including all living lizards and snakes

SYNAPSE Structure where an (electrical or chemical) signal is passed from one neuron to another cell (neural or otherwise)

SYNONYMOUS SUBSTITUTION A nucleotide substitution in a protein-encoding sequence that does not alter the amino-acid sequence

SYNAPSE An area in a nerve fiber responsible for transmission of an electrical or chemical signal from one neuron to the next, or from a neuron to a muscle cell

QUARANTINE An isolation period for individual animals that are new to a collection where the animals are observed and tested for illness

SYSTEMIC SYMPTOMS Symptoms associated with whole-body physiological systems, as opposed to local symptoms (restricted to a single part of the body)

TOXICOFERA Reptile clade of the lizard radiations Anguimorpha (anguids, helodermatids, lanthanotids, shinosaurids, varanids, and xenosaurids) and Iguania (agamids, chameleonids, and iguanids) plus snakes characterized by oral glands that produce bioactive proteins that act as venoms

TOXIN A biologically produced molecule used by one organism to disrupt endophysiological or biochemical processes in another in function of feeding, defense or competition

TOXIN PRECURSOR A precursor protein from which the functional toxin (a smaller peptide or protein) is cleaved at some point after translation

TOXIN RECRUITMENT EVENT Process by which a gene encoding for a normal body tissue is duplicated and the new gene copy preferentially expressed in a venom gland

TRANSCRIPTOME (TRANSCRIPTOMICS) The collective of all RNA molecules, including mRNA, rRNA, tRNA, and other noncoding RNA produced by one cell, a tissue, or an organism

TYPE II DIABETES MELLITUS A group of metabolic diseases characterized by chronic high blood sugar levels due to resistance of cells against the hormone insulin, and often due to insufficient secretion of the latter

VENOM A secretion produced in specialized cells in one animal, delivered to a target animal through the infliction of a wound, that disrupts endophysiological or biochemical processes in the receiving animal to facilitate feeding, defense, or competition by/of the producing animal

REFERENCES

Abbas, A. K., A. H. Lichtman, and S. Pillai (2011). "*Cellular and molecular immunology 7th edition.*" Philadelphia: Elsevier Science.

Abrahamson, M., M. Alvarez-Fernandez, and C. M. Nathanson (2003). "Cystatins." *Biochem Soc Symp* (70): 179–199.

Ackermann, E. J., E. T. Ang, J. R. Kanter, I. Tsigelny, and P. Taylor (1998). "Identification of pairwise interactions in the alpha-neurotoxin-nicotinic acetylcholine receptor complex through double mutant cycles." *J Biol Chem* 273(18): 10958–10964.

Ackermann, E. J., and P. Taylor (1997). "Nonidentity of the alpha-neurotoxin binding sites on the nicotinic acetylcholine receptor revealed by modification in alpha-neurotoxin and receptor structures." *Biochemistry* 36(42): 12836–12844.

ACMT-AACT-AAPCC-EAPCC-IST-APAMT (2011). "Position statement: Pressure immobilization after North American Crotalinae snake envenomation." *J Med Toxicol* 7: 322–323.

Adermann, K., F. Wattler, S. Wattler, G. Heine, M. Meyer, W. G. Forssmann, and M. Nehls (1999). "Structural and phylogenetic characterization of human SLURP-1, the first secreted mammalian member of the Ly-6/uPAR protein superfamily." *Protein Sci* 8(4): 810–819.

Agarwala, K. L., S. Kawabata, M. Hirata, M. Miyagi, S. Tsunasawa, and S. Iwanaga (1996). "A cysteine protease inhibitor stored in the large granules of horseshoe crab hemocytes: Purification, characterization, cDNA cloning and tissue localization." *J Biochem* 119(1): 85–94.

Ahn, M. Y., B. M. Lee, and Y. S. Kim (2007). "Characterization and cytotoxicity of L-amino acid oxidase from the venom of king cobra (*Ophiophagus hannah*)." *Int J Biochem Cell Biol* 29(6): 911–919.

Aird, S. D. (2002). "Ophidian envenomation strategies and the role of purines." *Toxicon* 40(4): 335–393.

Aird, S. D., I. I. Kaiser, R. V. Lewis, and W. G. Kruggel (1985). "Rattlesnake presynaptic neurotoxins: Primary structure and evolutionary origin of the acidic subunit." *Biochemistry* 24(25): 7054–7058.

Aird, S. D., W. G. Kruggel, and I. I. Kaiser (1991). "Multiple myotoxin sequences from the venom of a single prairie rattlesnake (*Crotalus viridis viridis*)." *Toxicon* 29(2): 265–268.

Alagón, A. (2002). "Anticuerpos seguros y eficaces: La revolución de los nuevos antivenenos." *Central Dossier, Revista de la Universidad de México*: 617.

Alagón, A., L. D. Possani, J. Smart, and W. D. Schleuning (1986). "Helodermatine, a kallikrein-like, hypotensive enzyme from the venom of *Heloderma horridum horridum* (Mexican beaded lizard)." *J Exp Med* 164(6): 1835–1845.

Alam, M. I., B. Auddy, and A. Gomes (1994). "Isolation, purification and partial characterization of viper venom inhibiting factor from the root extract of the Indian medicinal plant sarsaparilla (*Hemidesmus indicus R. Br*)." *Toxicon* 32(12): 1551–1557.

Alape-Giron, A., L. Sanz, J. Escolano, M. Flores-Diaz, M. Madrigal, M. Sasa, and J. J. Calvete (2008). "Snake venomics of the lancehead pitviper *Bothrops asper*: Geographic, individual, and ontogenetic variations." *J Proteome Res* 7(8): 3556–3571.

Alberts, M. B., M. Shalit, and F. LoGalbo (2004). "Suction for venomous snakebite: A study of 'mock venom' extraction in a human model." *Ann Emerg Med* 43(2): 181–186.

Albrand, J. P., M. J. Blackledge, F. Pascaud, M. Hollecker, and D. Marion (1995). "NMR and restrained molecular dynamics study of the three-dimensional solution structure of toxin FS2, a specific blocker of the L-type calcium channel, isolated from black mamba venom." *Biochemistry* 34(17): 5923–5937.

Aleksiev, B., and B. Tchorbanov (1976). "Action on phosphatidylcholine of the toxic phospholipase A_2 from the venom of bulgarian viper (*Vipera ammodytes ammodytes*)." *Toxicon* 14(6): 477–485.

Ali, S. A., K. Baumann, T. N. Jackson, K. Wood, S. Mason, E. A. Undheim, A. Nouwens, et al. (2013a). "Proteomic comparison of *Hypnale hypnale* (hump-nosed pit-viper) and *Calloselasma rhodostoma* (Malayan pit-viper) venoms." *J Proteomics* 91C: 338–343.

Ali, S. A., S. Stoeva, A. Abbasi, J. M. Alam, R. Kayed, M. Faigle, B. Neumeister, and W. Voelter (2000). "Isolation, structural, and functional characterization of an apoptosis-inducing L-amino acid oxidase from leaf-nosed viper (*Eristocophis macmahoni*) snake venom." *Arch Biochem Biophys* 384(2): 216–226.

Ali, S. A., D. C. Yang, T. N. Jackson, E. A. Undheim, I. Koludarov, K. Wood, A. Jones, et al. (2013b). "Venom proteomic characterization and relative antivenom neutralization of two medically important Pakistani elapid snakes (*Bungarus sindanus* and *Naja naja*)." *J Proteomics* 89: 15–23.

Allen, H. R., M. L. Merchant, R. K. Tucker, J. W. Fox, and C. R. Geren (1996). "Characterization and chemical modification of E toxin isolated from timber rattlesnake (*Crotalus horridus horridus*) venom." *J Nat Toxins* 5: 409–427.

Allen, H. R., R. K. Tucker, and C. R. Geren (1986). "Potentiation of the toxicity of basic peptides from rattlesnake venoms by sodium acetate." *Toxicon* 24(6): 553–558.

Allender, M. C., M. Dreslik, C. A. Phillips, and C. Maddox (2012). "Facial disfiguration syndrome in free-ranging snakes throughout the Eastern US: An emerging pathogen associated with *Chrysosporium*." *Proceedings*. AAZV, Oakland, CA.

Allender, M. C., M. Dreslik, S. Wylie, C. Phillips, D. B. Wylie, C. Maddox, M. A. Delaney, and M. J. Kinsel (2011). "*Chrysosporium* sp. infection in eastern massasauga rattlesnakes." *Emerg Infect Dis* 17(12): 2383–2384.

Almeida, A. P., A. B. Andrade, A. J. Ferreira, A. C. Pires, D. D. Damasceno, M. N. Alves, E. R. Gomes, et al. (2011). "Antiarrhythmogenic effects of a neurotoxin from the spider *Phoneutria nigriventer*." *Toxicon* 57(2): 217–224.

Aloe, L., E. Alleva, A. Bohm, and R. Levimontalcini (1986). "Aggressive behavior induces release of nerve growth factor from mouse salivary gland into the blood-stream." *Proc Natl Acad Sci USA* 83(16): 6184–6187.

Aloe, L., L. Braccilaudiero, E. Alleva, A. Lambiase, A. Micera, and P. Tirassa (1994). "Emotional stress induced by parachute jumping enhances blood nerve growth factor levels and the distribution of nerve growth factor receptors in lymphocytes." *Proc Natl Acad Sci USA* 91(22): 10440–10444.

Alper, C. A., and D. Balavitch (1976). "Cobra venom factor: Evidence for its being altered cobra C3 (the third component of complement)." *Science* 191(4233): 1275–1276.

Alpert, A. J. (1988). "Hydrophobic interaction chromatography of peptides as an alternative to reversed-phase chromatography." *J Chromatogr* 444: 269–274.

——— (1990). "Hydrophilic-interaction chromatography for the separation of peptides, nucleic acids and other polar compounds." *J Chromatogr* 499: 177–196.

Alpert, A. J. (1999). "Size exclusion high-performance liquid chromatography of small solutes." In *Column handbook for size exclusion chromatography*, edited by C. Wu. New York: Academic Press.

Alpert, A. J., and P. C. Andrews (1988). "Cation-exchange chromatography of peptides on poly(2-sulfoethyl aspartamide)-silica." *J Chromatogr* 443: 85–96.

Altman, L. K. (1986). "New shock therapy for snakebites." *New York Times*, August 5.

Altschul, S. F., W. Gish, W. Miller, E. W. Myers, and D. J. Lipman (1990). "Basic local alignment search tool." *J Mol Biol* 215(3): 403–410.

Altschul, S. F., T. L. Madden, A. A. Schaffer, J. Zhang, Z. Zhang, W. Miller, and D. J. Lipman (1997). "Gapped BLAST and PSI-BLAST: A new generation of protein database search programs." *Nucleic Acids Res* 25(17): 3389–3402.

Alvarez-Fernandez, M., A. J. Barrett, B. Gerhartz, P. M. Dando, J. Ni, and M. Abrahamson (1999). "Inhibition of mammalian legumain by some cystatins is due to a novel second reactive site." *J Biol Chem* 274(27): 19195–19203.

Alves, R. M., G. A. Antonucci, H. H. Paiva, A. C. Cintra, J. J. Franco, E. P. Mendonca-Franqueiro, D. J. Dorta, et al. (2008). "Evidence of caspase-mediated apoptosis induced by L-amino acid oxidase isolated from *Bothrops atrox* snake venom." *Comp Biochem Physiol A Mol Integr Physiol* 151(4): 542–550.

Alves, R. M., P. R. Feliciano, S. V. Sampaio, and M. C. Nonato (2011). "A rational protocol for the successful crystallization of L-amino-acid oxidase from *Bothrops atrox*." *Acta Crystallogr F Struct Biol Cryst Commun* 67 (pt. 4): 475–478.

Amaral, C. F. S., D. Campolina, M. B. Dias, C. M. Bueno, and N. A. Rezende (1998). "Tourniquet ineffectiveness to reduce the severity of envenoming after *Crotalus durissus* snake bite in Belo Horizonte, Minas Gerais, Brazil." *Toxicon* 36(5): 805–808.

Amarant, T., W. Burkhart, H. Levine, C. L. Arochapinango, and I. Parikh (1991). "Isolation and complete amino-acid-sequence of 2 fibrinolytic proteinases from the toxic saturnid caterpillar *Lonomia achelous*." *Biochim Biophys Acta* 1079(2): 214–221.

Ameri, M., X. Wang, M. J. Wilkerson, M. R. Kanost, and A. B. Broce (2008). "An immunoglobulin binding protein (antigen 5) of the stable fly (Diptera: Muscidae) salivary gland stimulates bovine immune responses." *J Med Entomol* 45: 94–101.

Amiconi, G., A. Amoresano, G. Boumis, A. Brancaccio, R. De Cristofaro, A. De Pascalis, S. Di Girolamo, B. Maras, and A. Scaloni (2000). "A novel venombin B from *Agkistrodon contortrix contortrix*: Evidence for recognition properties in the surface around the primary specificity pocket different from thrombin." *Biochemistry* 39(33): 10294–10308.

Aminetzach, Y. T., J. R. Srouji, C. Y. Kong, and H. E. Hoekstra (2009). "Convergent evolution of novel protein function in shrew and lizard venom." *Curr Biol* 19(22): 1925–1931.

Amininasab, M., M. M. Elmi, N. Endlich, K. Endlich, N. Parekh, H. Naderi-Manesh, J. Schaller, et al. (2004). "Functional and structural characterization of a novel member of the natriuretic family of peptides from the venom of *Pseudocerastes persicus*." *FEBS Lett* 557(1–3): 104–108.

Ande, S. R., P. R. Kommoju, S. Draxl, M. Murkovic, P. Macheroux, S. Ghisla, and E. Ferrando-May (2006). "Mechanisms of cell death induction by L-amino acid oxidase, a major component of ophidian venom." *Apoptosis* 11(8): 1439–1451.

Andersen, J. F., D. E. Champagne, A. Weichsel, J. M. C. Ribeiro, C. A. Balfour, V. Dress, and W. R. Montfort (1997). "Nitric oxide binding and crystallization of recombinant nitrophorin I, a nitric oxide transport protein from the blood-sucking bug *Rhodnius prolixus*." *Biochemistry* 36(15): 4423–4428.

Andersen, J. F., I. M. B. Francischetti, J. G. Valenzuela, P. Schuck, and J. M. C. Ribeiro (2003). "Inhibition of hemostasis by a high affinity biogenic amine-binding protein from the saliva of a blood-feeding insect." *J Biol Chem* 278(7): 4611–4617.

Andersen, J. F., N. P. Gudderra, I. M. B. Francischetti, and J. M. C. Ribeiro (2005). "The role of salivary lipocalins in blood feeding by *Rhodnius prolixus*." *Arch Insect Biochem Physiol* 58(2): 97–105.

Andersen, J. F., N. P. Gudderra, I. M. B. Francischetti, J. G. Valenzuela, and J. M. C. Ribeiro (2004). "Recognition of anionic phospholipid membranes by an antihemostatic protein from a blood-feeding insect." *Biochemistry* 43(22): 6987–6994.

Andersen, J. F., B. J. Hinnebusch, D. A. Lucas, T. P. Conrads, T. D. Veenstra, V. M. Pham, and J. M. C. Ribeiro (2007). "An insight into the sialome of the oriental rat flea, *Xenopsylla cheopis* (*Rots*)." *BMC Genomics* 8: 102.

Andreini, C., L. Banci, I. Bertini, S. Elmi, and A. Rosato (2005). "Comparative analysis of the ADAM and ADAMTS families." *J Proteome Res* 4(3): 881–888.

Andrews, R. K., W. J. Booth, J. J. Gorman, P. A. Castaldi, and M. C. Berndt (1989). "Purification of botrocetin from *Bothrops jararaca* venom: Analysis of the botrocetin-mediated interaction between von Willebrand factor and the human platelet membrane glycoprotein Ib-IX complex." *Biochemistry* 28(21): 8317–8326.

Andrews, R. K., M. H. Kroll, C. M. Ward, J. W. Rose, R. M. Scarborough, A. I. Smith, J. A. Lopez, and M. C. Berndt (1996). "Binding of a novel 50-kilodalton alboaggregin from *Trimeresurus albolabris* and related viper venom proteins to the platelet membrane glycoprotein Ib-IX-V complex: Effect on platelet aggregation and glycoprotein Ib-mediated platelet activation." *Biochemistry* 35(38): 12629–12639.

Andriao-Escarso, S. H., A. M. Soares, V. M. Rodrigues, Y. Angulo, C. Diaz, B. Lomonte, J. M. Gutiérrez, and J. R. Giglio (2000). "Myotoxic phospholipases A(2) in bothrops snake venoms: Effect of chemical modifications on the enzymatic and pharmacological properties of bothropstoxins from *Bothrops jararacussu*." *Biochimie* 82(8): 755–763.

Angulo, Y., J. Escolano, B. Lomonte, J. M. Gutiérrez, L. Sanz, and J. J. Calvete (2008). "Snake venomics of Central American pitvipers: Clues for rationalizing the distinct envenomation profiles of *Atropoides nummifer* and *Atropoides picadoi*." *J Proteome Res* 7(2): 708–719.

Angulo, Y., J. M. Gutiérrez, A. M. Soares, W. Cho, and B. Lomonte (2005). "Myotoxic and cytolytic activities of dimeric Lys49 phospholipase A_2 homologues are reduced, but not abolished, by a pH-induced dissociation." *Toxicon* 46(3): 291–296.

Anonymous, B. P. D. C. L. (1938). "Post-graduate medical journal," 338.

Anthony, J. (1955). "Essai sur l'évolution anatomique de l'appareil venimeux des ophidiens [Essay on the anatomical evolution of the venomous apparatus of ophidians]." *Ann Sci Nat Zool* 11: 7–53.

Antil, S., D. Servent, and A. Ménez (1999). "Variability among the sites by which curaremimetic toxins bind to torpedo acetylcholine receptor, as revealed by identification of the functional residues of alpha-cobratoxin." *J Biol Chem* 274(49): 34851–34858.

Antuch, W., K. D. Berndt, M. A. Chavez, J. Delfin, and K. Wuthrich (1993). "The nmr solution structure of a kunitz-type proteinase-inhibitor from the sea-anemone *Stichodactyla helianthus*." *Eur J Biochem* 212 (3): 675–684.

Anusiewicz, I., M. Jasionowski, P. Skurski, and J. Simons (2005). "Backbone and side-chain cleavages in electron detachment dissociation (EDD)." *J Phys Chem A* 109(49): 11332–11337.

Apte, S. S. (2004). "A disintegrin-like and metalloprotease (reprolysin type) with thrombospondin type 1 motifs: The ADAMTS family." *Int J Biochem Cell Biol* 36(6): 981–985.

Aquilina, J. A. (2009). "The major toxin from the Australian Common Brown Snake is a hexamer with unusual gas-phase dissociation properties." *Proteins* 75(2): 478–485.

Arai, H., S. Hori, I. Aramori, H. Ohkubo, and S. Nakanishi (1990). "Cloning and expression of a cDNA encoding an endothelin receptor." *Nature* 348(6303): 730–732.

Arcà, B., F. Lombardo, J. G. Valenzuela, I. M. B. Francischetti, O. Marinotti, M. Coluzzi, and J. M. C. Ribeiro (2005). "An updated catalogue of salivary gland transcripts in the adult female mosquito, *Anopheles gambiae*." *J Exp Biol* 208: 3971–3986.

Arias, H. R. (2000). "Localization of agonist and competitive antagonist binding sites on nicotinic acetylcholine receptors." *Neurochem Int* 36(7): 595–645.

Arinami, T., M. Ishikawa, A. Inoue, M. Yanagisawa, T. Masaki, M. C. Yoshida, and H. Hamaguchi (1991). "Chromosomal assignments of the human endothelin family genes: The endothelin-1 gene (EDN1) to 6p23-p24, the endothelin-2 gene (EDN2) to 1p34, and the endothelin-3 gene (EDN3) to 20q13.2-q13.3." *Am J Hum Genet* 48(5): 990–996.

Arita, H., K. Hanasaki, T. Nakano, S. Oka, H. Teraoka, and K. Matsumoto (1991). "Novel proliferative effect of phospholipase A_2 in Swiss 3T3 cells via specific binding site." *J Biol Chem* 266(29): 19139–19141.

Arlinghaus, F. T., and J. A. Eble (2013). "The collagen-binding integrin alpha2beta1 is a novel interaction partner of the *Trimeresurus flavoviridis* venom protein flavocetin-A." *J Biol Chem* 288(2): 947–955.

Arlinghaus, F. T., T. Momic, N. A. Ammar, E. Shai, G. Spectre, D. Varon, C. Marcinkiewicz, H. Heide, P. Lazarovici, and J. A. Eble (2013). "Identification of alpha2beta1 integrin inhibitor VP-i with anti-platelet properties in the venom of *Vipera palaestinae*." *Toxicon* 64: 96–105.

Arnaout, M. A., B. Mahalingam, and J. P. Xiong (2005). "Integrin structure, allostery, and bidirectional signaling." *Annu Rev Cell Dev Biol* 21: 381–410.

Arnold, K., L. Bordoli, J. Kopp, and T. Schwede (2006). "The SWISS-MODEL Workspace: A web-based environment for protein structure homology modelling." *Bioinformatics* 22: 195–201.

Arredondo, J., A. I. Chernyavsky, D. L. Jolkovsky, R. J. Webber, and S. A. Grando (2006). "SLURP-2: A novel cholinergic signaling peptide in human mucocutaneous epithelium." *J Cell Physiol* 208(1): 238–245.

Asakawa, D., N. Smargiasso, L. Quinton, and E. De Pauw (2013). "Peptide backbone fragmentation initiated by side-chain loss at cysteine residue in matrix-assisted laser desorption/ionization in-source decay mass spectrometry." *J Mass Spectrom* 48(3): 352–360.

Asakawa, D., and M. Takayama (2011). "Cα-C bond cleavage of the peptide backbone in MALDI in-source decay using salicylic acid derivative matrices." *J Am Soc Mass Spectrom* 22(7): 1224–1233.

Ascenzi, P., A. Bertollini, M. Bolognesi, M. Guarneri, E. Menegatti, and G. Amiconi (1986). "Primary specificity of ancrod, the coagulating serine proteinase from the Malayan pit viper (*Agkistrodon rhodostoma*) venom." *Biochim Biophys Acta* 871(2): 225–228.

Asgari, S., G. M. Zhang, R. Zareie, and O. Schmidt (2003). "A serine proteinase homolog venom protein from an endoparasitoid wasp inhibits melanization of the host hemolymph." *Insect Biochem Mol Biol* 33(10): 1017–1024.

Assakura, M. T., A. P. Reichl, and F. R. Mandelbaum (1986). "Comparison of immunological, biochemical and biophysical properties of three hemorrhagic factors isolated from the venom of Bothrops jararaca (jararaca)." *Toxicon* 24(9): 943–946.

——— (1994). "Isolation and characterization of five fibrin(ogen)olytic enzymes from the venom of *Philodryas olfersii* (green snake)." *Toxicon* 32(7): 819–831.

Assakura, M. T., M. G. Salomao, G. Puorto, and F. R. Mandelbaum (1992). "Hemorrhagic, fibrinogenolytic and edema-forming activities of the venom of the colubrid snake *Philodryas olfersii* (green snake)." *Toxicon* 30(4): 427–438.

Atkins, A. R., R. C. Martin, and R. Smith (1995). "1H NMR studies of sarafotoxin SRTb, a nonselective endothelin receptor agonist, and IRL 1620, an ETB receptor-specific agonist." *Biochemistry* 34(6): 2026–2033.

Atoda, H., M. Hyuga, and T. Morita (1991). "The primary structure of coagulation factor IX/factor X-binding protein isolated from the venom of *Trimeresurus flavoviridis*: Homology with asialoglycoprotein receptors, proteoglycan core protein, tetranectin, and lymphocyte Fc epsilon receptor for immunoglobulin E." *J Biol Chem* 266(23): 14903–14911.

Atoda, H., M. Ishikawa, H. Mizuno, and T. Morita (1998). "Coagulation factor X-binding protein from *Deinagkistrodon acutus* venom is a Gla domain-binding protein." *Biochemistry* 37(50): 17361–17370.

Atoda, H., M. Ishikawa, E. Yoshihara, F. Sekiya, and T. Morita (1995). "Blood coagulation factor IX-binding protein from the venom of *Trimeresurus flavoviridis*: Purification and characterization." *J Biochem* 118(5): 965–973.

Aubert, J. D., and L. Juillerat-Jeanneret (2009). "Therapeutic potential of endothelin receptor modulators: Lessons from human clinical trials." *Expert Opin Ther Targets* 13(9): 1069–1084.

Audebert, F., M. Urtizberea, A. Sabouraud, J. M. Scherrmann, and C. Bon (1994). "Pharmacokinetics of *Vipera aspis* venom after experimental envenomation in rabbits." *J Pharmacol Exp Ther* 268(3): 1512–1517.

Auffenberg, W. (1981). *Behavioral ecology of the Komodo monitor.* Gainesville: University Press of Florida, 245–247, 269–270, 288.

Baggio, L. L., Q. Huang, X. Cao, and D. J. Drucker (2008). "An albumin-exendin-4 conjugate engages central and peripheral circuits regulating murine energy and glucose homeostasis." *Gastroenterology* 134(4): 1137–1147.

Bagnato, A., and L. Rosano (2008). "The endothelin axis in cancer." *Int J Biochem Cell Biol* 40(8): 1443–1451.

Bagnato, A., F. Spinella, and L. Rosano (2008). "The endothelin axis in cancer: The promise and the challenges of molecularly targeted therapy." *Can J Physiol Pharmacol* 86(8): 473–484.

Bahrenberg, G., A. Brauers, H. G. Joost, and G. Jakse (2000). "Reduced expression of PSCA, a member of the LY-6 family of cell surface antigens, in bladder, esophagus, and stomach tumors." *Biochem Biophys Res Commun* 275(3): 783–788.

Bajwa, S. S., H. Kirakossian, K. N. Reddy, and F. S. Markland (1982). "Thrombin-like and fibrinolytic enzymes in the venoms from the Gaboon viper (*Bitis gabonica*), eastern cottonmouth moccasin (*Agkistrodon p. piscivorus*) and southern copperhead (*Agkistrodon c. contortrix*) snakes." *Toxicon* 20(2): 427–432.

Bakker, H. M., G. Tans, L. Y. Yukelson, T. W. Janssen-Claessen, R. M. Bertina, H. C. Hemker, and J. Rosing (1993). "Protein C activation by an activator purified from the venom of *Agkistrodon halys halys*." *Blood Coagul Fibrin* 4(4): 605–614.

Baldwin, M. (1995). "The snakestone experiments—an early-modern medical debate." *Isis* 86(3): 394–418.

Balsinde, J., M. A. Balboa, P. A. Insel, and E. A. Dennis (1999). "Regulation and inhibition of phospholipase A_2." *Annu Rev Pharmacol Toxicol* 39: 175–189.

Balsinde, J., R. Perez, and M. A. Balboa (2006). "Calcium-independent phospholipase A_2 and apoptosis." *Biochim Biophys Acta* 1761(11): 1344–1350.

Bandeira, N., K. R. Clauser, and P. A. Pevzner (2007). "Shotgun protein sequencing: Assembly of peptide tandem mass spectra from mixtures of modified proteins." *Mol Cell Proteomics* 6(7): 1123–1134.

Bando, M., Y. Ishii, Y. Sugiyama, and S. Kitamura (1999). "Elevated plasma brain natriuretic peptide levels in chronic respiratory failure with cor pulmonale." *Resp Med* 93(7): 507–514.

Banerjee, Y., R. Lakshminarayanan, S. Vivekanandan, G. S. Anand, S. Valiyaveettil, and R. M. Kini (2007). "Biophysical characterization of anticoagulant hemextin AB complex from the venom of snake *Hemachatus haemachatus*." *Biophys J* 93(11): 3963–3976.

Banerjee, Y., J. Mizuguchi, S. Iwanaga, and R. M. Kini (2005). "Hemextin AB complex, a unique anticoagulant protein complex from *Hemachatus haemachatus* (African Ringhals cobra) venom that inhibits clot initiation and factor VIIa activity." *J Biol Chem* 280(52): 42601–42611.

Bantscheff, M., M. Schirle, G. Sweetman, J. Rick, and B. Kuster (2007). "Quantitative mass spectrometry in proteomics: A critical review." *Anal Bioanal Chem* 389(4): 1017–1031.

Banumathi, S., K. R. Rajashankar, C. Notzel, B. Aleksiev, T. P. Singh, N. Genov, and C. Betzel (2001). "Structure of the neurotoxic complex vipoxin at 1.4 A resolution." *Acta Crystallogr D Biol Crystallogr* 57 (pt. 11): 1552–1559.

Bao, S., Y. Li, X. Lei, M. Wohltmann, W. Jin, A. Bohrer, C. F. Semenkovich, S. Ramanadham, I. Tabas, and J. Turk (2007). "Attenuated free cholesterol loading-induced apoptosis but preserved phospholipid composition of peritoneal macrophages from mice that do not express group VIA phospholipase A_2." *J Biol Chem* 282(37): 27100–27114.

Barber, C. M., G. K. Isbister, and W. C. Hodgson (2013). "Alpha neurotoxins." *Toxicon* 66: 47–58.

Barbouche, R., N. Marrakchi, K. Mabrouk, M. N. Krifi, J. Van Rietschoten, E. Fenouillet, M. El Ayeb, and H. Rochat (1998). "Anti-platelet activity of the peptides composing the Lebetin 1 family, a new class of inhibitors of platelet aggregation." *Toxicon* 36(12): 1939–1947.

Barbouche, R., N. Marrakchi, P. Mansuelle, M. Krifi, E. Fenouillet, H. Rochat, and M. el Ayeb (1996). "Novel anti-platelet aggregation polypeptides from *Vipera lebetina* venom: Isolation and characterization." *FEBS Lett* 392(1): 6–10.

Bargnoux, A. S., K. Klouche, J. Fareh, I. Barazer, S. Villard-Saussine, A. M. Dupuy, H. Leray-Moragues, I. Giuliani, B. Canaud, and J. P. Cristol (2008). "Prohormone brain natriuretic peptide (proBNP), BNP and N-terminal-proBNP circulating levels in chronic hemodialysis patients: Correlation with ventricular function, fluid removal and effect of hemodiafiltration." *Clin Chem Lab Med* 46(7): 1019–1024.

Barker, S., N. P. Charlton, and C. P. Holstege (2010). "Accuracy of Internet recommendations for prehospital care of venomous snake bites." *Wilderness Environ Med* 21(4): 298–302.

Barlow, A., C. E. Pook, R. A. Harrison, and W. Wüster (2009). "Coevolution of diet and prey-specific venom activity supports the role of selection in snake venom evolution." *Proc Biol Sci* 276(1666): 2443–2449.

Barrett, A. J., and N. D. Rawlings (1995). "Families and clans of serine peptidases." *Arch Biochem Biophys* 318(2): 247–250.

Barrett, G. L. (2000). "The p75 neurotrophin receptor and neuronal apoptosis." *Prog Neurobiol* 61(2): 205–229.

Barrett, J. S., G. Murphy, K. Peerlinck, I. De Lepeleire, R. J. Gould, D. Panebianco, E. Hand, et al. (1994). "Pharmacokinetics and pharmacodynamics of MK-383, a selective non-peptide platelet glycoprotein-IIb/IIIa receptor antagonist, in healthy men." *Clin Pharmacol Ther* 56(4): 377–388.

Barry, P. H. (1994). "JPCalc, a software package for calculating liquid junction potential corrections in patch-clamp, intracellular, epithelial and bilayer measurements and for correcting junction potential measurements." *J Neurosci Methods* 51(1): 107–116.

Barton, M., and M. Yanagisawa (2008). "Endothelin: 20 years from discovery to therapy." *Can J Physiol Pharmacol* 86(8): 485–498.

Basavarajappa, B. S., and T. V. Gowda (1992). "Comparative characterization of two toxic phospholipases A_2 from Indian cobra (*Naja naja naja*) venom." *Toxicon* 30(10): 1227–1238.

Bates, E. E., W. H. Fridman, and C. G. Mueller (2002). "The ADAMDEC1 (decysin) gene structure: Evolution by duplication in a metalloprotease gene cluster on chromosome 8p12." *Immunogenetics* 54(2): 96–105.

Bates, P. A., L. A. Kelley, R. M. MacCallum, and M. J. Sternberg (2001). "Enhancement of protein modeling by human intervention in applying the automatic programs 3D-JIGSAW and 3D-PSSM." *Proteins* supp. 5: 39–46.

Batista, C. V. F., S. A. Roman-Gonzalez, S. P. Salas-Castillo, F. Z. Zamudio, F. Gomez-Lagunas, and L. D. Possani (2007). "Proteomic analysis of the venom from the scorpion *Tityus stigmurus*: Biochemical and physiological comparison with other *Tityus* species." *Comp Biochem Physiol C Toxicol Pharmacol* 146(1–2): 147–157.

Batuwangala, T., M. Leduc, J. M. Gibbins, C. Bon, and E. Y. Jones (2004). "Structure of the snake-venom toxin convulxin." *Acta Crystallogr D Biol Crystallogr* 60(pt. 1): 46–53.

Bax, B., T. L. Blundell, J. Murray-Rust, and N. Q. McDonald (1997). "Structure of mouse 7S NGF: A complex of nerve growth factor with four binding proteins." *Structure* 5(10): 1275–1285.

Bayrhuber, M., V. Vijayan, M. Ferber, R. Graf, J. Korukottu, J. Imperial, J. E. Garrett, et al. (2005). "Conkunitzin-S1 is the first member of a new Kunitz-type neurotoxin family—structural and functional characterization." *J Biol Chem* 280(25): 23766–23770.

Bazaa, A., N. Marrakchi, M. El Ayeb, L. Sanz, and J. J. Calvete (2005). "Snake venomics: Comparative analysis of the venom proteomes of the Tunisian snakes *Cerastes cerastes, Cerastes vipera* and *Macrovipera lebetina*." *Proteomics* 5(16): 4223–4235.

Bazaa, A., E. Pasquier, C. Defilles, I. Limam, R. Kessentini-Zouari, O. Kallech-Ziri, A. El Battari, et al. (2010). "MVL-PLA_2, a snake venom phospholipase A_2, inhibits angiogenesis through an increase in microtubule dynamics and disorganization of focal adhesions." *PLoS One* 5(4): e10124.

Bdolah, A. (2010). "Sarafotoxins, the snake venom homologs of the endothelins." In *Handbook of venom and toxins of reptiles*, edited by S. P. Mackessy. Boca Raton, FL: CRC Press, 303–315.

Bdolah, A., E. Kochva, M. Ovadia, S. Kinamon, and Z. Wollberg (1997). "Resistance of the Egyptian mongoose to sarafotoxins." *Toxicon* 35(8): 1251–1261.

Bdolah, A., Z. Wollberg, I. Ambar, Y. Kloog, M. Sokolovsky, and E. Kochva (1989). "Disturbances in the cardiovascular system caused by endothelin and sarafotoxin." *Biochem Pharmacol* 38(19): 3145–3146.

Becchetti, A. (2011). "Ion channels and transporters in cancer. 1. Ion channels and cell proliferation in cancer." *Am J Physiol Cell Physiol* 301(2): C255–265.

Becker, A., E. B. Dowdle, U. Hechler, K. Kauser, P. Donner, and W. D. Schleuning (1993). "Bibrotoxin, a novel member of the endothelin/sarafotoxin peptide family, from the venom of the burrowing asp *Atractaspis bibroni*." *FEBS Lett* 315(1): 100–103.

Bénard-Valle, M. (2009). *Caracterización bioquímica e inmunoquímica de venenos de coralillos norteamericanos*. Universidad Autónoma del Estado de Morelos, Mexico.

Bendtsen, J. D., H. Nielsen, G. von Heijne, and S. Brunak (2004). "Improved prediction of signal peptides: SignalP 3.0." *J Mol Biol* 340(4): 783–795.

Benishin, C. G. (1990). "Potassium channel blockade by the B subunit of β-bungarotoxin." *Mol Pharmacol* 38(2): 164–169.

Bennett, T. L., and J. C. Rose (1991). "Atrial natriuretic factor responses to volume expansion in pregnant and nonpregnant sheep." *Am J Obstet Gynecol* 165(6 pt. 1): 1627–1634.

Beress, L. (1982). "Biologically-active compounds from coelenterates." *Pure Appl Chem* 54(10): 1981–1994.

Bergeron, Z. L., and J. P. Bingham (2012). "Scorpion toxins specific for potassium (K+) channels: A historical overview of peptide bioengineering." *Toxins (Basel)* 4(11): 1082–1119.

Bergmeier, W., D. Bouvard, J. A. Eble, R. Mokhtari-Nejad, V. Schulte, H. Zirngibl, C. Brakebusch, R. Fassler, and B. Nieswandt (2001). "Rhodocytin (aggretin) activates platelets lacking alpha(2) beta(1) integrin, glycoprotein VI, and the ligand-binding domain of glycoprotein Ibalpha." *J Biol Chem* 276(27): 25121–25126.

Bergthorsson, U., D. I. Andersson, and J. R. Roth (2007). "Ohno's dilemma: Evolution of new genes under continuous selection." *Proc Natl Acad Sci USA* 104(43): 17004–17009.

Berman, H. M., J. Westbrook, Z. Feng, G. Gilliland, T. N. Bhat, H. Weissig, I. N. Shindyalov, and P. E. Bourne (2000). "The Protein Data Bank." *Nucleic Acids Res* 28(1): 235–242.

Bernardes, C. P., D. L. Menaldo, E. Camacho, J. C. Rosa, T. Escalante, A. Rucavado, B. Lomonte, J. M. Gutiérrez, and S. V. Sampaio (2013). "Proteomic analysis of *Bothrops pirajai* snake venom and characterization of BpirMP, a new P-I metalloproteinase." *J Proteomics* 80: 250–267.

Bernett, M. J., S. I. Blaber, I. A. Scarisbrick, P. Dhanarajan, S. M. Thompson, and M. Blaber (2002). "Crystal structure and biochemical characterization of human kallikrein 6 reveals that a trypsin-like kallikrein is expressed in the central nervous system." *J Biol Chem* 277(27): 24562–24570.

Bernheimer, A. W., R. Linder, S. A. Weinstein, and K. S. Kim (1987). "Isolation and characterization of a phospholipase B from venom of Collett's snake, *Pseudechis colletti*." *Toxicon* 25(5): 547–554.

Bernstein, D., G. Fajardo, and M. Zhao (2011). "The role of beta-adrenergic receptors in heart failure: Differential regulation of cardiotoxicity and cardioprotection." *Prog Pediatr Cardiol* 31(1): 35–38.

Bhakta, N. S., D. L. Morgan, and D. J. Borys (2009). "Copperhead (*Agkistrodon contortrix*) snakebites in the United States: 2000–2007." *Ann Emerg Med* 54(3): S19.

Bhat, R. N. (1974). "Viperine snake bite poisoning in Jammu." *J Indian Med Assoc* 63(12): 383–392.

Bhoola, K. D., C. D. Figueroa, and K. Worthy (1992). "Bioregulation of kinins: Kallikreins, kininogens, and kininases." *Pharmacol Rev* 44(1): 1–80.

Bieber, A. L., R. H. McParland, and R. R. Becker (1987). "Amino acid sequences of myotoxins from *Crotalus viridis concolor* venom." *Toxicon* 25(6): 677–680.

Bieber, A. L., and D. Nedelkov (1997). "Structural, biological and biochemical studies of myotoxin a and homologous myotoxins." *J Toxicol Toxin Rev* 16: 33–52.

Bieber, A. L., T. Tu, and A. T. Tu (1975). "Studies of an acidic cardiotoxin isolated from the venom of Mojave rattlesnake (*Crotalus scutulatus*)." *Biochim Biophys Acta* 400(1): 178–188.

Bielanska, J., J. Hernandez-Losa, T. Moline, R. Somoza, S. R. Cajal, E. Condom, J. C. Ferreres, and A. Felipe (2012). "Increased voltage-dependent K(+) channel Kv1.3 and Kv1.5 expression correlates with leiomyosarcoma aggressiveness." *Oncol Lett* 4(2): 227–230.

Bilwes, A., B. Rees, D. Moras, R. Ménez, and A. Ménez (1994). "X-ray structure at 1.55 A of toxin gamma, a cardiotoxin from *Naja nigricollis* venom: Crystal packing reveals a model for insertion into membranes." *J Mol Biol* 239(1): 122–136.

Birrell, G. W., S. Earl, P. P. Masci, J. de Jersey, T. P. Wallis, J. J. Gorman, and M. F. Lavin (2006). "Molecular diversity in venom from the Australian Brown snake, *Pseudonaja textilis*." *Mol Cell Proteomics* 5: 379–389.

Birrell, G. W., S. T. Earl, T. P. Wallis, P. P. Masci, J. de Jersey, J. J. Gorman, and M. F. Lavin (2007). "The diversity of bioactive proteins in Australian snake venoms." *Mol Cell Proteomics* 7: 973–986.

Bjarnason, J. B., A. Barish, G. S. Direnzo, R. Campbell, and J. W. Fox (1983). "Kallikrein-like enzymes from *Crotalus atrox* venom." *J Biol Chem* 258(20): 12566–12573.

Bjoern, S., D. C. Foster, L. Thim, F. C. Wiberg, M. Christensen, Y. Komiyama, A. H. Pedersen, and W. Kisiel (1991). "Human plasma and recombinant factor VII: Characterization of O-glycosylations at serine residues 52 and 60 and effects of site directed mutagenesis at serine 52 to alanine." *J Biol Chem* 266: 11051–11057.

Bjorck, L., A. Grubb, and L. Kjellen (1990). "Cystatin C, a human proteinase inhibitor, blocks replication of herpes simplex virus." *J Virol* 64(2): 941–943.

Black, R. A., C. T. Rauch, C. J. Kozlosky, J. J. Peschon, J. L. Slack, M. F. Wolfson, B. J. Castner, et al. (1997). "A metalloproteinase disintegrin that releases tumour-necrosis factor-alpha from cells." *Nature* 385(6618): 729–733.

Blacklow, B., P. Escoubas, and G. M. Nicholson (2010). "Characterisation of the heterotrimeric presynaptic phospholipase A(2) neurotoxin complex from the venom of the common death adder (*Acanthophis antarcticus*)." *Biochem Pharmacol* 80(2): 277–287.

Blacklow, B., R. Kornhauser, P. G. Hains, R. Loiacono, P. Escoubas, A. Graudins, and G. M. Nicholson (2011). "Alpha-Elapitoxin-Aa2a, a long-chain snake alpha-neurotoxin with potent actions on muscle (alpha1)(2)betagammadelta nicotinic receptors, lacks the classical high affinity for neuronal alpha7 nicotinic receptors." *Biochem Pharmacol* 81(2): 314–325.

Blankenvoorde, M. F., W. van't Hof, E. Walgreen-Weterings, T. J. van Steenbergen, H. S. Brand, E. C. Veerman, and A. V. Nieuw Amerongen (1998). "Cystatin and cystatin-derived peptides have antibacterial activity against the pathogen *Porphyromonas gingivalis*." *Biol Chem* 379(11): 1371–1375.

Blaylock, R. S. (2001). "Normal oral bacterial flora from some southern African snakes." *Onderstepoort J Vet Res* 68(3): 175–182.

Blobel, C. P. (2005). "ADAMs: Key components in EGFR signalling and development." *Nat Rev Mol Cell Biol* 6(1): 32–43.

Bode, W., R. Engh, D. Musil, U. Thiele, R. Huber, A. Karshikov, J. Brzin, J. Kos, and V. Turk (1988). "The 2.0 A X-ray crystal structure of chicken egg white cystatin and its possible mode of interaction with cysteine proteinases." *EMBO J* 7(8): 2593–2599.

Bode, W., F. X. Gomis-Ruth, and W. Stockler (1993). "Astacins, serralysins, snake venom and matrix metalloproteinases exhibit identical zinc-binding environments (HEXXHXXGXXH and Met-turn) and topologies and should be grouped into a common family, the 'metzincins.'" *FEBS Lett* 331(1–2): 134–140.

Boeckman, F. A., and E. Aizenman (1996). "Pharmacological properties of acquired excitotoxicity in Chinese hamster ovary cells transfected with N-methyl-D-aspartate receptor subunits." *J Pharmacol Exp Ther* 279(2): 515–523.

Boersema, P. J., S. Mohammed, and A. J. Heck (2008). "Hydrophilic interaction liquid chromatography (HILIC) in proteomics." *Anal Bioanal Chem* 391(1): 151–159.

Boffa, G. A., M. C. Boffa, and J. J. Winchenne (1976). "A phospholipase A_2 with anticoagulant activity, I: Isolation from *Vipera berus* venom and properties." *Biochim Biophys Acta* 429(3): 828–838.

Boffa, M. C., and G. A. Boffa (1976). "A phospholipase A_2 with anticoagulant activity, II: Inhibition of the phospholiped activity in coagulation." *Biochim Biophys Acta* 429(3): 839–852.

Bogdan, G. M., R. C. Dart, S. C. Falbo, J. McNally, and D. Spaite (2000). "Recurrent coagulopathy after antivenom treatment of crotalid snakebite." *South Med J* 93(6): 562–566.

Bogert, C. M. (1943). "Dentitional phenomena in cobras and other elapids with notes on adaptive modifications of the fangs." *Bull Am Mus Nat Hist* 131: 285–360.

Bogert, C. M., and R. M. del Campo (1956). "The gila monster and its allies: The relationships, habits, and behavior of the lizards of the family Helodermatidae." *Bull Am Mus Nat Hist* 109: 1–238.

Bohlen, C. J., A. T. Chesler, R. Sharif-Naeini, K. F. Medzihradszky, S. Zhou, D. King, E. E. Sanchez, A. L. Burlingame, A. I. Basbaum, and D. Julius (2011). "A heteromeric Texas coral snake toxin targets acid-sensing ion channels to produce pain." *Nature* 479(7373): 410–414.

Boisbouvier, J., J. P. Albrand, M. Blackledge, M. Jaquinod, H. Schweitz, M. Lazdunski, and D. Marion (1998). "A structural homologue of colipase in black mamba venom revealed by NMR floating disulfide bridge analysis." *J Mol Biol* 283(1): 205–219.

Boldrini-Franca, J., C. Correa-Netto, M. M. Silva, R. S. Rodrigues, P. De La Torre, A. Perez, A. M. Soares, et al. (2010). "Snake venomics and antivenomics of *Crotalus durissus* subspecies from Brazil: Assessment of geographic variation and its implication on snakebite management." *J Proteomics* 73(9): 1758–1776.

Bon, C. (1997). "Multicomponent neurotoxic phospholipases A 2." In *Phospholipase A 2 enzymes: Structure, function and mechanism*, edited by R. M. Kini. Chichester: John Wiley, 269–285.

Bon, C., J. P. Changeux, T. W. Jeng, and H. Fraenkel-Conrat (1979). "Postsynaptic effects of crotoxin and of its isolated subunits." *Eur J Biochem* 99(3): 471–481.

Booth, W., C. F. Smith, P. H. Eskridge, S. K. Hoss, J. R. Mendelson III, and G. W. Schuett (2012). "Facultative parthenogenesis discovered in wild vertebrates." *Biol Lett* 8(6): 983–985.

Borgheresi, R., M. S. Palma, F. Ducancel, A. C. M. Camargo, and E. Carmona (2001). "Expression and processing of recombinant sarafotoxins precursor in *Pichia pastoris*." *Toxicon* 39(8): 1211–1218.

Borgono, C. A., I. P. Michael, and E. P. Diamandis (2004). "Human tissue kallikreins: Physiologic roles and applications in cancer." *Mol Cancer Res* 2(5): 257–280.

Bos, M. H., M. Boltz, L. St Pierre, P. P. Masci, J. de Jersey, M. F. Lavin, and R. M. Camire (2009). "Venom factor V from the common brown snake escapes hemostatic regulation through procoagulant adaptations." *Blood* 114(3): 686–692.

Boss, E. A., R. Maciel Jr., and E. C. Vasco de Toledo (2004). "Freeze drying process: Real time model and optimization." *Chem Eng Process* 43: 1475–1485.

Bossi, E., M. S. Fabbrini, and A. Ceriotti (2007). "Exogenous protein expression in *Xenopus* oocytes: Basic procedures." *Methods Mol Biol* 375: 107–131.

Botto, M. (2000). "C3." In *The complement facts book*, edited by B. J. Morley and M. J. Walport. New York: Academic Press, 87–94.

Bouabboud, C. F., and D. G. Kardassakis (1988). "Acute myocardial-infarction following a gila monster (*Heloderma suspectum cinctum*) bite." *West J Med* 148(5): 577–579.

Bouchier, C., J. C. Boulain, C. Bon, and A. Ménez (1991). "Analysis of cDNAs encoding the two subunits of crotoxin, a phospholipase A_2 neurotoxin from rattlesnake venom: The acidic non enzymatic subunit derives from a phospholipase A_2-like precursor." *Biochim Biophys Acta* 1088(3): 401–408.

Bouchier, C., F. Ducancel, G. Guignery-Frelat, C. Bon, J. C. Boulain, and A. Ménez (1988). "Cloning and sequencing of cDNAs encoding the two subunits of Crotoxin." *Nucleic Acids Res* 16(18): 9050.

Bougis, P. E., A. Khelif, and H. Rochat (1989). "On the inhibition of [Na+,K+]-ATPases by the components of *Naja mossambica mossambica* venom: Evidence for two distinct rat brain [Na+,K+]-ATPase activities." *Biochemistry* 28(7): 3037–3043.

Boulenger, G. A. (1893). "Catalogue of snakes in the British Museum (Natural History)." 1.

Boutemy, L. S., S. R. F. King, J. Win, R. K. Hughes, T. A. Clarke, T. M. A. Blumenschein, S. Kamoun, and M. J. Banfield (2011). "Structures of phytophthora rxlr effector proteins a conserved but adaptable fold underpins functional diversity." *J Biol Chem* 286(41): 35834–35842.

Bowie, D., and T. G. Smart (1993). "Interplay between expressed non-NMDA receptors and endogenous calcium-activated chloride currents in *Xenopus laevis* oocytes." *Neurosci Lett* 151(1): 4–8.

Bowie, J. H., and M. J. Tyler (2006). "Host defence peptides from Australian amphibians: Caerulein and other neuropeptides." In *The handbook of biologically active peptides*, edited by A. J. Kaplin. San Diego, CA: Elsevier.

Bowie, J. U., R. Luthy, and D. Eisenberg (1991). "A method to identify protein sequences that fold into a known three-dimensional structure." *Science* 253(5016): 164–170.

Boyd, J. J., G. Agazzi, D. Svajda, A. J. Morgan, S. Ferrandis, and R. L. Norris (2007). "Venomous snakebite in mountainous terrain: Prevention and management." *Wilderness Environ Med* 18(3): 190–202.

Boyer, L., J. Degan, A. M. Ruha, J. Mallie, E. Mangin, and A. Alagón (2013a). "Safety of intravenous equine F(ab')2: Insights following clinical trials involving 1534 recipients of scorpion antivenom." *Toxicon* 76: 386–393.

Boyer, L., S. Seifert, R. Clark, J. McNally, S. Williams, S. Nordt, F. Walter, and R. Dart (1999). "Recurrent and persistent coagulopathy following pit viper envenomation." *Arch Intern Med* 159(7): 706–710.

Boyer, L. V., P. B. Chase, J. A. Degan, G. Figge, A. Buelna-Romero, C. Luchetti, and A. Alagón (2013b). "Subacute coagulopathy in a randomized, comparative trial of Fab and F(ab')2 antivenoms." *Toxicon* 74: 101–108.

Boyer, L. V., S. A. Seifert, and J. S. Cain (2001). "Recurrence phenomena after immunoglobulin therapy for snake envenomations: Part 2. Guidelines for clinical management with crotaline Fab antivenom." *Ann Emerg Med* 37(2): 196–201.

Braasch, I., J. N. Volff, and M. Schartl (2009). "The endothelin system: Evolution of vertebrate-specific ligand-receptor interactions by three rounds of genome duplication." *Mol Biol Evol* 26(4): 783–799.

Bracht, T., F. Figueiredo de Rezende, J. Stetefeld, L. M. Sorokin, and J. A. Eble (2011). "Monoclonal antibodies reveal the alteration of the rhodocetin structure upon alpha2beta1 integrin binding." *Biochem J* 440(1): 1–11.

Bradley, K. N. (2000). "Muscarinic toxins from the green mamba." *Pharmacol Ther* 85(2): 87–109.

Braga, M. D., A. M. Martins, D. N. Amora, D. B. de Ménezes, M. H. Toyama, D. O. Toyama, S. Marangoni, et al. (2008). "Purification and biological effects of L-amino acid oxidase isolated from *Bothrops insularis* venom." *Toxicon* 51(2): 199–207.

Braganca, B. M., and Y. M. Sambray (1967). "Multiple forms of cobra venom phospholipase A." *Nature* 216(5121): 1210–1211.

Braud, S., C. Bon, and A. Wisner (2000). "Snake venom proteins acting on hemostasis." *Biochimie* 82(9–10): 851–859.

Brazil, O. V., J. Prado-Franceschi, and C. J. Laure (1979). "Repetitive muscle responses induced by crotamine." *Toxicon* 1(1): 61–67.

Briggs, J. J., M. H. Haugen, H. T. Johansen, A. I. Riker, M. Abrahamson, O. Fodstad, G. M. Maelandsmo, and R. Solberg (2010). "Cystatin E/M suppresses legumain activity and invasion of human melanoma." *BMC Cancer* 10: 17.

Brillard-Bourdet, M., V. Nguyen, M. Ferrer-di Martino, F. Gauthier, and T. Moreau (1998). "Purification and characterization of a new cystatin inhibitor from Taiwan cobra (*Naja naja atra*) venom." *Biochem J* 331(pt 1): 239–244.

Broad, A. J., S. K. Sutherland, and A. R. Coulter (1979). "Lethality in mice of dangerous Australian and other snake venom." *Toxicon* 17(6): 661–664.

Brodie, E. D. III (2010). "Convergent evolution: Pick your poison carefully." *Curr Biol* 20(4): R152–R154.

Bronstein, A. C., F. E. Russell, and J. B. Sullivan (1986). "Negative-pressure suction in the field treatment of rattlesnake bite victims." *Vet Hum Toxicol* 28(5): 485–485.

Brooks, D. E., A. R. Means, E. J. Wright, S. P. Singh, and K. K. Tiver (1986). "Molecular cloning of the cDNA for androgen-dependent sperm-coating glycoproteins secreted by the rat epididymis." *Eur J Biochem* 161: 13–18.

Brown, A. M., A. Yatani, A. E. Lacerda, G. B. Gurrola, and L. D. Possani (1987). "Neurotoxins that act selectively on voltage-dependent cardiac calcium channels." *Circ Res* 61(4 pt. 2): I6–9.

Brown, D. E., and N. B. Carmony (1999). *Gila monster: Facts and folklore of America's Aztec lizard.* Salt Lake City: University of Utah Press.

Brown, G. B., J. A. Johnston, and L. C. Tolbert (1983). "The effect of toxin from *Leiurus quinquestriatus* scorpion venom on the polymerization and stability of microtubules." *Toxicon* 21(5): 699–708.

Brown, J. R., C. A. Bishop, and R. J. Brooks (2009). "Effectiveness of short-distance translocation and its effects on western rattlesnakes." *J Wildl Manage* 73(3): 419–425.

Brown, M. C., I. Staniszewska, L. Del Valle, G. P. Tuszynski, and C. Marcinkiewicz (2008). "Angiostatic activity of obtustatin as alpha1beta1 integrin inhibitor in experimental melanoma growth." *Int J Cancer* 123(9): 2195–2203.

Brown, R. L., T. L. Haley, K. A. West, and J. W. Crabb (1999). "Pseudechetoxin: A peptide blocker of cyclic nucleotide-gated ion channels." *Proc Natl Acad Sci USA* 96: 754–759.

Brown, R. L., L. L. Lynch, T. L. Haley, and R. Arsanjani (2003). "Pseudechetoxin binds to the pore turret of cyclic nucleotide-gated ion channels." *J Gen Physiol* 122: 749–760.

Brown, R. S., and J. J. Lennon (1995). "Sequence-specific fragmentation of matrix-assisted laser-desorbed protein/peptide ions." *Anal Chem* 67(21): 3990–3999.

Brummel, K. E., S. G. Paradis, S. Butenas, and K. G. Mann (2002). "Thrombin functions during tissue factor-induced blood coagulation." *Blood* 100(1): 148–152.

Brunner, H. R., H. Gavras, and J. H. Laragh (1974). "Specific inhibition of the renin-angiotensin system: A key to understanding blood pressure regulation." *Prog Cardiovasc Dis* 17(2): 87–98.

Brust, A., K. Sunagar, E. A. Undheim, I. Vetter, D. C. Yang, N. R. Casewell, T. N. Jackson, et al. (2013). "Differential evolution and neofunctionalization of snake venom metalloprotease domains." *Mol Cell Proteomics* 12(3): 651–663.

Bucknall, N. C. (1991). "Electrical treatment of venomous bites and stings." *Toxicon* 29(4–5): 397–400.

Bull, J. J., T. S. Jessop, and M. Whiteley (2010). "Deathly drool: Evolutionary and ecological basis of septic bacteria in Komodo dragon mouths." *PLoS One* 5(6): e11097.

Burbrink, F. T., and B. I. Crother (2011). "Evolution and taxonomy of snakes." In *Reproductive biology and phylogeny of snakes*, edited by D. M. Sever. Boca Raton, FL: CRC Press, 19–54.

Burcham, G. N., M. A. Miller, and T. S. Hickok (2011). "Pathology in practice: Mycotic dermatitis, cellulitis, and myositis." *J Am Vet Med Assoc* 239(10): 1305–1307.

Burdett-Smith, P. (1997). "Stun gun injury." *J Accid Emerg Med* 4(6): 402–404.

Burnouf, T., E. Griffiths, A. Padilla, S. Seddik, M. A. Stephano, and J. M. Gutiérrez (2004). "Assessment of the viral safety of antivenoms fractionated from equine plasma." *Biologicals* 32(3): 115–128.

Buse, J. B., D. C. Klonoff, L. L. Nielsen, X. Guan, C. L. Bowlus, J. H. Holcombe, D. G. Maggs, and M. E. Wintle (2007). "Metabolic effects of two years of exenatide treatment on diabetes, obesity, and hepatic biomarkers in patients with type 2 diabetes: An interim analysis of data from the open-label, uncontrolled extension of three double-blind, placebo-controlled trials." *Clin Ther* 29: 139–153.

Bush, S. P. (2004). "Snakebite suction devices don't remove venom: They just suck." *Ann Emerg Med* 43(2): 187–188.

Bush, S. P., and D. L. Hardy (2001). "Immediate removal of extractor is recommended." *Ann Emerg Med* 38(5): 607–608.

Bush, S. P., K. G. Hegewald, S. M. Green, M. D. Cardwell, and W. K. Hayes (2000). "Effects of a negative pressure venom extraction device (Extractor) on local tissue injury after artificial rattlesnake envenomation in a porcine model." *Wilderness Environ Med* 11(3): 180–188.

Bush, S. P., S. A. Seifert, J. Oakes, S. D. Smith, T. H. Phan, S. R. Pearl, and E. T. Reibling (2013). "Continuous IV Crotalidae polyvalent immune fab (ovine) (FabAV) for selected North American rattlesnake bite patients." *Toxicon* 69: 29–37.

Butenas, S., K. E. Brummel, R. F. Branda, S. G. Paradis, and K. G. Mann (2002). "Mechanism of factor VIIa-dependent coagulation in hemophilia blood." *Blood* 99(3): 923–930.

Butenas, S., C. vantVeer, and K. G. Mann (1997). "Evaluation of the initiation phase of blood coagulation using ultrasensitive assays for serine proteases." *J Biol Chem* 272(34): 21527–21533.

Butler, H., B. Malone, and N. Clemann (2005a). "Activity patterns and habitat preferences of translocated and resident tiger snakes (*Notechis scutatus*) in a suburban landscape." *Wildlife Res* 32(2): 157–163.

——— (2005b). "The effects of translocation on the spatial ecology of tiger snakes (*Notechis scutatus*) in a suburban landscape." *Wildlife Res* 32(2): 165–171.

Butt, R. H., and J. R. Coorssen (2013). "Coomassie blue as a near-infrared fluorescent stain: A systematic comparison with Sypro Ruby for in-gel protein detection." *Mol Cell Proteomics* 12(12): 3834–3850.

Cal, S., A. J. Obaya, M. Llamazares, C. Garabaya, V. Quesada, and C. Lopez-Otin (2002). "Cloning, expression analysis, and structural characterization of seven novel human ADAMTSs, a family of metalloproteinases with disintegrin and thrombospondin-1 domains." *Gene* 283(1–2): 49–62.

Calmette, A. (1894a). "Contribution à l'étude du venin des serpents: Immunisation des animaux et traitement de l'envenimation [Contribution to the study of snake venoms : Immunization of animals and treatment of envenomation]." *Ann Inst Pasteur* 8: 275–291.

——— (1894b). "L'immunisation artificielle des animaux contre le venin des serpents, et la thérapeutic expérimentale des morsures venimeuses. [The artificial immunization of animals against the venom of snakes, and the experimental therapy of venomous bites]." *C R Soc Biol* 46: 120–124.

——— (1897). "Sur le venin des serpents et sur l'emploi du serum antivenimeux dans latherapeutique des morsures venimeuses chez l'homme et chez les animaux. [On the venom of snakes and the use of antivenom in the therapy of venomous bites in human and animals]." *Ann Inst Pasteur* 12: 214–237.

Calvete, J. (2011a). "Brief history and molecular determinants of snake venom disintegrin evolution." In *Toxins and hemostasis*, edited by R. M. Kini, K. J. Clemetson, F. S. Markland, M. A. McLane, and T. Morita. Netherlands: Springer, 285–300.

——— (2011b). "Proteomic tools against the neglected pathology of snake bite envenoming." *Expert Rev Proteomics* 8(6): 739–758.

Calvete, J. J., A. Borges, A. Segura, M. Flores-Diaz, A. Alape-Giron, J. M. Gutiérrez, N. Diez, et al. (2009a). "Snake venomics and antivenomics of *Bothrops colombiensis*, a medically important pit-viper of the *Bothrops atrox-asper* complex endemic to Venezuela: Contributing to its taxonomy and snakebite management." *J Proteomics* 72(2): 227–240.

Calvete, J. J., J. Escolano, and L. Sanz (2007). "Snake venomics of *Bitis* species reveals large intragenus venom toxin composition variation: Application to taxonomy of congeneric taxa." *J Proteome Res* 6(7): 2732–2745.

Calvete, J. J., E. Fasoli, L. Sanz, E. Boschetti, and P. G. Righetti (2009b). "Exploring the venom proteome of the western diamondback rattlesnake, *Crotalus atrox*, via snake venomics and combinatorial peptide ligand library approaches." *J Proteome Res* 8(6): 3055–3067.

Calvete, J. J., P. Juarez, and L. Sanz (2007). "Snake venomics. Strategy and applications." *J Mass Spectrom* 42(11): 1405–1414.

Calvete, J. J., C. Marcinkiewicz, D. Monleon, V. Esteve, B. Celda, P. Juarez, and L. Sanz (2005). "Snake venom disintegrins: Evolution of structure and function." *Toxicon* 45(8): 1063–1074.

Calvete, J. J., C. Marcinkiewicz, and L. Sanz (2007). "Snake venomics of *Bitis gabonica* gabonica. Protein family composition, subunit organization of venom toxins, and characterization of dimeric disintegrins bitisgabonin-1 and bitisgabonin-2." *J Proteome Res* 6(1): 326–336.

Calvete, J. J., A. Perez, B. Lomonte, E. E. Sanchez, and L. Sanz (2012). "Snake venomics of *Crotalus tigris*: The minimalist toxin arsenal of the deadliest Neartic rattlesnake venom: Evolutionary clues for generating a pan-specific antivenom against crotalid type II venoms." *J Proteome Res* 11(2): 1382–1390.

Calvete, J. J., L. Sanz, A. Perez, A. Borges, A. M. Vargas, B. Lomonte, Y. Angulo, et al. (2011). "Snake population venomics and antivenomics of *Bothrops atrox*: Paedomorphism along its transamazonian dispersal and implications of geographic venom variability on snakebite management." *J Proteomics* 74(4): 510–527.

Cambi, A., and C. G. Figdor (2003). "Dual function of C-type lectin-like receptors in the immune system." *Curr Opin Cell Biol* 15(5): 539–546.

Cameo, M. S., and J. Blaquier (1976). "Androgen-controlled specific proteins in rat epididymis." *J Endocrinol* 69: 47–55.

Cameron, D. L., and A. T. Tu (1977). "Characterization of myotoxin a from the venom of prairie rattlesnake (*Crotalus viridis viridis*)." *Biochemistry* 16(11): 2546–2553.

Campbell, C. L., K. A. Vandyke, G. J. Letchworth, B. S. Drolet, T. Hanekamp, and W. C. Wilson (2005). "Midgut and salivary gland transcriptomes of the arbovirus vector *Culicoides sonorensis* (Diptera: Ceratopogonidae)." *Insect Mol Biol* 14(2): 121–136.

Campbell, C. L., W. C. Wilson, and K. Manninen (2005). "Characterization of differentially expressed midge genes in orbivirus vector populations." *Am J Trop Med Hyg* 73(6): 144.

Canale, E., G. K. Isbister, and B. J. Currie (2009). "Investigating pressure bandaging for snakebite in a simulated setting: Bandage type, training and the effect of transport." *Emerg Med Australas* 21(3): 184–190.

Cantrell, F. L. (2003). "Envenomation by the Mexican beaded lizard: A case report." *J Toxicol Clin Toxicol* 41(3): 241–244.

Canutescu, A. A., A. A. Shelenkov, and R. L. Dunbrack Jr. (2003). "A graph-theory algorithm for rapid protein side-chain prediction." *Protein Sci* 12(9): 2001–2014.

Carbajal-Saucedo, A., E. Lopez-Vera, M. Bénard-Valle, E. N. Smith, F. Zamudio, A. R. de Roodt, and A. Olvera-Rodríguez (2013). "Isolation, characterization, cloning and expression of an alpha-neurotoxin from the venom of the Mexican coral snake *Micrurus laticollaris* (Squamata: Elapidae)." *Toxicon* 66: 64–74.

Carneiro, S. M., M. B. Zablith, C. M. Kerchove, A. M. Moura-da-Silva, D. O. Quissell, R. P. Markus, and N. Yamanouye (2006). "Venom production in long-term primary culture of secretory cells of the *Bothrops jararaca* venom gland." *Toxicon* 47(1): 87–94.

Carr, M. E. Jr., S. L. Carr, and P. E. Greilich (1996). "Heparin ablates force development during platelet mediated clot retraction." *Thromb Haemost* 75(4): 674–678.

Casewell, N. R. (2012). "On the ancestral recruitment of metalloproteinases into the venom of snakes." *Toxicon* 60(4): 449–454.

Casewell, N. R., D. A. Cook, S. C. Wagstaff, A. Nasidi, N. Durfa, W. Wüster, and R. A. Harrison (2010). "Pre-clinical assays predict pan-African *Echis* viper efficacy for a species-specific antivenom." *PLoS Negl Trop Dis* 4(10): e851.

Casewell, N. R., R. A. Harrison, W. Wüster, and S. C. Wagstaff (2009). "Comparative venom gland transcriptome surveys of the saw-scaled vipers (Viperidae: *Echis*) reveal substantial intra-family gene diversity and novel venom transcripts." *BMC Genomics* 10: 564.

Casewell, N. R., S. C. Wagstaff, R. A. Harrison, C. Renjifo, and W. Wüster (2011a). "Domain loss facilitates accelerated evolution and neofunctionalization of duplicate snake venom metalloproteinase toxin genes." *Mol Biol Evol* 28(9): 2637–2649.

Casewell, N. R., S. C. Wagstaff, R. A. Harrison, and W. Wüster (2011b). "Gene tree parsimony of multilocus snake venom protein families reveals species tree conflict as a result of multiple parallel gene loss." *Mol Biol Evol* 28(3): 1157–1172.

Casewell, N. R., S. C. Wagstaff, W. Wüster, D. A. Cook, F. M. Bolton, S. I. King, D. Pla, L. Sanz, J. J. Calvete, and R. A. Harrison (2014). "Medically important differences in snake venom composition are dictated by distinct postgenomic mechanisms." *Proc Natl Acad Sci USA* 111(25): 9205–9210. doi: 10.1073/pnas.1405484111.

Casewell, N. R., W. Wüster, F. J. Vonk, R. A. Harrison, and B. G. Fry (2013). "Complex cocktails: The evolutionary novelty of venoms." *Trends Ecol Evol* 28(4): 219–229.

Castino, R., D. Pace, M. Demoz, M. Gargiulo, C. Ariatta, E. Raiteri, and C. Isidoro (2002). "Lysosomal proteases as potential targets for the induction of apoptotic cell death in human neuroblastomas." *Int J Cancer* 97(6): 775–779.

Castro, E. N., B. Lomonte, M. Del Carmen Gutiérrez, A. Alagón, and J. M. Gutiérrez (2013). "Intraspecies variation in the venom of the rattlesnake *Crotalus simus* from Mexico: Different expression of crotoxin results in highly variable toxicity in the venoms of three subspecies." *J Proteomics* 87: 103–121.

Castro, O., J. M. Gutiérrez, M. Barrios, I. Castro, M. Romero, and E. Umaña (1999). "Neutralización del efecto hemorrágico inducido por veneno de *Bothrops asper* (Serpentes: Viperidae) por extractos de plantas tropicales." *Rev Biol Trop* 47: 605–616.

Chacon, D., S. Rodriguez, J. Arias, G. Solano, F. Bonilla, and A. Gomez (2012). "Maintaining coral snakes (*Micrurus nigrocinctus*, Serpentes: Elapidae) for venom production on an alternative fish-based diet." *Toxicon* 60(3): 249–253.

Chacur, M., I. Longo, G. Picolo, J. M. Gutiérrez, B. Lomonte, J. L. Guerra, C. F. Teixeira, and Y. Cury (2003). "Hyperalgesia induced by Asp49 and Lys49 phospholipases A_2 from *Bothrops asper* snake venom: Pharmacological mediation and molecular determinants." *Toxicon* 41(6): 667–678.

Chacur, M., G. Picolo, J. M. Gutiérrez, C. F. Teixeira, and Y. Cury (2001). "Pharmacological modulation of hyperalgesia induced by *Bothrops asper* (terciopelo) snake venom." *Toxicon* 39(8): 1173–1181.

Chagot, B., S. Diochot, C. Pimentel, M. Lazdunski, and H. Darbon (2005). "Solution structure of APETx1 from the sea anemone *Anthopleura elegantissima*: A new fold for an HERG toxin." *Proteins* 59(2): 380–386.

Chaisakul, J., G. K. Isbister, N. Konstantakopoulos, M. Tare, H. C. Parkington, and W. C. Hodgson (2012). "*In vivo* and *in vitro* cardiovascular effects of Papuan taipan (*Oxyuranus scutellatus*) venom: Exploring 'sudden collapse.'" *Toxicol Lett* 213(2): 243–248.

Chaisakul, J., G. K. Isbister, S. Kuruppu, N. Konstantakopoulos, and W. C. Hodgson (2013). "An examination of cardiovascular collapse induced by eastern brown snake (*Pseudonaja textilis*) venom." *Toxicol Lett* 221(3): 205–211.

Champagne, D. E., R. H. Nussenzveig, and J. M. C. Ribeiro (1995). "Purification, partial characterization, and cloning of nitric oxide-carrying heme-proteins (nitrophorins) from salivary-glands of the bloodsucking insect *Rhodnius prolixus*." *J Biol Chem* 270(15): 8691–8695.

Chang, C. C., and C. Y. Lee (1963). "Isolation of neurotoxins from the venom of *Bungarus multicinctus* and their modes of neuromuscular blocking action." *Arch Int Pharmacod Ther* 144: 241–257.

Chang, C. C., and K. H. Tseng (1978). "Effect of crotamine, a toxin of South American rattlesnake venom, on the sodium channel of murine skeletal muscle." *Br J Pharmacol* 63(3): 551–559.

Chang, C. H., C. H. Chung, C. C. Hsu, T. Y. Huang, and T. F. Huang (2010). "A novel mechanism of cytokine release in phagocytes induced by aggretin, a snake venom C-type lectin protein, through CLEC-2 ligation." *J Thromb Haemost* 8(11): 2563–2570.

Chang, L. S., C. Chung, B. N. Wu, and C. C. Yang (2002). "Characterization and gene organization of Taiwan banded krait (*Bungarus multicinctus*) gamma-bungarotoxin." *J Protein Chem* 21(4): 223–229.

Chang, M. S., D. G. Lowe, M. Lewis, R. Hellmiss, E. Chen, and D. V. Goeddel (1989). "Differential activation by atrial and brain natriuretic peptides of two different receptor guanylate cyclases." *Nature* 341(6237): 68–72.

Chao, M. V., and M. Bothwell (2002). "Neurotrophins: To cleave or not to cleave." *Neuron* 33(1): 9–12.

Charlab, R., J. G. Valenzuela, E. D. Rowton, and J. M. C. Ribeiro (1999). "Toward an understanding of the biochemical and pharmacological complexity of the saliva of a hematophagous sand fly *Lutzomyia longipalpis*." *Proc Natl Acad Sci USA* 96(26): 15155–15160.

Chatrath, S. T., A. Chapeaurouge, Q. Lin, T. K. Lim, N. Dunstan, P. Mirtschin, P. P. Kumar, and R. M. Kini (2011). "Identification of novel proteins from the venom of a cryptic snake *Drysdalia coronoides* by a combined transcriptomics and proteomics approach." *J Proteome Res* 10(2): 739–750.

Chattopadhyay, S., N. K. Sinha, S. Banerjee, D. Roy, D. Chattopadhyay, and S. Roy (2006). "Small cationic protein from a marine turtle has beta-defensin-like fold and antibacterial and antiviral activity." *Proteins* 64(2): 524–531.

Chaves, F., G. Leon, V. H. Alvarado, and J. M. Gutiérrez (1998). "Pharmacological modulation of edema induced by Lys-49 and Asp-49 myotoxic phospholipases A_2 isolated from the venom of the snake *Bothrops asper* (terciopelo)." *Toxicon* 36(12): 1861–1869.

Chemical Solutions (2014). "Snake Repel: Don't get bitten. Buyer's guide." http://www.chemicalsolutions-sa.co.za/product_data/snake_repel_buyers_guide.pdf.

Chen, C. P., C. H. Hsu, N. Y. Su, Y. C. Lin, S. H. Chiou, and S. H. Wu (2001). "Solution structure of a Kunitz-type chymotrypsin inhibitor isolated from the elapid snake *Bungarus fasciatus*." *J Biol Chem* 276(48): 45079–45087.

Chen, D. S., R. M. Kini, R. Yuen, and H. E. Khoo (1997). "Haemolytic activity of stonustoxin from stonefish (*Synanceja horrida*) venom: Pore formation and the role of cationic amino acid residues." *Biochem J* 325: 685–691.

Chen, H. S., J. M. Chen, C. W. Lin, K. H. Khoo, and I. H. Tsai (2008a). "New insights into the functions and N-glycan structures of factor X activator from Russell's viper venom." *FEBS J* 275(15): 3944–3958.

Chen, H. S., Y. M. Wang, W. T. Huang, K. F. Huang, and I. H. Tsai (2012b). "Cloning, characterization and mutagenesis of Russell's viper venom L-amino acid oxidase: Insights into its catalytic mechanism." *Biochimie* 94(2): 335–344.

Chen, J., M. Deng, Q. He, E. Meng, L. Jiang, Z. Liao, M. Rong, and S. Liang (2008b). "Molecular diversity and evolution of cystine knot toxins of the tarantula *Chilobrachys jingzhao*." *Cell Mol Life Sci* 65(15): 2431–2444.

Chen, J., T. G. Diacovo, D. G. Grenache, S. A. Santoro, and M. M. Zutter (2002). "The alpha(2) integrin subunit-deficient mouse: A multifaceted phenotype including defects of branching morphogenesis and hemostasis." *Am J Pathol* 161(1): 337–344.

Chen, K. S., and F. H. Gage (1995). "Somatic gene-transfer of NGF to the aged brain—behavioral and morphological amelioration." *J Neurosci* 15(4): 2819–2825.

Chen, P. C., M. A. Hayashi, E. B. Oliveira, and R. L. Karpel (2012a). "DNA-interactive properties of crotamine, a cell-penetrating polypeptide and a potential drug carrier." *PLoS One* 7(11): e48913.

Chen, L. W., P. H. Kao, Y. S. Fu, W. P. Hu, and L. S. Chang (2011a). "Bactericidal effect of *Naja nigricollis* toxin gamma is related to its membrane-damaging activity." *Peptides* 32(8): 1755–1763.

Chen, L. W., P. H. Kao, Y. S. Fu, S. R. Lin, and L. S. Chang (2011b). "Membrane-damaging activity of Taiwan cobra cardiotoxin 3 is responsible for its bactericidal activity." *Toxicon* 58(1): 46–53.

Chen, R. Q., Y. Jin, J. B. Wu, X. D. Zhou, Q. M. Lu, W. Y. Wang, and Y. L. Xiong (2003a). "A new protein structure of P-II class snake venom metalloproteinases: It comprises metalloproteinase and disintegrin domains." *Biochem Biophys Res Comm* 310(1): 182–187.

Chen, T. B., S. Farragher, A. J. Bjourson, D. F. Orr, P. Rao, and C. Shaw (2003b). "Granular gland transcriptomes in stimulated amphibian skin secretions." *Biochem J* 371: 125–130.

Chen, T. B., Y. Z. Xue, M. Zhou, and C. Shaw (2005). "Molecular cloning of mRNA from toad granular gland secretion and lyophilized skin: Identification of Bo8—a novel prokineticin from *Bombina orientalis*." *Peptides* 26(3): 377–383.

Chen, Y., S. M. Pitzenberger, V. M. Garsky, P. K. Lumma, G. Sanyal, and J. Baum (1991). "Proton NMR assignments and secondary structure of the snake venom protein echistatin." *Biochemistry* 30(50): 11625–11636.

Chen, Y. H., Y. M. Wang, M. J. Hseu, and I. H. Tsai (2004). "Molecular evolutionand structure-function relationships of crotoxin-like and asparagine-6-containing phospholipases A_2 in pit viper venoms." *Biochem J* 381: 25–34.

Chen, Y. L., and I. H. Tsai (1996). "Functional and sequence characterization of coagulation factor IX/factor X-binding protein from the venom of *Echis carinatus leucogaster*." *Biochemistry* 35(16): 5264–5271.

Chen, Y. Q. E., and D. J. Drucker (1997). "Tissue-specific expression of unique mRNAs that encode proglucagon-derived peptides or exendin 4 in the lizard." *J Biol Chem* 272(7): 4108–4115.

Chen, Z., L. Lin, Q. Huai, and M. Huang (2009). "Challenges for drug discovery—a case study of urokinase receptor inhibition." *Comb Chem High Throughput Screen* 12(10): 961–967.

Cheng, B. C., X. P. Zhou, Q. Zhu, S. Gong, Z. H. Qin, P. F. Reid, L. N. Raymond, Q. Z. Yin, and X. H. Jiang (2009). "Cobratoxin inhibits pain-evoked discharge of neurons in thalamic parafascicular nucleus in rats: Involvement of cholinergic and serotonergic systems." *Toxicon* 54(3): 224–232.

Cherdchu, C., N. Poopyruchpong, R. Adchariyasucha, and K. Ratanabanangkoon (1977). "The absence of antagonism between extracts of *Clinacanthus nutans Burm.* and *Naja naja siamensis* venom." *Southeast Asian J Trop Med Public Health* 8(2): 249–254.

Chernyavsky, A. I., J. Arredondo, V. Galitovskiy, J. Qian, and S. A. Grando (2009). "Structure and function of the nicotinic arm of acetylcholine regulatory axis in human leukemic T cells." *Int J Immunopathol Pharmacol* 22(2): 461–472.

Chetty, N., A. Du, W. C. Hodgson, K. Winkel, and B. G. Fry (2004). "The *in vitro* neuromuscular activity of Indo-Pacific sea-snake venoms: Efficacy of two commercially available antivenoms." *Toxicon* 44(2): 193–200.

Chevreux, B., T. Pfisterer, B. Drescher, A. J. Driesel, W. E. G. Müller, T. Wetter, and S. Suhai (2004). "Using the miraEST assembler for reliable and automated mRNA transcript assembly and SNP detection in sequenced ESTs." *Genome Res* 14(6): 1147–1159.

Cheymol, J., J. M. Goncalves, F. Bourillet, and M. Roch-Arveiller (1971). "A comparison of the neuromuscular action of crotamine and the venom of *Crotalus durissus terrificus* var. *crotaminicus*: 2. Isolated preparations." *Toxicon* 9(3): 287–289.

Chiappinelli, V. A., J. B. Cohen, and R. E. Zigmond (1981). "The effects of alpha- and beta-neurotoxins from the venoms of various snakes on transmission in autonomic ganglia." *Brain Res* 211(1): 107–126.

Chijiwa, T., E. Tokunaga, R. Ikeda, K. Terada, T. Ogawa, N. Oda-Ueda, S. Hattori, M. Nozaki, and M. Ohno (2006). "Discovery of novel [Arg49]phospholipase A_2 isozymes from *Protobothrops elegans* venom and regional evolution of Crotalinae snake venom phospholipase A_2 isozymes in the southwestern islands of Japan and Taiwan." *Toxicon* 48(6): 672–682.

Chimienti, F., R. C. Hogg, L. Plantard, C. Lehmann, N. Brakch, J. Fischer, M. Huber, D. Bertrand, and D. Hohl (2003). "Identification of SLURP-1 as an epidermal neuromodulator explains the clinical phenotype of Mal de Meleda." *Hum Mol Genet* 12(22): 3017–3024.

Ching, A. T., A. F. Paes Leme, A. Zelanis, M. M. Rocha, F. Furtado Mde, D. A. Silva, M. R. Trugilho, et al. (2012). "Venomics profiling of *Thamnodynastes strigatus* unveils matrix metalloproteinases and other novel proteins recruited to the toxin arsenal of rear-fanged snakes." *J Proteome Res* 11(2): 1152–1162.

Ching, A. T., M. M. Rocha, A. F. Paes Leme, D. C. Pimenta, D. F. M. de Fatima, S. M. Serrano, P. L. Ho, and I. L. Junqueira de Azevedo (2006). "Some aspects of the venom proteome of the Colubridae snake *Philodryas olfersii* revealed from a Duvernoy's (venom) gland transcriptome." *FEBS Lett* 580(18): 4417–4422.

Chioato, L., E. A. Aragao, T. Lopes Ferreira, A. I. Medeiros, L. H. Faccioli and R. J. Ward (2007). "Mapping of the structural determinants of artificial and biological membrane damaging activities of a Lys49 phospholipase A_2 by scanning alanine mutagenesis." *Biochim Biophys Acta* 1768(5): 1247–1257.

Chioato, L., A. H. de Oliveira, R. Ruller, J. M. Sa, and R. J. Ward (2002). "Distinct sites for myotoxic and membrane-damaging activities in the C-terminal region of a Lys49-phospholipase A_2." *Biochem J* 366 (pt. 3): 971–976.

Chioato, L., and R. J. Ward (2003). "Mapping structural determinants of biological activities in snake venom phospholipases A_2 by sequence analysis and site directed mutagenesis." *Toxicon* 42(8): 869–883.

Chippaux, J. P. (1998). "Snake-bites: Appraisal of the global situation." *Bull WHO* 76(5): 515–524.

Chippaux, J. P. (2008). "Estimating the global burden of snakebite can help to improve management." *PLoS Med* 5(11): e221.

Chippaux, J. P. (2009). "Snakebite in Africa: Current situation and urgent needs." In *Handbook of venoms and toxins of reptiles*, edited by S. P. Mackessy. Boca Raton, FL: CRC Press, 453–473.

Chippaux, J. P., and M. Goyffon (1991). "Production and use of snake antivenin." In *Reptile venoms and toxins*, edited by A. T. Tu. New York: Marcel Dekker, Vol. 5: 529–555.

——— (1998). "Venoms, antivenoms and immunotherapy." *Toxicon* 36(6): 823–846.

Chippaux, J. P., B. Ramos-Cerrillo, and R. P. Stock (2007). "Study of the efficacy of the black stone on envenomation by snake bite in the murine model." *Toxicon* 49(5): 717–720.

Chippaux, J. P., R. P. Stock, and A. Massougbodji (2010). "Methodology of clinical studies dealing with the treatment of envenomation." *Toxicon* 55(7): 1195–1212.

Chippaux, J. P., V. Williams, and J. White (1991). "Snake venom variability: Methods of study, results and interpretation." *Toxicon* 29(11): 1279–1303.

Chiszar, D., G. H. Rodda, and H. M. Smith (1997). "Experiments on chemical control of behavior in Brown Tree Snakes." In *Repellents in wildlife management*, edited by J. R. Mason. Fort Collins, CO: US Department of Agriculture National Wildlife Research Center, 121–127.

Chitty, J. (2011). "Hospitalization of birds and reptiles." *J ExotPet Med* 20(2): 98–106.

Choe, L., M. D'Ascenzo, N. R. Relkin, D. Pappin, P. Ross, B. Williamson, S. Guertin, P. Pribil, and K. H. Lee (2007). "8-plex quantitation of changes in cerebrospinal fluid protein expression in subjects undergoing intravenous immunoglobulin treatment for Alzheimer's disease." *Proteomics* 7(20): 3651–3660.

Choo, Y. M., B. H. Lee, K. S. Lee, B. Y. Kim, J. Li, J. G. Kim, J. H. Lee, H. D. Sohn, S. Y. Nah, and B. R. Jin (2008). "Pr-lynx1, a modulator of nicotinic acetylcholine receptors in the insect." *Mol Cell Neurosci* 38(2): 224–235.

Chothia, C., and A. M. Lesk (1986). "The relation between the divergence of sequence and structure in proteins." *EMBO J* 5(4): 823–826.

Chow, G., S. Subburaju, and R. M. Kini (1998). "Purification, characterization, and amino acid sequence determination of acanthins, potent inhibitors of platelet aggregation from *Acanthophis antarcticus* (common death adder) venom." *Arch Biochem Biophys* 354(2): 232–238.

Chu, C. C., S. T. Chu, S. W. Chen, and Y. H. Chen (1994). "The non-phospholipase-A_2 subunit of β-bungarotoxin plays an important role in the phospholipase-A_2 independent neurotoxic effect—characterization of three isotoxins with a common phospholipase-A_2 subunit." *Biochem J* 303: 171–176.

Chung, C., B. N. Wu, C. C. Yang, and L. S. Chang (2002). "Muscarinic toxin-like proteins from Taiwan banded krait (*Bungarus multicinctus*) venom: Purification, characterization and gene organization." *Biol Chem* 383(9): 1397–1406.

Chung, C. H., L. C. Au, and T. F. Huang (1999). "Molecular cloning and sequence analysis of aggretin, a collagen-like platelet aggregation inducer." *Biochem Biophys Res Commun* 263(3): 723–727.

Chung, C. H., K. T. Lin, C. H. Chang, H. C. Peng, and T. F. Huang (2009). "The integrin alpha2beta1 agonist, aggretin, promotes proliferation and migration of VSMC through NF-kB translocation and PDGF production." *Br J Pharmacol* 156(5): 846–856.

Chung, C. H., H. C. Peng, and T. F. Huang (2001). "Aggretin, a C-type lectin protein, induces platelet aggregation via integrin alpha(2)beta(1) and GPIb in a phosphatidylinositol 3-kinase independent pathway." *Biochem Biophys Res Commun* 285(3): 689–695.

Chung, C. H., W. B. Wu, and T. F. Huang (2004). "Aggretin, a snake venom-derived endothelial integrin alpha 2 beta 1 agonist, induces angiogenesis via expression of vascular endothelial growth factor." *Blood* 103(6): 2105–2113.

Cidade, D. A., T. A. Simao, A. M. Davila, G. Wagner, I. L. Junqueira de Azevedo, P. L. Ho, C. Bon, R. B. Zingali, and R. M. Albano (2006). "*Bothrops jararaca* venom gland transcriptome: Analysis of the gene expression pattern." *Toxicon* 48(4): 437–461.

Ciscotto, P., R. A. Machado de Avila, E. A. Coelho, J. Oliveira, C. G. Diniz, L. M. Farias, M. A. de Carvalho, et al. (2009). "Antigenic, microbicidal and antiparasitic properties of an L-amino acid oxidase isolated from *Bothrops jararaca* snake venom." *Toxicon* 53(3): 330–341.

Clemetson, K. J. (2010). "Snaclecs (snake C-type lectins) that inhibit or activate platelets by binding to receptors." *Toxicon* 56(7): 1236–1246.

Clemetson, K. J., and J. M. Clemetson (2001). "Platelet collagen receptors." *Thromb Haemost* 86(1): 189–197.

——— (2008). "Platelet GPIb complex as a target for anti-thrombotic drug development." *Thromb Haemost* 99(3): 473–479.

Clemetson, K. J., Q. M. Lu, and J. M. Clemetson (2005). "Snake C-type lectin-like proteins and platelet receptors." *Pathophysiol Haemost Thromb* 34(4–5): 150–155.

Clemetson, K. J., A. Navdaev, D. Dormann, X. Y. Du, and J. M. Clemetson (2001). "Multifunctional snake C-type lectins affecting platelets." *Haemostasis* 31(3–6): 148–154.

Cohen, S., and R. Levi-Montalcini (1956). "A nerve growth-stimulating factor isolated from snake venom." *Proc Natl Acad Sci USA* 42: 571–574.

Cohen, W. R., W. Wetzel, and A. Kadish (1992). "Local heat and cold application after Eastern cottonmouth moccasin (*Agkistrodon piscivorus*) envenomation in the rat—effect on tissue-injury." *Toxicon* 30(11): 1383–1386.

Collins, E., M. P. Bracamonte, J. C. Burnett Jr., and V. M. Miller (2000). "Mechanism of relaxations to dendroaspis natriuretic peptide in canine coronary arteries." *J Cardiovasc Pharmacol* 35(4): 614–618.

Colman, R. W. (2006). "Are hemostasis and thrombosis two sides of the same coin?" *J Exp Med* 203(3): 493–495.

Colucci, W. S., U. Elkayam, D. P. Horton, W. T. Abraham, R. C. Bourge, A. D. Johnson, L. E. Wagoner, et al. (2000). "Intravenous nesiritide, a natriuretic peptide, in the treatment of decompensated congestive heart failure: Nesiritide Study Group." *N Engl J Med* 343(4): 246–253.

Compeau, P. E., P. A. Pevzner, and G. Tesler (2011). "How to apply de Bruijn graphs to genome assembly." *Nat Biotechnol* 29 (11): 987–991.

Condrea, E. (1976). "Hemolytic disorders associated with a primary red cell membrane defect." *Experientia* 32(5): 537–542.

Conner, J. M., M. A. Darracq, J. Roberts, and M. H. Tuszynski (2001). "Nontropic actions of neurotrophins: Subcortical nerve growth factor gene delivery reverses age-related degeneration of primate cortical cholinergic innervation." *Proc Natl Acad Sci USA* 98(4): 1941–1946.

Cook, D. A., T. Owen, S. C. Wagstaff, J. Kinne, U. Wernery, and R. A. Harrison (2010a). "Analysis of camelid antibodies for antivenom development: Neutralisation of venom-induced pathology." *Toxicon* 56(3): 373–380.

——— (2010b). "Analysis of camelid IgG for antivenom development: Serological responses of venom-immunised camels to prepare either monospecific or polyspecific antivenoms for West Africa." *Toxicon* 56(3): 363–372.

Cope, E. D. (1900). "The crocodilians, lizards, and snakes of North America." *Rep US Natl Mus*: 153–1270.

Correa, P. G., and N. Oguiura (2013). "Phylogenetic analysis of β-defensin-like genes of Bothrops, *Crotalus* and *Lachesis* snakes." *Toxicon* 69: 65–74.

Correa-Netto, C., I. L. Junqueira de Azevedo, D. A. Silva, P. L. Ho, M. Leitao de Araujo, M. L. Alves, L. Sanz, et al. (2011). "Snake venomics and venom gland transcriptomic analysis of Brazilian coral snakes, *Micrurus altirostris* and *M. corallinus*." *J Proteomics* 74(9): 1795–1809.

Costa de Almeida, C. M., C. L. da Silva, H. Pena Couto, R. d. C. Mothé Escocard, D. Gitirana da Rocha, L. d. P. Sentinelli, T. Liberman Kipnis, and W. Dias da Silva (1998). "Development of snake antivenom antibodies in chickens and their purification from yolk." *Vet Rec* 143(21): 579–584.

Costa, J. O., K. C. Fonseca, C. C. Mamede, M. E. Beletti, N. A. Santos Jr., A. M. Soares, E. C. Arantes, et al. (2010). "Bhalternin: Functional and structural characterization of a new thrombin-like enzyme from *Bothrops alternatus* snake venom." *Toxicon* 55(7): 1365–1377.

Costa, L. A., H. Miles, C. E. Araujo, S. Gonzalez, and V. G. Villarrubia (1998). "Tumor regression of advanced carcinomas following intra- and/or peri-tumoral inoculation with VRCTC-310 in humans: Preliminary report of two cases." *Immunopharmacol Immunotoxicol* 20(1): 15–25.

Costa Torres, A. F., R. T. Dantas, M. H. Toyama, E. Diz Jr., F. J. Zara, M. G. Rodrigues de Queiroz, N. A. Pinto Nogueira, et al. (2010). "Antibacterial and antiparasitic effects of *Bothrops marajoensis* venom and its fractions: Phospholipase A_2 and L-amino acid oxidase." *Toxicon* 55(4): 795–804.

Cousin, X., S. Bon, N. Duval, J. Massoulié, and C. Bon (1996a). "Cloning and expression of acetylcholinesterase from *Bungarus fasciatus* venom: A new type of COOH-terminal domain; involvement of a positively charged residue in the peripheral site." *J Biol Chem* 271(25): 15099–15108.

Cousin, X., S. Bon, J. Massoulié, and C. Bon (1998). "Identification of a novel type of alternatively spliced exon from the acetylcholinesterase gene of *Bungarus fasciatus*: Molecular forms of acetylcholinesterase in the snake liver and muscle." *J Biol Chem* 273(16): 9812–9820.

Cousin, X., C. Creminon, J. Grassi, K. Meflah, G. Cornu, B. Saliou, S. Bon, J. Massoulié, and C. Bon (1996b). "Acetylcholinesterase from Bungarus venom: A monomeric species." *FEBS Lett* 387(2–3): 196–200.

Couturier, S., D. Bertrand, J. M. Matter, M. C. Hernandez, S. Bertrand, N. Millar, S. Valera, T. Barkas, and M. Ballivet (1990). "A neuronal nicotinic acetylcholine receptor subunit (alpha 7) is developmentally regulated and forms a homo-oligomeric channel blocked by alpha-BTX." *Neuron* 5(6): 847–856.

Cox, J. C., and A. R. Coulter (1997). "Adjuvants: A classification and review of their modes of action." *Vaccine* 15(3): 246–256.

Craig, R. and R. C. Beavis (2004). "TANDEM: Matching proteins with tandem mass spectra." *Bioinformatics* 20(9): 1466–1467.

Creer, S., A. Malhotra, R. S. Thorpe, R. S. Stocklin, P. S. Favreau, and W. S. Hao Chou (2003). "Genetic and ecological correlates of intraspecific variation in pitviper venom composition detected using matrix-assisted laser desorption time-of-flight mass spectrometry (MALDI-TOF-MS) and isoelectric focusing." *J Mol Evol* 56(3): 317–329.

Criste, A., B. Georgescu, F. Criste, B. Anca, and O. Negrea (2008). "The use of the anti-venom specific antibodies isolated from duck eggs for inactivation of the viper venom." *Lucrări Stiintifice Zootehnie si Biotehnologii* 41(1): 49–52.

Crocker, P. R., J. C. Paulson, and A. Varki (2007). "Siglecs and their roles in the immune system." *Nat Rev Immunol* 7(4): 255–266.

Crouser, E. D., D. A. Culver, K. S. Knox, M. W. Julian, G. Shao, S. Abraham, S. Liyanarachchi, et al. (2009). "Gene expression profiling identifies MMP-12 and ADAMDEC1 as potential pathogenic mediators of pulmonary sarcoidosis." *Am J Respir Crit Care Med* 179(10): 929–938.

Cui, J. S., K. S. Oshea, A. Purkayastha, T. L. Saunders, and D. Ginsburg (1996). "Fatal haemorrhage and incomplete block to embryogenesis in mice locking coagulation factor V." *Nature* 384(6604): 66–68.

Cull-Candy, S. G., J. Fohlman, D. Gustavsson, R. Lullmann-Rauch, and S. Thesleff (1976). "The effects of taipoxin and notexin on the function and fine structure of the murine neuromuscular junction." *Neuroscience* 1(3): 175–180.

Cummins, P. M., O. Dowling, and B. F. O'Connor (2011). "Ion-exchange chromatography: Basic principles and application to the partial purification of soluble mammalian prolyl oligopeptidase." *Methods Mol Biol* 681: 215–228.

Cumpston, K. L. (2011). "Is there a role for fasciotomy in Crotalinae envenomations in North America?" *Clin Toxicol* 49(5): 351–365.

Cunningham, A. A. (1996). "Disease risks of wildlife translocations." *Conserv Biol* 10(2): 349–353.

Currie, B. J. (2006). "Treatment of snakebite in Australia: The current evidence base and questions requiring collaborative multicentre prospective studies." *Toxicon* 48 (7): 941–956.

Currie, B. J., E. Canale, and G. K. Isbister (2008). "Effectiveness of pressure-immobilization first aid for snakebite requires further study." *Emerg Med Australas* 20(3): 267–270.

Currie, J. R., B. A. Oxon, C. B. Glas, M. A. Edin, and D. P. H. Birm (1907). "On the supersensitation of persons suffering from diphtheria by repeated injections of horse serum." *J Hyg* 8: 35–60.

Cushman, D. W., H. S. Cheung, E. F. Sabo, and M. A. Ondetti (1977). "Design of potent competitive inhibitors of angiotensin-converting enzyme: Carboxyalkanoyl and mercaptoalkanoyl amino acids." *Biochemistry* 16(25): 5484–5491.

Cushman, D. W., and M. A. Ondetti (1999). "Design of angiotensin converting enzyme inhibitors." *Nat Med* 5(10): 1110–1113.

D'Amore, D. C., K. Moreno, C. R. McHenry, and S. Wroe (2011). "The effects of biting and pulling on the forces generated during feeding in the Komodo dragon (*Varanus komodoensis*)." *PLoS One* 6(10): e26226.

Da Silva, I. R., R. Lorenzetti, A. L. Renno, L. Baldissera Jr., A. Zelanis, S. M. Serrano, and S. Hyslop (2012a). "BJ-PI2, a non-hemorrhagic metalloproteinase from *Bothrops jararaca* snake venom." *Biochim Biophys Acta Bioenergetics* 1820(11): 1809–1821.

Da Silva, J. A., P. Spencer, M. A. Camillo, and V. M. de Lima (2012b). "Gyroxin and its biological activity: Effects on CNS basement membranes and endothelium and protease-activated receptors." *Curr Med Chem* 19(2): 281–291.

Da Silva, N. J., S. D. Aird, C. Seebart, and I. I. Kaiser (1989). "A gyroxin analog from the venom of the bushmaster (*Lachesis muta muta*)." *Toxicon* 27(7): 763–771.

Dada, A., S. O. Giwa, W. Yinusa, M. Ugbeye, and S. Gbadegesin (2009). "Complications of treatment of musculoskeletal injuries by bone setters." *W Afr J Med* 28(1): 43–47.

Daltry, J. C., W. Wüster, and R. S. Thorpe (1996). "Diet and snake venom evolution." *Nature* 379(6565): 537–540.

Daly, J. W., H. M. Garraffo, T. F. Spande, V. C. Clark, J. Ma, H. Ziffer, and J. F. Cover Jr. (2003). "Evidence for an enantioselective pumiliotoxin 7-hydroxylase in dendrobatid poison frogs of the genus *Dendrobates*." *Proc Natl Acad Sci USA* 100(19): 11092–11097.

Danse, J. M., E. G. Rowan, S. Gasparini, F. Ducancel, H. Vatanpour, L. C. Young, G. Poorheidari, et al. (1994). "On the site by which α-dendrotoxin binds to voltage-dependent potassium channels—site directed mutagenesis reveals that the lysine triplet 28-30 is not essential for binding." *FEBS Lett* 356(2–3): 153–158.

Darriba, D., G. L. Taboada, R. Doallo, and D. Posada (2012). "jModelTest 2: more models, new heuristics and parallel computing." *Nat Methods* 9(8): 772.

Dart, R. C., and R. A. Gustafson (1991). "Failure of electric-shock treatment for rattlesnake envenomation." *Ann Emerg Med* 20(6): 659–661.

Dart, R. C., K. M. Hurlbut, R. Garcia, and J. Boren (1996). "Validation of a severity score for the assessment of crotalid snakebite." *Ann Emerg Med* 27(3): 321–326.

Dart, R. C., and J. McNally (2001). "Efficacy, safety, and use of snake antivenoms in the United States." *Ann Emerg Med* 37(2): 181–188.

Dascal, N. (1987). "The use of *Xenopus* oocytes for the study of ion channels." *CRC Crit Rev Biochem* 22(4): 317–387.

Datta, G., and A. T. Tu (1997). "Structure and other chemical characterizations of gila toxin, a lethal toxin from lizard venom." *J Pept Res* 50(6): 443–450.

Datyner, M. E., and P. W. Gage (1973). "Presynaptic and postsynaptic effects of the venom of the Australian tiger snake at the neuromuscular junction." *Br J Pharmacol* 49(2): 340–354.

David, P., and G. Vogel (2010). *Venomous snakes of Europe, northern, central and western Asia*. Frankfurt am Main, Germany: Chimaira Buchhandelgesellschaft.

Davies, K. T. J., J. A. Cotton, J. D. Kirwan, E. C. Teeling, and S. J. Rossiter (2012). "Parallel signatures of sequence evolution among hearing genes in echolocating mammals: An emerging model of genetic convergence." *Heredity* 108(5): 480–489.

Davis, D., K. Branch, N. B. Egen, F. E. Russell, K. Gerrish, and P. S. Auerbach (1992). "The effect of an electrical-current on snake-venom toxicity." *J Wilderness Med* 3(1): 48–53.

Davis, I. W., L. W. Murray, J. S. Richardson, and D. C. Richardson (2004). "MOLPROBITY: Structure validation and all-atom contact analysis for nucleic acids and their complexes." *Nucleic Acids Res* 32 (Web server issue): W615–619.

De, S. S. (1944). "Physicochemical studies on hemolysin." *J Indian Chem Soc* 21: 290.

De Albuquerque Modesto, J. C., P. J. Spencer, M. Fritzen, R. C. Valenca, M. L. Oliva, M. B. da Silva, A. M. Chudzinski-Tavassi, and M. C. Guarnieri (2006). "BE-I-PLA$_2$, a novel acidic phospholipase A$_2$ from *Bothrops erythromelas* venom: Isolation, cloning and characterization as potent anti-platelet and inductor of prostaglandin I2 release by endothelial cells." *Biochem Pharmacol* 72(3): 377–384.

De Bruijn, M. H., and G. H. Fey (1985). "Human complement component C3: cDNA coding sequence and derived primary structure." *Proc Natl Acad Sci USA* 82(3): 708–712.

De Faria, L., E. Antunes, C. Bon, and A. L. de Araujo (2001). "Pharmacological characterization of the rat paw edema induced by *Bothrops lanceolatus* (Fer de lance) venom." *Toxicon* 39(6): 825–830.

De la Rosa, G. (2011). "Producción de toxoides de serpientes africanas mediante radiación Gamma." Thesis, Instituto de Biotecnología. Cuernavaca, México: Universidad Nacional Autónoma de México.

De la Vega, R. C. R., E. Merino, B. Becerril, and L. D. Possani (2003). "Novel interactions between K+ channels and scorpion toxins." *Trends Pharmacol Sci* 24(5): 222–227.

De Nooij-van Dalen, A. G., G. A. van Dongen, S. J. Smeets, E. J. Nieuwenhuis, M. Stigter-van Walsum, G. B. Snow, and R. H. Brakenhoff (2003). "Characterization of the human Ly-6 antigens, the newly annotated member Ly-6K included, as molecular markers for head-and-neck squamous cell carcinoma." *Int J Cancer* 103(6): 768–774.

De Oliveira, A. H., J. R. Giglio, S. H. Andriao-Escarso, A. S. Ito, and R. J. Ward (2001). "A pH-induced dissociation of the dimeric form of a lysine 49-phospholipase A$_2$ abolishes Ca2+-independent membrane damaging activity." *Biochemistry* 40(23): 6912–6920.

De Oliveira, L., C. Jared, A. L. da Costa Prudente, H. Zaher, and M. M. Antoniazzi (2008). "Oral glands in dipsadine 'goo-eater' snakes: Morphology and histochemistry of the infralabial glands in *Atractus reticulatus, Dipsas indica*, and *Sibynomorphus mikanii*." *Toxicon* 51(5): 898–913.

De Plater, G. M., R. L. Martin, and P. J. Milburn (1998). "A C-type natriuretic peptide from the venom of the platypus (*Ornithorhynchus anatinus*): Structure and pharmacology." *Comp Biochem Physiol C ToxicolPharmacol* 120(1): 99–110.

De Roodt, A. R. (2002). "Estudio inmunobiológico del veneno de serpientes de importancia sanitaria de la Argentina." Thesis, Facultad de Farmacia y Bioquímica. Buenos Aires: Universidad de Buenos Aires.

De Roodt, A. R., E. De Titto, J. boyer A. Dolab, and J. P. Chippaux (2013). "Envenoming by coral snakes (*Micrurus*) in Argentina, during the period between 1979–2003." *Rev Inst Med Trop São Paulo* 55(1): 13–18.

De Roodt, A. R., L. C. Lanari, V. C. de Oliveira, R. D. Laskowicz, and R. P. Stock (2011). "Neutralization of *Bothrops alternatus* regional venom pools and individual venoms by antivenom: A systematic comparison." *Toxicon* 57(7–8): 1073–1080.

De Weille, J. R., H. Schweitz, P. Maes, A. Tartar, and M. Lazdunski (1991). "Calciseptine, a peptide isolated from black mamba venom, is a specific blocker of the L-type calcium channel." *Proc Natl Acad Sci USA* 88(6): 2437–2440.

Debnath, A., A. Saha, A. Gomes, S. Biswas, P. Chakrabarti, B. Giri, A. K. Biswas, and S. D. Gupta (2010). "A lethal cardiotoxic-cytotoxic protein from the Indian monocellate cobra (*Naja kaouthia*) venom." *Toxicon* 56(4): 569–579.

Delev, D., A. Pavlova, S. Heinz, M. C. Blaise, T. Chandra, B. Poetsch, E. Seifried, and J. Oldenburg (2008). "Modelling and expression studies of two novel mutations causing factor V deficiency." *Thromb Haemost* 100(5): 766–772.

Delpietro, H. A., and R. G. Russo (2009). "Acquired resistance to saliva anticoagulants by prey previously fed upon by vampire bats (*Desmodus rotundus*): Evidence for immune response." *J Mammal* 90(5): 1132–1138.

Dennis, E. A. (1994). "Diversity of group types, regulation, and function of phospholipase A_2." *J Biol Chem* 269(18): 13057–13060.

Dennis, E. A., S. G. Rhee, M. M. Billah, and Y. A. Hannun (1991). "Role of phospholipase in generating lipid second messengers in signal transduction." *FASEB J* 5(7): 2068–2077.

Deolindo, P., A. S. Teixeira-Ferreira, R. A. DaMatta, and E. W. Alves (2010). "L-amino acid oxidase activity present in fractions of *Bothrops jararaca* venom is responsible for the induction of programmed cell death in *Trypanosoma cruzi*." *Toxicon* 56(6): 944–955.

Deregnaucourt, C., and J. Schrevel (2000). "Bee venom phospholipase A_2 induces stage-specific growth arrest of the intraerythrocytic Plasmodium falciparum via modifications of human serum components." *J Biol Chem* 275(51): 39973–39980.

Despotis, G. J., J. H. Joist, and L. T. Goodnough (1997). "Monitoring of hemostasis in cardiac surgical patients: Impact of point-of-care testing on blood loss and transfusion outcomes." *Clin Chem* 43(9): 1684–1696.

Dessaud, E., D. Salaun, O. Gayet, M. Chabbert, and O. deLapeyriere (2006). "Identification of lynx2, a novel member of the ly-6/neurotoxin superfamily, expressed in neuronal subpopulations during mouse development." *Mol Cell Neurosci* 31(2): 232–242.

Devedjiev, Y., A. Popov, B. Atanasov, and H. D. Bartunik (1997). "X-ray structure at 1.76 A resolution of a polypeptide phospholipase A_2 inhibitor." *J Mol Biol* 266(1): 160–172.

Devi, C. M., M. V. Bai, A. V. Lal, P. R. Umashankar, and L. K. Krishnan (2002). "An improved method for isolation of anti-viper venom antibodies from chicken egg yolk." *J Biochem Biophys Methods* 51(2): 129–138.

Dewan, J. C., G. A. Grant, and J. C. Sacchettini (1994). "Crystal structure of kappa-bungarotoxin at 2.3-A resolution." *Biochemistry* 33(44): 13147–13154.

DeYoung, M. B., L. MacConell, V. Sarin, M. Trautmann, and P. Herbert (2011). "Encapsulation of exenatide in poly-(D,L-lactide-co-glycolide) microspheres produced an investigational long-acting once-weekly formulation for type 2 diabetes." *Diabetes Technol Ther* 13(11): 1145–1154.

Di Scipio, R. G., M. A. Hermodson, and E. W. Davie (1977). "Activation of human factor X (Stuart factor) by a protease from Russell's viper venom." *Biochemistry* 16(24): 5253–5260.

Diaz, C., J. M. Gutiérrez, B. Lomonte, and J. A. Gene (1991). "The effect of myotoxins isolated from *Bothrops* snake venoms on multilamellar liposomes: Relationship to phospholipase A_2, anticoagulant and myotoxic activities." *Biochim Biophys Acta* 1070(2): 455–460.

Diaz, C., L. Valverde, O. Brenes, A. Rucavado, and J. M. Gutiérrez (2005). "Characterization of events associated with apoptosis/anoikis induced by snake venom metalloproteinase BaP1 on human endothelial cells." *J Cell Biochem* 94(3): 520–528.

Dickey, D. M., J. C. Burnett Jr., and L. R. Potter (2008). "Novel bifunctional natriuretic peptides as potential therapeutics." *J Biol Chem* 283(50): 35003–35009.

Dickinson, D. P. (2002). "Salivary (SD-type) cystatins: Over one billion years in the making—but to what purpose?" *Crit Rev Oral Biol Med* 13(6): 485–508.

Diochot, S., A. Baron, M. Salinas, D. Douguet, S. Scarzello, A. S. Dabert-Gay, D. Debayle, et al. (2012). "Black mamba venom peptides target acid-sensing ion channels to abolish pain." *Nature* 490(7421): 552–555.

Diochot, S., E. Loret, T. Bruhn, L. Beress, and M. Lazdunski (2003). "APETx1, a new toxin from the sea anemone *Anthopleura elegantissima*, blocks voltage-gated human ether-a-go-go-related gene potassium channels." *Mol Pharmacol* 64(1): 59–69.

Dixon, R. W., and J. B. Harris (1999). "Nerve terminal damage by beta-bungarotoxin: Its clinical significance." *Am J Pathol* 154(2): 447–455.

Dobiey, M., and G. Vogel (2007). *Venomous snakes of Africa*. Frankfurt am Main, Germany: Chimaira Buchhandelgesellschaft.

Dodd, C. K. Jr., and R. A. Seigel (1991). "Relocation, repatriation, and translocation of amphibians and reptiles: Are they conservation strategies that work?" *Herpetologica*: 336–350.

Doley, R., and R. M. Kini (2009). "Protein complexes in snake venom." *Cell Mol Life Sci* 66(17): 2851–2871.

Doley, R., N. B. T. Nguyen, M. A. Reza, and R. M. Kini (2008). "Unusual accelerated rate of deletions and insertions in toxin genes in the venom glands of the pygmy copperhead (*Austrelaps labialis*) from Kangaroo island." *BMC Evol Biol* 8(70): 1–13.

Domingues Passero, L., and K. de Sousa (2010). "Snake venoms—a perspective for development of drugs against microbes." In *Venoms: Sources, toxicity and therapeutic uses. Advances in biology and medicine,* edited by J. Gjersoe and S. Hundstad. Hauppauge, New York: Nova Science Publishers, 77–95.

Donoghue, S. (2006). "Nutrition." In *Reptile medicine and surgery*, edited by D. R. Mader. St. Louis, MO: Saunders Elsevier, 280.

Dos Santos, M. C., L. Morhy, L. C. L. Ferreira, and E. B. Oliveira (1993). "Purification and properties of a crotamine analog from Crotalus durissus ruruima venom." *Toxicon* 31: 166.

Draper, C. S., R. D. Walker, and H. E. Lawler (1981). "Patterns of oral bacterial infection in captive snakes". *J Am Vet Med Assoc* 179: 1223–1226.

Drickamer, K. (1999). "C-type lectin-like domains." *Curr Opin Struct Biol* 9(5): 585–590.

Du, X. Y., and K. J. Clemetson (2010). "Reptile C-type lectins." In *Handbook of venoms and toxins of reptiles*, edited by S. P. Mackessy. Boca Raton, FL: CRC Press, 359–376.

——— (2002). "Snake venom L-amino acid oxidases." *Toxicon* 40(6): 659–665.

Du, X. Y., J. M. Clemetson, A. Navdaev, E. M. Magnenat, T. N. Wells, and K. J. Clemetson (2002). "Ophioluxin, a convulxin-like C-type lectin from *Ophiophagus hannah* (King cobra) is a powerful platelet activator via glycoprotein VI." *J Biol Chem* 277(38): 35124–35132.

Du, X. Y., A. Navdaev, J. M. Clemetson, E. Magnenat, T. N. Wells, and K. J. Clemetson (2001). "Bilinexin, a snake C-type lectin from *Agkistrodon bilineatus* venom agglutinates platelets via GPIb and alpha2beta1." *Thromb Haemost* 86(5): 1277–1283.

Ducancel, F. (2005). "Endothelin-like peptides." *Cell Mol Life Sci* 62(23): 2828–2839.

Ducancel, F., G. Guignery-Frelat, C. Bouchier, A. Ménez, and J. C. Boulain (1988). "Complete amino acid sequence of a PLA_2 from the tiger snake *Notechis scutatus* scutatus as deduced from a complementary DNA." *Nucleic Acids Res* 16(18): 9049.

Ducancel, F., V. Matre, C. Dupont, E. Lajeunesse, Z. Wollberg, A. Bdolah, E. Kochva, J. C. Boulain, and A. Ménez (1993). "Cloning and sequence analysis of cDNAs encoding precursors of sarafotoxins: Evidence for an unusual 'rosary-type' organization." *J Biol Chem* 268(5): 3052–3055.

Duckert, P., S. Brunak, and N. Blom (2004). "Prediction of proprotein convertase cleavage sites." *Protein Eng Des Sel* 17(1): 107–112.

Dufton, M. J. (1984). "Classification of elapid snake neurotoxins and cytotoxins according to chain-length—evolutionary implications." *J Mol Evol* 20(2): 128–134.

Duke, J. A. (1985). *Handbook of medicinal herbs*. Boca Raton, FL: CRC Press.

Dunkelberger, J. R., and W. C. Song (2010). “Complement and its role in innate and adaptive immune responses.” *Cell Res* 20(1): 34–50.

Duran, R., C. Cervenansky, and E. Karlsson (1996). “Effect of fasciculin on hydrolysis of neutral and choline esters by butyrylcholinesterase, cobra venom and chicken acetylcholinesterases.” *Toxicon* 34(8): 959–963.

Durban, J., P. Juarez, Y. Angulo, B. Lomonte, M. Flores-Diaz, A. Alape-Giron, M. Sasa, et al. (2011). “Profiling the venom gland transcriptomes of Costa Rican snakes by 454 pyrosequencing.” *BMC Genomics* 12(1): 259.

Durban, J., A. Perez, L. Sanz, A. Gomez, F. Bonilla, S. Rodriguez, D. Chacon, et al. (2013). “Integrated ‘omics’ profiling indicates that miRNAs are modulators of the ontogenetic venom composition shift in the Central American rattlesnake, *Crotalus simus simus*.” *BMC Genomics* 14(1): 234.

Dutertre, S., A. H. Jin, Q. Kaas, A. Jones, P. F. Alewood, and R. J. Lewis (2013a). “Deep venomics reveals the mechanism for expanded peptide diversity in cone snail venom.” *Mol Cell Proteomics* 12(2): 312–329.

——— (2013b). “Deep venomics reveals the mechanism for expanded peptide diversity in cone snail venom.” *Mol Cell Proteomics* 12(2): 312–329.

Earl, S. T. H., G. W. Birrell, T. P. Wallis, L. D. St Pierre, P. P. Masci, J. de Jersey, J. J. Gorman, and M. F. Lavin (2006). “Post-translational modification accounts for the presence of varied forms of nerve growth factor in Australian elapid snake venoms.” *Proteomics* 6(24): 6554–6565.

Eble, J. A., B. Beermann, H. J. Hinz, and A. Schmidt-Hederich (2001). “Alpha 2beta 1 integrin is not recognized by rhodocytin but is the specific, high affinity target of rhodocetin, an RGD-independent disintegrin and potent inhibitor of cell adhesion to collagen.” *J Biol Chem* 276(15): 12274–12284.

Eble, J. A., S. Niland, T. Bracht, M. Mormann, J. Peter-Katalinic, G. Pohlentz, and J. Stetefeld (2009). “The alpha2beta1 integrin-specific antagonist rhodocetin is a cruciform, heterotetrameric molecule.” *FASEB J* 23(9): 2917–2927.

Eble, J. A., S. Niland, A. Dennes, A. Schmidt-Hederich, P. Bruckner, and G. Brunner (2002). “Rhodocetin antagonizes stromal tumor invasion *in vitro* and other alpha2beta1 integrin-mediated cell functions.” *Matrix Biol* 21(7): 547–558.

Eble, J. A., and D. S. Tuckwell (2003). “The alpha2beta1 integrin inhibitor rhodocetin binds to the A-domain of the integrin alpha2 subunit proximal to the collagen-binding site.” *Biochem J* 376(pt. 1): 77–85.

Edgar, R. C. (2004). “MUSCLE: A multiple sequence alignment method with reduced time and space complexity.” *BMC Bioinformatics* 5: 113.

Eggertsen, G., P. Lind, and J. Sjoquist (1981). “Molecular characterization of the complement activating protein in the venom of the Indian cobra (*Naja N. siamensis*).” *Mol Immunol* 18(2): 125–133.

Eggertsen, G., A. Lundwall, U. Hellman, and J. Sjoquist (1983). “Antigenic relationships between human and cobra complement factors C3 and cobra venom factor (CVF) from the Indian cobra (*Naja naja*).” *J Immunol* 131(4): 1920–1923.

Ehara, T., S. Kitajima, N. Kanzawa, T. Tamiya, and T. Tsuchiya (2002). “Antimicrobial action of achacin is mediated by L-amino acid oxidase activity.” *FEBS Lett* 531(3): 509–512.

Eichhorn, E. J., and M. R. Bristow (2001). “The Carvedilol Prospective Randomized Cumulative Survival (COPERNICUS) trial.” *Curr Control Trials Cardiovasc Med* 2(1): 20–23.

Eissele, R., E. Bothesandfort, B. Goke, J. Eng, R. Arnold, and H. Koop (1994). “Rat gastric somatostatin and gastrin-release—interactions of exendin-4 and truncated glucagon-like peptide-1 (GLP-1) amide.” *Life Sci* 55(8): 629–634.

Elashoff, M., A. V. Matveyenko, B. Gier, R. Elashoff, and P. C. Butler (2011). “Pancreatitis, pancreatic, and thyroid cancer with glucagon-like peptide-1-based therapies.” *Gastroenterology* 141(1): 150–156.

Eldor, R., M. Kidron, Y. Greenberg-Shushlav, and E. Arbit (2010). "Novel glucagon-like peptide-1 analog delivered orally reduces postprandial glucose excursions in porcine and canine models." *J Diabetes Sci Technol* 4(6): 1516–1523.

Emerich, D. F., S. R. Winn, J. Harper, J. P. Hammang, E. E. Baetge, and J. H. Kordower (1994). "Implants of polymer-encapsulated human NGF-secreting cells in the nonhuman primate—rescue and sprouting of degenerating cholinergic basal forebrain neurons." *J Comp Neurol* 349(1): 148–164.

Emsley, J., C. G. Knight, R. W. Farndale, M. J. Barnes, and R. C. Liddington (2000). "Structural basis of collagen recognition by integrin alpha2beta1." *Cell* 101(1): 47–56.

Eng, J. (1992). "Exendin peptides." *Mt Sinai J Med* 59(2): 147–148.

Eng, J., P. C. Andrews, W. A. Kleinman, L. Singh, and J. P. Raufman (1990). "Purification and structure of exendin-3, a new pancreatic secretagogue isolated from *Heloderma horridum* venom." *J Biol Chem* 265(33): 20259–20262.

Eng, J., W. A. Kleinman, L. Singh, G. Singh, and J. P. Raufman (1992). "Isolation and characterization of exendin-4, an exendin-3 analog, from *Heloderma suspectum* venom—further evidence for an exendin receptor on dispersed acini from guinea-pig pancreas." *J Biol Chem* 267(11): 7402–7405.

Eng, J. K., B. C. Searle, K. R. Clauser, and D. L. Tabb (2011). "A face in the crowd: Recognizing peptides through database search." *Mol Cell Proteomics* 10(11).

Erbacher, A., F. Gieseke, R. Handgretinger, and I. Muller (2009). "Dendritic cells: Functional aspects of glycosylation and lectins." *Hum Immunol* 70(5): 308–312.

Escalante, T., A. Rucavado, J. W. Fox, and J. M. Gutiérrez (2011). "Key events in microvascular damage induced by snake venom hemorrhagic metalloproteinases." *J Proteomics* 74(9): 1781–1794.

Escoubas, P. (2006). "Mass spectrometry in toxinology: A 21st-century technology for the study of biopolymers from venoms." *Toxicon* 47 (6): 609–613.

Escoubas, P., and L. Rash (2004). "Tarantulas: Eight-legged pharmacists and combinatorial chemists." *Toxicon* 43(5): 555–574.

Evans, H. J., R. Franson, G. D. Qureshi, and W. F. Moo-Penn (1980). "Isolation of anticoagulant proteins from cobra venom (*Naja nigricollis*): Identity with phospholipases A_2." *J Biol Chem* 255(8): 3793–3797.

Ezawa, K., S. Oota, and N. Saitou (2006). "Genome-wide search of gene conversions in duplicated genes of mouse and rat." *Mol Biol Evol* 23(5): 927–940.

Fadel, V., P. Bettendorff, T. Herrmann, W. F. de Azevedo Jr., E. B. Oliveira, T. Yamane, and K. Wuthrich (2005). "Automated NMR structure determination and disulfide bond identification of the myotoxin crotamine from *Crotalus durissus terrificus*." *Toxicon* 46(7): 759–767.

Fang, K. S. Y., M. Vitale, P. Fehlner, and T. P. King (1988). "Cdna cloning and primary structure of a white-face hornet venom allergen, antigen-5." *Proc Natl Acad Sci USA* 85(3): 895–899.

Fantini, E., P. Athias, R. Tirosh, and A. Pinson (1996). "Effect of Taicatoxin (TCX) on the electrophysiological, mechanical and biochemical characteristics of spontaneously beating ventricular cardiomyocytes." *Mol Cell Biochem* 160–161(1): 61–66.

Farre, C., and N. Fertig (2012). "HTS techniques for patch clamp-based ion channel screening—advances and economy." *Expert Opin Drug Discov* 7(6): 515–524.

Faull, K. F., G. J. Feistner, K. Conklin, P. Roepstorff, and P. C. Andrews (1998). "Revised primary structures of rat pituitary gamma-lipotrophin and beta-endorphin." *Neuropeptides* 32(4): 339–349.

Faure, G., and C. Bon (1988). "Crotoxin, a phospholipase A_2 neurotoxin from the South American rattlesnake *Crotalus durissus terrificus*: Purification of several isoforms and comparison of their molecular structure and of their biological activities." *Biochemistry* 27(2): 730–738.

Faure, G., V. T. Gowda, and R. C. Maroun (2007). "Characterization of a human coagulation factor Xa-binding site on Viperidae snake venom phospholipases A_2 by affinity binding studies and molecular bioinformatics." *BMC Struct Biol* 7: 82.

Faure, G., J. L. Guillaume, L. Camoin, B. Saliou, and C. Bon (1991). "Multiplicity of acidic subunit isoforms of crotoxin, the phospholipase A_2 neurotoxin from *Crotalus durissus terrificus* venom, results from posttranslational modifications." *Biochemistry* 30(32): 8074–8083.

Faure, G., H. J. Xu, and F. A. Saul (2011). "Crystal structure of crotoxin reveals key residues involved in the stability and toxicity of this potent heterodimeric β-Neurotoxin." *J Mol Biol* 412: 176–191.

Fenard, D., G. Lambeau, E. Valentin, J. C. Lefebvre, M. Lazdunski, and A. Doglio (1999). "Secreted phospholipases A(2), a new class of HIV inhibitors that block virus entry into host cells." *J Clin Invest* 104(5): 611–618.

Feng, W., L. Pan, and M. Zhang (2011). "Combination of NMR spectroscopy and X-ray crystallography offers unique advantages for elucidation of the structural basis of protein complex assembly." *Sci China Life Sci* 54(2): 101–111.

Feofanov, A. V., G. V. Sharonov, M. V. Astapova, D. I. Rodionov, Y. N. Utkin, and A. S. Arseniev (2005). "Cancer cell injury by cytotoxins from cobra venom is mediated through lysosomal damage." *Biochem J* 390 (pt. 1): 11–18.

Feofanov, A. V., G. V. Sharonov, M. A. Dubinnyi, M. V. Astapova, I. A. Kudelina, P. V. Dubovskii, D. I. Rodionov, Y. N. Utkin, and A. S. Arseniev (2004). "Comparative study of structure and activity of cytotoxins from venom of the cobras *Naja oxiana, Naja kaouthia*, and *Naja haje*." *Biochemistry (Moscow)* 69(10): 1148–1157.

Fergusson, D. A., P. C. Hébert, C. D. Mazer, S. Fremes, C. MacAdams, J. M. Murkin, K. Teoh, et al. (2008). "A comparison of aprotinin and lysine analogues in high-risk cardiac surgery." *N Engl J Med* 358(22): 2319–2331.

Fernandez, A. Z., A. Tablante, S. Beguin, H. C. Hemker, and R. Apitz-Castro (1999). "Draculin, the anticoagulant factor in vampire bat saliva, is a tight-binding, noncompetitive inhibitor of activated factor X." *Biochim Biophys Acta* 1434(1): 135–142.

Fernandez, J., A. Alape-Giron, Y. Angulo, L. Sanz, J. M. Gutiérrez, J. J. Calvete, and B. Lomonte (2011). "Venomic and antivenomic analyses of the Central American coral snake, *Micrurus nigrocinctus* (Elapidae)." *J Proteome Res* 10(4): 1816–1827.

Fernandez, J., P. Caccin, G. Koster, B. Lomonte, J. M. Gutiérrez, C. Montecucco, and A. D. Postle (2013). "Muscle phospholipid hydrolysis by *Bothrops asper* Asp49 and Lys49 phospholipase A(2) myotoxins—distinct mechanisms of action." *FEBS J* 280(16): 3878–3886.

Ferreira, S. H. (1965). "A bradykinin-potentiating factor (Bpf) present in the venom of *Bothrops jararaca*." *Br J Pharmacol Chemother* 24: 163–169.

Fertig, N., R. H. Blick, and J. C. Behrends (2002). "Whole cell patch clamp recording performed on a planar glass chip." *Biophys J* 82(6): 3056–3062.

Filippovich, I., N. Sorokina, P. P. Masci, J. de Jersey, A. N. Whitaker, D. J. Winzor, P. J. Gaffney, and M. F. Lavin (2002). "A family of textilinin genes, two of which encode proteins with antihaemorrhagic properties." *Brit J Haematol* 119(2): 376–384.

Finkel, A., A. Wittel, N. Yang, S. Handran, J. Hughes, and J. Costantin (2006). "Population patch clamp improves data consistency and success rates in the measurement of ionic currents." *J Biomol Screen* 11(5): 488–496.

Finney, D. J. (1971). *Probit analysis*. London: Cambridge University Press.

Flandry, F., E. J. Lisecki, G. J. Domingue, R. L. Nichols, D. L. Greer, and R. J. Haddad Jr. (1989). "Initial antibiotic therapy for alligator bites: Characterization of the oral flora of *Alligator mississippiensis*." *South Med J* 82(2): 262–266.

Fleer, E. A., H. M. Verheij, and G. H. de Haas (1981). "Modification of carboxylate groups in bovine pancreatic phospholipase A_2: Identification of aspartate-49 as Ca2+-binding ligand." *Eur J Biochem* 113(2): 283–288.

Fletcher, J. E., and P. Rosenberg (1997). "The cellular effects and mechanisms of action of presynaptically acting phospholipase A 2 enzymes." In *Venom phospholipase A_2 enzymes: structure, function, and mechanism*, edited by R. M. Kini. Chichester: John Wiley, 413–454.

Fletcher, J. E., H. S. Selistre de Araujo, and C. L. Ownby (1997). "Phospholipase A_2 myotoxins from *Bothrops* snake venoms." In *Venom phospholipase A_2 enzymes: Structure, function, and mechanism*, edited by R. M. Kini.Chichester: John Wiley, 455–497.

Flight, S., L. Johnson, M. Trabi, P. Gaffney, M. Lavin, J. de Jersey, and P. Masci (2005). "Comparison of textilinin-1 with aprotinin as serine protease inhibitors and as antifibrinolytic agents." *Pathophysiol Haemost Thromb* 34(4–5): 188–193.

Flight, S. M., L. A. Johnson, Q. S. Du, R. L. Warner, M. Trabi, P. J. Gaffney, M. F. Lavin, J. de Jersey, and P. P. Masci (2009). "Textilinin-1, an alternative anti-bleeding agent to aprotinin: Importance of plasmin inhibition in controlling blood loss." *Br J Haematol* 145(2): 207–211.

Fohlman, J. (1979). "Comparison of 2 highly toxic australian snake-venoms—taipan (*Oxyuranus s. scutellatus*) and the fierce snake (*Parademansia microlepidotus*)." *Toxicon* 17(2): 170–172.

Fohlman, J., D. Eaker, E. Karlsson, and S. Thesleff (1976). "Taipoxin, an extremely potent presynaptic neurotoxin from venom of australian snake taipan (*Oxyuranus s. scutellatus*)—isolation, characterization, quaternary structure and pharmacological properties." *Eur J Biochem* 68(2): 457–469.

Fohlman, J., P. Lind, and D. Eaker (1977). "Taipoxin, an extremely potent presynaptic snake-venom neurotoxin—elucidation of primary structure of acidic carbohydrate-containing taipoxin-subunit, a prophospholipase homolog." *FEBS Lett* 84(2): 367–371.

Forgey, W. W. (1993). "More on snake-venom and insect-venom extractors." *N Engl J Med* 328(7): 516–516.

Fornelli, L., J. Parra, R. Hartmer, C. Stoermer, M. Lubeck, and Y. O. Tsybin (2013). "Top-down analysis of 30-80 kDa proteins by electron transfer dissociation time-of-flight mass spectrometry." *Anal Bioanal Chem* 405(26): 8505–8514.

Forouhar, F., W. N. Huang, J. H. Liu, K. Y. Chien, W. G. Wu, and C. D. Hsiao (2003). "Structural basis of membrane-induced cardiotoxin A3 oligomerization." *J Biol Chem* 278(24): 21980–21988.

Forst, S., J. Weiss, P. Elsbach, J. M. Maraganore, I. Reardon, and R. L. Heinrikson (1986). "Structural and functional properties of a phospholipase A_2 purified from an inflammatory exudate." *Biochemistry* 25(26): 8381–8385.

Forstner, M., R. Hilsenbeck, and J. Scudday (1997). "Geographic variation in whole venom profiles from the mottled rock rattlesnake (*Crotalus lepidus lepidus*) in Texas." *J Herpetol*: 277–287.

Fox, J. W., and J. B. Bjarnason (1995). "Atrolysins: Metalloproteinases from *Crotalus atrox* venom." *Methods Enzymol* 248: 368–387.

Fox, J. W., M. Elzinga, and A. T. Tu (1979). "Amino acid sequence and disulfide bond assignment of myotoxin a isolated from the venom of Prairie rattlesnake (*Crotalus viridis viridis*)." *Biochemistry* 18(4): 678–684.

Fox, J. W., and S. M. Serrano (2005). "Structural considerations of the snake venom metalloproteinases, key members of the M12 reprolysin family of metalloproteinases." *Toxicon* 45(8): 969–985.

Fox, J. W., and S. M. T. Serrano (2008). "Insights into and speculations about snake venom metalloproteinase (SVMP) synthesis, folding and disulfide bond formation and their contribution to venom complexity." *FEBS J* 275(12): 3016–3030.

Frade, J. M., and Y. A. Barde (1998). "Microglia-derived nerve growth factor causes cell death in the developing retina." *Neuron* 20(1): 35–41.

——— (1999). "Genetic evidence for cell death mediated by nerve growth factor and the neurotrophin receptor p75 in the developing mouse retina and spinal cord." *Development* 126(4): 683–690.

Frade, J. M., A. Rodriguez-Tebar, and Y. A. Barde (1996). "Induction of cell death by endogenous nerve growth factor through its p75 receptor." *Nature* 383(6596): 166–168.

Franca, S. C., S. Kashima, P. G. Roberto, M. Marins, F. K. Ticli, J. O. Pereira, S. Astolfi Jr., et al. (2007). "Molecular approaches for structural characterization of *Bothrops* L-amino acid oxidases with antiprotozoal activity: cDNA cloning, comparative sequence analysis, and molecular modeling." *Biochem Biophys Res Commun* 355(2): 302–306.

Francis, B., A. Bdolah and I. I. Kaiser (1995). "Amino acid sequences of a heterodimeric neurotoxin from the venom of the false horned viper (*Pseudocerastes fieldi*)." *Toxicon* 33(7): 863–874.

Francis, B., E. S. Williams, C. Seebart, and I. I. Kaiser (1993). "Proteins isolated from the venom of the common tiger snake (*Notechis scutatus scutatus*) promote hypotension and hemorrhage." *Toxicon* 31(4): 447–458.

Francis, B. R., N. J. da Silva Jr., C. Seebart, L. L. Casais e Silva, J. J. Schmidt, and I. I. Kaiser (1997). "Toxins isolated from the venom of the Brazilian coral snake (*Micrurus frontalis frontalis*) include hemorrhagic type phospholipases A_2 and postsynaptic neurotoxins." *Toxicon* 35(8): 1193–1203.

Francischetti, I. M., T. C. Assumpcao, D. Ma, Y. Li, E. C. Vicente, W. Uieda, and J. M. Ribeiro (2013). "The 'Vampirome': Transcriptome and proteome analysis of the principal and accessory submaxillary glands of the vampire bat *Desmodus rotundus*, a vector of human rabies." *J Proteomics* 82: 288–319.

Francischetti, I. M., V. My-Pham, J. Harrison, M. K. Garfield, and J. M. Ribeiro (2004). "*Bitis gabonica* (Gaboon viper) snake venom gland: Toward a catalog for the full-length transcripts (cDNA) and proteins." *Gene* 337: 55–69.

Francischetti, I. M., B. Saliou, M. Leduc, C. R. Carlini, M. Hatmi, J. Randon, A. Faili, and C. Bon (1997). "Convulxin, a potent platelet-aggregating protein from *Crotalus durissus terrificus* venom, specifically binds to platelets." *Toxicon* 35(8): 1217–1228.

Francischetti, I. M. B., J. F. Andersen, and J. M. C. Ribeiro (2002). "Biochemical and functional characterization of recombinant *Rhodnius prolixus* platelet aggregation inhibitor 1 as a novel lipocalin with high affinity for adenosine diphosphate and other adenine nucleotides." *Biochemistry* 41(11): 3810–3818.

Francischetti, I. M. B., J. M. C. Ribeiro, D. Champagne, and J. Andersen (2000). "Purification, cloning, expression, and mechanism of action of a novel platelet aggregation inhibitor from the salivary gland of the blood-sucking bug, *Rhodnius prolixus*." *J Biol Chem* 275(17): 12639–12650.

Francischetti, I. M. B., J. G. Valenzuela, J. F. Andersen, T. N. Mather, and J. M. C. Ribeiro (2002). "Ixolaris, a novel recombinant tissue factor pathway inhibitor (TFPI) from the salivary gland of the tick, *Ixodes scapularis*: Identification of factor X and factor Xa as scaffolds for the inhibition of factor VIIa/tissue factor complex." *Blood* 99(10): 3602–3612.

Frank, H. A. (1971). "Snakebite or frostbite: What are we doing? An evaluation of cryotherapy for envenomation." *West J Med* 114(5): 25–27.

French, W. J., W. K. Hayes, S. P. Bush, M. D. Cardwell, J. O. Bader, and E. D. Rael (2004). "Mojave toxin in venom of *Crotalus helleri* (Southern Pacific Rattlesnake): Molecular and geographic characterization." *Toxicon* 44(7): 781–791.

Fritzinger, D. C., R. Bredehorst, and C. W. Vogel (1994). "Molecular cloning and derived primary structure of cobra venom factor." *Proc Natl Acad Sci USA* 91(26): 12775–12779.

Fritzinger, D. C., E. C. Petrella, M. B. Connelly, R. Bredehorst, and C. W. Vogel (1992). "Primary structure of cobra complement component C3." *J Immunol* 149(11): 3554–3562.

Frobert, Y., C. Creminon, X. Cousin, M. H. Remy, J. M. Chatel, S. Bon, C. Bon, and J. Grassi (1997). "Acetylcholinesterases from Elapidae snake venoms: Biochemical, immunological and enzymatic characterization." *Biochim Biophys Acta* 1339(2): 253–267.

Froy, O., T. Sagiv, M. Poreh, D. Urbach, N. Zilberberg, and M. Gurevitz (1999). "Dynamic diversification from a putative common ancestor of scorpion toxins affecting sodium, potassium, and chloride channels." *J Mol Evol* 48(2): 187–196.

Fry, B. G. (2005). "From genome to 'venome': Molecular origin and evolution of the snake venom proteome inferred from phylogenetic analysis of toxin sequences and related body proteins." *Genome Res* 15(3): 403–420.

Fry, B. G., N. R. Casewell, W. Wüster, N. Vidal, B. Young, and T. N. Jackson (2012a). "The structural and functional diversification of the Toxicofera reptile venom system." *Toxicon* 60(4): 434–448.

Fry, B. G., N. G. Lumsden, W. Wüster, J. C. Wickramaratna, W. C. Hodgson, and R. M. Kini (2003a). "Isolation of a neurotoxin (alpha-colubritoxin) from a nonvenomous colubrid: Evidence for early origin of venom in snakes." *J Mol Evol* 57(4): 446–452.

Fry, B. G., K. Roelants, D. E. Champagne, H. Scheib, J. D. Tyndall, G. F. King, T. J. Nevalainen, et al. (2009a). "The toxicogenomic multiverse: Convergent recruitment of proteins into animal venoms." *Annu Rev Genomics Hum Genet* 10: 483–511.

Fry, B. G., K. Roelants, K. Winter, W. C. Hodgson, L. Griesman, H. F. Kwok, D. Scanlon, et al. (2010a). "Novel venom proteins produced by differential domain-expression strategies in beaded lizards and gila monsters (genus *Heloderma*)." *Mol Biol Evol* 27(2): 395–407.

Fry, B. G., H. Scheib, I. L. M. Junqueira de Azevedo, D. A. Silva, and N. R. Casewell (2012b). "Novel transcripts in the maxillary venom glands of advanced snakes." *Toxicon* 59(7–8): 696–708.

Fry, B. G., H. Scheib, L. van der Weerd, B. Young, J. McNaughtan, S. F. Ramjan, N. Vidal, R. E. Poelmann, and J. A. Norman (2008). "Evolution of an arsenal: Structural and functional diversification of the venom system in the advanced snakes (Caenophidia)." *Mol Cell Proteomics* 7(2): 215–246.

Fry, B. G., E. A. Undheim, S. A. Ali, T. N. Jackson, J. Debono, H. Scheib, T. Ruder, et al. (2013). "Squeezers and leaf-cutters: Differential diversification and degeneration of the venom system in toxicoferan reptiles." *Mol Cell Proteomics* 12(7): 1881–1899.

Fry, B. G., N. Vidal, J. A. Norman, F. J. Vonk, H. Scheib, S. F. Ramjan, S. Kuruppu, et al. (2006). "Early evolution of the venom system in lizards and snakes." *Nature* 439(7076): 584–588.

Fry, B. G., J. C. Wickramaratana, S. Lemme, A. Beuve, D. Garbers, W. C. Hodgson, and P. Alewood (2005). "Novel natriuretic peptides from the venom of the inland taipan (*Oxyuranus microlepidotus*): Isolation, chemical and biological characterisation." *Biochem Biophys Res Commun* 327(4): 1011–1015.

Fry, B. G., J. C. Wickramaratna, W. C. Hodgson, P. F. Alewood, R. M. Kini, H. Ho, and W. Wüster (2002). "Electrospray liquid chromatography/mass spectrometry fingerprinting of *Acanthophis* (death adder) venoms: Taxonomic and toxinological implications." *Rapid Commun Mass Spectrom* 16(6): 600–608.

Fry, B. G., K. Winter, J. A. Norman, K. Roelants, R. J. Nabuurs, M. J. van Osch, W. M. Teeuwisse, et al. (2010b). "Functional and structural diversification of the Anguimorpha lizard venom system." *Mol Cell Proteomics* 9(11): 2369–2390.

Fry, B. G., S. Wroe, W. Teeuwisse, M. J. van Osch, K. Moreno, J. Ingle, C. McHenry, et al. (2009b). "A central role for venom in predation by *Varanus komodoensis* (Komodo Dragon) and the extinct giant *Varanus (Megalania) priscus*." *Proc Natl Acad Sci USA* 106(22): 8969–8974.

Fry, B. G., and W. Wüster (2004). "Assembling an arsenal: Origin and evolution of the snake venom proteome inferred from phylogenetic analysis of toxin sequences." *Mol Biol Evol* 21(5): 870–883.

Fry, B. G., W. Wüster, R. M. Kini, V. Brusic, A. Khan, D. Venkataraman, and A. P. Rooney (2003b). "Molecular evolution and phylogeny of elapid snake venom three-finger toxins." *J Mol Evol* 57(1): 110–129.

Fry, B. G., W. Wüster, S. F. R. Ramjan, T. Jackson, P. Martelli, and R. M. Kini (2003c). "Analysis of Colubroidea snake venoms by liquid chromatography with mass spectrometry: Evolutionary and toxinological implications." *Rapid Commun Mass Spectrom* 17(18): 2047–2062.

Fry, M. R., S. S. Ghosh, J. M. East, and R. C. Franson (1992). "Role of human sperm phospholipase A_2 in fertilization: Effects of a novel inhibitor of phospholipase A_2 activity on membrane perturbations and oocyte penetration." *Biol Reprod* 47(5): 751–759.

Fujikawa, K., M. E. Legaz, and E. W. Davie (1972). "Bovine factor X 1 (Stuart factor): Mechanism of activation by protein from Russell's viper venom." *Biochemistry* 11(26): 4892–4899.

Fujisawa, D., Y. Yamazaki, B. Lomonte, and T. Morita (2008). "Catalytically inactive phospholipase A_2 homologue binds to vascular endothelial growth factor receptor-2 via a C-terminal loop region." *Biochem J* 411(3): 515–522.

Fujisawa, D., Y. Yamazaki, and T. Morita (2009). "Re-evaluation of M-LAO, L-amino acid oxidase, from the venom of Gloydius blomhoffi as an anticoagulant protein." *J Biochem* 146(1): 43–49.

Fukagawa, T., T. Nose, Y. Shimohigashi, T. Ogawa, N. Oda, K. Nakashima, C. C. Chang, and M. Ohno (1993). "Purification, sequencing and characterization of single amino acid-substituted phospholipase A_2 isozymes from *Trimeresurus gramineus* (green Habu snake) venom." *Toxicon* 31(8): 957–967.

Fukata, Y., H. Adesnik, T. Iwanaga, D. S. Bredt, R. A. Nicoll, and M. Fukata (2006). "Epilepsy-related ligand/receptor complex LGI1 and ADAM22 regulate synaptic transmission." *Science* 313(5794): 1792–1795.

Fukuda, K., T. A. Doggett, L. A. Bankston, M. A. Cruz, T. G. Diacovo, and R. C. Liddington (2002). "Structural basis of von Willebrand factor activation by the snake toxin botrocetin." *Structure* 10(7): 943–950.

Fukuda, K., H. Mizuno, H. Atoda, and T. Morita (1999). "Crystallization and preliminary X-ray studies of flavocetin-A, a platelet glycoprotein Ib-binding protein from the habu snake venom." *Acta Crystallogr D Biol Crystallogr* 55 (pt. 11): 1911–1913.

——— (2000). "Crystal structure of flavocetin-A, a platelet glycoprotein Ib-binding protein, reveals a novel cyclic tetramer of C-type lectin-like heterodimers." *Biochemistry* 39(8): 1915–1923.

Fuly, A. L., S. Calil-Elias, A. M. Martinez, P. A. Melo, and J. A. Guimaraes (2003). "Myotoxicity induced by an acidic Asp-49 phospholipase A(2) isolated from *Lachesis muta* snake venom: Comparison with lysophosphatidylcholine." *Int J Biochem Cell Biol* 35(10): 1470–1481.

Fuly, A. L., S. Calil-Elias, R. B. Zingali, J. A. Guimaraes, and P. A. Melo (2000). "Myotoxic activity of an acidic phospholipase A_2 isolated from *Lachesis muta* (Bushmaster) snake venom." *Toxicon* 38(7): 961–972.

Fuly, A. L., O. L. Machado, E. W. Alves, and C. R. Carlini (1997). "Mechanism of inhibitory action on platelet activation of a phospholipase A_2 isolated from *Lachesis muta* (Bushmaster) snake venom." *Thromb Haemost* 78(5): 1372–1380.

Funk, C., J. Gmur, R. Herold, and P. W. Straub (1971). "Reptilase-R—a new reagent in blood coagulation." *Br J Haematol* 21(1): 43–52.

Funk, C. D. (2001). "Prostaglandins and leukotrienes: Advances in eicosanoid biology." *Science* 294(5548): 1871–1875.

Furukawa, S., and K. Hayashi (1978). "Purification and characterization of nerve growth factor from venom of *Bungarus multicinctus*." *Biochim Biophys Acta* 533(2): 383–395.

Gabe, M., and H. Saint Girons (1969). "Données histologiques sur les glandes salivaires des lépidosauriens [Histological data on salivary glands of lepidosaurians]." *Mém Mus Natl Hist Nat Paris* 58: 1–118.

Galat, A. (2008). "The three-fingered protein domain of the human genome." *Cell Mol Life Sci* 65(21): 3481–3493.

Galat, A., G. Gross, P. Drevet, A. Sato, and A. Ménez (2008). "Conserved structural determinants in three-fingered protein domains." *FEBS J* 275(12): 3207–3225.

Gao, J., and Y. Xie (1999). "[CD59-blocked erythrocytes and hemolysis tests of paroxysmal nocturnal hemoglobinuria]." *Zhonghua Xue Ye Xue Za Zhi* 20(4): 187–190.

Gao, R., R. M. Kini, and P. Gopalakrishnakone (2002). "A novel prothrombin activator from the venom of *Micropechis ikaheka*: Isolation and characterization." *Arch Biochem Biophys* 408(1): 87–92.

Gao, R., Y. Zhang, Q. X. Meng, W. H. Lee, D. S. Li, Y. L. Xiong, and W. Y. Wang (1998). "Characterization of three fibrinogenolytic enzymes from Chinese green tree viper (*Trimeresurus stejnegeri*) venom." *Toxicon* 36(3): 457–467.

Gao, W., V. G. Starkov, Z. X. He, Q. H. Wang, V. I. Tsetlin, Y. N. Utkin, Z. J. Lin, and R. C. Bi (2009). "Functions, structures and Triton X-100 effect for the catalytic subunits of heterodimeric phospholipases A_2 from *Vipera nikolskii* venom." *Toxicon* 54(6): 709–716.

Garb, J. E., and C. Y. Hayashi (2013). "Molecular evolution of alpha-latrotoxin, the exceptionally potent vertebrate neurotoxin in black widow spider venom." *Mol Biol Evol* 30(5): 999–1014.

Garbuzynskiy, S. O., B. S. Melnik, M. Y. Lobanov, A. V. Finkelstein, and O. V. Galzitskaya (2005). "Comparison of X-ray and NMR structures: Is there a systematic difference in residue contacts between X-ray- and NMR-resolved protein structures?" *Proteins* 60(1): 139–147.

Garcia, M., M. Monge, G. Leon, S. Lizano, E. Segura, G. Solano, G. Rojas, and J. M. Gutiérrez (2002). "Effect of preservatives on IgG aggregation, complement-activating effect and hypotensive activity of horse polyvalent antivenom used in snakebite envenomation." *Biologicals* 30(2): 143–151.

Gasiorowska, J., A. Teisseyre, A. Uryga, and K. Michalak (2012). "The influence of 8-prenylnaringenin on the activity of voltage-gated Kv1.3 potassium channels in human Jurkat T cells." *Cell Mol Biol Lett* 17(4): 559–570.

Gasparini, S., J. M. Danse, A. Lecoq, S. Pinkasfeld, S. Zinn-Justin, L. C. Young, C. C. L. de Medeiros, E. G. Rowan, A. L. Harvey, and A. Ménez (1998). "Delineation of the functional site of α-dendrotoxin: The functional topographies of dendrotoxins are different but share a conserved core with those of other Kv1 potassium channel-blocking toxins." *J Biol Chem* 273(39): 25393–25403.

Gatineau, E., F. Toma, T. Montenay-Garestier, M. Takechi, P. Fromageot, and A. Ménez (1987). "Role of tyrosine and tryptophan residues in the structure-activity relationships of a cardiotoxin from *Naja nigricollis* venom." *Biochemistry* 26(25): 8046–8055.

Gauthier, J. A., M. Kearney, J. A. Maisano, O. Rieppel, and A. Behlke (2012). "Assembling the squamate tree of life: Perspectives from the phenotype and the fossil record." *Bull Peabody Mus Nat Hist* 53(1): 1–306.

Gavel, Y., and G. von Heijne (1990). "Sequence differences between glycosylated and non-glycosylated Asn-X Thr/Ser acceptor sites: Implications for protein engineering." *Protein Eng* 3: 433–442.

Gavras, H., H. R. Brunner, J. H. Laragh, J. E. Sealey, I. Gavras, and R. A. Vukovich (1974). "An angiotensin converting-enzyme inhibitor to identify and treat vasoconstrictor and volume factors in hypertensive patients." *N Engl J Med* 291(16): 817–821.

Gedulin, B. R., P. A. Smith, C. M. Jodka, K. Chen, S. Bhavsar, L. L. Nielsen, D. G. Parkes, and A. A. Young (2008). "Pharmacokinetics and pharmacodynamics of exenatide following alternate routes of administration." *Int J Pharm* 356(1–2): 231–238.

Gellert, G. A. (1992). "Snake-venom and insect-venom extractors—an unproved therapy." *N Engl J Med* 327(18): 1322–1322.

Georgieva, D., M. Murakami, M. Perband, R. Arni, and C. Betzel (2011). "The structure of a native L-amino acid oxidase, the major component of the *Vipera ammodytes ammodytes* venomic, reveals dynamic active site and quaternary structure stabilization by divalent ions." *Mol Biosyst* 7(2): 379–384.

German, B. T., J. B. Hack, K. Brewer, and W. J. Meggs (2005). "Pressure-immobilization bandages delay toxicity in a porcine model of eastern coral snake (*Micrurus fulvius fulvius*) envenomation." *Ann Emerg Med* 45(6): 603–608.

Geyer, A., T. B. Fitzpatrick, P. D. Pawelek, K. Kitzing, A. Vrielink, S. Ghisla, and P. Macheroux (2001). "Structure and characterization of the glycan moiety of L-amino-acid oxidase from the Malayan pit viper *Calloselasma rhodostoma*." *Eur J Biochem* 268 (14): 4044–4053.

Gibbins, J. M. (2004). "Platelet adhesion signalling and the regulation of thrombus formation." *J Cell Sci* 117(pt 16): 3415–3425.

Gibbs, G. M., M. J. Scanlon, J. Swarbrick, S. Curtis, E. Gallant, A. F. Dulhunty, and M. K. O'Bryan (2006). "The cysteine-rich secretory protein domain of Tpx-1 is related to ion channel toxins and regulates ryanodine receptor Ca2+ signaling." *J Biol Chem* 281: 4156–4163.

Gibbs, H. L., and S. P. Mackessy (2009). "Functional basis of a molecular adaptation: Prey-specific toxic effects of venom from *Sistrurus* rattlesnakes." *Toxicon* 53(6): 672–679.

Giese, A., R. Jude, H. Kuiper, F. Piumi, A. Schambony, G. Guérin, O. Distl, E. Töpfer-Petersen, and T. Leeb (2002). "Molecular characterization of the equine AEG1 locus." *Gene* 292: 65–72.

Gil-Alarcón, G., M. C. Sánchez-Villegas, and V. H. Reynoso (2011). "Tratamiento prehospitalario del accidente ofídico: revisión, actualización y problemática actual." *Gac Med Mex* 147: 195–208.

Gilar, M., P. Olivova, A. E. Daly, and J. C. Gebler (2005). "Orthogonality of separation in two-dimensional liquid chromatography." *Anal Chem* 77(19): 6426–6434.

Gillespie, D., F. L. Frye, S. L. Stockham, and T. Fredeking (2000). "Blood values in wild and captive Komodo dragons (*Varanus komodoensis*)." *Zoo Biol* 19(6): 495–509.

Gilquin, B., A. Lecoq, F. Desné, M. Guenneugues, S. Zinn-Justin, and A. Ménez (1999). "Conformational and functional variability supported by the BPTI fold: Solution structure of the Ca^{2+} channel blocker calcicludine." *Proteins Struct Funct Bioinf* 34(4): 520–532.

Ginsborg, B. L., and J. Warriner (1960). "The isolated chick biventer cervicis nerve-muscle preparation." *Br J Pharmacol Chemother* 15: 410–411.

Giovannucci, E., D. M. Harlan, M. C. Archer, R. M. Bergenstal, S. M. Gapstur, L. A. Habel, M. Pollak, J. G. Regensteiner, and D. Yee (2010). "Diabetes and cancer: A consensus report." *CA Cancer J Clin* 60(4): 207–221.

Girish, K. S., D. K. Jagadeesha, K. B. Rajeev, and K. Kemparaju (2002). "Snake venom hyaluronidase: An evidence for isoforms and extracellular matrix degradation." *Mol Cell Biochem* 240(1–2): 105–110.

Girolami, A., P. Simioni, B. Girolami, A. Marchiori, D. S. Millar, P. Bignell, V. V. Kakkar, and D. N. Cooper (1993). "A novel dysfunctional protein C (protein C Padua 2) associated with a thrombotic tendency: Substitution of Cys for Arg-1 results in a strongly reduced affinity for binding of Ca++." *Br J Haematol* 85(3): 521–527.

Glass, R. E., and D. V. Banthorpe (1975). "Properties of nerve growth factor from venom of *Bothrops atrox*." *Biochim Biophys Acta* 405(1): 23–26.

Glaudas, X., and J. A. Rodriguez-Robles (2011). "Vagabond males and sedentary females: spatial ecology and mating system of the speckled rattlesnake (*Crotalus mitchellii*)." *Biol J Linn Soc* 103(3): 681–695.

Glenn, J. L., and R. Straight (1978). "Mojave rattlesnake *Crotalus scutulatus scutulatus* venom: Variation in toxicity with geographical origin." *Toxicon* 16(1): 81–84.

Glenn, T. C. (2011). "Field guide to next-generation DNA sequencers." *Mol Ecol Resour* 11(5): 759–769.

Glish, G. L., and D. J. Burinsky (2008). "Hybrid mass spectrometers for tandem mass spectrometry." *J Am Soc Mass Spectrom* 19(2): 161–172.

Gmachl, M., and G. Kreil (1993). "Bee venom hyaluronidase is homologous to a membrane-protein of mammalian sperm." *Proc Natl Acad Sci USA* 90(8): 3569–3573.

Gocheva, V., and J. A. Joyce (2007). "Cysteine cathepsins and the cutting edge of cancer invasion." *Cell Cycle* 6(1): 60–64.

Goke, R., H. C. Fehmann, T. Linn, H. Schmidt, M. Krause, J. Eng, and B. Goke (1993). "Exendin-4 is a high potency agonist and truncated exendin-(9-39)-amide an antagonist at the glucagon-like peptide 1-(7-36)-amide receptor of insulin-secreting beta-cells." *J Biol Chem* 268(26): 19650–19655.

Gold, B. S., R. A. Barish, R. C. Dart, R. P. Silverman, and G. V. Bochicchio (2003). "Resolution of compartment syndrome after rattlesnake envenomation utilizing non-invasive measures." *J Emerg Med* 24(3): 285–288.

Goldin, A. L. (1992). "Maintenance of *Xenopus laevis* and oocyte injection." *Methods Enzymol* 207: 266–279.

Goldman, N., and Z. Yang (1994). "A codon-based model of nucleotide substitution for protein-coding DNA sequences." *Mol Biol Evol* 11(5): 725–736.

Goldstein, E. J., K. L. Tyrrell, D. M. Citron, C. R. Cox, I. M. Recchio, B. Okimoto, J. Bryja, and B. G. Fry (2013). "Anaerobic and aerobic bacteriology of the saliva and gingiva from 16 captive Komodo dragons (*Varanus komodoensis*): New implications for the 'bacteria as venom' model." *J Zoo Wildl Med* 44(2): 262–272.

Goldstein, E. J. C., E. O. Agyare, A. E. Vagvolgyi, and M. Halpern (1981). "Aerobic bacterial oral flora of garter snakes—development of normal flora and pathogenic potential for snakes and humans." *J Clin Microbiol* 13(5): 954–956.

Goldstein, E. J. C., D. M. Citron, B. Wield, U. Blachman, V. L. Sutter, T. A. Miller, and S. M. Finegold (1978). "Bacteriology of human and animal bite wounds." *J Clin Microbiol* 8(6): 667–672.

Gomez, F., A. Vandermeers, M. C. Vandermeers-Piret, R. Herzog, J. Rathe, M. Stievenart, J. Winand, and J. Christophe (1989). "Purification and characterization of five variants of phospholipase A_2 and complete primary structure of the main phospholipase A_2 variant in *Heloderma suspectum* (Gila monster) venom." *Eur J Biochem* 186(1–2): 23–33.

Gonçalves, J. M., and A. Polson (1947). "The electrophoretic analysis of snake venoms." *Arch Biochem Biophys* 13(2): 253–259.

Gong, E., L. D. Martin, D. A. Burnham, and A. R. Falk (2010). "The birdlike raptor *Sinornithosaurus* was venomous." *Proc Natl Acad Sci USA* 107(2): 766–768.

Gong, N., A. N. Ma, L. J. Zhang, X. S. Luo, Y. H. Zhang, M. Xu, and Y. X. Wang (2011). "Site-specific PEGylation of exenatide analogues markedly improved their glucoregulatory activity." *Br J Pharmacol* 163(2): 399–412.

Gong, W., X. Zhu, S. Liu, M. Teng, and L. Niu (1998). "Crystal structures of acutolysin A, a three-disulfide hemorrhagic zinc metalloproteinase from the snake venom of *Agkistrodon acutus*." *J Mol Biol* 283(3): 657–668.

Gonzalez, J. E., K. Oades, Y. Leychkis, A. Harootunian, and P. A. Negulescu (1999). "Cell-based assays and instrumentation for screening ion-channel targets." *Drug Discov Today* 4(9): 431–439.

Gopalakrishnakone, P., and E. Kochva (1990). "Venom glands and some associated muscles in sea snakes." *J Morph* 206: 85–96.

Gow, A. S. F., and A. F. Scholfield, eds. (1953). "The poems and poetical fragments." *Nicander of Colophon*. Cambridge: Cambridge University Press, xii–247.

Gowda, D. C., J. Glushka, H. Halbeek, R. N. Thotakura, R. Bredehorst, and C. W. Vogel (2001). "N-linked oligosaccharides of cobra venom factor contain novel alpha(1-3)galactosylated Le(x) structures." *Glycobiology* 11(3): 195–208.

Gowda, D. C., C. M. Jackson, P. Hensley, and E. A. Davidson (1994a). "Factor X-activating glycoprotein of Russell's viper venom: Polypeptide composition and characterization of the carbohydrate moieties." *J Biol Chem* 269(14): 10644–10650.

Gowda, D. C., E. C. Petrella, T. T. Raj, R. Bredehorst, and C. W. Vogel (1994b). "Immunoreactivity and function of oligosaccharides in cobra venom factor." *J Immunol* 152(6): 2977–2986.

Grabherr, M. G., B. J. Haas, M. Yassour, J. Z. Levin, D. A. Thompson, I. Amit, X. Adiconis, et al. (2011). "Full-length transcriptome assembly from RNA-Seq data without a reference genome." *Nat Biotechnol* 29(7): 644–652.

Graham, R. L., C. Graham, S. McClean, T. Chen, M. O'Rourke, D. Hirst, D. Theakston, and C. Shaw (2005). "Identification and functional analysis of a novel bradykinin inhibitory peptide in the venoms of New World Crotalinae pit vipers." *Biochem Biophys Res Comm* 338(3): 1587–1592.

Grando, S. A. (2008). "Basic and clinical aspects of non-neuronal acetylcholine: Biological and clinical significance of non-canonical ligands of epithelial nicotinic acetylcholine receptors." *J Pharmacol Sci* 106(2): 174–179.

Grant, G. A., and V. A. Chiappinelli (1985). "Kappa-bungarotoxin: Complete amino acid sequence of a neuronal nicotinic receptor probe." *Biochemistry* 24(6): 1532–1537.

Green, D. A. II, and C. G. Extavour (2012). "Convergent evolution of a reproductive trait through distinct developmental mechanisms in *Drosophila*." *Dev Biol* 372(1): 120–130.

Greenwald, J., W. H. Fischer, W. W. Vale, and S. Choe (1999). "Three-finger toxin fold for the extracellular ligand-binding domain of the type II activin receptor serine kinase." *Nat Struct Biol* 6(1): 18–22.

Greer, J. (1981). "Comparative model-building of the mammalian serine proteases." *J Mol Biol* 153(4): 1027–1042.

Greig, N. H., H. W. Holloway, K. A. De Ore, D. Jani, Y. Wang, J. Zhou, M. J. Garant, and J. M. Egan (1999). "Once daily injection of exendin-4 to diabetic mice achieves long-term beneficial effects on blood glucose concentrations." *Diabetologia* 42(1): 45–50.

Griffin, P. R., and S. D. Aird (1990). "A new small myotoxin from the venom of the prairie rattlesnake (*Crotalus viridis viridis*)." *FEBS Lett* 274(1–2): 43–47.

Grinevich, V. P., S. R. Letchworth, K. A. Lindenberger, J. Menager, V. Mary, K. A. Sadieva, L. M. Buhlman, et al. (2005). "Heterologous expression of human (alpha)6(beta)4(beta)3(alpha)5 nicotinic acetylcholine receptors: Binding properties consistent with their natural expression require quaternary subunit assembly including the (alpha)5 subunit." *J Pharmacol Exp Ther* 312(2): 619–626.

Gronroos, J. O., J. H. Salonen, M. Viander, T. J. Nevalainen, and V. J. Laine (2005). "Roles of group IIA phospholipase A_2 and complement in killing of bacteria by acute phase serum." *Scand J Immunol* 62(4): 413–419.

Grosset, C., A. Villeneuve, A. Brieger, and S. Lair (2011). "Cryptosporidiosis in juvenile bearded dragons (*Pogona vitticeps*): Effects of treatment with paromomycin." *J Herpetol Med Surg* 21(1): 10–15.

Grunclova, L., M. Horn, M. Vancova, D. Sojka, Z. Franta, M. Mares, and P. Kopacek (2006). "Two secreted cystatins of the soft tick *Ornithodoros moubata*: Differential expression pattern and inhibitory specificity." *Biol Chem* 387(12): 1635–1644.

Gu, L. C., Y. Q. Shen, S. J. Xu, Y. Y. Shu, T. Jiang, and Z. J. Lin (2002). "Crystallization and preliminary X-ray diffraction studies of cobra venom beta-nerve growth factor." *Biochem Biophys Res Commun* 297(4): 1008–1010.

Gu, Z., G. Thomas, J. Yamashiro, I. P. Shintaku, F. Dorey, A. Raitano, O. N. Witte, J. W. Said, M. Loda, and R. E. Reiter (2000). "Prostate stem cell antigen (PSCA) expression increases with high gleason score, advanced stage and bone metastasis in prostate cancer." *Oncogene* 19(10): 1288–1296.

Guasch, J. F., S. Cannegieter, P. H. Reitsma, E. T. Van't Veer-Korthof, and R. M. Bertina (1998). "Severe coagulation factor V deficiency caused by a 4 bp deletion in the factor V gene." *Br J Haematol* 101(1): 32–39.

Gudderra, N. P., J. M. C. Ribeiro, and J. F. Andersen (2005). "Structural determinants of factor IX(a) binding in nitrophorin 2, a lipocalin inhibitor of the intrinsic coagulation pathway." *J Biol Chem* 280(26): 25022–25028.

Guderian, R. H., C. D. Mackenzie, and I. F. Williams (1986). "High voltage shock treatment for snakebite." *Lancet* 2:229.

Guex, N., and M. C. Peitsch (1997). "SWISS-MODEL and the Swiss-PdbViewer: An environment for comparative protein modeling." *Electrophoresis* 18(15): 2714–2723.

Guidlolin, R., R. M. Marcelino, H. H. Gondo, J. F. Morais, R. A. Ferreira, C. L. Silva, T. L. Kipnis, J. A. Silva, J. Fafetine, and W. D. da Silva (2010). "Polyvalent horse F(ab')2 snake antivenom: Development of process to produce polyvalent horse F(ab')2 antibodies anti-African snake venom." *Afr J Biotechnol* 9(16): 2446–2455.

Guindon, S., J. F. Dufayard, V. Lefort, M. Anisimova, W. Hordijk, and O. Gascuel (2010). "New algorithms and methods to estimate maximum-likelihood phylogenies: Assessing the performance of PhyML 3.0." *Syst Biol* 59(3): 307–321.

Gumley, T. P., I. F. McKenzie, and M. S. Sandrin (1995). "Tissue expression, structure and function of the murine Ly-6 family of molecules." *Immunol Cell Biol* 73(4): 277–296.

Gunning, A. P., S. Chambers, C. Pin, A. L. Man, V. J. Morris, and C. Nicoletti (2008). "Mapping specific adhesive interactions on living human intestinal epithelial cells with atomic force microscopy." *FASEB J* 22(7): 2331–2339.

Guo, M., M. Teng, L. Niu, Q. Liu, Q. Huang, and Q. Hao (2005). "Crystal structure of the cysteine-rich secretory protein stecrisp reveals that the cysteine-rich domain has a K+ channel inhibitor-like fold." *J Biol Chem* 280: 12405–12412.

Guo, X. X., L. Zeng, W. H. Lee, Y. Zhang, and Y. Jin (2007). "Isolation and cloning of a metalloproteinase from king cobra snake venom." *Toxicon* 49(7): 954–965.

Gustavson, S. M., D. Chen, V. Somayaji, K. Hudson, D. J. Baltrukonis, J. Singh, T. L. Boyden, and R. A. Calle (2011). "Effects of a long-acting GLP-1 mimetic (PF-04603629) on pulse rate and diastolic blood pressure in patients with type 2 diabetes mellitus." *Diabetes Obes Metab* 13(11): 1056–1058.

Guthals, A., K. R. Clauser, A. M. Frank, and N. Bandeira (2013). "Sequencing-grade *de novo* analysis of MS/MS triplets (CID/HCD/ETD) from overlapping peptides." *J Proteome Res* 12(6): 2846–2857.

Gutiérrez, J. M., O. Arroyo, and R. Bolaños (1980). "Mionecrosis, hemorragia y edema inducidos por el veneno de *Bothrops asper* en ratón blanco." *Toxicon* 18(5–6): 603–610.

Gutiérrez, J. M., H. W. Fan, C. L. Silvera, and Y. Angulo (2009). "Stability, distribution and use of antivenoms for snakebite envenomation in Latin America: Report of a workshop." *Toxicon* 53(6): 625–630.

Gutiérrez, J. M., J. A. Gene, G. Rojas, and L. Cerdas (1985). "Neutralization of proteolytic and hemorrhagic activities of Costa Rican snake venoms by a polyvalent antivenom." *Toxicon* 23(6): 887–893.

Gutiérrez, J. M., G. Leon, and T. Burnouf (2011). "Antivenoms for the treatment of snakebite envenomings: The road ahead." *Biologicals* 39(3): 129–142.

Gutiérrez, J. M., G. Leon, and B. Lomonte (2003). "Pharmacokinetic-pharmacodynamic relationships of immunoglobulin therapy for envenomation." *Clin Pharmacokinet* 42(8): 721–741.

Gutiérrez, J. M., and B. Lomonte (1995). "Phospholipase A_2 myotoxins from *Bothrops* snake venoms." *Toxicon* 33(11): 1405–1424.

——— (1997). "Phospholipase A_2 myotoxins from *Bothrops* snake venoms." In *Venom phospholipase A_2 enzymes: Structure, function, and mechanism*, edited by R. M. Kini. Chichester: John Wiley, 321–352.

Gutiérrez, J. M., J. Nuñez, T. Escalante, and A. Rucavado (2006). "Blood flow is required for rapid endothelial cell damage induced by a snake venom hemorrhagic metalloproteinase." *Microvasc Res* 71(1): 55–63.

Gutiérrez, J. M., C. L. Ownby, and G. V. Odell (1984). "Isolation of a myotoxin from Bothrops asper venom: Partial characterization and action on skeletal muscle." *Toxicon* 22(1): 115–128.

Gutiérrez, J. M., M. Romero, C. Diaz, G. Borkow, and M. Ovadia (1995). "Isolation and characterization of a metalloproteinase with weak hemorrhagic activity from the venom of the snake *Bothrops asper* (terciopelo)." *Toxicon* 33(1): 19–29.

Gutiérrez, J. M., A. Rucavado, and T. Escalante (2009). "Snake venom metalloproteinases: Biological roles and participation in the pathophysiolgy of envenomation." In *Handbook of venoms and toxins of reptiles*, edited by S. P. Mackessy. Boca Raton, FL: CRC Press, 116–138.

Gutiérrez, J. M., A. Rucavado, T. Escalante, and C. Diaz (2005). "Hemorrhage induced by snake venom metalloproteinases: Biochemical and biophysical mechanisms involved in microvessel damage." *Toxicon* 45(8): 997–1011.

Gutiérrez, J. M., L. Sanz, J. Escolano, J. Fernandez, B. Lomonte, Y. Angulo, A. Rucavado, D. A. Warrell, and J. J. Calvete (2008). "Snake venomics of the Lesser Antillean pit vipers *Bothrops caribbaeus* and *Bothrops lanceolatus*: Correlation with toxicological activities and immunoreactivity of a heterologous antivenom." *J Proteome Res* 7(10): 4396–4408.

Gutiérrez, J. M., G. Solano, D. Pla, M. Herrera, A. Segura, M. Villalta, M. Vargas, et al. (2013). "Assessing the preclinical efficacy of antivenoms: From the lethality neutralization assay to antivenomics." *Toxicon* 69: 168–179.

Habermann, E. (1972). "Bee and wasp venoms." *Science* 177(4046): 314–322.

Habib, A. G. (2011). "Effect of pre-medication on early adverse reactions following antivenom use in snakebite: A systematic review and meta-analysis." *Drug Safety* 34(10): 869–880.

Haendler, B., J. Kratzschmar, F. Theuring, and W. D. Schleuning (1993). "Transcripts for cysteine-rich secretory protein-1 (CRISP-1; DE/AEG) and the novel related CRISP-3 are expressed under androgen control in the mouse salivary gland." *Endocrinology* 133(1): 192–198.

Hagiwara, H., H. Sakaguchi, M. Itakura, T. Yoshimoto, M. Furuya, S. Tanaka, and S. Hirose (1994a). "Autocrine regulation of rat chondrocyte proliferation by natriuretic peptide C and its receptor, natriuretic peptide receptor-B." *J Biol Chem* 269(14): 10729–10733.

Hagiwara, H., H. Sakaguchi, K. M. Lodhi, K. Suda, and S. Hirose (1994b). "Subtype switching of natriuretic peptide receptors in rat chondrocytes during *in vitro* culture." *J Biochem* 116(3): 606–609.

Halkier, T. (1991). *Mechanisms in blood coagulation fibrinolysis and the complement system.* Cambridge: Cambridge University Press.

Hall, E. L. (2001). "Role of surgical intervention in the management of crotaline snake envenomation." *Ann Emerg Med* 37(2): 175–180.

Hall, N. G., P. Klenotic, B. Anand-Apte, and S. S. Apte (2003). "ADAMTSL-3/punctin-2, a novel glycoprotein in extracellular matrix related to the ADAMTS family of metalloproteases." *Matrix Biol* 22(6): 501–510.

Hamako, J., T. Matsui, S. Nishida, S. Nomura, Y. Fujimura, M. Ito, Y. Ozeki, and K. Titani (1998). "Purification and characterization of kaouthiagin, a von Willebrand factor-binding and -cleaving metalloproteinase from *Naha kaouthia* cobra venom." *Thromb Haemost* 80(3): 499–505.

Hamako, J., T. Matsui, M. Suzuki, M. Ito, K. Makita, Y. Fujimura, Y. Ozeki, and K. Titani (1996). "Purification and characterization of bitiscetin, a novel von Willebrand factor modulator protein from *Bitis arietans* snake venom." *Biochem Biophys Res Commun* 226(1): 273–279.

Hamill, O. P., A. Marty, E. Neher, B. Sakmann, and F. J. Sigworth (1981). "Improved patch-clamp techniques for high-resolution current recording from cells and cell-free membrane patches." *Pflugers Arch* 391(2): 85–100.

Hardy, D., H. Greene, B. Tomberlin, and M. Webster (2001). "Relocation of nuisance rattlesnakes: Problems using short-distance translocation in a small rural community." *Sonoran Herpetologist* 14: 1–3.

Hardy, D. L. (1988). "Appropriate first aid measures for venomous snakebite should not come as a shock." *Tucson Herpetol Soc Newsl* 1: 12–13.

Hare, T., and J. McNally (1997). "Evaluation of a rattlesnake relocation program in the Tucson, Arizona, area." *Sonoran Herpetologist* 10: 26–31.

Hargreaves, A. D., M. T. Swain, M. J. Hegarty, D. W. Logan, and J. F. Mulley (2014a). "Restriction and recruitment—gene duplication and the origin and evolution of snake venom toxins." *Genome Biol Evol* 6: 2088–2095.

Hargreaves, A. D., M. T. Swain, D. W. Logan, and J. F. Mulley (2014b). "Testing the toxicofera: Comparative transcriptomics casts doubt on the single, early evolution of the reptile venom system." *Toxicon* 92: 140–156.

Hargrove, D. M., E. S. Kendall, J. M. Reynolds, A. N. Lwin, J. P. Herich, P. A. Smith, B. R. Gedulin, et al. (2007). "Biological activity of AC3174, a peptide analog of exendin-4." *Regul Pept* 141(1–3): 113–119.

Hariprasad, G., B. Singh, U. Das, A. S. Ethayathulla, P. Kaur, T. P. Singh, and A. Srinivasan (2007). "Cloning, sequence analysis and homology modeling of a novel phospholipase A_2 from *Heterometrus fulvipes* (Indian black scorpion)." *DNA Seq* 18(3): 242–246.

Harper, C. V., C. L. R. Barratt, and S. J. Publicover (2004). "Stimulation of human spermatozoa with progesterone gradients to simulate approach to the oocyte. Induction of [Ca(2+)](i) oscillations and cyclical transitions in flagellar beating." *J Biol Chem* 279: 46315–46325.

Harris, J. B. (1985). "Phospholipases in snake venoms and their effects on nerve and muscle." *Pharmacol Ther* 31(1–2): 79–102.

Harris, J. B. and A. Goonetilleke (2004). "Animal poisons and the nervous system: what the neurologist needs to know." *J Neurol Neurosurg Psychiatry* 75 suppl 3: iii40–46.

Harris, J. B., B. D. Grubb, C. A. Maltin, and R. Dixon (2000). "The neurotoxicity of the venom phospholipases A(2), notexin and taipoxin." *Exp Neurol* 161(2): 517–526.

Harrison, R. A., S. S. Hasson, M. Harmsen, G. D. Laing, K. Conrath, and R. D. Theakston (2006). "Neutralisation of venom-induced haemorrhage by IgG from camels and llamas immunised with viper venom and also by endogenous, non-IgG components in camelid sera." *Toxicon* 47(3): 364–368.

Harrison, R. A., F. Ibison, D. Wilbraham, and S. C. Wagstaff (2007). "Identification of cDNAs encoding viper venom hyaluronidases: Cross-generic sequence conservation of full-length and unusually short variant transcripts." *Gene* 392(1–2): 22–33.

Hart, A. J., G. K. Isbister, P. O'Donnell, N. A. Williamson, and W. C. Hodgson (2013). "Species differences in the neuromuscular activity of post-synaptic neurotoxins from two Australian black snakes (*Pseudechis porphyriacus* and *Pseudechis colletti*)." *Toxicol Lett* 219(3): 262–268.

Hartmann, S., and R. Lucius (2003). "Modulation of host immune responses by nematode cystatins." *Int J Parasitol* 33(11): 1291–1302.

Harvey, A. L. (1985). "Cardiotoxins from cobra venoms: Possible mechanisms of action." *Toxin Rev* 4(1): 41–69.

Harvey, A. L. (2001). "Twenty years of dendrotoxins." *Toxicon* 39(1): 15–26.

Harvey, A. L., and A. J. Anderson (1985). "Dendrotoxins: Snake toxins that block potassium channels and facilitate neurotransmitter release." *Pharmacol Ther* 31(1–2): 33–55.

Harvey, A. L., A. Barfaraz, E. Thomson, A. Faiz, S. Preston, and J. B. Harris (1994). "Screening of snake venoms for neurotoxic and myotoxic effects using simple *in vitro* preparations from rodents and chicks." *Toxicon* 32(3): 257–265.

Harvey, A. L., and E. Karlsson (1980). "Dendrotoxin from the venom of the green mamba, *Dendroaspis angusticeps*: A neurotoxin that enhances acetylcholine release at neuromuscular junctions." *Naunyn Schmiedebergs Arch Pharmacol* 312(1): 1–6.

Harvey, A. L., E. Kornisiuk, K. N. Bradley, C. Cervenansky, R. Duran, M. Adrover, G. Sanchez, and D. Jerusalinsky (2002). "Effects of muscarinic toxins MT1 and MT2 from green mamba on different muscarinic cholinoceptors." *Neurochem Res* 27(11): 1543–1554.

Harvey, A. L., R. J. Marshall, and E. Karlsson (1982). "Effects of purified cardiotoxins from the Thailand cobra (*Naja naja siamensis*) on isolated skeletal and cardiac muscle preparations." *Toxicon* 20(2): 379–396.

Harvey, A. L., and B. Robertson (2004). "Dendrotoxins: Structure-activity relationships and effects on potassium ion channels." *Curr Med Chem* 11(23): 3065–3072.

Harvey, A. L., E. G. Rowan, H. Vatanpour, A. Engstrom, B. Westerlund, and E. Karlsson (1997). "Changes to biological activity following acetylation of dendrotoxin I from *Dendroaspis polylepis* (black mamba)." *Toxicon* 35(8): 1263–1273.

Harwig, S. S., K. M. Swiderek, V. N. Kokryakov, L. Tan, T. D. Lee, E. A. Panyutich, G. M. Aleshina, O. V. Shamova, and R. I. Lehrer (1994). "Gallinacins: Cysteine-rich antimicrobial peptides of chicken leukocytes." *FEBS Lett* 342(3): 281–285.

Hase, S., S. Kawabata, H. Nishimura, H. Takeya, T. Sueyoshi, T. Miyata, S. Iwanaga, T. Takao, Y. Shimonishi, and T. Ikenaka (1988). "A new trisaccharide sugar chain linked to a serine residue in bovine blood coagulation factors VII and IX." *J Biochem* 104: 867–868.

Hati, A. K., M. Mandal, M. K. De, H. Mukherjee, and R. N. Hati (1992). "Epidemiology of snake bite in the district of Burdwan, West Bengal." *J Indian Med Assoc* 90(6): 145–147.

Hati, R. N., M. Mandal, and A. K. Hati (1990). "Active immunization of rabbit with gamma irradiated Russell's viper venom toxoid." *Toxicon* 28(8): 895–902.

Hausenloy, D. J., and D. M. Yellon (2012). "Taking lizard saliva to heart." *Eur Heart J* 33(12): 1426–1430.

Havugimana, P. C., P. Wong, and A. Emili (2007). "Improved proteomic discovery by sample pre-fractionation using dual-column ion-exchange high performance liquid chromatography." *J Chromatogr B* 847(1): 54–61.

Hawgood, B. J., and C. Bon (1991). "Snake venom presynaptic toxins." In *Handbook of natural toxins, V: Reptile venoms and toxins*," edited by A. T. Tu. New York: Marcell Dekker, 3.

Hayashi, M. A., C. Ligny-Lemaire, Z. Wollberg, M. Wery, A. Galat, T. Ogawa, B. H. Muller, et al. (2004). "Long-sarafotoxins: Characterization of a new family of endothelin-like peptides." *Peptides* 25(8): 1243–1251.

Hayashi, M. A., F. D. Nascimento, A. Kerkis, V. Oliveira, E. B. Oliveira, A. Pereira, G. Rádis-Baptista, et al. (2008). "Cytotoxic effects of crotamine are mediated through lysosomal membrane permeabilization." *Toxicon* 52(3): 508–517.

Hayashi, M. A., E. B. Oliveira, I. Kerkis, and R. L. Karpel (2012). "Crotamine: A novel cell-penetrating polypeptide nanocarrier with potential anti-cancer and biotechnological applications." *Methods Mol Biol* 906: 337–352.

Hayward, C. P. M., E. M. Cramer, W. H. Kane, S. L. Zheng, M. Bouchard, J. M. Masse, and G. E. Rivard (1997). "Studies of a second family with the Quebec platelet disorder: Evidence that the degradation of the alpha-granule membrane and its soluble contents are not secondary to a defect in targeting proteins to alpha-granules." *Blood* 89(4): 1243–1253.

He, X. L., A. Dukkipati, and K. C. Garcia (2006). "Structural determinants of natriuretic peptide receptor specificity and degeneracy." *J Mol Biol* 361(4): 698–714.

He, Y.-Y., S.-B. Liu, W.-H. Lee, J.-Q. Qian, and Y. Zhang (2008). "Isolation, expression and characterization of a novel dual serine protease inhibitor, OH-TCI, from king cobra venom." *Peptides* 29(10): 1692–1699.

Heidtmann, H. H., D. M. Nettelbeck, A. Mingels, R. Jager, H. G. Welker, and R. E. Kontermann (1999). "Generation of angiostatin-like fragments from plasminogen by prostate-specific antigen." *Br J Cancer* 81(8): 1269–1273.

Heinrikson, R. L., E. T. Krueger, and P. S. Keim (1977). "Amino acid sequence of phospholipase A_2-alpha from the venom of *Crotalus adamanteus*: A new classification of phospholipases A_2 based upon structural determinants." *J Biol Chem* 252(14): 4913–4921.

Hejnar, P., J. Bardon, P. Sauer, and M. Kolar (2007). "*Stenotrophomonas maltophilia* as a part of normal oral bacterial flora in captive snakes and its susceptibility to antibiotics." *Vet Microbiol* 121(3–4): 357–362.

Hendon, R. A., and A. T. Tu (1981). "Biochemical characterization of the lizard toxin gilatoxin." *Biochemistry* 20(12): 3517–3522.

Henry, R. R., D. Logan, T. Alessi, and M. A. Baron (2013). "A randomized, open-label, multicenter, 4-week study to evaluate the tolerability and pharmacokinetics of ITCA 650 in patients with type 2 diabetes." *Clin Ther* 35(5): 634–645, e631.

Hensley, P., M. C. O'Keefe, C. J. Spangler, J. C. Osborne Jr., and C. W. Vogel (1986). "The effects of metal ions and temperature on the interaction of cobra venom factor and human complement factor B." *J Biol Chem* 261(24): 11038–11044.

Herbert, S. P., and J. H. Walker (2006). "Group VIA calcium-independent phospholipase A_2 mediates endothelial cell S phase progression." *J Biol Chem* 281(47): 35709–35716.

Hernández, R. M., A. C. Bello, and M. E. Mávil (2007). "Etnobotánica y ecología de plantas utilizadas por tres curanderos contra la mordedura de serpiente en la región de Acayucán, Veracruz, México." *Bol Soc Bot Méx* 81: 89–100.

Hernandez-Divers, S. J. (2006). "Appendix A: Reptile viral diseases—summary table." In *Reptile medicine and surgery*. St. Louis, MO: Elsevier, 1147–1158.

Herrera, M., G. Leon, A. Segura, F. Meneses, B. Lomonte, J. P. Chippaux, and J. M. Gutiérrez (2005). "Factors associated with adverse reactions induced by caprylic acid-fractionated whole IgG preparations: Comparison between horse, sheep and camel IgGs." *Toxicon* 46(7): 775–781.

Herve-Grepinet, V., S. Rehault-Godbert, V. Labas, T. Magallon, C. Derache, M. Lavergne, J. Gautron, A. C. Lalmanach, and Y. Nys (2010). "Purification and characterization of avian beta-defensin 11, an antimicrobial peptide of the hen egg." *Antimicrob Agents Chemother* 54(10): 4401–4409.

Herzig, V., M. Kohler, K. F. Grund, S. Reeve, A. I. Smith, and W. C. Hodgson (2013). "Analysis of intraspecific variation in venoms of *Acanthophis antarcticus* death adders from South Australia." *J Venom Res* 4: 13–20.

Heyborne, W. H., and S. P. Mackessy (2013). "Identification and characterization of a taxon-specific three-finger toxin from the venom of the Green Vinesnake (*Oxybelis fulgidus*; family Colubridae)." *Biochimie* 95(10): 1923–1932.

Hill, J. H., and P. A. Ward (1971). "The phlogistic role of C3 leukotactic fragments in myocardial infarcts of rats." *J Exp Med* 133(4): 885–900.

Hill, R. E., G. M. Bogdan, and R. C. Dart (2001). "Time to reconstitution: Purified Fab antivenom vs. unpurified IgG antivenom." *Toxicon* 39(5): 729–731.

Hill, R. E., and S. P. Mackessy (2000). "Characterization of venom (Duvernoy's secretion) from twelve species of colubrid snakes and partial sequence of four venom proteins." *Toxicon* 38(12): 1663–1687.

Hirotsu, S., H. Mizuno, K. Fukuda, M. C. Qi, T. Matsui, J. Hamako, T. Morita, and K. Titani (2001). "Crystal structure of bitiscetin, a von Willebrand factor-dependent platelet aggregation inducer." *Biochemistry* 40(45): 13592–13597.

Hisada, M., H. Satake, K. Masuda, M. Aoyama, K. Murata, T. Shinada, T. Iwashita, Y. Ohfune, and T. Nakajima (2005). "Molecular components and toxicity of the venom of the solitary wasp, *Anoplius samariensis*." *Biochem Biophys Res Commun* 330(4): 1048–1054.

Hite, L. A., L. G. Jia, J. B. Bjarnason, and J. W. Fox (1994). "cDNA sequences for four snake venom metalloproteinases: Structure, classification, and their relationship to mammalian reproductive proteins." *Arch Biochem Biophys* 308(1): 182–191.

Hite, L. A., J. D. Shannon, J. B. Bjarnason, and J. W. Fox (1992). "Sequence of a cDNA clone encoding the zinc metalloproteinase hemorrhagic toxin e from *Crotalus atrox*: Evidence for signal, zymogen, and disintegrin-like structures." *Biochemistry* 31(27): 6203–6211.

Ho, C. L., and C. Y. Lee (1981). "Presynaptic actions of Mojave toxin isolated from Mojave rattlesnake (*Crotalus scutulatus*) venom." *Toxicon* 19(6): 889–892.

Hoch, J. C., M. W. Maciejewski, M. Mobli, A. D. Schuyler, and A. S. Stern (2014). "Nonuniform sampling and maximum entropy reconstruction in multidimensional NMR." *Acc Chem Res* 47(2): 708–717.

Hockin, M. F., K. M. Cawthern, M. Kalafatis, and K. G. Mann (1999). "A model describing the inactivation of factor Va by APC: Bond cleavage, fragment dissociation, and product inhibition." *Biochemistry* 38(21): 6918–6934.

Hockin, M. F., K. C. Jones, S. J. Everse, and K. G. Mann (2002). "A model for the stoichiometric regulation of blood coagulation." *J Biol Chem* 277(21): 18322–18333.

Hockin, M. F., M. Kalafatis, M. Shatos, and K. G. Mann (1997). "Protein C activation and factor Va inactivation on human umbilical vein endothelial cells." *Arterioscler Thromb Vasc Biol* 17(11): 2765–2775.

Hocknull, S. A., P. J. Piper, G. D. van den Bergh, R. A. Due, M. J. Morwood, and I. Kurniawan (2009). "Dragon's paradise lost: Palaeobiogeography, evolution and extinction of the largest-ever terrestrial lizards (Varanidae)." *PLoS One* 4(9): e7241.

Hodgson, W. C., C. A. Dal Belo, and E. G. Rowan (2007). "The neuromuscular activity of paradoxin: A presynaptic neurotoxin from the venom of the inland taipan (*Oxyuranus microlepidotus*)." *Neuropharmacology* 52(5): 1229–1236.

Hodgson, W. C., and J. C. Wickramaratna (2002). "*In vitro* neuromuscular activity of snake venoms." *Clin Exp Pharmacol Physiol* 29(9): 807–814.

Hofmann, H., and C. Bon (1987). "Blood coagulation induced by the venom of *Bothrops atrox*, 2: Identification, purification, and properties of two factor X activators." *Biochemistry* 26(3): 780–787.

Holstege, C. P., M. B. Miller, M. Wermuth, B. Furbee, and S. C. Curry (1997). "Crotalid snake envenomation." *Crit Care Clin* 13(4): 889.

Holstege, C. P., and E. M. Singletary (2006). "Images in emergency medicine: Skin damage following application of suction device for snakebite." *Ann Emerg Med* 48(1): 113–105, 113.

Holz, P., I. K. Barker, J. P. Burger, G. J. Crawshaw, and P. D. Conlon (1997). "The effect of the renal portal system on pharmacokinetic parameters in the red-fared slider (*Trachemys scripta elegans*)." *J Zoo Wildl Med* 28(4): 386–393.

Holz, P. H., J. P. Burger, R. Baker, and S. Young (2002). "Effect of injection site on carbenicillin pharmacokinetics in the carpet python, *Morelia spilota*." *J Herpetol Med Surg* 12(4): 12–16.

Homsi-Brandeburgo, M. I., L. S. Queiroz, H. Santo-Neto, L. Rodrigues-Simioni, and J. R. Giglio (1988). "Fractionation of *Bothrops jararacussu* snake venom: Partial chemical characterization and biological activity of bothropstoxin." *Toxicon* 26(7): 615–627.

Hong, P., S. Koza, and E. S. Bouvier (2012). "Size-exclusion chromatography for the analysis of protein biotherapeutics and their aggregates." *J Liq Chromatogr Relat Technol* 35(20): 2923–2950.

Hong, S. Y., H. Lee, W. K. You, K. H. Chung, D. S. Kim, and K. Song (2003). "The snake venom disintegrin salmosin induces apoptosis by disassembly of focal adhesions in bovine capillary endothelial cells." *Biochem Biophys Res Comm* 302(3): 502–508.

Honma, T., and K. Shiomi (2006). "Peptide toxins in sea anemones: Structural and functional aspects." *Mar Biotechnol* 8(1): 1–10.

Hooft, R. W., G. Vriend, C. Sander, and E. E. Abola (1996). "Errors in protein structures." *Nature* 381(6580): 272.

Hook, V., L. Funkelstein, D. Lu, S. Bark, J. Wegrzyn, and S. R. Hwang (2008). "Proteases for processing proneuropeptides into peptide neurotransmitters and hormones." *Annu Rev Pharmacol Toxicol* 48: 393–423.

Hook, V., T. Toneff, M. Bogyo, D. Greenbaum, K. F. Medzihradszky, J. Neveu, W. Lane, G. Hook, and T. Reisine (2005). "Inhibition of cathepsin B reduces beta-amyloid production in regulated secretory vesicles of neuronal chromaffin cells: Evidence for cathepsin B as a candidate beta-secretase of Alzheimer's disease." *Biol Chem* 386(9): 931–940.

Hooker, K. R., and E. M. Caravati (1994). "Gila monster envenomation." *Ann Emerg Med* 24(4): 731–735.

Hooley, E., E. Papagrigoriou, A. Navdaev, A. V. Pandey, J. M. Clemetson, K. J. Clemetson, and J. Emsley (2008). "The crystal structure of the platelet activator aggretin reveals a novel (alphabeta)2 dimeric structure." *Biochemistry* 47(30): 7831–7837.

Horie, M., K. Okutomi, Y. Taniguchi, Y. Ohbuchi, M. Suzuki, and E. Takahashi (1998). "Isolation and characterization of a new member of the human Ly6 gene family (LY6H)." *Genomics* 53(3): 365–368.

Horii, K., D. Okuda, T. Morita, and H. Mizuno (2003). "Structural characterization of EMS16, an antagonist of collagen receptor (GPIa/IIa) from the venom of *Echis multisquamatus*." *Biochemistry* 42(43): 12497–12502.

Houghton, P. J., M. J. Howes, C. C. Lee, and G. Steventon (2007). "Uses and abuses of *in vitro* tests in ethnopharmacology: Visualizing an elephant." *J Ethnopharmacol* 110(3): 391–400.

Howard, J. W., R. G. Kay, S. Pleasance, and C. S. Creaser (2012). "UHPLC for the separation of proteins and peptides." *Bioanalysis* 4(24): 2971–2988.

Howe, N. R., and J. L. Meisenheimer (1988). "Electric-shock does not save snakebitten rats." *Ann Emerg Med* 17(3): 254–256.

Howie, P. W., C. R. Prentice, and G. P. McNicol (1973). "A method of antithrombin estimation using plasma defibrinates with Ancrod." *Br J Haematol* 25(1): 101–110.

Hristova, M., and L. Aloe (2006). "Metabolic syndrome—neurotrophic hypothesis." *Med Hypotheses* 66(3): 545–549.

Hruska, M., J. Keefe, D. Wert, A. B. Tekinay, J. J. Hulce, I. Ibanez-Tallon, and R. Nishi (2009). "Prostate stem cell antigen is an endogenous lynx1-like prototoxin that antagonizes alpha7-containing nicotinic receptors and prevents programmed cell death of parasympathetic neurons." *J Neurosci* 29(47): 14847–14854.

Hu, H., P. Bandyopadhyay, B. Olivera, and M. Yandell (2011). "Characterization of the *Conus bullatus* genome and its venom-duct transcriptome." *BMC Genomics* 12(1): 60.

——— (2012). "Elucidation of the molecular envenomation strategy of the cone snail *Conus geographus* through transcriptome sequencing of its venom duct." *BMC Genomics* 13(1): 284.

Huai, Q., A. P. Mazar, A. Kuo, G. C. Parry, D. E. Shaw, J. Callahan, Y. Li, et al. (2006). "Structure of human urokinase plasminogen activator in complex with its receptor." *Science* 311(5761): 656–659.

Huang, L. F., H. B. Zheng, Y. Xu, H. T. Song, and C. X. Yu (2008). "A snake venom phospholipase A(2) with high affinity for muscarinic acetylcholine receptors acts on guinea pig ileum." *Toxicon* 51(6): 1008–1016.

Huang, M. Z., and P. Gopalakrishnakone (1996). "Pathological changes induced by an acidic phospholipase A_2 from *Ophiophagus hannah* venom on heart and skeletal muscle of mice after systemic injection." *Toxicon* 34(2): 201–211.

Huang, M. Z., P. Gopalakrishnakone, M. C. Chung, and R. M. Kini (1997). "Complete amino acid sequence of an acidic, cardiotoxic phospholipase A_2 from the venom of *Ophiophagus hannah* (King Cobra): A novel cobra venom enzyme with 'pancreatic loop.'" *Arch Biochem Biophys* 338(2): 150–156.

Huang, M. Z., Q. C. Wang, and G. F. Liu (1993). "Effects of an acidic phospholipase A_2 purified from *Ophiophagus hannah* (king cobra) venom on rat heart." *Toxicon* 31(5): 627–635.

Huang, P., and S. P. Mackessy (2004). "Biochemical characterization of phospholipase A_2 (trimorphin) from the venom of the Sonoran Lyre Snake *Trimorphodon biscutatus lambda* (family Colubridae)." *Toxicon* 44(1): 27–36.

Huang, T. F., and H. S. Chiang (1994). "Effect on human platelet-aggregation of phospholipase a(2) purified from *Heloderma horridum* (bearded lizard) venom." *Biochim Biophys Acta Lipids Lipid Metab* 1211(1): 61–68.

Huang, T. F., J. C. Holt, H. Lukasiewicz, and S. Niewiarowski (1987). "Trigramin: A low molecular weight peptide inhibiting fibrinogen interaction with platelet receptors expressed on glycoprotein IIb-IIIa complex." *J Biol Chem* 262(33): 16157–16163.

Huang, T. F., C. Z. Liu, and S. H. Yang (1995). "Aggretin, a novel platelet-aggregation inducer from snake (*Calloselasma rhodostoma*) venom, activates phospholipase C by acting as a glycoprotein Ia/IIa agonist." *Biochem J* 309(pt 3): 1021–1027.

Huang, T. T., J. B. Lynch, D. L. Larson, and S. R. Lewis (1974). "Use of excisional therapy in management of snakebite." *Ann Surg* 179(5): 598–607.

Huber, R., D. Kukla, A. Ruhlmann, O. Epp, and H. Formanek (1970). "The basic trypsin inhibitor of bovine pancreas, I: Structure analysis and conformation of the polypeptide chain." *Naturwissenschaften* 57(8): 389–392.

Huey, R., C. M. Bloor, M. S. Kawahara, and T. E. Hugli (1983). "Potentiation of the anaphylatoxins *in vivo* using an inhibitor of serum carboxypeptidase N (SCPN), I: Lethality and pathologic effects on pulmonary tissue." *Am J Pathol* 112(1): 48–60.

Hughes, A. L. (1994). "The evolution of functionally novel proteins after gene duplication." *Proc Biol Sci* 256(1346): 119–124.

Hultgren, H. N. (1994). "Rattlesnake bite—editorial comment." *J Wilderness Med* 5(2): 220–221.

Hung, C. C., and S. H. Chiou (2001). "Fibrinogenolytic proteases isolated from the snake venom of Taiwan habu: Serine proteases with kallikrein-like and angiotensin-degrading activities." *Biochem Biophys Res Commun* 281(4): 1012–1018.

Hussein, A. S., M. R. Chacon, A. M. Smith, R. Tosado-Acevedo, and M. E. Selkirk (1999). "Cloning, expression, and properties of a nonneuronal secreted acetylcholinesterase from the parasitic nematode *Nippostrongylus brasiliensis*." *J Biol Chem* 274(14): 9312–9319.

Hussein, A. S., M. E. Grigg, and M. E. Selkirk (1999). "*Nippostrongylus brasiliensis*: Characterisation of a somatic amphiphilic acetylcholinesterase with properties distinct from the secreted enzymes." *Exp Parasitol* 91(2): 144–150.

Hutchinson, D. A., A. Mori, A. H. Savitzky, G. M. Burghardt, X. Wu, J. Meinwald, and F. C. Schroeder (2007). "Dietary sequestration of defensive steroids in nuchal glands of the Asian snake *Rhabdophis tigrinus*." *Proc Natl Acad Sci USA* 104(7): 2265–2270.

Hutchinson, D. A., A. H. Savitzky, A. Mori, G. M. Burghardt, J. Meinwald, and F. C. Schroeder (2012). "Chemical investigations of defensive steroid sequestration by the Asian snake *Rhabdophis tigrinus*." *Chemoecology* 22(3): 199–206.

Hutton, R. A., and D. A. Warrell (1993). "Action of snake venom components on the haemostatic system." *Blood Rev* 7(3): 176–189.

Huxley-Jones, J., T. K. Clarke, C. Beck, G. Toubaris, D. L. Robertson, and R. P. Boot-Handford (2007). "The evolution of the vertebrate metzincins; insights from *Ciona intestinalis* and *Danio rerio*." *BMC Evol Biol* 7: 63.

Hyndman, T. H., C. M. Shilton, R. J. Doneley, and P. K. Nicholls (2012). "Sunshine virus in Australian pythons." *Vet Microbiol* 161(1–2): 77–87.

Hyndman, T. H., C. M. Shilton, and R. E. Marschang (2013). "Paramyxoviruses in reptiles: A review." *Vet Microbiol* 165(3–4): 200–213.

Ibanez-Tallon, I., J. M. Miwa, H. L. Wang, N. C. Adams, G. W. Crabtree, S. M. Sine, and N. Heintz (2002). "Novel modulation of neuronal nicotinic acetylcholine receptors by association with the endogenous prototoxin lynx1." *Neuron* 33(6): 893–903.

Iino, M., D. C. Foster, and W. Kisiel (1998). "Functional consequences of mutations in Ser-52 and Ser-60 in human blood coagulation factor VII." *Arch Biochem Biophys* 352: 182–192.

Illig, K. A., and K. Ouriel (1996). "Ancrod: Understanding the agent." *Semin Vasc Surg* 9(4): 303–314.

Imai, K., T. Nikai, H. Sugihara, and C. L. Ownby (1989). "Hemorrhagic toxin from the venom of *Agkistrodon bilineatus* (common cantil)." *Int J Biochem* 21(6): 667–673.

Inoue, K., and T. Morita (1993). "Identification of O-linked oligosaccharide chains in the activation peptides of blood coagulation factor X: The role of the carbohydrate moieties in the activation of factor X." *Eur J Biochem* 218(1): 153–163.

Isbister, G. K. (2009). "Procoagulant snake toxins: Laboratory studies, diagnosis, and understanding snakebite coagulopathy." *Semin Thromb Hemost* 35(01): 093–103.

Isbister, G. K., S. G. Brown, E. MacDonald, J. White, and B. J. Currie (2008). "Current use of Australian snake antivenoms and frequency of immediate-type hypersensitivity reactions and anaphylaxis." *Med J Aust* 188(8): 473–476.

Isbister, G. K., S. G. Brown, C. B. Page, D. L. McCoubrie, S. L. Greene, and N. A. Buckley (2013). "Snakebite in Australia: A practical approach to diagnosis and treatment." *Med J Aust* 199(11): 763–768.

Isbister, G. K., S. B. Duffull, and S. G. Brown (2009). "Failure of antivenom to improve recovery in Australian snakebite coagulopathy." *Q J Med* 102(8): 563–568.

Isbister, G. K., F. E. Scorgie, M. A. O'Leary, M. Seldon, G. A. Brown, and L. F. Lincz (2010). "Factor deficiencies in venom-induced consumption coagulopathy resulting from Australian elapid envenomation: Australian Snakebite Project (ASP-10)." *J Thromb Haemost* 8: 2504–2513.

Ishikawa, M., M. Kumashiro, Y. Yamazaki, H. Atoda, and T. Morita (2009). "Anticoagulant mechanism of factor IX/factor X-binding protein isolated from the venom of *Trimeresurus flavoviridis*." *J Biochem* 145(1): 123–128.

Ito, M., J. Hamako, Y. Sakurai, M. Matsumoto, Y. Fujimura, M. Suzuki, K. Hashimoto, K. Titani, and T. Matsui (2001). "Complete amino acid sequence of kaouthiagin, a novel cobra venom metalloproteinase with two disintegrin-like sequences." *Biochemistry* 40(14): 4503–4511.

Ito, N., M. Mita, Y. Takahashi, A. Matsushima, Y. G. Watanabe, S. Hirano, and S. Odani (2007). "Novel cysteine-rich secretory protein in the buccal gland secretion of the parasitic lamprey, *Lethehteron japonicum*." *Biochem Biophys Res Comm* 358(1): 35–40.

Itoh, N., N. Tanaka, I. Funakoshi, T. Kawasaki, S. Mihashi, and I. Yamashina (1988). "Organization of the gene for batroxobin, a thrombin-like snake venom enzyme: Homology with the trypsin/kallikrein gene family." *J Biol Chem* 263(16): 7628–7631.

Iwanaga, S., H. Nishimure, S. Kawabata, W. Kisiel, S. Hase, and T. Ikenaka (1991). "A new trisaccharide sugar chain linked to a serine residue in the first EGF-like domain of clotting factors VII and IX and protein Z." In *Fibrinogen thrombosis coagulation and fibrinolysis*, edited by C. Y. Liu and S. Chien. New York: Plenum, 121–125.

Iyengar, N. K., H. B. Sehra, B. Mukeiji, and R. N. Chopra (1938). "Choline esterase in cobra venom." *Curr Sci India* 7: 51–53.

Izidoro, L. F., M. C. Ribeiro, G. R. Souza, C. D. Sant'Ana, A. Hamaguchi, M. I. Homsi-Brandeburgo, L. R. Goulart, et al. (2006). "Biochemical and functional characterization of an L-amino acid oxidase isolated from *Bothrops pirajai* snake venom." *Bioorg Med Chem* 14(20): 7034–7043.

Jackson, C. G. Jr., and M. M. Jackson (1971). "The frequency of *Salmonella* and *Arizona* microorganisms in zoo turtles." *J Wildl Dis* 7(2): 130–132.

Jackson, K. (2003). "The evolution of venom-delivery systems in snakes." *Zool J Linnean Soc* 137: 337–354.

Jackson, T. N., K. Sunagar, E. A. Undheim, I. Koludarov, A. H. Chan, K. Sanders, S. A. Ali, I. Hendrikx, N. Dunstan, and B. G. Fry (2013). "Venom down under: Dynamic evolution of Australian elapid snake toxins." *Toxins (Basel)* 5(12): 2621–2655.

Jacobson, E. (2007a). "Parasites and parasitic diseases of reptiles." In *Infectious diseases and pathology of reptiles: A color atlas and text*, 1st ed., edited by E. R. Jacobson. Boca Raton, FL: CRC Press, 571–665.

Jacobson, E. R. (2007b). "Serodiagnostics." In *Infectious diseases and pathology of reptiles: A color atlas and text*, edited by E. R. Jacobson. Boca Raton, FL: CRC Press, 381–394.

Jalkanen, J., I. Huhtaniemi, and M. Poutanen (2005). "Mouse cysteine-rich secretory protein 4 (CRISP4): A member of the Crisp family exclusively expressed in the epididymis in an androgen-dependent manner." *Biol Reprod* 72: 1268–1274.

Jan, V., R. C. Maroun, A. Robbe-Vincent, L. De Haro, and V. Choumet (2002). "Toxicity evolution of *Vipera aspis aspis* venom: Identification and molecular modeling of a novel phospholipase A(2) heterodimer neurotoxin." *FEBS Lett* 527(1–3): 263–268.

Janatova, J. (1986). "Detection of disulfide bonds and localization of interchain linkages in the third (C3) and the fourth (C4) components of human complement." *Biochem J* 233(3): 819–825.

Jandera, P. (2011). "Stationary and mobile phases in hydrophilic interaction chromatography: A review." *Anal Chim Acta* 692(1–2): 1–25.

Jandrot-Perrus, M., A. H. Lagrue, M. Okuma, and C. Bon (1997). "Adhesion and activation of human platelets induced by convulxin involve glycoprotein VI and integrin alpha2beta1." *J Biol Chem* 272(43): 27035–27041.

Janes, D. E., C. L. Organ, M. K. Fujita, A. M. Shedlock, and S. V. Edwards (2010). "Genome evolution in Reptilia, the sister group of mammals." *Annu Rev Genomics Hum Genet*11: 239–264.

Janeway, C. A., P. Travers, M. Walport, and M. Shlomchick (2001). *Immunobiology: The immune system in health and disease*. New York: Garland.

Janeway, C. M., G. E. Rivard, P. B. Tracy, and K. G. Mann (1996). "*Factor V Quebec revisited*." *Blood* 87(9): 3571–3578.

Janssen, B. J., L. Gomes, R. I. Koning, D. I. Svergun, A. J. Koster, D. C. Fritzinger, C. W. Vogel, and P. Gros (2009). "Insights into complement convertase formation based on the structure of the factor B-cobra venom factor complex." *EMBO J* 28(16): 2469–2478.

Janssen, B. J. and P. Gros (2006). "Conformational complexity of complement component C3." *Adv Exp Med Biol* 586: 291–312.

Janssen, B. J., E. G. Huizinga, H. C. Raaijmakers, A. Roos, M. R. Daha, K. Nilsson-Ekdahl, B. Nilsson, and P. Gros (2005). "Structures of complement component C3 provide insights into the function and evolution of immunity." *Nature* 437(7058): 505–511.

Jayaraman, G., T. K. Kumar, C. C. Tsai, S. Srisailam, S. H. Chou, C. L. Ho, and C. Yu (2000). "Elucidation of the solution structure of cardiotoxin analogue V from the Taiwan cobra (*Naja naja atra*)—identification of structural features important for the lethal action of snake venom cardiotoxins." *Protein Sci* 9(4): 637–646.

Jebali, J., A. Bazaa, S. Sarray, K. Benhaj, A. Karboul, M. El Ayeb, N. Marrakchi, and A. Gargouri (2009). "C-type lectin protein isoforms of *Macrovipera lebetina*: cDNA cloning and genetic diversity." *Toxicon* 53(2): 228–237.

Jelinek, G. A., C. Tweed, D. Lynch, T. Celenza, B. Bush, and N. Michalopoulos (2004). "Cross reactivity between venomous, mildly venomous, and non-venomous snake venoms with the Commonwealth Serum Laboratories venom detection kit." *Emerg Med Australas* 16: 459–464.

Jenny, R. J., D. D. Pittman, J. J. Toole, R. W. Kriz, R. A. Aldape, R. M. Hewick, R. J. Kaufman, and K. G. Mann (1987). "Complete cDNA and derived amino-acid-sequence of human factor-V." *Proc Natl AcadSci USA* 84(14): 4846–4850.

Jeon, O. H., and D. S. Kim (1999a). "Cloning, expression, and characterization of a cDNA encoding snake venom metalloprotease." *Biochem Mol Biol Int* 47(3): 417–425.

——— (1999b). "Molecular cloning and functional characterization of a snake venom metalloprotease." *Eur J Biochem* 263(2): 526–533.

Jerusalinsky, D., and A. L. Harvey (1994). "Toxins from mamba venoms: Small proteins with selectivities for different subtypes of muscarinic acetylcholine receptors." *Trends Pharmacol Sci* 15(11): 424–430.

Jerusalinsky, D., E. Kornisiuk, R. Bernabeu, I. Izquierdo, and C. Cervenansky (1995). "Muscarinic toxins from the venom of *Dendroaspis* snakes with agonist-like actions." *Toxicon* 33(4): 389–397.

Jesupret, C., K. Baumann, T. N. Jackson, S. A. Ali, D. C. Yang, L. Greisman, L. Kern, et al. (2014). "Vintage venoms: Proteomic and pharmacological stability of snake venoms stored for up to eight decades." *J Proteomics*.

Jha, D., R. Mishra, S. Gottschalk, K. H. Wiesmuller, K. Ugurbil, M. E. Maier, and J. Engelmann (2011). "CyLoP-1: A novel cysteine-rich cell-penetrating peptide for cytosolic delivery of cargoes." *Bioconjug Chem* 22(3): 319–328.

Jia, L. G., K. Shimokawa, J. B. Bjarnason, and J. W. Fox (1996). "Snake venom metalloproteinases: Structure, function and relationship to the ADAMs family of proteins." *Toxicon* 34(11–12): 1269–1276.

Jia, L. G., X. M. Wang, J. D. Shannon, J. B. Bjarnason, and J. W. Fox (1997). "Function of disintegrin-like/cysteine-rich domains of atrolysin A: Inhibition of platelet aggregation by recombinant protein and peptide antagonists." *J Biol Chem* 272(20): 13094–13102.

Jiang, Y., Y. Li, W. Lee, X. Xu, Y. Zhang, R. Zhao, and W. Wang (2011). "Venom gland transcriptomes of two elapid snakes (*Bungarus multicinctus* and *Naja atra*) and evolution of toxin genes." *BMC Genomics* 12: 1.

Jiao, H. M., L. X. Yang, B. Lu, Y. Q. Wu, and Y. C. Zhou (2005). "Shedaoenase, a novel fibrinogenase from the venom of *Agkistrodon shedaoenthesis Zhao*." *Acta Biochim Biophys Sin (Shanghai)* 37(12): 835–842.

Jin, C. H., S. Y. Chae, S. Son, T. H. Kim, K. A. Um, Y. S. Youn, S. Lee, and K. C. Lee (2009). "A new orally available glucagon-like peptide-1 receptor agonist, biotinylated exendin-4, displays improved hypoglycemic effects in db/db mice." *J Control Release* 133(3): 172–177.

Jin, Y., W. H. Lee, L. Zeng, and Y. Zhang (2007). "Molecular characterization of L-amino acid oxidase from king cobra venom." *Toxicon* 50(4): 479–485.

Jin, Y., Q. Lu, X. Zhou, S. Zhu, R. Li, W. Wang, and Y. Xiong (2003). "Purification and cloning of cysteine-rich proteins from *Trimeresurus jerdonii* and *Naja atra* venoms." *Toxicon* 42: 539–547.

Johns, D. G., Z. Ao, B. J. Heidrich, G. E. Hunsberger, T. Graham, L. Payne, N. Elshourbagy, Q. Lu, N. Aiyar, and S. A. Douglas (2007). "Dendroaspis natriuretic peptide binds to the natriuretic peptide clearance receptor." *Biochem Biophys Res Comm* 358(1): 145–149.

Johnson, E. K., K. V. Kardong and S. P. Mackessy (1987). "Electric shocks are ineffective in treatment of lethal effects of rattlesnake envenomation in mice." *Toxicon* 25(12): 1347–1349.

Johnson, E. K., K. V. Kardong, and C. L. Ownby (1987). "Observations on white and yellow venoms from an individual southern Pacific rattlesnake (*Crotalus viridis helleri*)." *Toxicon* 25(11): 1169–1180.

Johnson, I., and M. T. Z. Spence (2010). *Molecular probes handbook: A guide to fluorescent probes and labeling technologies*. Carlsbad, CA: Invitrogen.

Johnson, M. E. (1988). "Shock cures dog." *Outdoor Life* 182(3): 8.

Johnson, R. (2011). "Clinical technique: Handling and treating venomous snakes." *J Exot Pet Med* 20(2): 124–130.

Johnson, R., and T. Hyndman (2014). "Blisters and blood—cutaneous and haematological manifestations of an emerging disease." *Proceedings of the Combined Meeting of the Unusual Pet and Avian Veterinarians Special Interest Group, the Association of Avian Veterinarians Australasian Committee, and the Association of Reptilian and Amphibian Veterinarians.* Cairns, Australia, 117–121.

Johnson, R. G. (1956). "The origin and evolution of the venomous snakes." *Evolution* 10(1): 56–65.

Johnson, R. S. (2012). "Health and diseases." In *A guide to Australian snakes in captivity: Elapids and colubrids*, edited by S. Eipper. Brisbane: ABK Publications.

Jones, B. F. (1996). "Cloning and characterization of ancylostoma-secreted protein." *J Biol Chem* 271: 6672–6678.

Jorge da Silva, N., and S. D. Aird (2001). "Prey specificity, comparative lethality and compositional differences of coral snake venoms." *Comp Biochem Physiol C Toxicol Pharmacol* 128: 425–456.

Jorge, M. T., J. S. de Mendoça, L. A. Ribeiro, M. I. R. d. Silva, K. E. J. U, and C. L. d. S. Cordero (1990). "Flora bacteriana da cavidade oral, presas e veneno de *Bothrops jararaca*: Possível fonte de infecçao no local da picada." *Rev Inst Med Trop São Paulo* 32(1): 6–10.

Joseph, J. S., M. C. Chung, K. Jeyaseelan, and R. M. Kini (1999). "Amino acid sequence of trocarin, a prothrombin activator from *Tropidechis carinatus* venom: Its structural similarity to coagulation factor Xa." *Blood* 94: 621–631.

Joseph, J. S., S. Thirumangalathu, F. Tsang, F. W. Wong, and R. M. Kini (2003a). "Trocarin, a blood coagulation factor Xa homologue from snake venom, causes inflammation and mitogenesis." *Toxicon* 42: 769–776.

Joseph, J. S., M. Valiyaveettil, D. C. Gowda, and R. M. Kini (2003b). "Occurrence of O-linked Xyl-GlcNAc and Xyl-Glc disaccharides in trocarin, a factor Xa homolog from snake venom." *J Thromb Haemost* 1: 545–550.

Joseph, R., S. Pahari, W. C. Hodgson, and R. M. Kini (2004). "Hypotensive agents from snake venoms." *Curr Drug Targets Cardiovasc Haematol Disord* 4(4): 437–459.

Juarez, P., I. Comas, F. Gonzalez-Candelas, and J. J. Calvete (2008). "Evolution of snake venom disintegrins by positive Darwinian selection." *Mol Biol Evol* 25(11): 2391–2407.

Juarez, P., S. C. Wagstaff, L. Sanz, R. A. Harrison, and J. J. Calvete (2006). "Molecular cloning of *Echis ocellatus* disintegrins reveals non-venom-secreted proteins and a pathway for the evolution of ocellatusin." *J Mol Evol* 63(2): 183–193.

Junqueira de Azevedo, I. L., and P. L. Ho (2002). "A survey of gene expression and diversity in the venom glands of the pitviper snake *Bothrops insularis* through the generation of expressed sequence tags (ESTs)." *Gene* 299(1–2): 279–291.

Junqueira de Azevedo, I. L., A. T. Ching, E. Carvalho, F. Faria, M. Y. Nishiyama Jr., P. L. Ho, and M. R. Diniz (2006). "*Lachesis muta* (Viperidae) cDNAs reveal diverging pit viper molecules and scaffolds typical of cobra (Elapidae) venoms: Implications for snake toxin repertoire evolution." *Genetics* 173(2): 877–889.

Junqueira de Azevedo, I. L., P. Ho, L. Milene, N. Yamanouye, and N. R. Casewell (2014). "Venom-related transcripts from *Bothrops jararaca* tissues provide novel molecular insights into the production and evolution of snake venom." *Mol Biol Evol.* In-press.

Kakinuma, Y., Y. Endo, M. Takahashi, M. Nakata, M. Matsushita, S. Takenoshita, and T. Fujita (2003). "Molecular cloning and characterization of novel ficolins from *Xenopus laevis*." *Immunogenetics* 55(1): 29–37.

Kalafatis, M., R. J. Jenny, and K. G. Mann (1990). "Identification and characterization of a phospholipid-binding site of bovine factor-Va." *J Biol Chem* 265(35): 21580–21589.

Kalafatis, M., and K. G. Mann (1993). "Role of the membrane in the inactivation of factor-Va by activated protein-C." *J Biol Chem* 268(36): 27246–27257.

Kalafatis, M., M. D. Rand, and K. G. Mann (1994a). "Factor-Va membrane interaction is mediated by 2 regions located on the light-chain of the cofactor." *Biochemistry* 33(2): 486–493.

Kalafatis, M., M. D. Rand, and K. G. Mann (1994b). "The mechanism of inactivation of human factor-V and human factor Va by activated protein-C." *J Biol Chem* 269(50): 31869–31880.

Kalafatis, M., J. C. Xue, C. M. Lawler, and K. G. Mann (1994). "Contribution of the heavy and light-chains of factor Va to the interaction with factor Xa." *Biochemistry* 33(21): 6538–6545.

Kaltashov, I. A., C. E. Bobst, and R. R. Abzalimov (2013). "Mass spectrometry-based methods to study protein architecture and dynamics." *Protein Sci* 22(5): 530–544.

Kamiguti, A. S., C. R. Hay, and M. Zuzel (1996). "Inhibition of collagen-induced platelet aggregation as the result of cleavage of alpha 2 beta 1-integrin by the snake venom metalloproteinase jararhagin." *Biochem J* 320(pt 2): 635–641.

Kamiguti, A. S., G. D. Laing, G. M. Lowe, M. Zuzel, D. A. Warrell, and R. D. Theakston (1994). "Biological properties of the venom of the Papuan black snake (*Pseudechis papuanus*): Presence of a phospholipase A_2 platelet inhibitor." *Toxicon* 32(8): 915–925.

Kamiguti, A. S., R. D. Theakston, N. Sherman, and J. W. Fox (2000). "Mass spectrophotometric evidence for P-III/P-IV metalloproteinases in the venom of the Boomslang (*Dispholidus typus*)." *Toxicon* 38(11): 1613–1620.

Kane, W. H., and P. W. Majerus (1981). "Purification and characterization of human coagulation factor-V." *J Biol Chem* 256(2): 1002–1007.

Kang, T. S., D. Georgieva, N. Genov, M. T. Murakami, M. Sinha, R. P. Kumar, P. Kaur, et al. (2011). "Enzymatic toxins from snake venom: Structural characterization and mechanism of catalysis." *FEBS J* 278(23): 4544–4576.

Kao, P. H., S. R. Lin, and L. S. Chang (2009). "Differential binding to phospholipid bilayers modulates membrane-damaging activity of *Naja naja atra* cardiotoxins." *Toxicon* 54(3): 321–328.

Kao, P. H., M. J. Wu, and L. S. Chang (2009). "Membrane-bound conformation of *Naja nigricollis* toxin gamma affects its membrane-damaging activity." *Toxicon* 53(3): 342–348.

Karbovs'kyi, V. L., O. M. Savchuk, H. L. Volkov, N. V. Zaichko, and T. Bukhan (2007). "[Effect of proteins from the *Agkistrodon blomhoffii ussuriensis* snake venom on platelets in the hemostasis system]." *Ukr Biokhim Zh* 79(4): 82–89.

Kardong, K. V. (1980). "Evolutionary patterns in advanced snakes." *Am Zool* 20: 269–282.

——— (1982). "The evolution of the venom apparatus in snakes from colubrids to viperids and elapids." *Memoirs of the Institute of Butanan* 46: 105–118.

——— (2012). "Replies to Fry et al. (*Toxicon* 2012, 60/4, 434–448). Part B. Properties and biological roles of squamate oral products: The 'venomous lifestyle' and preadaptation." *Toxicon* 60(5): 964–966.

Karlsson, E., M. Jolkkonen, E. Mulugeta, P. Onali, and A. Adem (2000). "Snake toxins with high selectivity for subtypes of muscarinic acetylcholine receptors." *Biochimie* 82(9–10): 793–806.

Karlsson, E., P. M. Mbugua, and D. Rodriguez-Ithurralde (1984). "Fasciculins, anticholinesterase toxins from the venom of the green mamba *Dendroaspis angusticeps*." *J Physiol (Paris)* 79(4): 232–240.

Karne, S., C. K. Jayawickreme, and M. R. Lerner (1993). "Cloning and characterization of an endothelin-3 specific receptor (ETC receptor) from *Xenopus laevis* dermal melanophores." *J Biol Chem* 268(25): 19126–19133.

Kashima, S., A. M. Soares, P. G. Roberto, J. O. Pereira, S. Astolfi, A. O. Cintra, M. R. M. Fontes, J. R. Giglio, and S. D. Franca (2002). "cDNA sequence and molecular modeling of a nerve growth factor from *Bothrops jararacussu* venomous gland." *Biochimie* 84(7): 675–680.

Kasturiratne, A., A. R. Wickremasinghe, N. de Silva, N. K. Gunawardena, A. Pathmeswaran, R. Premaratna, L. Savioli, D. G. Lalloo, and H. J. de Silva (2008). "The global burden of snakebite: A literature analysis and modelling based on regional estimates of envenoming and deaths." *PLoS Med* 5(11): e218.

Katagiri, C., H. H. Ishikawa, M. Ohkura, O. Nakagawasai, T. Tadano, K. Kisara, and Y. Ohizumi (1998). "Properties of specific binding site of myotoxin a, a powerful convulsant, in brain microsomes." *Can J Physiol Pharmacol* 76(4): 395–400.

Kato, H., J. M. Anderson, S. Kamhawi, F. Oliveira, P. G. Lawyer, V. M. Pham, C. S. Sangare, et al. (2006). "High degree of conservancy among secreted salivary gland proteins from two geographically distant *Phlebotomus duboscqi* sandflies populations (Mali and Kenya)." *BMC Genomics* 7: 226.

Katoh, K., and D. M. Standley (2013). "MAFFT multiple sequence alignment software version 7: Improvements in performance and usability." *Mol Biol Evol* 30(4): 772–780.

Katzir, I., J. Shani, G. Goshen, J. Sela, E. Ninary, A. M. Dogonovski, D. Shabashov, et al. (2003). "Characterization of nerve growth factors (NGFs) from snake venoms by use of a novel, quantitative bioassay utilizing pheochromocytoma (PC12) cells overexpressing human trkA receptors." *Toxicon* 42(5): 481–490.

Kawanabe, Y., and S. M. Nauli (2011). "Endothelin." *Cell Mol Life Sci* 68(2): 195–203.

Keane, T. M., C. J. Creevey, M. M. Pentony, T. J. Naughton, and J. O. McLnerney (2006). "Assessment of methods for amino acid matrix selection and their use on empirical data shows that ad hoc assumptions for choice of matrix are not justified." *BMC Evol Biol* 6: 29.

Kelly, C. M., N. P. Barker, and M. H. Villet (2003). "Phylogenetics of advanced snakes (Caenophidia) based on four mitochondrial genes." *Syst Biol* 52(4): 439–459.

Kemparaju, K., K. S. Girish, and S. Nagaraju (2010). "Hyaluronidases, a neglected class of glycosidases from snake venom: Beyond a spreading factor." In *Handbook of venoms and toxins of reptiles*, edited by S. P. Mackessy. Boca Raton, FL: CRC Press, 237–268.

Kendall, D. M., M. C. Riddle, J. Rosenstock, D. Zhuang, D. D. Kim, M. S. Fineman, and A. D. Baron (2005). "Effects of exenatide (exendin-4) on glycemic control over 30 weeks in patients with type 2 diabetes treated with metformin and a sulfonylurea." *Diabetes Care* 28(5): 1083–1091.

Kerfelec, B., K. S. LaForge, P. Vasiloudes, A. Puigserver, and G. A. Scheele (1990). "Isolation and sequence of the canine pancreatic phospholipase A_2 gene." *Eur J Biochem* 190(2): 299–304.

Kerkis, A., M. A. Hayashi, T. Yamane, and I. Kerkis (2006). "Properties of cell penetrating peptides (CPPs)." *IUBMB Life* 58(1): 7–13.

Kerkis, A., I. Kerkis, G. Rádis-Baptista, E. B. Oliveira, A. M. Vianna-Morgante, L. V. Pereira, and T. Yamane (2004). "Crotamine is a novel cell-penetrating protein from the venom of rattlesnake *Crotalus durissus terrificus*." *FASEB J* 18(12): 1407–1409.

Kerkis, I., S. Silva Fde, A. Pereira, A. Kerkis, and G. Rádis-Baptista (2010). "Biological versatility of crotamine—a cationic peptide from the venom of a South American rattlesnake." *Expert Opin Investig Drugs* 19(12): 1515–1525.

Kessentini-Zouari, R., J. Jebali, S. Taboubi, N. Srairi-Abid, M. Morjen, O. Kallech-Ziri, S. Bezzine, et al. (2010). "CC-PLA_2-1 and CC-PLA_2-2, two *Cerastes cerastes* venom-derived phospholipases A_2, inhibit angiogenesis both *in vitro* and *in vivo*." *Lab Invest* 90(4): 510–519.

Khan, M. F., M. R. Movahed, and J. Jung (2012). "*Pasteurella multocida* endocarditis." *J Heart Valve Dis* 21: 260–262.

Kheyri, H., B. W. Cribb, and D. J. Merritt (2013). "Comparing the secretory pathway in honeybee venom and hypopharyngeal glands." *Arthropod Struct Dev* 42(2): 107–114.

Khimji, A. K., and D. C. Rockey (2010). "Endothelin-biology and disease." *Cell Signal* 22(11): 1615–1625.

Kiehn, J., A. E. Lacerda, B. Wible, and A. M. Brown (1996). "Molecular physiology and pharmacology of HERG: Single-channel currents and block by dofetilide." *Circulation* 94(10): 2572–2579.

Kierszenbaum, A. L., O. Lea, P. Petrusz, F. S. French, and L. L. Tres (1981). "Isolation, culture, and immunocytochemical characterization of epididymal epithelial cells from pubertal and adult rats." *Proc Natl Acad Sci USA* 78(3): 1675–1679.

Kihara, H., R. Uchikawa, S. Hattori, and M. Ohno (1992). "Myotoxicity and physiological effects of three *Trimeresurus flavoviridis* phospholipases A_2." *Biochem Int* 28(5): 895–903.

Kim, M.-S., and A. Pandey (2012). "Electron transfer dissociation mass spectrometry in proteomics." *Proteomics* 12(4–5): 530–542.

King, G. F. (2011). "Venoms as a platform for human drugs: Translating toxins into therapeutics." *Expert Opin Biol Ther* 11(11): 1469–1484.

King, G. F., and M. Mobli (2010). "Determination of peptide and protein structures using NMR spectroscopy." In *Comprehensive natural products II: Chemistry and biology*, edited by L. Mander and H. Liu. Oxford: Elsevier Science & Technology, 279–325.

King, R., C. Berg, and B. Hay (2004). "A repatriation study of the eastern massasauga (*Sistrurus catenatus catenatus*) in Wisconsin." *Herpetologica* 60(4): 429–437.

King, T. P., G. Lu, M. Gonzalez, N. F. Qian, and L. Soldatova (1996). "Yellow jacket venom allergens, hyaluronidase and phospholipase: Sequence similarity and antigenic cross-reactivity with their hornet and wasp homologs and possible implications for clinical allergy." *Journal Allerg Clin Immunol* 98(3): 588–600.

King, T. P., and M. D. Spangfort (2000). "Structure and biology of stinging insect venom allergens." *Int Arch Allergy Immunol* 123: 99–106.

Kini, R. M. (1997). *Venom phospholipase* A_2 *enzymes: Structure, function and mechanism.* Chichester: John Wiley, 511.

——— (2003). "Excitement ahead: Structure, function and mechanism of snake venom phospholipase A_2 enzymes." *Toxicon* 42(8): 827–840.

——— (2005a). "Serine proteases affecting blood coagulation and fibrinolysis from snake venoms." *Pathophysiol Haemost Thromb* 34(4–5): 200–204.

——— (2005b). "Structure-function relationships and mechanism of anticoagulant phospholipase A_2 enzymes from snake venoms." *Toxicon* 45(8): 1147–1161.

Kini, R. M., and Y. M. Chan (1999). "Accelerated evolution and molecular surface of venom phospholipase A_2 enzymes." *J Mol Evol* 48(2): 125–132.

Kini, R. M., and R. Doley (2010). "Structure, function and evolution of three-finger toxins: Mini proteins with multiple targets." *Toxicon* 56(6): 855–867.

Kini, R. M., and H. J. Evans (1989a). "A common cytolytic region in myotoxins, hemolysins, cardiotoxins and antibacterial peptides." *Int J Pept Protein Res* 34(4): 277–286.

——— (1989b). "Role of cationic residues in cytolytic activity: Modification of lysine residues in the cardiotoxin from *Naja nigricollis* venom and correlation between cytolytic and antiplatelet activity." *Biochemistry* 28(23): 9209–9215.

——— Kini, R. M., and H. J. Evans (1997). "Effects of phospholipase A 2 enzymes on platelet aggregation." In*Venom phospholipase A 2 enzymes: Structure, function and mechanism*, edited by R. M. Kini. Chichester: John Wiley, 369–387.

Kini, R. M., and S. Iwanaga (1986). "Structure-function relationships of phospholipases I: Prediction of presynaptic neurotoxicity." *Toxicon* 24(6): 527–541.

Kirby, E. P., S. Niewiarowski, K. Stocker, C. Kettner, E. Shaw, and T. M. Brudzynski (1979). "Thrombocytin, a serine protease from *Bothrops atrox* venom, 1: Purification and characterization of the enzyme." *Biochemistry* 18(16): 3564–3570.

Kirschbaum, A., R. Junker, H. G. Koch, H. Vielhaber, and U. Nowak-Gottl (1999). "Anticoagulant response to *Agkistrodon contortrix* venom (ACV) in infants and children with genetic defects in the protein C anticoagulant pathway." *Eur J Pediatr* 158 suppl 3: S203–204.

Kishimoto, M., and T. Takahashi (2002). "Molecular cloning of HR1a and HR1b, high molecular hemorrhagic factors, from *Trimeresurus flavoviridis* venom." *Toxicon* 40(9): 1369–1375.

Kishore, U., T. J. Greenhough, P. Waters, A. K. Shrive, R. Ghai, M. F. Kamran, A. L. Bernal, K. B. Reid, T. Madan, and T. Chakraborty (2006). "Surfactant proteins SP-A and SP-D: Structure, function and receptors." *Mol Immunol* 43(9): 1293–1315.

Kisiel, W., M. A. Hermodson, and E. W. Davie (1976). "Factor X activating enzyme from Russell's viper venom: Isolation and characterization." *Biochemistry* 15(22): 4901–4906.

Kisiel, W., S. Kondo, K. J. Smith, B. A. McMullen, and L. F. Smith (1987). "Characterization of a protein C activator from *Agkistrodon contortrix contortrix* venom." *J Biol Chem* 262(26): 12607–12613.

Kiss, T. L. (2012). "Critical care for frostbite." *Crit Care Nurs Clin North Am* 24(4): 581.

Kita, M., Y. Nakamura, Y. Okumura, S. D. Ohdachi, Y. Oba, M. Yoshikuni, H. Kido, and D. Uemura (2004). "Blarina toxin, a mammalian lethal venom from the short-tailed shrew *Blarina brevicauda*: Isolation and characterization." *Proc Natl Acad Sci USA* 101(20): 7542–7547.

Kita, M., Y. Okumura, S. D. Ohdachi, Y. Oba, M. Yoshikuni, Y. Nakamura, H. Kido, and D. Uemura (2005). "Purification and characterisation of blarinasin, a new tissue kallikrein-like protease from the short-tailed shrew *Blarina brevicauda*: Comparative studies with blarina toxin." *Biol Chem* 386(2): 177–182.

Kitani, Y., N. Kikuchi, G. Zhang, S. Ishizaki, K. Shimakura, K. Shiomi, and Y. Nagashima (2008). "Antibacterial action of L-amino acid oxidase from the skin mucus of rockfish *Sebastes schlegelii*." *Comp Biochem Physiol B Biochem Mol Biol* 149(2): 394–400.

Kitchens, C. and T. Eskin (2008). "Fatality in a case of envenomation by Crotalus adamanteus initially successfully treated with polyvalent ovine antivenom followed by recurrence of defibrinogenation syndrome." *J Med Toxicol* 4(3): 180–183.

Klauber, L. M. (1956). *Rattlesnakes: Their habits, life history and influence on mankind.* Berkeley: University of California Press.

Klein, D. E., S. E. Stayrook, F. Shi, K. Narayan, and M. A. Lemmon (2008). "Structural basis for EGFR ligand sequestration by Argos." *Nature* 453(7199): 1271–1275.

Klenner, F. R. (1971). "Observations on the dose and administration of ascorbic acid when employed beyond the range of a vitamin in human pathology." *J Appl Nutr* 23(3–4): 61–87.

Klint, J. K., S. Senff, N. J. Saez, R. Seshadri, H. Y. Lau, N. S. Bende, E. A. B. Undheim, L. D. Rash, M. Mobli, and G. F. King (2013). "Production of recombinant disulfide-rich venom peptides for structural and functional analysis via expression in the periplasm of *E. coli*." *PLoS One* 8(5): e63865.

Kloog, Y., I. Ambar, M. Sokolovsky, E. Kochva, Z. Wollberg, and A. Bdolah (1988). "Sarafotoxin, a novel vasoconstrictor peptide: Phosphoinositide hydrolysis in rat heart and brain." *Science* 242(4876): 268–270.

Kloog, Y., and M. Sokolovsky (1989). "Similarity in mode and sites of action of sarafotoxins and endothelins." *Trends Pharmacol Sci* 10: 212–214.

Kochva, E. (1963a). "Development of the venom gland and trigeminal muscles in *Vipera palaestinae*." *Acta Anatomica* 52: 49–89.

——— (1963b). "The phylogenetic significance of the venom apparatus in snakes." *Am Zool* 3(4): 487–487.

——— (1974). "Glandes spécialisées de la mâchoire inférieure chez les Anguimorphes [Specialized glands of the inferior jaw in Anguimorphans]." In *Recherches biologiques contemporaines, dédiées à la mémoire du Dr. Manfred Gabe (1916–1973)* [In contemporary biological research, dedicated to the memory of Dr. Manfred Gabe], edited by L. Arvy. Nancy, France: Imprimerie Vagner, 281–286.

——— (1978). "Oral glands of the *Reptilia*." *Physiology B* 8: 43–162.

——— (1987). "The origin of snakes and evolution of the venom apparatus." *Toxicon* 25(1): 65–106.

Kochva, E., A. Bdolah, and Z. Wollberg (1993). "Sarafotoxins and endothelins: Evolution, structure and function." *Toxicon* 31(5): 541–568.

Kochva, E., and C. Gans (1965). "Venom gland of *Vipera palaestinae* with comments on glands of some other viperines." *Acta Anatomica* 62(3): 365–401.

Kochva, E., Z. Wollberg, and I. Golani (1985). "The burrowing asps genus *Atractaspis*, a tropical group of venomous snakes." *Toxicon* 23(4): 623–623.

Koduri, R. S., J. O. Gronroos, V. J. Laine, C. Le Calvez, G. Lambeau, T. J. Nevalainen, and M. H. Gelb (2002). "Bactericidal properties of human and murine groups I, II, V, X, and XII secreted phospholipases A(2)." *J Biol Chem* 277(8): 5849–5857.

Koehl, P., and M. Delarue (1994). "Application of a self-consistent mean field theory to predict protein side-chains conformation and estimate their conformational entropy." *J Mol Biol* 239(2): 249–275.

——— (1995). "A self consistent mean field approach to simultaneous gap closure and side-chain positioning in homology modelling." *Nat Struct Biol* 2(2): 163–170.

Kogan, A. E., G. V. Bashkov, I. D. Bobruskin, E. P. Romanova, V. A. Makarov, and S. M. Strukova (1993). "Protein C activator from the venom of *Agkistrodon blomhoffi ussuriensis* retards thrombus formation in the arterio-venous shunt in rats." *Thromb Res* 70(5): 385–393.

Koh, C. Y., and R. M. Kini (2012). "From snake venom toxins to therapeutics—cardiovascular examples." *Toxicon* 59(4): 497–506.

Koh, D. C. I., A. Armugam, and K. Jeyaseelan (2004). "Sputa nerve growth factor forms a preferable substitute to mouse 7S-beta nerve growth factor." *Biochem J* 383: 149–158.

Koh, G., A. Low, D. Poh, Y. Yao, S. K. Ng, V. V. Wong, V. Vagenende, K. P. Lam, and D. Y. Lee (2011). "Integrative analysis workflow for the structural and functional classification of C-type lectins." *BMC Bioinformatics* 12 suppl 14: S5.

Koh, G. Y., D. R. Nussenzveig, J. Okolicany, D. A. Price, and T. Maack (1992). "Dynamics of atrial natriuretic factor-guanylate cyclase receptors and receptor-ligand complexes in cultured glomerular mesangial and renomedullary interstitial cells." *J Biol Chem* 267(17): 11987–11994.

Koludarov, I., K. Sunagar, E. A. Undheim, T. N. Jackson, T. Ruder, D. Whitehead, A. C. Saucedo, et al. (2012). "Structural and molecular diversification of the *Anguimorpha* lizard mandibular venom gland system in the arboreal species *Abronia graminea*." *J Mol Evol* 75(5–6): 168–183.

Komori, Y., T. Nikai, A. Ohara, S. Yagihashi, and H. Sugihara (1993). "Effect of bilineobin, a thrombin-like proteinase from the venom of common cantil (*Agkistrodon bilineatus*)." *Toxicon* 31(3): 257–270.

Komori, Y., T. Nikai, and H. Sugihara (1988). "Purification and characterization of a lethal toxin from the venom of *Heloderma horridum horridum*." *Biochem Biophys Res Commun* 154(2): 613–619.

Kondo, H., S. Kondo, H. Ikezawa, and R. Murata (1960). "Studies on the quantitative method for determination of hemorrhagic activity of Habu snake venom." *Jpn J Med Sci Biol* 13: 43–52.

Kondo, K., K. Narita, and C. Y. Lee (1978a). "Amino acid sequences of the two polypeptide chains in β1-bungarotoxin from the venom of *Bungarus multicinctus*." *J Biochem (Tokyo)* 83(1): 101–115.

——— (1978b). "Chemical properties and amino acid composition of β1-bungarotoxin from venom *Bungarus multicinctus* (Formosan banded krait)." *J Biochem (Tokyo)* 83(1): 91–99.

Kondo, K., H. Toda, K. Narita, and C. Y. Lee (1982). "Amino acid sequences of three β-bungarotoxins (β3-, β4-, and β5-bungarotoxins) from *Bungarus multicinctus* venom: Amino acid substitutions in the A chains." *J Biochem (Tokyo)* 91(5): 1531–1548.

Konno, K., G. Picolo, V. P. Gutiérrez, P. Brigatte, V. O. Zambelli, A. C. Camargo, and Y. Cury (2008). "Crotalphine, a novel potent analgesic peptide from the venom of the South American rattlesnake *Crotalus durissus terrificus*." *Peptides* 29(8): 1293–1304.

Kopaciewicz, W., and F. E. Regnier (1983). "Mobile phase selection for the high-performance ion-exchange chromatography of proteins." *Anal Biochem* 133(1): 251–259.

Kordis, D., and F. Gubenek (2000). "Adaptive evolution of animal toxin multigene families." *Gene* 261(1): 43–52.

Kordis, D., and V. Turk (2009). "Phylogenomic analysis of the cystatin superfamily in eukaryotes and prokaryotes." *BMC Evol Biol* 9: 266–287.

Kornalik, F., and B. Blomback (1975). "Prothrombin activation induced by Ecarin—a prothrombin converting enzyme from *Echis carinatus* venom." *Thromb Res* 6(1): 57–63.

Kornfeld, R., and S. Kornfeld (1985). "Assembly of asparagine-linked oligosaccharides." *Annu Rev Biochem* 54(631–664).

Kornisiuk, E., D. Jerusalinsky, C. Cervenansky, and A. L. Harvey (1995). "Binding of muscarinic toxins MTx1 and MTx2 from the venom of the green mamba *Dendroaspis angusticeps* to cloned human muscarinic cholinoceptors." *Toxicon* 33(1): 11–18.

Korsching, S. (1993). "The neurotrophic factor concept—a reexamination." *J Neurosci* 13(7): 2739–2748.

Korsching, S., G. Auburger, R. Heumann, J. Scott, and H. Thoenen (1985). "Levels of nerve growth factor and its messenger RNA in the central nervous system of the rat correlate with cholinergic innervation." *EMBO J* 4(6): 1389–1393.

Kosakovsky Pond, S. L., and S. D. W. Frost (2005). "Not so different after all: A comparison of methods for detecting amino acid sites under selection." *Mol Biol Evol* 22(5): 1208–1222.

Kostanski, L. K., D. M. Keller, and A. E. Hamielec (2004). "Size-exclusion chromatography: A review of calibration methodologies." *J Biochem Biophys Methods* 58(2): 159–186.

Kostiza, T., and J. Meier (1996). "Nerve growth factors from snake venoms: Chemical properties, mode of action and biological significance." *Toxicon* 34(7): 787–806.

Kotsyfakis, M., A. Sa-Nunes, I. M. Francischetti, T. N. Mather, J. F. Andersen, and J. M. Ribeiro (2006). "Antiinflammatory and immunosuppressive activity of sialostatin L, a salivary cystatin from the tick *Ixodes scapularis*." *J Biol Chem* 281(36): 26298–26307.

Kourie, J. I. (1999a). "Calcium dependence of C-type natriuretic peptide-formed fast K(+) channel." *Am J Physiol* 277(1 pt 1): C43–50.

——— (1999b). "Characterization of a C-type natriuretic peptide (CNP-39)-formed cation-selective channel from platypus (*Ornithorhynchus anatinus*) venom." *J Physiol* 518(pt 2): 359–369.

——— (1999c). "A component of platypus (*Ornithorhynchus anatinus*) venom forms slow-kinetic cation channels." *J Membr Biol* 172(1): 37–45.

Kowalska, M. A., L. Tan, J. C. Holt, M. Peng, J. Karczewski, J. J. Calvete, and S. Niewiarowski (1998). "Alboaggregins A and B: Structure and interaction with human platelets." *Thromb Haemost* 79(3): 609–613.

Koyama, J., S. Inoue, K. Ikeda, and K. Hayashi (1992). "Purification and amino acid sequence of a nerve growth factor from the venom of *Vipera russelli russelli*." *Biochim Biophys Acta* 1160(3): 287–292.

Kratzschmar, J., B. Haendler, G. Langer, W. Boidol, P. Bringmann, A. Alagón, P. Donner, and W. D. Schleuning (1991). "The plasminogen-activator family from the salivary-gland of the vampire bat *Desmodus rotundus*—cloning and expression." *Gene* 105(2): 229–237.

Krieger, E. M., H. C. Salgado, and C. J. Assan (1971). "Greene LL, Ferreira SH: Potential screening test for detection of overactivity of renin-angiotensin system." *Lancet* 1(7693): 269–271.

Krishnan, V., K. Ponnuraj, Y. Xu, K. Macon, J. E. Volanakis, and S. V. Narayana (2009). "The crystal structure of cobra venom factor, a cofactor for C3- and C5-convertase CVFBb." *Structure* 17(4): 611–619.

Krishnaswamy, S. and K. G. Mann (1988). "The binding of factor-Va to phospholipid-vesicles." *J Biol Chem* 263(12): 5714–5723.

Krishnaswamy, S., K. G. Mann, and M. E. Nesheim (1986). "The prothrombinase-catalyzed activation of prothrombin proceeds through the intermediate meizothrombin in an ordered, sequential reaction." *J Biol Chem* 261(19): 8977–8984.

Krishnaswamy, S., G. D. Russell, and K. G. Mann (1989). "The reassociation of factor Va from its isolated subunits." *J Biol Chem* 264(6): 3160–3168.

*Krishnaswamy, S., E. B. Williams, and K. G. Mann (1986). "The binding of activated protein-C to factor-V and factor-Va." *J Biol Chem* 261(21): 9684–9693.

Kritzik, M., B. Savage, D. J. Nugent, S. Santoso, Z. M. Ruggeri, and T. J. Kunicki (1998). "Nucleotide polymorphisms in the alpha2 gene define multiple alleles that are associated with differences in platelet alpha2 beta1 density." *Blood* 92(7): 2382–2388.

Krizaj, I., A. L. Bieber, A. Ritonja, and F. Gubensek (1991). "The primary structure of ammodytin L, a myotoxic phospholipase A_2 homologue from *Vipera ammodytes* venom." *Eur J Biochem* 202(3): 1165–1168.

Kubista, H., R. A. Mafra, Y. Chong, G. M. Nicholson, P. S. Beirao, J. S. Cruz, S. Boehm, W. Nentwig, and L. Kuhn-Nentwig (2007). "CSTX-1, a toxin from the venom of the hunting spider *Cupiennius salei*, is a selective blocker of L-type calcium channels in mammalian neurons." *Neuropharmacology* 52(8): 1650–1662.

Kudo, I., M. Murakami, S. Hara, and K. Inoue (1993). "Mammalian non-pancreatic phospholipases A_2." *Biochim Biophys Acta* 1170(3): 217–231.

Kukhtina, V. V., C. Weise, T. A. Muranova, V. G. Starkov, P. Franke, F. Hucho, S. Wnendt, C. Gillen, V. I. Tsetlin, and Y. N. Utkin (2000). "Muscarinic toxin-like proteins from cobra venom." *Eur J Biochem* 267(23): 6784–6789.

Kumar, R., S. Beguin, and H. C. Hemker (1994). "The influence of fibrinogen and fibrin on thrombin generation—evidence for feedback activation of the clotting system by clot bound thrombin." *Thromb Haemost* 72(5): 713–721.

Kumar, S., J. R. Miranda-Massari, M. J. Gonzalez, and H. D. Riordan (2004). "Intravenous ascorbic acid as a treatment for severe jellyfish stings." *P R Health Sci J* 23(2): 125–126.

Kumar, T. K., C. S. Lee, and C. Yu (1996). "A case study of cardiotoxin III from the Taiwan cobra (*Naja naja atra*): Solution structure and other physical properties." *Adv Exp Med Biol* 391: 115–129.

Kunitz, M., and J. H. Northrop (1936). "Isolation from beef pancreas of crystalline trypsinogen, trypsin, a trypsin inhibitor, and an inhibitor-trypsin compound." *J Gen Physiol* 19(6): 991–1007.

Kuniyoshi, A. K., M. Rocha, D. Cajado Carvalho, M. A. Juliano, L. Juliano Neto, D. V. Tambourgi, and F. C. Portaro (2012). "Angiotensin-degrading serine peptidase: A new chymotrypsin-like activity in the venom of *Bothrops jararaca* partially blocked by the commercial antivenom." *Toxicon* 59(1): 124–131.

Kunkler, G. A. (1855). "On the bite of a copper snake." *Weekly Medical Gazette* 31: 481–483.

Kuruppu, S., S. Reeve, Y. Banerjee, R. M. Kini, A. I. Smith, and W. C. Hodgson (2005a). "Isolation and pharmacological characterization of cannitoxin, a presynaptic neurotoxin from the venom of the Papuan Taipan (*Oxyuranus scutellatus canni*)." *J Pharmacol Exp Ther* 315(3): 1196–1202.

Kuruppu, S., S. Reeve, A. I. Smith, and W. C. Hodgson (2005b). "Isolation and pharmacological characterisation of papuantoxin-1, a postsynaptic neurotoxin from the venom of the Papuan black snake (*Pseudechis papuanus*)." *Biochem Pharmacol* 70(5): 794–800.

Kuruppu, S., S. Robinson, W. C. Hodgson, and B. G. Fry (2007). "The *in vitro* neurotoxic and myotoxic effects of the venom from the *Suta* genus (curl snakes) of elapid snakes." *Basic Clin Pharmacol Toxicol* 101(6): 407–410.

Kwan, A. H., M. Mobli, P. R. Gooley, G. F. King, and J. P. Mackay (2011). "Macromolecular NMR spectroscopy for the nonspectroscopist." *FEBS J* 278(5): 687–703.

Kwok, H. F., T. Chen, M. O'Rourke, C. Ivanyi, D. Hirst, and C. Shaw (2008). "Helokinestatin: A new bradykinin B-2 receptor antagonist decapeptide from lizard venom." *Peptides* 29(1): 65–72.

Kwong, P. D., N. Q. McDonald, P. B. Sigle, and W. A. Hendrickson (1995). "Structure of β_2-bungarotoxin—potassium channel binding by Kunitz modules and targeted phospholipase action." *Structure* 3(10): 1109–1119.

Labarca, P., J. Lindstrom, and M. Montal (1984). "Acetylcholine receptor in planar lipid bilayers: Characterization of the channel properties of the purified nicotinic acetylcholine receptor from *Torpedo californica* reconstituted in planar lipid bilayers." *J Gen Physiol* 83(4): 473–496.

Laemmli, U. K. (1970). "Cleavage of structural proteins during the assembly of the head of bacteriophage T4." *Nature* 227(5259): 680–685.

Lah, T. T., M. B. Duran Alonso, and C. J. Van Noorden (2006). "Antiprotease therapy in cancer: Hot or not?" *Expert Opin Biol Ther* 6(3): 257–279.

Laing, G. D., A. S. Kamiguti, M. C. Wilkinson, G. M. Lowe, and R. D. Theakston (1995). "Characterisation of a purified phospholipase A_2 from the venom of the Papuan black snake (*Pseudechis papuanus*)." *Biochim Biophys Acta* 1250(2): 137–143.

Lakshmanan, J. (1987). "Nerve growth-factor levels in mouse serum—variations due to stress." *Neurochem Res* 12(4): 393–397.

Lakshminarayanan, R., S. Valiyaveettil, V. S. Rao, and R. M. Kini (2003). "Purification, characterization, and *in vitro* mineralization studies of a novel goose eggshell matrix protein, ansocalcin." *J Biol Chem* 278(5): 2928–2936.

Lakshminarayanan, R., S. Vivekanandan, R. P. Samy, Y. Banerjee, E. O. Chi-Jin, K. W. Teo, S. D. Jois, R. M. Kini, and S. Valiyaveettil (2008). "Structure, self-assembly, and dual role of a beta-defensin-like peptide from the Chinese soft-shelled turtle eggshell matrix." *J Am Chem Soc* 130(14): 4660–4668.

Lalloo, D. G., and R. D. G. Theakston (2003). "Snake antivenoms." *J Toxicol Clin Toxicol* 41(3): 277–290.

Lambeau, G., and M. H. Gelb (2008). "Biochemistry and physiology of mammalian secreted phospholipases A_2." *Annu Rev Biochem* 77: 495–520.

Lambeau, G., and M. Lazdunski (1999). "Receptors for a growing family of secreted phospholipases A_2." *Trends Pharmacol Sci* 20(4): 162–170.

Lamthanh, H., A. Bdolah, C. Creminon, J. Grassi, A. Ménez, Z. Wollberg, and E. Kochva (1994). "Biological activities of [Thr2]sarafotoxin-b, a synthetic analogue of sarafotoxin-b." *Toxicon* 32(9): 1105–1114.

Lanari, L. C., A. Olvera, V. Costa de Oliveira, R. D. Lascowicz, L. Boyer, N. R. Lago, A. Alagón, and A. R. de Roodt (2014). "Intraspecific differences in the immunochemical reactivity and neutralization of venom from Argentinean *Bothrops* (*Rhinocerophis*) *alternatus* by specific experimental antivenoms." *Toxicon* 85 : 31–45.

Landan, G., A. Bdolah, Z. Wollberg, E. Kochva, and D. Graur (1991a). "Evolution of the sarafotoxin/endothelin superfamily of proteins." *Toxicon* 29(2): 237–244.

——— (1991b). "The evolutionary history of the sarafotoxin/endothelin/endothelin-like superfamily." *J Cardiovasc Pharmacol* 17 suppl 7: S517–519.

Landon, C., C. Thouzeau, H. Labbe, P. Bulet, and F. Vovelle (2004). "Solution structure of spheniscin, a beta-defensin from the penguin stomach." *J Biol Chem* 279(29): 30433–30439.

Landucci, E. C., R. C. Castro, M. F. Pereira, A. C. Cintra, J. R. Giglio, S. Marangoni, B. Oliveira, G. Cirino, E. Antunes, and G. De Nucci (1998). "Mast cell degranulation induced by two phospholipase A_2 homologues: Dissociation between enzymatic and biological activities." *Eur J Pharmacol* 343(2–3): 257–263.

Landucci, E. C., R. C. Castro, M. Toyama, J. R. Giglio, S. Marangoni, G. De Nucci, and E. Antunes (2000). "Inflammatory oedema induced by the lys-49 phospholipase A(2) homologue piratoxin-i in the rat and rabbit: Effect of polyanions and p-bromophenacyl bromide." *Biochem Pharmacol* 59(10): 1289–1294.

Large, T. H., S. C. Bodary, D. O. Clegg, G. Weskamp, U. Otten, and L. F. Reichardt (1986). "Nerve growth factor gene expression in the developing rat brain." *Science* 234(4774): 352–355.

Laskowski, R. A., J. A. Rullmannn, M. W. MacArthur, R. Kaptein, and J. M. Thornton (1996). "AQUA and PROCHECK-NMR: Programs for checking the quality of protein structures solved by NMR." *J Biomol NMR* 8(4): 477–486.

Latallo, Z. S., and E. Teisseyre (1971). "Evaluation of Reptilase R and thrombin clotting time in the presence of fibrinogen degradation products and heparin." *Scand J Haematol* suppl 13: 261–266.

Lavergne, V., S. Dutertre, A. H. Jin, R. J. Lewis, R. J. Taft, and P. F. Alewood (2013). "Systematic interrogation of the *Conus marmoreus* venom duct transcriptome with ConoSorter reveals 158 novel conotoxins and 13 new gene superfamilies." *BMC Genomics* 14: 708.

Lawrence, M. G., J. Lai, and J. A. Clements (2010). "Kallikreins on steroids: Structure, function, and hormonal regulation of prostate-specific antigen and the extended kallikrein locus." *Endocr Rev* 31(4): 407–446.

Lawson, R., J. B. Slowinski, B. I. Crother, and F. T. Burbrink (2005). "Phylogeny of the Colubroidea (Serpentes): New evidence from mitochondrial and nuclear genes." *Mol Phylogenet Evol* 37(2): 581–601.

Le, T. N. M., M. Abu Reza, S. Swarup, and R. M. Kini (2005). "Gene duplication of coagulation factor V and origin of venom prothrombin activator in *Pseudonaja textilis* snake." *Thromb Haemost* 93(3): 420–429.

Leanza, L., B. Henry, N. Sassi, M. Zoratti, K. G. Chandy, E. Gulbins, and I. Szabo (2012). "Inhibitors of mitochondrial Kv1.3 channels induce Bax/Bak-independent death of cancer cells." *EMBO Mol Med* 4(7): 577–593.

Leao, L. I., P. L. Ho, and I. L. Junqueira de Azevedo (2009). "Transcriptomic basis for an antiserum against *Micrurus corallinus* (coral snake) venom." *BMC Genomics* 10: 112.

Lecompte, M. F., G. Bouix, and K. G. Mann (1994). "Electrostatic and hydrophobic interactions are involved in factor-Va binding to membranes containing acidic phospholipids." *J Biol Chem* 269(3): 1905–1910.

Leduc, M., and C. Bon (1998). "Cloning of subunits of convulxin, a collagen-like platelet-aggregating protein from *Crotalus durissus terrificus* venom." *Biochem J* 333(pt 2): 389–393.

Lee, C. Y., S. L. Chang, S. T. Kau, and S. H. Luh (1972). "Chromatographic separation of the venom of *Bungarus multicinctus* and characterization of its components." *J Chromatogr* 72(1): 71–82.

Lee, C. Y., H. H. Chen, O. Lisy, S. Swan, C. Cannon, H. D. Lieu, and J. C. Burnett Jr. (2009). "Pharmacodynamics of a novel designer natriuretic peptide, CD-NP, in a first-in-human clinical trial in healthy subjects." *J Clin Pharmacol* 49(6): 668–673.

Lee, J. W., J. H. Seu, I. K. Rhee, I. Jin, Y. Kawamura, and W. Park (1999). "Purification and characterization of brevinase, a heterogeneous two-chain fibrinolytic enzyme from the venom of Korean snake, *Agkistrodon blomhoffii brevicaudus*." *Biochem Biophys Res Commun* 260(3): 665–670.

Lee, M. L., N. H. Tan, S. Y. Fung, and D. S. Sekaran (2011). "Antibacterial action of a heat-stable form of L-amino acid oxidase isolated from king cobra (*Ophiophagus hannah*) venom." *Comp Biochem Physiol C Toxicol Pharmacol* 153(2): 237–242.

Lehrer, R. I., and T. Ganz (2002). "Defensins of vertebrate animals." *Curr Opin Immunol* 14(1): 96–102.

Leon, G., M. Herrera, A. Segura, M. Villalta, M. Vargas, and J. M. Gutiérrez (2013). "Pathogenic mechanisms underlying adverse reactions induced by intravenous administration of snake antivenoms." *Toxicon* 76: 63–76.

Leonardi, A., J. W. Fox, A. Trampus-Bakija, and I. Krizaj (2007). "Ammodytase, a metalloprotease from *Vipera ammodytes ammodytes* venom, possesses strong fibrinolytic activity." *Toxicon* 49(6): 833–842.

Levi-Montalcini, R., and R. Angeletti (1968). "Nerve growth factor." *Physiol Rev* 48(3): 534-569.

Levi-Montalcini, R., S. D. Skaper, R. DalToso, L. Petrelli, and A. Leon (1996). "Nerve growth factor: From neurotrophin to neurokine." *Trends Neurosci* 19(11): 514–520.

Levine, M., C. Conry-Cantilena, Y. H. Wang, R. W. Welch, P. W. Washko, K. R. Dhariwal, J. B. Park, et al. (1996). "Vitamin C pharmacokinetics in healthy volunteers: Evidence for a recommended dietary allowance." *Proc Natl Acad Sci USA* 93(8): 3704–3709.

Lewis, R. J., and M. L. Garcia (2003). "Therapeutic potential of venom peptides." *Nat Rev Drug Discov* 2(10): 790–802.

Lewis, R. L., and L. Gutmann (2004). "Snake venoms and the neuromuscular junction." *Semin Neurol* 24(2): 175–179.

Ley, A. C., W. Markland, and R. C. Ladner (1996). "Obtaining a family of high-affinity, high-specificity protein inhibitors of plasmin and plasma kallikrein." *Mol Divers* 2(1–2): 119–124.

Leyton-Ovando, R. (2001). Los culebreros: Medicina traditional viva. Mexico. Ed. Conaculta.

Li, M., C. M. Bullock, D. J. Knauer, F. J. Ehlert, and Q. Y. Zhou (2001). "Identification of two prokineticin cDNAs: Recombinant proteins potently contract gastrointestinal smooth muscle." *Mol Pharmacol* 59(4): 692–698.

Li, M., B. G. Fry, and R. M. Kini (2005a). "Eggs-only diet: Its implications for the toxin profile changes and ecology of the marbled sea snake (*Aipysurus eydouxii*)." *J Mol Evol* 60(1): 81–89.

——— (2005b). "Putting the brakes on snake venom evolution: The unique molecular evolutionary patterns of *Aipysuras eydouxii* (Marbled sea snake) phospholipase A(2) toxins." *Mol Biol Evol* 22(4): 934–941.

Li, S., J. Kwon, and S. Aksoy (2001). "Characterization of genes expressed in the salivary glands of the tsetse fly, *Glossina morsitans morsitans*." *Insect Mol Biol* 10: 69–76.

Li, X., L. Zheng, C. Kong, P. R. Kolatkar, and M. C. Chung (2004). "Purpureotin: A novel di-dimeric C-type lectin-like protein from *Trimeresurus purpureomaculatus* venom is stabilized by noncovalent interactions." *Arch Biochem Biophys* 424(1): 53–62.

Li, Z. Y., T. F. Yu, and E. C. Lian (1994). "Purification and characterization of L-amino acid oxidase from king cobra (*Ophiophagus hannah*) venom and its effects on human platelet aggregation." *Toxicon* 32(11): 1349–1358.

Liang, S. P. (2004). "An overview of peptide toxins from the venom of the Chinese bird spider *Selenocosmia huwena Wang* (= *Ornithoctonus huwena* [*Wang*])." *Toxicon* 43(5): 575–585.

Ligabue-Braun, R., H. Verli, and C. R. Carlini (2012). "Venomous mammals: A review." *Toxicon* 59(7–8): 680–695.

Lillywhite, H. B. (1996). "Husbandry of the little file snake, *Acrochordus granulatus*." *Zoo Biol* 15(3): 315–327.

Lillywhite, H. B., L. S. Babonis, C. M. Sheehy 3rd, and M. C. Tu (2008) "Sea snakes (*Laticauda* spp.) require fresh drinking water: Implication for the distribution and persistence of populations." *Physiol Biochem Zool* 81(6): 785–796.

Lin, K. L., C. M. Chien, C. Y. Hsieh, P. C. Tsai, L. S. Chang, and S. R. Lin (2012). "Antimetastatic potential of cardiotoxin III involves inactivation of PI3K/Akt and p38 MAPK signaling pathways in human breast cancer MDA-MB-231 cells." *Life Sci* 90(1–2): 54–65.

Lin, W. W., L. A. Smith, and C. Y. Lee (1995). "A study on the cause of death due to waglerin-I, a toxin from *Trimeresurus wagleri*." *Toxicon* 33(1): 111–114.

Lind, P. (1982). "Amino-acid-sequence of the beta-1 isosubunit of taipoxin, an extremely potent pre-synaptic neurotoxin from the australian snake taipan (*Oxyuranus s. scutellatus*)." *Eur J Biochem* 128(1): 71–75.

Lind, P., and D. Eaker (1981). "Amino acid sequence of a lethal myotoxic phospholipase A_2 from the venom of the common sea snake (*Enhydrina schistosa*)." *Toxicon* 19(1): 11–24.

Lindholm, D., R. Heumann, M. Meyer, and H. Thoenen (1987). "Interleukin-1 regulates synthesis of nerve growth factor in non-neuronal cells of rat sciatic nerve." *Nature* 330(6149): 658–659.

Lipman, D. J., and W. R. Pearson (1985). "Rapid and sensitive protein similarity searches." *Science* 227(4693): 1435–1441.

Lisy, O., B. K. Huntley, D. J. McCormick, P. A. Kurlansky, and J. C. Burnett Jr. (2008). "Design, synthesis, and actions of a novel chimeric natriuretic peptide: CD-NP." *J Am Coll Cardiol* 52(1): 60–68.

Liu, C. S., T. C. Wu, and T. B. Lo (1983). "Complete amino acid sequences of two protease inhibitors in the venom of *Bungarus fasciatus*." *Int J Pept Protein Res* 21(2): 209–215.

Liu, C. Z., H. Y. Huang, P. J. Tsai, and M. H. Shih (2005). "Blockade of glycoprotein IIb/IIIa by crotavirin, a member of disintegrins, prevents platelet from activation and aggregation by *Staphylococcus aureus* bacteria." *Thromb Res* 116(2): 145–155.

Liu, Q., W. Xu, X. Cheng, G. Jin, X. Shen, H. Lou, and J. Liu (1999). "Molecular cloning and sequence analysis of cDNA encoding haemorrhagic toxin acutolysin A from *Agkistrodon acutus*." *Toxicon* 37(11): 1539–1548.

Liu, S. Y., K. Yoshizumi, N. Oda, M. Ohno, F. Tokunaga, S. Iwanaga, and H. Kihara (1990). "Purification and amino acid sequence of basic protein II, a lysine-49-phospholipase A_2 with low activity, from *Trimeresurus flavoviridis* venom." *J Biochem* 107(3): 400–408.

Llinas, P., M. H. Le Du, H. Gardsvoll, K. Dano, M. Ploug, B. Gilquin, E. A. Stura, and A. Ménez (2005). "Crystal structure of the human urokinase plasminogen activator receptor bound to an antagonist peptide." *EMBO J* 24(9): 1655–1663.

Lock, B. and J. F. Wellehan (2014). "Ophidia (Snakes)." In *Fowler's zoo and wild animal medicine vol. 8*, edited by M. E. Fowler. St. Louis, MO: Elsevier Saunders, 60–74.

Lohse, M., A. M. Bolger, A. Nagel, A. R. Fernie, J. E. Lunn, M. Stitt, and B. Usadel (2012). "RobiNA: A user-friendly, integrated software solution for RNA-Seq-based transcriptomics." *Nucleic Acids Res* 40 (Web server issue): W622–627.

Lomonte, B., Y. Angulo, and L. Calderon (2003). "An overview of lysine-49 phospholipase A_2 myotoxins from crotalid snake venoms and their structural determinants of myotoxic action." *Toxicon* 42(8): 885–901.

Lomonte, B., Y. Angulo, S. Rufini, W. Cho, J. R. Giglio, M. Ohno, J. J. Daniele, P. Geoghegan, and J. M. Gutiérrez (1999). "Comparative study of the cytolytic activity of myotoxic phospholipases A_2 on mouse endothelial (tEnd) and skeletal muscle (C2C12) cells *in vitro*." *Toxicon* 37(1): 145–158.

Lomonte, B., Y. Angulo, M. Sasa, and J. M. Gutiérrez (2009). "The phospholipase A_2 homologues of snake venoms: Biological activities and their possible adaptive roles." *Protein Pept Lett* 16(8): 860–876.

Lomonte, B., J. Escolano, J. Fernandez, L. Sanz, Y. Angulo, J. M. Gutiérrez, and J. J. Calvete (2008). "Snake venomics and antivenomics of the arboreal neotropical pitvipers *Bothriechis lateralis* and *Bothriechis schlegelii*." *J Proteome Res* 7(6): 2445–2457.

Lomonte, B., and J. M. Gutiérrez (1989). "A new muscle damaging toxin, myotoxin II, from the venom of the snake *Bothrops asper* (terciopelo)." *Toxicon* 27(7): 725–733.

Lomonte, B., E. Moreno, A. Tarkowski, L. A. Hanson, and M. Maccarana (1994). "Neutralizing interaction between heparins and myotoxin II, a lysine 49 phospholipase A_2 from *Bothrops asper* snake venom: Identification of a heparin-binding and cytolytic toxin region by the use of synthetic peptides and molecular modeling." *J Biol Chem* 269(47): 29867–29873.

Lomonte, B., and J. Rangel (2012). "Snake venom Lys49 myotoxins: From phospholipases A(2) to non-enzymatic membrane disruptors." *Toxicon* 60(4): 520–530.

Lomonte, B., A. Tarkowski, and L. A. Hanson (1993). "Host response to *Bothrops asper* snake venom: Analysis of edema formation, inflammatory cells, and cytokine release in a mouse model." *Inflammation* 17(2): 93–105.

——— (1994). "Broad cytolytic specificity of myotoxin II, a lysine-49 phospholipase A_2 of *Bothrops asper* snake venom." *Toxicon* 32(11): 1359–1369.

Lonborg, J., H. Kelbaek, N. Vejlstrup, H. E. Botker, W. Y. Kim, L. Holmvang, E. Jorgensen, et al. (2012a). "Exenatide reduces final infarct size in patients with ST-segment-elevation myocardial infarction and short-duration of ischemia." *Circ Cardiovasc Interv* 5(2): 288–295.

Lonborg, J., N. Vejlstrup, H. Kelbaek, H. E. Botker, W. Y. Kim, A. B. Mathiasen, E. Jorgensen, et al. (2012b). "Exenatide reduces reperfusion injury in patients with ST-segment elevation myocardial infarction." *Eur Heart J* 33(12): 1491–1499.

Lopes-Ferreira, M., G. S. Magalhaes, J. H. Fernandez, I. L. Junqueira de Azevedo, P. L. Ho, C. Lima, R. H. Valente, and A. M. Moura-da-Silva (2011). "Structural and biological characterization of Nattectin, a new C-type lectin from the venomous fish *Thalassophryne nattereri*." *Biochimie* 93(6): 971–980.

López, N., C. Lopera, and A. Ramírez (2008). "Characteristics of patients with ophidic accidents (snakebites) and infectious complications at the Pablo Tobon Uribe Hospital between the years 2000 and 2006." *Acta Med Colomb* 33(3): 127–130.

Lopez-Lozano, J. L., M. V. de Sousa, C. A. Ricart, C. Chavez-Olortegui, E. Flores Sanchez, E. G. Muniz, P. F. Buhrnheim, and L. Morhy (2002). "Ontogenetic variation of metalloproteinases and plasma coagulant activity in venoms of wild *Bothrops atrox* specimens from Amazonian rain forest." *Toxicon* 40(7): 997–1006.

Loria, G. D., A. Rucavado, A. S. Kamiguti, R. D. Theakston, J. W. Fox, A. Alape, and J. M. Gutiérrez (2003). "Characterization of 'basparin A,' a prothrombin-activating metalloproteinase, from the venom of the snake *Bothrops asper* that inhibits platelet aggregation and induces defibrination and thrombosis." *Arch Biochem Biophys* 418(1): 13–24.

Losos, J. B. (2011). "Convergence, adaptation, and constraint." *Evolution* 65(7): 1827–1840.

Lourenco, A. Jr., C. F. Zorzella Creste, L. C. de Barros, L. Delazari dos Santos, D. C. Pimenta, B. Barraviera, and R. S. Ferreira, Jr. (2013). "Individual venom profiling of *Crotalus durissus terrificus* specimens from a geographically limited region: Crotamine assessment and captivity evaluation on the biological activities." *Toxicon* 69: 75–81.

Low, D. H., K. Sunagar, E. A. Undheim, S. A. Ali, A. C. Alagón, T. Ruder, T. N. Jackson, et al. (2013). "Dracula's children: Molecular evolution of vampire bat venom." *J Proteomics* 89: 95–111.

Low, K. S. Y., M. C. E. Gwee, R. Yuen, P. Gopalakrishnakone, and H. E. Khoo (1994). "Stonustoxin—effects on neuromuscular function *in-vitro* and *in-vivo*." *Toxicon* 32(5): 573–581.

Lowe, D. G., M. S. Chang, R. Hellmiss, E. Chen, S. Singh, D. L. Garbers, and D. V. Goeddel (1989). "Human atrial natriuretic peptide receptor defines a new paradigm for second messenger signal transduction." *EMBO J* 8(5): 1377–1384.

Lu, X., J. A. Williams, J. J. Deadman, G. P. Salmon, V. V. Kakkar, J. M. Wilkinson, D. Baruch, K. S. Authi, and S. Rahman (1994). "Preferential antagonism of the interactions of the integrin alpha IIb beta 3 with immobilized glycoprotein ligands by snake-venom RGD (Arg-Gly-Asp) proteins: Evidence supporting a functional role for the amino acid residues flanking the tripeptide RGD in determining the inhibitory properties of snake-venom RGD proteins." *Biochem J* 304(pt 3): 929–936.

Luetje, C. W., K. Wada, S. Rogers, S. N. Abramson, K. Tsuji, S. Heinemann, and J. Patrick (1990). "Neurotoxins distinguish between different neuronal nicotinic acetylcholine receptor subunit combinations." *J Neurochem* 55(2): 632–640.

Lumsden, N. G., B. G. Fry, R. M. Kini, and W. C. Hodgson (2004a). "*In vitro* neuromuscular activity of 'colubrid' venoms: Clinical and evolutionary implications." *Toxicon* 43(7): 819–827.

Lumsden, N. G., B. G. Fry, S. Ventura, R. M. Kini, and W. C. Hodgson (2004b). "The *in vitro* and *in vivo* pharmacological activity of *Boiga dendrophila* (mangrove catsnake) venom." *Auton Autacoid Pharmacol* 24(4): 107–113.

——— (2005). "Pharmacological characterisation of a neurotoxin from the venom of *Boiga dendrophila* (mangrove catsnake)." *Toxicon* 45(3): 329–334.

Lundwall, A., U. Hellman, G. Eggertsen, and J. Sjoquist (1984). "Chemical characterization of cyanogen bromide fragments from the beta-chain of human complement factor C3." *FEBS Lett* 169(1): 57–62.

Lunger, P. D., and H. F. Clark (1979). "Morphogenesis of fer-de-lance virus (FDLV) cultured at optimal (30 degrees C) cell growth temperature." *J Comp Pathol* 89(2): 265–279.

Luthy, R., J. U. Bowie and D. Eisenberg (1992). "Assessment of protein models with three-dimensional profiles." *Nature* 356(6364): 83–85.

Lyukmanova, E. N., Z. O. Shenkarev, M. A. Shulepko, K. S. Mineev, D. D'Hoedt, I. E. Kasheverov, S. Y. Filkin, et al. (2011). "NMR structure and action on nicotinic acetylcholine receptors of water-soluble domain of human LYNX1." *J Biol Chem* 286(12): 10618–10627.

Ma, C. B., H. Wang, Y. X. Wu, M. Zhou, G. Lowe, L. Wang, Y. Q. Zhang, T. B. Chen, and C. Shaw (2012). "Helokinestatin-7 peptides from the venoms of Heloderma lizards." *Peptides* 35(2): 300–305.

MacFarlane, R. G., and B. Barnett (1934). "The haemostatic possibilities of snake venom." *Lancet*: 985–987.

Macheroux, P., O. Seth, C. Bollschweiler, M. Schwarz, M. Kurfurst, L. C. Au, and S. Ghisla (2001). "L-amino-acid oxidase from the Malayan pit viper *Calloselasma rhodostoma*: Comparative sequence analysis and characterization of active and inactive forms of the enzyme." *Eur J Biochem* 268(6): 1679–1686.

Mackessy, S. P. (1988). "Venom ontogeny in the Pacific rattlesnakes *Crotalus viridis helleri* and *C. v. oreganus*." *Copeia* (1): 92–101.

——— (1991). "Morphology and ultrastructure of the venom glands of the northern pacific rattlesnake *Crotalus viridis oreganus*." *J Morph* 208(1): 109–128.

——— (2010). "Evolutionary trends in venom composition in the western rattlesnakes (*Crotalus viridis sensu lato*): Toxicity vs. tenderizers." *Toxicon* 55(8): 1463–1474.

Mackessy, S. P., and L. M. Baxter (2006). "Bioweapons synthesis and storage: The venom gland of front-fanged snakes." *Zool Anz* 245: 147–159.

Mackessy, S. P., N. A. Sixberry, W. H. Heyborne, and T. Fritts (2006). "Venom of the Brown Treesnake, *Boiga irregularis*: Ontogenetic shifts and taxa-specific toxicity." *Toxicon* 47(5): 537–548.

Mackey, C. (1988). "Shocks for first aid." *Outdoor Life* 182: 3.

Maclean, D., M. C. Fishbein, E. Braunwald, and P. R. Maroko (1978). "Long-term preservation of ischemic myocardium after experimental coronary artery occlusion." *J Clin Invest* 61(3): 541–551.

Madsbad, S., U. Kielgast, M. Asmar, C. F. Deacon, S. S. Torekov, and J. J. Holst (2011). "An overview of once-weekly glucagon-like peptide-1 receptor agonists—available efficacy and safety data and perspectives for the future." *Diabetes Obes Metab* 13(5): 394–407.

Maeda, N., N. Tamiya, T. R. Pattabhiraman, and F. E. Russell (1978). "Some chemical properties of the venom of the rattlesnake, *Crotalus viridis helleri*." *Toxicon* 16(5): 431–441.

Maeda, T., M. Sakashita, Y. Ohba, and Y. Nakanishi (1998). "Molecular cloning of the rat Tpx-1 responsible for the interaction between spermatogenic and Sertoli cells." *Biochem Biophys Res Commun* 248: 140–146.

Magklara, A., A. A. Mellati, G. A. Wasney, S. P. Little, G. Sotiropoulou, G. W. Becker, and E. P. Diamandis (2003). "Characterization of the enzymatic activity of human kallikrein 6: Autoactivation, substrate specificity, and regulation by inhibitors." *Biochem Biophys Res Commun* 307(4): 948–955.

Maita, N., K. Nishio, E. Nishimoto, T. Matsui, Y. Shikamoto, T. Morita, J. E. Sadler, and H. Mizuno (2003). "Crystal structure of von Willebrand factor A1 domain complexed with snake venom, bitiscetin: Insight into glycoprotein Ibalpha binding mechanism induced by snake venom proteins." *J Biol Chem* 278(39): 37777–37781.

Malhotra, R., L. Singh, J. Eng, and J. P. Raufman (1992). "Exendin-4, a new peptide from *Heloderma suspectum* venom, potentiates cholecystokinin-induced amylase release from rat pancreatic acini." *Regul Pept* 41(2): 149–156.

Mancheva, I., T. Kleinschmidt, B. Aleksiev, and G. Braunitzer (1984). "Sequence homology between phospholipase and its inhibitor in snake venom: The primary structure of the inhibitor of vipoxin from the venom of the Bulgarian viper (*Vipera ammodytes ammodytes*, Serpentes)." *Hoppe Seylers Z Physiol Chem* 365(8): 885–894.

——— (1987). "Sequence homology between phospholipase and its inhibitor in snake-venom—the primary structure of phospholipase-A_2 of vipoxin from the venom of the bulgarian viper (*Vipera ammodytes ammodytes*, Serpentes)." *Biol Chem Hoppe-Seyler* 368(4): 343–352.

Mancin, A. C., A. M. Soares, S. H. Andriao-Escarso, V. M. Faca, L. J. Greene, S. Zuccolotto, I. R. Pela, and J. R. Giglio (1998). "The analgesic activity of crotamine, a neurotoxin from *Crotalus durissus terrificus* (South American rattlesnake) venom: A biochemical and pharmacological study." *Toxicon* 36(12): 1927–1937.

Mann, K. G., M. F. Hockin, K. J. Bean, and M. Kalafatis (1997). "Activated protein C cleavage of factor Va leads to dissociation of the A2 domain." *J Biol Chem* 272(33): 20678–20683.

Mann, K. G., and M. Kalafatis (2003). "Factor V: A combination of Dr. Jekyll and Mr. Hyde." *Blood* 101(1): 20–30.

Mans, B. J., J. F. Andersen, T. G. Schwan, and J. M. C. Ribeiro (2008). "Characterization of anti-hemostatic factors in the argasid, *Argas monolakensis*: Implications for the evolution of blood-feeding in the soft tick family." *Insect Biochem Mol Biol* 38(1): 22–41.

Mans, B. J., A. I. Louw, and A. W. H. Neitz (2003). "The major tick salivary gland proteins and toxins from the soft tick, *Ornithodoros savignyi*, are part of the tick lipocalin family: Implications for the origins of tick toxicoses." *Mol Biol Evol* 20(7): 1158–1167.

Mans, B. J., and J. M. C. Ribeiro (2008). "Function, mechanism and evolution of the moubatin-clade of soft tick lipocalins." *Insect Biochem Mol Biol* 38(9): 841–852.

Maraganore, J. M., G. Merutka, W. Cho, W. Welches, F. J. Kezdy, and R. L. Heinrikson (1984). "A new class of phospholipases A_2 with lysine in place of aspartate 49: Functional consequences for calcium and substrate binding." *J Biol Chem* 259(22): 13839–13843.

Marchot, P., and A. Chatonnet (2012). "Enzymatic activity and protein interactions in alpha/beta hydrolase fold proteins: Moonlighting versus promiscuity." *Protein Pept Lett* 19(2): 132–143.

Marchot, P., C. N. Prowse, J. Kanter, S. Camp, E. J. Ackermann, Z. Radic, P. E. Bougis, and P. Taylor (1997). "Expression and activity of mutants of fasciculin, a peptidic acetylcholinesterase inhibitor from mamba venom." *J Biol Chem* 272(6): 3502–3510.

Marcinkiewicz, C., J. J. Calvete, M. M. Marcinkiewicz, M. Raida, S. Vijay-Kumar, Z. Huang, R. R. Lobb, and S. Niewiarowski (1999a). "EC3, a novel heterodimeric disintegrin from *Echis carinatus* venom, inhibits alpha4 and alpha5 integrins in an RGD-independent manner." *J Biol Chem* 274(18): 12468–12473.

Marcinkiewicz, C., J. J. Calvete, S. Vijay-Kumar, M. M. Marcinkiewicz, M. Raida, P. Schick, R. R. Lobb, and S. Niewiarowski (1999b). "Structural and functional characterization of EMF10, a heterodimeric disintegrin from *Eristocophis macmahoni* venom that selectively inhibits alpha 5 beta 1 integrin." *Biochemistry* 38(40): 13302–13309.

Marcinkiewicz, C., R. R. Lobb, M. M. Marcinkiewicz, J. L. Daniel, J. B. Smith, C. Dangelmaier, P. H. Weinreb, D. A. Beacham, and S. Niewiarowski (2000). "Isolation and characterization of EMS16, a C-lectin type protein from *Echis multisquamatus* venom, a potent and selective inhibitor of the alpha2beta1 integrin." *Biochemistry* 39(32): 9859–9867.

Marcinkiewicz, C., S. Vijay-Kumar, M. A. McLane, and S. Niewiarowski (1997). "Significance of RGD loop and C-terminal domain of echistatin for recognition of alphaIIb beta3 and alpha(v) beta3 integrins and expression of ligand-induced binding site." *Blood* 90(4): 1565–1575.

Marcinkiewicz, C., P. H. Weinreb, J. J. Calvete, D. G. Kisiel, S. A. Mousa, G. P. Tuszynski, and R. R. Lobb (2003). "Obtustatin: A potent selective inhibitor of alpha1beta1 integrin *in vitro* and angiogenesis *in vivo*." *Cancer Res* 63(9): 2020–2023.

Marcon, F., M. Leblanc, I. Vetter, R. J. Lewis, P. Escoubas, and G. M. Nicholson (2012). "Pharmacological characterization of alpha-elapitoxin-Al2a from the venom of the Australian pygmy copperhead (*Austrelaps labialis*): An atypical long-chain alpha-neurotoxin with only weak affinity for alpha7 nicotinic receptors." *Biochem Pharmacol* 84(6): 851–863.

Marcon, F., L. Purtell, J. Santos, P. G. Hains, P. Escoubas, A. Graudins, and G. M. Nicholson (2013). "Characterization of monomeric and multimeric snake neurotoxins and other bioactive proteins from the venom of the lethal Australian common copperhead (*Austrelaps superbus*)." *Biochem Pharmacol* 85(10): 1555–1573.

Marcussi, S., C. D. Sant'Ana, C. Z. Oliveira, A. Q. Rueda, D. L. Menaldo, R. O. Beleboni, R. G. Stabeli, J. R. Giglio, M. R. Fontes, and A. M. Soares (2007). "Snake venom phospholipase A_2 inhibitors: Medicinal chemistry and therapeutic potential." *Curr Top Med Chem* 7(8): 743–756.

Margres, M. J., K. Aronow, J. Loyacano, and D. R. Rokyta (2013). "The venom-gland transcriptome of the eastern coral snake (*Micrurus fulvius*) reveals high venom complexity in the intragenomic evolution of venoms." *BMC Genomics* 14: 531.

Margulies, M., M. Egholm, W. E. Altman, S. Attiya, J. S. Bader, L. A. Bemben, J. Berka, et al. (2005). "Genome sequencing in microfabricated high-density picolitre reactors." *Nature* 437(7057): 376–380.

Maritz, B., and G. J. Alexander (2012). "Movement patterns in the smallest viper, *Bitis schneideri*." *Copeia* 4: 732–737.

Markland, F. S. (1996). "Fibrolase, an active thrombolytic enzyme in arterial and venous thrombosis model systems." *Adv Exp Med Biol* 391: 427–438.

——— (1998). "Snake venoms and the hemostatic system." *Toxicon* 36(12): 1749–1800.

Markland, F. S., C. Kettner, S. Schiffman, E. Shaw, S. S. Bajwa, K. N. Reddy, H. Kirakossian, G. B. Patkos, I. Theodor, and H. Pirkle (1982). "Kallikrein-like activity of crotalase, a snake venom enzyme that clots fibrinogen." *Proc Natl Acad Sci USA* 79(6): 1688–1692.

Markland, F. S., and H. Pirkle (1977). "Thrombin-like enzyme from the venom of *Crotalus adamanteus* (Eastern diamondback rattlesnake)." *Thromb Res* 10(3): 487–494.

Maroko, P. R., C. B. Carpenter, M. Chiariello, M. C. Fishbein, P. Radvany, J. D. Knostman, and S. L. Hale (1978). "Reduction by cobra venom factor of myocardial necrosis after coronary artery occlusion." *J Clin Invest* 61(3): 661–670.

Marquez, B., G. Ignotz, and S. S. Suarez (2007). "Contributions of extracellular and intracellular Ca2+ to regulation of sperm motility: Release of intracellular stores can hyperactivate CatSper1 and CatSper2 null sperm." *Dev Biol* 303(1): 214–221.

Marrakchi, N., R. B. Zingali, H. Karoui, C. Bon, and M. el Ayeb (1995). "Cerastocytin, a new thrombin-like platelet activator from the venom of the Tunisian viper *Cerastes cerastes*." *Biochim Biophys Acta* 1244(1): 147–156.

Marsh, N., and V. Williams (2005). "Practical applications of snake venom toxins in haemostasis." *Toxicon* 45(8): 1171–1181.

Marsh, N. A. (2001). "Diagnostic uses of snake venom." *Haemostasis* 31(3-6): 211–217.

Marsh, R. E. (1993). "Test results of a new snake repellent." *Great Plains Wildlife Damage Control Workshop Proceedings*: 344.

Marshall, R. D. (1972). "Glycoproteins." *Annu Rev Biochem* 41: 673–702.

Martelli, P., S. Luz, and W. Meyer (2001). "Information on euthanasia of snakes—with recommendations for procedures causing the least distress possible." *Kleintierpraxis* 46(2): 105–109.

Martin, A., and V. Orgogozo (2013). "The loci of repeated evolution: A catalog of genetic hotspots of phenotypic variation." *Evolution* 67(5): 1235–1250.

Martin, M. (2011). "Cutadapt removes adapter sequences from high-throughput sequencing reads." *EMBnet.journal* 17(1): 10–12.

Martinez-Jimenez, D., and S. J. Hernandez-Divers (2007). "Emergency care of reptiles." *Vet Clin North Am Exot Anim Pract* 10(2): 557–585.

Martz, W. (1992). "Plants with a reputation against snakebite." *Toxicon* 30(10): 1131–1142.

Marx, H., and G. B. Rabb (1972). "Phyletic analysis of fifty characters of advanced snakes." *Fieldiana Zool* 63: 1–321.

Masci, P. P. (2007). *Patent WO2009079690A1: Haemostasis-modulating compositions and uses therefor.* Brisbane: University of Queensland.

Masci, P. P., A. N. Whitaker, and J. de Jersey (1988). "Purification and characterization of a prothrombin activator from the venom of the Australian brown snake, *Pseudonaja textilis textilis*." *Biochem Int* 17: 825–835.

Masci, P. P., A. N. Whitaker, L. G. Sparrow, J. de Jersey, D. J. Winzor, D. J. Watters, M. F. Lavin, and P. J. Gaffney (2000). "Textilinins from *Pseudonaja textilis textilis*: Characterization of two plasmin inhibitors that reduce bleeding in an animal model." *Blood Coagul Fibrin* 11(4): 385–393.

Massoulié, J., A. Anselmet, S. Bon, E. Krejci, C. Legay, N. Morel, and S. Simon (1998). "Acetylcholi nesterase: C-terminal domains, molecular forms and functional localization." *J Physiol (Paris)* 92(3–4): 183–190.

——— (1999). "The polymorphism of acetylcholinesterase: Post-translational processing, quaternary associations and localization." *Chem Biol Interact* 119–120: 29–42.

Massoulié, J., and S. Bon (2006). "The C-terminal T peptide of cholinesterases: Structure, interactions, and influence on protein folding and secretion." *J Mol Neurosci* 30(1–2): 233–236.

Massoulié, J., L. Pezzementi, S. Bon, E. Krejci, and F. M. Vallette (1993). "Molecular and cellular biology of cholinesterases." *Prog Neurobiol* 41(1): 31–91.

Mastrangeli, R., S. Donini, C. A. Kelton, C. He, A. Bressan, F. Milazzo, V. Ciolli, et al. (2003). "ARS Component B: Structural characterization, tissue expression and regulation of the gene and protein (SLURP-1) associated with Mal de Meleda." *Eur J Dermatol* 13(6): 560–570.

Masuda, S., H. Hayashi, and S. Araki (1998). "Two vascular apoptosis-inducing proteins from snake venom are members of the metalloprotease/disintegrin family." *Eur J Biochem* 253(1): 36–41.

Masuda, S., H. Hayashi, H. Atoda, T. Morita, and S. Araki (2001). "Purification, cDNA cloning and characterization of the vascular apoptosis-inducing protein, HV1, from *Trimeresurus flavoviridis*." *Eur J Biochem* 268(11): 3339–3345.

Masuda, S., T. Ohta, K. Kaji, J. W. Fox, H. Hayashi, and S. Araki (2000). "cDNA cloning and characterization of vascular apoptosis-inducing protein 1." *Biochem Biophys Res Comm* 278(1): 197–204.

Matavel, A. C., D. L. Ferreira-Alves, P. S. Beirao, and J. S. Cruz (1998). "Tension generation and increase in voltage-activated Na+ current by crotamine." *Eur J Pharmacol* 348(2–3): 167–173.

Matsui, T., and J. Hamako (2005). "Structure and function of snake venom toxins interacting with human von Willebrand factor." *Toxicon* 45(8): 1075–1087.

Matsui, T., J. Hamako, T. Matsushita, T. Nakayama, Y. Fujimura, and K. Titani (2002). "Binding site on human von Willebrand factor of bitiscetin, a snake venom-derived platelet aggregation inducer." *Biochemistry* 41(25): 7939–7946.

Matsui, T., Y. Sakurai, Y. Fujimura, I. Hayashi, S. Oh-Ishi, M. Suzuki, J. Hamako, et al. (1998). "Purification and amino acid sequence of halystase from snake venom of *Agkistrodon halys blomhoffii*, a serine protease that cleaves specifically fibrinogen and kininogen." *Eur J Biochem* 252(3): 569–575.

McArdle, J. J., T. L. Lentz, V. Witzemann, H. Schwarz, S. A. Weinstein, and J. J. Schmidt (1999). "Waglerin-1 selectively blocks the epsilon form of the muscle nicotinic acetylcholine receptor." *J Pharmacol Exp Ther* 289(1): 543–550.

McCarthy, C. J. (1987). "Adaptations of sea snakes that eat fish eggs; with a note on the throat musculature of *Aipysurus eydouxi* (Gray, 1894)." *J Nat Hist* 1: 1119–1128.

McDonald, K. L., and R. I. Webb (2011). "Freeze substitution in 3 hours or less." *J Microsc* 243(3): 227–233.

McDonald, T. F., S. Pelzer, W. Trautwein, and D. J. Pelzer (1994). "Regulation and modulation of calcium channels in cardiac, skeletal, and smooth muscle cells." *Physiol Rev* 74(2): 365–507.

McDowell, R. S., M. S. Dennis, A. Louie, M. Shuster, M. G. Mulkerrin, and R. A. Lazarus (1992). "Mambin, a potent glycoprotein IIb-IIIa antagonist and platelet aggregation inhibitor structurally related to the short neurotoxins." *Biochemistry* 31(20): 4766–4772.

McDowell, S. B. (1986). "The architecture of the corner of the mouth of colubroid snakes." *J Herpetol* 20: 353–407.

McDuffie, F. C., C. Giffin, R. Niedringhaus, K. G. Mann, C. A. J. Owen, E. J. W. Bowie, J. Peterson, G. Clark, and G. G. Hunder (1979). "Prothrombin, thrombin and prothrombin fragments in plasma of normal individuals and of patients with laboratory evidence of disseminated intravascular coagulation." *Thromb Res* 16(5–6): 759–773.

McEver, R. P. (2002). "Selectins: Lectins that initiate cell adhesion under flow." *Curr Opin Cell Biol* 14(5): 581–586.

McGinn, O. J., W. R. English, S. Roberts, A. Ager, P. Newham, and G. Murphy (2011). "Modulation of integrin alpha4beta1 by ADAM28 promotes lymphocyte adhesion and transendothelial migration." *Cell Biol Int* 35(10): 1043–1053.

McLelland, D. J., L. Johnson, and R. Reuter (2010). "Fatal cutaneous mycosis in a broad-headed snake (*Hoplocephalus bungaroides*) caused by the *Chrysosporium* anamorph of *Nannizziopsis vriesii*." *Proc Wildl Dis Assoc*: 55.

McMullen, B. A., K. Fujikawa, W. Kisiel, T. Sasagawa, W. N. Howald, E. Y. Kwa, and B. Weinstein (1983). "Complete amino acid sequence of the light chain of human coagulation factor X: Evidence for identification of residue 63 as beta-hydroxyaspartic acid." *Biochem Int* 22: 2875–2884.

Meakin, S. O., and E. M. Shooter (1992). "The nerve growth factor family of receptors." *Trends Neurosci* 15(9): 323–331.

Mebs, D., M. Ehrenfeld, and Y. Samejima (1983). "Local necrotizing effect of snake venoms on skin and muscle: Relationship to serum creatine kinase." *Toxicon* 21(3): 393–404.

Mebs, D., U. Kuch, F. I. Coronas, C. V. Batista, A. Gumprecht, and L. D. Possani (2006). "Biochemical and biological activities of the venom of the Chinese pitviper *Zhaoermia mangshanensis*, with the complete amino acid sequence and phylogenetic analysis of a novel Arg49 phospholipase A_2 myotoxin." *Toxicon* 47(7): 797–811.

Meenatchisundaram, S., G. Parameswari, A. Michael, and S. Ramalingam (2008). "Neutralization of the pharmacological effects of Cobra and Krait venoms by chicken egg yolk antibodies." *Toxicon* 52(2): 221–227.

Meinkoth, J. L., A. S. Alberts, W. Went, D. Fantozzi, S. S. Taylor, M. Hagiwara, M. Montminy, and J. R. Feramisco (1993). "Signal transduction through the cAMP-dependent protein kinase." *Mol Cell Biochem* 127–128: 179–186.

Melo, F., and E. Feytmans (1997). "Novel knowledge-based mean force potential at atomic level." *J Mol Biol* 267(1): 207–222.

Ménezes, M. C., M. F. Furtado, S. R. Travaglia-Cardoso, A. C. Camargo, and S. M. Serrano (2006). "Sex-based individual variation of snake venom proteome among eighteen *Bothrops jararaca* siblings." *Toxicon* 47(3): 304–312.

Menin, L., A. Perchuc, P. Favreau, F. Perret, S. Michalet, R. Schoni, M. Wilmer, and R. Stocklin (2008). "High throughput screening of bradykinin-potentiating peptides in *Bothrops moojeni* snake venom using precursor ion mass spectrometry." *Toxicon* 51(7): 1288–1302.

Menozzi, A., P. A. Merlini, and D. Ardissino (2005). "Tirofiban in acute coronary syndromes." *Expert Rev Cardiovasc Ther* 3(2): 193–206.

Messenger, K. R., D. Rosenberg, K. K. Caldwell, and W. L. Sargent (2012). "*Rhabdophis subminiatus helleri* (Red-necked Keel-back) defensive behaviour." *Herpetol Rev* 43(3): 497–497.

Meyer, T. S., and B. L. Lamberts (1965). "Use of Coomassie brilliant blue R250 for the electrophoresis of microgram quantities of parotid saliva proteins on acrylamide-gel strips." *Biochim Biophys Acta* 107(1): 144–145.

Meyer, W. K., S. Zhang, S. Hayakawa, H. Imai, and M. Przeworski (2013). "The convergent evolution of blue iris pigmentation in primates took distinct molecular paths." *Am J Phys Anthropol* 151(3): 398–407.

Meyer-Hoffert, U. (2009). "Reddish, scaly, and itchy: How proteases and their inhibitors contribute to inflammatory skin diseases." *Arch Immunol Ther Exp (Warsz)* 57(5): 345–354.

Michel, G. H., N. Murayama, T. Sada, M. Nozaki, K. Saguchi, H. Ohi, Y. Fujita, H. Koike, and S. Higuchi (2000). "Two N-terminally truncated forms of C-type natriuretic peptide from habu snake venom." *Peptides* 21(5): 609–615.

Miledi, R., I. Parker, and K. Sumikawa (1982). "Properties of acetylcholine receptors translated by cat muscle mRNA in *Xenopus* oocytes." *EMBO J* 1(11): 1307–1312.

Miller, F. D., and D. R. Kaplan (2001). "On Trk for retrograde signaling." *Neuron* 32(5): 767–770.

Miller, J. R., S. Koren, and G. Sutton (2010). "Assembly algorithms for next-generation sequencing data." *Genomics* 95(6): 315–327.

Miller, M. F. (1995). "Gila monster envenomation." *Ann Emerg Med* 25(5): 720–720.

Miller, R. E. (1996). "Quarantine protocols and preventive medicine procedures for reptiles, birds and mammals in zoos." *Rev Sci Tech OIE* 15(1): 183–189.

Millers, E. K. I., P. P. Masci, M. F. Lavin, J. de Jersey, and L. W. Guddat (2006). "Crystallization and preliminary X-ray analysis of a Kunitz-type inhibitor, textilinin-1 from *Pseudonaja textilis textilis*." *Acta Crystallogr F Struct Biol Cryst Commun* 62: 642–645.

Millotti, G., A. Vetter, K. Leithner, F. Sarti, G. Shahnaz Bano, P. Augustijns, and A. Bernkop-Schnurch (2013). "Development of thiolated poly(acrylic acid) microparticles for the nasal administration of exenatide." *Drug Dev Ind Pharm* 40(12): 1677–1682.

Milne, T. J., G. Abbenante, J. D. A. Tyndall, J. Halliday, and R. J. Lewis (2003). "Isolation and characterization of a cone snail protease with homology to CRISP proteins of the pathogenesis-related protein superfamily." *J Biol Chem* 278: 31105–31110.

Minagawa, S., M. Ishida, K. Shimakura, Y. Nagashima, and K. Shiomi (1997). "Isolation and amino acid sequences of two Kunitz-type protease inhibitors from the sea anemone *Anthopleura* aff. *xanthogrammica*." *Comp Biochem Physiol B Biochem Mol Biol* 118(2): 381–386.

Mitchell, J. S., A. B. Heckert, and H. D. Sues (2010). "Grooves to tubes: Evolution of the venom delivery system in a Late Triassic 'reptile.'" *Naturwissenschaften* 97(12): 1117–1121.

Miwa, J. M., I. Ibanez-Tallon, G. W. Crabtree, R. Sanchez, A. Sali, L. W. Role, and N. Heintz (1999). "Lynx1, an endogenous toxin-like modulator of nicotinic acetylcholine receptors in the mammalian CNS." *Neuron* 23(1): 105–114.

Mizuno, H., Z. Fujimoto, H. Atoda, and T. Morita (2001). "Crystal structure of an anticoagulant protein in complex with the Gla domain of factor X." *Proc Natl Acad Sci USA* 98(13): 7230–7234.

Mizuno, H., Z. Fujimoto, M. Koizumi, H. Kano, H. Atoda, and T. Morita (1997). "Structure of coagulation factors IX/X-binding protein, a heterodimer of C-type lectin domains." *Nat Struct Biol* 4(6): 438–441.

Mobli, M., A. S. Stern, W. Bermel, G. F. King, and J. C. Hoch (2010). "A non-uniformly sampled 4D HCC(CO)NH-TOCSY experiment processed using maximum entropy for rapid protein sidechain assignment." *J Magn Reson* 204(1): 160–164.

Mochca-Morales, J., B. M. Martin, and L. D. Possani (1990). "Isolation and characterization of helothermine, a novel toxin from *Heloderma horridum horridum* (Mexican beaded lizard) venom." *Toxicon* 28: 299–309.

Mohamed, A., P. F. Reid, L. Raymond, and T. Dufan (2006). "Amelioration of acute and relapsing stages of the experimental allergic encephalomyelitis by cobra toxins." *Biomed Sci Instrum* 42: 399–404.

Mohr, W. J., K. Jenabzadeh, and D. H. Ahrenholz (2009). "Cold injury." *Hand Clinics* 25(4): 481.

Molina, H., V. M. Holers, B. Li, Y. Fung, S. Mariathasan, J. Goellner, J. Strauss-Schoenberger, R. W. Karr, and D. D. Chaplin (1996). "Markedly impaired humoral immune response in mice deficient in complement receptors 1 and 2." *Proc Natl Acad Sci USA* 93(8): 3357–3361.

Moll, S., P. Kenyon, L. Bertoli, J. De Maio, H. Homesley, and S. R. Deitcher (2006). "Phase II trial of alfimeprase, a novel-acting fibrin degradation agent, for occluded central venous access devices." *J Clin Oncol* 24(19): 3056–3060.

Mollay, C., C. Wechselberger, G. Mignogna, L. Negri, P. Melchiorri, D. Barra, and G. Kreil (1999). "Bv8, a small protein from frog skin and its homologue from snake venom induce hyperalgesia in rats." *Eur J Pharmacol* 374(2): 189–196.

Momic, T., G. Cohen, R. Reich, F. T. Arlinghaus, J. A. Eble, C. Marcinkiewicz, and P. Lazarovici (2012). "Vixapatin (VP12), a C-type lectin-protein from *Vipera xantina palestinae* venom: Characterization as a novel anti-angiogenic compound." *Toxins (Basel)* 4(10): 862–877.

Monks, S. A., A. R. Gould, P. E. Lumley, P. F. Alewood, W. R. Kem, N. H. Goss, and R. S. Norton (1994). "Limited proteolysis study of structure-function relationships in Sh I, a polypeptide neurotoxin from a sea anemone." *Biochim Biophys Acta* 1207(1): 93–101.

Montefusco, M. C., S. Duga, R. Asselta, E. Santagostino, G. Mancuso, M. Malcovati, P. M. Mannucci, and M. L. Tenchini (2000). "A novel two base pair deletion in the factor V gene associated with severe factor V deficiency." *Br J Haematol* 111(4): 1240–1246.

Monteiro, R. Q. (2005). "Targeting exosites on blood coagulation proteases." *An Acad Bras Cienc* 77(2): 275–280.

Montfort, W. R., A. Weichsel, and J. F. Andersen (2000). "Nitrophorins and related antihemostatic lipocalins from *Rhodnius prolixus* and other blood-sucking arthropods." *Biochim Biophys Acta Protein Struct Mol Enzymol* 1482(1–2): 110–118.

Montgomery, J. M., D. Gillespie, P. Sastrawan, T. M. Fredeking, and G. L. Stewart (2002). "Aerobic salivary bacteria in wild and captive Komodo dragons." *J Wildl Dis* 38(3): 545–551.

Moore, G. W. (2007). "Combining Taipan snake venom time/Ecarin time screening with the mixing studies of conventional assays increases detection rates of lupus anticoagulants in orally anticoagulated patients." *Thromb J* 5: 12.

Moore, G. W., M. P. Smith, and G. F. Savidge (2003). "The Ecarin time is an improved confirmatory test for the Taipan snake venom time in warfarinized patients with lupus anticoagulants." *Blood Coagul Fibrin* 14(3): 307–312.

Mor, A., N. Chartrel, H. Vaudry, and P. Nicolas (1994). "Skin peptide tyrosine-tyrosine, a member of the pancreatic-polypeptide family—isolation, structure, synthesis, and endocrine activity." *Proc Natl Acad Sci USA* 91(22): 10295–10299.

Mora, R., B. Valverde, C. Diaz, B. Lomonte, and J. M. Gutiérrez (2005). "A Lys49 phospholipase A(2) homologue from *Bothrops asper* snake venom induces proliferation, apoptosis and necrosis in a lymphoblastoid cell line." *Toxicon* 45 (5): 651–660.

Mordvintsev, D. Y., Y. L. Polyak, D. I. Rodionov, J. Jakubik, V. Dolezal, E. Karlsson, V. I. Tsetlin, and Y. N. Utkin (2009). "Weak toxin WTX from *Naja kaouthia* cobra venom interacts with both nicotinic and muscarinic acetylcholine receptors." *FEBS J* 276(18): 5065–5075.

Moreno, K., S. Wroe, P. Clausen, C. McHenry, D. C. D'Amore, E. J. Rayfield, and E. Cunningham (2008). "Cranial performance in the Komodo dragon (*Varanus komodoensis*) as revealed by high-resolution 3-D finite element analysis." *J Anat* 212(6): 736–746.

Mori, A., and G. M. Burghardt (2008). "Comparative experimental tests of natracine antipredator displays, with special reference to the apparently unique displays in the Asian genus, *Rhabdophis*." *J Ethol* 26: 61–68.

Mori, A., G. M. Burghardt, A. H. Savitzky, K. A. Roberts, D. A. Hutchinson, and R. C. Goris (2012). "Nuchal glands: A novel defensive system in snakes." *Chemoecology* 22(3): 187–198.

Mori, N., T. Nikai, H. Sugihara, and A. T. Tu (1987). "Biochemical characterization of hemorrhagic toxins with fibrinogenase activity isolated from *Crotalus ruber ruber* venom." *Arch Biochem Biophys* 253(1): 108–121.

Morita, A., H. Isawa, Y. Orito, S. Iwanaga, Y. Chinzei, and M. Yuda (2006). "Identification and characterization of a collagen-induced platelet aggregation inhibitor, triplatin, from salivary glands of the assassin bug, *Triatoma infestans*." *FEBS J* 273(13): 2955–2962.

Morita, T. (2005a). "Structure-function relationships of C-type lectin-related proteins." *Pathophysiol Haemost Thromb* 34(4–5): 156–159.

——— (2005b). "Structures and functions of snake venom CLPs (C-type lectin-like proteins) with anticoagulant-, procoagulant-, and platelet-modulating activities." *Toxicon* 45(8): 1099–1114.

Morita, T., S. Iwanaga, and T. Suzuki (1976a). "Activation of bovine prothrombin by an activator isolated from *Echis carinatus* venom." *Thromb Res* 8(2 suppl): 59–65.

——— (1976b). "The mechanism of activation of bovine prothrombin by an activator isolated from *Echis carinatus* venom and characterization of the new active intermediates." *J Biochem* 79(5): 1089–1108.

Morrison, J. J., J. H. Pearn, and A. R. Coulter (1982). "The mass of venom injected by two elapidae: The taipan (*Oxyuranus scutellatus*) and the Australian tiger snake (*Notechis scutatus*)." *Toxicon* 20(4): 739–745.

Morrison, J. J., I. Tesseraux, J. H. Pearn, J. Harris, and P. P. Masci (1984). "Venom of the Australian rough-scaled snake, *Tropidechis carinatus*: Lethal potency and electrophysiological actions." *Toxicon* 22(5): 759–765.

Morrissette, J., J. Kratzschmar, B. Haendler, R. el-Hayek, J. Mochca-Morales, B. M. Martin, J. R. Patel, R. L. Moss, W. D. Schleuning, and R. Coronado (1995). "Primary structure and properties of helothermine, a peptide toxin that blocks ryanodine receptors." *Biophys J* 68(6): 2280–2288.

Morton, J. F. (1981). *Atlas of medicinal plants of Middle America: Bahamas to Yucatan*. Springfield, IL: Charles C. Thomas.

Mosley, C. (2005). "Anesthesia and analgesia in reptiles." *Semin Avian Exot Pet Med* 14(4): 243–262.

Mouhat, S., N. Andreotti, B. Jouirou, and J. M. Sabatier (2008). "Animal toxins acting on voltage-gated potassium channels." *Curr Pharm Des* 14(24): 2503–2518.

Mouhat, S., B. Jouirou, A. Mosbah, M. De Waard, and J. M. Sabatier (2004). "Diversity of folds in animal toxins acting on ion channels." *Biochem J* 378: 717–726.

Moura-da-Silva, A. M., R. D. Theakston, and J. M. Crampton (1996). "Evolution of disintegrin cysteine-rich and mammalian matrix-degrading metalloproteinases: Gene duplication and divergence of a common ancestor rather than convergent evolution." *J Mol Evol* 43(3): 263–269.

Mourier, G., M. Hajj, F. Cordier, A. Zorba, X. Gao, T. Coskun, A. Herbet, et al. (2012). "Pharmacological and structural characterization of long-sarafotoxins, a new family of endothelin-like peptides: Role of the C-terminus extension." *Biochimie* 94(2): 461–470.

Moustafa, I. M., S. Foster, A. Y. Lyubimov, and A. Vrielink (2006). "Crystal structure of LAAO from *Calloselasma rhodostoma* with an L-phenylalanine substrate: Insights into structure and mechanism." *J Mol Biol* 364(5): 991–1002.

Muanpasitporn, C., and P. Rojnuckarin (2007). "Expression and characterization of a recombinant fibrinogenolytic serine protease from green pit viper (*Trimeresurus albolabris*) venom." *Toxicon* 49(8): 1083–1089.

Mueller, L. (1988a). "A shocking cure for snakebites." *Outdoor Life* 182(1): 45–47, 76–78.

——— (1988b). "A shocking cure for snakebites: Part I." *Outdoor Life* 181(6): 64–65, 110–112.

Mukherjee, S., H. Zheng, M. G. Derebe, K. M. Callenberg, C. L. Partch, D. Rollins, D. C. Propheter, et al. (2014). "Antibacterial membrane attack by a pore-forming intestinal C-type lectin." *Nature* 505(7481): 103–107.

Mukoyama, M., K. Nakao, K. Hosoda, S. Suga, Y. Saito, Y. Ogawa, G. Shirakami, M. Jougasaki, K. Obata, and H. Yasue (1991). "Brain natriuretic peptide as a novel cardiac hormone in humans: Evidence for an exquisite dual natriuretic peptide system, atrial natriuretic peptide and brain natriuretic peptide." *J Clin Invest* 87(4): 1402–1412.

Mulder, R., M. K. ten Kate, H. C. Kluin-Nelemans, and A. B. Mulder (2010). "Low cut-off values increase diagnostic performance of protein S assays." *Thromb Haemost* 104(3): 618–625.

Muller-Esterl, W. (1989). "Kininogens, kinins and kinships." *Thromb Haemost* 61(1): 2–6.

Mulligan, M. S., E. Schmid, B. Beck-Schimmer, G. O. Till, H. P. Friedl, R. B. Brauer, T. E. Hugli, et al. (1996). "Requirement and role of C5a in acute lung inflammatory injury in rats." *J Clin Invest* 98(2): 503–512.

Mundy, H. R., S. J. Jones, J. C. Hobart, M. G. Hanna, and P. J. Lee (2003). "A randomized controlled study of modified cobratoxin in adrenomyeloneuropathy." *Neurology* 61(4): 528–530.

Murakami, M., and I. Kudo (2002). "Phospholipase A_2." *J Biochem* 131(3): 285–292.

Murrell, B., S. Moola, A. Mabona, T. Weighill, D. Sheward, S. L. Kosakovsky Pond, and K. Scheffler (2013). "FUBAR: A fast, unconstrained bayesian approximation for inferring selection." *Mol Biol Evol* 30(5): 1196–1205.

Murrell, B., J. O. Wertheim, S. Moola, T. Weighill, K. Scheffler, and S. L. Kosakovsky Pond (2012). "Detecting individual sites subject to episodic diversifying selection." *PLoS Genet* 8(7): e1002764.

Murthy, K. S., and G. M. Makhlouf (1999). "Identification of the G protein-activating domain of the natriuretic peptide clearance receptor (NPR-C)." *J Biol Chem* 274(25): 17587–17592.

Nagarajan, N., and M. Pop (2013). "*Sequence assembly demystified.*" *Nat Rev Genet* 14(3): 157–167.

Nagaraju, S., S. Devaraja, and K. Kemparaju (2007). "Purification and properties of hyaluronidase from *Hippasa partita* (funnel web spider) venom gland extract." *Toxicon* 50(3): 383–393.

Nagele, E., M. Vollmer, P. Horth, and C. Vad (2004). "2D-LC/MS techniques for the identification of proteins in highly complex mixtures." *Expert Rev Proteomics* 1(1): 37–46.

Nair, D. G., B. G. Fry, P. Alewood, P. P. Kumar, and R. M. Kini (2007). "Antimicrobial activity of omwaprin, a new member of the waprin family of snake venom proteins." *Biochem J* 402(1): 93–104.

Nakagaki, T., A. L. Kazim, and W. Kisiel (1990). "Isolation and characterization of a protein C activator from tropical moccasin venom." *Thromb Res* 58(6): 593–602.

Nakajima, K., S. Kubo, S. Kumagaye, H. Nishio, M. Tsunemi, T. Inui, H. Kuroda, et al. (1989). "Structure-activity relationship of endothelin: Importance of charged groups." *Biochem Biophys Res Comm* 163(1): 424–429.

Nakajima, M., K. Hanasaki, M. Ueda, and H. Arita (1992). "Effect of pancreatic type phospholipase A_2 on isolated porcine cerebral arteries via its specific binding sites." *FEBS Lett* 309(3): 261–264.

Nakashima, S., K. Kitamoto, and M. Arioka (2004). "The catalytic activity, but not receptor binding, of $sPLA_2$s plays a critical role for neurite outgrowth induction in PC12 cells." *Brain Res* 1015(1-2): 207–211.

Nanthakumar, K., S. M. Peng, K. Umapathy, P. Dorian, E. Sevaptsidis, and M. Waxman (2008). "Cardiac stimulation with high voltage discharge from stun guns." *Can Med Assoc J* 178(11): 1451–1457.

Nascimento, F. D., M. A. Hayashi, A. Kerkis, V. Oliveira, E. B. Oliveira, G. Rádis-Baptista, H. B. Nader, T. Yamane, I. L. Tersariol, and I. Kerkis (2007). "Crotamine mediates gene delivery into cells through the binding to heparan sulfate proteoglycans." *J Biol Chem* 282(29): 21349–21360.

Nascimento, F. D., L. Sancey, A. Pereira, C. Rome, V. Oliveira, E. B. Oliveira, H. B. Nader, et al. (2012). "The natural cell-penetrating peptide crotamine targets tumor tissue *in vivo* and triggers a lethal calcium-dependent pathway in cultured cells." *Mol Pharmcol* 9(2): 211–221.

Natesh, R., S. L. Schwager, E. D. Sturrock, and K. R. Acharya (2003). "Crystal structure of the human angiotensin-converting enzyme-lisinopril complex." *Nature* 421(6922): 551–554.

Nature Notes (1870). *Nature* 1: 288.

Nature Notes (1871). *Nature* 4: 287.

Naumann, G. B., L. F. Silva, L. Silva, G. Faria, M. Richardson, K. Evangelista, M. Kohlhoff, et al. (2011). "Cytotoxicity and inhibition of platelet aggregation caused by an L-amino acid oxidase from *Bothrops leucurus* venom." *Biochim Biophys Acta* 1810(7): 683–694.

Navdaev, A., D. Dormann, J. M. Clemetson, and K. J. Clemetson (2001). "Echicetin, a GPIb-binding snake C-type lectin from *Echis carinatus*, also contains a binding site for IgM-kappa responsible for platelet agglutination in plasma and inducing signal transduction." *Blood* 97(8): 2333–2341.

Navdaev, A., G. Lochnit, and J. A. Eble (2011). "The rhodocetin alphabeta subunit targets GPIb and inhibits von Willebrand factor induced platelet activation." *Toxicon* 57(7–8): 1041–1048.

Negri, L., R. Lattanzi, E. Giannini, and P. Melchiorri (2007). "Bv8/prokineticin proteins and their receptors." *Life Sci* 81(14): 1103–1116.

Nelsestuen, G. L., and J. W. Suttie (1972a). "The carbohydrate of bovine prothrombin. Partial structural determination demonstrating the presence of -galactose residues." *J Biol Chem* 247(19): 6096–6102.

——— (1972b). "Mode of action of vitamin K: Calcium binding properties of bovine prothrombin." *Biochemistry* 11(26): 4961–4964.

——— (1972c). "The purification and properties of an abnormal prothrombin protein produced by dicumarol-treated cows: A comparison to normal prothrombin." *J Biol Chem* 247(24): 8176–8182.

Nelson, M. E., A. Kuryatov, C. H. Choi, Y. Zhou, and J. Lindstrom (2003). "Alternate stoichiometries of alpha4beta2 nicotinic acetylcholine receptors." *Mol Pharmacol* 63(2): 332–341.

Neri-Castro, E. E. (2013). *Thesis: Variación intraespecífica en el veneno de* Crotalus si*mus de México.* Universidad Autónoma del Estado de Morelos, Mexico.

Nesheim, M. E., W. M. Canfield, W. Kisiel, and K. G. Mann (1982). "Studies of the capacity of factor-Xa to protect factor-Va from inactivation by activated protein-C." *J Biol Chem* 257(3): 1443–1447.

Nesheim, M. E., W. B. Foster, R. Hewick, and K. G. Mann (1984). "Characterization of factor-V activation intermediates." *J Biol Chem* 259(5): 3187–3196.

Nesheim, M. E., W. L. Nichols, T. L. Cole, J. G. Houston, R. B. Schenk, K. G. Mann, and E. J. W. Bowie (1986). "Isolation and study of an acquired inhibitor of human coagulation factor-V." *J Clin Invest* 77(2): 405–415.

Nesheim, M. E., J. B. Taswell, and K. G. Mann (1979). "Contribution of bovine factor-V and factor-Va to the activity of prothrombinase." *J Biol Chem* 254(21): 952–962.

Neuhoff, V., N. Arold, D. Taube, and W. Ehrhardt (1988). "Improved staining of proteins in polyacrylamide gels including isoelectric focusing gels with clear background at nanogram sensitivity using Coomassie Brilliant Blue G-250 and R-250." *Electrophoresis* 9(6): 255–262.

Neuhoff, V., R. Stamm, and H. Eibl (1985). "Clear background and highly sensitive protein staining with Coomassie Blue dyes in polyacrylamide gels: A systematic analysis." *Electrophoresis* 6(9): 427–448.

Nevalainen, T. J., G. G. Graham, and K. F. Scott (2008). "Antibacterial actions of secreted phospholipases A_2: Review." *Biochim Biophys Acta* 1781(1–2): 1–9.

Newman, W. J., N. F. Moran, R. D. G. Theakston, D. A. Warrell, and D. Wilkinson (1997). "Traditional treatments for snake bite in a rural African community." *Ann Trop Med Parasitol* 91(8): 967–969.

Ng, H. C., S. Ranganathan, K. L. Chua, and H. E. Khoo (2005). "Cloning and molecular characterization of the first aquatic hyaluronidase, SFHYA1, from the venom of stonefish (*Synanceja horrida*)." *Gene* 346: 71–81.

Nicastro, G., L. Franzoni, C. de Chiara, A. C. Mancin, J. R. Giglio, and A. Spisni (2003). "Solution structure of crotamine, a Na+ channel affecting toxin from *Crotalus durissus terrificus* venom." *Eur J Biochem* 270(9): 1969–1979.

Niewiarowski, S., E. P. Kirby, T. M. Brudzynski, and K. Stocker (1979). "Thrombocytin, a serine protease from *Bothrops atrox* venom, 2: Interaction with platelets and plasma-clotting factors." *Biochemistry* 18(16): 3570–3577.

Nikai, T., M. Momose, Y. Okumura, A. Ohara, Y. Komori, and H. Sugihara (1993). "Kallidin-releasing enzyme from *Bitis arietans* (puff adder) venom." *Arch Biochem Biophys* 307(2): 304–310.

Nikai, T., K. Taniguchi, Y. Komori, K. Masuda, J. W. Fox, and H. Sugihara (2000). "Primary structure and functional characterization of bilitoxin-1, a novel dimeric P-II snake venom metalloproteinase from *Agkistrodon bilineatus* venom." *Arch Biochem Biophys* 378(1): 6–15.

Niland, S., B. Ditkowski, D. Parrandier, L. Roth, H. Augustin, and J. A. Eble (2013). "Rhodocetin-alphabeta-induced neuropilin-1-cMet association triggers restructuring of matrix contacts in endothelial cells." *Arterioscler Thromb Vasc Biol* 33(3): 544–554.

Nirthanan, S., E. Charpantier, P. Gopalakrishnakone, M. C. Gwee, H. E. Khoo, L. S. Cheah, D. Bertrand, and R. M. Kini (2002). "Candoxin, a novel toxin from *Bungarus candidus*, is a reversible antagonist of muscle (alphabetagammadelta) but a poorly reversible antagonist of neuronal alpha 7 nicotinic acetylcholine receptors." *J Biol Chem* 277(20): 17811–17820.

Nishida, S., Y. Fujimura, S. Miura, Y. Ozaki, Y. Usami, M. Suzuki, K. Titani, E. Yoshida, M. Sugimoto, and A. Yoshioka (1994). "Purification and characterization of bothrombin, a fibrinogen-clotting serine protease from the venom of *Bothrops jararaca*." *Biochemistry* 33(7): 1843–1849.

Nishida, S., T. Fujita, N. Kohno, H. Atoda, T. Morita, H. Takeya, I. Kido, et al. (1995). "cDNA cloning and deduced amino acid sequence of prothrombin activator (ecarin) from Kenyan *Echis carinatus* venom." *Biochemistry* 34(5): 1771–1778.

Nishimura, H., S. Kawabata, W. Kisiel, S. Hase, T. Ikenaka, Y. Shimonishi, and S. Iwanaga (1989). "Identification of a disaccharide (Xyl-Glc) and a trisaccharide (Xyl2-Glc) O-glycosidically linked to a serine residue in the first epidermal growth factor-like domain of human factors VII and IX and protein Z and bovine protein Z." *J Biol Chem* 264: 20320–20325.

Nishio, H., E. Katoh, T. Yamazaki, T. Inui, Y. Nishiuchi, and T. Kimura (1999). "Structure-activity relationships of calcicludine and dendrotoxin-I, homologous peptides acting on different targets, calcium and potassium channels." *Biochem Biophys Res Comm* 262(2): 319–321.

Nobile, M., V. Magnelli, L. Lagostena, J. Mochca-Morales, L. D. Possani, and G. Prestipino (1994). "The toxin helothermine affects potassium currents in newborn rat cerebellar granule cells." *J Membr Biol* 139(1): 49–55.

Nobile, M., F. Noceti, G. Prestipino, and L. D. Possani (1996). "Helothermine, a lizard venom toxin, inhibits calcium current in cerebellar granules." *Exp Brain Res* 110: 15–20.

Noeske-Jungblut, C., B. Haendler, P. Donner, A. Alagón, L. Possani, and W. D. Schleuning (1995). "Triabin, a highly potent exosite inhibitor of thrombin." *J Biol Chem* 270(48): 28629–28634.

Noeske-Jungblut, C., J. Kratzschmar, B. Haendler, A. Alagón, L. Possani, P. Verhallen, P. Donner, and W. D. Schleuning (1994). "An inhibitor of collagen-induced platelet-aggregation from the saliva of *Triatoma pallidipennis*." *J Biol Chem* 269(7): 5050–5053.

Nolan, C., L. S. Hall, and G. H. Barlow (1976). "Ancrod, the coagulating enzyme from Malayan pit viper (*Agkistrodon rhodostoma*) venom." *Methods Enzymol* 45: 205–213.

Nolan, M. A., L. Wu, H. J. Bang, S. A. Jelinsky, K. P. Roberts, T. T. Turner, G. S. Kopf, and D. S. Johnston (2006). "Identification of rat cysteine-rich secretory protein 4 (Crisp4) as the ortholog to human CRISP1 and mouse Crisp4." *Biol Reprod* 74: 984–991.

Nomura, F., N. Kurobe, Y. Mori, A. Hikita, M. Kawai, M. Suwa, and Y. Okutani (2008). "Multicenter prospective investigation on efficacy and safety of carperitide as a first-line drug for acute heart failure syndrome with preserved blood pressure: COMPASS, Carperitide Effects Observed through Monitoring Dyspnea in Acute Decompensated Heart Failure Study." *Circ J* 72(11): 1777–1786.

Norris, R. L., J. Ngo, K. Nolan, and G. Hooker (2005). "Physicians and lay people are unable to apply pressure immobilization properly in a simulated snakebite scenario." *Wilderness Environ Med* 16(1): 16–21.

Novakova, L., and H. Vlckova (2009). "A review of current trends and advances in modern bio-analytical methods: Chromatography and sample preparation." *Anal Chim Acta* 656(1–2): 8–35.

Nowak, E. M., T. Hare, and J. McNally (2002). "Management of 'nuisance' vipers: Effects of translocation on Western diamond-backed rattlesnakes (*Crotalus atrox*)." In *Biology of the vipers*, edited by G. W. Schuett, M. Hoggren, M. E. Douglas, and H. W. Greene. Eagle Mountain, UT: Eagle Mountain, 535–560.

Nowak, G. (2003). "The ecarin clotting time, a universal method to quantify direct thrombin inhibitors." *Pathophysiol Haemost Thromb* 33(4): 173–183.

Nuñez, C. E., Y. Angulo, and B. Lomonte (2001). "Identification of the myotoxic site of the Lys49 phospholipase A(2) from *Agkistrodon piscivorus piscivorus* snake venom: Synthetic C-terminal peptides from Lys49, but not from Asp49 myotoxins, exert membrane-damaging activities." *Toxicon* 39(10): 1587–1594.

Nuñez, V., P. Cid, L. Sanz, P. De La Torre, Y. Angulo, B. Lomonte, J. M. Gutiérrez, and J. J. Calvete (2009). "Snake venomics and antivenomics of *Bothrops atrox* venoms from Colombia and the Amazon regions of Brazil, Peru and Ecuador suggest the occurrence of geographic variation of venom phenotype by a trend towards paedomorphism." *J Proteomics* 73(1): 57–78.

Nutt, E., T. Gasic, J. Rodkey, G. J. Gasic, J. W. Jacobs, P. A. Friedman, and E. Simpson (1988). "The amino acid sequence of antistasin: A potent inhibitor of factor Xa reveals a repeated internal structure." *J Biol Chem* 263(21): 10162–10167.

Nuyttens, B. P., T. Thijs, H. Deckmyn, and K. Broos (2011). "Platelet adhesion to collagen." *Thromb Res* 127 suppl 2: S26–29.

O'Shea, J. C., and J. E. Tcheng (2002). "Eptifibatide: A potent inhibitor of the platelet receptor integrin glycoprotein IIb/IIIa." *Expert Opin Pharmacother* 3(8): 1199–1210.

Obermajer, N., Z. Jevnikar, B. Doljak, and J. Kos (2008). "Role of cysteine cathepsins in matrix degradation and cell signalling." *Connect Tissue Res* 49(3): 193–196.

Ogawa, T., T. Chijiwa, N. Oda-Ueda, and M. Ohno (2005). "Molecular diversity and accelerated evolution of C-type lectin-like proteins from snake venom." *Toxicon* 45(1): 1–14.

Ogawa, Y., Y. Mamura, N. Murayama, and R. Yanoshita (2006). "Characterization and cDNA cloning of dipeptidyl peptidase IV from the venom of *Gloydius blomhoffi brevicaudus*." *Comp Biochem Physiol B Biochem Mol Biol* 145(1): 35–42.

Ogawa, Y., N. Murayama, Y. Fujita, and R. Yanoshita (2007). "Characterization and cDNA cloning of aminopeptidase A from the venom of *Gloydius blomhoffi brevicaudus*." *Toxicon* 49(8): 1172–1181.

Ogawa, Y., K. Nakao, O. Nakagawa, Y. Komatsu, K. Hosoda, S. Suga, H. Arai, K. Nagata, N. Yoshida, and H. Imura (1992). "Human C-type natriuretic peptide: Characterization of the gene and peptide." *Hypertension* 19(6 pt 2): 809–813.

Oguiura, N., M. Boni-Mitake, R. Affonso, and G. Zhang (2011). "*In vitro* antibacterial and hemolytic activities of crotamine, a small basic myotoxin from rattlesnake *Crotalus durissus*." *J Antibiot (Tokyo)* 64(4): 327–331.

Oguiura, N., M. E. Camargo, A. R. da Silva, and D. S. Horton (2000). "Quantification of crotamine, a small basic myotoxin, in South American rattlesnake (*Crotalus durissus terrificus*) venom by enzyme-linked immunosorbent assay with parallel-lines analysis." *Toxicon* 38(3): 443–448.

Oh, J. S., C. Han, and C. Cho (2009). "ADAM7 is associated with epididymosomes and integrated into sperm plasma membrane." *Mol Cells* 28(5): 441–446.

Ohtani, Y., Y. Yabuki, M. Mimura, and H. Takahashi (1988). "Purification of a kininogenase from the venom of *Agkistrodon caliginosus* (Kankoku-Mamushi)." *Toxicon* 26(10): 891–901.

Okuda, D., K. Horii, H. Mizuno, and T. Morita (2003). "Characterization and preliminary crystallographic studies of EMS16, an antagonist of collagen receptor (GPIa/IIa) from the venom of *Echis multisquamatus*." *J Biochem* 134(1): 19–23.

Okuda, D., H. Koike, and T. Morita (2002). "A new gene structure of the disintegrin family: A subunit of dimeric disintegrin has a short coding region." *Biochemistry* 41(48): 14248–14254.

Olfa, K. Z., L. Jose, D. Salma, B. Amine, S. A. Najet, A. Nicolas, L. Maxime, et al. (2005). "Lebestatin, a disintegrin from *Macrovipera* venom, inhibits integrin-mediated cell adhesion, migration and angiogenesis." *Lab Invest* 85(12): 1507–1516.

Olianas, M. C., A. Ingianni, C. Maullu, A. Adem, E. Karlsson, and P. Onali (1999). "Selectivity profile of muscarinic toxin 3 in functional assays of cloned and native receptors." *J Pharmacol Exp Ther* 288(1): 164–170.

Oliveira, J. S., A. J. Zaharenko, W. A. Ferreira Jr., K. Konno, C. S. Shida, M. Richardson, A. D. Lucio, P. S. Beirao, and J. C. de Freitas (2006). "BcIV, a new paralyzing peptide obtained from the venom of the sea anemone *Bunodosoma caissarum*: A comparison with the Na+ channel toxin BcIII." *Biochim Biophys Acta* 1764(10): 1592–1600.

Oliveira-Carvalho, A. L., P. R. Guimaraes, P. A. Abreu, D. L. Dutra, I. L. Junqueira de Azevedo, C. R. Rodrigues, P. L. Ho, H. C. Castro, and R. B. Zingali (2008). "Identification and characterization of a new member of snake venom thrombin inhibitors from *Bothrops insularis* using a proteomic approach." *Toxicon* 51(4): 659–671.

Olivera, B. M., and R. W. Teichert (2011). "Neuroscience: Chemical ecology of pain." *Nature* 479(7373): 306–307.

Olsen, J. V., B. Macek, O. Lange, A. Makarov, S. Horning, and M. Mann (2007). "Higher-energy C-trap dissociation for peptide modification analysis." *Nat Meth* 4(9): 709–712.

Olsen, J. V., S. E. Ong, and M. Mann (2004). "Trypsin cleaves exclusively C-terminal to arginine and lysine residues." *Mol Cell Proteomics* 3(6): 608–614.

Olson, J. H., X. Xiang, T. Ziegert, A. Kittelson, A. Rawls, A. L. Bieber, and D. E. Chandler (2001). "Allurin, a 21-kDa sperm chemoattractant from *Xenopus* egg jelly, is related to mammalian sperm-binding proteins." *Proc Natl Acad Sci USA* 98: 11205–11210.

Omori-Satoh, T., and S. Sadahiro (1979). "Resolution of the major hemorrhagic component of *Trimeresurus flavoviridis* venom into two parts." *Biochim Biophys Acta Bioenergetics* 580(2): 392–404.

OmPraba, G., A. Chapeaurouge, R. Doley, K. R. Devi, P. Padmanaban, C. Venkatraman, D. Velmurugan, Q. Lin, and R. M. Kini (2010). "Identification of a novel family of snake venom proteins Veficolins from *Cerberus rynchops* using a venom gland transcriptomics and proteomics approach." *J Proteome Res* 9(4): 1882–1893.

Ondetti, M. A. (1988). "Structural relationships of angiotensin converting-enzyme inhibitors to pharmacologic activity." *Circulation* 77(6 pt 2): I74–78.

Ondetti, M. A., and D. W. Cushman (1981). "Inhibition of the renin-angiotensin system: A new approach to the therapy of hypertension." *J Med Chem* 24(4): 355–361.

——— (1982). "Enzymes of the renin-angiotensin system and their inhibitors." *Annu Rev Biochem* 51: 283–308.

Ondetti, M. A., N. J. Williams, E. F. Sabo, J. Pluscec, E. R. Weaver, and O. Kocy (1971). "Angiotensin-converting enzyme inhibitors from the venom of *Bothrops jararaca*: Isolation, elucidation of structure, and synthesis." *Biochemistry* 10(22): 4033–4039.

Ono, S., T. Kimura, and T. Kubo (2011). "Characterization of voltage-dependent calcium channel blocking peptides from the venom of the tarantula *Grammostola rosea*." *Toxicon* 58(3): 265–276.

Ookuma, S., M. Fukuda, and E. Nishida (2003). "Identification of a DAF-16 transcriptional target gene, scl-1, that regulates longevity and stress resistance in *Caenorhabditis elegans*." *Curr Biol* 13: 427–431.

Orfeo, T., N. Brufatto, M. E. Nesheim, H. Xu, S. Butenas, and K. G. Mann (2004). "The factor V activation paradox." *J Biol Chem* 279(19): 19580–19591.

Ortel, T. L., M. A. Quinnallen, L. A. Charles, D. Devorecarter, and W. H. Kane (1992). "Characterization of an acquired inhibitor to coagulation factor-V—antibody-binding to the 2nd C-type domain of factor-V inhibits the binding of factor-V to phosphatidylserine and neutralizes procoagulant activity." *J Clin Invest* 90(6): 2340–2347.

Ortel, T. L., M. A. Quinnallen, F. G. Keller, J. A. Peterson, D. Larocca, and W. H. Kane (1994). "Localization of functionally important epitopes within the 2nd C-type domain of coagulation-factor V using recombinant chimeras." *J Biol Chem* 269(22): 15898–15905.

Orthner, C. L., P. Bhattacharya, and D. K. Strickland (1988). "Characterization of a protein C activator from the venom of *Agkistrodon contortrix contortrix*." *Biochemistry* 27(7): 2558–2564.

Osaka, H., S. Malany, B. E. Molles, S. M. Sine, and P. Taylor (2000). "Pairwise electrostatic interactions between alpha-neurotoxins and gamma, delta, and epsilon subunits of the nicotinic acetylcholine receptor." *J Biol Chem* 275(8): 5478–5484.

Oshikawa, K., and S. Terada (1999). "Ussuristatin 2, a novel KGD-bearing disintegrin from *Agkistrodon ussuriensis* venom." *J Biochem* 125(1): 31–35.

Osipov, A. V., M. V. Astapova, V. I. Tsetlin, and Y. N. Utkin (2004). "The first representative of glycosylated three-fingered toxins: Cytotoxin from the *Naja kaouthia* cobra venom." *Eur J Biochem* 271(10): 2018–2027.

Osipov, A. V., I. E. Kasheverov, Y. V. Makarova, V. G. Starkov, O. V. Vorontsova, R. Ziganshin, T. V. Andreeva, et al. (2008). "Naturally occurring disulfide-bound dimers of three-fingered toxins: A paradigm for biological activity diversification." *J Biol Chem* 283(21): 14571–14580.

Osipov, A. V., M. Y. Levashov, V. I. Tsetlin, and Y. N. Utkin (2005). "Cobra venom contains a pool of cysteine-rich secretory proteins." *Biochem Biophys Res Comm* 328: 177–182.

Osipov, A.V., P. Rucktooa, I. E. Kasheverov, S. Y. Filkin, V. G. Starkov, T. V. Andreeva, T. K. Sixma, et al. (2012). "Dimeric α-cobratoxin X-ray structure: Localization of intermolecular disulfides and possible mode of binding to nicotinic acetylcholine receptors." *J Biol Chem* 287(9): 6725–6734.

Osman, O. H., M. Ismail, and M. F. El-Asmar (1973). "Pharmacological studies of snake (*Dendroaspis angusticeps*) venom." *Toxicon* 11(2): 185–192.

Otero, R., R. Fonnegra, S. L. Jimenez, V. Nuñez, N. Evans, S. P. Alzate, M. E. Garcia, et al. (2000a). "Snakebites and ethnobotany in the northwest region of Colombia: Part I: Traditional use of plants." *J Ethnopharmacol* 71(3): 493–504.

Otero, R., J. M. Gutiérrez, G. Rojas, V. Nuñez, A. Diaz, E. Miranda, A. F. Uribe, et al. (1999). "A randomized blinded clinical trial of two antivenoms, prepared by caprylic acid or ammonium sulphate fractionation of IgG, in *Bothrops* and *Porthidium* snake bites in Colombia: Correlation between safety and biochemical characteristics of antivenoms." *Toxicon* 37(6): 895–908.

Otero, R., V. Nuñez, S. L. Jiménez, R. Fonnegra, R. G. Osorio, M. E. García, and A. Díaz (2000b). "Snakebites and ethnobotany in the northwest region of Colombia: Part II: Neutralization of lethal and enzimatic effects of Bothrops atrox venom." *J Ethnopharmacol* 71: 505–511.

Ouadid-Ahidouch, H., and A. Ahidouch (2008). "K+ channel expression in human breast cancer cells: Involvement in cell cycle regulation and carcinogenesis." *J Membr Biol* 221(1): 1–6.

Ouwens, P. A. (1912). "On a large *Varanus* species from the island of Komodo." *Bull Jardin Bot Buitenzorg* 6: 1–3.

Ovadia, M. (1987). "Isolation and characterization of a hemorrhagic factor from the venom of the snake *Atractaspis engaddensis* (Atractaspididae)." *Toxicon* 25(6): 621–630.

Ownby, C. (1998). "Structure, function, and biophysical aspects of myotoxins from snake venoms." *J Toxicol Tox Rev* 17: 213–238.

Ownby, C. L., H. S. Selistre de Araujo, S. P. White, and J. E. Fletcher (1999). "Lysine 49 phospholipase A_2 proteins." *Toxicon* 37(3): 411–445.

Oyama, E., T. Fukuda, and H. Takahashi (2008). "Amino acid sequence of a kinin-releasing enzyme, KR-E-1, from the venom of *Agkistrodon caliginosus* (Kankoku-mamushi)." *Toxicon* 52(5): 651–654.

Oyama, E., and H. Takahashi (2003). "Purification and characterization of a thrombin like enzyme, elegaxobin II, with lys-bradykinin releasing activity from the venom of *Trimeresurus elegans* (Sakishima-Habu)." *Toxicon* 41(5): 559–568.

Paaventhan, P., C. Kong, J. S. Joseph, M. C. Chung, and P. R. Kolatkar (2005). "Structure of rhodocetin reveals noncovalently bound heterodimer interface." *Protein Sci* 14(1): 169–175.

Paes Leme, A. F., B. C. Prezoto, E. T. Yamashiro, L. Bertholim, A. K. Tashima, C. F. Klitzke, A. C. Camargo, and S. M. Serrano (2008). "*Bothrops* protease A, a unique highly glycosylated serine proteinase, is a potent, specific fibrinogenolytic agent." *J Thromb Haemost* 6(8): 1363–1372.

Paesen, G. C., P. L. Adams, K. Harlos, P. A. Nuttall, and D. I. Stuart (1999). "Tick histamine-binding proteins: Isolation, cloning, and three-dimensional structure." *Mol Cell* 3(5): 661–671.

Paesen, G. A., P. L. Adams, P. A. Nuttall, and D. L. Stuart (2000). "Tick histamine-binding proteins: Lipocalins with a second binding cavity." *Biochim Biophys Acta Protein Struct Mol Enzymol* 1482(1–2): 92–101.

Page, M. J., and E. Di Cera (2008). "Serine peptidases: Classification, structure and function." *Cell Mol Life Sci* 65(7–8): 1220–1236.

Pahari, S., S. P. Mackessy, and R. M. Kini (2007). "The venom gland transcriptome of the Desert Massasauga rattlesnake (*Sistrurus catenatus edwardsii*): Towards an understanding of venom composition among advanced snakes (Superfamily Colubroidea)." *BMC Mol Biol* 8: 115.

Paine, M. J., H. P. Desmond, R. D. Theakston, and J. M. Crampton (1992a). "Gene expression in *Echis carinatus* (carpet viper) venom glands following milking." *Toxicon* 30: 379–386.

——— (1992b). "Purification, cloning, and molecular characterization of a high molecular weight hemorrhagic metalloprotease, jararhagin, from *Bothrops jararaca* venom: Insights into the disintegrin gene family." *J Biol Chem* 267 (32): 22869–22876.

Paniagua, D., L. Jimenez, C. Romero, I. Vergara, A. Calderon, M. Bénard, M. J. Bernas, et al. (2012). "Lymphatic route of transport and pharmacokinetics of *Micrurus fulvius* (coral snake) venom in sheep." *Lymphology* 45(4): 144–153.

Páramo, L., B. Lomonte, J. Pizarro-Cerda, J. A. Bengoechea, J. P. Gorvel, and E. Moreno (1998). "Bactericidal activity of Lys49 and Asp49 myotoxic phospholipases A_2 from *Bothrops asper* snake venom—synthetic Lys49 myotoxin II-(115-129)-peptide identifies its bactericidal region." *Eur J Biochem* 253(2): 452–461.

Pare, A., K. A. Coyle, L. Sigler, A. K. Maas III, and R. L. Mitchell (2006). "Pathogenicity of the *Chrysosporium* anamorph of *Nannizziopsis vriesii* for veiled chameleons (*Chamaeleo calyptratus*)." *Med Mycol* 44(1): 25–31.

Pare, J. (2014). "Update on fungal infections in reptiles." In *Current therapy in reptile medicine and surgery*, edited by D. R. Mader and S. J. Divers. St. Louis, MO: Saunders Elsevier, 53–56.

Pare, J. A., and E. R. Jacobson (2007). "Mycotic diseases of reptiles." In *Infectious diseases and pathology of reptiles: A color atlas and text*," edited by E. R. Jacobson. Boca Raton, FL: CRC Press, 527–570.

Pare, J. A., and L. Sigler (2002). "Fungal diseases of reptiles." In *Reptile medicine and surgery*, 2nd ed., edited by D. R. Mader. Philadelphia: W. B. Saunders, 217–226.

Paré, J. A., L. Sigler, K. L. Rypien, and C. F. C. Gibas (2003). "Cutaneous mycobiota of captive squamate reptiles with notes on the scarcity of *Chrysosporium* anamorph of *Nannizziopsis vriesii*." *J Herpetol Med Surg* 13(4): 10–15.

Park, D., H. Kim, K. Chung, D. S. Kim, and Y. Yun (1998). "Expression and characterization of a novel plasminogen activator from *Agkistrodon halys* venom." *Toxicon* 36(12): 1807–1819.

Parry, M. A., U. Jacob, R. Huber, A. Wisner, C. Bon, and W. Bode (1998). "The crystal structure of the novel snake venom plasminogen activator TSV-PA: A prototype structure for snake venom serine proteinases." *Structure* 6(9): 1195–1206.

Pasmans, F., S. Blahak, A. Martel, and N. Pantchev (2008). "Introducing reptiles into a captive collection: The role of the veterinarian." *Vet J* 175(1): 53–68.

Patino, A. C., J. A. Pereanez, V. Nuñez, D. M. Benjumea, M. Fernandez, A. Rucavado, L. Sanz, and J. J. Calvete (2010). "Isolation and biological characterization of Batx-I, a weak hemorrhagic and fibrinogenolytic PI metalloproteinase from Colombian *Bothrops atrox* venom." *Toxicon* 56(6): 936–943.

Patthy, L. (2003). "Modular assembly of genes and the evolution of new functions." *Genetica* 118(2–3): 217–231.

Paul, K., J. Manjula, E. P. Deepa, Z. E. Selvanayagam, K. A. Ganesh, and P. V. Subba Rao (2007). "Anti-*Echis carinatus* venom antibodies from chicken egg yolk: Isolation, purification and neutralization efficacy." *Toxicon* 50(7): 893–900.

Pavlopoulou, A., G. Pampalakis, I. Michalopoulos, and G. Sotiropoulou (2010). "Evolutionary history of tissue kallikreins." *PLoS One* 5(11): e13781.

Pawlak, J., S. P. Mackessy, B. G. Fry, M. Bhatia, G. Mourier, C. Fruchart-Gaillard, D. Servent, et al. (2006). "Denmotoxin, a three-finger toxin from the colubrid snake *Boiga dendrophila* (Mangrove Catsnake) with bird-specific activity." *J Biol Chem* 281(39): 29030–29041.

Pawlak, J., S. P. Mackessy, N. M. Sixberry, E. A. Stura, M. H. Le Du, R. Ménez, C. S. Foo, A. Ménez, S. Nirthanan, and R. M. Kini (2009). "Irditoxin, a novel covalently linked heterodimeric three-finger toxin with high taxon-specific neurotoxicity." *FASEB J* 23(2): 534–545.

Payne, R. W., B. M. Murphy, and M. C. Manning (2011). "Product development issues for PEGylated proteins." *Pharm Dev Technol* 16(5): 423–440.

Payne, V., and P. C. Kam (2004). "Mast cell tryptase: A review of its physiology and clinical significance." *Anaesthesia* 59(7): 695–703.

Peach, M. J. (1977). "Renin-angiotensin system: Biochemistry and mechanisms of action." *Physiol Rev* 57(2): 313–370.

Pearce, F. L., B. E. Banks, D. V. Banthorpe, A. R. Berry, H. S. Davies, and C. A. Vernon (1972). "The isolation and characterization of nerve-growth factor from the venom of *Vipera russelli*." *Eur J Biochem* 29(3): 417–425.

Pearce, F. L., and H. L. Thompson (1986). "Some characteristics of histamine secretion from rat peritoneal mast cells stimulated with nerve growth factor." *J Physiol* 372: 379–393.

Pearson, J. A., M. I. Tyler, K. V. Retson, and M. E. Howden (1993). "Studies on the subunit structure of textilotoxin, a potent presynaptic neurotoxin from the venom of the Australian common brown snake (*Pseudonaja textilis*), 3: The complete amino-acid sequences of all the subunits." *Biochim Biophys Acta* 1161(2–3): 223–229.

Peigneur, S., B. Billen, R. Derua, E. Waelkens, S. Debaveye, L. Beress, and J. Tytgat (2011). "A bifunctional sea anemone peptide with Kunitz type protease and potassium channel inhibiting properties." *Biochem Pharmacol* 82(1): 81–90.

Peigneur, S., D. J. Orts, A. R. Prieto da Silva, N. Oguiura, M. Boni-Mitake, E. B. de Oliveira, A. J. Zaharenko, J. C. de Freitas, and J. Tytgat (2012a). "Crotamine pharmacology revisited: Novel insights based on the inhibition of KV channels." *Mol Pharmacol* 82(1): 90–96.

Peigneur, S., L. Béress, C. Möller, F. Marí, W. G. Forssmann, and J. Tytgat (2012b). "A natural point mutation changes both target selectivity and mechanism of action of sea anemone toxins." *FASEB J* 26(12): 5141–5151.

Pejkic, S., N. Savic, M. Paripovic, M. Sladojevic, P. Doric, and N. Ilic (2014). "Vascular graft thrombosis secondary to activated protein C resistance: A case report and literature review." *Vascular* 22(1): 71–76.

Peng, J., and J. Xu (2011). "RaptorX: Exploiting structure information for protein alignment by statistical inference." *Proteins* 79 suppl 10: 161–171.

Peng, M., F. A. Emig, A. Mao, W. Lu, E. P. Kirby, S. Niewiarowski, and M. A. Kowalska (1995). "Interaction of echicetin with a high affinity thrombin binding site on platelet glycoprotein GPIb." *Thromb Haemost* 74(3): 954–957.

Peng, M., J. C. Holt, and S. Niewiarowski (1994). "Isolation, characterization and amino acid sequence of echicetin beta subunit, a specific inhibitor of von Willebrand factor and thrombin interaction with glycoprotein Ib." *Biochem Biophys Res Commun* 205(1): 68–72.

Peng, M., W. Lu, L. Beviglia, S. Niewiarowski, and E. P. Kirby (1993). "Echicetin: A snake venom protein that inhibits binding of von Willebrand factor and alboaggregins to platelet glycoprotein Ib." *Blood* 81(9): 2321–2328.

Peng, M., W. Lu, and E. P. Kirby (1991). "Alboaggregin-B: A new platelet agonist that binds to platelet membrane glycoprotein Ib." *Biochemistry* 30(49): 11529–11536.

Pennisi, E. (2003). "Drafting a tree." *Science* 300(5626): 1694.

Perbandt, M., J. C. Wilson, S. Eschenburg, I. Mancheva, B. Aleksiev, N. Genov, P. Willingmann, W. Weber, T. P. Singh, and C. Betzel (1997). "Crystal structure of vipoxin at 2.0 A: An example of regulation of a toxic function generated by molecular evolution." *FEBS Lett* 412(3): 573–577.

Perchuc, A. M., and M. M. Wilmer (2010). "Diagnostic use of snake venom components in the coagulation laboratory." In *Toxins and hemostasis*, edited by K. J. Clemetson, R. M. Kini, F. S. Markland, M. A. McLane, and T. Morita. Netherlands: Springer, 747–766.

Pereira, A., A. Kerkis, M. A. Hayashi, A. S. Pereira, F. S. Silva, E. B. Oliveira, A. R. Prieto da Silva, T. Yamane, G. Rádis-Baptista, and I. Kerkis (2011). "Crotamine toxicity and efficacy in mouse models of melanoma." *Expert Opin Investig Drugs* 20(9): 1189–1200.

Perez-Polo, J. R., H. Bomar, C. Beck, and K. Hall (1978). "Nerve growth factor from *Crotalus adamenteus* snake venom." *J Biol Chem* 253(17): 6140–6148.

Perkins, D. N., D. J. C. Pappin, D. M. Creasy, and J. S. Cottrell (1999). "Probability-based protein identification by searching sequence databases using mass spectrometry data." *Electrophoresis* 20(18): 3551–3567.

Perona, J. J., and C. S. Craik (1995). "Structural basis of substrate specificity in the serine proteases." *Protein Sci* 4(3): 337–360.

Perumal Samy, R., P. Gopalakrishnakone, M. M. Thwin, T. K. Chow, H. Bow, E. H. Yap, and T. W. Thong (2007). "Antibacterial activity of snake, scorpion and bee venoms: A comparison with purified venom phospholipase A_2 enzymes." *J Appl Microbiol* 102(3): 650–659.

Petan, T., I. Krizaj, M. H. Gelb, and J. Pungercar (2005). "Ammodytoxins, potent presynaptic neurotoxins, are also highly efficient phospholipase A_2 enzymes." *Biochemistry* 44(37): 12535–12545.

Petan, T., I. Krizaj, and J. Pungercar (2007). "Restoration of enzymatic activity in a Ser-49 phospholipase A_2 homologue decreases its Ca(2+)-independent membrane-damaging activity and increases its toxicity." *Biochemistry* 46(44): 12795–12809.

Petras, D., L. Sanz, A. Segura, M. Herrera, M. Villalta, D. Solano, M. Vargas, et al. (2011). "Snake venomics of African spitting cobras: Toxin composition and assessment of congeneric cross-reactivity of the pan-African EchiTAb-Plus-ICP antivenom by antivenomics and neutralization approaches." *J Proteome Res* 10(3): 1266–1280.

Petrey, D., Z. Xiang, C. L. Tang, L. Xie, M. Gimpelev, T. Mitros, C. S. Soto, et al. (2003). "Using multiple structure alignments, fast model building, and energetic analysis in fold recognition and homology modeling." *Proteins* 53 suppl 6: 430–435.

Petrova, S. D., V. N. Atanasov, and K. Balashev (2012). "Vipoxin and its components: Structure-function relationship." *Adv Protein Chem Struct Biol* 87: 117–153.

Pettersson, A., S. Nordlander, G. Nylund, A. Khorram-Manesh, S. Nordgren, and D. S. Delbro (2008). "Expression of the endogenous, nicotinic acetylcholine receptor ligand, SLURP-1, in human colon cancer." *Auton Autacoid Pharmacol* 28(4): 109–116.

Philippe, G., L. Angenot, M. Tits, and M. Frederich (2004). "About the toxicity of some *Strychnos* species and their alkaloids." *Toxicon* 44(4): 405–416.

Phisalix, C., and G. Bertrand (1894a). "Propriétés antitoxique du sang des animaux vaccineé contre le venin de vipère: Contribution à l'étude du mécanisme de la vaccination contre ce venin [Antitoxical properties of blood from vaccinated animal against viper venom : Contribution to the study of the mechanism of vaccination against this venom]." *Archives de Physiologie* 6: 611–619.

——— (1894b). "Sur la propriété antitoxique du sang des animaux vaccinée contre le venin de vipére [On the antitoxical property of blood from vaccinated animal against viper venom]." *C R Soc Biol* 46: 111–113.

Phisalix, M., and R. Caius (1918). "L'extension de la fonction venimeuse dans l'ordre entière des ophidiens et son existence chez des familles ou elle n'avait pas été soupçonnée jusqu'içi [Expanding of the venomous function in the entire order of ophidians and its existence in families where it was not suspected previously]." *J Physiol Pathol Gen* 17: 923–964.

Pillet, L., O. Tremeau, F. Ducancel, P. Drevet, S. Zinn-Justin, S. Pinkasfeld, J. C. Boulain, and A. Ménez (1993). "Genetic engineering of snake toxins: Role of invariant residues in the structural and functional properties of a curaremimetic toxin, as probed by site-directed mutagenesis." *J Biol Chem* 268(2): 909–916.

Pils, B., and J. Schultz (2004). "Inactive enzyme-homologues find new function in regulatory processes." *J Mol Biol* 340(3): 399–404.

Pinckard, R. N., R. A. O'Rourke, M. H. Crawford, F. S. Grover, L. M. McManus, J. J. Ghidoni, S. B. Storrs, and M. S. Olson (1980). "Complement localization and mediation of ischemic injury in baboon myocardium." *J Clin Invest* 66(5): 1050–1056.

Piojan, M. (2008). "Antídotos tribales herencia milenaria." *Etnofarmacia* 27: 105–109.

Piskurek, O., C. C. Austin, and N. Okada (2006). "Sauria SINEs: Novel short interspersed retroposable elements that are widespread in reptile genomes." *J Mol Evol* 62(5): 630–644.

Pla, D., J. M. Gutiérrez, and J. J. Calvete (2012). "Second generation snake antivenomics: Comparing immunoaffinity and immunodepletion protocols." *Toxicon* 60(4): 688–699.

Planes, C., and G. H. Caughey (2007). "Regulation of the epithelial Na+ channel by peptidases." *Curr Top Dev Biol* 78: 23–46.

Plummer, M. V., and N. E. Mills (2000). "Spatial ecology and survivorship of resident and translocated hognose snakes (*Heterodon platirhinos*)." *Journal of Herpetology*: 565–575.

Plummer, T. H. Jr., and M. Y. Hurwitz (1978). "Human plasma carboxypeptidase N: Isolation and characterization." *J Biol Chem* 253(11): 3907–3912.

Poillot, C., K. Dridi, H. Bichraoui, J. Pecher, S. Alphonse, B. Douzi, M. Ronjat, H. Darbon, and M. De Waard (2010). "D-Maurocalcine, a pharmacologically inert efficient cell-penetrating peptide analogue." *J Biol Chem* 285(44): 34168–34180.

Polgar, J., E. M. Magnenat, M. C. Peitsch, T. N. Wells, and K. J. Clemetson (1996). "Asp-49 is not an absolute prerequisite for the enzymic activity of low-M(r) phospholipases A_2: Purification, characterization and computer modelling of an enzymically active Ser-49 phospholipase A_2, ecarpholin S, from the venom of *Echis carinatus sochureki* (saw-scaled viper)." *Biochem J* 319(pt 3): 961–968.

Pond, S. L. K., S. D. W. Frost, and S. V. Muse (2005). "HyPhy: Hypothesis testing using phylogenies." *Bioinformatics* 21(5): 676–679.

Pool, W. R., and A. L. Bieber (1981). "Fractionation of midget faded rattlesnake (*Crotalus viridis concolor*) venom: Lethal fractions and enzymatic activities." *Toxicon* 19(4): 517–527.

Porschke, D., C. Creminon, X. Cousin, C. Bon, J. Sussman, and I. Silman (1996). "Electrooptical measurements demonstrate a large permanent dipole moment associated with acetylcholinesterase." *Biophys J* 70(4): 1603–1608.

Possani, L. D., B. M. Martin, A. Yatani, J. Mochca-Morales, F. Z. Zamudio, G. B. Gurrola, and A. M. Brown (1992). "Isolation and physiological characterization of taicatoxin, a complex toxin with specific effects on calcium channels." *Toxicon* 30(11): 1343–1364.

Powell, R. H. (1853). "Two fatal cases of snake-bite: With remarks." *Assoc Med J* 1(35): 773–774.

Prasarnpun, S., J. Walsh, S. S. Awad, and J. B. Harris (2005). "Envenoming bites by kraits: The biological basis of treatment-resistant neuromuscular paralysis." *Brain* 128(pt 12): 2987–2996.

Pung, Y. F., P. T. H. Wong, P. P. Kumar, W. C. Hodgson, and R. M. Kini (2005). "Ohanin, a novel protein from king cobra venom, induces hypolocomotion and hyperalgesia in mice." *J Biol Chem* 280(13): 13137–13147.

Pyron, R. A., and F. T. Burbrink (2012). "Extinction, ecological opportunity, and the origins of global snake diversity." *Evolution* 66(1): 163–178.

Pyron, R. A., F. T. Burbrink, G. R. Colli, A. N. de Oca, L. J. Vitt, C. A. Kuczynski, and J. J. Wiens (2011). "The phylogeny of advanced snakes (Colubroidea), with discovery of a new subfamily and comparison of support methods for likelihood trees." *Mol Phylogenet Evol* 58(2): 329–342.

Pyron, R. A., F. T. Burbrink, and J. J. Wiens (2013). "A phylogeny and revised classification of *Squamata*, including 4161 species of lizards and snakes." *BMC Evol Biol* 13: 93.

Pyz, E., A. S. Marshall, S. Gordon, and G. D. Brown (2006). "C-type lectin-like receptors on myeloid cells." *Ann Med* 38(4): 242–251.

Qu, H., D. Ricklin, and J. D. Lambris (2009). "Recent developments in low molecular weight complement inhibitors." *Mol Immunol* 47(2–3): 185–195.

Quan, A. N., D. Quan, and S. C. Curry (2010). "Improving Crotalidae polyvalent immune Fab reconstitution times." *Am J Emerg Med* 28(5): 593–595.

Quinton, L., N. Gilles, N. Smargiasso, A. Kiehne, and E. De Pauw (2011). "An unusual family of glycosylated peptides isolated from *Dendroaspis angusticeps* venom and characterized by combination of collision induced and electron transfer dissociation." *J Am Soc Mass Spectrom* 22(11): 1891–1897.

Quinton, L., E. Girard, A. Maiga, M. Rekik, P. Lluel, G. Masuyer, M. Larregola, et al. (2010). "Isolation and pharmacological characterization of AdTx1, a natural peptide displaying specific insurmountable antagonism of the alpha1A-adrenoceptor." *Br J Pharmacol* 159(2): 316–325.

Quinton, L., J. P. Le Caer, G. Phan, C. Ligny-Lemaire, J. Bourdais-Jomaron, F. Ducancel, and J. Chamot-Rooke (2005). "Characterization of toxins within crude venoms by combined use of Fourier transform mass spectrometry and cloning." *Anal Chem* 77(20): 6630–6639.

Rabilloud, T., L. Vuillard, C. Gilly, and J. J. Lawrence (1994). "Silver-staining of proteins in polyacrylamide gels: A general overview." *Cell Mol Biol (Noisy-le-Grand)* 40(1): 57–75.

Rádis-Baptista, G., B. G. de la Torre, and D. Andreu (2008). "A novel cell-penetrating peptide sequence derived by structural minimization of a snake toxin exhibits preferential nucleolar localization." *J Med Chem* 51(22): 7041–7044.

——— (2012). "Insights into the uptake mechanism of NrTP, a cell-penetrating peptide preferentially targeting the nucleolus of tumour cells." *Chem Biol Drug Des* 79(6): 907–915.

Rádis-Baptista, G., and I. Kerkis (2011). "Crotamine, a small basic polypeptide myotoxin from rattlesnake venom with cell-penetrating properties." *Curr Pharm Des* 17(38): 4351–4361.

Rádis-Baptista, G., T. Kubo, N. Oguiura, A. R. Prieto da Silva, M. A. Hayashi, E. B. Oliveira, and T. Yamane (2004). "Identification of crotasin, a crotamine-related gene of *Crotalus durissus terrificus*." *Toxicon* 43(7): 751–759.

Rádis-Baptista, G., T. Kubo, N. Oguiura, M. Svartman, T. M. B. Almeida, R. F. Batistic, E. B. Oliveira, A. M. Vianna-Morgante, and T. Yamane (2003). "Structure and chromosomal localization of the gene for crotamine, a toxin from the South American rattlesnake, *Crotalus durissus terrificus*." *Toxicon* 42: 747–752.

Rádis-Baptista, G., N. Oguiura, M. A. Hayashi, M. E. Camargo, K. F. Grego, E. B. Oliveira, and T. Yamane (1999). "Nucleotide sequence of crotamine isoform precursors from a single South American rattlesnake (*Crotalus durissus terrificus*)." *Toxicon* 37(7): 973–984.

Rae, J. L., and R. A. Levis (1992). "Glass technology for patch clamp electrodes." *Methods Enzymol* 207: 66–92.

Raibekas, A. A., and V. Massey (1998). "Primary structure of the snake venom L-amino acid oxidase shows high homology with the mouse B cell interleukin 4-induced Figl protein." *Biochem Biophys Res Commun* 248(3): 476–478.

Rajagopalan, N., Y. F. Pung, Y. Z. Zhu, P. T. Wong, P. P. Kumar, and R. M. Kini (2007). "Beta-cardiotoxin: A new three-finger toxin from *Ophiophagus hannah* (king cobra) venom with beta-blocker activity." *FASEB J* 21(13): 3685–3695.

Rajeev, S., D. A. Sutton, B. L. Wickes, D. L. Miller, D. Giri, M. Van Meter, E. H. Thompson, et al. (2009). "Isolation and characterization of a new fungal species, *Chrysosporium ophiodiicola*, from a mycotic granuloma of a black rat snake (*Elaphe obsoleta obsoleta*)." *J Clin Microbiol* 47(4): 1264–1268.

Ramos-Cerrillo, B., A. R. de Roodt, J. P. Chippaux, L. Olguin, A. Casasola, G. Guzman, J. Paniagua-Solis, A. Alagón, and R. P. Stock (2008). "Characterization of a new polyvalent antivenom (*Antivipmyn Africa*) against African vipers and elapids." *Toxicon* 52(8): 881–888.

Ranawaka, U. K., D. G. Lalloo, and H. J. de Silva (2013). "Neurotoxicity in snakebite—the limits of our knowledge." *PLoS Negl Trop Dis* 7(10): e2302.

Randolph, A., S. H. Chamberlain, H. L. Chu, A. D. Retzios, F. S. Markland Jr., and F. R. Masiarz (1992). "Amino acid sequence of fibrolase, a direct-acting fibrinolytic enzyme from *Agkistrodon contortrix contortrix* venom." *Protein Sci* 1(5): 590–600.

Rankin, T. L., K. J. Tsuruta, M. K. Holland, M. D. Griswold, and M. C. Orgebin-Crist (1992). "Isolation, immunolocalization, and sperm-association of three proteins of 18, 25, and 29 kilodaltons secreted by the mouse epididymis." *Biol Reprod* 46: 747–766.

Rao, V. S., J. S. Joseph, and R. M. Kini (2003). "Group D prothrombin activators from snake venom are structural homologues of mammalian blood coagulation factor Xa." *Biochem J* 369(3): 635–642.

Rao, V. S., and R. M. Kini (2002). "Pseutarin C, a prothrombin activator from *Pseudonaja textilis* venom: Its structural and functional similarity to mammalian coagulation factor Xa-Va complex." *Thromb Haemost* 88(4): 611–619.

Rao, V. S., S. Swarup, and R. M. Kini (2003). "The nonenzymatic subunit of pseutarin C, a prothrombin activator from Eastern Brown snake (*Pseudonaja textilis*) venom, shows structural similarity to mammalian coagulation factor V." *Blood* 102(4): 1347–1354.

——— (2004). "The catalytic subunit of pseutarin C, a group C prothrombin activator from the venom of *Pseudonaja textilis*, is structurally similar to mammalian blood coagulation factor Xa." *Thromb Haemost* 92: 509–521.

Rasquinha, D. (1996). "Snake stone for snake envenomization." *Am J Emerg Med* 14(1): 112–113.

Raufman, J. P. (1996). "Bioactive peptides from lizard venoms." *Regul Pept* 61(1): 1–18.

Raufman, J. P., L. Singh, G. Singh, and J. Eng (1992). "Truncated glucagon-like peptide-1 interacts with exendin receptors on dispersed acini from guinea-pig pancreas—identification of a mammalian analog of the reptilian peptide exendin-4." *J Biol Chem* 267(30): 21432–21437.

Rawlings, N. D., A. J. Barrett, and A. Bateman (2010). "MEROPS: The peptidase database." *Nucleic Acids Res* 38: D227–233.

——— (2012). "MEROPS: The database of proteolytic enzymes, their substrates and inhibitors." *Nucleic Acids Res* 40: D343–350.

Rawlings, N. D., D. P. Tolle, and A. J. Barrett (2004). "MEROPS: The peptidase database." *Nucleic Acids Res* 32 (database issue): D160–164.

Read, M. R. (2004). "Evaluation of the use of anesthesia and analgesia in reptiles." *J Am Vet Med Assoc* 224(4): 547–552.

Read, M. S., S. V. Smith, M. A. Lamb, and K. M. Brinkhous (1989). "Role of botrocetin in platelet agglutination: Formation of an activated complex of botrocetin and von Willebrand factor." *Blood* 74(3): 1031–1035.

REAP (Rural Extension with Africa's Poor) (2013). "Natural Medicines Teaching Leaflets: Black Stone." http://reap-eastafrica.org/reap/?p=20.

Redi, F. (1671). "Experiences relating to various natural things, in particular those that are brought to us from the Indies." Letter to the Highly Revered Father Anastasion Chircher of the Society of Jesus. Florence.

Reichl, A. P., S. M. Serrano, C. A. Sampaio, and F. R. Mandelbaum (1993). "Hydrolytic specificity of three basic proteinases isolated from the venom of *Bothrops moojeni* for the B-chain of oxidized insulin." *Toxicon* 31(11): 1479–1482.

Reid, P. F. (2007). "Alpha-cobratoxin as a possible therapy for multiple sclerosis: A review of the literature leading to its development for this application." *Crit Rev Immunol* 27(4): 291–302.

Reinert, H. K., and R. R. Rupert Jr. (1999). "Impacts of translocation on behavior and survival of timber rattlesnakes, *Crotalus horridus*." *J Herpetol*: 45–61.

Reis, C. V., S. A. Andrade, O. H. P. Ramos, C. R. R. Ramos, P. L. Ho, I. D. C. Batista, and A. M. Chudzinski-Tavassi (2006). "Lopap, a prothrombin activator from *Lonomia obliqua* belonging to the lipocalin family: Recombinant production, biochemical characterization and structure-function insights." *Biochem J* 398: 295–302.

Reiter, R. E., Z. Gu, T. Watabe, G. Thomas, K. Szigeti, E. Davis, M. Wahl, et al. (1998). "Prostate stem cell antigen: A cell surface marker overexpressed in prostate cancer." *Proc Natl Acad Sci USA* 95(4): 1735–1740.

Rendón-Anaya, M., L. Delaye, L. D. Possani, and A. Herrera-Estrella (2012). "Global transcriptome analysis of the scorpion *Centruroides noxius*: New toxin families and evolutionary insights from an ancestral scorpion species." *PLoS One* 7(8): e43331.

Renetseder, R., S. Brunie, B. W. Dijkstra, J. Drenth, and P. B. Sigler (1985). "A comparison of the crystal structures of phospholipase A_2 from bovine pancreas and *Crotalus atrox* venom." *J Biol Chem* 260(21): 11627–11634.

Repnik, U., and B. Turk (2011). "Lysosomal-mitochondrial cross-talk during cell death." *Mitochondrion* 10(6): 662–669.

Reyes-Velasco, J., D. C. Card, A. L. Andrew, K. J. Shaney, R. J. Adams, D. R. Schield, N. R. Casewell, S. P. Mackessy, and T. A. Castoe (2014). "Expression of venom gene homologs in diverse python tissues suggests a new model for the evolution of snake venom." *Mol Biol Evol*. In press. doi: 10.1093/molbev/msu294.

Reza, M. A., T. N. Minh Le, S. Swarup, and R. M. Kini (2006). "Molecular evolution caught in action: Gene duplication and evolution of molecular isoforms of prothrombin activators in *Pseudonaja textilis* (brown snake)." *J Thromb Haemost* 4: 1346–1353.

Reza, M. A., S. Swarup, and R. M. Kini (2005). "Two parallel prothrombin activator systems in Australian rough-scaled snake, *Tropidechis carinatus*." *Thromb Haemost* 93: 40–47.

——— (2006). "Structure of two genes encoding parallel prothrombin activators in *Tropidechis carinatus* snake: Gene duplication and recruitment of factor X gene to the venom gland." *J Thromb Haemost* 5: 117–126.

Ribeiro, J. M. C., F. Alarcon-Chaidez, I. M. B. Francischetti, B. J. Mans, T. N. Mather, J. G. Valenzuela, and S. K. Wikel (2006). "An annotated catalog of salivary gland transcripts from *Ixodes scapularis* ticks." *Insect Biochem Mol Biol* 36(2): 111–129.

Ribeiro, J. M. C., J. Andersen, M. A. C. Silva-Neto, V. M. Pham, M. K. Garfield, and J. G. Valenzuela (2004a). "Exploring the sialome of the blood-sucking bug *Rhodnius prolixus*." *Insect Biochem Mol Biol* 34: 61–79.

Ribeiro, J. M. C., B. Arcà, F. Lombardo, E. Calvo, V. M. Phan, P. K. Chandra, and S. K. Wikel (2007). "An annotated catalogue of salivary gland transcripts in the adult female mosquito, *Aedes aegypti*." *BMC Genomics* 8: 6.

Ribeiro, J. M. C., R. Charlab, V. M. Pham, M. Garfield, and J. G. Valenzuela (2004b). "An insight into the salivary transcriptome and proteome of the adult female mosquito *Culex pipiens quinquefasciatus*." *Insect Biochem Mol Biol* 34(6): 543–563.

Ribeiro, J. M. C., M. Schneider, and J. A. Guimaraes (1995). "Purification and characterization of prolixin-s (nitrophorin-2), the salivary anticoagulant of the bloodsucking bug *Rhodnius prolixus*." *Biochem J* 308: 243–249.

Ribeiro, J. M. C., and F. A. Walker (1994). "High-affinity histamine-binding and antihistaminic activity of the salivary nitric oxide-carrying heme protein (nitrophorin) of *Rhodnius prolixus*." *J Exp Med* 180(6): 2251–2257.

Ricci-Silva, M. E., R. H. Valente, I. R. Leon, D. V. Tambourgi, O. H. P. Ramos, J. Perales, and A. M. Chudzinski-Tavassi (2008). "Immunochemical and proteomic technologies as tools for unravelling toxins involved in envenoming by accidental contact with *Lonomia obliqua* caterpillars." *Toxicon* 51(6): 1017–1028.

Richards, D. P., A. Barlow, and W. Wüster (2012). "Venom lethality and diet: Differential responses of natural prey and model organisms to the venom of the saw-scaled vipers (*Echis*)." *Toxicon* 59(1): 110–116.

Richards, R., L. St Pierre, M. Trabi, L. A. Johnson, J. de Jersey, P. P. Masci, and M. F. Lavin (2011). "Cloning and characterisation of novel cystatins from elapid snake venom glands." *Biochimie* 93(4): 659–668.

Rigoni, M., P. Caccin, S. Gschmeissner, G. Koster, A. D. Postle, O. Rossetto, G. Schiavo, and C. Montecucco (2005). "Equivalent effects of snake PLA_2 neurotoxins and lysophospholipid-fatty acid mixtures." *Science* 310(5754): 1678–1680.

Rigoni, M., G. Schiavo, A. E. Weston, P. Caccin, F. Allegrini, M. Pennuto, F. Valtorta, C. Montecucco, and O. Rossetto (2004). "Snake presynaptic neurotoxins with phospholipase A_2 activity induce punctate swellings of neurites and exocytosis of synaptic vesicles." *J Cell Sci* 117(pt 16): 3561–3570.

Rigopoulou, E. I., D. Roggenbuck, D. S. Smyk, C. Liaskos, M. G. Mytilinaiou, E. Feist, K. Conrad, and D. P. Bogdanos (2012). "Asialoglycoprotein receptor (ASGPR) as target autoantigen in liver autoimmunity: Lost and found." *Autoimmun Rev* 12(2): 260–269.

Ritchie, B. (2006). "Virology." In *Medicine and surgery of reptiles*. St. Louis, MO: Elsevier, 391–417.

Ritonja, A., and F. Gubensek (1985). "Ammodytoxin A, a highly lethal phospholipase A_2 from *Vipera ammodytes ammodytes* venom." *Biochim Biophys Acta* 828(3): 306–312.

Ritonja, A., B. Meloun, and F. Gubensek (1983). "The primary structure of Vipera ammodytes venom chymotrypsin inhibitor." *Biochim Biophys Acta Protein Struct M* 746(3): 138–145.

Riviere, G., V. Choumet, B. Saliou, M. Debray, and C. Bon (1998). "Absorption and elimination of viper venom after antivenom administration." *J Pharmacol Exp Ther* 285(2): 490–495.

Rizzi, C. T., J. L. Carvalho-de-Souza, E. Schiavon, A. C. Cassola, E. Wanke, and L. R. Troncone (2007). "Crotamine inhibits preferentially fast-twitching muscles but is inactive on sodium channels." *Toxicon* 50(4): 553–562.

Robert, A., V. Eschwege, H. Hameg, L. Drouet, and M. F. Aillaud (1996). "Anticoagulant response to *Agkistrodon contortrix* venom (ACV test): A new global test to screen for defects in the anticoagulant protein C pathway." *Thromb Haemost* 75(4): 562–566.

Roberts, K. P., K. M. Ensrud, J. L. Wooters, M. A. Nolan, D. S. Johnston, and D. W. Hamilton (2006). "Epididymal secreted protein Crisp-1 and sperm function." *Mol Cell Endocrinol* 250: 122–127.

Roberts, R. S., T. A. Csencsitz, and C. W. Heard (1985). "*Upper extremity compartment syndromes following pit viper envenomation.*" *Clin Orthop Relat Res* (193): 184–188.

Rochelle, M. J., and K. V. Kardong (1993). "Constriction vs. envenomation in prey capture by brown tree snakes (*Boiga irregularis*) (Squamata, Colubridae)." *Herpetologica* 49: 301–304.

Rodrigues, M., B. G. de la Torre, G. Rádis-Baptista, N. C. Santos, and D. Andreu (2011). "Efficient cellular delivery of beta-galactosidase mediated by NrTPs, a new family of cell-penetrating peptides." *Bioconjug Chem* 22(11): 2339–2344.

Rodrigues, R. S., J. F. da Silva, J. Boldrini Franca, F. P. Fonseca, A. R. Otaviano, F. Henrique Silva, A. Hamaguchi, et al. (2009). "Structural and functional properties of Bp-LAAO, a new L-amino acid oxidase isolated from *Bothrops pauloensis* snake venom." *Biochimie* 91(4): 490–501.

Rodrigues, V. M., A. M. Soares, A. C. Mancin, M. R. Fontes, M. I. Homsi-Brandeburgo, and J. R. Giglio (1998). "Geographic variations in the composition of myotoxins from *Bothrops neuwiedi* snake venoms: Biochemical characterization and biological activity." *Comp Biochem Physiol A Mol Integr Physiol* 121(3): 215–222.

Rogers, Y.-H., and J. C. Venter (2005). "Genomics: Massively parallel sequencing." *Nature* 437(7057): 326–327.

Rojas, G., M. Espinoza, B. Lomonte, and J. M. Gutiérrez (1990). "Effect of storage temperature on the stability of the liquid polyvalent antivenom produced in Costa Rica." *Toxicon* 28(1): 101–105.

Rojas, G., J. M. Jimenez, and J. M. Gutiérrez (1994). "Caprylic acid fractionation of hyperimmune horse plasma: Description of a simple procedure for antivenom production." *Toxicon* 32(3): 351–363.

Rokyta, D., K. Wray, and M. Margres (2013). "The genesis of an exceptionally lethal venom in the timber rattlesnake (*Crotalus horridus*) revealed through comparative venom-gland transcriptomics." *BMC Genomics* 14(1): 394.

Rokyta, D. R., K. P. Wray, A. R. Lemmon, E. M. Lemmon, and S. B. Caudle (2011). "A high-throughput venom-gland transcriptome for the Eastern Diamondback Rattlesnake (*Crotalus adamanteus*) and evidence for pervasive positive selection across toxin classes." *Toxicon* 57(5): 657–671.

Ronquist, F., M. Teslenko, P. van der Mark, D. L. Ayres, A. Darling, S. Hohna, B. Larget, L. Liu, M. A. Suchard, and J. P. Huelsenbeck (2012). "MrBayes 3.2: Efficient Bayesian phylogenetic inference and model choice across a large model space." *Syst Biol* 61(3): 539–542.

Ros, A., M. Faupel, H. Mees, J. v. Oostrum, R. Ferrigno, F. Reymond, P. Michel, J. S. Rossier, and H. H. Girault (2002). "Protein purification by Off-Gel electrophoresis." *Proteomics* 2(2): 151–156.

Rosen, P. B., J. I. Leiva, and C. P. Ross (2000). "Delayed antivenom treatment for a patient after envenomation by *Crotalus atrox*." *Ann Emerg Med* 35(1): 86–88.

Rosenberg, M. B., T. Friedmann, R. C. Robertson, M. Tuszynski, J. A. Wolff, X. O. Breakefield, and F. H. Gage (1988). "Grafting genetically modified cells to the damaged brain—restorative effects of NGF expression." *Science* 242(4885): 1575–1578.

Rosenberg, P. (1990). "Phospholipases." In *Handbook of toxinology*, edited by W. T. Shier and D. Mebs. New York: Marcel Dekker.

Rosengren, K. J., N. L. Daly, M. R. Plan, C. Waine, and D. J. Craik (2003). "Twists, knots, and rings in proteins: Structural definition of the cyclotide framework." *J Biol Chem* 278(10): 8606–8616.

Rosenow, F., R. Ossig, D. Thormeyer, P. Gasmann, K. Schluter, G. Brunner, J. Haier, and J. A. Eble (2008). "Integrins as antimetastatic targets of RGD-independent snake venom components in liver metastasis [corrected]." *Neoplasia* 10(2): 168–176.

Roshak, A. K., E. A. Capper, C. Stevenson, C. Eichman, and L. A. Marshall (2000). "Human calcium-independent phospholipase A_2 mediates lymphocyte proliferation." *J Biol Chem* 275(46): 35692–35698.

Ross, P. L., Y. N. Huang, J. N. Marchese, B. Williamson, K. Parker, S. Hattan, N. Khainovski, et al. (2004). "Multiplexed protein quantitation in *Saccharomyces cerevisiae* using amine-reactive isobaric tagging reagents." *Mol Cell Proteomics* 3(12): 1154–1169.

Rotenberg, D., E. S. Bamberger, and E. Kochva (1971). "Studies on ribonucleic acid synthesis in the venom glands of *Vipera palaestinae* (Ophidia, Reptilia)." *Biochem J* 121: 609–612.

Rouget, C., L. Quinton, A. Maiga, C. Gales, G. Masuyer, C. Malosse, J. Chamot-Rooke, et al. (2010). "Identification of a novel snake peptide toxin displaying high affinity and antagonist behaviour for the alpha2-adrenoceptors." *Br J Pharmacol* 161 (6): 1361–1374.

Roy, A., X. Zhou, M. Z. Chong, D. D'Hoedt, C. S. Foo, N. Rajagopalan, S. Nirthanan, D. Bertrand, J. Sivaraman, and R. M. Kini (2010). "Structural and functional characterization of a novel homodimeric three-finger neurotoxin from the venom of *Ophiophagus hannah* (king cobra)." *J Biol Chem* 285(11): 8302–8315.

Rubin, L. J., and S. Roux (2002). "Bosentan: A dual endothelin receptor antagonist." *Expert Opin Investig Drugs* 11(7): 991–1002.

Rucavado, A., B. Lomonte, M. Ovadia, and J. M. Gutiérrez (1995). "Local tissue damage induced by BaP1, a metalloproteinase isolated from *Bothrops asper* (Terciopelo) snake venom." *Exp Mol Pathol* 63(3): 186–199.

Rucavado, A., M. Soto, T. Escalante, G. D. Loria, R. Arni, and J. M. Gutiérrez (2005). "Thrombocytopenia and platelet hypoaggregation induced by *Bothrops asper* snake venom: Toxins involved and their contribution to metalloproteinase-induced pulmonary hemorrhage." *Thromb Haemost* 94(1): 123–131.

Ruder, T., S. A. Ali, K. Ormerod, A. Brust, M. L. Roymanchadi, S. Ventura, E. A. Undheim, et el. (2013a). "Functional characterization on invertebrate and vertebrate tissues of tachykinin peptides from octopus venoms." *Peptides* 47: 71–76.

Ruder, T., K. Sunagar, E. A. Undheim, S. A. Ali, T. C. Wai, D. H. Low, T. N. Jackson, G. F. King, A. Antunes, and B. G. Fry (2013b). "Molecular phylogeny and evolution of the proteins encoded by coleoid (cuttlefish, octopus, and squid) posterior venom glands." *J Mol Evol* 76(4): 192–204.

Rufini, S., P. Cesaroni, A. Desideri, R. Farias, F. Gubensek, J. M. Gutiérrez, P. Luly, R. Massoud, R. Morero, and J. Z. Pedersen (1992). "Calcium ion independent membrane leakage induced by phospholipase-like myotoxins." *Biochemistry* 31(49): 12424–12430.

Rühlmann, A., D. Kukla, P. Schwager, K. Bartels, and R. Huber (1973). "Structure of the complex formed by bovine trypsin and bovine pancreatic trypsin inhibitor: Crystal structure determination and stereochemistry of the contact region." *J Mol Biol* 77(3): 417–436.

Rusch, S. L., and D. A. Kendall (1995). "Protein transport via amino-terminal targeting sequences: Common themes in diverse systems." *Mol Membr Biol* 12(4): 295–307.

Sackner-Bernstein, J., and K. D. Aaronson (2005). "Nesiritide—not verified." *N Engl J Med* 353(14): 1525–1527.

Sackner-Bernstein, J. D., M. Kowalski, M. Fox, and K. Aaronson (2005). "Short-term risk of death after treatment with nesiritide for decompensated heart failure: A pooled analysis of randomized controlled trials." *JAMA* 293(15): 1900–1905.

Sackner-Bernstein, J. D., H. A. Skopicki, and K. D. Aaronson (2005). "Risk of worsening renal function with nesiritide in patients with acutely decompensated heart failure." *Circulation* 111(12): 1487–1491.

Saint Girons, H. (1988). "Les glandes céphaliques exocrines des reptiles. I. Données anatomiques et histologiques [The exocrine cephalic glands of reptiles. I. Anatomical and histological data]." *Ann Sci Nat Zool* 9(9): 221–255.

——— (1989). "Les glandes céphaliques exocrines des reptiles. II. Considérations fonctionnelles et évolutives. [The exocrine cephalic glands of reptiles. II. Functional and evolutionary considerations]" *Ann Sci Nat Zool* 10: 1–17.

Sainz, I. M., R. A. Pixley, and R. W. Colman (2007). "Fifty years of research on the plasma kallikrein-kinin system: From protein structure and function to cell biology and *in-vivo* pathophysiology." *Thromb Haemost* 98(1): 77–83.

Saito, Y. (2010). "Roles of atrial natriuretic peptide and its therapeutic use." *J Cardiol* 56(3): 262–270.

Sakai, F., S. M. Carneiro, and N. Yamanouye (2012). "Morphological study of accessory gland of *Bothrops jararaca* and its secretory cycle." *Toxicon* 59(3): 393–401.

Sakurai, T., M. Yanagisawa, Y. Takuwa, H. Miyazaki, S. Kimura, K. Goto, and T. Masaki (1990). "Cloning of a cDNA encoding a non-isopeptide-selective subtype of the endothelin receptor." *Nature* 348(6303): 732–735.

Sakurai, Y., M. Shima, T. Matsumoto, H. Takatsuka, K. Nishiya, S. Kasuda, Y. Fujimura, and A. Yoshioka (2003). "Anticoagulant activity of M-LAO, L-amino acid oxidase purified from *Agkistrodon halys blomhoffii*, through selective inhibition of factor IX." *Biochim Biophys Acta* 1649(1): 51–57.

Sakurai, Y., H. Takatsuka, A. Yoshioka, T. Matsui, M. Suzuki, K. Titani, and Y. Fujimura (2001). "Inhibition of human platelet aggregation by L-amino acid oxidase purified from *Naja naja kaouthia* venom." *Toxicon* 39(12): 1827–1833.

Salach, J. I., P. Turini, R. Seng, J. Hauber, and T. P. Singer (1971). "Phospholipase A of snake venoms, I: Isolation and molecular properties of isoenzymes from *Naja naja* and *Vipera russellii* venoms." *J Biol Chem* 246(2): 331–339.

Salazar, A. M., B. Guerrero, B. Cantu, E. Cantu, A. Rodríguez-Acosta, J. C. Pérez, J. A. Galán, A. Tao, and E. E. Sánchez (2009). "Venom variation in hemostasis of the southern Pacific rattlesnake (*Crotalus oreganus helleri*): Isolation of hellerase." *Comp Biochem Physiol C Toxicol Pharmacol* 149(3): 307–316.

Sali, A., and T. L. Blundell (1993). "Comparative protein modelling by satisfaction of spatial restraints." *J Mol Biol* 234(3): 779–815.

Salomão, M. d. G., and I. L. Laporta-Ferreira (1994). "The role of secretions from the supralabial, infralabial, and Duvernoy's glands of the slug-eating snake *Sibynomorphus mikani* (Colubridae: Dipsadinae) in the immobilization of molluscan prey." *J Herpetol* 28: 369–371.

Samejima, Y., Y. Aoki and D. Mebs (1991). "Amino acid sequence of a myotoxin from venom of the eastern diamondback rattlesnake (*Crotalus adamanteus*)." *Toxicon* 29(4–5): 461–468.

Samel, M., K. Tonismagi, G. Ronnholm, H. Vija, J. Siigur, N. Kalkkinen, and E. Siigur (2008). "L-amino acid oxidase from *Naja naja oxiana* venom." *Comp Biochem Physiol B Biochem Mol Biol* 149(4): 572–580.

Samel, M., H. Vija, G. Ronnholm, J. Siigur, N. Kalkkinen, and E. Siigur (2006). "Isolation and characterization of an apoptotic and platelet aggregation inhibiting L-amino acid oxidase from *Vipera berus berus* (common viper) venom." *Biochim Biophys Acta* 1764(4): 707–714.

Sampaio, S. C., S. Hyslop, M. R. Fontes, J. Prado-Franceschi, V. O. Zambelli, A. J. Magro, P. Brigatte, V. P. Gutiérrez, and Y. Cury (2010). "Crotoxin: Novel activities for a classic beta-neurotoxin." *Toxicon* 55(6): 1045–1060.

Samsa, G. P., D. B. Matchar, G. R. Williams, and D. E. Levy (2002). "Cost-effectiveness of ancrod treatment of acute ischaemic stroke: Results from the Stroke Treatment with Ancrod Trial (STAT)." *J Eval Clin Pract* 8(1): 61–70.

San Julian, G. J. (1985). "What you wanted to know about all you ever heard concerning snake repellents." *Second Eastern Wildlife Damage Control Conference*: 41.

Sanchez, E. F., A. Magalhes, F. R. Mandelbaum, and C. R. Diniz (1991). "Purification and characterization of the hemorrhagic factor II from the venom of the Bushmaster snake (*Lachesis muta muta*)." *Biochim Biophys Acta Bioenergetics* 1074 Brust (3): 347–356.

Sanchez, E. F., C. I. Santos, A. Magalhaes, C. R. Diniz, S. Figueiredo, J. Gilroy, and M. Richardson (2000). "Isolation of a proteinase with plasminogen-activating activity from *Lachesis muta muta* (bushmaster) snake venom." *Arch Biochem Biophys* 378(1): 131–141.

Sanchez, G., O. Alvares Lde, M. V. Oberholzer, B. Genro, J. Quillfeldt, J. C. da Costa, C. Cervenansky, D. Jerusalinsky, and E. Kornisiuk (2009). "M4 muscarinic receptors are involved in modulation of neurotransmission at synapses of Schaffer collaterals on CA1 hippocampal neurons in rats." *J Neurosci Res* 87(3): 691–700.

Sangamnatdej, S., G. C. Paesen, M. Slovak, and P. A. Nuttall (2002). "A high affinity serotonin- and histamine-binding lipocalin from tick saliva." *Insect Mol Biol* 11(1): 79–86.

Santamaria, C., S. Larios, Y. Angulo, J. Pizarro-Cerda, J. P. Gorvel, E. Moreno, and B. Lomonte (2005). "Antimicrobial activity of myotoxic phospholipases A_2 from crotalid snake venoms and synthetic peptide variants derived from their C-terminal region." *Toxicon* 45(7): 807–815.

Santoro, M. L., T. S. Vaquero, A. F. Leme, and S. M. Serrano (2009). "NPP-BJ, a nucleotide pyrophosphatase/phosphodiesterase from *Bothrops jararaca* snake venom, inhibits platelet aggregation." *Toxicon* 54(4): 499–512.

Santos, A., J. M. C. Ribeiro, M. J. Lehane, N. F. Gontijo, A. B. Veloso, M. R. V. Sant'anna, R. N. Araujo, E. C. Grisard, and M. H. Pereira (2007). "The sialotranscriptome of the blood-sucking bug *Triatoma brasiliensis* (*Hemiptera, Triatominae*)." *Insect Biochem Mol Biol* 37(7): 702–712.

Sanz, L., N. Ayvazyan, and J. J. Calvete (2008). "Snake venomics of the Armenian mountain vipers *Macrovipera lebetina obtusa* and *Vipera raddei*." *J Proteomics* 71(2): 198–209.

Sanz, L., J. Escolano, M. Ferretti, M. J. Biscoglio, E. Rivera, E. J. Crescenti, Y. Angulo, et al. (2008). "Snake venomics of the South and Central American Bushmasters: Comparison of the toxin composition of *Lachesis muta* gathered from proteomic versus transcriptomic analysis." *J Proteomics* 71(1): 46–60.

Sanz, L., H. L. Gibbs, S. P. Mackessy, and J. J. Calvete (2006). "Venom proteomes of closely related *Sistrurus* rattlesnakes with divergent diets." *J Proteome Res* 5(9): 2098–2112.

Sanz, L., R. A. Harrison, and J. J. Calvete (2012). "First draft of the genomic organization of a PIII-SVMP gene." *Toxicon* 60(4): 455–469.

Sanz-Soler, R., C. Lorente, B. Company, L. Sanz, P. Juarez, A. Perez, Y. Zhang, et al. (2012). "Recombinant expression of mutants of the Frankenstein disintegrin, RTS-ocellatusin: Evidence for the independent origin of RGD and KTS/RTS disintegrins." *Toxicon* 60(4): 665–675.

Saravanakumar, A., T. Balasubramanian, K. Raja, and J. P. Trilles (2012). "A massive infestation of sea snakes by cymothoid isopods." *Parasitol Res* 110(6): 2529–2531.

Saravia, P., E. Rojas, V. Arce, C. Guevara, J. C. Lopez, E. Chaves, R. Velasquez, G. Rojas, and J. M. Gutiérrez (2002). "Geographic and ontogenic variability in the venom of the neotropical rattlesnake *Crotalus durissus*: Pathophysiological and therapeutic implications." *Rev Biol Trop* 50(1): 337–346.

Sarray, S., E. Delamarre, J. Marvaldi, M. El Ayeb, N. Marrakchi, and J. Luis (2007). "Lebectin and lebecetin, two C-type lectins from snake venom, inhibit alpha5beta1 and alphaV-containing integrins." *Matrix Biol* 26(4): 306–313.

Saul, F. A., P. Prijatelj-Znidarsic, B. Vulliez-le Normand, B. Villette, B. Raynal, J. Pungercar, I. Krizaj, and G. Faure (2010). "Comparative structural studies of two natural isoforms of ammodytoxin, phospholipases A_2 from *Vipera* ammodytes ammodytes which differ in neurotoxicity and anticoagulant activity." *J Struct Biol* 169(3): 360–369.

Savarie, P. J., and R. L. Bruggers (1999). "Candidate repellents, oral and dermal toxicants, and fumigants for brown treesnake control." In *Problem snake management: The habu and the brown treesnake*, edited by Gordon H. Rodda, Yoshio Sawai, David Chiszar, and Hiroshi Tanaka. Ithaca, New York: Cornell University Press, 417–422.

Sawyer Products. (2014). http://sawyer.com/products/type/first-aid/.

Scarborough, R. M., J. W. Rose, M. A. Hsu, D. R. Phillips, V. A. Fried, A. M. Campbell, L. Nannizzi, and I. F. Charo (1991). "Barbourin: A GPIIb-IIIa-specific integrin antagonist from the venom of *Sistrurus m. barbouri*." *J Biol Chem* 266(15): 9359–9362.

Scarborough, R. M., J. W. Rose, M. A. Naughton, D. R. Phillips, L. Nannizzi, A. Arfsten, A. M. Campbell, and I. F. Charo (1993). "Characterization of the integrin specificities of disintegrins isolated from American pit viper venoms." *J Biol Chem* 268(2): 1058–1065.

Schagger, H., and G. von Jagow (1987). "Tricine-sodium dodecyl sulfate-polyacrylamide gel electrophoresis for the separation of proteins in the range from 1 to 100 kDa." *Anal Biochem* 166(2): 368–379.

Schaloske, R. H., and E. A. Dennis (2006). "The phospholipase A(2) superfamily and its group numbering system." *Biochim Biophys Acta Mol Cell Biol Lipids* 1761(11): 1246–1259.

Schaper, A., H. Desel, M. Ebbecke, L. De Haro, M. Deters, H. Hentschel, M. Hermanns-Clausen, and C. Langer (2009). "Bites and stings by exotic pets in Europe: An 11 year analysis of 404 cases from Northeastern Germany and Southeastern France." *Clin Toxicol* 47(1): 39–43.

Schatz, G., and B. Dobberstein (1996). "Common principles of protein translocation across membranes." *Science* 271(5255): 1519–1526.

Schechte, I., and A. Berger (1967). "On size of active site in proteases I. papain." *Biochem Biophys Res Comm* 27(2): 157–162.

Schenberg, S. (1959). "Geographical pattern of crotamine distribution in the same rattlesnake subspecies." *Science* 129(3359): 1361–1363.

Schepp, W., J. Schmidtler, T. Riedel, K. Dehne, V. Schusdziarra, J. J. Holst, J. Eng, J. P. Raufman, and M. Classen (1994). "Exendin-4 and exendin-(9-39)nh2—agonist and antagonist, respectively, at the rat parietal-cell receptor for glucagon-like peptide-1-(7-36)nh2." *Eur J Pharmacol Mol Pharmacol Sect* 269(2): 183–191.

Schiffman, S., I. Theodor, and S. I. Rapaport (1969). "Separation from Russell's viper venom of one fraction reacting with factor X and another reacting with factor V." *Biochemistry* 8(4): 1397–1405.

Schleuning, W. D., A. Alagón, W. Boidol, P. Bringmann, T. Petri, J. Kratzschmar, B. Haendler, et al. (1992). "Plasminogen activators from the saliva of *Desmodus rotundus* (common vampire bat)—unique fibrin specificity." *Ann NY Acad Sci* 667: 395–403.

Schmidt, J. J., S. A. Weinstein, and L. A. Smith (1992). "Molecular properties and structure-function relationships of lethal peptides from venom of Wagler's pit viper, *Trimeresurus wagleri*." *Toxicon* 30(9): 1027–1036.

Schroeder, C. I., L. D. Rash, X. Vila-Farres, K. J. Rosengren, M. Mobli, G. F. King, P. F. Alewood, D. J. Craik, and T. Durek (2014). "Chemical synthesis, 3D structure, and ASIC binding site of the toxin mambalgin-2." *Angew Chem Int Ed Engl* 53(4): 1017–1020.

Schulz, M. H., D. R. Zerbino, M. Vingron, and E. Birney (2012). "Oases: Robust *de novo* RNA-seq assembly across the dynamic range of expression levels." *Bioinformatics* 28(8): 1086–1092.

Schumacher, J. (2006). "Selected infectious diseases of wild reptiles and amphibians." *J Exot Pet Med* 15(1): 18–24.

Schumacher, J., and T. Yelen (2006). "Anesthesia and analgesia." In *Reptile medicine and surgery*, edited by D. R. Mader. St. Louis, MO: Saunders Elsevier, 442–452.

Schuster, S. C. (2008). "Next-generation sequencing transforms today's biology." *Nat Meth* 5(1): 16–18.

Schweitz, H., J. N. Bidard, and M. Lazdunski (1990). "Purification and pharmacological characterization of peptide toxins from the black mamba (*Dendroaspis polylepis*) venom." *Toxicon* 28(7): 847–856.

Schweitz, H., T. Bruhn, E. Guillemare, D. Moinier, J. M. Lancelin, L. Beress, and M. Lazdunski (1995). "Kalicludines and kaliseptine—2 different classes of sea-anemone toxins for voltage-sensitive k+ channels." *J Biol Chem* 270(42): 25121–25126.

Schweitz, H., C. Heurteaux, P. Bois, D. Moinier, G. Romey, and M. Lazdunski (1994). "Calcicludine, a venom peptide of the Kunitz-type protease inhibitor family, is a potent blocker of high threshold

Ca^{2+} channels with a high affinity for L-type channels in cerebellar granule neurons." *Proc Natl Acad Sci USA* 91(3): 878–882.

Schweitz, H., P. Pacaud, S. Diochot, D. Moinier, and M. Lazdunski (1999). "MIT1, a black mamba toxin with a new and highly potent activity on intestinal contraction." *FEBS Lett* 461(3): 183–188.

Schweitz, H., P. Vigne, D. Moinier, C. Frelin, and M. Lazdunski (1992). "A new member of the natriuretic peptide family is present in the venom of the green mamba (*Dendroaspis angusticeps*)." *J Biol Chem* 267(20): 13928–13932.

Scott, D. L. (1997). "Phospholipase A_2 structure and catalytic properties." In *Venom phospholipase A_2 enzymes: Structure, function and mechanism,* edited by R. M. Kini. Chichester: John Wiley, 97–128.

Scott, D. L., A. Achari, J. C. Vidal, and P. B. Sigler (1992). "Crystallographic and biochemical studies of the (inactive) Lys-49 phospholipase A_2 from the venom of *Agkistridon piscivorus piscivorus.*" *J Biol Chem* 267(31): 22645–22657.

Scott, D. L., S. P. White, J. L. Browning, J. J. Rosa, M. H. Gelb, and P. B. Sigler (1991). "Structures of free and inhibited human secretory phospholipase A_2 from inflammatory exudate." *Science* 254(5034): 1007–1010.

Scriven, T. A., and J. C. Burnett Jr. (1985). "Effects of synthetic atrial natriuretic peptide on renal function and renin release in acute experimental heart failure." *Circulation* 72(4): 892–897.

Sealy, J. (1997). "Short-distance translocations of timber rattlesnakes in a North Carolina state park: A successful conservation and management program." *Sonoran Herpetologist* 10: 94–99.

Seifert, S. A., L. V. Boyer, R. C. Dart, R. S. Porter, and L. Sjostrom (1997). "Relationship of venom effects to venom antigen and antivenom serum concentrations in a patient with *Crotalus atrox* envenomation treated with a Fab antivenom." *Ann Emerg Med* 30(1): 49–53.

Seigel, R. A., and C. K. Dodd (2002). "Translocations of amphibians: Proven management method or experimental technique?" *Conserv Biol* 16(2): 552–554.

Seilhamer, J. J., T. L. Randall, M. Yamanaka, and L. K. Johnson (1986). "Pancreatic phospholipase A_2: Isolation of the human gene and cDNAs from porcine pancreas and human lung." *DNA* 5(6): 519–527.

Sekiya, F., H. Atoda, and T. Morita (1993). "Isolation and characterization of an anticoagulant protein homologous to botrocetin from the venom of *Bothrops jararaca.*" *Biochemistry* 32(27): 6892–6897.

Selkirk, M. E., O. Lazari, and J. B. Matthews (2005). "*Functional genomics of nematode acetylcholinesterases.*" *Parasitology* 131 Suppl: S3–18.

Semple, F., and J. R. Dorin (2012). "Beta-defensins: Multifunctional modulators of infection, inflammation and more?" *J Innate Immun* 4(4): 337–348.

Serrano, S. M., Y. Hagiwara, N. Murayama, S. Higuchi, R. Mentele, C. A. Sampaio, A. C. Camargo, and E. Fink (1998). "Purification and characterization of a kinin-releasing and fibrinogen-clotting serine proteinase (KN-BJ) from the venom of *Bothrops jararaca*, and molecular cloning and sequence analysis of its cDNA." *Eur J Biochem* 251(3): 845–853.

Serrano, S. M., L. G. Jia, D. Wang, J. D. Shannon, and J. W. Fox (2005). "Function of the cysteine-rich domain of the haemorrhagic metalloproteinase atrolysin A: Targeting adhesion proteins collagen I and von Willebrand factor." *Biochem J* 391(pt 1): 69–76.

Serrano, S. M., J. Kim, D. Wang, B. Dragulev, J. D. Shannon, H. H. Mann, G. Veit, et al. (2006). "The cysteine-rich domain of snake venom metalloproteinases is a ligand for von Willebrand factor A domains: Role in substrate targeting." *J Biol Chem* 281(52): 39746–39756.

Serrano, S. M., and R. C. Maroun (2005). "Snake venom serine proteinases: Sequence homology vs. substrate specificity, a paradox to be solved." *Toxicon* 45(8): 1115–1132.

Serrano, S. M., R. Mentele, C. A. Sampaio, and E. Fink (1995). "Purification, characterization, and amino acid sequence of a serine proteinase, PA-BJ, with platelet-aggregating activity from the venom of *Bothrops jararaca.*" *Biochemistry* 34(21): 7186–7193.

Serrano, S. M., C. A. Sampaio, R. Mentele, A. C. Camargo, and E. Fink (2000). "A novel fibrinogen-clotting enzyme, TL-BJ, from the venom of the snake *Bothrops jararaca*: Purification and characterization." *Thromb Haemost* 83(3): 438–444.

Serrano, S. M., D. Wang, J. D. Shannon, A. F. Pinto, R. K. Polanowska-Grabowska, and J. W. Fox (2007). "Interaction of the cysteine-rich domain of snake venom metalloproteinases with the A1 domain of von Willebrand factor promotes site-specific proteolysis of von Willebrand factor and inhibition of von Willebrand factor-mediated platelet aggregation." *FEBS J* 274(14): 3611–3621.

Servent, D., and C. Fruchart-Gaillard (2009). "Muscarinic toxins: Tools for the study of the pharmacological and functional properties of muscarinic receptors." *J Neurochem* 109(5): 1193–1202.

Servent, D., V. Winckler-Dietrich, H. Y. Hu, P. Kessler, P. Drevet, D. Bertrand, and A. Ménez (1997). "Only snake curaremimetic toxins with a fifth disulfide bond have high affinity for the neuronal alpha7 nicotinic receptor." *J Biol Chem* 272(39): 24279–24286.

Sevcik, C., G. D'Suze, V. Salazar, P. Díaz, and H. Vázquez (2012). "Horse IgG- and ostrich IgY-F(ab')$_2$ groups have different affinities for mice erythrocytes and lymphocytes: Implications for avian immunoglobulin therapeutic usefulness." *Toxicon* 60: 1215—1221.

Sevcik, C., P. Diaz, and G. D'Suze (2008). "On the presence of antibodies against bovine, equine and poultry immunoglobulins in human IgG preparations, and its implications on antivenom production." *Toxicon* 51(1): 10–16.

Shah, A., and B. Bano (2009). "Cystatins in health and diseases." *Int J Pept Res Ther* 15: 43–48.

Shannon, J. D., E. N. Baramova, J. B. Bjarnason, and J. W. Fox (1989). "Amino acid sequence of a *Crotalus atrox* venom metalloproteinase which cleaves type IV collagen and gelatin." *J Biol Chem* 264(20): 11575–11583.

Shapiro, S. D. (2003). "Proteolysis in the lung." *Eur Respir J* suppl 44: 30s–32s.

Sharabi, O., Y. Peleg, E. Mashiach, E. Vardy, Y. Ashani, I. Silman, J. L. Sussman, and J. M. Shifman (2009). "Design, expression and characterization of mutants of fasciculin optimized for interaction with its target, acetylcholinesterase." *Protein Eng Des Sel* 22(10): 641–648.

Shargel, L., S. Wu-Pong, and A. B. C. Yu (2005). *Applied biopharmaceutics and pharmacokinetics.* New York: McGraw-Hill Medical.

Shebuski, R. J., D. R. Ramjit, G. R. Sitko, P. K. Lumma, and V. M. Garsky (1990). "Prevention of canine coronary artery thrombosis with echistatin, a potent inhibitor of platelet aggregation from the venom of the viper, *Echis carinatus*." *Thromb Haemost* 64(4): 576–581.

Shek, K. C., K. L. Tsui, K. K. Lam, P. Crow, K. H. Ng, G. Ades, K. T. Yip, et al. (2009). "Oral bacterial flora of the Chinese cobra (*Naja atra*) and bamboo pit viper (*Trimeresurus albolabris*) in Hong Kong SAR, China." *Hong Kong Med J* 15(3): 183–190.

Shi, G. N., Y. L. Liu, H. M. Lin, S. L. Yang, Y. L. Feng, P. F. Reid, and Z. H. Qin (2011). "Involvement of cholinergic system in suppression of formalin-induced inflammatory pain by cobratoxin." *Acta Pharmacol Sinica* 32(10): 1233–1238.

Shigetomi, H., A. Onogi, H. Kajiwara, S. Yoshida, N. Furukawa, S. Haruta, Y. Tanase, et al. (2010). "Anti-inflammatory actions of serine protease inhibitors containing the Kunitz domain." *Inflamm Res* 59(9): 679–687.

Shikamoto, Y., K. Suto, Y. Yamazaki, T. Morita, and H. Mizuno (2005). "Crystal structure of a CRISP family Ca2+-channel blocker derived from snake venom." *J Mol Biol* 350: 735–743.

Shilov, I. V., S. L. Seymour, A. A. Patel, A. Loboda, W. H. Tang, S. P. Keating, C. L. Hunter, L. M. Nuwaysir, and D. A. Schaeffer (2007). "The paragon algorithm, a next generation search engine that uses sequence temperature values and feature probabilities to identify peptides from tandem mass spectra." *Mol Cell Proteomics* 6(9): 1638–1655.

Shimohigashi, Y., A. Tani, H. Matsumoto, K. Nakashima, and Y. Yamaguchi (1995). "Lysine-49-phospholipases A$_2$ from *Trimeresurus flavoviridis* venom are membrane-acting enzymes." *J Biochem* 118(5): 1037–1044.

Shimokawa, K., L. G. Jia, J. D. Shannon, and J. W. Fox (1998). "Isolation, sequence analysis, and biological activity of atrolysin E/D, the non-RGD disintegrin domain from *Crotalus atrox* venom." *Arch Biochem Biophys* 354(2): 239–246.

Shimokawa, K., L. G. Jia, X. M. Wang, and J. W. Fox (1996). "Expression, activation, and processing of the recombinant snake venom metalloproteinase, pro-atrolysin E." *Arch Biochem Biophys* 335(2): 283–294.

Shimokawa, K., and H. Takahashi (1993a). "Purification of a capillary permeability-increasing enzyme-2 from the venom of *Agkistrodon caliginosus* (Kankoku-Mamushi)." *Toxicon* 31(10): 1213–1219.

——— (1993b). "Some properties of a capillary permeability-increasing enzyme-2 from the venom of *Agkistrodon caliginosus* (Kankoku-Mamushi)." *Toxicon* 31(10): 1221–1227.

Shin, Y., I. Okuyama, J. Hasegawa, and T. Morita (2000). "Molecular cloning of glycoprotein Ib-binding protein, flavocetin-A, which inhibits platelet aggregation." *Thromb Res* 99(3): 239–247.

Shine, R., and J. Koenig (2001). "Snakes in the garden: An analysis of reptiles 'rescued' by community-based wildlife carers." *Biol Conserv* 102(3): 271–283.

Shiu, J. H., C. Y. Chen, L. S. Chang, Y. C. Chen, Y. H. Lo, Y. C. Liu, and W. J. Chuang (2004). "Solution structure of gamma-bungarotoxin: The functional significance of amino acid residues flanking the RGD motif in integrin binding." *Proteins* 57(4): 839–849.

Shoemaker, C. F. (2009). Laboratory diagnosis in equine practice. In *AAEVT's equine manual for veterinary technicians*, edited by D. Reeder, S. Miller, D. Wilfong, M. Leitch, and D. Zimmel. Ames, IA: Wiley-Blackwell, 187–216.

Shridhar, R., J. Zhang, J. Song, B. A. Booth, C. G. Kevil, G. Sotiropoulou, B. F. Sloane, and D. Keppler (2004). "Cystatin M suppresses the malignant phenotype of human MDA-MB-435S cells." *Oncogene* 23(12): 2206–2215.

Shulov, A., D. Benshaul, R. Rosin, and M. Nitzan (1969). "Efficacy of suction as a method of treatment in snake venom poisoning." *Toxicon* 7(1): 15–18.

Siigur, E., A. Aaspollu, K. Trummal, K. Tonismagi, I. Tammiste, N. Kalkkinen, and J. Siigur (2004). "Factor X activator from *Vipera lebetina* venom is synthesized from different genes." *Biochim Biophys Acta Bioenergetics* 1702(1): 41–51.

Siigur, E., T. Neuman, V. Jarve, A. Tara, and J. Siigur (1985). "Isolation and characterization of nerve growth factor from *Vipera lebetina* (snake) venom." *Comp Biochem Physiol B Biochem Mol Biol* 81(1): 211–215.

Siigur, E., M. Samel, K. Tonismagi, J. Subbi, T. Reintamm, and J. Siigur (1998). "Isolation, properties and N-terminal amino acid sequence of a factor V activator from *Vipera lebetina* (Levantine viper) snake venom." *Biochim Biophys Acta* 1429(1): 239–248.

Siigur, E., K. Tonismagi, K. Trummal, M. Samel, H. Vija, A. Aaspollu, G. Ronnholm, J. Subbi, N. Kalkkinen, and J. Siigur (2011). "A new tyrosine-specific chymotrypsin-like and angiotensin-degrading serine proteinase from *Vipera lebetina* snake venom." *Biochimie* 93(2): 321–330.

Siigur, E., K. Tonismagi, K. Trummal, M. Samel, H. Vija, J. Subbi, and J. Siigur (2001). "Factor X activator from *Vipera lebetina* snake venom, molecular characterization and substrate specificity." *BiochimBiophys Acta Bioenergetics* 1568(1): 90–98.

Siigur, J., U. Arumae, T. Neuman, M. Samel, E. Siigur, and M. Saarma (1986). "Isolation and characterization of nerve growth factor from *Vipera berus berus* (common viper) venom." *Comp Biochem Physiol B Biochem Mol Biol* 83(3): 621–625.

Siigur, J., and E. Siigur (2010). "Activation of factor X by snake venom proteases." In *Toxins and hemostasis*, edited by R. M. Kini, K. J. Clemetson, F. S. Markland, M. A. McLane, and T. Morita. New York: Springer, 447–464.

Sikic, K., S. Tomic, and O. Carugo (2010). "Systematic comparison of crystal and NMR protein structures deposited in the protein data bank." *Open Biochem J* 4: 83–95.

Silva, C. A., J. P. Zuliani, M. T. Assakura, R. Mentele, A. C. Camargo, C. F. Teixeira, and S. M. Serrano (2004). "Activation of alpha(M)beta(2)-mediated phagocytosis by HF3, a P-III class metalloproteinase isolated from the venom of *Bothrops jararaca*." *Biochem Biophys Res Comm* 322(3): 950–956.

Silva, M. B., M. Schattner, C. R. Ramos, I. L. Junqueira de Azevedo, M. C. Guarnieri, M. A. Lazzari, C. A. Sampaio, et al. (2003). "A prothrombin activator from *Bothrops erythromelas* (jararaca-da-seca) snake venom: Characterization and molecular cloning." *Biochem J* 369(pt 1): 129–139.

Sim, K. L. (1998). "Purification and preliminary characterisation of praelongin phospholipases, antiplatelet agents from the snake venom of *Acanthophis praelongus*." *Biochim Biophys Acta* 1379(2): 198–206.

Simmer, J. P., A. S. Richardson, C. E. Smith, Y. Hu, and J. C. C. Hu (2011). "Expression of kallikrein-related peptidase 4 in dental and non-dental tissues." *EurJ Oral Sci* 119: 226–233.

Singh, G., J. Eng, and J. P. Raufman (1994). "Use of 125I-[Y39]exendin-4 to characterize exendin receptors on dispersed pancreatic acini and gastric chief cells from guinea pig." *Regul Pept* 53(1): 47–59.

Sippl, M. J. (1993). "Recognition of errors in three-dimensional structures of proteins." *Proteins* 17(4): 355–362.

Siqueira, A. M., N. F. Martins, M. E. De Lima, C. R. Diniz, A. Cartier, D. Brown, and B. Maigret (2002). "A proposed 3D structure for crotamine based on homology building, molecular simulations and circular dichroism." *J Mol Graph Model* 20(5): 389–398.

Sivertsen, J., J. Rosenmeier, J. J. Holst, and T. Vilsboll (2012). "The effect of glucagon-like peptide 1 on cardiovascular risk." *Nat Rev Cardiol* 9(4): 209–222.

Six, D. A., and E. A. Dennis (2000). "The expanding superfamily of phospholipase A(2) enzymes: Classification and characterization." *Biochim Biophys Acta* 1488(1–2): 1–19.

Sixma, J. J., G. H. van Zanten, E. G. Huizinga, R. M. van der Plas, M. Verkley, Y. P. Wu, P. Gros, and P. G. de Groot (1997). "Platelet adhesion to collagen: An update." *Thromb Haemost* 78(1): 434–438.

Sjostrom, L., I. H. al-Abdulla, S. Rawat, D. C. Smith, and J. Landon (1994). "A comparison of ovine and equine antivenoms." *Toxicon* 32(4): 427–433.

Skarzynski, T. (1992). "Crystal structure of α-dendrotoxin from the green mamba venom and its comparison with the structure of bovine pancreatic trypsin inhibitor." *J Mol Biol* 224(3): 671–683.

Skobe, M., and M. Detmar (2000). "Structure, function, and molecular control of the skin lymphatic system." *J Investig Dermatol Symp Proc* 5(1): 14–19.

Sleeman, J. (2013). "Snake fungal disease in the United States." *USGS—National Wildlife Health Center, Wildlife Health Bulletin*, April 22.

Sloan, D. J., M. J. Dedicoat, and D. G. Lalloo (2007). "Healthcare-seeking behaviour and use of traditional healers after snakebite in Hlabisa sub-district, KwaZulu Natal." *Tropical Medicine and International Health* 12(11): 1386–1390.

Smith, D. C., K. R. Reddi, G. Laing, R. G. Theakston, and J. Landon (1992a). "An affinity purified ovine antivenom for the treatment of *Vipera berus* envenoming." *Toxicon* 30(8): 865–871.

Smith, M. A., S. Makino, S. Y. Kim, and R. Kvetnansky (1995). "Stress increases brain-derived neurotropic factor messenger ribonucleic acid in the hypothalamus and pituitary." *Endocrinology* 136(9): 3743–3750.

Smith, P. J., W. F. Brandt, B. J. Stickells, and C. Vonholt (1992b). "*Bitis arietans* nerve growth factor is a disulfide-linked homodimer." *Comp Biochem Physiol B Biochem Mol Biol* 103(4): 975–980.

Soares, A. M., S. H. Andriao-Escarso, Y. Angulo, B. Lomonte, J. M. Gutiérrez, S. Marangoni, M. H. Toyama, R. K. Arni, and J. R. Giglio (2000). "Structural and functional characterization of myotoxin I, a Lys49 phospholipase A(2) homologue from *Bothrops moojeni* (Caissaca) snake venom." *Arch Biochem Biophys* 373(1): 7–15.

Soares, A. M., S. H. Andriao-Escarso, R. K. Bortoleto, L. Rodrigues-Simioni, R. K. Arni, R. J. Ward, J. M. Gutiérrez, and J. R. Giglio (2001). "Dissociation of enzymatic and pharmacological properties of piratoxins-I and -III, two myotoxic phospholipases A_2 from *Bothrops pirajai* snake venom." *Arch Biochem Biophys* 387(2): 188–196.

Soares, A. M., and J. R. Giglio (2003). "Chemical modifications of phospholipases A_2 from snake venoms: Effects on catalytic and pharmacological properties." *Toxicon* 42(8): 855–868.

Soares, A. M., V. M. Rodrigues, M. I. Homsi-Brandeburgo, M. H. Toyama, F. R. Lombardi, R. K. Arni, and J. R. Giglio (1998). "A rapid procedure for the isolation of the Lys-49 myotoxin II from *Bothrops moojeni* (caissaca) venom: Biochemical characterization, crystallization, myotoxic and edematogenic activity." *Toxicon* 36(3): 503–514.

Soares, M. R., A. L. Oliveira-Carvalho, L. S. Wermelinger, R. B. Zingali, P. L. Ho, I. D. M. Junqueira de Azevedo, and M. R. V. Diniz (2005). "Identification of novel bradykinin-potentiating peptides and C-type natriuretic peptide from *Lachesis muta venom*." *Toxicon* 46(1): 31–38.

Sobotka, A. K., R. M. Franklin, N. F. Adkinson Jr., M. Valentine, H. Baer, and L. M. Lichtenstein (1976). "Allergy to insect stings, II: Phospholipase A: the major allergen in honeybee venom." *J Allergy Clin Immunol* 57(1): 29–40.

Sokolovsky, M., and Z. Shraga-Levine (2001). "Sarafotoxins and their relationship to the endothelin family of peptides." In *Endothelin and its inhibitors*, edited by T. D. Warner. Berlin: Springer.

Solymoss, S., M. M. Tucker, and P. B. Tracy (1988). "Kinetics of inactivation of membrane-bound factor-Va by activated protein-C—protein-S modulates factor-Xa protection." *J Biol Chem* 263(29): 14884–14890.

Sommers, C. D., J. L. Bobbitt, K. G. Bemis, and D. W. Snyder (1992). "Porcine pancreatic phospholipase A_2 -induced contractions of guinea pig lung pleural strips." *Eur J Pharmacol* 216(1): 87–96.

Song, B., F. Wang, Y. Guo, Q. Sang, M. Liu, D. Li, W. Fang, and D. Zhang (2012). "Protein-protein interaction network-based detection of functionally similar proteins within species." *Proteins Struct Funct Bioinf* 80(7): 1736–1743.

Song, Y., P. Wilkins, W. Hu, K. S. Murthy, J. Chen, Z. Lee, R. Oyesanya, J. Wu, S. E. Barbour, and X. Fang (2007). "Inhibition of calcium-independent phospholipase A_2 suppresses proliferation and tumorigenicity of ovarian carcinoma cells." *Biochem J* 406(3): 427–436.

Sonne, D. P., T. Engstrom, and M. Treiman (2008). "Protective effects of GLP-1 analogues exendin-4 and GLP-1(9-36) amide against ischemia-reperfusion injury in rat heart." *Regul Pept* 146(1–3): 243–249.

Sorg, O. (2004). "Oxidative stress: A theoretical model or a biological reality?" *C R Biol* 327(7): 649–662.

Sotiropoulou, G., A. Anisowicz, and R. Sager (1997). "Identification, cloning, and characterization of cystatin M, a novel cysteine proteinase inhibitor, down-regulated in breast cancer." *J Biol Chem* 272(2): 903–910.

Sotiropoulou, G., V. Rogakos, T. Tsetsenis, G. Pampalakis, N. Zafiropoulos, G. Simillides, A. Yiotakis, and E. P. Diamandis (2003). "Emerging interest in the kallikrein gene family for understanding and diagnosing cancer." *Oncol Res* 13(6–10): 381–391.

Sousa, S. R., I. Vetter, L. Ragnarsson, and R. J. Lewis (2013). "Expression and pharmacology of endogenous Cav channels in SH-SY5Y human neuroblastoma cells." *PLoS One* 8(3): e59293.

Souza, D. H., L. M. Eugenio, J. E. Fletcher, M. S. Jiang, R. C. Garratt, G. Oliva, and H. S. Selistre-de-Araujo (1999). "Isolation and structural characterization of a cytotoxic L-amino acid oxidase from *Agkistrodon contortrix laticinctus* snake venom: Preliminary crystallographic data." *Arch Biochem Biophys* 368(2): 285–290.

Sriprapat, S., S. Aeksowan, S. Sapsutthipas, C. Chotwiwatthanakun, P. Suttijitpaisal, R. Pratanaphon, O. Khow, V. Sitprija, and K. Ratanabanangkoon (2003). "The impact of a low dose, low volume, multi-site immunization on the production of therapeutic antivenoms in Thailand." *Toxicon* 41(1): 57–64.

Srour, N., A. Lebel, S. McMahon, I. Fournier, M. Fugere, R. Day, and C. M. Dubois (2003). "TACE/ADAM-17 maturation and activation of sheddase activity require proprotein convertase activity." *FEBS Lett* 554(3): 275–283.

St Pierre, L., G. W. Birrell, S. T. H. Earl, T. P. Wallis, J. J. Gorman, J. de Jersey, P. P. Masci, and M. F. Lavin (2007a). "Diversity of toxic components from the venom of the evolutionarily distinct black whip snake, *Demansia vestigiata*." *J Proteome Res* 6: 3093–3107.

St Pierre, L., S. T. Earl, I. Filippovich, N. Sorokina, P. P. Masci, J. de Jersey, and M. F. Lavin (2008). "Common evolution of waprin and Kunitz-like toxin families in Australian venomous snakes." *Cell Mol Life Sci* 65(24): 4039–4054.

St Pierre, L., H. Fischer, D. J. Adams, M. Schenning, N. Lavidis, J. de Jersey, P. P. Masci, and M. F. Lavin (2007b). "Distinct activities of novel neurotoxins from Australian venomous snakes for nicotinic acetylcholine receptors." *Cell Mol Life Sci* 64(21): 2829–2840.

St Pierre, L., S. Flight, P. P. Masci, K. J. Hanchard, R. J. Lewis, P. F. Alewood, J. de Jersey, and M. F. Lavin (2006). "Cloning and characterisation of natriuretic peptides from the venom glands of Australian elapids." *Biochimie* 88(12): 1923–1931.

St Pierre, L., P. P. Masci, I. Filipovich, N. Sorokina, N. Marsh, D. J. Miller, and M. F. Lavin (2005). "Comparative analysis of prothrombin activators from the venom of Australian elapids." *Mol Biol Evol* 22(9): 1853–1864.

Stabeli, R. G., S. Marcussi, G. B. Carlos, R. C. Pietro, H. S. Selistre-de-Araujo, J. R. Giglio, E. B. Oliveira, and A. M. Soares (2004). "Platelet aggregation and antibacterial effects of an L-amino acid oxidase purified from *Bothrops alternatus* snake venom." *Bioorg Med Chem* 12(11): 2881–2886.

Stabeli, R. G., C. D. Sant'Ana, P. H. Ribeiro, T. R. Costa, F. K. Ticli, M. G. Pires, A. Nomizo, et al. (2007). "Cytotoxic L-amino acid oxidase from *Bothrops moojeni*: Biochemical and functional characterization." *Int J Biol Macromol* 41(2): 132–140.

Stahl, S. J. (2001). "Reptile production medicine." *Sem Avian Exot Pet Med* 10(3): 140–150.

Stamatakis, A. (2014). "RAxML version 8: a tool for phylogenetic analysis and post-analysis of large phylogenies." *Bioinformatics* 30(9): 1312–1313.

Staniszewska, I., E. M. Walsh, V. L. Rothman, A. Gaathon, G. P. Tuszynski, J. J. Calvete, P. Lazarovici, and C. Marcinkiewicz (2009). "Effect of VP12 and viperistatin on inhibition of collagen-receptor-dependent melanoma metastasis." *Cancer Biol Ther* 8(15): 1507–1516.

Stegemann, C., A. Kolobov Jr., Y. F. Leonova, D. Knappe, O. Shamova, T. V. Ovchinnikova, V. N. Kokryakov, and R. Hoffmann (2009). "Isolation, purification and *de novo* sequencing of TBD-1, the first beta-defensin from leukocytes of reptiles." *Proteomics* 9(5): 1364–1373.

Steuten, J., K. Winkel, T. Carroll, N. A. Williamson, V. Ignjatovic, K. Fung, A. W. Purcell, and B. G. Fry (2007). "The molecular basis of cross-reactivity in the Australian Snake Venom Detection Kit (SVDK)." *Toxicon* 50(8): 1041–1052.

Stiles, B. G. (1993). "Acetylcholine receptor binding characteristics of snake and cone snail venom postsynaptic neurotoxins: Further studies with a nonradioactive assay." *Toxicon* 31(7): 825–834.

Stiles, B. G., F. W. Sexton, and S. A. Weinstein (1991). "Antibacterial effects of different snake venoms: Purification and characterization of antibacterial proteins from *Pseudechis australis* (Australian king brown or mulga snake) venom." *Toxicon* 29(9): 1129–1141.

Stocker, K., and G. H. Barlow (1976). "The coagulant enzyme from *Bothrops atrox* venom (batroxobin)." *Methods Enzymol* 45: 214–223.

Stocker, K., H. Fischer, J. Meier, M. Brogli, and L. Svendsen (1987). "Characterization of the protein C activator Protac from the venom of the southern copperhead (*Agkistrodon contortrix*) snake." *Toxicon* 25(3): 239–252.

Stocker, K. F. (1998). "Diagnostic and medicinal uses of snake venom enzymes. In *Enzymes from snake venom*, edited by G. S. Bailey. Fort Collins, CO: Alaken, 705–736.

Stockwell, C. A., M. Mulvey, and G. L. Vinyard (1996). "Translocations and the preservation of allelic diversity." *Conserv Biol* 10(4): 1133–1141.

Stone, I. (1972). *The healing factor: Vitamin C against disease.* New York: Grosset & Dunlap.

Stoud, C., H. Amon, T. Wagner, and J. L. Falk (1989). "Effect of electric shock therapy on local tissue reaction to poisonous snake venom injection in rabbits." *Ann Emerg Med* 18(4): 447.

Straight, R. C., and J. L. Glenn (1988). "Isolation and characterization of basic phospholipase (PLA 2) and acidic subunits of canebrake toxin from *Crotalus horridus atricaudatus* venom using HPLC." *Toxicon* 27: 80.

Straight, R. C., J. L. Glenn, T. B. Wolt, and M. C. Wolfe (1991). "Regional differences in content of small basic peptide toxins in the venoms of *Crotalus adamanteus* and *Crotalus horridus*." *Comp Biochem Physiol B Biochem Mol Biol* 100(1): 51–58.

Strimple, P. D., A. J. Tomassoni, E. J. Otten, and D. Bahner (1997). "Report on envenomation by a Gila monster (*Heloderma suspectum*) with a discussion of venom apparatus, clinical findings, and treatment." *Wilderness Environ Med* 8(2): 111–116.

Strong, P. N., J. Goerke, S. G. Oberg, and R. B. Kelly (1976). "Beta-bungarotoxin, a pre-synaptic toxin with enzymatic activity." *Proc Natl Acad Sci USA* 73(1): 178–182.

Stühmer, W., and A. Parekh (1995). "Electrophysiological recordings from *Xenopus* oocytes." In *Single-channel recording*, edited by B. Sakmann and E. Neher. New York: Springer, 341–356.

Štulík, K., V. Pacáková, J. Suchánková, and H. A. Claessens (1997). "Stationary phases for peptide analysis by high performance liquid chromatography: A review." *Anal Chim Acta* 352(1–3): 1–19.

Su, M. J., and C. C. Chang (1984). "Presynaptic effects of snake venom toxins which have phospholipase A_2 activity (beta-bungarotoxin, taipoxin, crotoxin)." *Toxicon* 22(4): 631–640.

Su, M. J., A. R. Coulter, S. K. Sutherland, and C. C. Chang (1983). "The presynaptic neuromuscular blocking effect and phospholipase A_2 activity of textilotoxin, a potent toxin isolated from the venom of the Australian brown snake, *Pseudonaja textilis*." *Toxicon* 21(1): 143–151.

Suat, Z., A. Behcet, Y. Pinar, T. Seyithan, B. Hasan, Y. Cuma, and A. Nurten (2013). "Oxidant/antioxidant status in cases of snake bite." *J Emerg Med* 45(1): 39–45.

Subburaju, S., and R. M. Kini (1997). "Isolation and purification of superbins I and II from *Austrelaps superbus* (copperhead) snake venom and their anticoagulant and antiplatelet effects." *Toxicon* 35(8): 1239–1250.

Sudoh, T., N. Minamino, K. Kangawa, and H. Matsuo (1990). "C-type natriuretic peptide (CNP): A new member of natriuretic peptide family identified in porcine brain." *Biochem Biophys Res Comm* 168(2): 863–870.

Suga, S., K. Nakao, H. Itoh, Y. Komatsu, Y. Ogawa, N. Hama, and H. Imura (1992a). "Endothelial production of C-type natriuretic peptide and its marked augmentation by transforming growth factor-beta: Possible existence of 'vascular natriuretic peptide system.'" *J Clin Invest* 90(3): 1145–1149.

Suga, S., K. Nakao, I. Kishimoto, K. Hosoda, M. Mukoyama, H. Arai, G. Shirakami, Y. Ogawa, Y. Komatsu, and O. Nakagawa (1992b). "Phenotype-related alteration in expression of natriuretic peptide receptors in aortic smooth muscle cells." *Circ Res* 71(1): 34–39.

Sullivan, B. K., M. A. Kwiatkowski, and G. W. Schuett (2004). "Translocation of urban Gila Monsters: A problematic conservation tool." *Biol Conserv* 117(3): 235–242.

Sun, L. K., Y. Yoshii, A. Hyodo, H. Tsurushima, A. Saito, T. Harakuni, Y. P. Li, K. Kariya, M. Nozaki, and N. Morine (2003). "Apoptotic effect in the glioma cells induced by specific protein extracted from Okinawa Habu (*Trimeresurus flavoviridis*) venom in relation to oxidative stress." *Toxicol in vitro* 17(2): 169–177.

Sunagar, K., B. G. Fry, T. N. Jackson, N. R. Casewell, E. A. Undheim, N. Vidal, S. A. Ali, et al. (2013a). "Molecular evolution of vertebrate neurotrophins: Co-option of the highly conserved nerve growth factor gene into the advanced snake venom arsenal." *PLoS One* 8(11): e81827.

Sunagar, K., T. Jackson, E. Undheim, S. Ali, A. Antunes, and B. Fry (2013b). "Three-fingered RAVERs: Rapid accumulation of variations in exposed residues of snake venom toxins." *Toxins* 5(11): 2172–2208.

Sunagar, K., W. E. Johnson, S. J. O'Brien, V. Vasconcelos, and A. Antunes (2012). "Evolution of CRISPs associated with toxicoferan-reptilian venom and mammalian reproduction." *Mol Biol Evol* 29(7): 1807–1822.

Sunagar, K., E. A. Undheim, A. H. Chan, I. Koludarov, S. A. Munoz-Gomez, A. Antunes, and B. G. Fry (2013c). "Evolution stings: The origin and diversification of scorpion toxin peptide scaffolds." *Toxins (Basel)* 5(12): 2456–2487.

Sunagar, K., E. A. Undheim, H. Scheib, E. C. Gren, C. Cochran, C. E. Person, I. Koludarov, et al. (2014). "Intraspecific venom variation in the medically significant Southern Pacific Rattlesnake (*Crotalus oreganus helleri*): Biodiscovery, clinical and evolutionary implications." *J Proteomics* 99: 68–83.

Sung, J. M. L., K. S. Y. Low, and H. E. Khoo (2002). "Characterization of the mechanism underlying stonustoxin-mediated relaxant response in the rat aorta *in vitro*." *Biochem Pharmacol* 63(6): 1113–1118.

Suwa, M., Y. Seino, Y. Nomachi, S. Matsuki, and K. Funahashi (2005). "Multicenter prospective investigation on efficacy and safety of carperitide for acute heart failure in the 'real world' of therapy." *Circ J* 69(3): 283–290.

Suzuki, K., B. Dahlback, and J. Stenflo (1982). "Thrombin-catalyzed activation of human coagulation factor-V." *J Biol Chem* 257(11): 6556–6564.

Suzuki, N., Y. Shikamoto, Z. Fujimoto, T. Morita, and H. Mizuno (2005). "Crystallization and preliminary X-ray analysis of coagulation factor IX-binding protein from habu snake venom at pH 6.5 and 4.6." *Acta Crystallogr F Struct Biol Cryst Commun* 61(pt 1): 147–149.

Suzuki, N., Y. Yamazaki, R. L. Brown, Z. Fujimoto, T. Morita, and H. Mizuno (2008). "Structures of pseudechetoxin and pseudecin, two snake-venom cysteine-rich secretory proteins that target cyclic nucleotide-gated ion channels: Implications for movement of the C-terminal cysteine-rich domain." *Acta Crystallogr D Biol Crystallogr* 64: 1034–1042.

Suzuki-Inoue, K., G. L. Fuller, A. Garcia, J. A. Eble, S. Pohlmann, O. Inoue, T. K. Gartner, et al. (2006). "A novel Syk-dependent mechanism of platelet activation by the C-type lectin receptor CLEC-2." *Blood* 107(2): 542–549.

Sweet, S. S., and E. R. Pianka (2007). "Monitors, mammals and Wallace's line." *Mertensiella* 16: 79–99.

Swenson, S., C. F. Toombs, L. Pena, J. Johansson, and F. S. Markland Jr. (2004). "Alpha-fibrinogenases." *Curr Drug Targets Cardiovasc Haematol Disord* 4 Ouwens (4): 417–435.

Szabo, I., M. Zoratti, and E. Gulbins (2010). "Contribution of voltage-gated potassium channels to the regulation of apoptosis." *FEBS Lett* 584(10): 2049–2056.

Szepesy, L., and G. Rippel (1994). "Effect of the characteristics of the phase system on the retention of proteins in hydrophobic interaction chromatography." *J Chromatogr A* 668(2): 337–344.

Szeto, T. H., X. H. Wang, R. Smith, M. Connor, M. J. Christie, G. M. Nicholson, and G. F. King (2000). "Isolation of a funnel-web spider polypeptide with homology to mamba intestinal toxin 1 and the embryonic head inducer Dickkopf-1." *Toxicon* 38(3): 429–442.

Szewczyk, A., W. Jarmuszkiewicz, and W. S. Kunz (2009). *"Mitochondrial potassium channels." IUBMB Life* 61(2): 134–143.

Tabrizi, M. A., C. M. Tseng, and L. K. Roskos (2006). "Elimination mechanisms of therapeutic monoclonal antibodies." *Drug Discovery Today* 11(1–2): 81–88.

Takacs, Z., K. C. Wilhelmsen, and S. Sorota (2004). "Cobra (*Naja* spp.) nicotinic acetylcholine receptor exhibits resistance to Erabu sea snake (*Laticauda semifasciata*) short-chain alpha-neurotoxin." *J Mol Evol* 58(5): 516–526.

Takai, N., N. Kira, T. Ishii, T. Yoshida, M. Nishida, Y. Nishida, K. Nasu, and H. Narahara (2012). "Bufalin, a traditional oriental medicine, induces apoptosis in human cancer cells." *Asian Pac J Cancer Prev* 13(1): 399–402.

Takasaki, C., J. Suzuki, and N. Tamiya (1990). "Purification and properties of several phospholipases A_2 from the venom of Australian king brown snake (*Pseudechis australis*)." *Toxicon* 28(3): 319–327.

Takasaki, C., N. Tamiya, A. Bdolah, Z. Wollberg, and E. Kochva (1988). "Sarafotoxins S6: several isotoxins from *Atractaspis engaddensis* (burrowing asp) venom that affect the heart." *Toxicon* 26(6): 543–548.

Takatsuka, H., Y. Sakurai, A. Yoshioka, T. Kokubo, Y. Usami, M. Suzuki, T. Matsui, et al. (2001). "Molecular characterization of L-amino acid oxidase from *Agkistrodon halys blomhoffii* with special reference to platelet aggregation." *Biochim Biophys Acta* 1544(1–2): 267–277.

Takayama, M. (2001). "N-Cα bond cleavage of the peptide backbone via hydrogen abstraction." *J Am Soc Mass Spectrom* 12(9): 1044–1049.

Takeya, H., M. Arakawa, T. Miyata, S. Iwanaga, and T. Omori-Satoh (1989). "Primary structure of H2-proteinase, a non-hemorrhagic metalloproteinase, isolated from the venom of the habu snake, *Trimeresurus flavoviridis*." *J Biochem* 106(1): 151–157.

Takeya, H., S. Nishida, T. Miyata, S. Kawada, Y. Saisaka, T. Morita, and S. Iwanaga (1992). "Coagulation factor X activating enzyme from Russell's viper venom (RVV-X): A novel metalloproteinase with disintegrin (platelet aggregation inhibitor)-like and C-type lectin-like domains." *J Biolog Chem* 267(20): 14109–14117.

Takeya, H., A. Onikura, T. Nikai, H. Sugihara, and S. Iwanaga (1990). "Primary structure of a hemorrhagic metalloproteinase, HT-2, isolated from the venom of *Crotalus ruber ruber*." *J Biochem* 108(5): 711–719.

Tamaoki, H., Y. Kobayashi, S. Nishimura, T. Ohkubo, Y. Kyogoku, K. Nakajima, S. Kumagaye, T. Kimura, and S. Sakakibara (1991). "Solution conformation of endothelin determined by means of 1H-NMR spectroscopy and distance geometry calculations." *Protein Eng* 4(5): 509–518.

Tamkus, D., A. Al-Janadi, G. Fink, G. Krishnan, and N. V. Dimitrov (2009). "Endothelins as a potential target for anticancer therapy." *J Cancer Mol* 4: 163–167.

Tan, N. H. (1998). L-amino acid oxidases and lactate dehydrogenases. In *Enzymes from snake venom*, edited by G. S. Bailey. Fort Collins, CO: Alaken, 579–598.

Tan, N. H., and S. Y. Fung (2009). "Snake venom L-amino acid oxidases." In *Handbook of venoms and toxins of reptiles*, edited by S. P. Mackessy. Boca Raton, FL: CRC Press, 220–233.

Tan, N. H., and M. N. Saifuddin (1990). "Isolation and characterization of a hemorrhagin from the venom of *Ophiophagus hannah* (king cobra)." *Toxicon* 28(4): 385–392.

Tan, N. H., and C. S. Tan (1988a). "Biological properties of *Trimeresurus purpureomaculatus* (shore pit viper) venom and its fractions." *Toxicon* 26(11): 989–996.

——— (1988b). "Partial purification of acetylcholinesterase from the venom of the shore pit viper (*Trimeresurus purpureomaculatus*)." *Toxicon* 26(5): 505–508.

Taniuchi, Y., T. Kawasaki, and Y. Fujimura (2000). "The high molecular mass, glycoprotein Ib-binding protein flavocetin-A induces only small platelet aggregates *in vitro*." *Thromb Res* 97(2): 69–75.

Taniuchi, Y., T. Kawasaki, Y. Fujimura, M. Suzuki, K. Titani, Y. Sakai, S. Kaku, N. Hisamichi, N. Satoh, and T. Takenaka (1995). "Flavocetin-A and -B, two high molecular mass glycoprotein Ib binding proteins with high affinity purified from *Trimeresurus flavoviridis* venom, inhibit platelet aggregation at high shear stress." *Biochim Biophys Acta* 1244(2–3): 331–338.

Tanjoni, I., R. Weinlich, M. S. Della-Casa, P. B. Clissa, R. F. Saldanha-Gama, M. S. de Freitas, C. Barja-Fidalgo, G. P. Amarante-Mendes, and A. M. Moura-da-Silva (2005). "Jararhagin, a snake venom metalloproteinase, induces a specialized form of apoptosis (anoikis) selective to endothelial cells." *Apoptosis* 10(4): 851–861.

Tanos, P. P., G. K. Isbister, D. G. Lalloo, C. M. Kirkpatrick, and S. B. Duffull (2008). "A model for venom-induced consumptive coagulopathy in snake bite." *Toxicon* 52: 769–780.

Tashima, A. K., L. Sanz, A. C. Camargo, S. M. Serrano, and J. J. Calvete (2008). "Snake venomics of the Brazilian pitvipers *Bothrops cotiara* and *Bothrops fonsecai*: Identification of taxonomy markers." *J Proteomics* 71(4): 473–485.

Tashima, A. K., A. Zelanis, E. S. Kitano, D. Ianzer, R. L. Melo, V. Rioli, S. S. Sant'anna, A. C. Schenberg, A. C. Camargo, and S. M. Serrano (2012). "Peptidomics of three Bothrops snake venoms: Insights into the molecular diversification of proteomes and peptidomes." *Mol Cell Proteomics* 11(11): 1245–1262.

Taub, A. M. (1967). "Comparative histological studies on Duvernoy's gland of colubrid snakes." *Bull Am Mus Nat Hist* 138: 1–50.

Tchorbanov, B., E. Grishin, B. Aleksiev, and Y. Ovchinnikov (1978). "A neurotoxic complex from the venom of the Bulgarian viper (*Vipera ammodytes ammodytes*) and partial amino acid sequence of the toxic phospholipase A_2." *Toxicon* 16(1): 37–44.

Teixeira, C., Y. Cury, V. Moreira, G. Picolo, and F. Chaves (2009). "Inflammation induced by *Bothrops asper* venom." *Toxicon* 54(1): 67–76.

Teixeira, C. F., S. R. Zamuner, J. P. Zuliani, C. M. Fernandes, M. A. Cruz-Hofling, I. Fernandes, F. Chaves, and J. M. Gutiérrez (2003). "Neutrophils do not contribute to local tissue damage, but play a key role in skeletal muscle regeneration, in mice injected with *Bothrops asper* snake venom." *Muscle Nerve* 28(4): 449–459.

Tekinay, A. B., Y. Nong, J. M. Miwa, I. Lieberam, I. Ibanez-Tallon, P. Greengard, and N. Heintz (2009). "A role for LYNX2 in anxiety-related behavior." *Proc Natl Acad Sci USA* 106(11): 4477–4482.

Tempone, A. G., H. F. Andrade Jr., P. J. Spencer, C. O. Lourenco, J. R. Rogero, and N. Nascimento (2001). "*Bothrops moojeni* venom kills *Leishmania* spp. with hydrogen peroxide generated by its L-amino acid oxidase." *Biochem Biophys Res Commun* 280(3): 620–624.

Ten Kate, M. K., and J. van der Meer (2008). "Protein S deficiency: A clinical perspective." *Haemophilia* 14(6): 1222–1228.

Terrat, Y., D. Biass, S. Dutertre, P. Favreau, M. Remm, R. Stöcklin, D. Piquemal, and F. Ducancel (2012). "High-resolution picture of a venom gland transcriptome: Case study with the marine snail *Conus consors*." *Toxicon* 59(1): 34–46.

Terrat, Y., K. Sunagar, B. G. Fry, T. N. Jackson, H. Scheib, R. Fourmy, M. Verdenaud, G. Blanchet, A. Antunes, and F. Ducancel (2013). "*Atractaspis aterrima* toxins: The first insight into the molecular evolution of venom in side-stabbers." *Toxins (Basel)* 5(11): 1948–1964.

Theakston, R. D., R. E. Phillips, S. Looareesuwan, P. Echeverria, T. Makin, and D. A. Warrell (1990). "Bacteriological studies of the venom and mouth cavities of wild Malayan pit vipers (*Calloselasma rhodostoma*) in southern Thailand." *Trans R Soc Trop Med Hyg* 84(6): 875–879.

Theakston, R. D., and H. A. Reid (1983). "Development of simple standard assay procedures for the characterization of snake venom." *Bull WHO* 61(6): 949–956.

Theoharides, T. C. (1990). "Mast cells: The immune gate to the brain." *Life Sci* 46(9): 607–617.

Thiagarajan, P., V. V. Pengo, and S. S. Shapiro (1986). "The use of the dilute Russell viper venom time for the diagnosis of lupus anticoagulants." *Blood* 68: 869–874.

Thoenen, H., C. Bandtlow, and R. Heumann (1987). "The physiological function of nerve growth factor in the central nervous system—comparison with the periphery." *Rev Physiol Biochem Pharmacol* 109: 145–178.

Thompson, J. D., T. J. Gibson, F. Plewniak, F. Jeanmougin, and D. G. Higgins (1997). "The CLUSTAL_X windows interface: Flexible strategies for multiple sequence alignment aided by quality analysis tools." *Nucleic Acids Res* 25(24): 4876–4882.

Thouzeau, C., Y. Le Maho, G. Froget, L. Sabatier, C. Le Bohec, J. A. Hoffmann, and P. Bulet (2003). "Spheniscins, avian beta-defensins in preserved stomach contents of the king penguin, *Aptenodytes patagonicus*." *J Biol Chem* 278(51): 51053–51058.

Till, G. O., K. J. Johnson, R. Kunkel, and P. A. Ward (1982). "Intravascular activation of complement and acute lung injury: Dependency on neutrophils and toxic oxygen metabolites." *J Clin Invest* 69(5): 1126–1135.

Till, G. O., M. L. Morganroth, R. Kunkel, and P. A. Ward (1987). "Activation of C5 by cobra venom factor is required in neutrophil-mediated lung injury in the rat." *Am J Pathol* 129(1): 44–53.

Timmers, L., J. P. Henriques, D. P. de Kleijn, J. H. Devries, H. Kemperman, P. Steendijk, C. W. Verlaan, et al. (2009). "Exenatide reduces infarct size and improves cardiac function in a porcine model of ischemia and reperfusion injury." *J Am Coll Cardiol* 53(6): 501–510.

Tokunaga, F., K. Nagasawa, S. Tamura, T. Miyata, S. Iwanaga, and W. Kisiel (1988). "The factor V-activating enzyme (RVV-V) from Russell's viper venom: Identification of isoproteins RVV-V alpha, -V beta, and -V gamma and their complete amino acid sequences." *J Biol Chem* 263(33): 17471–17481.

Tonismagi, K., M. Samel, K. Trummal, G. Ronnholm, J. Siigur, N. Kalkkinen, and E. Siigur (2006). "L-amino acid oxidase from *Vipera lebetina* venom: Isolation, characterization, effects on platelets and bacteria." *Toxicon* 48(2): 227–237.

Toplon, D. E., S. P. Terrell, L. Sigler, and E. R. Jacobson (2013). "Dermatitis and cellulitis in leopard geckos (*Eublepharis macularius*) caused by the *Chrysosporium* anamorph of *Nannizziopsis vriesii*." *Vet Pathol* 50(4): 585–589.

Topol, E. J., T. V. Byzova, and E. F. Plow (1999). "Platelet GPIIb-IIIa blockers." *Lancet* 353(9148): 227–231.

Torbet, J. (1987). "Fibrin assembly after fibrinopeptide A release in model systems and human plasma studied with magnetic birefringence." *Biochem J* 244(3): 633–637.

Torii, S., K. Yamane, T. Mashima, N. Haga, K. Yamamoto, J. W. Fox, M. Naito, and T. Tsuruo (2000). "Molecular cloning and functional analysis of apoxin I, a snake venom-derived apoptosis-inducing factor with L-amino acid oxidase activity." *Biochemistry* 39(12): 3197–3205.

Torres, A. M., G. M. de Plater, M. Doverskog, L. C. Birinyi-Strachan, G. M. Nicholson, C. H. Gallagher, and P. W. Kuchel (2000). "Defensin-like peptide-2 from platypus venom: Member of a class of peptides with a distinct structural fold." *Biochem J* 348(pt 3): 649–656.

Torres, A. M., and P. W. Kuchel (2004). "The beta-defensin-fold family of polypeptides." *Toxicon* 44(6): 581–588.

Torres, A. M., I. Menz, P. F. Alewood, P. Bansal, J. Lahnstein, C. H. Gallagher, and P. W. Kuchel (2002). "D-amino acid residue in the C-type natriuretic peptide from the venom of the mammal, *Ornithorhynchus anatinus*, the Australian platypus." *FEBS Lett* 524(1–3): 172–176.

Toso, R., and R. M. Camire (2004). "Removal of B-domain sequences from factor V rather than specific proteolysis underlies the mechanism by which cofactor function is realized." *J Biol Chem* 279(20): 21643–21650.

Touqui, L. and M. Alaoui-El-Azher (2001). "Mammalian secreted phospholipases A_2 and their pathophysiological significance in inflammatory diseases." *Curr Mol Med* 1: 739–754.

Toutant, J. P., T. Rouaud, and G. H. Le Douarin (1981). "Histochemical properties of the biventer cervicis muscle of the chick: A relationship between multiple innervation and slow-tonic fibre types." *Histochem J* 13(3): 481–493.

Townsend, T., A. Larson, E. Louis, and J. R. Macey (2004). "Molecular phylogenetics of squamata: The position of snakes, amphisbaenians, and dibamids, and the root of the squamate tree." *Syst Biol* 53(5): 735–757.

Toyama, M. H., O. Toyama Dde, L. F. Passero, M. D. Laurenti, C. E. Corbett, T. Y. Tomokane, F. V. Fonseca, et al. (2006). "Isolation of a new L-amino acid oxidase from *Crotalus durissus cascavella* venom." *Toxicon* 47(1): 47–57.

Tracy, P. B., L. L. Eide, and E. J. Bowie (1982). "Radioimmunoassay of factor V in human plasma and platelets." *Blood* 60: 59–63.

Tracy, P. B., A. R. Giles, K. G. Mann, L. L. Eide, H. Hoogendoorn, and G. E. Rivard (1984). "Factor-V (Quebec)—a bleeding diathesis associated with a qualitative platelet factor-V deficiency." *J Clin Invest* 74(4): 1221–1228.

Trape, J. F., Y. Mané, and I. Ineich (2006). "*Atractaspis microlepidota, A. micropholis* et *A. watsoni* en Afrique occidentale et centrale." *Bull Soc Herpétologique de France* 199: 5–16.

Tremeau, O., C. Lemaire, P. Drevet, S. Pinkasfeld, F. Ducancel, J. C. Boulain, and A. Ménez (1995). "Genetic engineering of snake toxins: The functional site of Erabutoxin a, as delineated by site-directed mutagenesis, includes variant residues." *J Biol Chem* 270(16): 9362–9369.

Trikha, M., Y. A. De Clerck, and F. S. Markland (1994). "Contortrostatin, a snake venom disintegrin, inhibits beta 1 integrin-mediated human metastatic melanoma cell adhesion and blocks experimental metastasis." *Cancer Res* 54(18): 4993–4998.

Trinh, H. V., J. Grossmann, P. Gehrig, B. Roschitzki, R. Schlapbach, U. F. Greber, and S. Hemmi (2013). "iTRAQ-based and label-free proteomics approaches for studies of human adenovirus infections." *Int J Proteomics* 2013: 581862.

Triplett, D. A. (2000). "Use of the dilute Russell viper venom time (dRVVT): Its importance and pitfalls." *J Autoimmun* 15: 173–178.

Triplett, D. A., K. F. Stocker, G. A. Unger, and L. K. Barna (1993). "The Textarin/Ecarin ratio: A confirmatory test for lupus anticoagulants." *Thromb Haemost* 70(6): 925–931.

Tripodi, A. (2007). "Laboratory testing for lupus anticoagulants: A review of issues affecting results." *Clin Chem* 53(9): 1629–1635.

Trofimov, Y. (2007). "Snakes on a plain: Living with cobras outside Calcutta." *Wall Street Journal.*

Tsai, I. H., P. J. Lu, Y. M. Wang, C. L. Ho, and L. L. Liaw (1995). "Molecular cloning and characterization of a neurotoxic phospholipase A_2 from the venom of Taiwan habu (Trimeresurus mucrosquamatus)." *Biochem J* 311(pt 3): 895–900.

Tsai, I. H., Y. M. Wang, Y. H. Chen, T. S. Tsai, and M. C. Tu (2004). "Venom phospholipases A_2 of bamboo viper (*Trimeresurus stejnegeri*): Molecular characterization, geographic variations and evidence of multiple ancestries." *Biochem J* 377(pt 1): 215–223.

Tsai, I. H., Y. M. Wang, A. C. Cheng, V. Starkov, A. Osipov, I. Nikitin, Y. Makarova, R. Ziganshin, and Y. Utkin (2011a). "cDNA cloning, structural, and functional analyses of venom phospholipases A(2) and a Kunitz-type protease inhibitor from steppe viper *Vipera ursinii renardi*." *Toxicon* 57(2): 332–341.

Tsai, I. H., Y. M. Wang, and M. J. Hseu (2011b). "Mutagenesis analyses explore residues responsible for the neurotoxic and anticoagulant activities of Trimucrotoxin, a pit-viper venom Asn6-phospholipase A_2." *Biochimie* 93(2): 277–285.

Tsai, S. H., Y. C. Chen, L. Chen, Y. M. Wang, and I. H. Tsai (2007). "Binding of a venom Lys-49 phospholipase A(2) to LPS and suppression of its effects on mouse macrophages." *Toxicon* 50(7): 914–922.

Tseng, Y. L., H. C. Peng, and T. F. Huang (2004). "Rhodostomin, a disintegrin, inhibits adhesion of neutrophils to fibrinogen and attenuates superoxide production." *J Biomed Sci* 11(5): 683–691.

Tsetlin, V. (1999). "Snake venom alpha-neurotoxins and other 'three-finger' proteins." *Eur J Biochem* 264(2): 281–286.

Tsueshita, T., H. Onyukusel, V. Sethi, S. Gandhi, and I. Rubinstein (2004). "Helospectin I and II evoke vasodilation in the intact peripheral microcirculation." *Peptides* 25(1): 65–69.

Tsuji, H., K. Okamoto, Y. Matsuzaka, H. Iizuka, G. Tamiya, and H. Inoko (2003). "SLURP-2, a novel member of the human Ly-6 superfamily that is up-regulated in psoriasis vulgaris." *Genomics* 81(1): 26–33.

Tu, A. T., and R. R. Hendon (1983). "Characterization of lizard venom hyaluronidase and evidence for its action as a spreading factor." *Comp Biochem Physiol B Biochem Mol Biol* 76(2): 377–383.

Tun-Pe, Aye-Aye-Myint, H. Khin Ei, H. Thi, and Tin-Nu-Swe (1995). "Local compression pads as a first-aid measure for victims of bites by Russell's viper (*Daboia russelii siamensis*) in Myanmar." *Trans R Soc Trop Med Hyg* 89(3): 293–295.

Tun-Pe, Tin-Nu-Swe, Myint-Lwin, D. A. Warrell, and Than-Win (1987). "The efficacy of tourniquets as a 1st-aid measure for Russell's viper bites in Burma." *Trans R Soc Trop Med Hygiene* 81(3): 403–405.

Tunget-Johnson, C. L., P. D. Pearigen, M. J. McDermott, and B. L. Gillingham (1998). "Resolution of elevated compartment pressures after rattlesnake envenomation with antivenin." Abstract. Conference of Clinical Toxicology. Orlando, FL.

Tuszynski, G. P., T. B. Gasic, and G. J. Gasic (1987). "Isolation and characterization of antistasin: An inhibitor of metastasis and coagulation." *J Biol Chem* 262(20): 9718–9723.

Tuszynski, M. H. (2002). "Growth-factor gene therapy for neurodegenerative disorders." *Lancet Neurology* 1(1): 51–57.

Tuszynski, M. H., L. Thal, M. Pay, D. P. Salmon, U. H. Sang, R. Bakay, P. Patel, et al. (2005). "A phase 1 clinical trial of nerve growth factor gene therapy for Alzheimer disease." *Nat Med* 11(5): 551–555.

Tyler, M. I., D. Barnett, P. Nicholson, I. Spence, and M. E. Howden (1987). "Studies on the subunit structure of textilotoxin, a potent neurotoxin from the venom of the Australian common brown snake (*Pseudonaja textilis*)." *Biochim Biophys Acta* 915(2): 210–216.

Uddman, R., P. J. Goadsby, I. Jansen-Olesen, and L. Edvinsson (1999). "Helospectin-like peptides: Immunochemical localization and effects on isolated cerebral arteries and on local cerebral blood flow in the cat." *J Cereb Blood Flow Metab* 19(1): 61–67.

Ueberheide, B. M., D. Fenyo, P. F. Alewood, and B. T. Chait (2009). "Rapid sensitive analysis of cysteine rich peptide venom components." *Proc Natl Acad Sci USA* 106(17): 6910–6915.

Ullmann, J. F., G. Cowin, N. D. Kurniawan, and S. P. Collin (2010). "Magnetic resonance histology of the adult zebrafish brain: Optimization of fixation and gadolinium contrast enhancement." *NMR Biomed* 23(4): 341–346.

Underwood, G. (2002). "On the rictal structures of some snakes." *Herpetologica* 58: 1–17.

Underwood, G., and E. Kochva (1993). "On the affinities of the burrowing asps *Atractaspis* (Serpents: Atractaspididae)." *Zool J Linnean Soc* 107: 3–24.

Undheim, E., K. Sunagar, V. Herzig, L. Kely, D. Low, T. Jackson, A. Jones, et al. (2013). "A proteomics and transcriptomics investigation of the venom from the barychelid spider *Trittame loki* (brush-foot trapdoor)." *Toxins (Basel)* 5(12): 2488–2503.

Undheim, E. A., A. Jones, K. R. Clauser, J. W. Holland, S. S. Pineda, G. F. King, and B. G. Fry (2014). "Clawing through evolution: Toxin diversification and convergence in the ancient lineage *Chilopoda (Centipedes)*." *Mol Biol Evol* 31(8): 2124–2148.

Unuigbe, E. I., J. Ikhidero, A. O. Ogbemudia, A. Bafor, and A. O. Isah (2009). "Multiple digital gangrene arising from traditional therapy: A case report." *W Afr J Med* 28(6): 397–399.

Urano, T., H. Ihara, Y. Takada, M. Fujie, and A. Takada (2000). "The cleavage and inactivation of plasminogen activator inhibitor type 1 and alpha2-antiplasmin by reptilase, a thrombin-like venom enzyme." *Blood Coagul Fibrin* 11(2): 145–153.

Usami, Y., Y. Fujimura, M. Suzuki, Y. Ozeki, K. Nishio, H. Fukui, and K. Titani (1993). "Primary structure of two-chain botrocetin, a von Willebrand factor modulator purified from the venom of *Bothrops jararaca*." *Proc Natl Acad Sci USA* 90(3): 928–932.

Usami, Y., M. Suzuki, E. Yoshida, Y. Sakurai, K. Hirano, T. Kawasaki, Y. Fujimura, and K. Titani (1996). "Primary structure of alboaggregin-B purified from the venom of *Trimeresurus albolabris*." *Biochem Biophys Res Commun* 219(3): 727–733.

Utaisincharoen, P., S. P. Mackessy, R. A. Miller, and A. T. Tu (1993). "Complete primary structure and biochemical properties of gilatoxin, a serine protease with kallikrein-like and angiotensin-degrading activities." *J Biol Chem* 268(29): 21975–21983.

Utkin, Y. N., V. V. Kukhtina, E. V. Kryukova, F. Chiodini, D. Bertrand, C. Methfessel, and V. I. Tsetlin (2001). "'Weak toxin' from *Naja kaouthia* is a nontoxic antagonist of alpha 7 and muscle-type nicotinic acetylcholine receptors." *J Biol Chem* 276(19): 15810–15815.

Utkin, Y. N., C. Weise, I. E. Kasheverov, T. V. Andreeva, E. V. Kryukova, M. N. Zhmak, V. G. Starkov, et al. (2012). "Azemiopsin from *Azemiops feae* viper venom, a novel polypeptide ligand of nicotinic acetylcholine receptor." *J Biol Chem* 287(32): 27079–27086.

Vadas, P., J. Browning, J. Edelson, and W. Pruzanski (1993). "Extracellular phospholipase A_2 expression and inflammation: The relationship with associated disease states." *J Lipid Mediat* 8(1): 1–30.

Vaiyapuri, S., R. A. Harrison, A. B. Bicknell, J. M. Gibbins, and G. Hutchinson (2010). "Purification and functional characterisation of rhinocerase, a novel serine protease from the venom of *Bitis gabonica rhinoceros*." *PLoS One* 5(3): e9687.

Vaiyapuri, S., N. Thiyagarajan, E. G. Hutchinson, and J. M. Gibbins (2012). "Sequence and phylogenetic analysis of viper venom serine proteases." *Bioinformation* 8(16): 763–772.

Vaiyapuri, S., S. C. Wagstaff, R. A. Harrison, J. M. Gibbins, and E. G. Hutchinson (2011). "Evolutionary analysis of novel serine proteases in the venom gland transcriptome of *Bitis gabonica rhinoceros*." *PLoS One* 6(6): e21532.

Valdez-Cruz, N. A., C. V. Batista, and L. D. Possani (2004). "Phaiodactylipin, a glycosylated heterodimeric phospholipase A from the venom of the scorpion *Anuroctonus phaiodactylus*." *Eur J Biochem* 271(8): 1453–1464.

Valentin, E., and G. Lambeau (2000). "Increasing molecular diversity of secreted phospholipases A(2) and their receptors and binding proteins." *Biochim Biophys Acta* 1488(1–2): 59–70.

Van Asbeck, E. C., A. I. Hoepelman, J. Scharringa, B. L. Herpers, and J. Verhoef (2008). "Mannose binding lectin plays a crucial role in innate immunity against yeast by enhanced complement activation and enhanced uptake of polymorphonuclear cells." *BMC Microbiol* 8: 229.

Van de Locht, A., M. T. Stubbs, W. Bode, T. Friedrich, C. Bollschweiler, W. Hoffken, and R. Huber (1996). "The ornithodorin-thrombin crystal structure, a key to the TAP enigma?" *EMBO J* 15(22): 6011–6017.

Van den Berg, L. M., S. I. Gringhuis, and T. B. Geijtenbeek (2012). "An evolutionary perspective on C-type lectins in infection and immunity." *Ann NY Acad Sci* 1253: 149–158.

Van den Bergh, C. J., A. J. Slotboom, H. M. Verheij, and G. H. de Haas (1988). "The role of aspartic acid-49 in the active site of phospholipase A_2: A site-specific mutagenesis study of porcine pancreatic phospholipase A_2 and the rationale of the enzymatic activity of [lysine49]phospholipase A_2 from *Agkistrodon piscivorus piscivorus*' venom." *Eur J Biochem* 176(2): 353–357.

Van den Heuvel, R. H., E. van Duijn, H. Mazon, S. A. Synowsky, K. Lorenzen, C. Versluis, S. J. Brouns, et al. (2006). "Improving the performance of a quadrupole time-of-flight instrument for macromolecular mass spectrometry." *Anal Chem* 78(21): 7473–7483.

Van der Weyden, L., P. Hains, K. Broady, D. Shaw, and P. Milburn (2001). "Amino acid sequence of a neurotoxic phospholipase A_2 enzyme from common death adder (*Acanthophis antarcticus*) venom." *J Nat Toxins* 10(1): 33–42.

Van der Weyden, L., P. Hains, M. Morris, and K. Broady (1997). "Acanthoxin, a toxic phospholipase A_2 from the venom of the common death adder (*Acanthophis antarcticus*)." *Toxicon* 35(8): 1315–1325.

Van Loon, L. C., M. Rep, and C. M. J. Pieterse (2006). "Significance of inducible defense-related proteins in infected plants." *Annu Rev Phytopathol* 44: 135–162.

Vandermeers, A., M. C. Vandermeerspiret, L. Vigneron, J. Rathe, M. Stievenart, and J. Christophe (1991). "Differences in primary structure among 5 phospholipases-a2 from *Heloderma suspectum*." *Eur J Biochem* 196(3): 537–544.

Vargas, L. J., M. Londono, J. C. Quintana, C. Rua, C. Segura, B. Lomonte, and V. Nuñez (2012). "An acidic phospholipase A(2) with antibacterial activity from *Porthidium nasutum* snake venom." *Comp Biochem Physiol B Biochem Mol Biol* 161(4): 341–347.

Veenstra, J. A. (2000). "Mono- and dibasic proteolytic cleavage sites in insect neuroendocrine peptide precursors." *Arch Insect Biochem Physiol* 43(2): 49–63.

Veiga, A. B. G., J. M. C. Ribeiro, J. A. Guimaraes, and I. M. B. Francischetti (2005). "A catalog for the transcripts from the venomous structures of the caterpillar *Lonomia obliqua*: Identification of the proteins potentially involved in the coagulation disorder and hemorrhagic syndrome." *Gene* 355: 11–27.

Verheij, H. M., M. C. Boffa, C. Rothen, M. C. Bryckaert, R. Verger, and G. H. de Haas (1980a). "Correlation of enzymatic activity and anticoagulant properties of phospholipase A_2." *Eur J Biochem* 112(1): 25–32.

Verheij, H. M., J. J. Volwerk, E. H. Jansen, W. C. Puyk, B. W. Dijkstra, J. Drenth, and G. H. de Haas (1980b). "Methylation of histidine-48 in pancreatic phospholipase A_2: Role of histidine and calcium ion in the catalytic mechanism." *Biochemistry* 19(4): 743–750.

Vetter, I. (2012). "Development and optimization of FLIPR high throughput calcium assays for ion channels and GPCRs." *Adv Exp Med Biol* 740: 45–82.

Vetter, I., Z. Dekan, O. Knapp, D. J. Adams, P. F. Alewood, and R. J. Lewis (2012). "Isolation, characterization, and total regioselective synthesis of the novel muO-conotoxin MfVIA from *Conus magnificus* that targets voltage-gated sodium channels." *Biochem Pharmacol* 84(4): 540–548.

Vetter, I., and R. J. Lewis (2010). "Characterization of endogenous calcium responses in neuronal cell lines." *Biochem Pharmacol* 79(6): 908–920.

Vidal, N. (2002). "Colubroid systematics: Evidence for an early appearance of the venom apparatus followed by extensive evolutionary tinkering." *J Toxicol Tox Rev* 21(1–2): 21–41.

Vidal, N., W. R. Branch, O. S. G. Pauwels, S. B. Hedges, D. G. Broadley, M. Wink, C. Cruaud, U. Joger, and Z. T. Nagy (2008). "Dissecting the major African snake radiation: A molecular phylogeny of the Lamprophiidae Fitzinger (Serpentes, Caenophidia)." *Zootaxa* 1945: 51–66.

Vidal, N., and P. David (2004). "New insights into the early history of snakes inferred from two nuclear genes." *Mol Phylogenet Evol* 31(2): 783–787.

Vidal, N., A. S. Delmas, P. David, C. Cruaud, A. Couloux, and S. B. Hedges (2007). "The phylogeny and classification of caenophidian snakes inferred from seven nuclear protein-coding genes." *C R Biol* 330(2): 182–187.

Vidal, N., M. Dewynter, and D. J. Gower (2010). "Dissecting the major American snake radiation: A molecular phylogeny of the Dipsadidae Bonaparte (Serpentes, Caenophidia)." *C R Biol* 333(1): 48–55.

Vidal, N., and S. B. Hedges (2002). "Higher-level relationships of caenophidian snakes inferred from four nuclear and mitochondrial genes." *C R Biol* 325(9): 987–995.

——— (2004). "Molecular evidence for a terrestrial origin of snakes." *Proc Biol Sci* 271 suppl 4: S226–229.

——— (2005). "The phylogeny of squamate reptiles (lizards, snakes, and amphisbaenians) inferred from nine nuclear protein-coding genes." *C R Biol* 328(10–11): 1000–1008.

——— (2009). "The molecular evolutionary tree of lizards, snakes, and amphisbaenians." *C R Biol* 332(2–3): 129–139.

Vidal, N., S. G. Kindl, A. Wong, and S. B. Hedges (2000). "Phylogenetic relationships of xenodontine snakes inferred from 12S and 16S ribosomal RNA sequences." *Mol Phylogenet Evol* 14(3): 389–402.

Vidal, N., J. Marin, M. Morini, S. Donnellan, W. R. Branch, R. Thomas, M. Vences, A. Wynn, C. Cruaud, and S. B. Hedges (2010). "Blindsnake evolutionary tree reveals long history on Gondwana." *Biol Lett* 6(4): 558-561.

Vidal, N., J. C. Rage, A. Couloux, and S. B. Hedges (2009). Snakes (Serpentes). In *The timetree of life*, edited by S. B. Hedges and S. Kumar. New York: Oxford University Press, 390–397.

Vik, D. P., P. Amiguet, G. J. Moffat, M. Fey, F. Amiguet-Barras, R. A. Wetsel, and B. F. Tack (1991). "Structural features of the human C3 gene: Intron/exon organization, transcriptional start site, and promoter region sequence." *Biochemistry* 30(4): 1080–1085.

Vikrant, S., and B. S. Verma (2014). "Monitor lizard bite-induced acute kidney injury—a case report." *Ren Fail* 36(3): 444–446.

Violette, A., D. Biass, S. Dutertre, D. Koua, D. Piquemal, F. Pierrat, R. Stöcklin, and P. Favreau (2012). "Large-scale discovery of conopeptides and conoproteins in the injectable venom of a fish-hunting cone snail using a combined proteomic and transcriptomic approach." *J Proteomics* 75(17): 5215–5225.

Vishwanath, B. S., R. M. Kini, and T. V. Gowda (1987). "Characterization of three edema-inducing phospholipase A_2 enzymes from habu (*Trimeresurus flavoviridis*) venom and their interaction with the alkaloid aristolochic acid." *Toxicon* 25(5): 501–515.

Visser, L. E., S. Kyei-Faried, D. W. Belcher, D. W. Geelhoed, J. S. van Leeuwen, and J. van Roosmalen (2008). "Failure of a new antivenom to treat *Echis ocellatus* snake bite in rural Ghana: The importance of quality surveillance." *Trans R Soc Trop Med Hyg* 102(5): 445–450.

Vogel, C. W. (1991). "Cobra venom factor, the complement-activating protein of cobra venom." In *Handbook of natural toxins: Reptile and amphibian venoms*, edited by A. T. Tu. New York: Marcel Dekker, 147–188.

Vogel, C. W., R. Bredehorst, D. C. Fritzinger, T. Grunwald, P. Ziegelmuller, and M. A. Kock (1996). "Structure and function of cobra venom factor, the complement-activating protein in cobra venom." *Adv Exp Med Biol* 391: 97–114.

Vogel, C. W., and D. C. Fritzinger (2007). "Humanized cobra venom factor: Experimental therapeutics for targeted complement activation and complement depletion." *Curr Pharm Des* 13(28): 2916–2926.

Vogel, C. W., and H. J. Muller-Eberhard (1982). "The cobra venom factor-dependent C3 convertase of human complement: A kinetic and thermodynamic analysis of a protease acting on its natural high molecular weight substrate." *J Biol Chem* 257(14): 8292–8299.

Vogel, C. W., C. A. Smith, and H. J. Muller-Eberhard (1984). "Cobra venom factor: structural homology with the third component of human complement." *J Immunol* 133(6): 3235–3241.

Vogt, W., G. Schmidt, B. Von Buttlar, and L. Dieminger (1978). "A new function of the activated third component of complement: Binding to C5, an essential step for C5 activation." *Immunology* 34(1): 29–40.

Von Ahsen, N., P. Lewczuk, E. Schütz, M. Oellerich, and H. Ehrenreich (2000). "Prothrombin activity and concentration in healthy subjects with and without the prothrombin G20210A mutation." *Thromb Res* 99(6): 549–556.

Von Reumont, B. M., A. Blanke, S. Richter, F. Alvarez, C. Bleidorn, and R. A. Jenner (2014). "The first venomous crustacean revealed by transcriptomics and functional morphology: Remipede venom glands express a unique toxin cocktail dominated by enzymes and a neurotoxin." *Mol Biol Evol* 31(1): 48–58.

Von Segesser, L. K., X. Mueller, B. Marty, J. Horisberger, and A. Corno (2001). "Alternatives to unfractionated heparin for anticoagulation in cardiopulmonary bypass." *Perfusion* 16(5): 411–416.

Vonk, F. J., J. F. Admiraal, K. Jackson, R. Reshef, M. A. G. de Bakker, K. Vanderschoot, I. van den Berge, et al. (2008). "Evolutionary origin and development of snake fangs." *Nature* 454(7204): 630–633.

Vonk, F. J., N. R. Casewell, C. V. Henkel, A. M. Heimberg, H. J. Jansen, R. J. McCleary, H. M. Kerkkamp, et al. (2013). "The king cobra genome reveals dynamic gene evolution and adaptation in the snake venom system." *Proc Natl Acad Sci USA* 110(51): 20651–20656.

Voris, H. K., and H. H. Voris (1983). "Feeding strategies in marine snakes: An analysis of evolutionary, morphological, behavioral and ecological relationships." *Am Zool* 23: 411–425.

Vouldoukis, I., Y. Shai, P. Nicolas, and A. Mor (1996). "Broad spectrum antibiotic activity of skin-PYY." *FEBS Lett* 380(3): 237–240.

Vulfius, C. A., E. V. Gorbacheva, V. G. Starkov, A. V. Osipov, I. E. Kasheverov, T. V. Andreeva, M. E. Astashev, V. I. Tsetlin, and Y. N. Utkin (2011). "An unusual phospholipase A(2) from puff adder *Bitis arietans* venom—a novel blocker of nicotinic acetylcholine receptors." *Toxicon* 57(5): 787–793.

Wagner, E., and M. M. Frank (2010). "Therapeutic potential of complement modulation." *Nat Rev Drug Discov* 9(1): 43–56.

Wagstaff, S. C., P. Favreau, O. Cheneval, G. D. Laing, M. C. Wilkinson, R. L. Miller, R. Stocklin, and R. A. Harrison (2008). "Molecular characterisation of endogenous snake venom metalloproteinase inhibitors." *Biochem Biophys Res Commun* 365(4): 650–656.

Wagstaff, S. C., and R. A. Harrison (2006). "Venom gland EST analysis of the saw-scaled viper, *Echis ocellatus*, reveals novel alpha(9)beta(1) integrin-binding motifs in venom metalloproteinases and a new group of putative toxins, renin-like aspartic proteases." *Gene* 377: 21–32.

Wagstaff, S. C., G. D. Laing, R. D. Theakston, C. Papaspyridis, and R. A. Harrison (2006). "Bioinformatics and multiepitope DNA immunization to design rational snake antivenom." *PLoS Med* 3(6): e184.

Wagstaff, S. C., L. Sanz, P. Juarez, R. A. Harrison, and J. J. Calvete (2009). "Combined snake venomics and venom gland transcriptomic analysis of the ocellated carpet viper, *Echis ocellatus*." *J Proteomics* 71(6): 609–623.

Wajima, T., G. K. Isbister, and S. B. Duffull (2009). "A comprehensive model for the humoral coagulation network in humans." *Clin Pharmacol Ther* 86: 290–298.

Walport, M. J. (2001). "Complement: First of two parts." *N Engl J Med* 344(14): 1058–1066.

Wang, C., M. Eufemi, C. Turano, and A. Giartosio (1996). "Influence of the carbohydrate moiety on the stability of glycoproteins." *Biochemistry* 35(23): 7299–7307.

Wang, C. H., J. H. Liu, S. C. Lee, C. D. Hsiao, and W. G. Wu (2006a). "Glycosphingolipid-facilitated membrane insertion and internalization of cobra cardiotoxin: The sulfatide-cardiotoxin complex structure in a membrane-like environment suggests a lipid-dependent cell-penetrating mechanism for membrane binding polypeptides." *J Biol Chem* 281(1): 656–667.

Wang, C. H., and W. G. Wu (2005). "Amphiphilic beta-sheet cobra cardiotoxin targets mitochondria and disrupts its network." *FEBS Lett* 579 (14): 3169–3174.

Wang, F., H. Li, M. N. Liu, H. Song, H. M. Han, Q. L. Wang, C. C. Yin, et al. (2006b). "Structural and functional analysis of natrin, a venom protein that targets various ion channels." *Biochem Biophys Res Commun* 351(2): 443–448.

Wang, J., B. Shen, M. Guo, X. Lou, Y. Duan, X. P. Cheng, M. Teng, et al. (2005a). "Blocking effect and crystal structure of natrin toxin, a cysteine-rich secretory protein from Naja atra venom that targets the BKCa channel." *Biochemistry* 44: 10145–10152.

Wang, N., A. Orr-Urtreger, J. Chapman, Y. Ergun, R. Rabinowitz, and A. D. Korczyn (2005b). "Hidden function of neuronal nicotinic acetylcholine receptor beta2 subunits in ganglionic transmission: Comparison to alpha5 and beta4 subunits." *J Neurol Sci* 228(2): 167–177.

Wang, R., R. M. Kini, and M. C. Chung (1999). "Rhodocetin, a novel platelet aggregation inhibitor from the venom of *Calloselasma rhodostoma* (Malayan pit viper): Synergistic and noncovalent interaction between its subunits." *Biochemistry* 38(23): 7584–7593.

Wang, W. J., and T. F. Huang (2002). "Purification and characterization of a novel metalloproteinase, acurhagin, from *Agkistrodon acutus* venom." *Thromb Haemost* 87(4): 641–650.

Wang, X., L. Du, and B. Z. Peterson (2007). "Calcicludine binding to the outer pore of L-type calcium channels is allosterically coupled to dihydropyridine binding." *Biochemistry* 46(25): 7590–7598.

Wang, Y. W., G. H. Cui, M. Zhao, J. Yang, C. Wang, R. W. Giese, and S. Q. Peng (2008). "Bioassay-directed purification of an acidic phospholipase A(2) from *Agkistrodon halys pallas* venom." *Toxicon* 51(7): 1131–1139.

Ward, R. J., L. Chioato, A. H. de Oliveira, R. Ruller, and J. M. Sa (2002). "Active-site mutagenesis of a Lys49-phospholipase A_2: Biological and membrane-disrupting activities in the absence of catalysis." *Biochem J* 362(pt 1): 89–96.

Wardi, A. H., and G. A. Michos (1972). "Alcian blue staining of glycoproteins in acrylamide disc electrophoresis." *Anal Biochem* 49(2): 607–609.

Warner, R. L., S. D. McClintock, A. G. Barron, and F. de la Iglesia (2007). "Hemostatic properties of a venomic protein in rodent dermal injuries." *Exp Mol Pathol* 83: 241–248.

——— (2009). "Hemostatic properties of a venom protein in rat organ trauma." *Exp Mol Pathol* 87: 204–211.

Warrell, D. A. (2008). "Unscrupulous marketing of snake bite antivenoms in Africa and Papua New Guinea: Choosing the right product—'what's in a name?'" *Trans R Soc Trop Med Hyg* 102(5): 397–399.

Warrell, D. A. (2009). "Commissioned article: Management of exotic snakebites." *Q J Med* 102(9): 593–601.

Warrell, D. A., B. M. Greenwood, N. M. Davidson, L. D. Ormerod, and C. R. Prentice (1976). "Necrosis, haemorrhage and complement depletion following bites by the spitting cobra (*Naja nigricollis*)." *Q J Med* 45(177): 1–22.

Warrell, D. A., S. Looareesuwan, R. D. Theakston, R. E. Phillips, P. Chanthavanich, C. Viravan, W. Supanaranond, J. Karbwang, M. Ho, and R. A. Hutton (1986). "Randomized comparative trial of three monospecific antivenoms for bites by the Malayan pit viper (*Calloselasma rhodostoma*) in southern Thailand: Clinical and laboratory correlations." *Am J Trop Med Hyg* 35(6): 1235–1247.

Watson, A. A., and C. A. O'Callaghan (2011). "Molecular analysis of the interaction of the snake venom rhodocytin with the platelet receptor CLEC-2." *Toxins (Basel)* 3(8): 991–1003.

Watt, G., B. D. Meade, R. D. Theakston, L. P. Padre, M. L. Tuazon, C. Calubaquib, E. Santiago, and C. P. Ranoa (1989). "Comparison of Tensilon and antivenom for the treatment of cobra-bite paralysis." *Trans R Soc Trop Med Hyg* 83(4): 570–573.

Watt, G., R. D. Theakston, C. G. Hayes, M. L. Yambao, R. Sangalang, C. P. Ranoa, E. Alquizalas, and D. A. Warrell (1986). "Positive response to edrophonium in patients with neurotoxic envenoming by cobras (*Naja naja philippinensis*): A placebo-controlled study." *N Engl J Med* 315(23): 1444–1448.

Waxman, L., and T. M. Connolly (1993). "Isolation of an inhibitor selective for collagen-stimulated platelet-aggregation from the soft tick *Ornithodoros moubata*." *J Biol Chem* 268(8): 5445–5449.

Waxman, L., D. E. Smith, K. E. Arcuri, and G. P. Vlasuk (1990). "Tick anticoagulant peptide (TAP) is a novel inhibitor of blood coagulation factor Xa." *Science* 248(4955): 593–596.

Weber, D. J., J. S. Wolfson, M. N. Swartz, and D. C. Hooper (1984). "Pasteurella multocida infections. Report of 34 cases and review of the literature." *Medicine (Baltimore)* 63: 133–154.

Wei, A. Z., R. S. Alexander, J. Duke, H. Ross, S. A. Rosenfeld, and C. H. Chang (1998). "Unexpected binding mode of tick anticoagulant peptide complexed to bovine factor Xa." *J Mol Biol* 283(1): 147–154.

Wei, J. F., X. L. Wei, Q. Y. Chen, and S. H. He (2010). "Induction of inflammatory cell accumulation by TM-N49 and promutoxin, two novel phospholipase A(2)." *Toxicon* 56(4): 580–588.

Wei, J. F., H. W. Yang, X. L. Wei, L. Y. Qiao, W. Y. Wang, and S. H. He (2009). "Purification, characterization and biological activities of the L-amino acid oxidase from *Bungarus fasciatus* snake venom." *Toxicon* 54(3): 262–271.

Wei, X. L., J. F. Wei, T. Li, L. Y. Qiao, Y. L. Liu, T. Huang, and S. H. He (2007). "Purification, characterization and potent lung lesion activity of an L-amino acid oxidase from *Agkistrodon blomhoffii ussurensis* snake venom." *Toxicon* 50(8): 1126–1139.

Weinstein, S. A., D. E. Keyler, and J. White (2012). "Replies to Fry et al. (*Toxicon* 2012, 60/4, 434–448). Part A. Analyses of squamate reptile oral glands and their products: A call for caution in formal assignment of terminology designating biological function." *Toxicon* 60(5): 954–963.

Weiser, E., Z. Wollberg, E. Kochva, and S. Y. Lee (1984). "Cardiotoxic effects of the venom of the burrowing asp, *Atractaspis engaddensis* (Atractaspididae, Ophidia)." *Toxicon* 22(5): 767–774.

Welch, E. B., and B. J. Gales (2001). "Use of stun guns for venomous bites and stings: a review." *Wilderness EnvironMed* 12(2): 111–117.

Weldon, C. L., and S. P. Mackessy (2010). "Biological and proteomic analysis of venom from the Puerto Rican Racer (*Alsophis portoricensis:* Dipsadidae)." *Toxicon* 55(2–3): 558–569.

Wellehan, J. F., and S. J. Stahl (2013). "Cryptosporidiosis." In *Clinical veterinary advisor: Birds and exotic pets*, 1st ed., edited by J. Mayer and T. M. Donnelly. St. Louis, MO: Saunders, 89.

Welton, R. E., and J. N. Burnell (2005). "Full length nucleotide sequence of a Factor V-like subunit of oscutarin from *Oxyuranus scutellatus scutellatus* (coastal Taipan)." *Toxicon* 46(3): 328–336.

Wen, S. P., D. T. R. Wilson, S. Kuruppu, M. L. J. Korsinczky, J. Hedrick, L. Pang, T. Szeto, et al. (2005). "Discovery of an MIT-like atracotoxin family: Spider venom peptides that share sequence homology but not pharmacological properties with AVIT family proteins." *Peptides* 26(12): 2412–2426.

White, J. M. (2003). "ADAMs: Modulators of cell-cell and cell-matrix interactions." *Curr Opin Cell Biol* 15(5): 598–606.

Whiting, M. (1997). "Translocation of non-threatened species for 'humanitarian' reasons: What are the conservation risks?" 93(5): 217–218.

Whittington, C. M., A. T. Papenfuss, P. Bansal, A. M. Torres, E. S. Wong, J. E. Deakin, T. Graves, et al. (2008). "Defensins and the convergent evolution of platypus and reptile venom genes." *Genome Res* 18(6): 986–994.

WHO (2010). "WHO guidelines for the production, control and regulation of snake antivenom immunoglobulins." *Bull WHO*: 1–34.

Wickramaratna, J. C., B. G. Fry, M. I. Aguilar, R. M. Kini, and W. C. Hodgson (2003). "Isolation and pharmacological characterisation of a phospholipase A(2) myotoxin from the venom of the Irian Jayan death adder (*Acanthophis rugosus*)." *Br J Pharmacol* 138(2): 333–342.

Wickramaratna, J. C., B. G. Fry, R. E. Loiacono, M. I. Aguilar, P. F. Alewood, and W. C. Hodgson (2004). "Isolation and characterization at cholinergic nicotinic receptors of a neurotoxin from the venom of the *Acanthophis* sp. Seram death adder." *Biochem Pharmacol* 68(2): 383–394.

Wiederstein, M., and M. J. Sippl (2007). "ProSA-web: Interactive Web service for the recognition of errors in three-dimensional structures of proteins." *Nucleic Acids Res* 35 (Web server issue): W407–410.

Wiens, J. J., C. R. Hutter, D. G. Mulcahy, B. P. Noonan, T. M. Townsend, J. W. Sites Jr., and T. W. Reeder (2012). "Resolving the phylogeny of lizards and snakes (*Squamata*) with extensive sampling of genes and species." *Biol Lett* 8(6): 1043–1046.

Wierzbicka-Patynowski, I., S. Niewiarowski, C. Marcinkiewicz, J. J. Calvete, M. M. Marcinkiewicz, and M. A. McLane (1999). "Structural requirements of echistatin for the recognition of alpha(v) beta(3) and alpha(5)beta(1) integrins." *J Biol Chem* 274(53): 37809–37814.

Wiesmann, C., K. J. Katschke, J. Yin, K. Y. Helmy, M. Steffek, W. J. Fairbrother, S. A. McCallum, et al. (2006). "Structure of C3b in complex with CRIg gives insights into regulation of complement activation." *Nature* 444(7116): 217–220.

Wijeyewickrema, L. C., E. E. Gardiner, E. L. Gladigau, M. C. Berndt, and R. K. Andrews (2010). "Nerve growth factor inhibits metalloproteinase-disintegrins and blocks ectodomain shedding of platelet glycoprotein VI." *J Biol Chem* 285(16): 11793–11799.

Wijeyewickrema, L. C., E. E. Gardiner, M. Moroi, M. C. Berndt, and R. K. Andrews (2007a). "Snake venom metalloproteinases, crotarhagin and alborhagin, induce ectodomain shedding of the platelet collagen receptor, glycoprotein VI." *Thromb Haemost* 98(6): 1285–1290.

Wijeyewickrema, L. C., E. E. Gardiner, Y. Shen, M. C. Berndt, and R. K. Andrews (2007b). "Fractionation of snake venom metalloproteinases by metal ion affinity: a purified cobra metalloproteinase, Nk, from *Naja kaouthia* binds Ni2+-agarose." *Toxicon* 50(8): 1064–1072.

Wilhelm, A. J. (2009). "Next-generation DNA sequencing techniques." *New Biotech* 25(4): 195–203.

Williams, A., and L. G. Baird (2003). "DX-88 and HAE: A developmental perspective." *Transfus Apher Sci* 29(3): 255–258.

Williams, D. J., J. M. Gutiérrez, J. J. Calvete, W. Wüster, K. Ratanabanangkoon, O. Paiva, N. I. Brown, et al. (2011). "Ending the drought: New strategies for improving the flow of affordable, effective antivenoms in Asia and Africa." *J Proteomics* 74(9): 1735–1767.

Willis, T. W., and A. T. Tu (1988). "Purification and biochemical characterization of atroxase, a non-hemorrhagic fibrinolytic protease from western diamondback rattlesnake venom." *Biochemistry* 27(13): 4769–4777.

Wilson, H. I., G. M. Nicholson, M. I. Tyler, and M. E. Howden (1995a). "Induction of giant miniature end-plate potentials during blockade of neuromuscular transmission by textilotoxin." *Naunyn Schmiedebergs Arch Pharmacol* 352(1): 79–87.

——— (1995b). "Induction of giant miniature end-plate potentials during blockade of neuromuscular-transmission by textilotoxin." *Naunyn Schmiedebergs Arch Pharmacol* 352(1): 79–87.

Wingert, W. A., and L. Chan (1988). "Rattlesnake bites in Southern California and rationale for recommended treatment." *West J Med* 148(1): 37–44.

Winnik, W. M. (2005). "Continuous pH/salt gradient and peptide score for strong cation exchange chromatography in 2D-nano-LC/MS/MS peptide identification for proteomics." *Anal Chem* 77(15): 4991–4998.

Wollberg, M., E. Kochva, and G. Underwood (1998). "On the rictal glands of some atractaspid snakes." *Herpetol J* 8: 137–143.

Wollberg, Z., A. Bdolah, and E. Kochva (1989). "Vasoconstrictor effects of sarafotoxins in rabbit aorta: Structure-function relationships." *Biochem Biophys Res Commun* 162(1): 371–376.

Wollberg, Z., D. Bousso-Mittler, A. Bdolah, Y. Kloog, E. Kochva, and M. Sokolovsky (1992). "Endothelins and sarafotoxins: Effects on motility, binding properties and phosphoinositide hydrolysis during the estrous cycle of the rat uterus." *J Basic Clin Physiol Pharmacol* 3(1): 41–57.

Wollberg, Z., A. Matz, S. Kinamon, and A. Bdolah (1998). "SRTX-d/e inhibits sarafotoxin-b-induced contraction of guinea pig intestine." *Toxicon* 36: 1252.

Wollberg, Z., R. Shabo-Shina, N. Intrator, A. Bdolah, E. Kochva, G. Shavit, Y. Oron, B. A. Vidne, and S. Gitter (1988). "A novel cardiotoxic polypeptide from the venom of *Atractaspis engaddensis* (burrowing asp): Cardiac effects in mice and isolated rat and human heart preparations." *Toxicon* 26(6): 525–534.

Wonderlin, W. F., and J. S. Strobl (1996). "Potassium channels, proliferation and G1 progression." *J Membr Biol* 154(2): 91–107.

Wong, E. S., and K. Belov (2012). "Venom evolution through gene duplications." *Gene* 496(1): 1–7.

Wong, E. S., M. C. Hardy, D. Wood, T. Bailey, and G. F. King (2013). "SVM-based prediction of propeptide cleavage sites in spider toxins identifies toxin innovation in an Australian tarantula." *PLoS One* 8(7): e66279.

Wong, E. S., D. Morgenstern, E. Mofiz, S. Gombert, K. M. Morris, P. Temple-Smith, M. B. Renfree, et al. (2012). "Proteomics and deep sequencing comparison of seasonally active venom glands in the platypus reveals novel venom peptides and distinct expression profiles." *Mol Cell Proteomics* 11(11): 1354–1364.

Wong, J. W., and G. Cagney (2010). "An overview of label-free quantitation methods in proteomics by mass spectrometry." *Methods Mol Biol* 604: 273–283.

Woo, P. C., S. K. Lau, P. Martelli, S. W. Hui, C. C. Lau, R. Y. Fan, J. M. Groff, E. W. Tam, K. H. Chan, and K. Y. Yuen (2014). "Fatal systemic necrotizing infections associated with a novel paramyxovirus, anaconda paramyxovirus, in green anaconda juveniles." *J Clin Microbiol* 52(10): 3614–3623.

Woolf, C. J. (1996). "Phenotypic modification of primary sensory neurons: The role of nerve growth factor in the production of persistent pain." *Philos Trans R Soc Lond B Biol Sci* 351(1338): 441–448.

Wu, C., F. Wu, J. Pan, J. Morser, and Q. Wu (2003). "Furin-mediated processing of Pro-C-type natriuretic peptide." *J Biol Chem* 278(28): 25847–25852.

Wu, J., Y. Jin, S. Zhong, R. Chen, S. Zhu, W. Wang, Q. Lu, and Y. Xiong (2008). "A unique group of inactive serine protease homologues from snake venom." *Toxicon* 52(2): 277–284.

Wu, P. F., and L. S. Chang (2001). "Expression of A chain and B chain of β-bungarotoxin from Taiwan banded krait: The functional implication of the interchain disulfide bond between A chain and B chain." *J Protein Chem* 20(5): 413–421.

Wu, P. F., S. N. Wu, C. C. Chang, and L. S. Chang (1998). "Cloning and functional expression of B chains of β-bungarotoxins from *Bungarus multicinctus* (Taiwan banded krait)." *Biochem J* 334(1): 87–92.

Wu, P. L., S. C. Lee, C. C. Chuang, S. Mori, N. Akakura, W. G. Wu, and Y. Takada (2006). "Non-cytotoxic cobra cardiotoxin A5 binds to alpha(v)beta3 integrin and inhibits bone resorption: Identification of cardiotoxins as non-RGD integrin-binding proteins of the Ly-6 family." *J Biol Chem* 281(12): 7937–7945.

Wu, W. B., S. C. Chang, M. Y. Liau, and T. F. Huang (2001). "Purification, molecular cloning and mechanism of action of graminelysin I, a snake-venom-derived metalloproteinase that induces apoptosis of human endothelial cells." *Biochem J* 357(pt 3): 719–728.

Wu, Y., Z. F. Wang, and Y. L. Shi (2002). "Beta-agkistrodotoxin inhibits large-conductance calcium-activated potassium channels in rat hippocampal CA1 pyramidal neurons." *Brain Res* 940(1–2): 21–26.

Wüthrich, K. (1986). *NMR of proteins and nucleic acids.* New York: John Wiley.

——— (1989). "The development of nuclear magnetic resonance spectroscopy as a technique for protein structure determination." *Acc Chem Res* 22(1): 36–44.

Xu, J., X. Zhang, Z. Guo, J. Yan, L. Yu, X. Li, X. Xue, and X. Liang (2012). "Short-chain peptides identification of scorpion *Buthus martensi* Karsch venom by employing high orthogonal 2D-HPLC system and tandem mass spectrometry." *Proteomics* 12(19–20): 3076–3084.

Xu, J. M., S. T. Song, F. Y. Feng, F. L. Huang, Y. Yang, G. R. Xie, L. G. Xu, C. Z. Zhang, M. Bruno, and A. Paradiso (2006). "Cobrotoxin-containing analgesic compound to treat chronic moderate to severe cancer pain: Results from a randomized, double-blind, cross-over study and from an open-label study." *Oncol Rep* 16(5): 1077–1084.

Xu, S., L. Gu, Q. Wang, Y. Shu, and Z. Lin (2002). "Preliminary crystallographic study of an acidic phospholipase A_2 from *Ophiophagus hannah* (king cobra)." *Acta Crystallogr D Biol Crystallogr* 58 pt 10/pt 2): 1836–1837.

Xu, Y., J. F. Bruno, and B. J. Luft (2005). "Identification of novel tick salivary gland proteins for vaccine development." *Biochem Biophys Res Commun* 326(4): 901–904.

Xu, Y., M. Ma, G. C. Ippolito, H. W. Schroeder Jr., M. C. Carroll, and J. E. Volanakis (2001). "Complement activation in factor D-deficient mice." *Proc Natl Acad Sci USA* 98(25): 14577–14582.

Yaffe, D., and O. Saxel (1977). "Serial passaging and differentiation of myogenic cells isolated from dystrophic mouse muscle." *Nature* 270(5639): 725–727.

Yamada, D., F. Sekiya, and T. Morita (1996). "Isolation and characterization of carinactivase, a novel prothrombin activator in *Echis carinatus* venom with a unique catalytic mechanism." *J Biol Chem* 271(9): 5200–5207.

Yamada, K., K. Hino, S. Tomoyasu, Y. Honma, and N. Tsuruoka (1998). "Enhancement by bufalin of retinoic acid-induced differentiation of acute promyelocytic leukemia cells in primary culture." *Leuk Res* 22(7): 589–595.

Yamaguchi, Y. (2000). "Lecticans: Organizers of the brain extracellular matrix." *Cell Mol Life Sci* 57(2): 276–289.

Yamaji, N., M. J. Little, H. Nishio, B. Billen, E. Villegas, Y. Nishiuchi, J. Tytgat, G. M. Nicholson, and G. Corzo (2009). "Synthesis, solution structure, and phylum selectivity of a spider delta-toxin that slows inactivation of specific voltage-gated sodium channel subtypes." *J Biol Chem* 284(36): 24568–24582.

Yamamoto, C., D. Tsuru, N. Oda-Ueda, M. Ohno, S. Hattori, and S. T. Kim (2002). "Flavoxobin, a serine protease from *Trimeresurus flavoviridis* (habu snake) venom, independently cleaves Arg726-Ser727 of human C3 and acts as a novel, heterologous C3 convertase." *Immunology* 107(1): 111–117.

Yamanaka, H., X. He, K. Matsumoto, S. Shiosaka, and S. Yoshida (1999). "Protease M/neurosin mRNA is expressed in mature oligodendrocytes." *Brain Res Mol Brain Res* 71(2): 217–224.

Yamane, E. S., F. C. Bizerra, E. B. Oliveira, J. T. Moreira, M. Rajabi, G. L. Nunes, A. O. de Souza, et al. (2013). "Unraveling the antifungal activity of a South American rattlesnake toxin crotamine." *Biochimie* 95(2): 231–240.

Yamazaki, Y., R. L. Brown, and T. Morita (2002). "Purification and cloning of toxins from elapid venoms that target cyclic nucleotide-gated ion channels." *Biochemistry* 41(38): 11331–11337.

Yamazaki, Y., H. Koike, Y. Sugiyama, K. Motoyoshi, T. Wada, S. Hishinuma, M. Mita, and T. Morita (2002). "Cloning and characterization of novel snake venom proteins that block smooth muscle contraction." *Eur J Biochem* 269(11): 2708–2715.

Yamazaki, Y., Y. Matsunaga, Y. Tokunaga, S. Obayashi, M. Saito, and T. Morita (2009). "Snake venom Vascular Endothelial Growth Factors (VEGF-Fs) exclusively vary their structures and functions among species." *J Biol Chem* 284(15): 9885–9891.

Yamazaki, Y., and T. Morita (2004). "Structure and function of snake venom cysteine-rich secretory proteins." *Toxicon* 44: 227–231.

——— (2007). "Snake venom components affecting blood coagulation and the vascular system: Structural similarities and marked diversity." *Curr Pharm Des* 13(28): 2872–2886.

——— (2009). "Structures and functions of snake venom toxins affecting thrombosis and hemostasis." *Tanpakushitsu Kakusan Koso* 54(5): 628–634.

Yamazaki, Y., K. Takani, H. Atoda, and T. Morita (2003). "Snake venom vascular endothelial growth factors (VEGFs) exhibit potent activity through their specific recognition of KDR (VEGF receptor 2)." *J Biol Chem* 278(52): 51985–51988.

Yanagisawa, M., H. Kurihara, S. Kimura, Y. Tomobe, M. Kobayashi, Y. Mitsui, Y. Yazaki, K. Goto, and T. Masaki (1988). "A novel potent vasoconstrictor peptide produced by vascular endothelial cells." *Nature* 332(6163): 411–415.

Yang, C. C. (1994). "Structure-function relationship of phospholipase A_2 from snake venoms." *J Toxicol Tox Rev* 13: 125–177.

Yang, S. H., C. M. Chien, M. C. Lu, Y. J. Lu, Z. Z. Wu, and S. R. Lin (2005). "Cardiotoxin III induces apoptosis in K562 cells through a mitochondrial-mediated pathway." *Clin Exp Pharmacol Physiol* 32(7): 515–520.

Yang, Y., S. Cun, L. Peng, X. Xie, J. Wei, W. Yang, and A. Xu (2003). "cDNA cloning, identification and characterization of a novel cystatin from the tentacle of *Cyanea capillata*." *Biochimie* 85(10): 1033–1039.

Yang, Y., J. D. Dignam, and L. E. Gentry (1997). "Role of carbohydrate structures in the binding of beta1-latency-associated peptide to ligands." *Biochemistry* 36(39): 11923–11932.

Yang, Z. (1998). "Likelihood ratio tests for detecting positive selection and application to primate lysozyme evolution." *Mol Biol Evol* 15(5): 568–573.

Yang, Z. (2007). "PAML 4: Phylogenetic analysis by maximum likelihood." *Mol Biol Evol* 24(8): 1586–1591.

Yang, Z., and R. Nielsen (2002). "Codon-substitution models for detecting molecular adaptation at individual sites along specific lineages." *Mol Biol Evol* 19(6): 908–917.

Yang, Z. H., R. Nielsen, N. Goldman, and A. M. K. Pedersen (2000). "Codon-substitution models for heterogeneous selection pressure at amino acid sites." *Genetics* 155(1): 431–449.

Yang, Z. H., W. S. W. Wong, and R. Nielsen (2005). "Bayes empirical Bayes inference of amino acid sites under positive selection." *Mol Biol Evol* 22(4): 1107–1118.

Yang, Z. M., Q. Guo, Z. R. Ma, Y. Chen, Z. Z. Wang, X. M. Wang, Y. M. Wang, and I. H. Tsai (2015). "Structures and functions of crotoxin-like heterodimers and acidic phospholipases A_2 from *Gloydius intermedius* venom: Insights into the origin of neurotoxic-type rattlesnakes." *J Proteomics* 112: 210–223.

Yeh, C. H., H. C. Peng, R. S. Yang, and T. F. Huang (2001). "Rhodostomin, a snake venom disintegrin, inhibits angiogenesis elicited by basic fibroblast growth factor and suppresses tumor growth by a selective alpha(v)beta(3) blockade of endothelial cells." *Mol Pharmacol* 59(5): 1333–1342.

Yennamalli, R. M., A. J. Rader, J. D. Wolt, and T. Z. Sen (2011). "Thermostability in endoglucanases is fold-specific." *BMC Struct Biol* 11: 10.

Yoshida, E., Y. Fujimura, S. Miura, M. Sugimoto, H. Fukui, N. Narita, Y. Usami, M. Suzuki, and K. Titani (1993). "Alboaggregin-B and botrocetin, two snake venom proteins with highly homologous amino acid sequences but totally distinct functions on von Willebrand factor binding to platelets." *Biochem Biophys Res Commun* 191(3): 1386–1392.

Yoshimura, M., H. Yasue, E. Morita, N. Sakaino, M. Jougasaki, M. Kurose, M. Mukoyama, Y. Saito, K. Nakao, and H. Imura (1991). "Hemodynamic, renal, and hormonal responses to brain natriuretic peptide infusion in patients with congestive heart failure." *Circulation* 84(4): 1581–1588.

You, W. K., H. J. Seo, K. H. Chung, and D. S. Kim (2003). "A novel metalloprotease from *Gloydius halys* venom induces endothelial cell apoptosis through its protease and disintegrin-like domains." *J Biochem* 134(5): 739–749.

Young, B. A., M. Blair, K. Zahn, and J. Marvin (2001). "Mechanics of venom expulsion in *Crotalus*, with special reference to the role of the fang sheath." *Anat Rec* 264(4): 415–426.

Young, B. A., K. Dunlap, K. Koenig, and M. Singer (2004). "The buccal buckle: The functional morphology of venom spitting in cobras." *J Exp Biol* 207(pt 20): 3483–3494.

Young, B. A., F. Herzog, P. Friedel, S. Rammensee, A. Bausch, and J. L. van Hemmen (2011). "Tears of venom: Hydrodynamics of reptilian envenomation." *Phys Rev Lett* 106(19): 198103.

Young, B. A., and K. V. Kardong (1991). "Dentitional surface-features in snakes." *Am Zool* 31(5): A51

——— (1996). "Dentitional surface features in snakes (Reptilia: Serpentes)." *Amphibia-Reptilia* 17(3): 261–276.

——— (2007). "Mechanisms controlling venom expulsion in the western diamondback rattlesnake, *Crotalus atrox*." *J Exp Zool A Ecol Genet Physiol* 307(1): 18–27.

Young, B. A., and K. Zahn (2001). "Venom flow in rattlesnakes: Mechanics and metering." *J Exp Biol* 204(pt 24): 4345–4351.

Young, P. M., and T. E. Wheat (1990). "Optimization of high-performance liquid chromatographic peptide separations with alternative mobile and stationary phases." *J Chromatogr* 512: 273–281.

Yount, N. Y., D. Kupferwasser, A. Spisni, S. M. Dutz, Z. H. Ramjan, S. Sharma, A. J. Waring, and M. R. Yeaman (2009). "Selective reciprocity in antimicrobial activity versus cytotoxicity of hBD-2 and crotamine." *Proc Natl Acad Sci USA* 106(35): 14972–14977.

Yount, N. Y., and M. R. Yeaman (2004). "Multidimensional signatures in antimicrobial peptides." *Proc Natl Acad Sci USA* 101(19): 7363–7368.

Yousef, G. M., and E. P. Diamandis (2001). "The new human tissue kallikrein gene family: Structure, function, and association to disease." *Endocr Rev* 22(2): 184–204.

Yu, B. Z., O. G. Berg, and M. K. Jain (1993). "The divalent cation is obligatory for the binding of ligands to the catalytic site of secreted phospholipase A_2." *Biochemistry* 32(25): 6485–6492.

Yu, C., R. Bhaskaran, L. C. Chuang, and C. C. Yang (1993). "Solution conformation of cobrotoxin: A nuclear magnetic resonance and hybrid distance geometry-dynamical simulated annealing study." *Biochemistry* 32(9): 2131–2136.

Yuan, C. H., Q. Y. He, K. Peng, J. B. Diao, L. P. Jiang, X. Tang, and S. P. Liang (2008). "Discovery of a distinct superfamily of kunitz-type toxin (KTT) from tarantulas." *PLoS One* 3(10): e3414.

Yuliang, X., Z. Rujin, Y. Datong, K. Fukiren, Z. Ermi, F. Liang, J. Yueming, M. Yunxiang, and Y. Hengchu (1979). "Experimental studies on therapy for venomous snakebite with Yunnan snakebite drug." *Acta Pharm Sinica* 14: 557–560.

Zaganelli, G. L., M. G. Zaganelli, A. Magalhaes, C. R. Diniz, and M. E. de Lima (1996). "Purification and characterization of a fibrinogen-clotting enzyme from the venom of jararacucu (*Bothrops jararacussu*)." *Toxicon* 34(7): 807–819.

Zaharenko, A. J., W. A. Ferreira Jr., J. S. Oliveira, M. Richardson, D. C. Pimenta, K. Konno, F. C. Portaro, and J. C. de Freitas (2008). "Proteomics of the neurotoxic fraction from the sea anemone *Bunodosoma cangicum* venom: Novel peptides belonging to new classes of toxins." *Comp Biochem Physiol D Genomics Proteomics* 3(3): 219–225.

Zamudio, F. Z., R. Conde, C. Arevalo, B. Becerril, B. M. Martin, H. H. Valdivia, and L. D. Possani (1997). "The mechanism of inhibition of ryanodine receptor channels by imperatoxin I, a heterodimeric protein from the scorpion *Pandinus imperator*." *J Biol Chem* 272(18): 11886–11894.

Zeeuwen, P. L., I. M. van Vlijmen-Willems, W. Hendriks, G. F. Merkx, and J. Schalkwijk (2002). "A null mutation in the cystatin M/E gene of ichq mice causes juvenile lethality and defects in epidermal cornification." *Hum Mol Genet* 11(23): 2867–2875.

Zeeuwen, P. L., I. M. van Vlijmen-Willems, D. Olthuis, H. T. Johansen, K. Hitomi, I. Hara-Nishimura, J. C. Powers, et al. (2004). "Evidence that unrestricted legumain activity is involved in disturbed epidermal cornification in cystatin M/E deficient mice." *Hum Mol Genet* 13(10): 1069–1079.

Zelensky, A. N., and J. E. Gready (2005). "The C-type lectin-like domain superfamily." *FEBS J* 272(24): 6179–6217.

Zhang, B., D. He, Y. Fan, N. Liu, and Y. Chen (2014). "Oral delivery of exenatide via microspheres prepared by cross-linking of alginate and hyaluronate." *PLoS One* 9(1): e86064.

Zhang, D., I. Botos, F. X. Gomis-Ruth, R. Doll, C. Blood, F. G. Njoroge, J. W. Fox, W. Bode, and E. F. Meyer (1994a). "Structural interaction of natural and synthetic inhibitors with the venom metalloproteinase, atrolysin C (form d)." *Proc Natl Acad Sci USA* 91(18): 8447–8451.

Zhang, H., M. Teng, L. Niu, Y. Wang, Q. Liu, Q. Huang, Q. Hao, Y. Dong, and P. Liu (2004a). "Purification, partial characterization, crystallization and structural determination of AHP-LAAO, a novel L-amino-acid oxidase with cell apoptosis-inducing activity from *Agkistrodon halys pallas* venom." *Acta Crystallogr D Biol Crystallogr* 60(pt 5): 974–977.

Zhang, H., Q. Yang, M. Sun, M. Teng, and L. Niu (2004b). "Hydrogen peroxide produced by two amino acid oxidases mediates antibacterial actions." *J Microbiol* 42(4): 336–339.

Zhang, H. L., R. Han, Z. X. Chen, Z. L. Gu, P. F. Reid, L. N. Raymond, and Z. H. Qin (2006a). "Analgesic effects of receptin, a chemically modified cobratoxin from Thailand cobra venom." *Neurosci Bull* 22(5): 267–273.

Zhang, H. L., R. Han, Z. L. Gu, Z. X. Chen, B. W. Chen, P. F. Reid, L. N. Raymond, and Z. H. Qin (2006b). "A short-chain alpha-neurotoxin from *Naja naja atra* produces potent cholinergic-dependent analgesia." *Neurosci Bull* 22(2): 103–109.

Zhang, J., R. Shridhar, Q. Dai, J. Song, S. C. Barlow, L. Yin, B. F. Sloane, et al. (2004c). "Cystatin M: A novel candidate tumor suppressor gene for breast cancer." *Cancer Res* 64(19): 6957–6964.

Zhang, J. H., T. D. Chung, and K. R. Oldenburg (1999). "A simple statistical parameter for use in evaluation and validation of high throughput screening assays." *J Biomol Screen* 4(2): 67–73.

Zhang, J. Z., R. Nielsen, and Z. H. Yang (2005). "Evaluation of an improved branch-site likelihood method for detecting positive selection at the molecular level." *Mol Biol Evol* 22(12): 2472–2479.

Zhang, L., and L. Cui (2007). "A cytotoxin isolated from *Agkistrodon acutus* snake venom induces apoptosis via Fas pathway in A549 cells." *Toxicol in Vitro* 21(6): 1095–1103.

Zhang, L., and L. J. Wei (2007). "ACTX-8, a cytotoxic L-amino acid oxidase isolated from *Agkistrodon acutus* snake venom, induces apoptosis in Hela cervical cancer cells." *Life Sci* 80(13): 1189–1197.

Zhang, P., J. Shi, B. Shen, X. Li, Y. Gao, Z. Zhu, Z. Zhu, Y. Ji, M. Teng, and L. Niu (2009). "Stejnihagin, a novel snake metalloproteinase from *Trimeresurus stejnegeri* venom, inhibited L-type Ca2+ channels." *Toxicon* 53(2): 309–315.

Zhang, S., B. Ma, J. Sakai, H. Shiono, T. Matsui, I. Sugie, and T. Okada (2001). "Characterization of a thrombin-like serine protease, Kangshuanmei, isolated from the venom of a Chinese snake, *Agkistrodon halys brevicaudus stejneger*." *J Nat Toxins* 10(3): 221–238.

Zhang, X. H., C. Zhao, K. Seleznev, K. Song, J. J. Manfredi, and Z. A. Ma (2006c). "Disruption of G1-phase phospholipid turnover by inhibition of Ca2+-independent phospholipase A_2 induces a p53-dependent cell-cycle arrest in G1 phase." *J Cell Sci* 119(pt 6): 1005–1015.

Zhang, Y., B. R. Fonslow, B. Shan, M. C. Baek, and J. R. Yates III (2013). "Protein analysis by shotgun/bottom-up proteomics." *Chem Rev* 113(4): 2343–2394.

Zhang, Y., W. H. Lee, Y. L. Xiong, W. Y. Wang, and S. W. Zu (1994b). "Characterization of OhS1, an arginine/lysine amidase from the venom of king cobra (*Ophiophagus hannah*)." *Toxicon* 32(5): 615–623.

Zhang, Y., A. Wisner, Y. Xiong, and C. Bon (1995). "A novel plasminogen activator from snake venom: Purification, characterization, and molecular cloning." *J Biol Chem* 270(17): 10246–10255.

Zhang, Y. J., J. H. Wang, W. H. Lee, Q. Wang, H. Liu, Y. T. Zheng, and Y. Zhang (2003). "Molecular characterization of *Trimeresurus stejnegeri* venom L-amino acid oxidase with potential anti-HIV activity." *Biochem Biophys Res Commun* 309(3): 598–604.

Zhao, G., X. Y. Dong, and Y. Sun (2009). "Ligands for mixed-mode protein chromatography: Principles, characteristics and design." *J Biotechnol* 144(1): 3–11.

Zhong, S. R., Y. Jin, J. B. Wu, Y. H. Jia, G. L. Xu, G. C. Wang, Y. L. Xiong, and Q. M. Lu (2009). "Purification and characterization of a new L-amino acid oxidase from *Daboia russellii siamensis* venom." *Toxicon* 54(6): 763–771.

Zhou, J. L., M. Ueda, R. Umemiya, B. Battsetseg, D. Boldbaatar, X. A. Xuan, and K. Fujisaki (2006). "A secreted cystatin from the tick *Haemaphysalis longicornis* and its distinct expression patterns in relation to innate immunity." *Insect Biochem Mol Biol* 36(7): 527–535.

Zhou, Q., C. Dangelmaier, and J. B. Smith (1996). "The hemorrhagin catrocollastatin inhibits collagen-induced platelet aggregation by binding to collagen via its disintegrin-like domain." *Biochem Biophys Res Comm* 219(3): 720–726.

Zhou, Q., J. B. Smith, and M. H. Grossman (1995). "Molecular cloning and expression of catrocollastatin, a snake-venom protein from *Crotalus atrox* (western diamondback rattlesnake) which inhibits platelet adhesion to collagen." *Biochem J* 307(pt 2): 411–417.

Zhu, S., B. Gao, M. Deng, Y. Yuan, L. Luo, S. Peigneur, Y. Xiao, S. Liang, and J. Tytgat (2010). "Drosotoxin, a selective inhibitor of tetrodotoxin-resistant sodium channels." *Biochem Pharmacol* 80(8): 1296–1302.

Zhu, S., S. Peigneur, B. Gao, X. Lu, C. Cao, and J. Tytgat (2012). "Evolutionary diversification of *Mesobuthus* alpha-scorpion toxins affecting sodium channels." *Mol Cell Proteomics* 11(1): M111.012054.

Zhu, S. Y., I. Huys, K. Dyason, F. Verdonck, and J. Tytgat (2004). "Evolutionary trace analysis of scorpion toxins specific for K-channels." *Proteins Struct Funct Genet* 54(2): 361–370.

Zingali, R. B., M. L. Bianconi, and R. Q. Monteiro (2001). "Interaction of bothrojaracin with prothrombin." *Haemostasis* 31(3-6): 273–278.

Zingali, R. B., M. S. Ferreira, M. Assafim, F. S. Frattani, and R. Q. Monteiro (2005). "Bothrojaracin, a *Bothrops jararaca* snake venom-derived (pro)thrombin inhibitor, as an anti-thrombotic molecule." *Pathophysiol Haemost Thromb* 34(4–5): 160–163.

Zouari-Kessentini, R., J. Luis, A. Karray, O. Kallech-Ziri, N. Srairi-Abid, A. Bazaa, E. Loret, S. Bezzine, M. El Ayeb, and N. Marrakchi (2009). "Two purified and characterized phospholipases A_2 from *Cerastes cerastes* venom that inhibit cancerous cell adhesion and migration." *Toxicon* 53(4): 444–453.

Zuliani, J. P., A. M. Kayano, K. D. Zaqueo, A. C. Neto, S. V. Sampaio, A. M. Soares, and R. G. Stabeli (2009). "Snake venom L-amino acid oxidases: Some consideration about their functional characterization." *Protein Pept Lett* 16(8): 908–912.

Zuraw, B., U. Yasothan, and P. Kirkpatrick (2010). "Ecallantide." *Nat Rev Drug Discov* 9(3): 189–190.

INDEX